The New York
Botanical Garden
Illustrated Encyclopedia
of Horticulture

ERRATA

Volumes 1–3
Page xiii, line 15:
"Barnaby" should read "Barneby"

Volume 3
Page 742, column 1, line 6:
"*buckleyi*" should read "*bridgesii*"

Page 798, column 2, line 45:
Add "and Pellaea."

Page 807, column 2, picture caption:
"*Coccoloba diversifolia*" should read "*Cocculus
   laurifolius*"

Page 860, column 1, line 4:
"*Catystegia*" should read "*Calystegia*"

Page 861, column 2, line 22:
"**ORA**" should read "**ORO**"

Page 945, column 2, picture caption:
"*Cupressus*, unidentified species in Scotland" should read
   "*Chamaecyparis nootkatensis*"

Page 983, column 1, lines 14, 15:
"eighty genera and" should read "ninety genera and
   4,000"

Volume 3 (Continued)
Page 986, lower right picture caption:
"*acaule*" should read "*reginae*"

Page 1010, column 2, line 14:
Insert "The flowers are yellow." after "eight."

The New York Botanical Garden Illustrated Encyclopedia of Horticulture

Thomas H. Everett

Volume 3
Cha-Di

Garland Publishing, Inc.
New York & London

15 14 13 12 11 10 9 8 7 6 5 4 3 2 1

Library of Congress Cataloging in Publication Data

Everett, Thomas H
    The New York Botanical Garden illustrated
encyclopedia of horticulture.

    1. Horticulture—Dictionaries. 2. Gardening—
Dictionaries. 3. Plants, Ornamental—Dictionaries.
4. Plants, Cultivated—Dictionaries. I. New York
(City). Botanical Garden. II. Title.
SB317.58.E94   635.9'03'21   80-65941
ISBN 0-8240-7233-2

PHOTO CREDITS

Black and White

George J. Ball: Chrysanthemums from cuttings (part b), p. 751. W. Atlee Burpee Company: *Cynoglossum amabile*, p. 981. Colonial Williamsburg, Va.: Colonial gardens as restored at Colonial Williamsburg, pp. 830–832. Desert Botanical Garden, Phoenix, Ariz.: Desert garden views (part a), p. 1044. A. B. Graf: *Chironia baccifera*, p. 733; *Chrysocoma coma-aurea*, p. 758. The Horticultural Society of New York, Inc.: Greenhouse chrysanthemums exhibited at the flower show in 1916, p. 748; First-prize winner in the one-specimen class at the flower show in 1916, p. 749. Malak, Ottawa, Canada: *Chionodoxa luciliae tmoli*, p. 730; Dutch crocuses, three varieties, p. 920; Dutch crocus 'Pickwick', p. 920; Dutch crocus 'Pickwick', p. 924. National Chrysanthemum Society, Inc.: Classes of chrysanthemums, pp. 749–750. The New York Botanical Garden: *Chamaecyparis pisifera*, p. 709; *Chimaphila maculata*, p. 725; Norway spruce and balsam fir, p. 742; Annual chrysanthemum, garden variety, p. 746; Hardy chrysanthemums, p. 747; Greenhouse chrysanthemums, p. 747; Outdoor chrysanthemums, disbudded to one bloom on each stem, p. 753; *Cladrastis lutea* (flowers), p. 779; *Coccothrinax* species, p. 808; Fruit of coconut with one-half of the outer fibrous husk removed, p. 813; *Copernicia hospita*, p. 861; *Cornus canadensis*, in fruit, p. 877; *Crassula falcata*, p. 903; *Crocus chrysanthus* 'Canary Bird', p. 921; *Crocus etruscus*, p. 922; *Crocus medius*, p. 923; *Crocus asturicus*, p. 924; *Chryptanthus beuckeri*, p. 929; Black currants, p. 949; Rooting cuttings in water (part a), p. 954; Rooted bulb cuttings (part a), p. 960; Evergreen cuttings (part b), p. 960; Hardwood cuttings (part b), p. 962; *Cymophyllus fraseri*, p. 978; *Cypella herbertii*, p. 982; *Cypripedium passerinum*, p. 986; *Cyrtanthus ochroleucus*, p. 988; Modern hybrid delphiniums (part a), p. 1030; *Deutzia scabra*, p. 1049; *Deutzia lemoinei*, p. 1050; *Dianthus arenarius*, p. 1052; *Dianthus gratianopolitanus*, p. 1052; *Dianthus knappi*, p. 1053; *Dianthus alpinus*, p. 1053. New York State College of Agriculture: Chlorosis in apple leaves caused by iron deficiency, p. 739. Other photographs by Thomas H. Everett.

Color

Malak, Ottawa, Canada: *Chionodoxa luciliae*; Dutch crocuses, garden varieties. The New York Botanical Garden: *Clarkia purpurea viminea, Claytonia lanceolata, Cliftonia monophylla, Clintonia borealis* (fruit), *Colchicum autumnale, Corallorhiza maculata, Corylopsis sinensis, Dalea* species, *Dasylirion,* unidentified species, *Dendrobium chrysotoxum, Dentaria diphylla.* Other photographs by Thomas H. Everett.

Published by Garland Publishing, Inc.
136 Madison Avenue, New York, New York 10016

Printed in the United States of America

This work is dedicated to the honored memory of the distinguished horticulturists and botanists who most profoundly influenced my professional career: Allan Falconer of Cheadle Royal Gardens, Cheshire, England; William Jackson Bean, William Dallimore, and John Coutts of the Royal Botanic Gardens, Kew, England; and Dr. Elmer D. Merrill and Dr. Henry A. Gleason of The New York Botanical Garden.

Foreword

According to Webster, an encyclopedia is a book or set of books giving information on all or many branches of knowledge generally in articles alphabetically arranged. To the horticulturist or grower of plants, such a work is indispensable and one to be kept close at hand for frequent reference.

The appearance of *The New York Botanical Garden Illustrated Encyclopedia of Horticulture* by Thomas H. Everett is therefore welcomed as an important addition to the library of horticultural literature. Since horticulture is a living, growing subject, these volumes contain an immense amount of information not heretofore readily available. In addition to detailed descriptions of many thousands of plants given under their generic names and brief description of the characteristics of the more important plant families, together with lists of their genera known to be in cultivation, this Encyclopedia is replete with well-founded advice on how to use plants effectively in gardens and, where appropriate, indoors. Thoroughly practical directions and suggestions for growing plants are given in considerable detail and in easily understood language. Recommendations about what to do in the garden for all months of the year and in different geographical regions will be helpful to beginners and will serve as reminders to others.

The useful category of special subject entries (as distinct from the taxonomic presentations) consists of a wide variety of topics. It is safe to predict that one of the most popular will be Rock and Alpine Gardens. In this entry the author deals helpfully and adequately with a phase of horticulture that appeals to a growing group of devotees, and in doing so presents a distinctly fresh point of view. Many other examples could be cited.

The author's many years as a horticulturist and teacher well qualify him for the task of preparing this Encyclopedia. Because he has, over a period of more than a dozen years, written the entire text (submitting certain critical sections to specialists for review and suggestions) instead of farming out sections to a score or more specialists to write, the result is remarkably homogeneous and cohesive. The Encyclopedia is fully cross referenced so that one may locate a plant by either its scientific or common name.

If, as has been said, an encyclopedia should be all things to all people, then the present volumes richly deserve that accolade. Among the many who call it "friend" will be not only horticulturists ("gardeners," as our author likes to refer to them), but growers, breeders, writers, lecturers, arborists, ecologists, and professional botanists who are frequently called upon to answer questions to which only such a work can provide answers. It seems safe to predict that it will be many years before lovers and growers of plants will have at their command another reference work as authoritative and comprehensive as T. H. Everett's Encyclopedia.

John M. Fogg, Jr.
Director Emeritus, Arboretum of the Barnes Foundation
Emeritus Professor of Botany, University of Pennsylvania

Preface

The primary objective of *The New York Botanical Garden Illustrated Encyclopedia of Horticulture* is a comprehensive description and evaluation of horticulture as it is known and practiced in the United States and Canada by amateurs and by professionals, including those responsible for botanical gardens, public parks, and industrial landscapes. Although large-scale commercial methods of cultivating plants are not stressed, much of the content of the Encyclopedia is as basic to such operations as it is to other horticultural enterprises. Similarly, although landscape design is not treated on a professional level, landscape architects will find in the Encyclopedia a great deal of importance and interest to them. Emphasis throughout is placed on the appropriate employment of plants both outdoors and indoors, and particular attention is given to explaining in considerable detail the how-and when-to-do-it aspects of plant growing.

It may be useful to assess the meanings of two words I have used. Horticulture is simply gardening. It derives from the Latin *hortus,* garden, and *cultura,* culture, and alludes to the intensive cultivation in gardens and nurseries of flowers, fruits, vegetables, shrubs, trees, and other plants. The term is not applicable to the extensive field practices that characterize agriculture and forestry. Amateur, as employed by me, retains its classic meaning of a lover from the Latin *amator*; it refers to those who garden for pleasure rather than for financial gain or professional status. It carries no implication of lack of knowledge or skills and is not to be equated with novice, tyro, or dabbler. In truth, amateurs provide the solid basis upon which American horticulture rests; without them the importance of professionals would diminish. Numbered in millions, amateur gardeners are devotees of the most widespread avocation in the United States. This avocation is serviced by a great complex of nurseries, garden centers, and other suppliers; by landscape architects and landscape contractors; and by garden writers, garden lecturers, Cooperative Extension Agents, librarians, and others who dispense horticultural information. Numerous horticultural societies, garden clubs, and botanical gardens inspire and promote interest in America's greatest hobby and stand ready to help its enthusiasts.

Horticulture as a vocation presents a wide range of opportunities which appeal equally to women and men. It is a field in which excellent prospects still exist for capable entrepreneurs. Opportunities at professional levels occur too in nurseries and greenhouses, in the management of landscaped grounds of many types, and in teaching horticulture.

Some people confuse horticulture with botany. They are not the same. The distinction becomes more apparent if the word gardening is substituted for horticulture. Botany is the science that encompasses all systematized factual knowledge about plants, both wild and cultivated. It is only one of the several disciplines upon which horticulture is based. To become a capable gardener or a knowledgeable plantsman or plantswoman (I like these designations for gardeners who have a wide, intimate, and discerning knowledge of plants in addition to skill in growing them) it is not necessary to study botany formally, although such study is likely to add greatly to one's pleasure. In the practice of gardening many botanical truths are learned from experience. I have known highly competent gardeners without formal training in botany and able and indeed distinguished botanists possessed of minimal horticultural knowledge and skills.

Horticulture is primarily an art and a craft, based upon science, and at some levels perhaps justly regarded as a science in its own right. As an art it calls for an appreciation of beauty and form as expressed in three-dimensional spatial relationships and an ability to translate aesthetic concepts into reality. The chief materials used to create gardens are living plants, most of which change in size and form with the passing of time and often show differences in color and texture and in other ways from season to season. Thus it is important that designers of gardens have a wide familiarity with the sorts of plants that lend themselves to their purposes and with plants' adaptability to the regions and to the sites where it is proposed to plant them.

As a craft, horticulture involves special skills often derived from ancient practices passed from generation to generation by word of mouth and apprenticeship-like contacts. As a technology it relies on this backlog of empirical knowledge supplemented by that acquired by scientific experiment and investigation, the results of which often serve to explain rather than supplant old beliefs and practices but sometimes point the way to more expeditious methods of attaining similar results. And from time to time new techniques are developed that add dimensions to horticultural practice; among such of fairly recent years that come to mind are the manipulation of blooming season by artificial day-length, the propagation of orchids and some other plants by meristem tissue culture, and the development of soilless growing mixes as substitutes for soil.

One of the most significant developments in American horticulture in recent decades is the tremendous increase in the number of different kinds of plants that are cultivated by many more people than formerly. This is particularly true of indoor plants or houseplants, the sorts grown in homes, offices, and other interiors, but is by no means confined to that group. The relative affluence of our society and the freedom and frequency of travel both at home and abroad has contributed to this expansion, a phenomenon that will surely continue as avid collectors of the unusual bring into cultivation new plants from the wild and promote wider interest in sorts presently rare. Our garden flora is also constantly and beneficially expanded as a result of the work of both amateur and professional plant breeders.

It is impracticable in even the most comprehensive encyclopedia to describe or even list all plants that somewhere within a territory as large as the United States and Canada are grown in gardens. In this Encyclopedia the majority of genera known to be in cultivation are described, and descriptions and often other pertinent information about a complete or substantial number of their species and lesser categories are given. Sorts likely to be found only in collections of botanical gardens or in those of specialists may be omitted.

The vexing matter of plant nomenclature inevitably presents itself when an encyclopedia of horticulture is contemplated. Conflicts arise chiefly between the very understandable desire of gardeners and others who deal with cultivated plants to retain long-familiar names and the need to reflect up-to-date botanical interpretations. These points of view are basically irreconcilable and so accommodations must be reached.

As has been well demonstrated in the past, it is unrealistic to attempt to standardize the horticultural usage of plant names by decree or edict. To do so would negate scientific progress. But it is just as impracticable to expect gardeners, nurserymen, arborists, seedsmen, dealers in bulbs, and other amateur and professional horticulturists to keep current with the interpretations and recommendations of plant taxonomists; particularly as these sometimes fail to gain the acceptance even of other botanists and it is not unusual for scientists of equal stature and competence to prefer different names for the same plant.

In practice time is the great leveler. Newly proposed plant names accepted in botanical literature are likely to filter gradually into horticultural usage and eventually gain currency value, but this sometimes takes several years. The complete up-to-dateness and niceties of botanical naming are less likely to bedevil horticulturists than uncertainties concerned with correct plant identification. This is of prime importance. Whether a tree is labeled *Pseudotsuga douglasii, P. taxifolia,* or *P. menziesii* is of less concern than that the specimen so identified is indeed a Douglas-fir and not some other conifer.

After reflection I decided that the most sensible course to follow in *The New York Botanical Garden Illustrated Encyclopedia of Horticulture* was to accept almost in its entirety the nomenclature adopted in *Hortus Third* published in 1976. By doing so, much of the confusion that would result from two major comprehensive horticultural works of the late twentieth century using different names for the same plant is avoided, and it is hoped that for a period of years a degree of stability will be attained. Always those deeply concerned with critical groups of plants can adopt the recommendations of the latest monographers. Exceptions to the parallelism in nomenclature in this Encyclopedia and *Hortus Third* are to be found in the CACTACEAE for which, with certain reservations but for practical purposes, as explained in the Encyclopedia entry Cactuses, the nomenclature of Curt Backeburg's *Die Cactaceae,* published in 1958–62, is followed; and the ferns, where I mostly accepted the guidance of Dr. John T. Mickel of The New York Botanical Garden. The common or colloquial names employed are those deemed to have general acceptance. Cross references and synonomy are freely provided.

The convention of indicating typographically whether or not plants of status lesser than species represent entities that propagate and persist in the wild or are sorts that persist only in cultivation is not followed. Instead, as explained in the Encyclopedia entry Plant Names, the word variety is employed for all entities below specific rank and if in Latin form the name is written in italic, if in English or other modern language, in Roman type, with initial capital letter, and enclosed in single quotation marks.

Thomas H. Everett
Senior Horticulture Specialist
The New York Botanical Garden

Acknowledgments

I am indebted to many people for help and support generously given over the period of more than twelve years it has taken to bring this Encyclopedia to fruition. Chief credit belongs to four ladies. They are Lillian M. Weber and Nancy Callaghan, who besides accepting responsibility for the formidable task of filing and retrieving information, typing manuscript, proofreading, and the management of a vast collection of photographs, provided much wise council; Elizabeth C. Hall, librarian extraordinary, whose superb knowledge of horticultural and botanical literature was freely at my disposal; and Ellen, my wife, who displayed a deep understanding of the demands on time called for by an undertaking of this magnitude, and with rare patience accepted inevitable inconvenience. I am also obliged to my sister, Hette Everett, for the valuable help she freely gave on many occasions.

Of the botanists I repeatedly called upon for opinions and advice and from whom I sought elucidation of many details of their science abstruse to me, the most heavily burdened have been my friends and colleagues at The New York Botanical Garden, Dr. Rupert C. Barnaby, Dr. Arthur Cronquist, and Dr. John T. Mickel. Other botanists and horticulturists with whom I held discussions or corresponded about matters pertinent to my text include Dr. Theodore M. Barkley, Dr. Lyman Benson, Dr. Ben Blackburn, Professor Harold Davidson, Dr. Otto Degener, Harold Epstein, Dr. John M. Fogg, Jr., Dr. Alwyn H. Gentry, Dr. Alfred B. Graf, Brian Halliwell, Dr. David R. Hunt. Dr. John P. Jessop, Dr. Tetsuo Koyama, Dr. Bassett Maguire, Dr. Roy A. Mecklenberg, Everitt L. Miller, Dr. Harold N. Moldenke, Dr. Dan H. Nicolson, Dr. Pascal P. Pirone, Dr. Ghillean Prance, Don Richardson, Stanley J. Smith, Ralph L. Snodsmith, Marco Polo Stufano, Dr. Bernard Verdcourt, Dr. Edgar T. Wherry, Dr. Trevor Whiffin, Dr. Richard P. Wunderlin, Dr. John J. Wurdack, Yuji Yoshimura, and Rudolf Ziesenhenne.

Without either exception or stint these conferees and correspondents shared with me their knowledge, thoughts, and judgments. Much of the bounty so gleaned is reflected in the text of the Encyclopedia but none other than I am responsible for interpretations and opinions that appear there. To all who have helped, my special thanks are due and are gratefully proferred.

I acknowledge with much pleasure the excellent cooperation I have received from the Garland Publishing Company and most particularly from its President, Gavin Borden. To Ruth Adams, Nancy Isaac, Carol Miller, and Melinda Wirkus, I say thank you for working so understandingly and effectively with me and for shepherding my raw typescript through the necessary stages.

How to Use This Encyclopedia

A vast amount of information about how to use, propagate, and care for plants both indoors and outdoors is contained in the thousands of entries that compose the *New York Botanical Garden Illustrated Encyclopedia of Horticulture*. Some understanding of the Encyclopedia's organization is necessary in order to find what you want to know.

Arrangement of the Entries

Genera

The entries are arranged in alphabetical order. Most numerous are those that deal with taxonomic groups of plants. Here belong approximately 3,500 items entered under the genus name, such as ABIES, DIEFFENBACHIA, and JUGLANS. If instead of referring to these names you consult their common name equivalents of FIR, DUMB CANE, and WALNUT, you will find cross references to the genus names.

Bigeneric Hybrids & Chimeras

Hybrids between genera that have names equivalent to genus names—most of these belonging in the orchid family—are accorded separate entries. The same is true for the few chimeras or graft hybrids with names of similar status. Because bigeneric hybrids frequently have characteristics similar to those of their parents and require similar care, the entries for them are often briefer than the regular genus entries.

Families

Plant families are described under their botanical names, with their common name equivalents also given. Each description is followed by a list of the genera accorded separate entries in this Encyclopedia.

Vegetables, Fruits, Herbs, & Ornamentals

Vegetables and fruits that are commonly cultivated, such as broccoli, cabbage, potato, tomato, apple, peach, and raspberry; most culinary herbs, including basil, chives, parsley, sage, and tarragon; and a few popular ornamentals, such as azaleas, carnations, pansies, and poinsettias, are treated under their familiar names, with cross references to their genera. Discussions of a few herbs and some lesser known vegetables and fruits are given under their Latin scientific names with cross references to the common names.

Other Entries

The remaining entries in the Encyclopedia are cross references, definitions, and more substantial discussions of many subjects of interest to gardeners and others concerned with plants. For example, a calendar of gardening activity, by geographical area, is given under the names of the months and a glossary of frequently applied species names (technically, specific epithets) is provided in the entry Plant Names. A list of these general topics, which may provide additional information about a particular plant, is provided at the beginning of each volume of the Encyclopedia (see pp. xvii–xx).

Cross References & Definitions

The cross references are of two chief types: those that give specific information, which may be all you wish to know at the moment:
Boojam Tree is *Idria columnaris.*
Cobra plant is *Darlingtonia californica.*
and those that refer to entries where fuller explanations are to be found:
Adhatoda. See Justicia.
Clubmoss. See Lycopodium and Selaginella.

Additional information about entries of the former type can, of course, be found by looking up the genus to which the plant belongs—*Idria* in the case of the boojam tree and *Darlingtonia* for the cobra plant.

ORGANIZATION OF THE GENUS ENTRIES

Pronunciation

Each genus name is followed by its pronunciation in parentheses. The stressed syllable is indicated by the diacritical mark ´ if the vowel sound is short as in man, pet, pink, hot, and up; or by ` if the vowel sound is long as in mane, pete, pine, home, and fluke.

Genus Common Names
Family Common Names
General Characteristics

Following the pronunciation, there may be one or more common names applicable to the genus as a whole or to certain of its kinds. Other names may be introduced later with the descriptions of the species or kinds. Early in the entry you will find the common and botanical names of the plant family to which the genus belongs, the number of species the genus contains, its natural geographical distribution, and the derivation of its name. A description that stresses the general characteristics of the genus follows, and this may be supplemented by historical data, uses of some or all of its members, and other pertinent information.

Identification of Plants

Descriptions of species, hybrids, and varieties appear next. The identification of unrecognized plants is a fairly common objective of gardeners; accordingly, in this Encyclopedia various species have been grouped within entries in ways that make their identification easier. The groupings may bring into proximity sorts that can be adapted for similar landscape uses or that require the same cultural care, or they may emphasize geographical origins of species or such categories as evergreen and deciduous or tall and low members of the same genus. Where the description of a species occurs, its name is designated in *bold italic.* Under this plan, the description of a particular species can be found by referring to the group to which it belongs, scanning the entry for the species name in bold italic, or referring to the opening sentences of paragraphs which have been designed to serve as lead-ins to descriptive groupings.

Gardening & Landscape Uses
Cultivation
Pests & Diseases

At the end of genus entries, subentries giving information on garden and landscape uses, cultivation, and pests or diseases or both are included, or else reference is made to other genera or groupings for which these are similar.

xvi

General Subject Listings

The lists below organize some of the encyclopedia entries into topics which may be of particular interest to the reader. They are also an aid in finding information other than Latin or common names of plants.

PLANT ANATOMY AND TERMS USED IN PLANT DESCRIPTIONS

All-America Selections
Alternate
Annual Rings
Anther
Apex
Ascending
Awl-shaped
Axil, Axillary
Berry
Bloom
Bracts
Bud
Bulb
Bulbils
Bulblet
Bur
Burl
Calyx
Cambium Layer
Capsule
Carpel
Catkin
Centrals
Ciliate
Climber
Corm
Cormel
Cotyledon
Crown
Deciduous
Disk or Disc
Double Flowers
Drupe
Florets
Flower
Follicle
Frond
Fruit
Glaucous
Gymnosperms
Head
Hips
Hose-in-Hose

Inflorescence
Lanceolate
Leader
Leaf
Leggy
Linear
Lobe
Midrib
Mycelium
Node
Nut and Nutlet
Oblanceolate
Oblong
Obovate
Offset
Ovate
Palmate
Panicle
Pedate
Peltate
Perianth
Petal
Pinnate
Pip
Pistil
Pit
Pod
Pollen
Pompon
Pseudobulb
Radials
Ray Floret
Rhizome
Runners
Samara
Scion or Cion
Seeds
Sepal
Set
Shoot
Spore
Sprigs
Spur
Stamen
Stigma
Stipule

Stolon
Stool
Style
Subshrub
Taproot
Tepal
Terminal
Whorl

GARDENING TERMS AND INFORMATION

Acid and Alkaline Soils
Adobe
Aeration of the Soil
Air and Air Pollution
Air Drainage
Air Layering
Alpine Greenhouse or Alpine House
Amateur Gardener
April, Gardening Reminders For
Aquarium
Arbor
Arboretum
Arch
Asexual or Vegetative Propagation
Atmosphere
August, Gardening Reminders For
Balled and Burlapped
Banks and Steep Slopes
Bare-Root
Bark Ringing
Baskets, Hanging
Bed
Bedding and Bedding Plants
Bell Jar
Bench, Greenhouse
Blanching
Bleeding
Bog
Bolting
Border
Bottom Heat
Break, Breaking
Broadcast
Budding
Bulbs or Bulb Plants

Gardening Terms and Information (Continued)

State Agricultural Experimental Stations
Stock or Understock
Straightedge
Strawberry Jars
Strike
Stunt
Succession Cropping
Sundials
Syringing
Thinning or Thinning Out
Tillage
Tilth
Tools
Top-Dressing
Topiary Work
Training Plants
Tree Surgery
Tree Wrapping
Trenching
Trowels
Tubs
Watering
Weeds and Their Control
Window Boxes

FERTILIZERS AND OTHER SUBSTANCES RELATED TO GARDENING

Algicide
Aluminum Sulfate
Ammonium Nitrate
Ammonium Sulfate
Antibiotics
Ashes
Auxins
Basic Slag
Blood Meal
Bonemeal
Bordeaux Mixture
Calcium Carbonate
Calcium Chloride
Calcium Metaphosphate
Calcium Nitrate
Calcium Sulfate
Carbon Disulfide
Chalk
Charcoal
Coal Cinders
Cork Bark
Complete Fertilizer
Compost and Composting
Cottonseed Meal
Creosote
DDT
Dormant Sprays
Dried Blood
Fermate or Ferbam
Fertilizers
Fishmeal
Formaldehyde
Fungicides
Gibberellic Acid
Green Manuring
Growth Retardants
Guano
Herbicides or Weed-Killers
Hoof and Horn Meal

Hormones
Humus
Insecticide
John Innes Composts
Lime and Liming
Liquid Fertilizer
Liquid Manure
Manures
Mulching and Mulches
Muriate of Potash
Nitrate of Ammonia
Nitrate of Lime
Nitrate of Potash
Nitrate of Soda
Nitrogen
Orchid Peat
Organic Matter
Osmunda Fiber or Osmundine
Oyster Shells
Peat
Peat Moss
Permagnate of Potash
Potassium
Potassium Chloride
Potassium-Magnesium Sulfate
Potassium Nitrate
Potassium Permagnate
Potassium Sulfate
Pyrethrum
Rock Phosphate
Rotenone
Salt Hay or Salt Marsh Hay
Sand
Sawdust
Sodium Chloride
Sprays and Spraying
Sulfate
Superphosphate
Trace Elements
Urea
Urea-Form Fertilizers
Vermiculite
Wood Ashes

TECHNICAL TERMS

Acre
Alternate Host
Annuals
Antidessicant or Antitranspirant
Biennals
Binomial
Botany
Chromosome
Climate
Clone
Composite
Conservation
Cross or Crossbred
Cross Fertilization
Cross Pollination
Cultivar
Decumbent
Dicotyledon
Division
Dormant
Endemic
Environment
Family

Fasciation
Fertility
Fertilization
Flocculate
Floriculture
Genus
Germinate
Habitat
Half-Hardy
Half-Ripe
Hardy Annual
Hardy Perennial
Heredity
Hybrid
Indigenous
Juvenile Forms
Juvenility
Legume
Monocotyledon
Monoecious
Mutant or Sport
Mycorrhiza or Mycorhiza
Nitrification
Perennials
pH
Plant Families
Photoperiodism
Photosynthesis
Pollination
Pubescent
Saprophyte
Self-Fertile
Self-Sterile
Species
Standard
Sterile
Strain
Terrestrial
Tetraploid
Transpiration
Variety

TYPES OF GARDENS AND GARDENING

Alpine Garden
Artificial Light Gardening
Backyard Gardens
Biodynamic Gardening
Bog Gardens
Botanic Gardens and Arboretums
Bottle Garden
City Gardening
Colonial Gardens
Conservatory
Container Gardening
Cutting Garden
Desert Gardens
Dish Gardens
Flower Garden
Fluorescent Light Gardening
Formal and Semiformal Gardens
Greenhouses and Conservatories
Heath or Heather Garden
Herb Gardens
Hydroponics or Nutriculture
Indoor Lighting Gardening
Japanese Gardens
Kitchen Garden
Knot Gardens

Types of Gardens and Gardening (Continued)

Miniature Gardens
Native Plant Gardens
Naturalistic Gardens
Nutriculture
Organic Gardening
Rock and Alpine Gardens
Roof and Terrace Gardening
Salads or Salad Plants
Seaside Gardens
Shady Gardens
Sink Gardening
Terrariums
Vegetable Gardens
Water and Waterside Gardens
Wild Gardens

PESTS, DISEASES, AND OTHER TROUBLES

Ants
Aphids
Armyworms
Bagworms
Bees
Beetles
Billbugs
Biological Control of Pests
Birds
Blight
Blindness
Blotch
Borers
Budworms and Bud Moths
Bugs
Butterflies
Canker
Cankerworms or Inchworms
Casebearers
Caterpillars
Cats
Centipede, Garden
Chinch Bugs
Chipmunks
Club Root
Corn Earworm
Crickets
Cutworms
Damping Off
Deer
Die Back
Diseases of Plants
Downy Mildew
Earthworms
Earwigs
Edema
Fairy Rings
Fire Blight
Flies
Fungi or Funguses
Galls
Gas Injury

Gophers
Grasshoppers
Grubs
Gummosis
Hornworms
Inchworms
Insects
Iron Chelates
Iron Deficiency
Lace Bugs
Lantana Bug
Lantern-Flies
Larva
Leaf Blight
Leaf Blister
Leaf Blotch
Leaf Curl
Leaf Cutters
Leaf Hoppers
Leaf Miners
Leaf Mold
Leaf Rollers
Leaf Scorch
Leaf Skeletonizer
Leaf Spot Disease
Leaf Tiers
Lightening Injury
Maggots
Mantis or Mantid
Mealybugs
Mice
Midges
Milky Disease
Millipedes
Mites
Mold
Moles
Mosaic Diseases
Moths
Muskrats
Needle Cast
Nematodes or Nemas
Parasite
Pests of Plants
Plant Hoppers
Plant Lice
Praying Mantis
Psyllids
Rabbits
Red Spider Mite
Rootworms
Rots
Rust
Sawflies
Scab Diseases
Scale Insects
Scorch or Sunscorch
Scurf
Slugs and Snails
Smut and White Smut Diseases
Sowbugs or Pillbugs
Spanworms

Spittlebugs
Springtails
Squirrels
Stunt
Suckers
Sun Scald
Thrips
Tree Hoppers
Virus
Walking-Stick Insects
Wasps
Webworms
Weevils
Wilts
Witches' Brooms
Woodchucks

GROUPINGS OF PLANTS

Accent Plants
Aquatics
Aromatic Plants
Bedding and Bedding Plants
Berried Trees and Shrubs
Bible Plants
Broad-leaved and Narrow-leaved Trees and Shrubs
Bulbs or Bulb Plants
Bush Fruits
Carnivorous or Insectivorous Plants
Dried Flowers, Foliage, and Fruits
Edging Plants
Epiphyte or Air Plant
Evergreens
Everlastings
Fern Allies
Filmy Ferns
Florists' Flowers
Foliage Plants
Fragrant Plants and Flowers
Gift Plants
Graft Hybrids
Grasses, Ornamental
Hard-Wooded Plants
Houseplants or Indoor Plants
Japanese Dwarfed Trees
Medicinal or Drug Plants
Night-Blooming Plants
Ornamental-Fruited Plants
Pitcher Plants
Poisonous Plants
Shrubs
State Flowers
State Trees
Stone Fruits
Stone or Pebble Plants
Stove Plants
Succulents
Tender Plants
Trees
Windowed Plants

The New York
Botanical Garden
Illustrated Encyclopedia
of Horticulture

CHAMAECEREUS (Chamae-cèreus)—Peanut Cactus. The only member of this genus, the peanut cactus (*Chamaecereus silvestri*) is a popular and delightful little plant, as easy to grow by the window gardener as by the skilled fancier of succulents with splendid greenhouse facilities. Native of Argentina, its name comes from the Greek *chamai*, dwarf or on the ground, and *Cereus*, another genus of the cactus family CACTACEAE.

The peanut cactus (***Chamaecereus sylvestri***) is so called because its short, cylindrical stem segments, up to 2 inches or so long by about ½ inch wide, suggest peanuts in size and shape. They have six to nine ribs furnished with short, soft-white spines. The flowers, open by day, are 2 to 2½ inches long and have narrow funnel-shaped tubes and spreading, glowing red perianth segments (petals). Their styles are yellowish-green. The peanut cactus forms dense clusters, not more than about 3 inches high, of easily detachable stem segments. These root readily and give rise to new plants. The blooms appear quite freely in late winter or early spring. The fruits are woolly, hairy, small, round, and dry. A variety with yellow rather than green stems and one with cristate (crested) stem segments are cultivated. Hybrids between this species and other cactuses, chiefly *Lobivia famatimensis*, have less easily detachable stem segments and usually larger blooms.

Garden Uses and Cultivation. The peanut cactus and its hybrids are properly included in collections of succulents and are also charming window garden plants. In

Chamaecereus sylvestri, grafted on *Hylocereus undatus*

Chamaecereus sylvestri

addition to being easy to grow as low specimens on their own roots, they can be had as miniature "trees" by grafting them onto stems of *Pereskia, Selenicereus, Hylocereus,* and other upright-growing cactuses. This is the best way to grow the yellow and crested varieties. This procedure is simple. The stem of the understock is beheaded at the desired height and cleft in two, three, or more places to admit the bases, cut in wedge-shaped fashion, of the joints of the *Chamaecereus*. These are pinned into place with long cactus spines or are held in position by rubber bands stretched around them and the understock. In a few weeks the cut surfaces knit and the union is complete. The peanut cactus can be raised from seeds, but cuttings root so freely that when available they provide all needed increase. General culture is described under Cactuses.

CHAMAECYPARIS (Chamaecý-paris) — False-Cypress. In gardens members of the genus *Chamaecyparis* are often called cypresses, a name more correctly reserved for related *Cupressus*, from which chamaecyparises or false-cypresses differ in each of their cone scales usually having two, and rarely up to five seeds. In *Cupressus*, with the exception of *C. funebris*, each has six to twenty. Also, the branchlets of false-cypresses are distinctly flattened, which is rarely true of true cypresses. The fruiting cones of false-cypresses, except for those of *C. nootkatensis*, mature in their first season, those of *Cupressus* in their second. The two genera closely resemble each other, and it often takes a botanical eye to determine which is which. Important to northern gardeners is the comforting knowledge that false-cypresses as a group are much hardier than cypresses.

Belonging to the cypress family CUPRESSACEAE, false-cypresses are conifers. Their name comes from the Greek *chamai*, dwarf, and *kyparissos*, cypress. It alludes to the botanical relationship and habit of growth of some kinds. Seven species are recognized, natives of North America, Japan, and Taiwan. In addition there are numerous horticultural varieties, some as difficult to identify as are varieties of the related arbor-vitaes (*Thuja*).

As they occur in the wild, false-cypresses are pyramidal trees of impressive size, but many horticultural varieties remain bushy and compact, and some are quite dwarf. These evergreens have two types of foliage, juvenile, characteristic of young plants raised from seeds and which in some horticultural varieties is the only kind developed for an indefinite period, and adult which normally gradually replaces the juvenile foliage as seedlings attain maturity. Juvenile leaves are needle-shaped and radiate all around the shoots. Adult leaves are in four rows along the shoots and are tiny, scalelike, and more or less pressed against the shoots. Those of the two lateral rows are boat-shaped and spread at their tips, those of the other two rows are smaller and have less spreading tips. The branchlets are commonly in frondlike sprays. Flowers of both sexes appear on the same tree, the males usually yellow, rarely red or brown. The fruiting cones, small and rounded, mostly ripen their seeds the first year, but those of *C. nootkatensis* not until their second year. Each cone consists of six to twelve scales joined to the axes on which they are carried at the insides of their margins, like those of the cypresses (*Cupressus*), but quite differently from arbor-vitaes (*Thuja*). There are five or fewer seeds to each scale.

Some kinds are called *Retinispora* by gardeners, but the name is not acceptable botanically. Its use dates back to an early misunderstanding of certain horticultural varieties of *Chamaecyparis* and *Thuja*. As explained above, and under *Thuja*, young seedlings of both genera have foliage entirely different in appearance from that of more mature plants. Oriental horticulturists, by propagating from juvenile plants and then for many generations repropagating before adult foliage developed, obtained stable forms that retain juvenile-type or intermediate-type foliage indefinitely even in large plants. They have more the aspect of certain junipers than false-cypresses or arbor-vitaes, but have softer foliage and leaves not sharp-pointed, and often have on their undersides whitish or grayish markings that are never present on the undersides of juniper leaves. When these were first brought to Europe and America botanists thought they represented a new genus and applied to them the name *Retinispora*, but this was discarded once their true origin became known. They are dealt with here as horticultural varieties of *Chamaecyparis* and *Thuja* according to the genus from which they are derived.

The native American species of false-cypress are three, two from the west and one from the east. The latter, the white-cedar (**C. thyoides**), is a coastal tree that occurs from Maine to Florida and Mississippi, usually in swamps. It is the least ornamental of the genus, but has the virtue of being the hardiest. It succeeds in cultivation north of its natural range, well into Canada and is a lumber producer of some importance. The white-cedar has a slender, often rather ragged, spirelike head up to 80 or 90 feet in height, with a trunk sometimes more than 3 feet in diameter covered with fissured, reddish-brown bark, and ascending or spreading branches. The slender, irregularly-arranged, flattened branchlets are in tufts. They do not droop. The adult foliage is light green or glaucous-bluish-green and consists of small scalelike leaves each with a resin gland on its back. The male cones are minute. The fruiting cones have six scales and are about ¼ inch in diameter. Like the species, the varieties of this tree generally appeal only to gardeners in climates where other, and more ornamental, kinds will not grow. The chief varieties are *C. t. andelyensis*, the best of the white-cedar variants, a bushy, pyramidal kind that bears both juvenile and adult foliage and is slightly less hardy and is slower-growing than the typical species. *C. t. andelyensis nana*, a dwarfer, rounded or flat-topped shrub, grows even more slowly than the last variety. *C. t. ericoides* is a dense, stiff, pyramidal or columnar shrub with erect branches and all bright green, juvenile foliage that becomes bronzy-red or purplish-brown in winter. The undersides of its leaves are marked with two bluish lines. *C. t. glauca* is a compact plant with silvery or light glaucous-blue foliage and reddish twigs. *C. t. hoveyi* is slender and has its foliage in dense tufts. *C. t. nana* is very low and compact and has glaucous foliage. *C. t. variegata* has branchlets variegated with yellow.

The Nootka false-cypress (**C. nootkatensis**) inhabits northwest America from Oregon to Alaska. It is a good ornamental,

Chamaecyparis nootkatensis

less hardy than the white-cedar, but much more resistant to cold than the third American species *C. lawsoniana*. It is hardy in New England. A slender tree up to 120 feet in height, it has a trunk up to 6 feet in diameter, brownish-gray bark, and pendulous, quadrangular branchlets in flattened sprays and without white markings on the undersides of its dark green leaves. The foliage when crushed smells strongly of turpentine, and the leaves remain attached to the branches for a long time after they die. The fruiting cones, about ½ inch across, mature in the late spring or early summer of their second year. Its more outstanding varieties include *C. n. lutea* (syn. *C. n. aurea*), with shoots and leaves at first yellowish. *C. n. compacta*, a dwarf, dense shrub. *C. n. glauca*, with very glaucous foliage. *C. n. pendula*, very distinct with

Chamaecyparis lawsoniana (foliage)

Chamaecyparis nootkatensis pendula

spreading branches and pendulous branchlets up to several feet long.

Lawson's false-cypress (***C. lawsoniana***) is one of the most magnificent of America's conifers, and one of the most useful, for it is an important lumber tree. Towering up to 200 feet and slender, when not crowded it retains its branches and foliage to the ground throughout its life. It is the source of about two hundred horticultural varieties, some of which are among the most lovely of evergreens. Unfortunately neither the typical species nor its variants are as hardy as other American false-cypresses, and they prosper only in humid climates. Lawson's false-cypress is limited in its natural distribution to a strip not over forty miles wide and extending from Eureka, California to Coos Bay, Oregon. Its trunks are covered with very thick, spongy, furrowed, reddish bark, and its branches droop as do also its flat, fan-shaped branchlets. The leaves are streaked indistinctly with white on their under surfaces. When crushed they have what has been described as a resinous, parsley-like odor. The pink or crimson male cones appear in spring. The fruiting cones, about ⅓ inch in diameter, at first purplish or glaucous-green become reddish-brown when mature at the end of their first season.

The horticultural varieties of Lawson's false-cypress include kinds showing a wide range of color variations, others that are markedly dissimilar in habit from the typical species include narrowly-columnar or fastigiate forms, weeping varieties, and dwarf kinds. A selection of the best and most commonly cultivated varieties follows: *C. l. albo-spica* is densely-pyramidal and has its branchlets tipped with white, but is not one of the most attractive kinds. *C. l. allumii*, one of the finest varieties, with glaucous-blue foliage, is compact and narrow-pyramidal. *C. l. argentea* is a name sometimes applied to *C. l.* 'Silver Queen'. *C. l. ellwoodii*, dwarf, dense, conical and slow-growing, has glaucous-blue juvenile

Chamaecyparis lawsoniana

foliage. *C. l. erecta* (syn. *C. l. erecta viridis*) forms a dense narrow column of erect branches and bright green foliage and has phases with grayish green, golden, and glaucous-green foliage known as *C. l. e. alba*, *C. l. e. aurea*, and *C. l. e. glauca*, respectively. Other forms of this variety are *C. l. e. filiformis*, with short, stiff branches and longer branchlets than those of *C. erecta*, and *C. l. e. witzeliana*, a slender variety. *C. l. fletcheri* grows slowly, is conical, and has dense, light glaucous-green foliage tinted brown or purplish in winter. *C. l. forsteckensis*, one of the best dwarfs, has grayish-green leaves on dense cocks-comb-like branchlets. *C. l. glauca*, a vigorous variety with steel-blue foliage exists in a number of phases differing slightly in color (here belongs *C. l.* 'Triompf van Boskoop'). *C. l. gracilis* (syn. *C. l. gracilis pendula*) is elegant and has long drooping branches; *C. l. g. aurea*, with golden foliage, and *C. l. g. glauca*, with blue foliage, are variants. *C. l. intertexta* is a handsome, vigorous,

pyramidal tree with arching branches and widely-spaced branchlets. *C. l.* 'Kilmacurragh' is erect, narrowly-columnar, and bright green. *C. l. minima*, broadly-conical, dwarf and compact, with branchlets in vertical planes, has two variants, *C. l. m. aurea*, with golden foliage, and *C. l. m. glauca*, a steel-blue kind. *C. l. nana* is dark green, dwarf, and globose, usually broader than tall, and has the variant *C. l. n. alba*, which is yellowish-white. *C. l. n. glauca*, is a glaucous-foliaged plant. *C. l. pendula*, a beautiful variety with horizontal branches and drooping branchlets also exists as a variant called *C. l. p. vera*, in which both branches and branchlets droop emphatically. *C. l. pottenii*, rigid, conical, and slow-growing, has partly juvenile foliage. *C. l.* 'Silver Queen' has silvery-white foliage. *C. l. stewartii* is attractive, with shoots when young deep yellow. *C. l. westermannii* is an exceptionally vigorous grower with wide-spreading branches, drooping branchlets, and shoots and foliage that are light golden-yellow when young. *C. l. wisselii* is glaucous-foliaged and columnar, with crowded, tufted branchlets.

Chamaecyparis lawsoniana stewartii

Chamaecyparis lawsoniana wisselii

Chamaecyparis lawsoniana 'Kilmacurragh'

Chamaecyparis lawsoniana erecta

Chamaecyparis lawsoniana pottenii

Chamaecyparis lawsoniana fletcheri

The Sawara false-cypress (*C. pisifera*) and its variations are hardy throughout New England. The species is a native of Japan where it attains a height of 150 feet and a trunk diameter of 5 feet. It is narrowly-pyramidal and has horizontal branches, and rather smooth, reddish-brown bark that peels in long strips. Its branchlets, flattened and two-ranked, are in horizontal planes. The leaves have obscure glands on their backs, are lustrous green above, and have white areas on their undersides. The fruiting cones, at the ends of short branchlets, are glaucous-green through the summer, but change to dark brown when they mature. About ¼ inch in diameter, they have ten or sometimes twelve scales, each with one or two seeds. The typical Sawara false-cypress grows fairly rapidly, but as it ages it is likely to lose its lower branches and to become thin and open. It is less satisfactory as an ornamental than the Hinoki false-cypress. Its varieties are numerous and mostly very attractive, and some have only juvenile foliage. Among the better and most popular are *C. p. aurea*, with bright yellow foliage in early summer that fades later and is of poor appearance through the winter. *C. p. filifera*, a tall shrub, is usually broader than

Chamaecyparis pisifera

Chamaecyparis pisifera squarrosa

Chamaecyparis pisifera aurea

Chamaecyparis pisifera plumosa (foliage)

Chamaecyparis pisifera cyanoviridis (foliage)

high, with mostly undivided, long whip-like branchlets. *C. p. f. nana* and *C. p. f. aurea* are forms of *C. p. filifera*, the first a dwarf, the other with golden foliage. *C. p. plumosa* and its variants have foliage intermediate between juvenile and adult. *C. p. plumosa* itself is dense and pyramidal. *C. p. p. albo-picta* has white tips to its branchlets. *C. p. p. argentea* is similar to the last named. *C. p. p. flavescens* has yellowish young branchlet tips. *C. p. squarrosa* is distinguished by its loose, feathery, glaucous-blue-green juvenile foliage, white-marked on both surfaces, and not arranged in flat, frondlike sprays. *C. p. s. minima* and *C. p. s. sulphurea* are phases of *C. p. squarrosa*, the former dwarf and compact, the latter with pale yellow foliage. Attractive *C. p. cyanoviridis* is broad-pyramidal and dark green. The smallest of the *C. pisifera* varieties, extremely slow growing *C. p. plumosa compressa* forms a compact, dense mound a few inches high with miniature foliage that seems to vary in color depending on the situation and season.

Chamaecyparis pisifera cyanoviridis

Chamaecyparis pisifera plumosa compressa

The Hinoki false-cypress (**C. obtusa**) is one of the most ornamental of the genus. Slow-growing and available in both tree and shrubby varieties, it is hardy throughout New England and into southern Canada. In its typical form it is a broadly-pyramidal tree that in its native Japan attains a height of 120 feet and a trunk diameter of 4 feet, but is much smaller as known in cultivation in America and Europe, although a specimen more than 70 feet tall is reported from England and, as it was not brought into cultivation in that country until 1861, it may eventually grow as big as in its native land. The Hinoki false-cypress has a straight trunk, usually buttressed at its base, and smoothish, reddish-brown bark that peels in long, slender strips. The branches are spreading and the flattened branchlets are in very definite horizontal planes with their nonglandular leaves dark green above and with waxy, whitish, X-shaped marks on their undersides. The cones, larger than those of other false-cypresses, are up to ½ inch in diameter and orange-brown. They have eight or rarely ten scales, each with two to five seeds.

In Japan the Hinoki false-cypress is of religious significance and is commonly planted near Shinto Temples. It is also an important commercial timber tree, its lumber being greatly favored for general construction and building, to make interior trim and furniture, and especially as a base for fine lacquer ware. It is an important component of Japanese forests, both natural and planted.

There are numerous horticultural varieties of the Hinoki false-cypress. Among the better known and most attractive are these: C. o. albo-spica has white young shoots. C. o. aurea, has golden young shoots and foliage, sometimes with green leaves intermixed. C. o. breviramea, is narrow-pyramidal and has slender branchlets. C. o. compacta is a broad, compact dwarf kind, slow-growing and adapted for rock gardens. C. o. crippsii is especially beautiful, narrow-pyramidal, and has golden-yellow foliage. C. o. erecta is a narrow variety with upright branches and bright green foliage. C. o. filicoides has ferny sprays of short branchlets of nearly equal length and forms an irregular, slow-growing, dense bush. C. o. gracilis is bright green and pyramidal, its branches slightly pendulous at their tips. C. o. g. aurea is similar to the last, but its young foliage is golden. C. o. lycopodioides, a dwarf kind with irregularly-spreading branches, has its branchlets crowded and not in one plane, and its foliage is mosslike. C. o. l. aurea is similar to the last, but has golden foliage. C. o. magnifica is a vigorous kind with bright green foliage. C. o. nana is very slow growing, dark, dull green, and makes a low, flat-topped specimen suitable for rock gardens. C. o. n. aurea has golden foliage, but otherwise is similar to the last. C. o. n. gracilis resembles C. o. nana, but has brighter green foliage. C. o. pygmaea is extremely dwarf, has prostrate or nearly prostrate branches, short, crowded branchlets, and bluish-green foliage. C. o. sanderi (syn. C. o. ericoides) is very dwarf, pyramidal or nearly globose, and has bluish-gray juvenile-type foliage that resembles that of certain junipers. C. o. tetragona, rare in cultivation, is a dwarf with erect quadrangular branchlets. C. o. t. aurea, similar to the last, but with portions of its foliage yellow, is much commoner.

Chamaecyparis obtusa

Chamaecyparis obtusa (foliage)

Chamaecyparis obtusa crippsii

Chamaecyparis obtusa gracilis

Chamaecyparis obtusa nana

Chamaecyparis obtusa juniperoides

Other varieties of *C. obtusa* include very distinctive *C. o. coralliformis*, which forms a low shrub with curiously contorted and twisted branches and branchlets, fancifully suggestive of coral, instead of normal sprays of foliage. A compact, dark green pygmy, *C. o. intermedia* is considered by the English authority Humphrey J. Welch to be the smallest form practicable for the open garden, except in sheltered districts. Another extremely dwarf variety, *C. o. juniperoides* has fan-shaped sprays of dark green foliage. Prostrate *C. o. repens* has green foliage without a trace of bronze. Dwarf, conical *C. o. sanderi*, often grown as *C. o. ericoides*, has bluish-green leaves of juvenile type. It is considered by Humphrey J. Welch to be a variety of *Thuja orientalis*.

Chamaecyparis obtusa repens

Called mourning-cypress, although mourning false-cypress would be a more appropriate name, **C. funebris** (syn. *Chamaecyparis funebris*) is a lovely native of China. From 60 to 70 feet in height, it has pendulous branches and spreading, flat sprays of gray-green foliage. In many ways this sort is intermediate between *Chamaecyparis* and *Cupressus*. In its homeland it is much favored for planting near temples, monasteries, and tombs.

Garden and Landscape Uses. False-cypresses are highly regarded for landscape adornment. It is practical to use many kinds in climates much too cold for true cypresses (*Cupressus*), and they serve very much the same purposes. They offer a wide variety of colors and forms from which to select, and grow without difficulty if they are given a fair chance. They need a reasonably fertile, moist soil, sufficiently well drained to prevent water accumulating in it (*C. thyoides* will stand wet soil, but it is the least attractive species), and good light. False-cypresses may be grouped attractively or planted as individual specimens as accents. The more vigorous kinds make splendid screens and hedges. The dwarf ones are appropriate for rock gardens. They may also be used, especially the Hinoki false-cypress and its varieties in Japanese gardens, and the less vigorous growers lend themselves to training as bonsai.

Cultivation. No special problems attend the cultivation of false-cypresses. They are among the easiest of evergreens to transplant successfully, and they need little or no pruning beyond, possibly, an occasional shearing of horticultural varieties that tend to grow loosely or must be restricted to size, and the regular shearing or

Chamaecyparis obtusa coralliformis

Chamaecyparis obtusa intermedia

knife-pruning of hedges. The best time to prune is just before new growth begins in spring, but pruning can also be successfully carried out in late fall. Shearing should not be so severe that old wood is cut into. All that should be shortened, almost to their bases if necessary, are shoots of the last summer's growth. A permanently maintained mulch of compost, peat moss, or other suitable organic material is beneficial, and for specimens in poor soil or that have exhausted the earth of most available nutrients, it is very helpful to apply a dressing of complete fertilizer each spring. Periodic deep watering in dry weather promotes vigor and health. Such soakings are especially likely to be needed by specimens planted near buildings where free access of rain to the soil is impeded by overhanging roofs or by walls and where the soil may dry excessively because of free subsurface drainage. Inexperienced gardeners are often unduly concerned because in late summer or fall the older branchlets, those farthest away from the tips of the branches, turn brown and are quite conspicuous as they hang for a period on the branches before they drop. This is no cause for worry. It is quite natural. But if younger branchlets near the branch tips shrivel and die it is a matter for concern. It may be caused by root damage during transplanting, excessive dryness, infestations with red spider mites or scale, or other unfavorable circumstances. Propagation of the natural species of *Chamaecyparis* is best done by seeds sown in sandy peaty soil in a cold frame or protected bed outdoors in fall or spring or in a cool greenhouse in winter or spring. Varieties, and species if seeds are not available, may be rooted from cuttings taken in late summer or fall and inserted in a propagating bench under mist or in a very humid cold frame or cool greenhouse. They may also be increased by veneer grafting onto seedling understocks in a greenhouse in late winter or spring.

Diseases and Pests. Juniper twig blight sometimes infects false-cypresses and causes dying back of branchlets. Its control is as for junipers. Fungus-caused witches' brooms develop on some kinds. The alternate hosts of the fungus are *Myrica* and *Comptonia*. Removal of the brooms and segregation of the alternate hosts are the only recommended control measures. Insect pests most likely to infest false-cypresses are scale insects, red spider mites, and bag-worms. For further information see Conifers.

CHAMAEDAPHNE (Chamae-dáphne)—Leather Leaf. One loosely-branched evergreen shrub rather similar to *Leucothoe* is the only species of *Chamaedaphne*. It belongs in the heath family ERICACEAE, and is indigenous in bogs and wet soils throughout the cooler parts of North

America, Europe, and Asia. From *Leucothoe* it is easily distinguished by the dense coating of brown scales on the shoots and undersides of the leaves. Its name is from the Greek *chamai*, on the ground, and *daphne*, the ancient Greek name of the laurel (*Laurus nobilis*).

Leather leaf (*C. calyculata*), up to 4½ feet tall, has alternate, short-stalked, sometimes minutely-toothed, oblong, elliptic, or rarely obovate, bluntish leaves up to 2 inches long. The short-stalked flowers are in terminal, one-sided, leafy-bracted racemes 2 to 5 inches long. They have deeply-five-cleft calyxes and white, nearly cylindrical corollas about ¼ inch long, slightly constricted at their throats, and with five short, recurved lobes. There are ten included, flat-stalked stamens and a long style. The fruits are small, flattened-globular, five-angled capsules. Variety *C. c. angustifolia* has narrower, wavy-edged leaves. Variety *C. c. nana* is up to 1 foot tall.

Chamaedaphne calyculata

Garden and Landscape Uses. The leather leaf is hardy and pretty and is esteemed for its evergreen foliage and its sprays of blooms in evidence early in spring. It is well deserving of planting in borders, to "front down" other evergreens, and in rock gardens and similar locations. For its success it needs moist or even wet soil with an acid reaction and dappled or part-day shade.

Cultivation. No great problems attend the cultivation of this species provided soil and site reasonably suit its needs. Routine care is minimal. Propagation is by seeds sown very shallowly or merely pressed into the surface of sandy peaty soil in pans (shallow pots) or flats in a greenhouse or cold frame. The soil is kept uniformly moist. Cuttings taken in late summer can be rooted in a greenhouse propagating bench. Layering and suckers provide other methods of increase.

CHAMAEDOREA (Chamae-dòrea). A hundred species of tropical American palms with solitary or clustered, ringed or jointed, reedy stems belong here. Chiefly they are shade-loving forest plants, usually unisexual, but some kinds bisexual. Most are low. A few are climbers. The group is not well understood botanically. Several kinds are cultivated, but their correct identifications are not always easy because often plants of one sex only are available for study and usually fruits are absent. Also, there are some hybrids. Belonging in the palm family PALMAE, the genus takes its name from the Greek *chamai*, dwarf, and *dorea*, a gift. This refers to the fruits being at convenient heights for picking.

Most species of *Chamaedorea* have leaves that are pinnate or sometimes pinnatisect (divided in pinnate-fashion, but not cleft deeply enough to form separate leaflets). More rarely they are undivided with their veins arranged pinnately. The branched or branchless flower clusters arise from among or below the foliage. They have usually prominent spathes. The male blooms have six stamens. The pea-sized fruits are red, orange, purplish, or black. The juice of ripe fruits irritates the skin. Where these palms are native the young flower clusters of some kinds, notably *C. elegans*, *C. sartori*, and *C. tepejilote*, are cooked and eaten as are the fruits of *C. elegans* and perhaps others. The irritant principle is destroyed in cooking.

A clump-forming kind, *C. erumpens*, native to Guatemala and Honduras, has bamboo-like stems up to 10 to 12 feet tall and rather short, pinnate leaves with five to fifteen pairs of leaflets. In plants with the fewest pairs of leaflets the terminal ones are much broader than the others. The flowers are yellow, the fruits black. This kind produces suckers freely. Native of Guatemala, *C. graminifolia* is a clump-former with erect stems and pinnate leaves of about forty leaflets 1 foot long or longer and set closely together. Its male flower clusters are erect. One of the most orna-

mental sorts, Mexican *C. seifrizii* is admirable for planting outdoors and in containers. It forms dense clusters of slender canes. Another multiple-stemmed kind, *C. stolonifera*, of southern Mexico, is about 3 feet tall and has globose fruits. A popular solitary-stemmed kind is the parlor palm (*C. elegans* syns. *Collinia elegans* and *Neanthe bella*). Native to Mexico and Guatemala, it attains an eventual height of 3 to 6 feet, but blooms when very much smaller. Its leaves have twelve pairs or more of slender leaflets. The flower clusters are 3 feet or less, often very much less, long and have pale yellow flowers. The fruits are small and globular.

Other kinds likely to be cultivated include these: *C. cataractarum*, of Mexico, sends up several stems from a much-branching horizontal rhizome. Its leaves have many slender leaflets. *C. ernesti-angustii*, native from Mexico to Honduras, has a solitary stem with aerial brace roots at its base and leaves deeply-cleft at their apexes and undivided or with three to five leaflets on each side of the midrib. The flowers are orange, the females in branchless, erect spikes, the males in freely-branched clusters. *C. elatior*, of Mexico, has slender stems up to 15 feet tall, and

Chamaedorea elegans

Chamaedorea elegans, with flowers

Chamaedorea erumpens

about 1 inch thick, that need support such as may be offered by a shrub or tree among the branches of which they can intrude. This has pinnate leaves up to 7 feet long with many long-pointed, lanceolate leaflets. The flowers are in more or less drooping, branched clusters. **C. martiana** is a low, creeping clump-forming kind from Mexico. Its pinnate leaves have many drooping leaflets. **C. metallica** is Mexican. It is low and solitary-trunked and has normally undivided leaves cupped and deeply-cleft at their ends. Very rarely the leaves are divided into separate leaflets, they are dark green with a metallic sheen. The branchless female flower spikes spread upward at an angle of about forty-five degrees. **C. sartori,** native from Mexico to Honduras, has a single trunk and 3-foot-long leaves with usually about nine leaflets on either side of the midrib, but sometimes with fewer and the terminal portion then much broader than the other leaflets. The female inflorescence has more than two branches. **C. tenella,** of Mexico, has a solitary, slender, erect trunk up to 2 feet long

Chamaedorea tenella, young specimen

or perhaps longer. Its dark bluish-green leaves, except for a deep, wide cleft at the apex, are undivided. About 1 foot long and 6 inches wide, they have a prominent midrib from each side of which angle a dozen veins. The upper leaf margins are shallowly-toothed. The black fruits are ⅓ inch in diameter. **C. tepejilote** (syn. *C. wendlandiana,* of Mexico to Central America, has solitary stems up to 12 feet tall and pinnate leaves up to 4 feet long with twenty to thirty leaflets. The flower clusters are branched, those of male flowers long and drooping.

Garden and Landscape Uses. Chamaedoreas are useful for planting outdoors in shaded locations in warm regions where little or no frost is experienced, such as Hawaii, southern California, and southern Florida, and are excellent for pot and tub

cultivation. They are good houseplants and provided the atmosphere is not excessively dry tolerate low light intensities and other unfavorable conditions to which houseplants are subjected better than many other kinds. Outdoors they can be used effectively under trees and on the shaded sides of buildings and the clump-formers make good informal hedges and screens. They grow in any garden soil that is reasonably moist, but thrive best in one that contains an abundance of leaf mold, peat moss, or other decayed organic material.

Cultivation. These are among the easiest palms to grow. They come readily from fresh seeds sown in sandy, peaty soil in a temperature of 75 to 90°F, and the suckering, clump-forming kinds can be increased by careful division in late winter or spring just before new growth begins. At least some sorts, including *C. elegans* and probably others that produce adventitious roots from their stems such as *C. tepejilote,* can be increased by air layering.

In greenhouses chamaedoreas need a minimum winter night temperature of 55 to 60°F and night temperatures at other seasons several degrees higher. Day temperatures should be five to ten degrees higher than those maintained at night. A humid atmosphere and shade from strong sun are necessary, as are coarse, fertile, porous soil and well-drained containers. Water is applied freely from spring through fall, more moderately in winter, without allowing the soil to be ever really dry. After the plants are well rooted they benefit from biweekly applications of dilute liquid fertilizer from spring through fall. For more information see Palms.

CHAMAELAUCIUM. See Chamelaucium.

CHAMAELIRIUM (Chamae-lírium)—Blazing Star or Fairy Wand or Devil's Bit. To the lily family LILIACEAE belongs the only member of this genus, a native of bogs and moist woodlands from Massachusetts to Ontario, Michigan, Florida, Mississippi, and Arkansas. Its name comes from the Greek *chamai,* on the ground, and *lirion,* a lily. The implication is a little misleading for, far from ground-hugging, the blazing star, fairy wand, or Devil's bit (*Chamaelirium luteum*) has erect stems, 1 foot to 3 feet tall, with feathery masses of small flowers in early summer. True, its chief foliage is basal. The lower leaves form rosettes and are 3½ to 6 inches long and spoon-shaped or obovate. The stem leaves are smaller; they diminish in size and become narrower from the base upward.

The blazing star (**C. luteum**) is a hardy, deciduous perennial with unisexual flowers, the sexes on separate plants. The flowers, many together in spikelike racemes, are ¼ inch wide. The spikes of male plants are 2 to 5 inches long and approximately ½ inch wide, those of females proportion-

ately slenderer and up to 1 foot long. Each flower has six narrow, white, perianth parts commonly called petals, but correctly tepals. The males have six stamens with slender, flattened stalks (filaments) and yellow anthers responsible for the slightly yellowish appearance of the male spikes. The females have six rudimentary, nonfunctional stamens and three stigmas. The fruits are roundish or egg-shaped capsules.

Garden and Landscape Uses and Cultivation. For shaded locations in moderately acid, moist, fertile soils this attractive American wild flower is well suited. It transplants readily and is easily raised from seeds sown as soon as they are ripe or kept cool and moist and sown in a cold frame in fall. They germinate in sandy peaty soil kept evenly moist. Established plants need little care. An annual mulch of rich compost placed around them in fall or early spring is beneficial. This is a good plant for naturalizing in semiwild places and for including in native plant gardens.

CHAMAEMELUM (Chamae-mèlum)—Camomile. Three species of herbaceous perennials constitute *Chamaemelum,* a genus of the daisy family COMPOSITAE, formerly included in *Anthemis.* The name, alluding to a supposed melon-like fragrance of the foliage of some kind, derives from the Greek *chamai,* low or on the ground, and *melon,* a melon.

Natives of Europe and the Mediterranean region, these plants have leafy stems with solitary flower heads with both disk and ray florets terminating the branches. The leaves are alternate and once to thrice pinnately-cleft. The fruits are seedlike achenes, laterally flattened and without ribs.

Camomile or chamomile (**C. nobile** syn. *Anthemis nobilis*), the flower heads of which are used medicinally, is a procumbent-stemmed, more or less spreading or creeping, freely-branched hardy perennial, native of western Europe, North Africa, and

Chamaemelum nobile

the Azores, and naturalized to some extent in North America. From 6 inches to 1 foot tall, it has pleasantly-scented, finely-twice-pinnate or pinnately-lobed leaves, the final segments linear or awl-shaped. Its white-rayed flower heads are ¾ to 1 inch in diameter. The bracts of the involucre are blunt. The achenes are top-shaped and bluntly three-angled. Double flowers, or more correctly flower heads, the disk florets replaced by ray florets, are borne by *C. n. plena*.

Garden Uses and Cultivation. The sorts described are chiefly planted in gardens devoted to herbs and medicinal plants. They grow readily in well-drained soil in sunny locations and are easily propagated by division in spring or early fall and the typical species, but not the variety, by seeds sown in spring.

CHAMAERANTHEMUM (Chamaer-ánthe-mum). The name of this genus of eight species of low, evergreen, herbaceous plants of the acanthus family ACANTHACEAE is sometimes spelled *Chameranthemum*. The name comes from the Greek *chamai*, dwarf, and *anthos*, a flower.

Entirely tropical American, *Chamaeranthemum* differs from related *Eranthemum* in its blooms having four instead of two stamens. Its leaves are opposite and undivided. The asymmetrical flowers have five-parted calyxes and tubular corollas with five spreading lobes (petals). The fruits are capsules.

Native to Brazil, *C. gaudichaudii* has creeping stems, and short-stalked, oblong-elliptic leaves with blades 2 to 4 inches long and silvery-gray toward their centers. The small lavender blooms, each with a central white star, are in erect, terminal racemes 3 to 6 inches long. Also, from Brazil, *C. venosum* is a low plant with firm, short-stalked, broad-ovate leaves, clearly veined with silver, 2 to 2½ inches long. Believed to be a variant of the last, *C. v.* 'India Plant', introduced from Brazil, has leathery, broad-elliptic, dark gray-green leaves about 5 inches in length and charmingly veined with silver. The flowers are lavender with white centers. The plant grown as *C. igneum* is *Xantheranthemum igneum*.

Garden and Landscape Uses and Cultivation. Attractive foliage plants, the sorts described here are suitable for rock gardens and similar locations in the humid tropics, and for tropical greenhouses and terrariums. They are easily increased by division and cuttings and grow enthusiastically in porous, well-drained soil that contains an abundance of leaf mold, peat moss, or other similar organic material. Shade from strong sun is needed, and indoor temperatures in the 60 to 70°F range at night, and warmer by day. The soil should be always reasonably moist, but not constantly soggy. High atmospheric humidity is desirable. Pans (shallow pots) and similar containers are suitable, or they may be grown in beds of soil in terrariums and greenhouses.

CHAMAEROPS (Chamaè-rops)—European Fan Palm. This, the only genus of palms native in Europe, consists of one, or according to some authorities, two species. It is a member of the palm family PALMAE. The name from the Greek *chamai*, dwarf, and *rhops*, a bush, alludes to its low height as compared with that of many tropical palms.

A fan-leaved species native to the Mediterranean region, *Chamaerops humilis* usually forms a clump or cluster of several trunks, but sometimes is single-stemmed. Typically, its trunks are not more than 5 feet in height and often are lower, but sometimes they are three to five times as

Chamaerops humilis, as a clump

tall. They are usually clothed with dark brown or blackish fibers and old leaf bases. The rigid, glaucous-blue or green leaves, mostly 2 to 3 feet in diameter, are deeply-divided into numerous, one-veined, bifid segments that stand out stiffly in the plane of the body of leaf and do not droop as do those of *Trachycarpus fortunei*. The long, slender leafstalks are furnished with many long spines angled toward the leaf blade. The short, erect, dense clusters of flowers are axillary among the leaves. They consist of yellow blooms with the sexes usually on separate plants. The fleshy, yellow to brown fruits resemble dates in size and shape. This palm exhibits considerable variation. A glaucous-blue-leaved form is sometimes distinguished as *C. argentea*, a name without botanical standing.

Garden and Landscape Uses. The European fan palm, one of the hardiest of palms, is commonly grown for ornament, especially in its lower-growing, clump-making forms, in subtropical and tropical countries as well as in more temperate regions. It is hardy in sheltered locations as far north as Victoria, British Columbia, Norfolk, Virginia, and Edinburgh, Scotland. Comparatively slow-growing, it is well suited for sites where space is limited. It is effective as a lawn specimen and for plantings in which it is combined with other plants. Often it is used to good advantage with architectural features such as masonry steps and in foundation plantings. An interesting use for it is as a hedge.

Chamaerops humilis, a single-trunked specimen

Plants set 4 to 6 feet apart and kept fertilized and watered to encourage maximum growth, fill out fairly quickly and eventually make an impenetrable evergreen barrier 8 to 10 feet tall.

Cultivation. The European fan palm thrives in sun or part-shade in any ordinary garden soil. It withstands dry conditions quite well, but grows faster where moisture is not too restricted. Although it is sometimes possible to secure increase by dividing old specimens, this is not a reliable method, and it is better to rely upon fresh seeds sown in sandy, peaty soil at a temperature of 70 to 80°F.

Chamaerops humilis, a tub specimen

The European fan palm is excellent for growing in large tubs for the decoration of terraces and other outdoor areas. The containers must be well drained, and the soil used should be coarse, porous, and fertile. Potting or tubbing is done in early spring. Specimens in large tubs require retubbing at intervals of several years only. In the intervening years they should be topdressed each spring and fed liberally with dilute liquid fertilizer at weekly or two-weekly intervals from spring to fall. At no time should the soil be permitted to dry out, but in winter it may become much drier between waterings than at other seasons. Where winters are too cold for the plants to live outdoors, they may be stored in a light cellar, garage, or other building where the temperature ranges between 35 and 50°F through that season. This palm is also excellent for planting in conservatories. It does well where the atmosphere is fairly moist and the night temperature from fall through spring is held at 40 to 50°F with a rise of about ten degrees permitted during the day. For further information see Palms.

CHAMBEYRONIA (Chambeyrò-nia). Two handsome species of New Caledonia constitute *Chambeyronia* of the palm family PALMAE. The generic name honors A. M. Chambeyron, a French ship captain of the nineteenth century. These palms are sometimes cultivated under the names *Kentia* and *Kentiopsis,* but they differ in technical details from those genera. Chambeyronias

Chamelaucium uncinatum

are tall, feather-leaved, unisexual trees little known in cultivation except occasionally as greenhouse plants. For this latter purpose they are especially attractive because their young leaves are red.

Up to 60 feet or more in height, *C. macrocarpa* has large leaves with numerous leaflets. Its flower clusters arise from below the crown of foliage. Their blooms normally are in groups of three. The ovoid-elliptic fruits, about 1½ inches long, are seated in leafy cuplike receptacles. From the last, *C. hookeri* differs in having long-ellipsoid fruits about 2 inches in length, larger male flowers, and leaves that are paler on their undersides.

Garden and Landscape Uses and Cultivation. Little experience with the cultivation of these palms is recorded. They may be expected to be hardy outdoors in southern Florida and Hawaii. Essentially they are plants for collectors of the unusual and should be afforded the same general culture as *Howeia.* For further information see Palms.

CHAMELAUCIUM (Cham-elàucium) — Geraldton Wax Flower. The Geraldton wax flower is the best known of this Australian genus of a dozen species of evergreen, heathlike shrubs of the myrtle family MYRTACEAE. By no means obvious or apt in its application to all kinds, the name *Chamelaucium* comes from the Greek *chamai,* low, and *leukos,* white.

Chamelauciums have usually opposite, stalkless leaves with the flowers originating in the axils of the upper ones or on short terminal racemes. The bell- to top-shaped calyxes have five spreading lobes.

There are five petals and ten stamens, which alternate with the same number of staminodes (nonfunctional stamens). The fruits are capsules.

Geraldton wax flower (*C. uncinatum*) is a broad, slender-stemmed shrub up to 12 feet tall or sometimes taller. Its white-glandular leaves, ¾ inch to 1¼ inches long, are triangular in section and hooked at their tips. Profusely produced in few-flowered clusters from the leaf axils, the showy blooms form large, long-stemmed, airy trusses, well suited for cutting for use in flower arrangements. In color they vary considerably, some plants having blooms of rich rosy-purple, others of pink or lavender-pink shades, and some with white flowers. The flowers have broad, smooth-margined sepals, and nearly circular petals. They are about ½ inch in diameter. This species is often misidentified as *C. ciliatum,* which kind differs from the Geraldton wax flower in being 3 to 4 feet tall and having blunt, cylindrical, or three-angled leaves about ¼ inch long and pink or white, nearly tubular flowers with fringed calyx lobes from the axils of the upper leaves. The blooms are ⅛ to ¼ inch in diameter.

Garden and Landscape Uses and Cultivation. Chamelauciums have somewhat the appearance of loose-growing heaths, and are well adapted for associating with that popular group of plants on sunny banks and in other well-drained locations in California and elsewhere where summers are warm and dry, and winter frost is absent or light and rare. The Geraldton wax flower is well worthy of inclusion in gardens devoted to raising flowers for cutting. Its blooms last well. Chamelauciums are readily raised from seeds and cuttings. Only those seedlings of the Geraldton wax flower that show superior flower color should be chosen for permanent plantings. Seedlings usually first bloom when about two years old. Routine care of these shrubs is not exacting. Pruning, either by cutting flowering branches for flower arrangements or by shortening the branches as soon as blooming is over, keeps the plants shapely and reasonably compact. The maintenance of a mulch around the plants is beneficial as is occasional fertilization. Watering must be done with care. Occasional deep applications during dry summer periods are helpful, but too much is likely to cause trouble.

CHAMISO or GREASEWOOD is *Adenostoma fasciculatum.*

CHAMOMILE. See Camomile.

CHANAR. See Gourliea.

CHAPARRAL-BROOM is *Baccharis pilularis consanguinea.*

CHAPARRAL-PEA is *Pickeringia montana.*

CHAPTALIA (Chap-tàlia). Natives of the warmer parts of the Americas, there are twenty-five species of *Chaptalia* of the daisy family COMPOSITAE. Their name honors the French chemist Comte Jean Antonin Claude Chaptal de Chanteloup, who died in 1832.

Chaptalias are stemless, fibrous-rooted herbaceous perennials with basal rosettes of obovate or elliptic to oblanceolate leaves that narrow gradually to their bases, are covered on their undersides with whitish hairs, and are toothless, toothed, or shallowly-pinnately-lobed. The flower heads are stalked, solitary, and mostly nodding. They have centers of male or bisexual disk-type florets encircled by female, petal-like ray florets. The fruits are seedlike achenes.

Native from North Carolina to Florida and Texas, **C. tomentosa** is 6 inches to 1 foot tall. It has thickish, elliptic to oblanceolate, remotely-toothed or toothless leaves 3 to 4 inches long and ½ inch to a little over 1 inch wide. Their undersides are densely-white-hairy. Their top surfaces, except when very young, are hairless. The slender, bractless flower stalks carry heads, about ¾ inch across, that nod in bud and after they are past their best, but at full maturity look upward. The about twenty, raylike outer florets are white with purplish undersides.

Garden and Landscape Uses and Cultivation. The species described is sometimes planted in its home territory and elsewhere. Its exact hardiness is not determined, but it undoubtedly can be grown north of its native range. It can be used in flower gardens, rock gardens, and informal areas and is satisfied with porous, fairly fertile, somewhat acid soil and open situations. Propagation is by seed and division.

CHARACEAE—Stonewort Family. This family of algae consists of about 215 species of nonflowering green plants botanists consider to have characteristics intermediate between those of algae and mosses. They have branched stemlike parts with whorls (circles) of slender, branched, leaflike processes from the nodes. The plants are sometimes crusted with lime (calcium carbonate). The only genus cultivated is *Nitella.*

CHARAS. See Cannabis.

CHARCOAL. The best type of charcoal for horticultural purposes is produced from hard woods by charring or by partially burning them under conditions that limit access of air. Such charcoal, broken to peanut size or smaller, is an excellent additive to potting soils containing abundant organic matter such as leaf mold or peat moss. Charcoal has the ability to absorb gases and some other products of decay. In gardeners' language it helps to keep the soil sweet.

CHARD or SWISS CHARD. One of the easiest vegetables to grow, this is a variety, *Beta vulgaris cicla,* of beet grown for its

Swiss chard

leaves. These are cooked and eaten like spinach. Unlike beets, chard does not have tuberous roots. It belongs in the goosefoot family CHENOPODIACEAE and succeeds in cool and warm climates, producing continuously from early summer to fall and in mild climates in winter and spring.

Not very particular about soil, chard thrives best in well-drained, fertile, slightly acid to neutral earth that does not lack for moisture. The best results are had in full sun. The preparation of the soil is as for beets.

Sow as soon as the ground can be made ready in spring, and again in July, in rows 1½ to 2 feet apart. Cover the seeds to a depth of ½ to 1 inch. Thin the seedlings first to about 4 inches apart and when they begin to crowd again to 8 or 9 inches. The later thinnings can be used in the kitchen. Subsequent care involves shallow surface cultivation, or mulching to control weeds, and generous watering during dry spells. In early August cut all but the tiniest leaves off the spring-sown crop and apply a light dressing of nitrate of soda or urea to invigorate the plants and stimulate new growth.

Harvesting is a continuous process. Even if not wanted for the table do not allow the leaves to become full sized. Keep them picked to encourage the production of young, tender ones. These are best for kitchen use when partially grown. Favorite varieties are 'Fordhook Giant', with thick white stalks and crinkled leaves, 'Rhubarb', the stalks of which have brilliant red stalks and midribs, and 'Lucellus'. Chard is subject to the same pests and diseases as beets.

CHARIEIS (Chár-ieis). Two species of annuals endemic to South Africa are the only representatives of *Charieis* of the daisy family COMPOSITAE. The name comes from the Greek *charieis,* elegant, and refers to the appearance of the plants. This genus has undivided, opposite or alternate leaves and solitary, daisy-like flower heads with a central eye of nearly always bisexual disk florets, surrounded by a single row of strap-shaped, female ray florets. The seedlike fruits are achenes.

Fairly commonly cultivated, **C. heterophylla** (syn. *Kaulfussia amelloides*) is attractive, bushy, softly-hairy, and up to 1 foot

Charieis heterophylla

in height. It has gray-green, oblong leaves up to 2½ inches long, on the lower parts of the plant opposite, alternate above. The long-stalked, upturned flower heads have blue rays and yellow or violet-blue centers. Faintly fragrant, about 1 inch in diameter, they open by day and close at night. Variety *C. h. atroviolacea* has very dark blue blooms. Those of *C. h. kermesiana* are deep violet-red.

Garden Uses and Cultivation. In places where summers are not excessively hot this is charming for edging paths and flower beds and for porch and window boxes. It is also pleasing in pots in greenhouses. For outdoor display sow seeds in early spring where the plants are to bloom in a sunny location in ordinary, fertile garden earth. Cover with soil to a depth of about ¼ inch. Thin out the young plants before they crowd so that they stand about 6 inches apart. No care other than the removal of weeds, and watering before the ground becomes excessively dry, is needed. Blooming begins in ten to twelve weeks from seed sowing and continues for a month to six weeks or more.

For winter and spring flowering in greenhouses sow seeds from September to February in pots in a temperature of 55 to 60°F. When the young plants are big enough to handle transplant them into well-drained, 6-inch pans (shallow pots) or pots, five plants in each. Use porous, fertile soil. Water sparingly at first, more copiously as roots penetrate the earth. Grow in full sun in a night temperature of 50 to 55°F with a rise of five to ten degrees in the day permitted. Ventilate freely whenever the weather is favorable.

CHARITY is *Polemonium caeruleum*.

CHARLESWORTHARA. This is the name of trigeneric orchid hybrids the parents of which are *Cochlioda*, *Miltonia*, and *Oncidium*.

CHARLOCK is *Sinapis arvensis*.

CHASMANTHE (Chas-mànthe). This group of seven species of African plants related to *Gladiolus* was formerly included in the once ambiguous genus *Antholyza*. It belongs in the iris family IRIDACEAE and contains several interesting summer-blooming plants of which only one is much cultivated. The name comes from the Greek *chasme*, gaping, and *anthe*, a flower, and alludes to the open mouth of the bloom.

Chasmanthes have underground storage organs that look like and are often called bulbs. They are corms. They differ in not being formed of concentric scales like an onion or lily bulb, but being solid throughout. The usually branchless or sometimes few-branched stems arise from the corms. The leaves are erect and sword-shaped, the basal ones in fans. Asymmetrical and tubular-funnel-shaped, the flowers are in spikes that, unlike those of *Antholyza*, are terminal on the stems. The blooms differ from those of *Gladiolus* in having their curved perianth tubes suddenly constricted at or below their middles into slender basal portions. There are six perianth lobes (petals) of which the upper hooded one is spatula-shaped and considerably longer than the others, which spread

and are of equal length. The three stamens nestle under the projecting long petal and are about as long. The style is three-branched. The fruits are capsules.

Most cultivated is *C. aethiopica* (syn. *Antholyza aethiopica*), of South Africa. This

Chasmanthe aethiopica

has comparatively large, more or less spherical corms and is 3 to 4 feet tall. Its rigid leaves are 1 foot to 2 feet long by about 1 inch wide and have conspicuous midribs. The slender, wiry stems, commonly branched, carry erect, dense spikes of many red and yellow flowers with almost 2-inch-long perianth tubes. The upper petal, 1 inch long and curved, is much bigger than the other, spreading ones and is just exceeded by the stamens. Toward its tip the style forks into three threadlike branches. Very similar to *C. aethiopica* but only about one-half as tall and with flowers that have the lower parts of their perianth tubes and lower petals yellow or greenish, is *C. bicolor* (syn. *Antholyza bicolor*), also of South Africa. Another not very different kind is *C. floribunda.* The chief differences between it and *C. aethiopica* are that its leaves are wider, and the broad upper portions of its corolla tubes slope gradually rather than being constricted suddenly, to the slender basal parts. This species also is South African.

Often misnamed *C. intermedia* in gardens, *C. caffra* (syn. *Antholyza caffra*) has usually branchless stems up to 2 feet tall.

Its leaves are linear-lanceolate and mostly not over 1 foot long. The brilliant red blooms are two-ranked and up to twenty together in loose spikes. They have curved, funnel-shaped perianth tubes, a little more than 1 inch long, that broaden gradually rather than suddenly. The upper petal is approximately 1 inch long, about equal to the stamens. The lower petal is strongly deflexed, a characteristic that distinguishes *C. caffra* from true *C. intermedia*. Both species are South African.

Garden and Landscape Uses. For summer flower beds and borders and for cutting, chasmanthes are decidedly useful. They afford attractive color and a grace that complements many other plants as does their swordlike foliage. In cool greenhouses these plants can be had in bloom in late winter and spring.

Cultivation. Full sun and agreeable, well-drained soil such as suits the majority of vegetables and flower garden plants is satisfying to chasmanthes. The corms are planted 3 to 4 inches deep, spaced 3 to 4 inches apart, as soon as the soil becomes warm in spring. No staking or other special attention is ordinarily needed, but in dry weather it is important that they do not dry out. When flowers are cut, as much as practicable of the foliage should be left to build strength in the corms for the following year's bloom. After the first frosts of fall, or before if the foliage has died naturally, the corms are dug, cleaned, and stored over winter in a dry place at a temperature of 40 to 50°F. In areas where the soil does not freeze to a depth of more than about 1 inch, or where it can be prevented from freezing by covering, chasmanthes may remain in the ground all winter. Propagation is by natural multiplication of the corms and by seeds.

For early blooming in greenhouses corms may be planted shallowly, 2 to 3 inches apart, in 7- or 8-inch pots in January or February. They need fertile, porous soil. Until good root development is established water moderately, afterwards copiously. When the containers become filled with roots applications of dilute liquid fertilizer aid growth. A night temperature of 45 to 50°F with a daytime rise of five to ten, or on very sunny days fifteen degrees gives good results. Full sun is needed. After blooming every effort should be made to retain the foliage as long as possible; only after it begins to die naturally should water be gradually withheld. After the foliage is completely brown the corms are taken from the soil and stored in a cool, dry, shady place until the next planting season.

CHASMANTHIUM (Chas-mánthium). Wild-Oats or Spike Grass. One of the five species of *Chasmanthium*, a North American genus of the grass family GRAMINEAE, is cultivated as an ornamental. The name, alluding to the form of the flowers, is derived from the Greek *chasme,* gaping, and *anthemon,* a flower.

Chasmanthiums are perennial grasses with flat leaves and panicles of flat spikelets of two to many florets.

Wild-oats or spike grass (*C. latifolium* syn. *Uniola latifolia*) is a native of moist woodlands from New Jersey to Ohio, Illinois, Indiana, Florida, and Texas. It grows from 2 to 5 feet tall, has clusters or usually branchless stems, and produces an abundance of loose flower panicles. The leaves, ½ to 1 inch wide, are up to 9 inches long. Except for their roughened margins they are hairless. The nodding flat green spikelets of flowers are pointed-elliptic to pointed-ovate.

Garden and Landscape Uses. This quite elegant grass is attractive for grouping in flower borders and for colonizing in light woodlands and other informal areas. Its flowers are quite charming when cut, and used fresh or dried.

Cultivation. If the soil is sufficiently moist this species prospers as well in sun as in part-shade, but in dry sunny locations it becomes stunted and unhappy-looking. A fertile, fairly deep soil gives the best results. It is easily raised from seeds, and with care, plants can be successfully divided in spring or early fall. Established clumps are benefited by a spring application of a complete fertilizer. Watering during dry weather promotes well-being.

CHASMATOPHYLLUM (Chasmáto-phyllum). This genus of six species belongs to the *Mesembryanthemum* relationship of the carpetweed family AIZOACEAE. Its name from the Greek *chasma,* an open mouth, and *phyllon,* a leaf, alludes to the jawlike appearance of the leaves.

Endemic to arid parts of South Africa, *Chasmatophyllum* consists of subshrubby, succulent plants with clustered, short or long, branched stems, at first erect, later prostrate. The leaves, in section semicircular or keeled, are in pairs alternating at right angles to each other, with or without portions of stem visible between neighbor pairs. Toward their tips, on the leaf margins and underneath, there are frequently a few blunt teeth. The under surfaces, and toward the leaf apexes, the upper surfaces, are covered with tiny, whitish protuberances (tubercles). Short-stalked and terminal, the yellow flowers are daisy-like in appearance, but not in structure. Unlike daisies, each flower head of which consists of numerous florets, the flowers of *Chasmatophyllum* are single blooms. The fruits are capsules.

Its matted stems crowded with foliage, *C. musculinum* (syns. *Stomatium musculinum, Mesembryanthemum musculinum*) is an attractive kind. Curved slightly inward, and with slightly hollowed upper surfaces, the usually blunt-toothed, ½- to ¾-inch-long leaves are semicylindrical to approximately triangular in section. Their upper surfaces are gray-green and have many minute, translucent dots. A little more than ½ inch wide, the flowers, with petals somewhat reddish toward the apexes of their undersides, open in the afternoon. As known in cultivation *C. nelii* has stems at first erect, later prostrate, and toothless, spatula-shaped leaves, the pairs spaced about 1 inch apart, up to ½ inch long. Grayish-green, they have flat or slightly convex upper surfaces and semicylindrical under surfaces covered with white tubercles, and with, at their apexes, short keels. In the wild this kind has more congested, smaller, ovate leaves, the stems between neighbor pairs barely visible. The golden-yellow blooms are about ½ inch in diameter.

Garden Uses and Cultivation. Appropriate for collections of succulents and for rock gardens and similar places in warm, desert and semidesert regions, and for greenhouses, chasmatophyllums are cultivated without difficulty. Porous, perfectly drained soil, kept on the dryish side, a sunny location, and indoors a winter night temperature of about 50°F, with an increase of five to fifteen degrees by day, are favorable. Dry atmospheric conditions are essential. Increase is easy by seeds and by cuttings.

CHASTE TREE is *Vitex agnus-castus.*

CHATHAM-ISLAND-FORGET-ME-NOT is *Myosotidium hortensia.*

CHAYOTE is *Sechium edule.*

CHECKER or CHECKERBLOOM is *Sidalcea malvaeflora.*

CHECKERBERRY is *Gaultheria procumbens.*

CHECKERED-LILY is *Fritillaria meleagris.*

CHEDDAR PINK is *Dianthus gratianopolitanus.*

CHEILANTHES (Cheil-ánthes)—Lip Fern, Lace Fern. The genus *Cheilanthes* consists of approximately 180 species of mostly small ferns of the pteris family PTERIDACEAE. As natives they are widely distributed throughout many temperate and warm regions of the Old and New worlds. The name, from the Greek *cheilos,* a lip, and *anthos,* a flower, alludes to the clusters of spore capsules and their location.

These are ground-growing and rock-inhabiting rather than epiphytic (tree-perching) plants. Many kinds are restricted to arid regions. They have abundantly scaly rhizomes, short-creeping or becoming erect. The fronds (leaves), generally clustered, are one, two, three, or sometimes four times pinnately-divided. Often hairy or scaly, they are sometimes smooth. The clusters of spore capsules are at the ends of veins, which do not extend to the leaf edges. They are covered or nearly covered by a continuous or an interrupted indusium formed of the scarcely modified, rolled-under leaf margin. For plants previously named *C. californica* and *C. siliquosa* see *Aspidotis.*

Hardy *C. lanosa* inhabits shale outcrops and cliffs from Connecticut to Wisconsin, Minnesota, Georgia, and Texas. Its fronds, scattered along creeping rhizomes, are up to 1 foot long, mostly shorter, with narrow-oblong to linear-lanceolate blades much longer than the stalks and 1 inch to 2 inches wide. They are twice-pinnate, with the final divisions lobed. There are ten to twenty pairs of ovate primary divisions, the lower ones opposite, those nearer the apex alternate. Each has seven to ten pairs of secondary divisions. The upper surfaces of the leaf blades are sparsely-hairy. Their undersides are clothed with shining, white hairs. Favoring limestone rocks and cliffs from Virginia to Missouri, Georgia, Oklahoma, and Texas, *C. alabamensis* has rhizomes clothed with bright orange scales. Its fronds, up to 1 foot long, have shining black stalks, hairy on their upper sides, hairless or nearly so beneath, not longer than the ¾- to 2-inch-wide, linear-lanceolate blades. They are twice-pinnate with the final segments lobed. There are eighteen to twenty-five pairs of primary divisions, the lower subopposite, those toward the apex alternate. The secondary divisions are oblong to linear-oblong. The covering of the spore clusters is continuous.

Lace fern (*C. gracillima*), of western North America, forms tufts of twice-pinnate or partly-thrice-pinnate leaves with blades 1 inch to 4 inches long and stalks about equaling them in length. Their upper surfaces soon lose any hairs they may have had when young, their undersides are hairy or woolly, but not scaly. There are nine to twenty pairs of primary divisions each of about nine tiny secondary ones usually longer than wide and reduced to almost beadlike proportions. The coverings of the spore clusters are generally continuous. Native from Texas and Arizona to tropical America, *C. myriophylla* has short, scaly rhizomes and fronds 3 to

Cheilanthes myriophylla

9 inches long and three to four times divided into numerous tiny ultimate segments clothed on their undersides with scales and wool. The coverings of the clusters of spore capsules are usually continuous.

Other sorts sometimes cultivated include *C. feei,* native from Illinois and Minnesota to British Columbia, Texas, and California. This grows in shaded crevices on rocks and cliffs most often, but not exclusively of limestone. A dainty miniature, it has short rhizomes. It forms crowded clusters of mostly twice-pinnate fronds 3 to 8 inches long with ovate-lanceolate blades 1 inch to 1½ inches wide, densely-hairy especially on their undersides. They mostly lie flat against rock faces. Native of the Rocky Mountain region and to the west and south and into Mexico, *C. fendleri* (syn. *Notholaena fendleri*) has somewhat the aspect of *C. feei,* but its fronds, in less compact tufts, are bigger and scaley, but not hairy. From 4 to 10 inches long, they have thrice-pinnate, ovate-lanceolate blades 1 inch to 1½ inches wide, the final segments cleft into oblong lobes.

South African *C. hirta* has rhizomes up to 6 inches long and crowded, erect, oblong-bladed fronds 4½ to 10 inches long by up to 1 inch wide. They are twice-pinnate and clothed with brown hairs. The primary divisions are rather distantly spaced. The secondary divisions are pinnately-lobed, and often up-curled. The spore clusters form continuous lines around the leaf edges. Also South African, *C. bergiana* (syn. *Hypolepis bergiana*) forms tufts of triangular fronds 1 to 2 feet long by 6 to 9 inches wide and four times pinnately-lobed. Its clusters of spore capsules are very small.

Cheilanthes bergiana

Garden Uses and Cultivation. These ferns are for the most part only of interest to collectors. They can to some extent be planted in rock gardens and similar locations in regions where they are native. Their hardiness reflects the climates of such regions. Generally they are not diffi-

cult to grow in well-drained soil. Care must be taken not to give too much water and to expose them to as strong light as they can tolerate without being scorched. In greenhouses a dryish rather than humid atmosphere is required and overhead watering must be strictly avoided. For more information see Ferns.

CHEIRANTHUS (Cheir-ánthus) — Wallflower, Gilliflower. Wallflowers are not much grown in America, but in Great Britain and other parts of northern Europe the English wallflower is grown in countless millions and in many varieties to brighten flower beds in parks, other public places, and home gardens. It is undoubtedly one of the most if not the most popular spring bedding plant. Bunches of its fragrant blooms are commonly sold in the flower markets. At one time splendid and extensive displays were featured each spring at The New York Botanical Garden. Because the winters at New York City are too severe for these plants to survive outdoors these displays were made possible by carrying them over in well-protected cold frames.

The genus *Cheiranthus* belongs in the mustard family CRUCIFERAE and consists of ten, usually somewhat subshrubby perennials (some commonly cultivated as biennials) indigenous to the Mediterranean region, the Himalayas, and Madeira and the Canary Islands. The generic name is thought to be formed of a modification of an Arabic one combined with the latinized form of the Greek *anthos,* a flower. From nearly related *Erysimum* this genus differs in having its seeds in its somewhat angled seed pods more or less in two rows and in its flowers being without median nectary glands. The seed pods of *Erysimum* have their seeds in a single row. They are usually more angular than those of *Cheiranthus,* and the flowers have median nectary glands as well as lateral ones at the bottoms of the stamens.

The English wallflower (*C. cheiri*), 1 foot to 2½ feet tall, is freely-branched and has toothless, lanceolate or pointed, narrowly-oblong leaves usually slightly grayish-green and up to about 3 inches long. The foliage is densest at the ends of non-blooming shoots and just below the flowers, which are in showy, terminal spikes and are up to 1 inch or more in diameter. They are deliciously spicy-fragrant, yellow to yellow-brown, and displayed in spring. The fruits are slender, erect, angular pods. Garden varieties of English wallflowers are numerous and vary in height and other characteristics, and most especially in flower color. They range from creamy-white to deepest oxblood-red with many intermediate hues including primrose-yellow, golden-yellow, orange-yellow, tan, brown, and chestnut-red. Some varieties have double flowers. There are annual strains of English wallflowers that bloom the first

Cheiranthus cheiri

summer from seeds sown in spring, but these are less beautiful and less showy than the spring-blooming biennial kinds. The plant previously called *C. senoneri,* native to Greece, is now accepted as a variety of *C. cheiri* named *C. c. senoneri.* It has oblong-lanceolate, sometimes short-toothed leaves and fragrant orange-yellow flowers. The colloquial name gilliflower is applied to the English wallflower as well as to stocks (*Mathiola*).

A hybrid wallflower, raised at the Royal Botanic Gardens, Kew, England in 1896 and named *C. kewensis* has as parents a red-flowered garden variety of English wallflower and a hybrid between *C. semperflorens* and a yellow-flowered variety of English wallflower. It has flowers reddish-brown on their outsides and on their faces brownish-orange changing to lilac-purple as they age. This is a winter-blooming kind adapted for growing in cool greenhouses. A native of Morocco, *C. semperflorens* (syn. *C. mutabilis*) is a shrubby species 2 to 3 feet tall and well branched. It has slightly hairy leaves and flowers that are cream-colored when they open, but soon become purple or purple-striped. Variety *C. s.* 'Wenlock Beauty' has large smoky-purple blooms. The Siberian-wallflower, often called *C. allionii,* is probably *Erysimum hieraciifolium.*

Garden and Landscape Uses. English wallflowers are beautiful plants for spring flower beds and borders and porch and window boxes. They can be used effectively with tulips, English daisies, forget-me-nots, pansies, polyanthus primroses, and other spring bedding plants. They are also good cut flowers. For that purpose and as decorative pot plants they are grown in greenhouses. The Kew wallflower (*C. kewensis*) is occasionally cultivated for winter bloom in greenhouses and, like its parent *C. semperflorens,* is probably suitable for growing outdoors in regions, such as parts of California, where winters are nearly free of frost. In regions of mild winters, where the temperature does not drop below 20°F, English wallflowers can be grown outdoors as easily as in Europe. With the protection of a cold frame they ordinarily survive temperatures ten or fifteen degrees lower and, if the cold frame is well insulated and is covered with mats on cold nights and during exceptionally cold days they are not damaged when the outside temperature drops as low as −10°F.

Cultivation. Double-flowered wallflowers and the original form of the hybrid *C. kewensis* are increased by cuttings made in spring from nonflowering shoots rooted in a propagating bed in a greenhouse or cold frame. Other kinds are ordinarily raised from seeds, and there is a seed-producing form of *C. kewensis* that is reported to be fairly true to type. Seeds of English wallflowers sown in outdoor beds or in cold frames, at such a time that the plants by fall will be compact, moderately-sized specimens 6 or 7 inches in diameter, produce the best results. If the plants are much larger than this when winter closes in they are more likely to suffer from winterkill or to be straggly, ungainly specimens difficult to transplant and not looking their best in their blooming locations. If they are seriously smaller than indicated the best floral display will not be realized. At New York City mid-July is the appropriate time to sow, the date must be adjusted where the growing season is longer or shorter. As soon as the seedlings are big enough to handle they are transplanted to cold frames or in regions of mild winters to outdoor nursery beds. Spacing in frames should be 8 inches each way between plants, outdoors they may be set in rows 1 foot apart with 8 inches between plants in the rows.

Wallflowers have a natural tendancy to develop long, rangy root systems instead of compact fibrous ones that would make for easier transplanting. To encourage the development of good root balls, the tips of the main roots of the seedlings should be nipped off at the first transplanting and the soil in which they are set should be firm and compact. Any organic material added, such as compost or peat moss, should be restricted to the upper four or five inches, and the earth must not be over-rich. The idea is to develop sturdy, compact, well-branched plants. To this end, it is helpful to mix in superphosphate with the soil at the rate of 2 ounces to 10 square feet. If the soil is acid a liberal application of lime is indicated. After-planting summer and fall care calls for nothing more than keeping down weeds and watering if long periods of dry weather occur. Wallflowers stand considerable dry-

ness, and excessive watering is to be avoided because it encourages rank growth.

Winter care of plants in cold frames consists of ventilating the structures as freely as possible whenever the outside temperature is above 32°F and of giving a little ventilation even if it is lower than this if the temperature inside the frame rises above 35°F. This is done to reduce the excessive humidity of the air in the frame as well as to control the temperature. Whenever the outside temperature is likely to drop below about 10°F the frame sash should be covered with heavy mats. It was the practice at The New York Botanical Garden, where winter temperatures occasionally drop to −10°F or lower, to surround the frames in fall with a framework of boards standing about 1 foot from the outside walls of the frame and to pack the space between with dry leaves or straw covered with tar paper or polyethylene plastic to keep out moisture. This additional insulation was not installed until late in fall, well after the ground had frozen, so that it would not attract mice looking for congenial winter quarters. In mild climates English wallflowers can be planted in fall in their flowering quarters, elsewhere as early in spring as it is possible to

Heavy mats are used in cold regions to protect English wallflowers in winter

Cold frame containing English wallflowers protected by a layer of straw packed within a wooden framework and to be covered with plastic

work the ground. They are spaced about 1 foot apart.

Wallflowers for greenhouse display are most conveniently had by treating them exactly as advised above for outdoor use until early fall. Then they are lifted and potted in any good soil in containers very little larger than their root balls (6-inch pots are usually satisfactory). After potting, the plants are watered well and put in cold frames. For a week or two they are kept covered with sash and shaded, and their foliage is sprayed lightly with water two or three times a day to help them recover from the shock of the root disturbance, but later throughout the fall and early winter they are grown as "hard" as possible without allowing the soil in the pots to actually freeze. Later, usually in late December or January, they are transferred to a sunny greenhouse with a night temperature of 40 to 50°F. Day temperatures are not permitted to rise more than five to ten degrees above the night minimum without the greenhouse ventilators being opened. At all times these plants should be "grown cool." Conditions appropriate for carnations suit wallflowers. Once new growth begins, weekly and later semiweekly applications of dilute liquid fertilizers are helpful.

Diseases and Pests. Diseases that affect wallflowers include gray mold or botrytis, mildews, white rust, bacterial wilt, and club root. They are sometimes damaged by nematodes, aphids, caterpillars, and fleabeetles.

CHEIRIDOPSIS (Cheirid-ópsis). Approximately 100 species of these fascinating, low, tufted, short-stemmed or stemless herbaceous perennials of the carpetweed family AIZOACEAE comprise *Cheiridopsis*. All are natives of southwestern Africa. The name, from the Greek *cheiris*, a sleeve, and *opsis*, resembling, alludes to the manner in which the parts of pairs of old leaves of many kinds wither to form papery sheaths around the young pairs.

In *Cheiridopsis* each shoot has one to three pairs of leaves with alternating pairs often differing markedly in size, shape, or proportions of their lengths from their united bases. During the dormant period with most, but not all species the basal portion of one pair of leaves, the tops of which have withered, remain as a sleeve or sheath around or enclosing the young succeeding pair. The firm or soft, usually dark-dotted, green, glaucous, or whitish leaves are nearly cylindrical to hatchet-shaped, boat-shaped, or spatula-shaped, frequently with one or two little teeth on the keel beneath or at the apex. Solitary, terminal, and generally stalked, the most often yellow, orange, or white, less commonly violet-red, blooms in aspect suggest daisies, but structurally are very different. Instead of being heads of many florets they are single blooms. They vary from ½ inch to 4 inches in diameter, and have four to five sepals, many petals in several rows, and numerous stamens. The fruits are capsules.

Kinds with leaves more than twice as long as wide include those now to be described. *C. acuminata* has thick stems, all similar leaves 2 to 2¾ inches long by about ½ inch wide, and long-stalked, pale yellow blooms up to 2¼ inches wide. It is without sheaths. *C. bibracteata,* stemless or nearly so, has dotted, gray-green leaves, minutely-hairy at their edges and apexes, 2 to 3 inches long by ⅓ inch wide or slightly wider, united below, the free portions flat above, their undersides rounded below, keeled toward the apex. One leaf of each pair is round-ended, the other pointed. The yellow flowers are 1½ inches in diameter. *C. bifida* has soft, dotted, gray-green leaves 2 inches long or a little longer by up to slightly more than ⅓ inch wide, one of each pair usually longer and slenderer than the other. Flattened on their upper sides, rounded beneath, they have somewhat keeled apexes. Long-stalked, the yellow flowers are 1½ inches wide. *C. candidissima* has beautiful, green-dotted, nearly white, slender boat-shaped leaves 3 to 4 inches long, nearly ½ inch wide and rather thicker, united for one-third or more of their lengths. The long-stalked, 2-inch-wide flowers are white or delicate pink. *C. cigarettifera* when dormant has its young leaves almost completely sheathed. Fully grown they are ½ inch to 1½ inches long by up to ⅓ inch wide. Glaucous and spotted, they have flat upper surfaces, rounded lower ones, and a keel minutely toothed at its tip. Long-stalked and golden-yellow, the blooms are 1 inch in diameter. *C. derenbergiana* is stemless and mat-forming. When dormant its young leaves are completely enclosed in long, white sheaths. The densely-hairy, gray-green leaves are 1¼ inches long, ¼ inch wide and thick, and sharply keeled. Long-stalked, 2¾ to 3¼ inches across, the flowers are pale yellow. *C. duplessii* has short stems with persistent dead foliage attached. The outer leaves are 1½ to 2 inches long or sometimes longer, ¼ to nearly ½ inch wide, the central ones up to 1 inch long by ⅙ inch wide and thick. They are bluish or whitish. About 2 inches across, the flowers are yellow. *C. lecta,* short-stemmed, has one of each pair of slightly gray-green, dotted leaves incurved, the other outcurved. They are up to 3¼ inches long by ⅓ inch wide, jointed for one-half their lengths, flat on their upper surfaces, rounded beneath, and keeled toward their apexes, the keel ending in a chin. Their blades spread. *C. marlothii* has dotted, gray-green, mealy, outcurving leaves up to 2¼ inches long by ⅓ inch wide, those of each pair joined for about one-third of their lengths. The undersides of the leaves are rounded toward

their bases and sharply keeled with the keel ending in a chin. When dormant the young growths have long sleeves. The lemon-yellow blooms are scarcely ½ inch wide. *C. peersii* has glaucous leaves 2½ to 3½ inches long by less than ½ inch wide, keeled beneath, the lower ones rounded toward their bases. Golden-yellow, the flowers are 2 inches across. *C. rostrata,* almost or quite stemless, forms mats of

Cheiridopsis rostrata

tufts of two or four gray-green leaves joined at their bases to form a cylindrical body. They are 2 to 3 inches long by approximately ½ inch wide and often curve to one side. Narrowed toward their tips, they have a flat upper surface and a rounded lower one with its upper part keeled. The longish-stalked flowers are 2 inches or more in width. *C. tuberculata* has dotted, bluish, gray, or green, sometimes reddish-tinged leaves 2½ to 4½ inches long, their lower parts conspicuously united. Their free portions narrow toward their apexes. They have flat uppersides and rounded undersides that become keeled toward their tips. The long-stalked, 1½-inch-wide blooms are yellow.

Kinds with leaves less than twice as long as wide include these: *C. herrei* has bluish to dull green, round-sided leaves ½ to ¾ inch long by somewhat less than ½ inch wide, with flat upper surfaces, bluntly keeled beneath. There is a large pustule at the base. The short-stalked, yellow flowers are up to 2¼ inches across. *C. peculiaris* is unique. One pair of short-pointed, broad-ovate, flat-topped leaves, each 1½ to 2 inches long and wide, spread horizontally. From their centers stand erectly a pair of completely united leaves. About 1½ inches across, the flowers are yellow. *C. pillansii* has dark-dotted, whitish-gray leaves joined for one-third of their lengths, only slightly separated, forming a fat plant body. They are nearly 2 inches long by 1 inch wide, flat on their upper or infacing sides, rounded beneath, and ending in a blunt chin. The yellow flowers are 2¼ inches wide. *C. verrucosa* also forms fat plant bodies, each of a pair of little-separated leaves united for one-half or more of their lengths. The leaves are ½ to ¾ inch long

by two-thirds as wide and nearly as thick. Their upper or inner surfaces are slightly convex, their lower ones rounded and obscurely keeled. The short-stalked flowers are yellow, ¾ to 1 inch across.

Garden Uses and Cultivation. Choice collectors' succulents, these are suited for growing in greenhouses under conditions suitable for *Lithops* and most other succulents of the *Mesembryanthemum* relationship. Some adapt quite well to cultivation, others are more recalcitrant. Early summer is the chief growing season for most kinds. From the time new growth begins until fall water should be given in moderation, then gradually reduced and finally withheld during the winter period of dormancy. Propagation is by seeds and cuttings. For further information see Succulents.

CHEIROSTEMON. See Chiranthodendron.

CHELIDONIUM (Cheli-dònium)—Greater Celandine. Except for its quite pretty, double-flowered variety, the greater celandine (*Chelidonium majus*) scarcely warrants inclusion among ornamentals. It is too weedy in appearance. Admittance of its single-flowered phase is justified in herb gardens and medicinal gardens on the basis of its medicinal virtues and uses. It contains several alkaloids and has been employed as a purgative, diuretic, diaphoretic, and expectorant. In Europe, country people believe that its yellow juice cures warts. The greater celandine belongs in the poppy family PAPAVERACEAE. A hardy native of Europe and Asia, it is naturalized freely in eastern North America. Its name is from the Greek *chelidon,* a swallow, and alludes to associations between the bird and this plant in folklore.

The genus *Chelidonium* includes only the species *C. majus,* which bears a remarkable resemblance, especially in its foliage, to *Stylophorum,* but differs in its flowers not having well-developed, elongated styles. The greater celandine is an erect, branched, slightly succulent biennial or perennial, 1

Chelidonium majus

foot to 3 feet tall, with basal and stem leaves, the latter alternate. The leaves are variously and deeply-pinnately-lobed and toothed, the terminal division broadly-obovate and three-lobed. Their undersides are bluish-glaucous. The flowers, which appear over a long period in summer, have two sepals that fall early, four yellow petals, and numerous slender stamens. They are about ⅔ inch across, in few-flowered, loose clusters, and are succeeded by slender, cylindrical, hairless seed pods. In addition to the typical species there is *C. m. flore-pleno*, with double flowers, and *C. m. laciniatum*, which has more finely-dissected foliage.

Chelone glabra

Chelidonium majus flore-pleno

Garden Uses and Cultivation. This is a plant for moist soils. It thrives in sun or light shade and needs little or no care. Usually it self-sows freely. Propagation is by seed or division in spring or early fall. The garden uses of the single-flowered greater celandine are mentioned at the beginning of this entry. The double-flowered kind is appropriate for semiwild parts of gardens.

CHELONE (Chelò-ne)—Turtlehead. There are four species of these North American hardy herbaceous perennials, members of the figwort family SCROPHULARIACEAE and close relatives of *Penstemon*. Their name is from the Greek *chelone*, a tortoise, and alludes to the shape of the flower, which is compared fancifully to that of a turtle's head. From penstemons chelones differ in having two or three large, sepal-like bracts present at the base of each flower. They have many erect, branchless or sparingly-branched, hairless stems, opposite, usually hairless, undivided, toothed leaves, and terminal dense spikes of creamy-white or pink blooms of quite distinctive shape, and indeed reminiscent of the head of a turtle. They are thick and lumpy with curved, broad corolla tubes, two-lipped at their mouths, the lower lip three-lobed, the upper arching and once-notched. There

Chelone glabra (flowers)

Chelone lyonii

are four hairy, stalked, fertile stamens and a fifth, shorter, hairless, sterile one. The fruits are capsules.

Chelones, 1 foot to 3 feet tall, chiefly inhabit wet woodlands. A common species, ranging from Newfoundland to Minnesota, Georgia, and Alabama, is *C. glabra.* This is highly variable in the shape of its leaves and in the color pattern of its flowers. The latter are about 1 inch long and may be white, white and greenish-yellow, white tinged with pink, or white with purplish tips. Its leaves, scarcely or not at all stalked, and tapered at their bases, are narrow to broad-lanceolate. The bracts of the flower spikes are prolonged at their apexes. From the above *C. obliqua* is distinguished by its flowers being entirely purplish-rose-red; they are 1 inch to 1½ inches long. It has leaves with definite short stalks and is indigenous from Maryland to Alabama

and from Indiana to Minnesota and Arkansas. Stalkless leaves with rounded bases distinguish **C. cuthbertii.** Native from Virginia to North Carolina, this has violet-purple flowers about 1 inch long. Indigenous only in the mountains of North Carolina, South Carolina, and Tennessee, **C. lyonii** may be identified by its broad-ovate leaves, rounded or wedge-shaped at their bases and having stalks ¾ inch to 1¼ inches long.

Garden Uses. The turtleheads are of secondary importance among hardy herbaceous perennials. Although they can not compare in display value with such standbys as coreopsis, phlox, or delphiniums so far as summer show is concerned, they are worth establishing in flower borders and are especially esteemed for moist and wet soils. They look at home and especially appropriate beside lakes, ponds, or streams where their roots can reach a wet substratum. They tolerate part-shade and flower for long periods in late summer and fall.

Cultivation. No plants are easier to grow. Once planted in a suitable site they pretty much take care of themselves, usually requiring no staking or other attention than the clearing away of the tops after they die in fall. Propagation is simple, by division in early fall or spring, by summer cuttings, and by seeds sown in constantly moist soil.

CHEMICULTURE. See Hydroponics or Nutriculture.

CHENAR TREE is *Platanus orientalis*.

CHENILLE PLANT is *Acalypha hispida*.

CHENOPODIACEAE — Goosefoot Family. Just in excess of 100 genera and some 1,400 species of dicotyledons compose this family, which has a very wide natural distribution especially in saline and dry soils. Besides many weedy members, the group includes beets, spinach, and a few ornamentals as well as the little known vegetable Good King Henry; also quinoa, its seeds an important food in parts of South America, and the weed lamb's quarters, sometimes gathered and eaten as greens. Its sorts are chiefly annuals, biennials, and herbaceous perennials, with a few shrubs or small trees included. Most have alternate, fewer have opposite leaves. A minority are leafless with succulent green stems that function as leaves. Chenopods, the group name for members of this family, have greenish, bisexual or unisexual flowers with persistent calyxes of one to five lobes or sepals or sometimes without calyxes. There are up to five stamens and two to five usually stalkless stigmas. The fruits are utricles. Cultivated genera are *Atriplex, Beta, Camphorosma, Chenopodium, Enchylaena, Eurotia, Kochia, Rhagodia, Sarcobatus, Spinacia,* and *Suaeda*.

CHENOPODIUM (Cheno-pòdium) — Goosefoot, Good King Henry, Jerusalem-Oak, Mexican-Tea. Few of the 100 or more species of this genus of the goosefoot family CHENOPODIACEAE are cultivated. Many are weeds of gardens and waste places. In the Andes of South America, quinoa is grown for its edible seeds, and in Europe and North America, Good King Henry or mercury and strawberry-blite are planted occasionally as pot-herbs or novelty vegetables. As edible greens for cooking lamb's quarters is to some extent collected from the wild. The name is from the Greek *chen*, a goose, and *pous*, a foot, in allusion to the fancied resemblance of the leaves to the webbed feet of geese. This genus is widely distributed as a native of temperate regions.

Chenopodiums are summer- and fall-blooming annuals, biennials, herbaceous perennials, and subshrubs. They have stems and foliage glandular or with the appearance of having been dusted lightly with white meal. Their leaves are alternate, linear to ovate, undivided, lobed, or toothed. The greenish or reddish flowers, of insignificant size and appearance, are in tight little heads in axillary or terminal spikes or panicles. Each has usually two, but up to five persistent calyx lobes, no petals, one to five stamens, and two or rarely three styles. The fruits, small and bladder-like, are sometimes enclosed by the fleshy calyxes so that the clusters resemble berries.

Good King Henry or mercury *(Chenopodium bonus-henricus)*, a native of Europe and naturalized in North America, is

Chenopodium bonus-henricus

a thick-rooted perennial. Up to 2½ feet tall and sparsely-mealy, it has many long-stalked, more or less triangular, arrow-shaped leaves 3 to 4½ inches long, and mostly terminal, slender flower panicles that are leafless in their upper parts.

Strawberry-blite *(C. capitatum)* is occasionally cultivated as a potherb. It is an erect or diffuse annual up to 1½ feet tall, with triangular-ovate, toothed leaves, and clusters of female flowers that ripen into fleshy, red, berry-like fruits in interrupted spikes. Native to Europe, this is naturalized in North America.

Lamb's quarters *(C. album)* is a highly variable annual native of North America, Europe, and Asia. It attains heights of 1 foot to 3 feet. Erect and branching freely, this sort has diamond-shaped to ovate or lanceolate, usually toothed, green or white-mealy leaves 1½ to 4 inches long that often assume reddish tints late in the season. The flowers are in terminal panicles.

Quinoa *(C. quinoa)*, native to the Andes, is an annual 4 to 5 feet tall. It has angularly-lobed, triangular-ovate, dull glaucous leaves, and mealy, terminal and axillary

Chenopodium quinoa

flower panicles. The seeds are large and after being washed in lime water to rid them of their natural bitter qualities are in Andean South America cooked and eaten like rice.

Jerusalem-oak or feather-geranium *(C. botrys)* is a strongly scented annual, slightly sticky because of short, glandular hairs. Native to Europe, Asia, and Africa, and a naturalized weed in waste places in North America, it is 9 inches to 2 feet tall and has stems without branches or that branch from near their bases. Most of the oblong to ovate leaves are pinnately-lobed in a manner that suggests those of oaks. Those

Chenopodium botrys

Chenopodium botrys, in bloom

toward the tops of the stems are coarsely-toothed or toothless and are much smaller than those below, which are up to 3½ inches long. The slender axillary flower panicles, from the upper leaf axils, are long-lasting and used in bouquets.

The red or purple-red coloring of their young shoots and foliage gives reason for *C. amaranticolor* and *C. purpurascens* being sometimes cultivated in flower gardens. The leaves and young shoots of the former and probably of the latter are sometimes cooked and eaten like spinach. From 6 to 8 feet tall, **C. amaranticolor** is a much-branched annual with young parts dusted with brightly colored meal and its older foliage red-striped. The triangular-ovate to diamond-shaped leaves are coarsely- and irregularly-toothed. They are 2 to 5 inches long. The flowers are in terminal and axillary panicles. The native country of this species is not known. It is naturalized in North America. Very similar, but not more than about 3 feet tall, annual **C. purpurascens** (syn. *C. atriplicis*), of China, has triangular-ovate, angular-toothed, long-stalked lower leaves, toothless upper ones, and flowers in dense, pyramidal clusters.

Mexican-tea or American wormseed **(C. ambrosioides)**, an annual or perennial native of tropical America and naturalized in

Chenopodium ambrosioides

parts of North America, has been used as a vermifuge. It is 3 to 3½ feet tall, strong-smelling, and has lobed or coarse-toothed leaves up to 5 inches long.

A native of northern Asia including Japan and Taiwan, *C. virgatum* is a vigorous annual of some interest because of the attractive, bright red, raspberry-like fruits ¼ to ⅓ inch long that festoon its branches. It has erect stems 1 to 2 feet long and ascending to spreading or trailing branches. The leaves have blades ¾ to 1½ inches long that are narrowly-lanceolate to ovate and have a few coarse teeth.

Garden Uses and Cultivation. The uses of cultivated chenopodiums are given above with their descriptions. Their cultivation

Chenopodium virgatum

presents no difficulties. All prefer fertile, well-drained soils and sunny locations, and are easily raised from seeds.

CHERIMOYA. See Annona.

CHERRY. In addition to orchard cherries, dealt with in the next entry in this Encyclopedia, and various other kinds of *Prunus* known as cherries, bush cherries, cherry-laurels, and sand-cherries discussed under the entry Prunus, there are other unrelated plants that have the word cherry included in their common names. These include African cherry-orange (*Citropsis*), Australian brush-cherry (*Syzygium paniculatum*), Barbados-cherry (*Malpighia glabra*), cherry-of-the-Rio Grande (*Eugenia aggregata*), cherry pie (*Heliotropium*), cherry tomato (*Lycopersicon esculentum cerasiforme*), Christmas- or Jerusalem-cherry (*Solanum pseudo-capsicum*), Cornelian-cherry (*Cornus mas*), False-Jerusalem-cherry (*Solanum cap-*

Cherry (fruits)

sicastrum), ground-cherry (*Physalis pruinosa*), Indian-cherry (*Rhamnus caroliniana*), madden-cherry (*Maddenia hypoleuca*), Spanish-cherry (*Mimusops elengi*), Surinam-cherry (*Eugenia uniflora*).

CHERRY. Here are considered cherries cultivated for their edible fruits, not ornamental flowering cherries, sometimes grouped as Japanese cherries, grown only for their blooms. The latter are discussed under Prunus. Because of the great attraction they have for birds, which quickly strip isolated trees and groups of few trees (they are less destructive in large orchards) and because of uncertainties of the fruits ripening perfectly cherries are less well adapted to home garden cultivation than many fruits.

Fruiting or orchard cherries are derivatives of species native to Transcaucasia and adjoining parts of Asia Minor and Iran.

Cherry tree, in bloom

Brought to America by the early settlers, they are now a major commercial crop.

Cherries fall into three groups, sweet cherries (varieties of *Prunus avium*), sour cherries (varieties of *P. cerasus*), and Duke cherries (believed to be hybrids between sweet and sour varieties). Sweet cherry orchards are mostly in California, Oregon, and Washington and in regions adjoining the Great Lakes in New York and Michigan, those of sour cherries in Colorado, New York, Michigan, Ohio, and Wisconsin. Duke cherries, less hardy than sweet and sour kinds, are chiefly grown in the Pacific Coast region.

In choosing locations for cherries remember they bloom early. Because of this, sites subject to late frosts are quite unsuitable. Avoid low places where cold air can pocket. If possible plant on sloping ground, but not on windswept tops of hills.

Soils of a variety of types are suitable for cherries, but they must be well drained. These trees are short-lived on wet ground. Prepare the soil for planting very adequately in the manner recommended for apples. Unless the soil is in good condition and planting is done well and at the right season, losses are likely to be high.

Understocks need consideration. These are of two types, mazzard (*Prunus avium*) and mahaleb (*P. mahaleb*). Because the latter is easier to grow and gives larger, more uniform young trees that come into bearing earlier it is often favored by nurserymen. But cherries on mazzard understocks generally perform much better as orchard trees and are much longer-lived. If possible obtain trees on this understock. To check which understock trees are budded on cut off a little bark below the union of understock and bud and chew it. That of mazzard is very bitter, that of mahaleb not. Another test is to scrape a root lightly and wet it with ferric chloride dissolved in water. This causes instant blackening of the tissues of mazzard roots, a much slower greenish-blackening of those of mahaleb understocks.

Another important factor to consider is that sweet and Duke cherries are not self-fertile. They must be cross-pollinated with others to ensure fruiting, and not all varieties fertilize all other varieties. Good pollinators for interplanting with other varieties are 'Black Tartarian', 'Giant', 'Seneca', and 'Lyons'. It is well to seek the advice of the State Agricultural Experiment Station of the state in which you propose to plant or other competent local authority before making final selections. Sour cherries are self-fertile. They set ample crops of fruit without cross-pollination.

Planting in early spring may in cold regions be better than fall planting. Elsewhere the latter is preferable. If done in spring plant before growth begins, otherwise losses may be high. Distances of 25 to 30 feet each way should be allowed between sweet and Duke cherries, 20 to 25 between sour varieties except 'English Morello', which may be spaced at 18 feet. Plant carefully, making sure to spread the roots and to work mellow soil between them and tread it firm. It is better not to prune the branches at planting time.

Routine care of established trees consists of cultivating the soil shallowly to keep down weeds during the early part of the season and after harvesting, of lightly fertilizing and sowing rye grass or winter rye as a cover crop to check erosion and supply organic material to turn under the following spring. As an alternative to cultivating and cover cropping, the orchard may be kept mulched with straw, hay, salt hay, or other suitable material.

Pruning, as compared with that needed for most fruits, is minimal. With young trees it is directed toward training them to assume the modified leader form. At planting time with sour cherries select for retention three or four scaffold or main framework branches, the lowest not under 1½ feet from the ground. They should be well spaced around the trunk and spread from it at a wide angle. Erect branches that form acute angles with the trunk are susceptible to breakage by storms, the weight of heavy crops, and ice accumulations. Remove all other branches. Two or three years later cut the leader (central shoot) back to a strong, out-pointing lateral and allow an additional two or three well-placed scaffold branches to develop, thus bringing the total to six. The only other pruning needed is any necessary to eliminate cross branches and to thin out weak branches and any crowded to the extent that they shade others. Pruning sweet and Duke cherries in their early years is similar to that for sour cherries except that the leader is usually topped at 4 feet and the vertical distance between scaffold branches should be 8 inches to 1 foot. The only pruning needed later is to cut back scaffold branches that tend to outgrow the leader and any light cutting necessary to take out weak branches from the center of the tree and broken or dead ones. Old trees may be renovated and tall ones lowered by severe heading back.

Varieties of sweet cherries include these: 'Bing', fruits large, dark red, tend to crack in wet weather, excellent for canning; 'Black Tartarian', fruits medium-sized, purplish-black to black, good for home use; 'Emperor Francis', fruits firm, of good quality, red over yellow; 'Geant d'Hedelfingen', fruits large, black, of high quality; 'Lambert', fruits large, black, tending to crack in wet weather; 'Lyons', fruits large, soft, purplish-red, ripen early; 'Napoleon' (called 'Royal Anne' on the Pacific Coast), fruits large, yellow blushed with red, the best for maraschino cherries; 'Windsor', fruits large, dark purplish-red, mature late; 'Yellow Spanish', fruits large, yellow with a pink cheek, used chiefly for maraschino cherries. Duke cherries most commonly grown are 'Reine Hortense', fruits light red, soft, of high quality; 'Royal Duke', fruits large, soft, dark red. Sour cherry varieties include 'Early Richmond', fruits small, red, maturing early; 'English Morello', fruits medium-sized, black, late maturing; 'Montmorency', fruits large, bright red, of superior quality.

Pests and Diseases. Where small numbers of cherry trees are grown birds are likely to be a serious pest. The only effective way of foiling them is to cover the trees when the fruits begin to ripen with cheesecloth or tobacco cloth (in tobacco-growing regions used cloth is sometimes available cheaply). Other pests include black aphids, larvae of the cherry fruit fly and plum curculio, sluglike larvae of the cherry sawfly, which skeletonize the foliage, and tent caterpillars. Brown rot disease affects the shoots and causes rotting of the fruits before they ripen. Shot-hole damage to the leaves results from bacterial leaf spot. Fungus leaf spots may also appear. Virus diseases, for which there are no known controls, are severe in some regions. For full current information concerning pests and diseases and recommendations for control measures best suited for particular areas consult County Cooperative Agents and State Agricultural Experiment Stations.

Other troubles may include winter injury. To minimize this, avoid pruning heavily or if this must be done do not do it following an exceptionally severe winter and delay it until after the period of severe cold has passed. Pest and disease control measures that enable the trees to retain their foliage without damage until late fall minimizes winter injury.

CHERVIL. Salad chervil is *Anthriscus cerefolium*. Turnip-rooted chervil is *Chaerophylum bulbosum*.

CHESTNUT. See Castanea. The Cape-chestnut is *Calodendrum capense*, the Chinese water-chestnut *Eleocharis dulcis*, the Guinea-chestnut *Pachira aquatica*, the horse-chestnut *Aesculus*, the Moreton-Bay-chestnut *Castanospermum australe*, and the water-chestnut *Trapa natans*.

CHEWING GUM TREE is *Manilkara zapota*.

CHIAN TURPENTINE is *Pistacia terebinthus*.

CHIAPASIA (Chiápas-ia). The genus *Chiapasia*, named after the Mexican state of Chiapas where it is native, consists of one desirable species related to *Epiphyllum* and by some authorities included in *Disocactus*. It belongs in the cactus family CACTACEAE.

Because it grows perched on trees, but takes no nourishment from them, *C. nelsonii*, like many orchids, bromeliads, and plants of like habit, is termed an epiphyte. It is a spineless cactus with slender stems, at first erect, later arching or pendulous.

They are 2 to 4 feet in length with flat, leaflike branches spirally arranged toward their ends. The branches are 4 inches to 1 foot long and 1¼ to 1½ inches wide. They have notched margins. The violet-scented, lily-shaped flowers remain open for several days and are borne freely from the upper parts of the branches. Three inches or more in diameter, they have green perianth tubes about ¾ inch long and about eight gracefully spreading or recurved carmine perianth segments (petals). The many stamens, white above and reddish below, and the light red, slender style terminated by a five-lobed, white stigma, protrude from the throat of the flower. Hybrids between *C. nelsonii* and *Epiphyllum* have been developed.

Garden Uses and Cultivation. These are as for *Epiphyllum*. An attractive species, *C. nelsonii* usually gives the best results when grafted onto *Hylocereus undatus* or other suitable understock. For further information see Cactuses.

CHIASTOPHYLLUM (Chiastó-phyllum). The one species of this genus is closely related to *Cotyledon* and is often grown under the names *Cotyledon oppositifolia* and *C. simplicifolia*. A hardy native of the Caucasus, where it grows at altitudes of more than 6,000 feet, its correct name is *Chiastophyllum oppositifolium*. It belongs in the orpine family CRASSULACEAE. The name is derived from the Greek *chiastos*, diagonally, and *phyllon*, a leaf, and alludes to the arrangement of the leaves, which are opposite, with alternate pairs at right angles to each other.

From creeping stems, this hairless, herbaceous perennial *(C. oppositifolium)* sends up shoots 2 or 3 inches to 1 foot in height, toward their bases with three or four pairs of thickish, stalked, coarsely-round-toothed, broad-elliptic to round-ovate, bright green leaves. The tiny, yellowish-white to bright yellow, bell-shaped flowers are in panicles with recurving or drooping branches. The fruits are follicles.

Garden Uses and Cultivation. Appropriate for rock gardens and dry walls, this species needs full sun and thoroughly drained soil, preferably containing limestone. It is very easily propagated by seeds and cuttings and needs no special care.

CHICHICASTE is *Urera baccifera*.

CHICK-PEA is *Cicer arietinum*.

CHICKASAW-LIMA-BEAN is *Canavalia ensiformis*.

CHICKEN-CORN is *Sorghum bicolor*.

CHICKWEED. See Stellaria. The chickweed-wintergreen is *Trientalis*.

CHICORY is *Chichorium intybus*. The chicory-grape is *Coccoloba venosa*.

CHICORY, FRENCH or BELGIAN ENDIVE, WHITLOOF CHICORY. The crop known by these names is a minor one of vegetable gardens. It is a garden variety of *Cichorium intybus* of the daisy family COMPOSITAE. A European native, this species is widely naturalized in North America. Its cultivated form is grown commercially for its roots, which are the source of chicory used in coffee, and as a salad plant. For this last purpose its young shoots are forced and blanched out of season. Although often called French or Belgian endive, it is quite distinct from true endive (*Cichorium endivia*).

To have plump, crisp heads of whitloof chicory select a place where the earth is deep and fertile. If manure is used apply it for a previous crop such as cabbage or celery rather than in immediate preparation for the whitloof. Make sure the soil is well and deeply pulverized and mix in before seeding a dressing of a complete garden fertilizer of a kind suitable for carrots, beets, and turnips.

Sow in early spring in drills 1¼ to 1½ feet apart, about ½ inch deep. Before they crowd, thin the seedlings to 7 to 9 inches apart. Throughout the long growing season cultivate between the rows and hand pull in them to keep down weeds. In dry weather water regularly. If needed to maintain continuous vigorous growth apply a side dressing of fertilizer about midsummer, but do not overdo this and beware of using excessive amounts of nitrogen. The objective is to build up heavy, carrot-like roots, not foliage at the expense of roots.

In fall dig the roots with care not to break them. Cut off the leaves and store the roots horizontally between layers of sand or fine soil outdoors, in a cold frame or root cellar, or in boxes in a garage or other unheated place.

Blanched heads may be had from about Christmas until well into spring by forcing successive batches. To do this, take the number of roots you want to cut from in a weekly or two-weekly period from storage and plant them closely together vertically in deep boxes with soil, sand, or peat moss, vermiculite, or similar moisture-holding medium packed between them. Water thoroughly and stand the boxes in a fairly warm place where it is completely dark, or cover them to exclude all light, leaving enough space above them of course for the heads to develop. At a temperature of 55°F they will be ready in about three weeks. In cooler environments four or five weeks will be needed.

Harvest by slicing the tight heads off with a tiny piece of root attached to each. Do this before the leaves begin to expand. After harvest the roots are of no further use. To maintain a continuous supply bring new batches of roots in to force every week or two throughout the winter season.

CHILE and CHILEAN. These words are used in the common names of a number of plants. Examples are: Chile bells (*Lapageria rosea*), Chile-hazel or Chilean nut (*Gevuina avellana*), Chilean-bellflower (*Lapageria rosea* and *Nolana paradoxa*), Chilean-cedar (*Austrocedrus*), Chilean-cranberry or Chilean-guava (*Ugni molinae*), Chilean-crocus (*Tecophilaea*), Chilean glory flower (*Eccremocarpus*), Chilean-jasmine (*Mandevilla*), Chilean-pine (*Araucaria araucana*), Chilean potato-tree (*Solanum crispum*), and Chilean wine palm (*Jubaea chilensis*).

CHILENIA. See Neoporteria.

CHILI. See Capsicum.

CHILICOTHE is *Marah macrocarpus*.

CHILOPSIS (Chi-lópsis)—Desert-Willow; Flowering-Willow. Native of dry parts of the southwestern United States and Mexico, often favoring stream sides and the vicinities of springs, the desert-willow or flowering-willow (*Chilopsis linearis*) belongs in the bignonia family BIGNONIACEAE. It is the only species of its genus. Its name comes from the Greek *cheilos*, a lip, and *opsis*, similar to, and alludes to the formation of the corolla. A deciduous shrub or tree up to 20 feet tall or occasionally taller, *C. linearis* has slender, opposite or alternate, lobeless, hairless, and often sticky leaves up to 1 foot long. Its beautiful, fragrant, trumpet- to bell-shaped blooms, in short terminal racemes, are 1 inch to 2 inches long and almost as wide. They have a deeply-two-lipped, five-toothed, inflated calyx. The corollas have jagged margins and are lavender or lilac with two yellow stripes in their throats. They are slightly two-lipped with the upper lip with two lobes, the lower, three. The lobes are crimped. There are four fertile stamens and one abortive (a staminode) enclosed in the corolla tube. Slender and 6 inches to 1 foot long, the seed pods contain numerous seeds with long hairs at their ends. Variety *C. l. alba* has white flowers.

Garden and Landscape Uses and Cultivation. As an ornamental this good-looking tree or shrub is useful for landscaping in warm dry climates and can be used effectively as a single specimen or grouped. It is not hardy in the north. It grows in ordinary soil in sun and is propagated by seed.

CHIMAPHILA (Chimá-phila)—Pipsissewa; Spotted-Wintergreen. Low, evergreen, subshrubby or herbaceous plants of North America, the West Indies, Asia, and northern Europe, make up *Chimaphila*. They belong in the heath family ERICACEAE; by some botanists they are segregated, together with *Pyrola* and *Moneses*, as the shinleaf family PYROLACEAE. Pipsissewas have running underground stems from which arise erect, branching, above-ground stems with thickish, short-stalked, toothed, glossy leaves, usually in irregular whorls (circles of more than two). The flowers are few together in terminal clusters atop long stalks. They have five sepals, five white or pink, wide-spreading, rounded petals, concave on their upper sides, and ten stamens with stalks (filaments) that broaden at their bases. The fruits are capsules containing numerous tiny seeds. There are eight species. The name is from the Greek *cheima*, winter, and *phileo*, to love, and like the common name spotted-wintergreen, alludes to the plants remaining green throughout the winter. There are eight species of Chimaphila.

The pipsissewas of North America include two varieties of the European species *C. umbellata.* One, *C. u. cisatlantica*, occurs from Quebec and Nova Scotia to Virginia, Ohio, and Indiana. It gradually merges westward into *C. u. occidentalis*, which occurs from Michigan to British Columbia, Alaska, Colorado, and California. These varieties differ in minor ways from the typical species, the eastern in having thicker, more conspicuously veined leaves, and usually recurved flower stalks, the western with thicker leaves than the European type, but without conspicuous veins, and with mostly erect flower stalks. These pipsissewas are 4 inches to 1 foot tall and have oblanceolate, toothed leaves, 1¼ to 2½ inches long, that taper to short stalks. There are four to eight blooms to each stalk, from slightly less to slightly more than ½ inch across. Another western American species, *C. menziesii*, occurs in the coastal mountains from California to British Columbia. It is up to 8 inches tall and differs from *C. u. occidentalis* in having its leaves not distinctly in whorls (circles) and having white instead of pink flowers, singly or in two or threes. Its leaves are sometimes variegated with white in the manner of *C. maculata*.

The striped pipsissewa or spotted-wintergreen (*C. maculata*) is native from Massachusetts to Michigan, Georgia, Alabama,

Chimaphila maculata

and Kentucky. It is a pretty plant with stems up to 8 inches tall and lanceolate, short-stalked, remotely-toothed leaves with conspicuous white stripes along the midveins; they are ¾ inch to 2 inches long. The flowers may be solitary or in clusters of two to five. They are ½ to ¾ inch wide.

Garden Uses and Cultivation. The chief uses of pipsissewas are as plants of interest for shaded places in wild gardens and rock gardens. They succeed best where soil conditions parallel those under which they grow in the wild, that is to say, in dryish, acid, sandy earth. They may be propagated by seeds sown in sandy peaty soil, by division in spring, and by cuttings taken in late summer and fall and planted in a mixture of peat moss and sand in a propagating bed in a cold frame or cool greenhouse.

CHIMERA. Chimeras or chimaeras are plants that have within their bodies tissues of two genetically distinct individuals growing in juxtaposition with neither affecting the hereditary makeup of the other. There is no fusion of cell contents or nuclei, as there is in hybrids. Not inaptly, such plants, of which many are known to science, are named for the fire-breathing monster of Greek mythology called a chimaera. This had the body of a goat, the head of a lion, and the tail of a serpent.

Chimeras are numerous among variegated-leaved pelargoniums (geraniums), citrus fruits, bouvardias, and occur in apples and many other cultivated plants. Among the most interesting chimeras are those produced artificially as a result of grafting. Although not hybrids in any true sense of the word, they are often known as graft-hybrids. The first to receive particular notice was *Laburnocytisus adamii*. For an account of its origin see Laburnocytisus. See also Crataegomespilus.

CHIMONANTHUS (Chimon-ánthus) — Winter Sweet. To an older generation of gardeners this genus was known as *Meratia*. It is still occasionally grown under that now botanically discarded name. It consists of two species, only one of which is in cultivation. The name is from the Greek *cheimon*, winter, and *anthos*, a flower, and alludes to the proclivity of *Chimonanthus praecox* for opening its blooms in winter if mild weather is experienced. In bearing flowers well before its leaves this deciduous member of the Chinese genus *Chimonanthus* differs from the closely related American sweet shrubs (*Calycanthus*). It also differs from *Calycanthus*, as does the other, evergreen species of *Chimonanthus*, in not having deep red blooms. The two genera are the only representatives of the calycanthus family CALYCANTHACEAE.

Winter sweet (*C. praecox* syn. *C. fragrans*) is a compact shrub 8 or 9 feet tall with opposite, short-stalked, toothless, elliptic-ovate to ovate-lanceolate, dark green

Chimonanthus praecox

leaves 2 to 5 inches long; their upper surfaces are rough. The leaves are hairless, except when very young along their chief veins. The exceedingly fragrant blooms are solitary on short stalks and are produced along the shoots of the previous year, in mild climates from November to March. The flowers are ¾ to 1 inch across with no differentiation between sepals and petals, so that all are often called by one name or the other. Whatever they are called, the outer ones are translucent greenish-yellow and the inner ones purplish-brown or striped with purplish-brown. There are five or six short stamens. The fruits are ellipsoid, 1½ inches in length, and constricted at the mouth. Variety *C. p. grandiflorus* has leaves up to 9 inches long, and flowers up to 1¾ inches across, of a clearer yellow color, but less fragrant than those of the typical species. Variety *C. p. luteus* has all its petal-like parts yellow.

Garden and Landscape Uses. Because this shrub is nearly unique in producing in even fairly cold climates (it is hardy about as far north as Philadelphia) deliciously fragrant flowers in winter, it is deserving of more attention from gardeners. Cut branches brought indoors and placed in water soon open their blooms and are especially pleasing. Although rightly classed as deciduous, in quite mild climates this shrub is often almost evergreen. Winter sweet needs a sheltered warm location. This is particularly true near the northern limits of the area in which it can be grown. It succeeds in any fairly good garden soil that is well drained, preferring one that is deep and fertile. Full sun or part-day shade are agreeable.

Cultivation. This shrub is notoriously difficult to propagate by cuttings, although this may be easier to accomplish with modern mist systems than heretofore. More usual methods of increase are by layering and by seeds sown in sandy peaty soil. No regular or systematic pruning is required.

CHIMONOBAMBUSA (Chimonobambù-sa). About fourteen species of bamboos of the grass family GRAMINEAE constitute *Chimonobambusa*, native from India to Japan. Its name is from the Greek *cheimon*, winter, and *Bambusa*, the name of another genus of bamboos. It refers to the season when new shoots begin to form. From *Arundinaria* this genus differs in that its flowers have two styles instead of one.

Chimonobambusas are evergreen perennials with slender running rhizomes from which sprout at intervals erect, nearly solid stems (canes) furnished with deciduous sheaths having conspicuous appendages. The branch sprout from the nodes in groups of several.

Native of Japan, *C. marmorea* (syn. *Arundinaria marmorea*) is indigenous to Japan. It has slender rhizomes and thin, nearly solid stems, at first green, but becoming purplish later, and mostly not branching until their second year. The branches are usually three from each node. The leaf sheaths are deciduous in fall or last perhaps until the following spring. At first they are purplish splashed with pinkish-gray. The leaves may be 6 inches long by ½ inch wide. Their apexes taper and end in a small tongue. They are bright green, moderately tessellated above, dull grayish-green on their undersides. One leaf margin is fringed with fine bristles, the other partially so. This bamboo is a vigorous spreader that needs ample room. It may be expected to be hardy about as far north as Washington, D.C. Variety *C. m. variegata* has white-variegated foliage. These are attractive tub plants.

Hailing from the Himalayan region of India and too tender to grow outdoors in the north, *C. falcata* (syn. *Arundinaria falcata*) is remarkable because of the deep red of its young shoots and the grayish-green of its more mature canes. They grow in tight clumps with the peripheral ones gracefully arching outward and are up to 20 feet in height. The branches are many from each node. The long leaf sheaths are deciduous. The papery-thin leaves, green above and dull gray-green beneath, are up to 6 inches long and up to 1 inch wide. They show no evident tessellation. Their margins are fringed with fine hairs, one side more conspicuously than the other. This species does not spread rapidly. It is attractive in tubs.

Indian *C. hookeriana* (syn. *Arundinaria hookeriana*) is decidedly ornamental, but is suitable for outdoor cultivation in mild climates only. It is also attractive for growing in tubs in greenhouses. A pleasing feature is the coloring of the canes, which are bright yellow with green, and when grown in the sun, pink streakings. Remains of the leaf sheaths stay attached to the stems for a considerable time. There are many branches from each node. The leaves may be 1 foot long by 3 inches wide, but often

are considerably smaller. Their margins are fringed with fine bristles. The upper leaf surface is light bluish-green, the undersurfaces paler.

A square-stemmed bamboo sounds an unlikely component of a group of plants that typically have canes that are round, or rarely round with one flattened side as in *Phyllostachys*, yet *C. quadrangularis* (syn. *Bambusa angulata*) is just that. Its mature stems are very decidedly four-cornered. Native to China, this curious plant ordinarily is up to about 10 feet tall, but specimens 30 feet in height are reported to grow in Japan. It does not form a tight clump, but from underground rhizomes sends up rather widely-spaced canes at first dark green, but later brownish-green. The leaf sheaths are early deciduous. The branches are many and comparatively long. Unlike those of most bamboos they can be easily broken from the cane without using knife or shears. The tapered leaves, mostly two to four at the end of each branchlet, are up to 9 inches long by 1¼ inches wide. Their margins are fringed throughout with bristly hairs. The upper leaf surface is dark olive-green and moderately tessellated. The undersides are duller and paler. This bamboo is not hardy in the north, but is suitable for regions of mild winters. It also is an attractive tub plant.

Garden and Landscape Uses and Cultivation. For information about these see Bamboos.

CHINA or CHINESE. These words are used as part of the names of many plants including these: China-aster (*Callistephus chinensis*), China-fir (*Cunninghamia*), China fleece vine (*Polygonum aubertii*), China-grass (*Boehmeria nivea candicans*), Chinaberry or China tree (*Melia azedarach*), China wood oil tree (*Aleurites fordii*), Chinese angelica tree (*Aralia chinensis*), Chinese chives (*Allium tuberosum*), Chinese evergreen (*Aglaeonema modestum*), Chinese-forget-me-not (*Cynoglossum amabile*), Chinese goddess bamboo (*Bambusa glaucescens rivsereorum*), Chinese-gooseberry (*Actinidia chinensis*), Chinese hat plant (*Holmskioldia sanguinea*), Chinese jujube (*Ziziphus jujuba*), Chinese jute (*Abutilon theophrasti*), Chinese lantern plant (*Physalis alkekengi*), Chinese parasol tree (*Firmiana simplex*), Chinese sacred-bamboo (*Nandina domestica*), Chinese sacred-lily (*Narcissus tazetta orientalis*), Chinese scholar tree (*Sophora japonica*), Chinese silk plant (*Boehmeria nivea*), Chinese tallow tree (*Sapium sebiferum*), Chinese water-chestnut (*Eleocharis dulcis*), Chinese watermelon (*Benincasa hispida*), Chinese yam (*Dioscorea opposita*), wild China tree (*Sapindus drummondii*), and yellow Chinese-poppy (*Meconopsis integrifolia*).

CHINCH BUGS. The group of true bugs identified as chinch bugs are minute, soft-bodied insects much longer than wide that

do immense damage to corn, grains, and grasses. They hibernate through the winter among dry leaves, grasses, or other vegetation. In spring each female lays up to 500 eggs at the bases or on the roots of host plants. Soon, the young, bright red or yellow marked with brown, hatch and begin sucking the plant juices. As a result, the stems shrink, become yellow, and often are killed. There are two types of adult chinch bugs, those with long wings that rarely fly, although they can, and a short-winged type that cannot fly. The common chinch bug, 1/16 inch long, and black with red legs and white wings, when crushed emits a foul odor. It is chiefly a pest of corn and grains. The hairy chinch bug favors lawns, causing brown patches of killed grasses. It is especially fond of bent and St. Augustine grasses. Unlike the grasses in the brown patches caused by beetle grubs, those in damaged areas resulting from chinch bug damage cannot be readily rolled up like a carpet. Chinch bugs are most active in hot weather and in sunny places. There are two, or in the south sometimes three broods each season. To detect chinch bugs in lawns part the grass near the edges of damaged areas and with the face close to the ground peer intently at the base of the grasses. The chinch bugs move rapidly. Control is by spraying with insecticides recommended by Cooperative Extension Agents and other reliable authorities.

CHINCHERINCHEE is *Ornithogalum thyrsoides*.

CHINESE CABBAGE. Very different from ordinary cabbage, the vegetables known as Chinese cabbage, pe-tsai, pak-choi, and celery cabbage, are varieties of Asian forms of rape (*Brassica rapa*) of the mustard family CRUCIFERAE. They are grown for their leafy heads, used in salads and cooked as greens,

Chinese cabbage

and are much more delicately flavored than cabbage. Those of pe-tsai (formerly *B. pekinensis*) have long, compact heads somewhat like those of a cos lettuce, but bigger. The pak-choi (formerly *B. chinensis*) varieties have much looser heads, more like those of Swiss chard.

Chinese cabbage goes quickly to seed in hot weather. It is important to sow early enough to have plants usable before sultry days arrive. In most sections this involves starting plants for early use indoors. Fall harvests are had from seeds sown outdoors in June, July, and in regions of mild winters, later. The soil and its preparation should be as for cabbage. High fertility and adequate moisture encourage rapid growth and good quality. Favorite varieties are 'Chihli', 'Michihli', and 'Wong Bok'.

Early sowings in greenhouses and hotbeds are made, and the seedlings managed, in the same way as for early cabbage. Because the plants resent root disturbance take a little extra care when transplanting. Set them in the open ground as soon as danger of frost is past, in rows 1½ to 2½ feet apart with 9 inches to 1½ feet between individuals. Where summers are cool sow thinly outdoors in spring as early as the ground can be made mellow. Space the rows as recommended above and thin the seedlings, in two or three steps, to the distances given. The later thinnings may be used in the kitchen. Late sowings for autumn harvesting are made and handled in similar fashion.

Care during the growing season consists of prompt elimination of weeds by shallow surface cultivation and if necessary hand pulling, drenching the soil with water at about five-day intervals in dry weather, and applying a dressing of nitrate of soda, urea, or other quick-acting high-nitrogen fertilizer when the plants are about half grown. Harvest for immediate use by cutting the heads off, as soon as they are firm and fully mature, just above the roots. Remove the outer leaves. Harvest for storing before severe frost by pulling up the plants with roots attached. They will keep for a few weeks in a cool, humid cellar or similar place. Diseases and pests are those common to cabbage and other brassica crops.

CHINQUAPIN. See Castanea and Castanopsis. The water-chinquapin is *Nelumbo lutea*, the chinquapin oak *Quercus prinoides*.

CHIOCOCCA (Chio-cócca) — Snowberry. This, a genus of twenty species confined to the tropics and subtropics of the Americas, is quite different from, and not related to, the familiar snowberry of northern gardens (*Symphoricarpos*). It belongs to the madder family RUBIACEAE and so has kinship with *Bouvardia* and *Ixora*, although in aspect it closely resembles neither. Its name, from the Greek *chion*, snow, and *kokkos*, a berry, refers to the white fruits.

Chiococcas are hairless, often climbing shrubs, with opposite, leathery leaves and small flowers in branched or branchless racemes from the leaf axils. The blooms have persistent calyxes with swollen tubes and five lobes or teeth. The corollas have funnel-shaped tubes and five spreading or reflexed lobes (petals). There are five stamens and a style ending in a shortly two-lobed or lobeless stigma. The fruits are berry-like drupes.

Widely dispersed in the wild from Florida to Mexico, Central America, and the West Indies, *Chiococca alba* (syn. *C. racemosa*) is a variable, more or less vinelike shrub with elliptic, ovate, or lanceolate, thin leaves 1½ to 3½ inches long, and greenish-white to yellow flowers ⅓ inch long and nearly as wide in racemes about as long as the leaves. The stamens do not protrude. The fruits are compressed, white, and approximately ⅕ inch in diameter.

Garden and Landscape Uses and Cultivation. Sometimes planted as a general purpose ornamental shrub of interest chiefly for its berries, this snowberry succeeds in warm, essentially frostless climates. It is propagated by seeds and by cuttings.

CHIOGENES. The genus previously known by this name is included in *Gaultheria*.

CHIONANTHUS (Chion-ánthus)—Fringe Tree. There are two species of fringe tree, one native of the eastern United States, the other of China. They are deciduous small trees or tall shrubs of the olive family OLEACEAE. The name, from the Greek *chion*, snow, and *anthos*, a flower, applies to their abundant white blooms.

Fringe trees have opposite, undivided, stalked leaves that, except sometimes on young plants, are toothless. The flowers, in showy panicles, are remarkable for their long, narrow, white petals, which give the fringe effect that inspired the common name. The blooms have small, persistent,

four-lobed calyxes, and usually four, more rarely five or six petals joined for a short distance at their bases. There are two or rarely three short stamens, and a short style with a two-lobed stigma. The fruits, produced only by female trees, and then only if a male is nearby for pollination, are more or less egg-shaped drupes (plumlike in structure), usually with one seed.

Common fringe tree (*Chionanthus virginicus*) is hardy in southern New England, although its natural range reaches only to New Jersey, from where it extends to Florida and Texas. It attains a height of about 30 feet, but often is lower. Its leaves are narrow-elliptic to obovate-oblong and 3½ to 8 inches long. Their undersides at first are hairy, at least on the veins, as are the leafstalks. The drooping panicles of slightly greenish, white blooms, about 6 inches long, develop from the upper parts of the previous year's shoots and are displayed in large, moplike clusters below the new growth. The petals are ¾ inch to 2¼ inches long and under ¹⁄₁₀ inch wide. The blooms and flower panicles of male plants are larger than those of females. The decorative dark blue fruits are up to ¾ inch long. This species blooms in the Deep South in March and April, in the vicinity of New York City in late May or early June.

Chinese fringe tree (*C. retusus*) is especially lovely. Broadly round-headed, grow-

Chionanthus retusus

Chionanthus virginicus

Chionanthus retusus (flowers)

ing in the same locality it blooms two to three weeks earlier than its American counterpart. Its flowers are whiter than those of the latter, and in cultivation the tree, as well as its foliage and blooms, is smaller. Ordinarily it does not exceed 20 feet in height, although in China it is 30 to 40 feet tall. The lustrous leaves, downy beneath and on the midribs of their upper sides, are 1 inch to 4 inches long by ¾ inch to 2 inches wide. From 2 to 4 inches long, the panicles of pure white blooms are produced in such profusion, on short shoots of the current season's growth, that the Greek equivalent of snow flower that is the botanical name of this species is well merited.

Garden and Landscape Uses and Cultivation. Both species of fringe tree are beautiful additions to landscape plantings. Much admired in bloom, they have foliage that turns yellow in fall. Used as lawn specimens and in other ways that allow space for their development and adequate display, they are very effective. Fringe trees do best in full sun, but stand part-day shade. For their most successful growth they need fertile, moist but well-drained soil. No regular pruning is required. Increase is by seeds, by cuttings made of young shoots planted under mist or in a greenhouse propagating bench, and by grafting or budding onto seedling ash trees. If seeds are sown outdoors in a protected place or in a cold frame in fall they usually do not germinate until the second spring. Faster results are had by stratifying the seeds in fall and sowing in spring. This is done by mixing them with slightly damp peat moss, sand, a mixture of peat moss and sand or vermiculite, sealing them in a polyethylene plastic bag, and storing them for three months in a temperature of 60 to 65°F and then for three months at 40°F before sowing in a greenhouse or cold frame. Seed stored in a dry, cool place for up to a year will germinate after being stratified.

CHIONODOXA (Chiono-dóxa)—Glory-of-the-Snow. The blooming of forsythias overlaps the flowering of glory-of-the-snows, members of the genus *Chionodoxa*. They belong in the lily family LILIACEAE and are natives of mountains from Crete to Asia Minor. There are about six species, not all cultivated. The name is from the Greek *chion*, snow, and *doxa*, glory. It alludes to the flowers appearing as the snows recede in spring. The plant known as *Chionodoxa allenii* is *Chionoscilla allenii*.

Glory-of-the-snows have small bulbs and linear to oblanceolate, flat leaves, two or three to each flower stalk. The flowers form short racemes of a few, stalked blooms with the perianth consisting of a short tube and six spreading lobes (commonly called petals). There are six separate stamens, which have broad, flattened stalks (fila-

ments) and are crowded in a cone in the center of the flower, and a single pistil. The seeds are in three-lobed or three-winged capsules. Chionodoxas differ from squills (*Scilla*) in that the perianth lobes are joined for an appreciable distance at their bases; those of *Scilla* are separate or essentially so. The blooms of glory-of-the-snows face unashamedly upward without any suggestion of the drooping posture of many squills.

As hardy bulbs go, glory-of-the-snows are comparatively newcomers to the horticultural world. The first discovered, but one not common in cultivation, was *C. nana*, an inhabitant of the mountains of Crete. It was found by the Austrian traveler Sieber in 1820. As late as 1879 the genus was scarcely known to gardeners. In that year the great English horticulturist and connoisseur of plants Henry J. Elwes advised his countrymen and others who read the "Gardeners' Chronicle" that *C. luciliae* "is one of the best, if not the very best of its class, for surpassing any of the squills, and apparently as hardy and as easy to increase as *Scilla siberica*." Note the uncertainty about the hardiness of *C. luciliae*. By then it had not been tried sufficiently for Elwes to be sure how cold resistant it was. Fortunately time and experience have dispelled all doubts, *C. luciliae* is indeed as hardy as the Siberian squill. One of the most commonly culti-

Chionodoxa luciliae

vated kinds, it has pear-shaped, small bulbs and rather broad, recurving leaves. Up to six or even seven or eight starry, sky-blue blooms that pale to white toward their centers and have purple ovaries, are carried on stems 4 to 6 inches or so tall. Three or four only of the flowers are open at one time. They are up to 1 inch in diameter. The variety *C. l. alba* is white-flowered. The blooms of *C. l. rosea* blush faintly on their upper sides, the reverses of their petals and their flower buds are more obviously *rosea*. A more robust variety is *C. l. 'Pink Giant'*. It has taller stalks with brighter pink blooms.

Chionodoxa sardensis

The deep gentian-blue flowers of *C. sardensis* lack the broad white centers of those of *C. luciliae*, instead they have tiny white disks in their throats. Also, their ovaries are olive-green and the blooms are slightly smaller, about ¾ inch in diameter. More venturesome than *C. luciliae*, in the vicinity of New York City this kind bows in well before March is gone, ahead of the first forsythia and pussy willow blooms. It is a native of mountains to the east of Smyrna.

The latest glory-of-the-snow to bloom is *C. l. tmoli*, named from its native locale, a mountain to the east of Smyrna called

Chionodoxa luciliae tmoli

Boz Dagh, the ancient name of which was Tmolus. There, it inhabits deep valleys and gorges where, in winter and well into spring, it sleeps under thick blankets of drifted snow. Untold centuries of life under such conditions must have established its propensity for not awakening too early, a habit that a few decades of cultivation has failed to change. Its blooms are larger than those of *C. luciliae* and appear later in the spring. They are blue with white centers.

Chamaecereus sylvestri

Chelidonium majus flore-pleno

Chamaecyparis obtusa aurea

Chionanthus virginicus

Cheiranthus cheiri, varieties

Chionodoxa luciliae

Chrysanthemum, annual

Chrysanthemum, disbudded decorative variety

Chrysanthemums, hardy garden varieties

Chrysanthemums, hardy garden variety
with blue salvia

Garden Uses. Glory-of-the-snows are not bulbs to be planted in niggardly fashion. Half a dozen or one dozen make no show worthy of the name. Fifty to a hundred may give some satisfaction, but to be seen at their loveliest they should be strewn in hundreds, or better still in thousands, and planted 2 or 3 inches apart, not with regularity, but spaced unevenly as though they have sprung from seeds self sown. As a matter of fact, under favorable conditions, they do increase from seeds they drop, and colonies gradually increase in numbers. This more often is true of *C. luciliae* and *C. sardensis* than of *C. l. tmoli.* The last-named needs more shade and damper soil than the others, which grow best in full sun although they tolerate light shade. Glory-of-the-snows are splendid for planting in places kept free of grass around the bases of trees and shrubs. Because they bloom at the same time, they can be used with great effectiveness beneath forsythias, pussy willows, star magnolias, and *Corylopsis.* They are also delightful plants for rock gardens and for the fringes of shrub borders and the fronts of perennial beds.

Cultivation. In agreeable soil glory-of-the-snows need little care. It is better that they be left undisturbed for as long as possible to multiply and colonize. When they become obviously overcrowded they may be taken up when their foliage fades and replanted. They are not very particular about the kind of earth in which they grow, but it must be porous enough to let water drain freely and not so compact that roots and emerging shoots are impeded. A soil of moderate fertility is adequate. The bulbs are planted in fall, as early as they can be obtained, with their bases about 3 inches below the surface. An early spring dressing of a complete fertilizer, applied every year or two, encourages vigor and perhaps multiplication. The seed pods should be allowed to mature as sources of possible increase. Propagation under controlled conditions can be done by sowing ripened seeds as soon as they are mature in a cold frame or outdoor bed, but bulbs are so inexpensive that it is the rare gardener who goes to this trouble, most depending upon self-sown seeds and natural multiplication by offsets.

CHIONOGRAPHIS (Chionó-graphis). Several species, possibly seven, of this Asian genus of the lily family LILIACEAE are native in Japan, southern Korea, and southern China. From their American close cousin *Chamaelirium* they differ in that two or three of the six perianth segments (petals or more properly tepals) usual in the lily family are absent or vestigial, so that the blooms have, or appear to have, only three or four petals. These are very narrowly-linear or even threadlike. There are six short stamens and three styles. The fruits

are capsules. Because the white flower spikes suggest slender brushes, implements used in the Orient for writing, the genus is named from the Greek *chion,* snow, and *graphis,* a pencil. They are herbaceous perennials.

Native of Japan and Korea, ***Chionographis japonica*** has a thick, short rhizome and a loose rosette of basal leaves, narrowly-obovate to oblongish, and with short or long stalks. They are up to 3 inches or slightly more in length and 1¼ inches in width. Their edges are slightly wavy. From 6 inches to 1½ feet tall, the erect flower stalks are furnished on their lower parts with stalkless, up-pointing, linear to linear-lanceolate leaves. The narrow spires of feathery, white blooms that terminate them are 2 to 8 inches in length. This species is hardy in sheltered locations in southern New York. It blooms there in early June.

Garden Uses and Cultivation. A choice plant for rock and woodland gardens, this needs a shaded place where the soil contains considerable leaf mold, peat moss, or other agreeable organic material, and is not excessively dry. It can be propagated by careful division in early spring or immediately after flowering and by seed.

CHIONOPHILA (Chionó-phila). Two alpine species of the figwort family SCROPHULARIACEAE are the only members of the genus *Chionophila.* One, *C. jamesii,* inhabits the southern reaches of the Rocky Mountains, the other, *C. tweedyi,* the mountains of Idaho and adjacent Montana. The name, referring to the habitats, comes from the Greek *chion,* snow, and *philos,* beloved.

Chinophilas are low, erect, herbaceous perennials with chiefly basal foliage and few, inconspicuous, opposite stem leaves. The leaves are undivided, spatula-shaped to lanceolate, lobeless and toothless. The flowers, in one-sided spikes or racemes with usually one bloom at each node, have calyxes with five teeth or lobes, a tubular, two-lipped corolla not cleft to its base, and four fertile stamens and one shorter, sterile one (staminode). The fruits are capsules.

Usually with one slender stem, *C. tweedyi* is 2 to 10 inches tall, and except for glandular hairs on the flowering portion of the stem, hairless. The oblanceolate basal leaves including their stalks are 1 inch to 3½ inches long by up to ½ inch wide. There are three or fewer pairs of leaves, rarely as long as ¾ inch, on the stems. The flower spikes carry ten or fewer pale lavender blooms ⅓ inch long or slightly longer.

From the last, *C. jamesii* differs in its creamy, approximately ½-inch-long flowers being in squatter, much denser, one-sided spikes. They have larger and more funnel-shaped calyxes than *C. tweedyi* and corolla tubes that are nearly or quite straight.

Garden Uses and Cultivation. Considered difficult, perhaps nearly impossible to

tame in cultivation, the members of this genus challenge cultivators of alpine plants. Success would be likeliest under conditions suitable for alpine penstemons and other high altitude species. Seed undoubtedly affords the best means of propagation.

CHIONOSCILLA (Chiono-scílla). This is the name of bigeneric hybrids between *Chionodoxa* and *Scilla.* They show characteristics intermediate between their parents. Known from the wild in Asia Minor and from stocks resulting from hybridization in gardens, ***Chionoscilla allenii*** has as its parents *Chionodoxa luciliae* and *Scilla bifolia.* A quite charming, spring-blooming bulb plant, it is much like a superior *Chionodoxa luciliae,* but with its perianth segments separated essentially to their bases, as in *Scilla bifolia,* rather than joined. There are two channeled, strap-shaped leaves to each flower stalk, and six or seven rich blue blooms that become somewhat violet with age, but are without white centers. The uses and cultural needs of this plant are those of its parents.

CHIPMUNKS. Gardens are rarely seriously harmed by these pretty, pert, friendly, small ground squirrels, but on occasion chipmunks become pests and must be controlled. They can cause much disturbance in rock gardens by burrowing and by their occasional appetites for small bulbs and succulent herbage. Their normal foods are nuts, seeds, and berries.

Chipmunks excavate underground dens of several chambers at depths of about 3 feet. Sloping tunnels lead to them. The soil removed is not left in mounds, but scattered. Each den is occupied by a family of several members. Certain chambers are used as storehouses for food, others for sleeping. During winter the chipmunks remain below ground sometimes torpid for short periods, sometimes awake. In spring litters of five or six are born. They are blind until they are about one month old. At about three months they are able to take care of themselves.

Trapping is the best method of dealing with chipmunks. Live-taking traps baited with corn, wheat, oats, or peanut butter and placed near the opening to their burrows work well. The ''catch'' may be released in woods or similar environments some considerable distance from the garden.

CHIRANTHODENDRON (Chirantho-déndron)—Mexican Handflower Tree. For long, inhabitants of the Valley of Mexico believed that there existed in the world only one specimen and some cultivated progeny of the tree they called *arbol de las manitas.* Their tree was an ancient one at Toluca near Mexico City. They did not know it had been brought, before European dis-

Chirita lavandulacea

Chiranthodendron pentadactylon

covery of America, from forests to the south, or that in parts of Guatemala its species was the most abundant element of the forest. Partly because of its supposed uniqueness, but largely because of the curious, and to the Mexicans significant, shape of the blooms, the tree at Toluca was and is venerated in religious observances. It is known by many names including *arbol de las manitas*, *mano de mico*, devil's hand, monkey's hand, and macpalxochiquahuitl. One of the most celebrated of Mexican trees, this species is commonly planted in that country.

The Mexican handflower tree (*C. pentadactylon* syns. *C. platanoides, Cheirostemon platanoides*), of the bombax family BOMBACACEAE, is the only member of its genus. Its name, alluding to the flowers, is derived from the Greek *cheir*, a hand, *anthos*, a flower, and *dendron*, a tree.

Related to *Fremontodendron*, the Mexican handflower tree (*C. pentadactylon*) attains a height of up to 50 feet. The long-stalked leaves, with blades up to 1 foot long, are usually shallowly three- to seven-lobed and are conspicuously clothed with stellate (star-shaped) hairs on their undersides. The most remarkable feature of this species is its petal-less blooms. From their red-streaked, green, bell-shaped, deeply-five-cleft calyxes 1½ to 2 inches long, and

containing in their bases five nectar-secreting glands, five bright red stamens about 3 inches in length extend in one plane. They suggest the fingers of an outstretched hand, a hand with but a tiny palm formed by the joining of the lower parts of the stamens, and with curved clawlike appendages for fingernails. The style, shorter than the stamens and pointing to one side of them, suggests a thumb. The leaves of the Mexican handflower tree are shaped somewhat like those of lindens. When crushed, they have the odor of fennel. The fruits are woody capsules up to about 5 inches long, with five prominent longitudinal ridges.

Garden and Landscape Uses and Cultivation. In California and other warm places where little or no frost is encountered this unusual tree is grown as something of a curiosity. It succeeds in well-drained, dryish soil and is propagated by seed.

CHIRITA (Chir-ìta). Native of warm parts of Asia, *Chirita*, of the gesneria family GESNERIACEAE, comprises some eighty species, including a few cultivated ones of considerable decorative merit. The name, from *cheryta*, Hindustani for a gentian, alludes to the blue flowers of some species. Chiritas are annuals and herbaceous perennials, mostly low or of moderate height, and

often with succulent stems. Their leaves are opposite or in whorls (circles of more than two) and generally are broad-elliptic to lanceolate, frequently with asymmetrical bases. The tubular flowers have calyxes of five, usually narrow lobes and funnel-shaped corollas with five rounded lobes (petals). There are two fertile stamens and a solitary style. The fruits are capsules.

Annual chiritas cultivated are *C. lavandulacea, C. elphinstonia,* and *C. micromusa;* the first, as its name indicates, has lavender blooms, the other two are yellow-flowered. All have more or less translucent, succulent stems. Under favorable conditions *C. lavandulacea* attains a height of 2 to 3 feet; the others are up to 1 foot tall. The pale green leaves of *C. lavandulacea* have broad-elliptic, clearly veined, softly-hairy blades up to 8 inches long by two-thirds as wide. The lower leaves are long-stalked and are the largest. Upward, they gradually diminish in size and stalk length. The blooms, 1 inch to 1¾ inches long, have glandular-hairy stalks and are few together in the leaf axils and from between pairs of sessile leaves that terminate the stems and branches. The corolla tubes are white with a yellow spot inside at their bases. The petals are a lovely lilac-blue. The style is glandular, the stigma two-lobed. The seed capsules are up to 2½ inches in length. This species is believed to be native of the Indo-Malaya region. Thailand is home to both *C. elphinstonia* and *C. micromusa*. The chief differences between these are that *C. micromusa* has flowers lacking the red-brown spots of those of the former and shorter seed capsules. Also, it is usually somewhat lower. Up to 1 foot tall, *C. elphinstonia* has lustrous, bright green, heart-shaped leaves, and deep yellow blooms about ¾ inch long, with a pair of red-brown spots in their throats.

Chirita micromusa

A native perennial of Thailand, *C. ana-choreta* attains a height of 2 to 3 feet and produces from its base several stems. Its thinnish, hairy, lanceolate leaves have asymmetrical bases. Lemon-yellow and 2½ inches long by 1½ inches across their faces, the flowers, from the leaf axils, are mostly in threes. A curious feature is the yellow knob at about the centers of the white stalks of the stamens.

Chiritas with blue to purple flowers more or less relieved by white, include perennial *C. asperifolia* (syn. *C. blumei*), of Indonesia. This has reddish stems and rough-hairy, bright green leaves with paler undersides. The waxy flowers, 1½ inches in length, have white petals and rich purple bases to the corolla tubes. A variety with all-white blooms occurs. Burmese *C. pumila* is up to 2 feet tall. Its hairy leaves, marked with patches of brown, are elliptic and about 5½ inches long by 2 inches wide. With white corolla tubes and light violet petals, the flowers are about 1¼ inches in length. Western China is home to *C. speciosa* (syn. *C. trailliana*), which has a short creeping stem and all basal, long-stalked, round-toothed, broad-ovate, hairless leaves with blades 4 to 8 inches long. The 2¾-inch-long flowers, pale mauve on their outsides, darker within, have their throats marked with two yellow lines.

Chirita pumila

Chirita sinensis

Most distinct, and very un-chirita-like in appearance, *C. sinensis* is native to China, and also Hong Kong. Its leaves, in flat, basal rosettes, are thick and fleshy. They may be unrelieved green or conspicuously and attractively variegated with patches of silvery-white. On long, erect stems from the leaf axils the large clusters of flowers are borne. The blooms are 1½ inches long and light lilac with yellow-orange marks in their throats. The stigma is tongue-shaped.

Garden Uses and Cultivation. Chiritas are of much interest to collectors of gesneriads (members of the gesneria family) and, the showier ones, for the embellishment of greenhouses and conservatories. Since many come from fairly high altitudes in the tropics, they find moderate temperatures, such as suit many begonias, agreeable and are satisfied with 55 to 60°F at night in winter, with an increase of five to fifteen degrees by day. In spring temperatures may be raised a few degrees, and in summer they will naturally often be higher. High humidity is important, as is good light with shade from strong sun. Soil that is coarse and fairly well supplied with organic matter and that permits the free passage of water is to the liking of chiritas. Except for *C. sinensis*, which needs exceedingly porous soil kept decidedly on the dry side in winter and moist at other times, it should be kept evenly, but not excessively moist. Care should be exercised not to wet the leaves, or at least not to do so if there is any possibility that they will remain wet for longer than about an hour. Long periods of wetness cause the foliage, noticeably that of *C. lavandulacea* and *C. sinensis*, to rot. Exposure to too strong sun also damages the foliage; but *C. sinensis* needs more light than most. The perennial kinds can be increased by cuttings, leaf cuttings, and sometimes division, and by seeds.

Annual chiritas are best propagated by seeds that are usually produced freely, but if necessary they can be multiplied and kept going by cuttings. To obtain good specimens of the annuals the seeds are sown in spring, and it is important to keep

the plants growing without check otherwise they become stunted and are likely to bloom prematurely and less satisfactorily than they would otherwise. With this end in view, they are repotted into larger containers each time their roots reasonably fill the ones they occupy until they are in their flowering pots. The final shift is given in late summer, and in the case of *C. lavandulacea* raised from early-sown seeds, may be into 6- or 7-inch pots. Smaller-growing kinds, such as *C. micromusa*, are finished in considerably smaller receptacles as are plants of *C. lavandulacea* from later sowings. Once the roots have filled the final containers, weekly applications of dilute liquid fertilizer will stimulate continued favorable growth. For further information see Gesneriads.

CHIRONIA (Chirò-nia). African and Madagascan annuals, perennial herbaceous plants, or rarely subshrubs totaling thirty species compose *Chironia* of the gentian family GENTIANACEAE. The name is that of one of the centaurs of Greek mythology. Chironias have opposite, undivided, toothless, stalkless leaves. Their pink, magenta, purplish, or rarely white flowers are solitary or in terminal clusters. They have a five-lobed calyx, a slender-tubed, five-lobed corolla with five wide-spreading lobes (petals), five stamens, and a single style. The fruits are capsules or rarely berries. In general aspect these plants resemble the American genus *Sabatia*.

An undershrub 6 inches to 1¼ feet tall, *C. baccifera* has linear to oblong leaves up to ¾ inch long, with margins slightly rolled under. Its flowers are ¾ to 1 inch wide and have corolla tubes as long as the spreading, blunt petals. Bright red berries, responsible for the common name Christmas berry applied to this species in South Africa, are its fruits.

Chironia baccifera

Chironia linoides

Chives

A herbaceous perennial, sometimes slightly woody below, **C. linoides** is 6 inches to 1 foot high. Its stems are more or less prostrate below, then erect. From 1 inch to 2 inches long, its leaves are flat and narrowly-linear. The blooms, 1½ to 2 inches in diameter, have corolla tubes slightly more than ½ inch long and pointed, ovate-lanceolate petals. The fruits are capsules. This species grows in the wild in South Africa in moistish soils.

Garden and Landscape Uses and Cultivation. Tender to frost, chironias are adopted for outdoor cultivation only where none or very little is experienced and for greenhouses. They are gay plants for rock gardens, fronts of flower beds, and similar places, and are attractive in pots. They need a sunny location and well-drained soil, preferably of a sandy, peaty nature and not excessively dry. Chironias are easily raised from cuttings and seeds. In sunny greenhouses they thrive where winter temperatures are 50°F at night and five to fifteen degrees higher by day. Pots rather small for the size of the plants suit better than larger ones. They must be well drained. Coarse, sandy peaty, fertile soil is best. It should be kept always just moist, with excessive wetness guarded against, especially in winter. During their early stages of growth the tips of the stems are pinched out occasionally to induce branching. Plants a year old or older are pruned moderately and are repotted in spring. They bloom in fall.

CHITTAMWOOD is *Bumelia lanuginosa.*

CHIVES. Among the easiest to grow of culinary herbs, chives (*Allium schoenoprasum*) is a hardy deciduous herbaceous perennial native to Europe and temperate Asia. Its variety *A. s. sibiricum* is indigenous in North America. A close relative of the onion, chives is pretty enough in foliage and flower to warrant a place in flower beds and rock gardens as well as vegetable and herb gardens. It is easy to grow in pots indoors on sunny window sills.

This herb forms crowded clumps of narrow bulbs from which sprout slender, dark green, hollow, cylindrical leaves 6 to 9 inches tall, onion-scented and -flavored. The flower stalks lift the small globular heads of lavender-purple blooms for display just above the foliage. Chinese-chives is *Allium tuberosum.*

Ordinary garden soil and a location in full sun suit chives. Plant small tufts in early spring 6 to 9 inches apart in rows to form edgings to beds or space them similarly in solid beds. Lift, separate, and replant every three or four years. Chives need no care other than that necessary to suppress weeds and, if the foliage is harvested regularly, the application once or twice a year of a little fertilizer.

Harvest repeatedly and regularly whether the produce is needed or not by cutting the "grass" about 1 inch above the bulbs with a sharp knife. This stimulates the growth of young, tender leaves. To assure a winter supply of fresh chives pot plants in 4- or 5-inch containers in early fall and transfer them to the kitchen window or other sunny place indoors. Keep them well watered and give them dilute liquid fertilizer every week or two. Those without gardens can cultivate chives in this way throughout the year.

Propagation is usually by planting small tufts had by pulling old clumps apart, but seeds can also be used. Sow outdoors in spring in rows 1 foot apart, and thin the seedlings to 3 inches apart in the rows. Do not harvest the first summer. The following spring transfer the young plants to their permanent locations.

CHLIDANTHUS (Chlid-ánthus). One South American species of *Chlidanthus* is the only member of this genus of the amaryllis family AMARYLLIDACEAE. Its name derives from the Greek *chlidanos*, delicate, and *anthos*, a flower. As the name indicates, its flowers are delicately fragrant, with something of the scent of narcissus and just a suggestion of lemon.

A bulbous plant, **C. fragrans** has about six glaucous, daffodil-like, narrow-linear basal leaves up to 8 inches long and a solid flower stalk 1 foot to 2 feet tall that bears at its top a few-flowered umbel of yellow, waxy, lily-shaped flowers 4 to 5 inches long by 3 inches across. The blooms have long perianth tubes, six segments or petals, the same number of stamens, and a protruding style. The fruits are capsules.

Garden Uses. In mild climates where little or no frost is experienced this delightful summer-blooming bulb can be accommodated in outdoor beds. It is also charming when grown in pots in greenhouses. It needs a warm, well-drained soil and a sunny location.

Cultivation. Plant the large, round bulbs out of doors in spring with their tips 3 inches beneath the soil surface. Watering to keep the soil moderately moist during the growing period is beneficial, but in late summer and fall, after flowering, it is better that the soil be dry.

For indoor cultivation plant the bulbs in well-drained pots of porous fertile soil in spring, with their tips 1 inch beneath the soil surface. Water thoroughly immediately after planting and put the containers in a cold frame or cool greenhouse. Until leaf growth begins keep them shaded from direct sun and water rather sparsely until the new roots permeate the soil. Later, increase the amount of water given to keep the soil always moist and apply dilute liquid fertilizer at weekly intervals to pots that are completely filled with healthy roots. After flowering, when the foliage begins to die down naturally, gradually reduce the water supply and finally, when the leaves are dead, withhold it completely and store the bulbs in the soil in which they grew in a cool, dry, frost-proof place until the following spring.

Greenhouse-grown specimens need a night temperature of 50 to 55°F and day temperatures a little higher at the beginning of their growing season. In summer, when outdoor temperatures are adequate, the greenhouse should be well ventilated and lightly shaded. Repot every three or four years. In intermediate years the upper surface of the old soil should be removed and replaced with a rich mixture. This top dressing or repotting is done in early spring just before new growth begins.

CHLORANTHACEAE—Chloranthus Family. Sixty-five species of dicotyledons contained in five tropical and subtropical genera compose this family. Trees, shrubs, and herbaceous plants, they have opposite leaves and spikes or clusters of small bisexual or unisexual flowers usually without sepals and petals, but sometimes with se-

pal-like petals. They have one to three stamens united to each other and the ovary,
and one pistil. The fruits are drupelike.
Genera sometimes cultivated are *Ascarina*
and *Chloranthus.*

CHLORANTHUS (Chlor-ánthus). This genus, like other members of the chloranthus
family CHLORANTHACEAE is mostly tropical
and subtropical. Its fifteen members are indigenous to eastern and southeast Asia.
The name derives from the Greek *chloros,*
green, and *anthos,* a flower, in allusion to
the color of the flowers of some kinds. Related to the pepper family PIPERACEAE,
these plants are aromatic, evergreen shrubs
and herbaceous perennials, with jointed
stems and opposite, undivided leaves. In
slender, terminal spikes, the minute, inconspicuous flowers have the calyx and
corolla reduced to a single small scale.
Usually there are three stamens, sometimes fewer.

Native of warm parts of China and
southeastern Asia, **Chloranthus spicatus**
(syn. *C. inconspicuus*) is a low evergreen
shrub with oblong-ovate to ovate-elliptic
leaves up to 3½ inches in length, about
one-half as wide as long, and spaced
evenly along the stems. Their margins are
shallowly and distantly-bluntly-toothed.
From the branch ends, the yellowish flower
spikes, ¾ inch to 1¼ inches long, arise in
clusters. This species is not hardy in the
north.

A hardy kind, *C. japonicus* is an attractive, deciduous herbaceous perennial that
prospers outdoors at least as far north as
New York City. Native of Japan, Korea,
and Manchuria, it has short, creeping rhizomes and erect, usually branchless stems
furnished on their lower parts with small,
scalelike leaves, and at their tops with
what looks like a whorl (circle) of three or
four leaves, but which really represent two
pairs of pointed short-stalked leaves set
closely together. They are ovate to elliptic,
2 to 4½ inches in length, about one-half
as wide as long, and sharply-toothed.
Branchless and usually solitary, the spikes
of tiny, white stalkless flowers are from a
little less to a little more than 1 inch long
and are at the ends of stalks 1 inch to 2
inches long.

Garden and Landscape Uses. The tender
kind described is an excellent low foliage
shrub for outdoors in the humid tropics
and subtropics. It thrives in ordinary soil
in partial shade and can be used with
pleasant effects in landscape plantings. It
also makes an attractive greenhouse and
conservatory foliage plant for pots or
ground beds. The other is suitable for
shady places in rock gardens, woodland
gardens, and beneath trees the roots of
which are not too concentrated near the
surface. It makes an attractive and distinctive foliage pattern and grows with little

trouble in somewhat acid, not excessively
dry, woodland soil.

Cultivation. Cuttings provide a simple
way of procuring young plants of *C. spicatus,* and *C. japonicus* can be increased by
carefully dividing the plants in early spring.
Both can be raised from seed. Once established, neither needs special care. In
greenhouses *C. spicatus* thrives in well-
drained fertile soil. Repotting when needed
should be done in late winter or early
spring. Enough water is given to keep the
soil uniformly moist from spring through
fall, rather drier, but never completely so,
in winter. Well-rooted specimens benefit
from regular applications of dilute liquid
fertilizer. Temperatures of 55 to 60°F at
night and a few degrees more by day are
satisfactory from fall through spring. A
fairly humid atmosphere and some shade
from summer sun are appreciated.

CHLORINE IN WATER. In many areas
water is chlorinated to reduce its population of harmful microorganisms and make
it fit for drinking and other human uses.
Often water so treated has a decided taste
or slight odor of chlorine. But it is not
harmful to plants. To be so it would have
to contain chlorine in amounts that it
would make it undrinkable. The chlorine
gas soon dissipates if the water is aerated
or left standing in an open container for a
few hours. Nervous growers, reluctant to
apply freshly drawn chlorinated tapwater
to their houseplants, may prefer to follow
one of the above procedures. They are not
necessary, but if they make you feel better,
why not?

CHLORIS (Chlòr-is)—Finger Grass. This
genus of fifty species of the grass family
GRAMINEAE inhabits warm regions of the
Old and New worlds, including the southern United States. Its name is that of the
Greek goddess of flowers *Chloris.* In the
tropics and subtropics Rhodes grass (*C.
gayana*) is esteemed for pastures. These annual and perennial grasses are tufted or
spread by stolons. They have slender stems
and folded or flat leaf blades. The one-
sided spikes of flowers are solitary, or
more frequently in umbel-like clusters, at
the tops of the stems. The stalkless, flattened, usually awned (bristle-tipped)
spikelets, arranged in two rows, each contain one fertile and one or more sterile
flowers.

Spontaneous in parts of the United States,
C. virgata (syn. *C. elegans*), a species widely
distributed in the tropics, is a tufted annual up to 2 feet tall. It has flat leaves with
blades up to ¼ inch wide and about 4
inches long. The silky or feathery, green
or purplish, bearded flower spikes, in umbels of six to fifteen, are 1½ to 3 inches in
length.

Other species include *C. barbata* and *C.*

truncata, the former widely distributed in
the tropics, the latter a native of Australia
and naturalized in North America. An annual, *C. barbata* (syn. *C. inflata*) forms
loose tufts of stems, sometimes prostrate
at their bases, 1 foot to 2½ feet tall. Up to
about 10 inches long and ¼ inch wide, the
leaf blades are flat. The umbels consist of
five to fifteen usually purplish or brownish, white-bristly spikes 1 inch to 3 inches
in length, with bristle-tipped spikelets.

Perennial and spreading by underground stolons, *C. truncata* is 6 inches to
1 foot tall. Its leaves, flat or folded, are up
to ¹⁄₁₀ inch wide. The slender stems are
topped by umbels of six to ten spikes 2 to
6 inches in length, not bearded and consisting of spikelets with short awns.

Also sometimes cultivated are *C. verticillata* and *C. distichophylla,* the latter South
American, the former North American.
Tufted and a few inches to more than 1
foot tall, *C. verticillata* has very slender
leaves crowded at the bases of its stems.
Its flower spikes, in umbels of seven to
ten, are spreading and in one to three
whorls (circles). The spikelets are awned.
From 3 to 9 feet in height, *C. distichophylla* has slender leaves up to 7 inches
long and brownish, awnless flower spikes
2 to 3 inches long and in clusters of eight
to fifteen.

**Garden and Landscape Uses and Culti-
vation.** As ornamentals and to supply material for fresh and dried flower arrangements, these grasses may be cultivated in
beds and borders in ordinary garden soil
in sunny locations. In mild climates the
perennials can be treated as such, but all
described may be grown as annuals. The
seeds are sown in spring where the plants
are to remain and the seedlings thinned so
that they do not crowd each other. Propagation of the perennials is by division as
well as by seed.

CHLOROCODON (Chloro-còdon)—Mundi
Root. Sometimes named *Mondia,* this African genus of two species belongs in the
milkweed family ASCLEPIADACEAE. The
name, from the Greek *chloros,* green, and
kodon, a bell, is descriptive of the blooms.

Chlorocodons are tuberous-rooted, tall
twining vines with opposite, heart-shaped
or oblong-ovate, toothless leaves, and axillary panicles of greenish or purplish,
somewhat bell-shaped blooms. The flowers have five-parted calyxes, corollas with
five lobes (petals), and a five-lobed crown
or corona. The fruits consist of two podlike
follicles.

Mundi root (*Chlorocodon whitei*), native
from tropical Africa to Natal, has twining
stems and pointed, ovate leaves about 7
inches long, with heart-shaped bases. The
pale green flowers, purplish toward the
bottoms of the petals, with white crowns,
are ¾ to 1 inch across.

Garden and Landscape Uses and Cultivation. Of no great display value, mundi root is sometimes planted for variety and interest in warm subtropical and tropical climates. It grows in ordinary soil and is increased by cuttings of firm shoots and by seeds.

CHLOROGALUM (Chloró-galum) — Soap Plant or Amole. The generic name is derived from the Greek *chloros*, green, and *gala*, milk, and refers to the color of the sap. Although none has great beauty, one or more of the five species of this genus of the lily family LILIACEAE are occasionally planted. The most likely to be grown is the soap plant or amole (*C. pomeridianum*), of interest because of the uses to which it was put by American Indians. They cooked and ate the bulbs, crushed them and used them to stupefy fish, and broke and rubbed in water the young bulbs to create lather (hence the name soap plant).

Natives of western North America, chiefly of California, chlorogalums have several linear basal leaves forming tufts and tall, almost leafless flower stalks. The leaves on the flower stalks are much smaller than the basal ones. The flowers have six white, pink, or blue quite separate, linear to oblong petals (actually tepals) that do not drop when they fade. There are six stamens and a long, slightly three-cleft style. The capsules containing the black seeds are nearly spherical. The flowers of *C. pomeridianum* are on stout, glaucous stems that branch freely above and are 2 to 8 feet tall. Each has a 1-inch-long stalk and narrow white petals with green or purplish mid-veins. The petals are up to 1 inch in length. The basal leaves may be 2 feet long or even longer and up to 1 inch wide; they are glaucous and very wavy. This species, native chiefly of dry open hills and plains, opens its blooms in the evening. This is true also of *C. grandiflorum,* a smaller species distinguished by the stalks of its individual flowers being up to ⅕ inch long, and *C. angustifolium,* also smaller and with nonwavy, very narrow leaves. Day-blooming kinds are *C. parviflorum* and *C. purpureum,* the former 1 foot to 3 feet tall and with white to pink blooms, the latter not more than 1¼ feet tall and with deep blue flowers.

Garden Uses and Cultivation. These plants are likely to appeal only to collectors of western American native flora and those interested in plants that were of importance to the Indians. Their cultivation is the same as for *Camassia.*

CHLOROPHORA (Chloró-phora)—Fustic. The dozen species of this genus are natives of tropical South America, the West Indies, tropical Africa, and Madagascar. They belong in the mulberry family MORACEAE. The name, from the Greek *chloros*, green, and *phoreo*, to bear, alludes to wood of the

Chlorophytum comosum variegatum

Chlorophytum comosum mandaianum

fustic tree being a source of yellow, khaki, and green dyes.

Milky-juiced, sometimes spiny trees, chlorophoras have alternate, toothed or toothless, stalked leaves. Their unisexual flowers are small, the males in slender, cylindrical spikes, the females in spherical to oblongish heads. They come from the leaf axils. The blooms have four-parted calyxes, but no petals. The males have four stamens, the females ovaries enclosed by the calyx. The fruits are compound, formed of many achenes.

The fustic tree (*Chlorophora tinctoria*), native of the West Indies and South America, is a nearly evergreen, broad-headed tree up to 60 feet in height. It has stalked, pointed, lanceolate to elliptic, slightly-toothed or toothless leaves up to 5 inches long. The flower clusters are stalked, the male spikes 1½ to 3½ inches long, the female heads spherical and less than ½ inch in diameter. The durable lumber of the fustic tree takes a high polish. It is used for furniture and other purposes.

Garden and Landscape Uses and Cultivation. In the humid tropics the fustic tree is planted for interest and ornament. It is sometimes grown in tropical greenhouses in which emphasis is placed on plants useful to man. It can be raised from seeds, and from cuttings of firm shoots planted in a greenhouse propagating bed with bottom heat. Ordinary soils and locations are suitable where conditions are warm and humid. In greenhouses pruning to restrict the sizes of specimens is done in late winter, and at that time repotting receives attention.

CHLOROPHYLL. This is the name of several closely similar green pigments that occur in plants and have the remarkable ability of being able to absorb energy from light and make it available to molecules active in photosynthesis. Without chlorophyll there is no photosynthesis, without photosynthesis no elaboration of food from simple elements, and without food no life

on earth. Associated with chlorophyll are always one or more accessory light-energy-absorbing pigments that function in part by passing on the energy they collect to the chlorophyll. Notable among these accessory pigments are carotenes and xanthophylls. In green foliage chlorophyll is sufficiently abundant to mask the colors of the accessory pigments, but these are dominant in the yellow, red, and other colored leaves of some plants and become clearly evident in much foliage that becomes colorful in autumn. Chlorophyll is chemically akin to the hemoglobin of blood except that the iron of blood is replaced by magnesium. In addition, chlorophyll contains carbon, hydrogen, oxygen, and nitrogen. It is a complex protein.

CHLOROPHYTUM (Chloróphyt-um) — Spider Plant. Very few of the more than 100 species of this rather bewildering genus have been cultivated, and with few exceptions those that have been grown have been restricted to botanical gardens and suchlike special collections. One exception is the well-known spider plant (*Chlorophytum comosum*), which finds favor as a houseplant and is a frequent inhabitant of barber shop windows and other places favored for the display of easy-to-grow, long-lasting pot plants. The genus belongs in the lily family LILIACEAE. Its name is from the Greek *chloros*, green, and *phyton*, a plant. The genus enjoys a wide distribution. Its chief concentration is in tropical and subtropical Africa, but it also appears in the native floras of the warmer parts of Asia, Australia, Tasmania, Brazil, and Peru.

Closely related to *Anthericum*, but differing in its deeply-three-lobed, sharply-angled or -winged seed capsules, and its flatish, disklike, black seeds, *Chlorophytum* consists of stemless plants with short rhizomes. They have fibrous or thick, fleshy roots and leaves either stalkless and linear or lanceolate to ovate; the leaves have parallel, longitudinal veins. The flowers are several to many on slender stalks and may be clustered tightly or distantly spaced, according to species; the lower part of the flower stalk is usually furnished with one or more leaflike bracts. Individual blooms are rather small and starry, white or greenish, and have six petal-like segments spreading like the spokes of a wheel, six stamens, and a single pistil ending in a small knob. The fruits are capsules, dry when ripe.

The most commonly cultivated species is the spider plant (*C. comosum*), a native of South Africa. It produces an abundance of evergreen, linear to linear-lanceolate leaves up to 1¼ feet long by ¼ to ¾ inch wide, that narrow toward their bases and apexes; they have twelve to sixteen longitudinal veins in addition to the midrib. The flower stalks, branched or branchless, exceed the leaves in length and have scattered racemes of white flowers, each bloom about ¾ inch across and the racemes 6 inches to 1 foot long. Young plants, consisting of rosettes of leaves and roots, almost always develop along the flower stalks and as they increase in size and weight cause them to arch and finally become pendant, producing a cascade-like effect of well-spaced spidery plantlets attached to slender drooping stalks. More familiar to gardeners than the plain, green-leaved species are varieties with variegated foliage. One such, *C. c. variegatum,* has leaves with yellow or cream-colored margins; *C. c. mandaianum* has a yellow central stripe to each leaf. Popular *C. c. vittatum* has recurved leaves with a white center stripe. The kinds long cultivated as *C. capense* (syn. *C. elatum*) and *C. capense variegatum* are rather strong-growing varieties of *C. comosum*. True **C. capense** differs in having glaucous leaves, flattened flower stalks, which never bear plantlets, and longer-than-wide seed capsules. Rather rare in its native South Africa, this species is probably not cultivated.

Smaller than either of the above, **C. bichetii** is becoming increasingly popular. Botanists seem not entirely agreed as to whether it should be classified as a *Chlorophytum* or an *Anthericum,* but a consensus favors the former. About 8 inches in height, it has fleshy roots and tufts of linear-lanceolate leaves 4 to 8 inches long and ⅓ to ⅔ inch wide, striped longitudinally, especially along their margins, with yellowish-white. The flower stalks, distinctly shorter than the foliage, are slender, lax, and usually branchless. They bear white flowers rather larger than those of other cultivated chlorophytums. This species, introduced to cultivation in 1900 is native to tropical Africa.

An attractive species of East Africa is *C. amaniense.* It is an inhabitant of rain forests in the general region of the home of the African-violet (*Saintpaulia*), the Usambara Mountains. Introduced to cultivation early in the twentieth century, this kind differs from others discussed here in having stout, erect, densely-flowered racemes. They are much shorter than the leaves, being less than 6 inches tall. The flowers are greenish. The leaves, which number six to ten, are broadly-lanceolate and glossy green. They have blades 5

inches to 1 foot long by 1¼ to 3 inches wide that taper into short channeled leaf-stalks. The latter, and often the bases of the leaves, are pinkish-buff-orange, a characteristic particularly well marked in specimens grown in good light. The blades have approximately twenty to twenty-six veins.

Somewhat glossy, lanceolate leaves, 1½ to 2½ feet long by 3 to 4 inches wide, narrowed to a short, deeply-channeled, stalk-like base, and with thirty to forty veins, are characteristic of tropical African *C. macrophyllum.* The white flowers with recurved petals about ¼ inch long have yellow anthers and are in slender racemes 9 inches to about 1 foot long terminating slender, erect, branchless stalks about as long.

Commonly cultivated in Europe, but less popular in America is *C. orchidastrum,* a native of tropical Africa of somewhat plantain-lily (*Hosta*) aspect that has six to nine spreading lustrous, oblong to ovate-lanceolate, wavy-edged leaves up to 1 foot long and 2 to 3 inches wide, which taper below into channeled leafstalks; they have fourteen to twenty-four longitudinal veins. The flower stalks, stiffly erect, have a solitary broad leaflike bract below their cen-

Chlorophytum bichetii

Chlorophytum macrophyllum

Chlorophytum comosum vittatum

Chlorophytum bichetii (flowers)

Chlorophytum macrophyllum (flowers)

Chlorophytum triflorum

ters and smaller, narrower ones above. Their upright branches bear somewhat distantly spaced clusters of small, nodding, green flowers with reflexed segments. The flowers open in the afternoon. Rare in cultivation, South African *C. triflorum* has thick black roots and from an underground rhizome develops rigid, erect stems 1¼ to 2½ feet tall. It has numerous two-ranked, folded leaves that are glaucous and, except for short hairs along their margins, glabrous. The erect, branchless flowering stalks bear blooms of starry aspect about ¾ inch wide that remain open throughout the day.

Garden Uses. Chlorophytums are chiefly employed as decorative plants for pots, hanging baskets, and other containers. In essentially frost-free climates they can be grown permanently outdoors and lend themselves for grouping or massing in borders, mainly for their foliage effects. Elsewhere they may be used in this way for summer display and be lifted before fall frosts and wintered indoors. The most popular are *C. comosum* and its varieties, which are very easily multiplied from the plantlets that develop on their flower stalks. These are likely to be passed by one amateur gardener to another and, as they easily "take hold" when set in any ordinary soil and show great tenacity to life, they are likely to remain favorites with those

who appreciate window plants that persist with minimum care.

Cultivation. Few plants are easier to grow than these. They are adaptable to a wide range of environments and can withstand considerable abuse. It is by no means unusual for neglected and sometimes rather woe-begone specimens to survive for years in store windows and other less-than-ideal environments with no attention other than sporadic watering. Such treatment is not, of course, conducive to the best results, and that chlorophytums survive it is indicative of their resistance to widely changeable and often adverse conditions of temperature, humidity, light, and moisture, and to their ability to persist without regular attention to such details as fertilizing and repotting. As previously mentioned, the most popular kinds are easily propagated by plantlets that develop spontaneously on their flower stalks. They, and others, are also readily increased by division and, when available, by seed. The latter soon germinate in a temperature of 60 to 70°F if sown in sandy soil kept moist, but not wet. Chlorophytums are not at all choosey about soil, but prefer one fairly porous. They should be planted in well-drained containers or, in subtropical and tropical climates, outdoor beds. Repotting should be done whenever the plants show by their gradual deterioration that they are

in need of this attention and, if they become too large or crowded with shoots, they may be divided at the same time. Spring is the most favorable time for these operations, but they may be undertaken at any time. When specimens are divided it is well to remove some of their lower leaves and to shorten others to compensate for the inevitable damage done to the roots. Established plants should be watered sufficiently often to keep the soil evenly moist, but not constantly saturated; it is often helpful to keep specimens that are heavily rooted and pot-bound with their containers standing in saucers of water, but this should not be done with less well-rooted plants. Occasional applications of dilute liquid fertilizer from spring through fall are helpful. The South African species and their varieties survive considerable cold, even a touch of frost, but ideally prefer rather cool, but frost-free conditions. Winter night temperatures of 45 to 50°F, give or take a few degrees, are satisfactory and day temperatures may be five to ten degrees higher. Excessively high temperatures, especially when accompanied by poor light make for soft, weak, and in the case of variegated-leaved kinds, poorly colored foliage. Tropical kinds, such as *C. bichetii*, respond well to warmer conditions. Chlorophytums are evergreens, but if kept cool and nearly dry in winter, *C. bichetii* loses its foliage then and puts forth a fresh crop when watering is resumed in spring; under warm, moist conditions it retains its leaves throughout the year.

CHLOROPICRIN. This is a water-insoluble, colorless to yellowish, poisonous liquid related to chloroform and employed as a soil fumigant especially to control nematodes in greenhouses and outdoors. It also kills weeds and their seeds. The gas given off by chloropicrim is a tear gas employed in warfare and for quelling riots and similar purposes. Because sophisticated techniques and special equipment are needed for its safe and successful use it is not practicable for amateurs to use it.

CHLOROSIS. As applied to plants chlorosis means the abnormal yellowing or whitening of foliage that results from partial or complete failure to develop normal amounts of chlorophyll. It may be caused by nutrient deficiencies, by viruses, or by lack of sufficient oxygen in the soil, the result of poor subsurface drainage.

Iron deficiency is a common cause of chlorosis in azaleas, rhododendrons, mountain-laurel, pin oaks, and many other acid-soil plants when they are grown on neutral or alkaline soils. It manifests itself by the leaves, except usually along the veins, turning yellow, and by reduced vigor. Its cause is the plant's inability to obtain iron in sufficient quantities, usually not because the soil lacks iron, but because

Chlorosis caused by iron deficiency, showing (from left to right) increasingly severe symptoms in apple leaves

it is there in forms unavailable to the plant. Relief is had quickly by spraying the foliage with an iron chelate, according to directions on the package, or with iron (ferrous) sulfate at ½ ounce to a gallon of water. Alternatively, a mixture of equal parts of iron sulfate and powdered sulfur applied to the soil at ¼ ounce to the square foot will give more slowly effective, but longer-lasting results.

Magnesium deficiency causes chlorosis in many plants in regions where this element is in short supply in the soil. Tomatoes and many flowering plants develop a yellowing of the leaf tissues between the veins, which remain green. Rate of growth is reduced. Puckering of the foliage often occurs. Tomatoes also develop small black spots next to the veins. Cabbages and other vegetables of that tribe, such as broccoli and cauliflower, evidence magnesium deficiency by their lower leaves puckering and the foliage becoming mottled and developing white patches and margins. Mature leaves of fruit trees develop tan-colored patches and drop successively from the base to the tips of the shoots. Correction of this trouble is had by applying dolomitic limestone to the soil or by using fertilizers containing magnesium. For lime-hating plants such as azaleas, Epsom salts (magnesium sulfate) can be used as a foliage spray at 1 pound to 6 gallons of water.

Chlorosis caused by viruses result in all-over yellowing as with aster yellows, mosaic effects as with dahlia mosaic and bean mosaic, ring spots as with tobacco ring spot disease, or other discolorations. Dwarfing usually occurs and sometimes proliferation of shoot growth and other symptoms. There are no cures. For preventive measures see Virus.

Wet soil deficient in oxygen as a cause of chlorosis can only be remedied by improving the subsurface drainage. The alternative is to select for wet places plants that grow satisfactorily under such circumstances.

CHOCOLATE. See Theobroma.

CHOISYA (Choísy-a) — Mexican-Orange. The popular Mexican-orange (*Choisya ternata*) is the only one of its genus commonly cultivated. Like the other five species, it is an evergreen shrub. The group is indigenous from Arizona to Mexico, the cultivated kind is Mexican. The name of the genus, which belongs in the rue family RUTACEAE, commemorates J. D. Choisy, a Swiss botanist, who died in 1859. Choisyas have opposite leaves of three to thirteen leaflets arranged like the extended fingers of a hand. Like those of most members of the rue family they are bespeckled with minute translucent dots, most easily seen when viewed against the light. The five-parted white blooms have ten stamens, a five-lobed ovary and a five-lobed stigma. The fruits are capsules.

The Mexican-orange (*C. ternata*) is a good-looking aromatic shrub 4 to 8 feet tall. Its young shoots are hairy, and it has stalked, thick leaves, of usually three narrowly-elliptic-oblong to obovate, blunt-ended leaflets, 1 inch to 3 inches long. Their edges are rolled under and are dotted with tiny glands. The sweetly fragrant, white blooms, about 1 inch in diameter, are in loose axillary clusters.

Garden and Landscape Uses. This attractive shrub can be grown outdoors in sheltered places about as far north as Virginia. It stands a little frost, but is at its best in regions of decidedly mild winters. It is also satisfactory for cool greenhouses. The Mexican-orange can be used effectively in foundation plantings, shrub borders, and other places where an evergreen flowering shrub of its stature is needed, and for screening. In large pots or tubs it makes a useful decoration for patios, terraces, and similar places.

Cultivation. The Mexican-orange thrives in any ordinary well-drained soil, in full sun. When grown in containers the earth should be coarse and fertile and kept evenly moist, a little drier in winter than summer. When the receptacles are well filled with roots regular applications of liquid fertilizer from spring through fall encourage good foliage color and health. In greenhouses a minimum night temperature in winter of 45 to 50°F is adequate; by day the temperature may be five to fifteen degrees higher. Any pruning needed, and some likely will be to keep the plant shapely and force the production of new growth from its lower parts, should be done as soon as flowering is through. Propagation is chiefly by cuttings made from the ends of shoots and planted in summer in a propagating bench, preferably one that is kept a few degrees warmer than the air above it, or by cuttings of firmer, older wood, inserted in late summer or early fall.

CHOKE CHERRY is *Prunus virginiana*.

CHOKEBERRY. See Aronia.

CHOLLA. Pronounced chol-ya or cho-ya, this is the name given to cylindrical-stemmed members of the genus *Opuntia*.

CHONDROBOLLEA. This is the name of a natural orchid hybrid the parents of which are *Bollea* and *Chondrorhyncha*.

CHONDROPETALUM. This is the name of bigeneric orchids the parents of which are *Chondrorhyncha* and *Zygopetalum*.

CHONEMORPHA (Chon-emòrpha). Here belong some fourteen species of the dogbane family APOCYNACEAE. Native from Malaya to Indonesia and India, *Chonemorpha* has a name that alludes to the shape of the blooms, derived from the Greek *chone*, a funnel, and *morpha*, form.

Chonemorphas are vigorous vines with milky sap and opposite, undivided leaves.

Choisya ternata

Their flowers, in branched clusters, are large, have a tubular calyx, a funnel-shaped corolla with five lobes (petals) twisted to the right, and five stamens. The fruits are a pair of long, hairy, podlike follicles.

Native from India to Malaya and Java, **C. fragrans** (syns. *Rhynchospermum fragrans*, *Trachelospermum fragrans*) is an attractive vine esteemed for its foliage and sweetly-scented blooms. It has ovate to elliptic, evergreen leaves 6 inches to over 1 foot long and up to 9 inches wide, densely clothed with short, fine hairs on their undersides, sparingly-hairy above. The plumeria-like flowers in hairy-stalked clusters of several to many, about 1½ inches long by 2 to 2½ inches across their faces, are white with yellow throats. The fruits may be 1 foot long or longer.

Garden and Landscape Uses and Cultivation. The species described is useful for clothing pillars and other supports in the tropics and warm subtropics and in tropical greenhouses. It does well in fertile, well-drained soil that has an abundant organic content and is moderately moist, but not wet. Some shade from strong sun is desirable. Propagation is easy by cuttings and seeds.

CHORISIA (Chor-ísia)—Floss-Silk Tree. Its name pronounced korísia, this genus belongs in the bombax family BOMBACACEAE. It consists of five species of spiny trees, natives of tropical and subtropical South America. The name honors Louis Choris, an artist, who died in 1828.

Chorisias have alternate leaves divided into five or seven toothed or toothless leaflets that spread, in hand-fashion, from the end of the leafstalk. The flowers, large and showy, come from the leaf axils and may be solitary or more or less in racemes. They have cup-shaped, irregularly two- to five-lobed calyxes, and five spreading or recurved, linear to oblong petals. The stamens form two tubes, one inside the other. The outer, shorter stamens have sterile anthers, those of the inner ring are fertile. The fruits are pear-shaped capsules containing many seeds with flossy hairs attached. The floss of the floss-silk tree is used for stuffing pillows and similar purposes. The trunks of *Chorisia insignis* are used for making dugout canoes, its inner bark cordage.

The floss-silk tree (**C. speciosa**) is spectacular in bloom and very variable. It is possible that plants in cultivation may involve more than one species. It may attain a height 50 or 60 feet and spread to a width equaling its height. Its trunk is usually conspicuously studded with thick-based, conical, sharp thorns, which sometimes fall away as the tree ages, and which in some individuals are not developed from the beginning. The flowers of *C. speciosa* are crimson, pink, or white, about 3 inches in diameter, and have long staminal tubes.

Rarer in cultivation, **C. insignis** is a variable, rather awkwardly shaped tree up to 50 feet or so tall, with a more open, irregular head, and an immense bellied trunk that at its widest is sometimes 6 feet in diameter. The trunk is generally studded with conelike spines, but these may be absent from old specimens. The flowers are yellow, paling with age, yellow with chestnut-brown blotches, or white.

Garden and Landscape Uses and Cultivation. Too large for small gardens, chorisias are excellent ornamentals for larger landscapes. The floss-silk tree is especially handsome. The other described above is sufficiently unusual to attract attention as something of a curiosity. Both need subtropical or tropical climates, but are able to survive occasional brief drops in temperature to below freezing. They succeed in sunny locations in ordinary, well-drained, reasonably fertile soil. Seed appears to afford the only practicable method of propagation.

CHORIZANTHE (Chorizán-the). The North American species of this genus of fifty to sixty include the only kinds likely to be cultivated. They are desert and semidesert annuals. Some South American species are perennials. Chorizanthes belong to the buckwheat family POLYGONACEAE. Their name, from the Greek *chorizo*, to divide, and *anthos*, a flower, alludes to the parted calyxes.

These are low, erect or prostrate plants, commonly with basal rosettes of foliage that under dry conditions shrivels early. The forking stems generally have bracts in place of leaves. The upper bracts are in twos or threes, the lower leaves are alternate and without stipules (basal appendages). The tiny stalked or nearly stalkless flowers, usually solitary, but sometimes few together, have cylindrical, urn-shaped, or triangular, stalkless involucres (collars of leafy bracts) and are grouped in heads or clusters. They have calyxes of six, or rarely five, colored lobes or sepals, no petals, three, six, or nine stamens, and three styles. The small, hairless fruits, technically achenes, are three-angled.

Natives of southern California the Turk's rug (*Chorizanthe staticoides*) and *C. palmeri* are cultivated. Both usually have erect stems, those of the former mostly about 8 inches tall, but sometimes taller, those of the latter up to 1 foot in height, sometimes prostrate. The Turk's rug (*C. staticoides*) has reddish or purplish stems, fragile at the joints, and forking above. Its leaves are basal, oblong to oblong-ovate, long-stalked, ¾ inch to 2¼ inches long, green or reddish above, and white-hairy on their under-

Chorisia speciosa, in fruit

Chorisia insignis

Chorisia insignis (trunk)

sides. The flowers are in solitary, cylindrical involucres, about ⅙ inch long in the leaf axils, or are clustered at the branchlet ends. They are rose-pink or paler, and have nine or rarely six stamens. Within this very variable species several botanical varieties are recognized. The leaves of *C. palmeri* are in basal tufts. They are oblanceolate to spatula-shaped, stalked, somewhat hairy on their under surfaces, slightly so above, and ½ inch to 1¼ inches long. The short-stalked, rose-pink flowers are in nearly cylindrical involucres in dense heads at the branchlet ends, and usually solitary at the forks of the branches. The flowers have nine stamens.

Garden Uses and Cultivation. These species are best suited for growing as annuals in warm, dry climates. They may be raised from seeds sown in porous, fertile soil where the plants are to remain and the seedlings thinned so that the plants do not become uncomfortably crowded.

CHORIZEMA (Chorí-zema)—Flame-Mock-Orange, Bush-Flame-Pea. An Australian group of the pea family LEGUMINOSAE, the genus *Chorizema* consists of fifteen species of delightful, evergreen, nonhardy shrubs and subshrubs. Its name, said to have been applied fancifully to express the joy

of the botanist who first discovered the genus, the original plant beside a fresh water spring following a protracted period during which he and his party had found only salt water wells, comes from the Greek *choros*, a dance, and *zema*, a drink.

Chorizemas have slender, wiry, somewhat straggling and frequently semitwining stems, and small, often spiny-toothed leathery leaves. In all species except *C. ilicifolium* the leaves are alternate. The pealike small blooms are orange, red, purplered, or combinations of these colors. They are in terminal or axillary racemes and have five-lobed calyxes and five petals narrowed at their bases into shafts or claws. The standard or banner petal is broadly-rounded or kidney-shaped. The wing petals are much smaller. The keels are short and inconspicuous. There are ten separate stamens. The fruits are short, ovoid pods not constricted between the seeds.

A shrub 4 to 6 feet tall, *C. varium* has downy shoots and broadly-ovate, heart-shaped leaves 1 inch to 2½ inches long by one-half or somewhat more as wide and downy on their lower surfaces. Their margins have a few prickly teeth or are toothless. The short-stalked flowers, in close terminal and axillary racemes 1 inch to 2 inches long, have kidney-shaped standard petals, orange-yellow with a purple-red blotch at the base, from which radiate lines of the same color, and reddish-purple wing petals. The blooms are about ½ inch wide.

Differing from the last in its laxer, weaker habit of growth, and in its ovate-lanceolate to lanceolate leaves being much more spiny-toothed and almost or quite hairless on their undersides, *C. ilicifolium* is low and rather sprawling. Its leaves ¾ inch to 2 inches long by approximately one-third as wide, often have heart-shaped bases. In terminal and axillary racemes up to 4 inches long, the few closely-arranged blooms are orange streaked with red. There is a yellow spot at the base of the standard petal.

A weak-stemmed shrub from 3 to, when supported, 10 feet tall, *C. cordatum* is nearly or quite hairless. It has ovate to ovate-lanceolate leaves with heart-shaped bases, 1 inch to 2½ inches in length and approximately one-half as wide as long. The spine-toothed leaf margins are scarcely or not wavy. Quite widely spaced along the up-to-6-inch long terminal and axillary racemes, the ten or fewer flowers, ½ inch wide or a little wider, have broadly-kidney-shaped, standard petals that are orange-red or scarlet, blotched with yellow at their bases. The wing petals and keel are purplish. In gardens this kind is sometimes misidentified as *C. varium* and *C. ilicifolium*.

Especially fine is *C. dicksonii*. From 1½ to 3 feet tall or perhaps sometimes taller, this has hairy shoots, and hairless or nearly

hairless, linear to oblong-lanceolate, spine-tipped leaves up to ¾ inch long by up to ¼ inch wide. The flowers, few together in terminal racemes 1 inch to 3 inches long, have velvety calyxes, and standard petals, ¾ inch wide, deeply notched at their apexes, red to apricot, with a small basal spot of yellow. The wing petals are darker than the standards.

Garden and Landscape Uses. Chorizemas are adaptable for outdoors in warm dry climates such as that of California, and for cool greenhouses. They may be grown as self-supporting shrubs, or trained in more or less vine fashion to clothe trellises, walls, pillars, and other supports. When used on walls or pillars, wires or trellises are needed to tie the shoots to, or for them to intertwine with. They grow satisfactorily in a variety of well-drained soils, preferring those of a peaty, loamy character that are moderately moist, but not wet. Very light shade from the strongest sun and sheltered locations rather than exposure to wind are advantageous.

Cultivation. Chorizemas can be raised from seeds, and quite readily from cuttings made from firm shoots planted in a mixture of sand and peat moss or other usual propagating medium in a bench in a cool greenhouse. During their young stages occasional pinching out of the tips of the young shoots is done to promote branching. Established plants benefit greatly from having the ground around them kept mulched with peat moss, compost, or similar organic material. Pruning to keep the plants shaped and tidy is needed and is best done as soon as flowering is through.

In greenhouses, chorizemas succeed where the night temperature in winter is 45 to 50°F and that by day five to fifteen degrees higher. A fairly airy rather than a heavy, humid atmosphere is appreciated, as is full sun except for some shade in summer. At that season the plants benefit from being removed to a lath house or a lightly shaded place outdoors. Repot immediately flowering is over. A well-drained, nourishing, sandy peaty, not-too-coarse soil mix gives the best results. It should be packed moderately firmly about the roots. Water to keep the earth always moderately moist without ever becoming stagnant. Specimens that have filled their containers with roots may be given dilute liquid fertilizer occasionally from spring to fall.

CHOROGI. See Artichoke, Chinese or Japanese.

CHRIST-THORN is *Paliurus spina-christi.*

CHRISTIEARA. This is the name of hybrid orchids the parents of which include *Aerides*, *Ascocentrum*, and *Vanda*.

CHRISTMAS. The word Christmas is used as part of the names of various plants including these: Christmas bells (*Blandfordia*), Christmas-berry (*Heteromeles arbutifolia* and *Lycium carolinianum*), Christmas berry tree (*Schinus terebinthifolius*), Christmas bush (*Cassia bicapsularis*), Christmas cactus (*Schlumbergera buckleyi*), Christmas-cherry (*Solanum pseudo-capsicum*), Christmas fern (*Polystichum acrostichoides*), Christmas-heather (*Erica canaliculata*), Christmas jewels (*Aechmea racinae*), Christmas orchid (*Cattleya trianaei*), Christmas palm (*Veitchia merrillii*), Christmas-rose (*Helleborus niger*), Christmas vine (*Porana paniculata*), New South Wales Christmas bush (*Ceratopetalum gummiferum*), and New Zealand Christmas tree (*Metrosideros excelsus*).

CHRISTMAS GREENS. See Greens.

CHRISTMAS TREES. The average citizen has little or no contact with Christmas trees except at the festive season when it is strong American tradition to install and decorate one. For most people this involves a trip to a local purveyor and the careful selection of a specimen that seems best to fit one's need and pocketbook. Fewer are in the happy position of being able to cut trees from the wild or plantations, and increasing numbers, with reluctance it is hoped if not with certain shame, accept artificial substitutes.

Just when the custom of decorating a tree at Christmas came to America is not certainly known. Hessian soldiers did so in December 1778 and by the late 1800s large cities such as New York, Boston, Chicago, Cleveland, and Detroit, were supplied with train and boat loads harvested from the wild.

The Christmas tree as we know it originated in western Germany more than 250 years ago and from there spread slowly throughout Europe and eventually much of the rest of the world. But its origin can be traced further. It is a line descendent of the "fir tree" hung with apples to represent the Garden of Eden, that in medieval Europe was the sole prop on the stage when the traditional Paradise play was presented during the Christmas season. There are those who go so far as to associate the custom with the use of evergreens at ancient pagan winter festivals. That may or may not be true.

The popularity of Christmas trees is unquestioned. In the United States interest in them supports a multimillion dollar business devoted to raising, harvesting, transporting, and merchandising them. The business of raising trees especially for the trade has grown tremendously in the last half century. One of the earliest publicized suggestions that such ventures might be worthwhile and profitable appeared in a bulletin of Michigan State Agricultural College in 1925. The Christmas tree industry is represented by the National Christmas Tree Growers' Association, which publishes a substantial monthly magazine, and by organizations at a regional level.

The cultivation of Christmas trees on a commercial basis is beyond the scope of this Encyclopedia. It is not impracticable, however, for those with land at their disposal to grow a few for home use. This involves selecting sorts suitable to the locality and planting a few young ones every two or three or four years to ensure succession. Trees for such use must be given ample space to develop without crowding and needed care to keep them growing vigorously and free of pests and diseases. A certain amount of pruning to shape may be desirable.

Qualities most desired in Christmas trees are that they be straight-trunked, well-foliaged and compact, and about two-thirds as wide at the base as tall. They should have four good face sides and be of a dark green or blue-green color. Light green or yellow-green is a disadvantage. The needles should look fresh and hold well without dropping. Additional virtues are that the tree have a good "handle" (basal part of the trunk free of branches) and that it have a pleasant aroma.

Kinds of evergreens most favored as Christmas trees vary somewhat in different parts of the country. Individuals are often strongly influenced by types they were accustomed to as children. Not all sorts offered for sale are equally as satisfactory. The red-cedar (*Juniperus virginiana*) and Arizona cypress (*Cupressus arizonica*), for example, have branches too limber, and the foliage of the first tends to become bronze in fall in the north. Norway spruce (*Picea abies*) and white spruce (*P. glauca*), although in other respects ideal, drop their

Used as Christmas trees, under identical conditions the Norway spruce sheds its needles and the balsam fir retains them

needles quickly, especially in warm rooms. Among the most popular and satisfactory Christmas trees are balsam fir (*Abies balsamea*), white fir (*Abies concolor*), Southern balsam fir (*Abies fraseri*), red fir (*Abies magnifica*), red pine (*Pinus resinosa*), white pine (*Pinus strobus*), Scots pine (*Pinus sylvestris*), and Douglas-fir (*Pseudotsuga menziesii*).

Living Christmas trees have gained favor in recent years. If specimens of sorts adaptable to the area, dug with good, unbroken root balls are obtainable, and if they are properly cared for after purchase they are satisfactory. Otherwise it is better to settle for cut trees, which are much less expensive than living ones of like quality and size. If you decide on a living tree obtain it from a reliable nursery and ask that it be sprayed with an anti-transpirant, such as Wiltpruf, or do this yourself. If the tree arrives a week or so before Christmas store it in a cool garage, cellar, or shaded place sheltered from wind outdoors, with its container or root ball wrapped around with straw, leaves, old blankets, or other insulation to keep it from freezing. Make sure the ball is kept moist.

Care of a living tree indoors should be based upon locating it in a place as cool as reasonably possible, not for longer than necessary, and keeping the soil well watered. When it has served its purpose indoors stand it in a sheltered place outdoors exposed neither to strong winds nor strong sun until the weather makes planting practicable. In the north, where the root ball is apt to freeze, pack around it a thick insulation layer of straw, leaves or other material secured inside a fence of chicken wire, boards, or a wrapping of polyethylene plastic film. Keep the ball moist by watering as necessary and in spring plant the tree in its permanent location.

Care of cut Christmas trees indoors calls for attention. To have them last in good condition saw at least 1 inch off the butt end of the trunk and during the entire time the tree is inside keep it standing in a container of water. As a fire retardant measure, spray the foliage with a mist of water in which has been dissolved, when the water is warm, boric acid crystals or powder (4 ounces to a gallon) or borax (9 ounces to a gallon).

Other precautions against fire, applicable to both cut and living trees, are to make sure the tree is supported securely and is not close to fireplaces, radiators, television sets, or other sources of heat, to make sure electrical circuits are not overloaded and that no worn or frayed wires are used, and to see that metal foil "icicles" and tinsel do not enter light sockets. Never permit lighted candles or other flames on or near Christmas trees, and do not use flammable or combustible decorations or reflectors for lights. Keep accumulations of wrapping papers and electrical toys from beneath the tree. Do not station Christmas trees where they block fire exists.

CHRISTOPHINE is *Sechium edule.*

CHROMOSOME. Chromosomes are specialized rodlike bodies within the nuclei of plant and animal cells that carry the genes that determine heredity characteristics. For each particular species the number in a complete set of chromosomes is normally constant and in plants may range from one to five hundred or more, but more often is between six and fifty.

When in the growth process a cell divides each chromosome splits lengthwise and one-half goes to each daughter cell. Each body cell normally has two complete sets of chromosomes and is said to be diploid. Such body cells differ from male and female reproductive cells (sperms and eggs or ovules), which are called gametes and each of which has only one set of chromosomes. Cells of this kind are termed haploid. The reduced number of chromosomes in haploid cells is the result of a complicated and beautifully carried-out process, called reduction division or meiosis.

When in the process of fertilization a male and female gamete unite the result is a diploid cell, called a zygote. This, the first diploid cell of a new individual under favorable circumstances divides and redivides to develop into the vast number of diploid body cells that constitute a complete individual.

Variations from the normal haploid pattern of two sets of chromosomes in body cells occur. Individuals with more than two sets, and some have as many as eight or even more, are called polyploids. The most frequent polyploids are tetraploids. These have in their body cells four sets of chromosomes. Polyploids occur naturally and are artificially induced by the use of colchicine and other techniques. Artificially developed polyploids are commonly larger and more vigorous than their diploid counterparts and may differ in other ways often horticulturally advantageous. They may or may not produce fertile seeds.

CHRYSALIDOCARPUS (Chrysalidocár-pus). Although *Chrysalidocarpus*, a genus of feather-leaved palms of Madagascar, Pemba Island, and the Comoro Islands, consists of twenty species, few are cultivated. By far the best known is *C. lutescens,* a favorite for growing indoors in pots and tubs and for planting outdoors in southern Florida, Hawaii, and elsewhere in the tropics. The genus belongs in the palm family PALMAE. Its name, derived from the Greek *chrysos,* gold, and *karpos,* fruit, refers to the color of the fruits of some kinds.

Chrysalidocarpuses, small to medium-sized, are without spines. Most species

form clumps of slender stems and gracefully arching leaves. A few are single-stemmed. Their branched flower clusters arise in the leaf axils and have two spathes of unequal size. The fruits are yellow, reddish, or violet-black.

Highly decorative *C. lutescens* (syn. *Areca lutescens*) forms a clump of sometimes branched, yellowish, canelike stems 10 to 25 feet tall with feathery, glossy, light

Chrysalidocarpus lutescens

Chrysalidocarpus lutescens, young specimen

green, arching leaves with yellowish stalks and midribs. The canes are not more than 3 inches in diameter and are tightly clustered near the ground. Above, the outer canes bend outward to produce a wide-spreading mass of foliage. The leaflets, ½ to ¾ inch in width and slender-pointed and notched at their ends, arch gracefully. Each has a single prominent central vein. There are thirty leaflets or more on each side of the leaf axis. The violet-black fruits are over ½ inch long and under ¼ inch thick. This species is native to Malagasy (Madagascar).

Other kinds include these: *C. cabadae,* a red-fruited tree up to 30 feet tall or taller, has several to many trunks. Its glossy, dark green leaves are 5 to 6 feet long and

have up to thirty leaflets on each side of the midrib. This species was described from cultivated plants in Cuba. Its nativity is unknown. *C. madagascariensis* has leaves with many more leaflets than those of *C. lutescens* and forms a denser head. It is a native of Madagascar. *C. m. lucubensis,* native to an island off the coast of Madagascar, is single-stemmed and attains a height of about 30 feet. *C. pembanus* is closely related to *C. cabadae,* and like it, has red fruits. Its several trunks attain heights of 60 feet and the leaves, with forty to fifty leaflets on each side of the midrib, are up to 7½ feet long. This is native to Pemba Island to the north of Madagascar.

Garden and Landscape Uses. Too tender to stand frost, *C. lutescens* is one of the most ornamental of the multistemmed species. It is especially effective as specimen clumps on lawns and for use in association with architectural features such as buildings, terraces, and steps. A popular employment for it in many tropical countries is for planting along driveways. It is also one of the most useful palms for screens and informal hedges. The other sorts described may be used similarly.

Cultivation. Outdoors *C. lutescens* thrives in sun or part-shade in ordinary, not excessively dry garden soil. Indoors it needs a minimum winter temperature of 60 to 65°F and a fairly humid atmosphere. From spring through fall the minimum indoor night temperature should be 65°F and at all seasons the day temperature should exceed the night by five to ten degrees. Shade from strong sun is necessary, and the soil should be kept always evenly moist, but not constantly saturated. From spring through fall biweekly applications of dilute liquid fertilizer are helpful. Seeds, which germinate in about a month if sown in sandy, peaty soil at a temperature of about 80°F, afford a ready means of propagation. Other kinds of *Chrysalidocarpus* may be expected to respond to similar care. For additional information see Palms.

CHRYSAMPHORA. See Darlingtonia.

CHRYSANTHEMUM (Chrys-ánthemum)— Costmary, Feverfew, Nippon-Daisy. To most people chrysanthemum means just one thing, the popular hardy perennials that decorate gardens in temperate regions so gloriously in fall and through the wonder of modern day-length control techniques are persuaded by florists to bloom in greenhouses throughout the year. Among them are the enormous snowball-like "mums" popular at fall football games as well as a wealth of other kinds. These garden and florists' chrysanthemums, as they are called, belong indeed in the genus *Chrysanthemum,* but they represent but a tiny fraction of the wide range of plants included there by botanists. Other well-known flowers that belong are shasta-daisies, pyrethrums, marguerites or Paris-daisies, and annual chrysanthemums. To prevent this treatment from becoming clumsy and confusing, these are discussed fully under the entries Chrysanthemums, Florists' or Garden; Marguerite; Pyrethrum; Shasta-Daisy; and Chrysanthemums, Annual. The name *Chrysanthemum,* derived from the Greek *chrysos,* gold, and *anthemon,* a flower, alludes to the yellow flowers of some sorts. For *C. tchihatchewii* see *Tripleurospermum tchihatchewii.*

The genus *Chrysanthemum* belongs to the daisy family COMPOSITAE. It consists of about 200 species of annuals, herbaceous perennials, and subshrubs mostly natives of the Old World. Often their foliage and other parts are strongly odoriferous when crushed. Generally they have erect, often much-branched stems, hairless, or hairy and sometimes slightly tacky. The leaves are alternate, lobed, toothed, or smooth-edged. The flower heads, as is common for their family, consist of many crowded florets some of which, except in a few double-flowered varieties, are disk florets such as compose the eyes of daisies. These are generally encircled by a row of toothed or toothless, petal-like ray florets. The disk florets are bisexual and capable of seed production. The ray florets, also mostly fertile, are female. The involucre (collar behind the flower head) is of usually several rows of overlapping, partially membranous bracts, whitish or brownish and more or less translucent except at their often green centers. The fruits are seedlike achenes. The genus *Tanacetum,* which includes the common tansy, is by some botanists included in *Chrysanthemum.* In this Encyclopedia it is treated separately.

The ox-eye-daisy or field-daisy (*C. leucanthemum*), native of Europe and temperate Asia, is widely naturalized in North America. A hairless or sparsely-hairy deciduous perennial, it has a rhomatous rootstock and erect, branchless or few-branched stems 1 to 2½ feet tall. Oblanceolate to spatula-shaped and stalked, the basal leaves are up to 6 inches long, round-toothed, and often more or less lobed. The stem leaves are smaller, blunt-toothed, and short-stalked or stalkless. Solitary at the ends of the stems and branches, the flower heads are up to 2 inches wide and have yellow centers and fifteen to thirty white ray florets.

Costmary or mint-geranium (*C. balsamita*) is a coarse, hardy herbaceous perennial 2 to 4 feet in height, a native of west-

Chrysanthemum balsamita

ern Asia, but naturalized to some extent along roadsides and waste places in parts of North America. At one time it was reputed to possess medicinal virtues and was used for flavoring foods. It has a bitter taste. Esteemed for the pleasant, sweet odor of its finely-hairy foliage, this kind has small-toothed, blunt leaves, lobeless except for sometimes a pair of small basal projections. They are oblong to elliptic, 2 to 6 inches, or the basal ones sometimes up to 10 inches long. The abundant flower heads, about ½ inch across, have yellowish disks and a few white ray florets. Variety *C. b. tanacetoides* has heads up to ⅓ inch wide, without ray florets. This is the most frequent kind.

Feverfew (*C. parthenium*), native from Europe to the Caucasus and naturalized in waste places in North America, is a hardy perennial, much branched, with powerfully-scented, nearly hairless foliage. It is 1 foot to 3 feet in height and has broad-ovate to broad-oblong leaves seldom more than 3 inches long and often smaller. Those below are pinnate with the leaflets pinnately-lobed. The upper ones are pinnately-divided into two or three pairs of toothed segments. The flower heads are typically of yellow disk florets encircled by ten to twenty white rays, or in variants, may be of all yellow disk florets or of all white ray florets. Known in gardens as golden feather, low *C. p. aureum* has yellow foliage. The taller, double-flowered feverfew is *C. p. flore-pleno.*

Chrysanthemum parthenium aureum

Chrysanthemum parthenium flore-pleno

Nippon-daisy (*C. nipponicum*) is a native of seashore habitats in Japan. It is an attractive, late-flowering, bushy, hairless or nearly hairless subshrub 1½ to 2½ feet tall with semiwoody or woody perennial stems, their lower parts soon bare of foliage. The stiffish, fleshy, lustrous, blunt-spatula-shaped leaves 2 to 3½ inches in length and ½ to ¾ inch wide are blunt-toothed near their apexes. The numerous, solitary, stalked, circular flower heads approximately 2¼ inches in diameter have yellow centers and thick-textured, pure white ray florets. This is a fine ornamental and a handsome garden plant.

Chrysanthemum leucanthemum

Chrysanthemum parthenium

Chrysanthemum parthenium flore-pleno (flowers)

Chrysanthemum nipponicum

Chrysanthemum alpinum

Chrysanthemum arcticum

Chrysanthemum serotinum

Chrysanthemum haradjanii

Other hardy perennial species cultivated include these: **C. alpinum**, a delightful native of mountains in Europe, is a low, tufted or matting plant with leaves seldom exceeding 1½ inches in length and often much smaller. They are pinnately-dissected into slender segments. Its solitary, attractive flower heads carried on nearly naked stalks to from 2 to 6 inches above the foliage are up to 1½ inches in diameter. They have orange-yellow disk florets and eight to twelve white or rarely lilac-pink rays little or not at all toothed at their apexes. **C. arcticum** is native to arctic and subarctic regions in Europe and Asia. Clump-forming, it has wedge-shaped leaves with, at their ends, three main lobes each again more or less lobed. The lower leaves are wider than they are long. Branchless or branched from their bases and 6 inches up to occasionally 1½ feet long, the stems are often more or less prostrate with their ends ascending. The daisy-like flower heads are solitary, 1 inch to 2 inches in diameter. They terminate erect stalks 3 to 4 inches long or sometimes longer furnished with a few small leaves or bracts. The ray florets are white usually tinged with pink or lav-

ender-pink. **C. cinerariaefolium**, the dalmatian pyrethrum, is a source of the insecticide pyrethrum. Native of Dalmatia and widely cultivated commercially elsewhere, this is a distinctly gray-hairy herbaceous perennial with slender, erect stems 1 foot to 2½ feet tall. Its oblong to elliptic leaves with stalks 6 inches to 1 foot long are pinnate with the leaflets pinnately-cleft into slender, ultimate segments. The solitary, long-stalked, decorative flower heads, about 1½ inches wide, have slender, white ray florets. **C. haradjanii**, of Syria, is an attractive, broad subshrub densely clothed with white hairs and up to 1½ feet tall. It has oblong-ovate to ovate leaves 1 inch to 2 inches long and deeply-twice- to thrice-pinnately-lobed. Rayless and ⅛ inch wide, the flower heads are in loose clusters. **C. indicum**, a native of Japan and China, has creeping rhizomes and erect or more or

less decumbent leafy stems. The chief leaves are ovate-oblong, 1¼ to 2 inches long by approximately three-quarters as wide. They are pinnately-lobed with the blunt terminal lobe the largest. The two pairs of side lobes are toothed. About 1 inch in diameter, the yellow flower heads are in loose, terminal clusters. **C. serotinum** (syn. *C. uliginosum*), of central Europe, is an admirable late-blooming ornamental. Its leafy stems, branched above, attain heights of 4 to 7 feet. Its long-lanceolate leaves are coarsely-sharp-toothed. Each stem carries several stalked, white-rayed, daisy-like flower heads 2 to 3 inches in diameter. As they age their yellow centers darken. **C. weyrichii** inhabits rocky seashores on Sakhalin Island near Japan. This mat-former has branched or branchless, purplish stems up to 1 foot tall. The lower leaves are long-stalked, have fleshy, nearly round blades cleft palmately (in hand-fashion) into five lobes or leaflets. The higher leaves are smaller, pinnately-lobed or linear and not cleft. The long-stalked flower heads are 1¼ to 1¾ inches wide, their ray florets white or pink. **C. yezoense**, of Japan, is allied to *C. arcticum* and appears to be the kind most commonly cultivated under that name. Somewhat more robust than *C. arcticum*, it differs from it most obviously in its flower heads being mostly three, sometimes more, on a stem, instead of being solitary. White fading to pink and about 1½ inches in diameter, they are on stems 9 inches to 1 foot tall. **C. zawadskii** (syn. *C. sibiricum*) has a wide natural range chiefly in mountainous regions in Japan, northern China, Manchuria, and Siberia and in the Carpathian mountains. It has creeping rhizomes and erect stems with lower leaves, like the basal ones, long-stalked and with ovate blades up to about 1½ inches long, often slightly wider than long. They are twice-pinnately-divided into narrow, toothed segments. The upper stem leaves are smaller, narrower, and upward progressively shorter-stalked to stalkless. The solitary, yellow-centered, white flower heads are 1¼ to 2¼ inches wide. Variety *C. z. latilobum* (syn. *C. rubellum*), native of Japan, northern China, Korea, and Manchuria, has upper leaves wedge-shaped and with usually four side lobes. This makes compact clumps of foliage and has 2- to 3 feet-tall, more or less branched stems ending in loose sprays of few to many pale to deep pink blooms 2½ to 3¼ inches in diameter. Their slender ray florets are well separated and not toothed or notched at their ends.

In addition to the sorts described above, these lesser known, hardy perennial species are well worth growing. *C. achilleifolium* of the Caucasus and Siberia is 6 inches to 2 feet tall. It has creeping stems and attractive gray-hairy foliage. Its oblong leaves are 1 to 1½ inches long and ¼ inch wide. They are pinnate with slender, short, finely-lobed leaflets. The clustered yellow flower heads usually have seven or eight very short ray florets. *C. argenteum* of western Asia forms attractive spreading

Chrysanthemum achilleifolium

Chrysanthemum argenteum

clumps 4 to 6 inches tall. Its much-divided, twice-pinnate, stalked ovate leaves, 1 to 1½ inches long, are clothed with silvery-white hairs. The solitary flower heads have white ray florets. *C. corymbosum* of Europe, adjacent Asia, and North Africa has much the aspect of an *Achillea*. From 1 to 1½ feet tall, it has erect stems branched near their tops and fernlike, twice-pinnately-cleft leaves. The tiny flower heads, up to slightly more than 1 inch across, are in loose, flattish clusters. Their ray florets are white.

Garden and Landscape Uses. The species described here are all hardy perennials and, except the Nippon-daisy and *C. haradjanii*, die to the ground in winter. The former is an excellent low, rounded shrub suitable for featuring as an accent in rock gardens, in perennial borders, at the fronts of shrub beds, and elsewhere where it can be displayed to advantage as a single specimen. Costmary is suitable for inclusion in herb gardens, the Dalmatian pyrethrum in collections of plants of economic importance to man. Low, compact varieties of feverfew are grown as annuals for temporary display in summer beds, window and porch boxes, and similar decorative containers. Other kinds presented here are useful for bringing summer or fall bloom to rock gardens, banks, and perennial borders. The Nippon-daisy does especially well near the sea. Some kinds, notably *C. zawadskii latilobum* and *C. serotinum* supply useful cut flowers. The ox-eye-daisy can be used to good effect in wild gardens and naturalistic areas.

Cultivation. No special care is demanded by these plants. They succeed in ordinary soil, well-drained, that neither lacks fertility nor is excessively rich in nitrogen. In the main they are lovers of sunshine, although some, including feverfew and costmary, tolerate slight shade. Propagation is easy by division, cuttings, and seeds. To grow golden feather and other low varieties of *C. parthenium* as annuals for summer flower beds, window and porch boxes, and similar uses sow seeds indoors about ten weeks before it is safe to set the young plants outside, which will ordinarily be after tulips have faded and tomato planting time has arrived. Transplant the seedlings about 2 inches apart in flats or separately in small pots and grow them in a sunny greenhouse or window where the night temperature is about 50°F, that by day five to fifteen degrees higher depending upon the brightness of the weather. Space the plants outdoors 7 to 9 inches apart.

CHRYSANTHEMUM-WEED is *Artemisia vulgaris*.

CHRYSANTHEMUMS, ANNUAL. Annual chrysanthemums have much more the aspect of large daisies than of what are generally thought of as chrysanthemums. Yet they belong to the genus *Chrysanthemum* and in botanical characteristics detailed in the previous entry in this Encyclopedia agree with other members of that group.

The kinds cultivated are horticultural varieties and hybrids of the corn-marigold (*C. segetum*), *C. carinatum*, and *C. coronarium*, all natives of the Mediterranean region. Naturalized to some extent in the United States, **C. segetum** is 1 foot to 2 feet tall, erect and freely-branched. It has shal-

Chrysanthemum segetum

lowly-pinnately-lobed or toothed leaves not cleft into long, narrow lobes. The flower heads are golden-yellow to white, 1 inch to 2 inches in diameter. Sometimes called tricolor chrysanthemum, bushy, hairless **C. carinatum** is 1 foot to 2 feet tall. It has somewhat succulent stems and rather fleshy, twice-pinnately-lobed, narrowly-segmented leaves. Its long-stalked, daisy-like flower heads 1½ to 2½ inches in diameter, have white, yellow, or red ray florets differently colored at their bases so that the flower head has a zone of a different hue around the central eye. Sometimes known as garland chrysanthemum or

Annual chrysanthemum, garden variety

crown-daisy, *C. coronarium* differs from the last in generally being taller and in having nonsucculent, thinner foliage. Also, its leaves are more deeply-lobed into broader segments. The flower heads, 1 inch to 2 inches wide, are more globular and have a tendency to doubleness. They are yellow to yellowish-white. A variety of this is used by the Chinese as edible greens.

Garden varieties of annual chrysanthemums have larger blooms than the wild species. They can be had with flowers in a splendid array of gay patterns and solid colors, including ivory-whites, pale and deep yellows, orange-yellows, chestnut-reds, and crimson-scarlets, sometimes zoned with purple, and white. There are single- and double-flowered kinds, tall and compact varieties. They are described in the catalogs of seedsmen.

Garden and Landscape Uses. Annual chrysanthemums are splendid garden ornamentals for beds and borders, for cut flowers, and for growing in greenhouses for late winter and spring blooms. Because they do not stand torrid weather well, in regions where summers are hot they fail comparatively early and if in places where continuous bloom is needed must be replaced. For the best results site them in full sun. Any reasonably good, well-drained soil that will produce satisfactory crops of vegetables and the run of garden annuals and perennials suits.

Cultivation. Sow outdoors as early in spring as the ground can be worked and, where summers are not uncomfortably hot, to maintain a succession of flowers for cutting make one or two additional sowings at about three-week intervals. In decorative beds of annuals or mixed annuals and perennials sow in patches of 1 to 3 square yards or for longer, ribbon effects in narrow borders. For cut flowers sow in rows 1½ to 2 feet apart. Cover the seeds to a depth of about ¼ inch. Thin the young seedlings, preferably in two stages, so that they ultimately stand about 1 foot apart in beds or borders, 9 inches apart in rows. Subsequent care involves surface cultivation to prevent the growth of weeds, light staking or the provision of other supports to prevent storm damage, soaking the ground with water at about weekly intervals in dry weather, and the prompt removal of faded blooms.

For earlier bloom than can be had from seeding outdoors in spring, gain a start on the season by sowing indoors about eight weeks before the young plants can be transferred to the open garden, which may be done as soon after the last frost as the ground can be gotten into suitable condition for planting. In regions of nearly frostless winters early crops are had by sowing outdoors in fall.

For winter and early spring bloom in greenhouses sow in September or October, and for smaller flowering specimens

Hardy chrysanthemums outdoors at The New York Botanical Garden

in January. Transplant the seedlings to small pots and later to benches filled with soil or ground beds or successively to bigger pots until they occupy those 6 to 8 inches in diameter. Use porous, nourishing soil and space the plants 8 inches apart each way. When they are about 4 inches tall pinch out the tips of the stems to induce branching. Grow throughout in full sun where the night temperature is 45 to 50°F, that by day five to ten degrees higher. Maintain a dryish atmosphere such as suits carnations, not an excessively humid one. Whenever the weather is favorable, open the greenhouse ventilators. When the final pots are filled with roots initiate a program of weekly applications of dilute liquid fertilizer. Earlier blooms can be had indoors by using artificial light to increase winter day length by four hours.

CHRYSANTHEMUMS, FLORISTS' or GARDEN. One of the chief glories of temperate-region fall gardens, and at one time only possible to have at that season, florists' or garden chrysanthemums are now brought to flower in greenhouses by commercial growers at all times of the year. This is done by manipulating the number of hours in each day of twenty-four that they are exposed to light, natural or artificial, and carefully regulating other conditions. Chrysanthemums, or "mums" as they are affectionately called, are short-day plants. To initiate flower buds they must have nights of ten to eleven hours of uninterrupted darkness according to variety. For precise results, control of day length must be accompanied by temperature regulation. In fall, normal outdoor conditions

Greenhouse chrysanthemums at The New York Botanical Garden

meet these needs, and the plants bloom then naturally.

Florists' or garden chrysanthemums are superb examples of breeders' skills, not so much those of scientists, but of practical gardeners and nurserymen who were dedicated to improving this great flower. The work was done in many parts of the world, notably in the Orient, in France, England, the United States, and Australia. So successful have results been that modern varieties bear little resemblance to the small-daisy-flowered species from which they are believed to have been derived.

The history of chrysanthemums as garden flowers goes back at least 1,500 years, very likely considerably longer. It begins with their cultivation in China, the homeland of the one or more species possibly no longer in existence, but *Chrysanthemum indicum* probably was involved; these are the presumed progenitors of the horticultural kinds now for convenience grouped as *C. morifolium* (syn. *C. hortorum*). About the middle of fourth century A.D. the cultivation of chrysanthemums became fashionable in China. The flowers were frequently portrayed in paintings and tapestries and on pottery and other works of art. The oldest book extant on the chrysanthemum is Chinese. Written about a thousand years ago, it makes reference to thirty diverse varieties. From China seeds were introduced to Japan in A.D. 366 and in the eighth century living plants of Chinese varieties were imported. In the year 910 the chrysanthemum was named the national flower of Japan, and in the twelfth century a sixteen-petaled bloom was proclaimed the Imperial crest. Like the Chinese, the Japanese reproduced chrysanthemums in a great variety of art forms. The earliest known Japanese catalog of these flowers, published in 1736, depicted 100 varieties.

The first chrysanthemums came to Europe from Japan to Holland in 1689, but strangely, the horticulturally capable Dutch failed to keep them alive, and all six varieties were lost to cultivation. A chrysanthemum received at Chelsea Physic garden in England from China in 1764 and named *Matricaria indica* suffered the same fate. Permanent establishment of chrysanthemums in European gardens did not begin until 1789 when three varieties were introduced from China to France. One purple-flowered sort survived. This, primitive by modern standards, was sent to Kew Gardens in England and from there was distributed and gained considerable renown as 'Old Purple'. Other sendings from China to France soon followed, and by 1826 the first breeding program undertaken outside the Orient was started in the south of France. This resulted in varieties that bloomed earlier than those imported.

In 1864 the great English plant hunter Robert Fortune sent from China to his homeland small-flowered chrysanthemums he called Chusan-daisies. Not well received in England, these, welcomed by French breeders, gave rise to the varieties we call pompons. Later, chrysanthemums from Japan came to Europe, among them some with large, informal blooms with reflexed petals sent by Robert Fortune and again unappreciated in England, but thought highly of in France.

England got away to a slow start in chrysanthemum growing and breeding, but once interest was aroused its growers and breeders excelled. For long they were entirely responsible for the development of hardy outdoor varieties. The first chrysanthemum society of importance, formed at Stoke Newington in 1846, held its initial flower show the following year, and in 1865 the first book in English on chrysanthemums, its author John Salter, appeared. The National Chrysanthemum Society, established in 1884, organized meetings and flower shows and in other ways stimulated an already strong interest in its name flower. Until nearly that time the varieties available bloomed too late for outdoor display in England. Now earlier-blooming ones began to appear. But it was not until the twentieth century that the great wealth of hardy, early flowering sorts in a wide variety of flower sizes, forms, and colors were developed.

American experience with chrysanthemums began with the introduction, by John Stevens of Hoboken, New Jersey, of a variety called 'Dark Purple'. This was in 1798. Surprisingly, none is listed in the detailed catalog published in 1811 of plants growing in the famous Elgin Botanic Garden in New York City, yet William Prince of Flushing, New York, in 1828 in 'A Short Treatise on Horticulture' lists as being grown in America nearly forty varieties and two years later, on November 20th, the first chrysanthemums recorded as exhibited in the United States were shown to the Massachusetts Horticultural Society. Well before the middle of the nineteenth century American breeders were producing new varieties, and many imports, some although late-flowering, but resistant to frost, were being made from Europe. Direct importations from Japan followed. Founded as the National Chrysanthemum Society, what soon became the Chrysanthemum Society of America was established in 1890. The following year the book *Chrysanthemum Culture for America* by James Morton appeared, and at least twenty-seven major specialty chrysanthemum flower shows were held in the United States and Canada.

A cornerstone significant in the development of chrysanthemums was unwittingly laid in 1892. The wife of a sea captain of Boston was on her husband's ship in that year when a young Japanese stowaway was discovered. Mrs. Alpheus Hardy persuaded her husband to allow the lad to go to Boston with them and eventually sent him to college. Later he returned to Japan and as an expression of gratitude sent his benefactor a shipment of chrysanthemum plants or cuttings. Although these arrived in poor condition, several were resuscitated by a local florist, among them one which when first exhibited caused quite a sensation. It had white, incurved, hairy florets and was appropriately named 'Mrs. Alpheus Hardy'. Other hairy-petaled seedlings raised from this, like 'Mrs. Alpheus Hardy', were extremely popular for many years.

In the early years of the twentieth century the art of growing immense specimens of formally trained chrysanthemums for exhibition and competition at fall flower shows was developed and became popular among wealthy estate owners who maintained staffs of highly qualified gardeners. In the United States this took place chiefly near Boston, New York, and Philadelphia. The Great Depression of 1932 effectively brought an end to these superb expressions of horticultural skill.

By 1914 hardy varieties that bloomed in the northeast before frost were advertised in America, but it was not until the 1920s, and more especially after the coming of the first Korean hybrids that more than very few, and those not very good in comparison to later varieties, were available. The most reliable of these fairly early-blooming

Greenhouse chrysanthemums, each a single plant, exhibited at the November 1916 flower show of the Horticultural Society of New York

First prize winner in the class for one-specimen plant at the November 1916 flower show of the Horticultural Society of New York

hardy chrysanthemums of the 1920s were the pompon varieties 'Pink Doty', 'Red Doty', and 'White Doty'.

Distribution by their raiser, Alex Cumming of Bristol, Connecticut, in 1932 of what were called Korean chrysanthemums created a great stir in the horticultural world and revitalized interest in outdoor chrysanthemums. The name was given in the mistaken belief that they were hybrids of *C. coreanum*. Actually, the species Cumming crossed with existing garden varieties to give rise to the hardy, vigorous handsome Korean strain, was *C. zawadskii*. The hybrids eventually became available in a bewildering variety of colors, forms, and statures. Practically all hardy chrysanthemums now in cultivation have ancestries stemming back to the Korean hybrids. They must therefore be regarded as representing *C. morifolium* reinforced and rejuvenated by admixture of genes from *C. zawadskii*.

Classification of modern chrysanthemums is based chiefly on the forms of their blooms and florets (commonly referred to as petals) with secondary groupings based on flower size recognized within the classes. A new American scheme for classifying chrysanthemums was established in 1978 by the National Chrysanthemum Society. It replaces a more complicated system formerly approved by the society, and it agrees more closely with the international classification scheme. From the latter, the American plan differs in providing for a greater number of varieties than are usually exhibited at international shows. The Japanese employ another even more elaborate system.

Below is a summary of the classes accepted as official by the National Chrysanthemum Society. More detailed descriptions are published by that organization.

The National Chrysanthemum Society's classification is based upon the form of the blooms when fully developed. At earlier stages the blooms may appear somewhat different from those typical of their class. This especially is true of varieties that belong in classes 1 to 5. In interpreting the descriptions remember that, basically, the structure of a chrysanthemum flower is that of a daisy. It is a head consisting of a center of disk florets (analogous to those that form the eye of a daisy) surrounded by ray florets (the equivalent of the petal-like ones of a daisy). In some kinds the ray florets are very numerous and may be greatly elongated. Often they curve inward and hide the much less obvious disk florets. In other kinds the disk florets form a prominent eyelike center to the chrysanthemum flower.

Class 1: Irregular incurve

Mature ray florets usually broad and incurving over tops of bloom in a regular or irregular manner to form a very large bloom with breadth and depth nearly equal. Lower florets generally loosely incurving. They may also reflex or open outward to form a skirted effect. No disk apparent.

Class 2: Reflex

Mature ray florets narrow to broad and long, gracefully overlapping and reflexing in a regular or irregular manner. Blooms globular to somewhat flattened with breadth and depth nearly equal. No disk apparent.

Class 3: Regular incurve

Mature ray florets narrow to broad and smooth and completely incurved in a regular or irregular manner to produce a globular bloom with breadth and depth equal, with no disk apparent.

Class 4: Decorative

Mature ray florets short and broad to narrow and pointed, most florets generally reflex at distal end. Center florets may tend to slightly incurve. Blooms more flattened than globular and not as deeply re-

Classes of chrysanthemums: (a) Irregular incurve

(c) Regular incurve

(b) Reflex

(d) Decorative

(e) Intermediate incurve

(h) Semi-double

(k) Quill

(f) Pompon

(i) Anemone

(l) Spider

(g) Single

(j) Spoon

(m) Brush and thistle

flexing as in the reflex. Floret ends may be laciniated. This class will include most cultivars formerly classified as laciniated. Size range from very small to large.

Class 5: Intermediate incurve

Mature ray florets broad to narrow and shorter than the regular incurve and only partially·incurving at the distal end. Centers often depressed and blooms present a more open appearance than the regular incurve with breadth and depth nearly equal or slightly flattened. Lower florets may be partially reflexed. Disk not apparent.

Class 6: Pompon

Mature ray florets short and incurved or reflexed in a regular or irregular manner

to form a globular to half round bloom somewhat flattened in an immature stage. No disk apparent. Size range from miniature button type to large disbudded bloom just under 4″.

Class 7: Single and semi-doubles

Mature ray florets in one or more rows at right angles to the stem, sometimes curving upward or downward at the tips, generally not containing more than 5 rows of florets. Disk prominent, flat and circular in outline.

Class 8: Anemone

Mature ray florets variable, from flattened, broad and equal in length to reflexing, pointed at the tip and unequal in length. Disk florets are numerous, tube-like and elongated so as to form a prominent disk which may range from flat to hemispherical in form.

Class 9: Spoon

Mature ray florets regular and tubular, usually straight, distal portion open, flattened and spoon like. Disk must be apparent and circular in outline.

Class 10: Quill

Mature ray florets tubular, straight, and not coiled or hooked, may be either closed at the tip and pointed or open and spatulate. Blooms fully double with no disk apparent.

Class 11: Spider

Mature ray florets long and tubular, very fine to coarse; and may assume a wide variety of direction. Distal portion of the florets may be closed or open and spatulate, tips show definite coils or hooks. Disk must not be apparent.

Class 12: Brush and thistle

Blooms small with tubular ray florets very fine to medium in size. Brush forms with florets nearly parallel to the stem as in saga types or flattened and twisted as in the thistle or Ise type. Disk nearly concealed; normally grown to sprays.

Class 13: Unclassified types

Bloom forms that do not fit descriptions of other classification. Blooms should be distinctive and unique, blooms may partially fit two or more classifications. Ray florets tubular to flat, and may swirl, twist or have hair like pubescence. May be open centered.

Garden and Landscape Uses. Chrysanthemums of practically all types, but most especially those with long-stemmed blooms, rank highly as cut flowers and are grown commercially in vast quantities for that purpose, in greenhouses and outdoors. They are also much esteemed as decorative pot plants, great numbers of which are produced for the florists' trade. For outdoor fall landscape displays they are unexcelled. No plants are more lavish in their production of flowers. Although the duration of the show is much shorter than, say, that of roses, the plentitude of the blooms is much greater and so too are the forms of the flowers. But chrysanthemums and roses are not competitors. Rather they supplement each other. The queen of fall bows in as the reign of summer's undisputed monarch wanes. Also, roses are permanent occupiers of garden beds, chrysanthemums are not. Because of this they lend themselves to far greater flexibility in placement and make possible changing locations and arrangements of displays from year to year. Exhibits of outdoor chrysanthemums can be extensive and appropriate for public parks, botanical gardens, and suchlike places or more modest ones suitable for home gardens. Extensive plantings and even less lavish ones can provide kaleidoscopes of bloom swirling across the landscape as eye-delighting, varicolored seas or can be displayed in more formal pattern beds as the last of the great outdoor display of the season. Chrysanthemums are also ideal for grouping in mixed flower beds. There, they may either be grown *in situ* or planted late from pots or nursery beds to fill gaps left by annuals that have ceased to be decorative. A few plants, a row or two perhaps, set out in a cut flower garden, nursery area, or vegetable garden will supply the home with a fine supply of blooms for cutting without degrading ornamental beds. For filling window boxes, porch boxes, decorative urns, and other containers, and to supply color after summer occupants are tiring, hardy chrysanthemums are unexcelled. For such purposes they may be grown in pots until ready for transfer to their display quarters, or alternatively in nursery beds.

Cultivation of Garden Chrysanthemums. Chrysanthemums, those that can be relied upon for outdoor displays in the north, are of necessity sorts that come into bloom early enough to flaunt themselves before killing frost. They stand with impunity light frosts. There are numerous varieties. New ones are added each year. They differ greatly in height and spread of plant, form and color of flowers, and other details. There are early, midseason, and late bloomers. Consult nurserymen's and specialist dealers' catalogs for descriptions and illustrations or visit garden centers.

Despite their general acceptance as hardy herbaceous perennials, not all varieties of chrysanthemums that bloom in time for garden display are sufficiently hardy to survive outdoors in cold sections, and none do as well when left undisturbed from year to year as when propagated and raised afresh from divisions or better still cuttings each year. These traits, as we shall see, pose no formidable problems.

Soil for outdoor chrysanthemums should be deep, fertile, and well-drained. One that will produce a variety of good vegetables is satisfactory. Spade it deeply or achieve the same purpose with rototiller or plow and mix in generous amounts of rotted manure, compost, or other agreeable, partially-decayed organic material. Do this as far in advance of planting as practicable. If the pH of the soil is below 6 or above 7 bring it to within this range, in the first case by applying lime, preferably in the form of ground limestone, in the second case by applying aluminum sulfate or pulverized sulfur. Your Cooperative Extension Agent can arrange for soil tests to be made and make recommendations as to the amounts to be used. A week or two prior to planting apply a complete general purpose garden fertilizer, and fork this into the upper 3 or 4 inches of soil.

Propagation can be by division, but where nematodes are a problem, and that is almost everywhere, cuttings are much more satisfactory because with care they give clean stock for planting. If you elect to divide, do so in early spring. Let each division consist of a single-rooted young shoot, not a small clump as is done with many hardy herbaceous perennials.

Take cuttings from stock plants that were dug up in fall, planted closely together in boxes, and put in a cold frame.

Chrysanthemums from cuttings: (a) Making the cuttings

(b) Rooted cuttings ready for potting

In January move the boxes to a sunny greenhouse where the night temperature is 45 to 50°F, that by day five to ten degrees higher. Water, taking great care not to wet the foliage. Allow the shoots to grow to a height of 9 or 10 inches before you remove their tops as cuttings. This, and the avoidance of wetting stems and foliage, is to foil leaf nematodes which swim up the stems through films of water and enter the young leaves. If no water is permitted on stems and foliage they are unable to ascend, and cuttings taken from high levels will produce nematode-free, young plants.

Make the cuttings 2 to 3 inches long. Remove their lower leaves and with a sharp knife or razor blade cut the stem cleanly across just beneath a node. Dip

Dividing chrysanthemums: (a) Cutting an old clump apart

(b) Single-shoot divisions ready for planting

them in a suitably diluted insectide and allow to dry before planting them in perlite, vermiculite, or compacted coarse sand, in a propagating bench, flats, or pots. Water well immediately after planting. Shade from strong sun. Maintain a moderately humid atmosphere and temperatures by night of 50°F, by day five to fifteen degrees higher. In bright weather mist the foliage lightly with water, but never so late in the day that it goes into the night wet. After a week or so, when the cuttings will stand exposure to sun without wilting, cease shading. Water to keep the rooting medium pleasantly moist, not constantly saturated. To do so encourages mildew disease and rotting. In three to four weeks the cuttings will have rooted. Then plant them 2 to 2½ inches apart in flats or if you prefer, individually in small pots, in nourishing, porous soil. Cuttings taken from plants brought indoors in January will give fine sturdy plants for setting out in April. If you are without a greenhouse or equivalent facility take later cuttings from long shoots that develop from stock plants left in the cold frame. These will give plants to set outdoors in May. As with those brought indoors, take great care to keep the stems and foliage of stock plants left in cold frames dry.

Seeds afford yet another means of obtaining stocks of young plants. This method is occasionally employed where mixed progeny rather than strictly uniform plants are not a disadvantage and, of course, is used for breeding new varieties. The vast majority of seedlings are to some degree inferior to good named varieties, yet they make gay showings and can sometimes be used advantageously. Sow the seeds indoors early. As soon as the young plants are big enough to handle, transplant them to flats or small pots and manage them as advised for rooted cuttings.

Care until planting outdoors consists of keeping the young plants growing in cool, dryish, airy conditions in full sun. Too high temperatures, especially at night, and poor light result in weak, spindling plants. Excessive humidity encourages diseases. As soon as the plants have taken hold in the flats or pots pinch out their tips to promote branching. Before transplanting outdoors harden them by standing the flats or pots in a sunny place outside for a week or two.

Planting outdoors may be done as soon as the ground is in suitable condition and there is no longer danger from severe frost, or it may be delayed if circumstances necessitate even until tulips and other spring flowers have faded. Late planting produces somewhat smaller, but otherwise no less satisfactory plants. Distances between plants may be varied somewhat according to the growth habits of varieties. Good averages are rows 2 to 2½ feet apart with about 1½ feet between individuals and 2 feet each way between plants in groups in beds.

After-planting care calls for regular, but not onerous attention. It is important to keep the upper inch of soil, but no more, loose by repeated use of a cultivator or scuffle hoe. This is to be preferred to

Pinching outdoor chrysanthemums to cause them to branch

mulching early in the season, but when hot weather arrives the latter can with advantage be substituted. Until the middle or end of July pinch the tips out of all stems and branches as soon as they are 6 inches long. Water so that the earth is drenched to a depth of 6 to 8 inches at whatever intervals are needed to keep the foliage from wilting. In dry weather this may mean as often as twice a week. One or two surface dressings of a complete garden fertilizer may be made during the summer to promote continuous growth, or this can be encouraged after the plants begin to crowd by making applications of dilute liquid fertilizer.

Supports of some kind are needed by most hardy chrysanthemums, although this is not needed for some of the bushier varieties. The neatest and most natural effect is had with groups in beds by sticking firmly into the ground among them, when they are about half grown, sharpened pieces of leafless brushwood. Branches of birch are most excellent, but almost any kind can be used. Supports of this type can be trimmed to height when the plants begin to form flower buds. They are then unobtrusive. Alternatively, stakes or canes can be used and the stems secured loosely to them with ties or Twist-ems. Use the

Planted outdoor chrysanthemums: (a) Flat of young plants raised from cuttings

(c) Press the soil firmly about the roots with the fingers

(b) Make holes of ample size to accommodate the roots

(d) Alternatively, tamp soil with the handle of a trowel

Staking outdoor chrysanthemums: (a) Brushwood stakes afford excellent support

Outdoor chrysanthemums, disbudded to one bloom on each stem

(b) The pieces of brushwood, their bottoms sharpened, are pushed vertically into the ground

fewest stakes that will do the job. Chrysanthemums grown in rows for cut blooms can be supported by sticking stakes firmly into the ground at intervals along both sides of the rows and stretching strings tautly between them.

Disbudding is not commonly done with hardy chrysanthemums, but superior blooms, sometimes important with cut flowers and for exhibition, are had from medium-sized and large-flowered ones by doing this. The procedure is as with greenhouse chrysanthemums. After blooming is over cut the tops off an inch or two above ground level. Kinds undoubtedly hardy may be left in the ground all winter, perhaps with a covering of salt hay, dry leaves, branches of evergreens, or similar protective material, but it is far better to dig up sufficient stock plants and plant them closely together in deep boxes to be wintered in a cold frame or similar place and discard the remaining plants. This makes it possible to spade or plow the soil in fall and leave it with a rough surface over winter to be improved by weathering, or if this is impracticable, the gaining of an early start in spring.

Where nematodes or other soil pests are troublesome, and it is not practicable to change the location of the chrysanthemum planting each year, sterilizing the soil with formaldehyde or other suitable material may be necessary. This must be done well in advance of planting. Clearing the ground in fall makes this practicable.

Cultivation of Greenhouse Chrysanthemums. Most, but not all varieties conveniently grouped as greenhouse chrysanthemums bloom too late to be relied upon under natural conditions to make satisfactory outdoor displays in the north and some are not winter hardy. Keen gardeners prepared to give them special attention, including perhaps provision of more or less improvised shelter in the fall, succeed in growing some varieties outdoors, but most commonly their cultivation is limited to greenhouses. These mums may be had in a plethora of varieties with flowers ranging from button-like pompons to enormous footballs, formal and informal, singles, doubles, semidoubles, and anemones, kinds suitable for disbudding to single blooms, others better adapted for growing as sprays, and all in a great range of colors. The plants may be tall and stately, of intermediate height and bushy, or low and compact. There are sorts for all purposes and tastes. New varieties are introduced each year. Consult catalogs of specialist dealers for descriptions and illustrations of those available.

In greenhouses chrysanthemums are grown in ground beds and benches for cut blooms, and as pot plants. By careful control of day length and temperature they can be had in bloom at any time, but first we shall consider their cultivation, without such manipulations, to flower at their normal season, according to variety, from Oc-

tober to January. Young plants are raised from cuttings as described for outdoor chrysanthemums. This is easily done even by amateurs, but more and more it has become the practice for greenhouse operators to purchase their stocks from specialists. This eliminates the need for wintering over stock plants and assures healthy starts. Cuttings are taken three weeks to a month before the young plants are to be ready for their first potting into 2½-inch pots from which they are soon planted in beds or benches or transferred to larger pots.

The soil must be fertile and friable, one fairly heavy, but porous. A good loamy topsoil mixed with well-rotted manure, compost, or peat moss, and sufficient coarse sand or perlite to maintain aeration is suitable. Fortify it by mixing in some complete, slow-acting fertilizer or some bonemeal. Let the mix be coarse rather than finely sifted. It is important that an adequate drainage layer underlie the soil.

Greenhouse bench of newly planted chrysanthemums

In benches and beds planting is done from 2- or 2½-inch pots from May to July, the earlier dates for early-blooming varieties, or rooted cuttings may be set directly in the beds or benches. Allow 1 foot each way between pompons and other small-flowered varieties to be grown as sprays, 6 to 8 inches each way between larger-flowered varieties to carry one stem disbudded to a single bloom, 8 inches each way if the plants are each to have two stems each with a disbudded bloom, and 9 inches each way if three single-bloomed stems are to be allowed to each plant.

Potted plants for normal-season blooming are managed similarly to those in beds or benches, except that the first transfer from 2- or 2½-inch pots is to ones 4 inches in diameter and later shifts to larger containers, the last one in June, are given as the increased size of the plants makes necessary. Specimens put into 4-inch pots in May will usually flower in 6-inch con-

Greenhouse chrysanthemum in a 2½-inch pot ready for repotting into a larger container

tainers. To have specimens to finish in 7-, 8-, or 9-inch pots, propagate earlier, in time to have plants ready for 4-inch pots in April. At the final potting leave the soil level 2 or 3 inches below the top of the rim of the pot to allow for a top dressing of rich soil later. Make sure plants to be potted or planted are well watered an hour or two previously. Do not water them immediately afterward, but keep their foliage misted with water, and water the soil before the leaves begin to wilt. Commercial growers follow the practice of potting four or five rooted cuttings directly into a 6-inch pot or pan, the container in which the plants are bloomed. This eliminates the need for repotting and gives full bushy specimens in a shorter time than if only one plant was set in each pot. Because of this, the first potting of the rooted cuttings can be done considerably later.

Summer care from the time of bedding, benching, or potting demands regular, but

Regular and adequate watering is needed by greenhouse chrysanthemums

not onerous attention. Chrysanthemums need full sun and a buoyant, airy atmosphere. Whenever outside weather permits, ventilate the greenhouse freely. Avoid dank, heavily humid conditions. Natural air circulation may with advantage be augmented by the use of fans. This is especially desirable during periods of heavy, humid weather. Water regularly and thoroughly, not so often that the soil is constantly sodden, because this will result in the roots rotting, but frequently enough to prevent wilting. Even though chrysanthemum foliage that droops from lack of water perks up quickly when the soil is wetted, plants subjected to this neglect suffer a setback that does not show up until later. On bright days when the moisture will dry within an hour, spray the foliage lightly with water several times, but never so late that it does not dry before dark. This does much to keep the stems soft and growing actively. Keep the surface soil shallowly cultivated until roots begin to invade it, then apply a shallow top dressing of rich earth or mulch. If the greenhouse is needed for other purposes in summer, stand pot chrysanthemums in a sunny place outdoors then, but generally it is better to keep them inside.

Fertilize on a regular basis beginning about a month after planting, or as soon as the containers of potted plants begin to be filled with healthy roots, and continue until the flowers begin to show color. Then cease. Weekly applications of dilute liquid fertilizer are needed or light dressings of slow-release solid fertilizers at longer intervals may be substituted. Fertilizer of relatively high nitrogen content such as a 10-6-4 is recommended, particularly during the early part of the season. Later one containing more potash may be desirable.

Support is needed. With pot plants supply this with stakes and ties. Do the job neatly to enhance the natural beauty of the plant and so that when it blooms the supports are concealed. In beds and benches large-flowered varieties may be held by se-

Support for greenhouse chrysanthemums grown for cut flowers: (a) Inserting a stick across the bench to keep the outer horizontal lengthwise wires spread

(b) Lacing strings across and to the horizontal lengthwise wires

(c) The finished job, showing the supporting squares of longitudinal wires and crosswise strings

(d) The crop, disbudded chrysanthemums, in bloom

curing each stem to a wire stake. With small-flowered kinds grown to have sprays of bloom it is more usual to install as the plants grow, first at a height of 1 foot, later at 2 feet, and eventually if needed at 3 feet above the soil, horizontal wires stretched tightly longitudinally with strings laced across them and twisted around each wire to form a gridiron with a square for each plant, or alternatively to use a special welded wire fabric favored by many commercial growers, installed so that it can be raised as the stems lengthen.

Pinching is important to control branching. Except with football-type mums and other large-flowered varieties the plants of which are being restricted to a single stem

(branches) from what are called break buds (early-produced flower buds that terminate the stems, but do not develop into blooms), but most do not and so the time pinch is necessary.

Disbudding is done to limit the number of flowers. With many varieties it is necessary to assure blooms of good form and large size. It is not needed with pompons and other small-flowered varieties and is done not at all or only to a limited extent if sprays of singles, anemones, and other kinds with medium-sized blooms are the objective. Disbudding involves two procedures, the removal while small of all or many of the side shoots from stems that

Small-flowered single chrysanthemums need no disbudding

Pot specimens disbudded to one flower to each stem (both figures, above)

Decorative type of chrysanthemum grown without disbudding to produce sprays of cut flowers

The last pinch, given long enough before flowering to ensure long stems

each with one bloom, the first pinch is given when the young plant is 3 or 4 inches tall. This, a soft pinch, is done by nipping out the upper ½ inch or less of the stem. Depending upon how early the plants were propagated, one or more later pinches will be needed. With plants grown as bush specimens pinch the tips out of the shoots each time they attain a length of about 6 inches. Very important is the last pinch or time pinch. With late-propagated plants this may also be the first pinch. It is done to make certain that the blooms will have satisfactory long stems. Just when to give the time pinch differs somewhat according to variety from ninety to one hundred and ten days before the flowering date. Suppliers of rooted chrysanthemum cuttings provide this information in their catalogs. A few varieties develop natural breaks

develop as a result of the time pinch and, where single large blooms instead of sprays of flowers are desired, the elimination of the tiny buds that surround the terminal one (the last bud of the season to be produced, which terminates the stem). Take off the buds from around the terminal one while they are tiny, before their individual stalks lengthen appreciably. Do this by supporting the top of the stem with your fingers, placing the ball of the thumb on the bud to be removed, and rolling it gently downward. Take great care not to damage the center bud in the process.

Management in the fall calls for the maintenance of a cool, airy environment without excessive humidity. Too much humidity invariably brings serious mildew disease. With the coming of September cease wetting the foliage, but supply water liberally to the roots and, until the flower buds show color, fertilize regularly. Bring plants summered outdoors inside before frost. On all occasions ventilate the greenhouse as much as possible to keep temperatures low and encourage a free circulation of air. Maintain night temperatures as close to the 45 to 50°F range as possible. After flowering, if plants are to be retained as propagating stock cut them down to a height of 8 or 9 inches. Pot any that have

Chrysanthemums trained in cascade fashion

One of the first chrysanthemums trained in the United States in cascade fashion. (grown by T. H. Everett in 1930)

been in beds or benches. Be sure each is labeled clearly with the name of the variety. Store these stock plants until cuttings are needed from them under good conditions in a deep, frost-proof cold frame or in a greenhouse where they are in good light and where the night temperature is 40 to 50°F. Do not, as is sometimes done, put them under a bench or in some other out-of-the-way place where light is poor, temperatures perhaps too high, and moisture conditions conducive to rots and other diseases. Such treatment cannot be ex-

pected to produce plants that will provide strong, healthy cuttings.

Trained chrysanthemums can be had in several attractive forms in addition to the familiar bush and single-stemmed plants. The most popular are cascades, tree-form or standards, and espaliers. The art of training cascade chrysanthemums was introduced from Japan, where it had long been practiced, to America and Europe early in the 1930s. A cascade is a specimen that in bloom displays itself as a floral cascade hanging down from its container. Only varieties with long, slender, pliable stems are adaptable to this mode of training. Here is how to do it. Beginning in late winter or early spring with a rooted cutting, allow one to three main stems to develop. Never pinch the tips out of these, the leaders. Do pinch the tips out of all lateral shoots, sublaterals, and later branches that develop up to the beginning of August as soon as they are 3 or 4 inches long. Then cease pinching. From the time of their first potting insert stakes in the pots at an angle of 45 degrees and keep the leaders tied to these. Make sure that the plants are positioned so that the tips of the stakes point north. This is important. It assures that all parts of the plant, which slopes downward to the south, receive the benefit of maximum sunshine. In late summer or fall when the flower buds first show, lower the stakes by two or three stages spaced a week or so apart, to the horizontal, still with their tips to the north. Finally, when the flower buds begin to show color, remove the stakes and allow the stems to cascade over the sides of the pot, which must of course then be raised by putting it on a stand, shelf, or other support. At this time, too, the plant must

be turned so that the cascade faces south, again so that it is assured full sun. After the buds begin to open cascade chrysanthemums can be planted in large hanging baskets, urns, and similar containers to achieve stunning effects.

Espaliered chrysanthemums are achieved in much the same way as cascades. The difference is that the stems at blooming time are erect instead of pendulous and are tied to a trellis. Varieties suitable as cascades are appropriate for espaliering. The only difference in procedure is that the supports, in the early stages stakes, and later trellis, and which like those for cascades are throughout the early part of the growing season kept at an angle of 45 degrees with their tips pointed to the north, are when the flower buds form in late summer, gradually raised to an upright position instead of being lowered. They then will be faced to the south.

Tree-form or standard-trained chrysanthemums are easy to obtain. Begin early in the year with a strong, rooted cutting. Keep this growing vigorously and tied to a stake. Do not pinch out its tip until it has reached the height you want the clear stem or trunk to be, but do pinch out all side shoots as soon as they appear. When the trunk reaches the desired height, usually 2½ to 3½ feet, nip out its apex and allow four to six shoots to develop just below. These are to become the main branches of the head. When they have developed four or five leaves, pinch their tips out and repeat this procedure with all other branches that develop until the beginning or middle of August. In addition to keeping the main stem tied to a stake, the branches of the head are likely to need some tying in, and if they are so numerous as to become unduly crowded, a little early thinning out. To encourage the development of symmetrical heads turn the plants around once a week so that all sides in turn are exposed to maximum light.

Novelty specimens of other types are sometimes produced by training flexible-stemmed varieties to wire frameworks of various shapes and by grafting more than one variety onto a single plant. By this last procedure it is possible to have tree-form specimens with different colored flowers on different segments of the head.

The American discovery in the 1920s that day length, or more correctly night length, is a critical factor in determining when chrysanthemums bloom, revolutionized their commercial production and made it possible for amateurs with but little effort to have garden chrysanthemums bloom earlier outdoors and to produce large-bloomed varieties in the open that without special treatment do not come into bloom before killing frost. It also made it possible to flower chrysanthemums in greenhouses at any time of the year.

Briefly the situation is this. When nights

(a) Wooden frames over beds of outdoor chrysanthemums in preparation for shading to induce early blooming

(b) A black cloth is placed over the plants to lengthen the hours of darkness

(c) A planting of outdoor chrysanthemums, some shaded, others not shaded

are less than nine and one-half to eleven hours long, the exact period depending upon the variety, chrysanthemums produce only shoot and leaf growth. Nights longer than these critical lengths cause initiation of flower buds. Under normal conditions natural changes in day length are responsible for flowers appearing at their appointed times in fall and early winter.

Time taken from the beginning of long nights to blooming varies with varieties from seven to fifteen weeks. Kinds taking no more than seven or eight weeks include early-flowering sorts used for outdoor displays in the north. Further south, varieties needing nine to eleven weeks from when the short days begin to blooming may be successfully flowered in the open. These also include the majority of greenhouse varieties. Those that require more than

eleven weeks are late-flowering greenhouse varieties that under normal conditions do not bloom before late December or January. Until the discovery of the effects of day length and the development of techniques to control it, chrysanthemums flowered only in their natural seasons, a succession of blooms being maintained through that period by growing a selection of from seven- to fifteen-week varieties.

Techniques for altering day length include shading to lengthen nights and artificial lighting to lengthen days. Only the first is applicable to mums grown outdoors. Shading is done by covering the plants with lightproof black cloth or black plastic draped over a suitable wood or metal framework so that it encloses the plants without resting on them. All light must be excluded. The shade is put in place each evening and taken off the next morning. Usually it is left in position for longer than the minimum period of darkness required. From 8 P.M. to 8 A.M. is satisfactory. For convenience, commercial florists often shade earlier, from 5 to 6 P.M., but this can result in too high temperatures inside the shade and also reduces unnecessarily the hours of light during which photosynthesis can take place. Begin shading the required number of weeks of short days required by the variety before the blooming date you wish to achieve. Thus, to bring a ten-week-response variety that normally flowers on November 5th into bloom by October 20th technically requires shading to begin August 11th. But night temperatures of more than 60°F, likely to occur that early in the season, have a delaying effect on bud initiation; therefore in this case it is advisable to begin shading four or five days earlier. A little experimentation and experience will provide information for precise scheduling of blooming. Once flower buds are well in evidence, approximately one month after shading begins, there is no need to continue it. From then on the buds will develop normally.

Lighting to delay blooming is done in greenhouses. Light sufficiently rich in red rays to produce the required response is produced by incandescent (Mazda-type) bulbs. If fluorescent tubes are used pink or red ones supply suitable illumination. Low intensities only are necessary. A minimum of ten foot-candles of light is usual, but lesser amounts are effective. A common arrangement is to space 60-watt incandescent lights 4 feet apart at a height of 2 feet above the plants and in such a way that the lights can be raised to maintain that distance as the plants grow. Because it is short periods of darkness rather than long periods of light that prevent the initiation of flower buds, it is more economical to break the night by introducing approximately midway through it a period of light

than, as is sometimes done, to extend the natural day by lighting from dusk until about nine hours before dawn. In practice it has been found that an interruption of darkness during the night of two hours' duration in August and May, three hours' in September, October, March, and April, and four hours' from November to February produces satisfactory results. Where the installation is extensive the load on the electric power line can be halved by lighting one-half the crop for the required period immediately before midnight and the other one-half for a similar time immediately after midnight. This can be done by an automatic time switch.

Temperature also plays a critical part in the initiation of chrysanthemum flower buds. Ordinarily at night it must not be under 60°F. With plants grown to bloom at their normal season this presents no problems. Ordinary outdoor temperatures at the time of bud initiation meet this minimum. The same or slightly higher levels must be maintained in greenhouses during the period when long nights are in effect to encourage bud formation to give out of season blooms. When flower buds are clearly visible the night temperature should be lowered to 55°F, a necessary procedure to favor the rapid and satisfactory development of blooms. From the time the flowers begin to open, out-of-season chrysanthemums may need shading very lightly from brilliant sun to prevent the blooms being scorched.

Growth retardants are used by commercial growers of pot chrysanthemums to supplement timing of taking cuttings, pinching, and exposure to maximum light, as height controls. Spraying with B-Nine at the rate of 6 ounces to a gallon two weeks after the young plants are pinched and perhaps again a week or two later is effective. Somewhat more uncertain as to results, but otherwise effective, Phosphon can be used as a soil treatment.

Pests and Diseases. Among common pests of chrysanthemums are aphids, cutworms, cyclamen mites, leaf miners, mealybugs, leaf nematodes, red spider mites, root knot nematodes, slugs, snails, tarnished plant bug, and whiteflys. A particular pest, chrysanthemum midge causes tiny galls on shoots and foliage inside which the larvae of small flies live. The chief diseases are botrytis blight, leaf spots, powdery mildew, rust, stem and root rots, verticillium wilt, virus yellows, and virus stunt.

CHRYSOBALANUS (Chryso-bálanus)—Coco-plum. This genus of four tropical American and African species is by some botanists placed in the rose family ROSACEAE, by others is included along with a few other genera in the chrysobalanus family, CHRYSOBALANACEAE. The name, derived from the Greek *chrysos,* gold, and *balanus*, an acorn, alludes to the appearance of the fruits of some kinds.

The sorts of *Chrysobalanus* are trees and shrubs with alternate, lobeless leaves and clusters of tiny white or greenish flowers with five each sepals and petals, fifteen or more stamens some of which may be sterile and thus technically staminodes, and one style. The fleshy fruits contain a single stone.

The coco-plum (**C. icaco**), native from southern Florida to West Indies and Brazil is an evergreen up to 30 feet tall, but often is only a low shrub. Its broadly-ovate to nearly-round, blunt, glossy leaves narrow to very short stalks and are up to 3 inches long. The white flowers, in erect, axillary clusters, are succeeded by round edible fruits about as big as plums, dark red to purplish or whitish, and much wrinkled when ripe. The fruits are very oily and in Mexico are sometimes strung on sticks as candles. In that country, too, the roots are used as an astringent, and a black dye is obtained from the leaves.

Garden and Landscape Uses and Cultivation. The coco-plum is quite useful for warm, frost-free climates and especially for seaside planting. Attractively foliaged, in the United States it rarely exceeds tall shrub size. It grows in ordinary soil, and requires little care. Most usually propagated by seeds; it can also be increased by cuttings.

CHRYSOCOMA (Chrysó-coma). Some fifty species of the daisy family COMPOSITAE, natives of Africa and South America, constitute this genus. The name, from the Greek *chrysos*, gold, and *kome*, hair, alludes to the fuzzy appearance of the flower heads.

Chrysocomas are small shrubs or subshrubs with alternate, linear leaves, and flower heads, solitary from the branch ends, of all disk florets (the type that compose the centers of daisies). The florets are tubular and five-lobed. The seedlike fruits are achenes.

The only kind at all commonly cultivated **Chrysocoma coma-aurea**, native of South Africa, is a hairless evergreen, in bloom up to 2½ feet tall. It has dark green, spreading or recurved, ½-inch-long leaves and pleasantly scented, dense, button-like, golden-yellow flower heads ¾ inch across.

Garden and Landscape Uses and Cultivation. Suitable for blooming in 5- or 6-inch pots in late winter and spring, *C. coma-aurea* is also well adapted for outdoor flower beds, rock gardens, and other locations in mild Mediterranean-type climates, such as that of California. Its requirements are simple, exposure to full sun and well-drained, moderately fertile earth. It may be sheared to shape, and pot specimens are repotted as soon as blooming is over. Increase is easy by cuttings and seed. Indoors a winter night temperature of 45 to 50°F is adequate. By day this may be increased by five to fifteen degrees. On all favorable occasions the greenhouse must be ventilated freely.

CHRYSOGONUM (Chrysóg-onum) — Golden Star. One species constitutes this genus of the daisy family COMPOSITAE, a native in woodlands from Pennsylvania and Ohio to Florida and Alabama. Its name, derived from the Greek *chrysos*, gold, and *gonu*, knee, has no obvious application.

Golden star (**Chrysogonum virginianum**) is a pretty, hardy, herbaceous perennial with a short, brittle rhizome, fibrous roots, and stems usually under 1 foot tall, but frequently considerably shorter. The stems are covered with soft, spreading, almost transparent, glandular hairs. The long-stalked, opposite leaves, 1 inch to 3½ inches long, are ovate to nearly round. To a greater or lesser degree they are hairy, at least on their undersides. Their margins are round-toothed. In the south the foliage remains green through the winter. In the north it dies. The flower heads, solitary or few together on slender terminal and axillary stalks, have a disk of sterile flowers and about five, widely spaced, golden-yellow fertile ray florets (often mistakenly called petals). They measure 1 inch to 1¼ inches across. There is variation among individuals as to height and compactness. Select dwarfer forms for planting. The fruits are seedlike achenes.

Garden Uses. Golden star blooms for many weeks in spring and early summer and sometimes more sparsely later. It is worthwhile for rock gardens, wild gardens, margins of woodland paths, and like locations and may be used as a choice groundcover in places given reasonable attention in the matter of weeding.

Cultivation. This plant thrives in any well-drained soil that contains a fair amount of organic matter and is not too dry. It stands considerable shade, but flowers best where it receives dappled sunlight filtered through the overhead canopy. A too-fertile soil promotes excessive stem length. Short stems and compact plants are most desirable. Propagation is easy by division in spring or early fall and by seed.

CHRYSOLARIX. See Pseudolarix.

CHRYSOLEPIS. See Castanopsis.

CHRYSOPHYLLUM (Chryso-phýllum) — Star Apple or Cainito or Caimito, Satin Leaf, Orleans-Plum. This genus of the sapodilla family SAPOTACEAE includes about 150 species of evergreen trees of the tropics and subtropics of the Old and the New Worlds and Australia, including one indigenous to Florida. Its name comes from the Greek *chrysos*, gold, and *phyllon*, a leaf, and has reference to the undersides of the foliage of some species.

Chrysophyllums have alternate, undivided leaves, with numerous parallel veins extending from the mid-veins to the margins. The small flowers are in clusters and usually have five-lobed calyxes and tubular-bell-shaped to somewhat wheel-shaped corollas, commonly with five lobes. There are five stamens. The fruits, technically berries, fleshy or leathery, are in some kinds edible.

The star-apple, cainito, or caimito (**Chrysophyllum cainito**), native of Central America and the West Indies, is planted there and in other warm places, including Florida and Hawaii, for its edible fruits and as an ornamental. It is a good-looking, round-topped tree that in Florida usually does not exceed 25 or 30 feet in height and may be lower, but in good soils elsewhere it sometimes attains 50 or even 65 feet. It has a narrow trunk, long, slender branches, and two-ranked, short-stalked, ovate to oblong leaves, 3 to 6 inches long by 1 inch to 3 inches wide, green above, and densely clothed with yellow or coppery, silky hairs on their undersides. The inconspicuous purplish flowers are borne in summer in

Chrysocoma coma-aurea

Chrysogonum virginianum

small clusters or sometimes solitary. They have five-lobed calyxes and corollas, five stamens, and a stalkless eight- to ten-parted stigma. The fruits ripen the following late spring. They are white or dull purplish, smooth, apple- or slightly pear-shaped, and 2 to 4 inches in diameter. When sliced horizontally they reveal a star-shaped core. They are sweet when fully ripe and are usually eaten without cooking.

The satin leaf (*C. oliviforme*) is an ever-green shrub or tree up to 35 feet tall. It inhabits southern Florida, including the Keys, and the West Indies. Its elliptic to ovate, leathery leaves, 1½ inches long, are lustrous green above and on their under-sides are covered with satiny, silvery to yellow hairs. The bell-shaped flowers are not decorative. They have five-lobed ca-lyxes and corollas, the latter white and up to ¼ inch wide. The fruits, about ¾ inch in diameter, are purple.

The Orleans-plum (*C. pruniferum*) is the only species native to Australia. It is usu-ally a small tree, but in its homeland oc-casionally may grow as high as 60 feet. Its leaves, russet on their undersides, are ob-long and 3 to 4 inches long. The small, clustered flowers are succeeded by dark blue, plumlike, succulent fruits 1 inch to 2 inches in diameter that contain one or two large seeds.

Garden and Landscape Uses. All the species discussed are attractive ornamen-tals, useful for planting singly or in groups. In addition, the star-apple is well worth growing as a source of edible fruit. It is a common door yard tree in the West Indies and Caribbean region.

Cultivation. For their successful growth these plants require humid tropical condi-tions. They grow in a variety of soils, but are most responsive to those deep and fer-tile. When young they are especially sub-ject to damage even by light frosts, later they are slightly more tolerant, but even then will not stand much cold. Propaga-tion is by seeds, which germinate readily in well-drained, sandy soil. Because the cropping and fruiting quality of seedling star-apples varies so much it would be ad-visable to increase desirable individuals by budding onto seedling understocks or by cuttings. The latter, made from firm shoots, can be rooted under greenhouse condi-tions with a little bottom heat provided in the cutting bench.

CHRYSOPSIS (Chrys-ópsis)—Golden-As-ter. Entirely North American, *Chrysopsis* is a genus of twenty species of the daisy fam-ily COMPOSITAE. Its name, calling attention to the color of the flower heads, comes from the Greek *chrysos*, gold, and *opsis*, re-sembling.

Golden-asters are summer- and fall-blooming herbaceous perennials, annuals, and biennials, with alternate, undivided leaves, all stalkless, or the lower ones stalked and those above without stalks. Except for their color mostly aster-like, the flower heads commonly have both disk and petal-like ray florets. The first are mostly bisexual, the rays female. The in-volucres (collars at the backs of the flower heads) are of bracts in several overlapping series. The hairy, seedlike fruits are achenes. Plants treated in this Encyclope-dia as *Pityopsis* are by some authorities in-cluded in *Chrysopsis*.

Perennial, and native of pine woods and sandy soils from New York to Ohio, Ken-tucky, Florida, and Louisiana, *C. mariana*, 8 inches to 2 feet in height, has stems woolly-hairy when young. Its lanceolate to oblong or elliptic, or the lower oblanceo-late to obovate, usually toothed, pinnately-veined leaves are up to 6 inches long. The flower heads, ¾ to 1 inch wide, are in crowded clusters at the ends of the branches. Their involucres are furnished with glandular, sticky hairs. From the last, *C. villosa* differs in the involucres of its flower heads not having glandular hairs. A highly variable perennial (botanists rec-ognize several varieties, including *C. v. bakeri*, which is sometimes grown as *C. bak-eri*), this is wild in dry, open places from the Middle West to the Pacific coast and from British Columbia to Texas and New Mexico. From 8 inches to 3 feet tall or taller, and hairy, it has many oblong-ellip-tic to linear-lanceolate, pinnate-veined leaves up to 3½ inches long by ½ inch wide, and toothed or not. The several flower heads are up to 1 inch wide.

Garden and Landscape Uses and Culti-vation. Golden-asters' chief appeals are for naturalizing in semiwild areas, for plant-ing in rock gardens and similar places, and for grouping at the fronts of flower beds. They are especially useful for sandy, dryish soils and in exposed, sunny locations. Al-though they get along with minimal care, they respond to reasonable fertilizing and watering. Propagation is easy by seed and by division.

CHRYSOSPLENIUM (Chrysos-plènium)—Golden-Saxifrage. Creeping or prostrate herbaceous plants of wet soils, the golden-saxifrages number about fifty species. Na-tives of temperate parts of the northern hemisphere, they belong in the saxifrage family SAXIFRAGACEAE. The name, from the Greek *chrysos*, gold, and *splen*, the spleen, is of obscure application.

Chryspleniums have alternate or op-posite leaves and solitary or clustered small flowers without petals. The blooms have four spreading sepals, and four to eight short-stalked stamens attached to an eight-lobed disk. The fruits are capsules.

A native American, *Chrysosplenium americanum* occurs in shaded places from southern Canada to Virginia and Indiana. Its branching stems, 2 to 8 inches long, have short-stalked, ovate to round leaves mostly under ½ inch in diameter and slightly toothed or toothless. The lower leaves are opposite, the upper ones alter-nate. The solitary greenish-yellow or red-tinged blooms are under ¼ inch across. They have stamens with red anthers.

Natives of Europe and Asia, *C. alterni-folium* and *C. oppositifolium* are sometimes cultivated. With underground creeping stems from which rise to a height of up to 8 inches numerous erect stems bearing alternate leaves, *C. alternifolium* has long-stalked basal leaves with rounded, shal-lowly-lobed blades up to ¾ inch in diam-eter. The upper leaf surfaces are hairy. The yellow blooms, under ¼ inch in diameter, are in terminal, flat-topped clusters. From the last, *C. oppositifolium* differs in its creeping stems being above ground, and in its rounded leaves being opposite.

Garden Uses and Cultivation. These rather insignificant hardy plants may be used to carpet bog gardens and other wet soils, preferably where there is some shade. They are easily increased by division and by seed.

CHRYSOTHAMNUS (Chryso-thámnus)—Rabbit Brush. Rabbit brushes are usually much-branched, erect shrubs and sub-shrubs typically of desert, semidesert, and other dry soils in western North America. There are thirteen species. They belong in the daisy family COMPOSITAE. Their botan-ical name, from the Greek *chrysos*, gold, and *thamnos*, a shrub, refers to their quite showy blooms.

These plants have alternate, or rarely opposite, undivided, sometimes glandu-lar-dotted, one- to three-veined leaves. Their flower heads, rarely including ray florets and usually consisting of five disk florets, have comparatively long and nar-row involucres (collars) of overlapping bracts, generally in five, but sometimes fewer, vertical rows. Although small, the heads are numerous and are usually as-sembled in racemes or panicles. Often the foliage, when crushed, is strongly scented. Some species contain rubber, but it has not proven practicable to exploit this commer-cially. The fruits are seedlike achenes.

Up to 7 feet in height, but often much lower, very variable *C. nauseosus* has branches permanently felted with white or greenish hairs. Its leaves, when bruised often ill-scented, are narrowly-linear to lin-ear-oblanceolate. They are ¾ inch to 2¾ inches long and hairy to nearly hairless. The flower heads are in terminal rounded clusters. This kind is native from Texas and adjacent Mexico to Colorado, Wyo-ming, Saskatchewan, California, and Brit-ish Columbia. Variety *C. n. graveolens* (syn. *C. graveolens*), 3 to 5 feet tall, has erect stems that are white-hairy when young, later becoming hairless. Its linear leaves 1 inch to 3 inches long, are hairless at maturity, and less strongly scented than those of *C. nauseosus*. The flower heads are in clusters 1¼ to 4 inches across. Highly variable *C. viscidiflorus* in its various forms ranges from California to New Mexico, the Rocky Mountains and British Columbia. Typically not more than 3 feet tall, it has hairless or somewhat hairy, linear to linear-lanceolate, flat or twisted leaves up to 2 inches long. The plant previously designated *C. pumilus* is *C. viscidiflorus pumilus*. It is up to 1½ feet tall.

Garden and Landscape Uses and Cultivation. In dry regions these shrubs are sometimes planted, but elsewhere they are of little or no horticultural importance. They succeed in full sun, in well-drained soil where rainfall is scant. Some kinds are highly tolerant of alkaline soils. They may be raised from seed.

CHRYSOTHEMIS (Chryso-thèmis). This is a genus of about six species of the gesneria family GESNERIACEAE that includes the plants previously named *Tussacia*. It ranges in the wild from Panama to the West Indies and Brazil and is closely related to *Alloplectus*. Its name is, according to Greek mythology, that of the daughter of Clytomnestra and Agamemnon.

The chief distinguishing feature of *Chrysothemis* is its bell- or cup-shaped calyx composed of five strongly-keeled or winged segments united almost to their tips. The corolla is trumpet-shaped and has five lobes (petals). There are four rounded stamens united in a square. The fruits are globose capsules containing seeds with threadlike attachments. Usually chrysothemises form substantial tubers, but under lush growing conditions they are less likely to do so than when subjected to seasonal drought in the wild or in cultivation when the roots are somewhat crowded in the pots. The thick, upright stems are furnished with opposite leaves in the upper axils of which the small clusters of flowers develop, and in the lower ones sometimes small tubers. Although individual blooms last only a few days, their brightly colored calyxes remain attractive for several months.

From 1 foot to 2 feet tall or taller, and with round stems, *C. friedrichsthaliana* is native to Central America and the West Indies. The largest of its hairy, bright green, pointed-ovate, toothed, thin leaves are up to 1 foot long by 5 inches wide, and much wrinkled between their prominent veins. The flowers, in clusters of three or more, have yellow-green, bell-shaped calyxes, with sharp-pointed tips to each of the joined segments. The orange-yellow corolla, ¾ inch long, scarcely exceeds the calyx in length. Its lobes, which spread to form a ½-inch-wide face to the flower, are longitudinally striped with crimson.

Similar to the last, *C. pulchella* (syns. *C. woodsonii*, *Tussacia pulchella*) differs in several details. Its stems are angled instead of

Chrysothemis friedrichsthaliana

Chrysothemis pulchella

round, and its leaves toothed, lustrous green and wrinkled on their upper sides, and up to 1¼ feet long by 5½ inches wide. In clusters of four to ten, the flowers have bright orange to orange-red calyxes, and ¾- to 1-inch-long corolla tubes that project farther than those of *C. friedrichsthaliana* from them. The blooms measure about ¾ inch across their faces. A variant of this species with narrower, olive-green leaves with red undersides and redder calyxes is known.

Garden and Landscape Uses. Because of the long-lasting attractiveness of the colorful calyxes, coupled with good-looking foliage, these easy-to-grow gesneriads are highly desirable. Only in the humid tropics and subtropics are they useful for outdoor cultivation. They need shade from strong sun and soil that is moist throughout the growing season, with abundant organic matter. More commonly they are grown in pots in greenhouses, and by gesneriad fanciers who lack greenhouses, indoors under conditions that suit gloxinias (*Sinningia*).

Cultivation. Soil for pot cultivation should be fertile, porous, sandy loam with which has been mixed a generous proportion of peat moss or humus. The containers must be well drained. When stems

and foliage are present the earth is maintained evenly moist, but not wet. When the foliage dies the tubers are kept dry in the soil in which they grew until signs of new growth appear, then they are shaken free of the old soil and repotted in new, and watering is resumed. A humid atmosphere and shade from strong sun are requisite, as is during the growing season a minimum night temperature of 55 to 60°F. By day the temperatures are increased by five to fifteen degrees. Higher temperatures in summer are appropriate. Once the pots in which the plants are to bloom (which may be 5 or 6 inches in diameter) are filled with roots, applications at weekly or ten-day intervals of dilute liquid fertilizer are helpful. Usually chrysothemises are grown without pinching out the tips of the stems, but this may be done when the plants are only a few inches tall to induce branching and the production of fuller, broader specimens. Propagation is very easy by cuttings made from young shoots, branches, or single leaves, and by seeds.

CHUFA is *Cyperus esculentus sativus*.

CHUPAROSA. As a common name this is applied to *Anisacanthus thurberi* and *Justicia californica*.

CHUSQUEA (Chusquè-a). This is a group of forty or possibly more species of bamboos of the grass family GRAMINEAE. They are natives of the New World, mostly of high plateaus in South America, but with the range of the genus extending from Mexico and the West Indies to southern Chile. These are bushy plants, or sometimes straggly climbers, with hollow or solid, cylindrical, jointed stems, leaves with flat, linear to lanceolate blades, and spikelets, each with one fertile flower with three stamens, in terminal clusters. The name is a native South American one.

From 4 to 18 feet or more tall, *Chusquea culeou* has solid canes 1 inch to 1½ inches in diameter at their bases and densely-clustered branches and branchlets. The

Chusquea culeou

canes, at first olive-green, change later to dull yellowish-brown. When young they have a waxy band beneath each node. Branches are not developed until the second year. The slender-pointed, five-veined leaves are up to 3½ inches in length and ⅓ inch wide, medium-green on their upper surfaces, and paler beneath. Bristly hairs fringe their margins along both sides. The leaf sheaths persist through the first season. This bamboo is of distinct appearance and makes a compact clump.

Garden and Landscape Uses and Cultivation. Chusqueas are little known in cultivation and it is unlikely that they will thrive where summers are very hot or winters very cold. At Edinburgh, Scotland, *C. culeou* prospers. It and perhaps others should be adaptable for cultivation in the Pacific Northwest. For more information see Bamboos.

CHYSIS (Chȳ-sis). The number of species into which this genus of the orchid family ORCHIDACEAE is divided depends upon individual concepts of what constitutes a species and varies from two to about six. Here the conservative viewpoint is adopted, and certain variants sometimes given specific rank are treated as varieties. The group, native from Mexico to South America, because of the fused appearance of its pollen masses was named from the Greek *chysis*, melting.

Chysises are deciduous or nearly deciduous epiphytes that grow perched on trees or more rarely on rocks. They have usually pendulous, club-shaped or spindle-shaped pseudobulbs approximately 1 foot in length and several pleated, lanceolate to elliptic-lanceolate leaves. Developing from the lower leaf axils of the young shoots, the thick, waxy, long-lasting, aromatically-fragrant blooms are in short racemes. Quite large and handsome, they have somewhat cupped sepals and petals, and a deeply-concave lip frequently of a different color.

Chiefly summer-blooming, *Chysis aurea* is native from Mexico to Venezuela and Peru. Its pseudobulbs are up to 1½ feet long, its leaves as long and approximately 2 inches wide. The long-stalked racemes, sometimes 1 foot in length, have five to twelve flowers up to 3 inches in diameter. Most commonly creamy- to lemon-yellow, they have whitish lips marked with red or brown and at their bases three to five velvety, yellowish ridges. In *C. a. bractescens* (syn. *C. bractescens*) the bracts from the axils of which the individual flowers come are more instead of less than 1 inch long. The blooms of *C. a. lemminghei* (syn. *C. lemminghei*) have purple-marked sepals and petals. The sepals and petals of *C. a. maculata* are marked with reddish-brown.

From *C. aurea* and its varieties *C. laevis* differs in having blooms usually under 3 inches across with sepals and petals yellow in their lower halves and with a conspic-

Chysis aurea bractescens

uous orange spot almost filling their tops. Their crimson-blotched, yellow lips have a wavy-margined, rounded center lobe and two-sickle-shaped side ones. Native from Mexico to Costa Rica, this flowers from spring to early summer.

Garden Uses and Cultivation. Of easy cultivation, chysises are worthy of inclusion in all orchid collections. They appreciate temperatures and environments suitable for cattleyas with, in winter, slightly cooler, drier, more airy conditions. They need bright light with only enough shade to prevent scorching of the foliage. Better suited for hanging baskets than pots, chysises succeed when planted in osmunda or tree fern fiber mixed with crocks (pieces of broken clay flower pots). From spring until the new pseudobulbs have matured in fall, generous watering is in order, but less water is needed during the resting season. When growth is active regular applications of dilute liquid fertilizer are beneficial. Because root disturbance is inimical, repotting is not done unless quite necessary. For further information see Orchids.

CIBOTIUM (Cibò-tium). To *Cibotium* of the dicksonia family DICKSONIACEAE, belong ten species of tree ferns. The group is indigenous to Mexico, Central America, the Hawaiian Islands, Malaysia, and southeastern Asia. Its name is from the Greek *kibotos*, a box, and alludes to the appearance of the spore cases.

Cibotiums have stout, usually erect trunks, but in comparatively young plants grown in greenhouses these are often not

well developed, and the leaves originate from not much above ground level. The trunks, especially at their apexes, and the leafstalks are covered with soft, limp hairs, which distinguishes them from all other cultivated tree ferns. A botanical diagnostic is that the sori (clusters of spore cases) are at the ends of veinlets at or near the margins of the leaf segments. The fronds (leaves) are arching and mostly three-times pinnately-lobed; they are ovate to triangular, often with their undersides covered with a waxy bloom.

The legend of the Scythian or Tartarian lamb owes its origin to *C. barometz*. The fabulous creature, so the story went, was a living lamb, about 3 feet tall, of flesh, blood, and wool; it was anchored to the ground by a root descending from its navel. The lamb could turn and feed on vegetation within reach, but when that was exhausted it pined away and died, leaving only its withered remains as evidence of its being. Examples were displayed in museums. In 1725 it was revealed that the Scythian lambs exhibited as scientific curiosities were actually the roots of a giant fern (*C. barometz*), covered with dense yellow hair and with the stumps of a few leafstalks left to suggest legs and horns.

The silky hairs, in Hawaii called pulu, that clothe the young parts of cibotium ferns have been used both there and in China for dressing wounds, embalming the dead, and other purposes. At one time it was important as a stuffing for mattresses and pillows, and considerable amounts of pulu were shipped from Hawaii to the continental United States for these purposes. In Hawaii the stems of the young fronds and the trunks were fed to hogs, and in times of famine the Hawaiians ate the starch from the trunks themselves. For a short period a laundry starch industry based on the use of Hawaiian cibotiums existed. Today bare-root trunks of Hawaiian species are imported to the mainland for growing into ornamental living specimens and pieces of the trunk are employed to support the growth of orchids, bromeliads, and other epiphytes.

Two Hawaiian Island species are quite commonly cultivated and one more rarely. The commonest is *C. glaucum,* often misidentified as *C. chamissoi,* a name rightfully belonging to another kind. With a trunk up to 15 feet tall, **C. glaucum** has fronds 6 to 7½ feet long. The two lowermost of the final sickle-shaped and toothed segments that compose the stalkless pinnules (leaflets) have one or two ears or lobes at their bases; these distinguish this kind from *C. splendens.* Also, the undersides of the leaves are usually bright-glaucous and hairless, whereas *C. splendens* has the lower sides of its leaves green and clothed with cobwebby hairs. The younger parts of the trunk are covered thickly with lustrous golden-brown hairs. Attaining about 35 feet in height, **C. chamissoi** has the younger portions of its trunk clothed with yellowbrown to black-brown hairs. The leafstalks and main midribs of the fronds are furnished with long, stiff, purplish-black hairs that, on the lower portions of the leafstalks, are intermixed with soft, yellowish hairs. The presence of these stiff hairs distinguishes this species from other cultivated cibotiums. The undersides of its final leaf segments sometimes have minute tufts of matted hairs. Rare in cultivation, **C. splendens** is similar to *C. glaucum,* but the bases of its pinnules are not lobed or eared and their undersurfaces are clothed with cobweb-like hairs. Also the hairs of the leafstalk are more matted and duller.

Mexican tree ferns of this genus include the very popular **C. schiedei,** which is

Cibotium schiedei

grown in considerable quantities in pots and tubs as a florist's decorative plant and is usually seen in its fairly juvenile, trunkless or nearly trunkless stage. It differs from other cultivated cibotiums in its lacier appearance and more membranous fronds with shorter pinnules, the latter stalkless, tapering, and up to 5 inches long and with the sori protruding slightly beyond their margins. The undersides of the fronds are glaucous. At its maximum the trunk may be 15 feet in height; its younger parts and the leafstalks are covered with lustrous, silky, yellow-brown hairs. Another Mexican species, **C. regale** is almost 30 feet in height and has fronds about 5 feet long. A characteristic feature that helps distinguish it from other kinds is that the veins are divided up to three times instead of not more than twice. Distinctions between this species and *C. schiedei* are that its sori do not protrude beyond the margins of the leaf segments, there are no hairs on the stalks of the fronds, and the fronds are more decidedly glaucous beneath.

The Scythian lamb (**C. barometz**) has a very short or prostrate trunk with its younger parts very densely covered with lustrous brown hairs; the leafstalks are similarly hairy. The fronds are up to 6 feet in length and have stalked leaflets and glaucous undersides. The veins are unforked or branch only once. Sori are borne only on the lower halves of the segments. This species is native of warm parts of Asia and Malaysia.

Garden and Landscape Uses. Cibotiums are among the most popular large ferns used to ornament office buildings, public places, and homes. They are used in tubs and planters as well as in ground beds in areas where they are shaded from strong sun. Kinds chiefly employed in this way are the Hawaiian species. The Mexican kinds, more especially *C. schiedei,* are popular with florists for using in the ways described above, but more commonly are employed in temporary decorations. Because in cultivation they are ordinarily low and essentially trunkless, their widespreading leaves take up considerable room and for this reason they are often less adapted to permanent locations where they are likely to be damaged by passersby. In Hawaii, southern California, Florida and other warm-climate regions cibotiums are grown outdoors as well as indoors; elsewhere their use is of necessity confined to greenhouses and other indoor areas.

Cultivation. Like most ferns, cibotiums are lovers of humidity, moist soil, and shade from strong sun. Yet they do not tolerate constantly saturated soil; the beds or containers in which they are planted must be well drained. They grow best in coarse, lumpy earth that contains a generous proportion of organic matter such as leaf mold, peat moss, or rich compost. The addition of gritty sand and some chopped

charcoal is helpful. Planting, potting, and repotting is best done in late winter or spring immediately before new growth begins. It is important not to put them in too large containers; they thrive best when the pots or tubs are snug about their roots. Established specimens that have filled their containers with roots benefit from weekly or biweekly applications of dilute fertilizer from spring through fall. When grown in greenhouses a minimum night temperature in winter of about 55°F is satisfactory; at other times it may be higher and day temperatures may always exceed those maintained at night by five or ten degrees. Cibotiums are propagated by spores. For additional information see Ferns.

CICER (Cì-cer)—Chick-Pea or Garbanzo. The chick-pea or garbanzo, native to western Asia, has long been cultivated there and elsewhere as a nutritious food. It was familiar to the ancient Hebrews, Egyptians, and Greeks and is popular in parts of Asia, southern Europe, and South America.

Of the pea family LEGUMINOSAE, the genus *Cicer* consists of twenty species of annual and perennial herbaceous plants and is indigenous from North Africa, Ethiopia, and the eastern Mediterranean region to Central Asia. Its name is the ancient Latin one for vetch (*Vicia*).

Cicers have leaves with an unequal number of toothed leaflets, or the terminal leaflet may be replaced by a tendril or spine, and small, mostly solitary, white, blue, violet, or reddish, pealike blooms. Their fruits are pods.

The chick-pea (**C. arietinum**) is a bushy, hairy annual about 2 feet tall with leaves of small, roundish leaflets, and solitary white to reddish, axillary flowers. The finely-pubescent pods contain flattish, somewhat ram's-head-shaped, black, red, or white seeds that are boiled, roasted, and used in soups.

Cultivation. After the weather warms, the seeds of chick-peas are sown thinly in rows 1½ to 2 feet apart in sunny locations in fertile, well-drained soil. The plants are thinned, if necessary, to 8 inches to 1 foot apart. This is a drought-resistant crop.

CICERBITA. See Lactuca.

CICHORIUM (Cich-òrium)—Endive, Chicory, Succory. Except some as salad crops, the nine species of this genus of the daisy family COMPOSITAE are seldom cultivated. They are Old World annuals, biennials, and herbaceous perennials, usually with deep, thick roots and sometimes quite pleasing flowers. The name *Cichorium* is a latinized version of an Arabic one.

Cichoriums have usually mostly basal foliage and flower heads that like those of the dandelion consist of all ray florets, but are blue, purple, white, or pink.

Chrysogonum virginianum

Chrysothemis pulchella

Cinerarias

Cissus adenopoda

Cissus discolor

Claytonia virginica

Clarkia purpurea viminea

Cleome hasslerana

Claytonia lanceolata

Cichorium intybus

Cimicifuga racemosa

Chicory (**C. intybus**), a blue-flowered native of Europe widely naturalized in North America, is a familiar roadside weed in many places. Curious **C. spinosum,** of Greece, is an interesting biennial, hardy where winters are not too severe. A low, branched plant with each branch ending in a formidable spine, this kind has rather fleshy leaves and clear blue flower heads about ½ inch in diameter. Appropriate for

Cichorium spinosum

rock gardens, it needs full sun and extremely well-drained soil. Propagation is by seed. For information about the cultivation of chicory and endive as salads consult the entries under those names.

CICUTA (Ci-cùta)—Water-Hemlock. Like their relatives the poison-hemlocks (*Conium*), plants of this genus are deadly poisonous and should under no circumstances be cultivated except under expert and careful supervision in botanical collections if they are needed for research or teaching. The group consists of about twenty species in the northern hemisphere, including North America. Characteristically they are natives of swamps and marshy places including tidal flats. They belong in the carrot family UMBELLIFERAE and are not related to the true hemlock (*Tsuga*). The name *Cicuta* is an ancient

Latin one, probably for the poison-hemlock.

Water-hemlocks are biennial and perennial herbaceous, hairless plants, with more or less tuberous roots and stout, erect, usually tall stems. Their principal leaves are twice- or thrice-pinnately-divided. Their small white flowers are in compound umbels. Several species are native to North America. One of the commonest is the musquash root or spotted cowbane (*C. maculata*), which has stems streaked with purple and tuberous roots that smell like parsnips and look like small sweet potatoes. They are deadly. Another widely distributed kind is the fine-leaved **C. bulbifera,** which develops bulblets in its upper leaf axils. In the west, the coarser-leaved **C. douglasii** kills livestock. It is reported that a piece of root of these plants will kill a cow or a man. Unlike that of the poison-hemlock (*Conium*) the very virulent poison of water-hemlocks causes violent convulsions. It behooves all who collect wild plants for eating to thoroughly familiarize themselves with these and other poisonous species.

CIDER GUM is *Eucalyptus gunnii.*

CIENFUEGOSIA. See Alyogyne.

CIGAR FLOWER is *Cuphea ignea.*

CILIATE. This botanical term means fringed with hairs.

CIMICIFUGA (Cimicí-fuga) — Bugbane, Black Snakeroot or Black Cohosh. A few quite attractive hardy ornamentals are among the fifteen species of *Cimicifuga* of the buttercup family RANUNCULACEAE. The genus inhabits temperate parts of North America, Europe, and Asia. Its name, de-

rived from the Latin *cimex,* a bug, and *fugare* to drive away, alludes to a use formerly made of an Asian species.

Bugbanes are tall, erect, deciduous herbaceous perennials with something of the aspect of baneberries (*Actaea*), from which they differ in their flowers being in long wandlike racemes or panicles rather than short racemes and in having as fruits dry, podlike follicles, not fleshy berries. The large leaves are divided two or three times. The little white flowers, often slightly ill-scented, have two to five petal-like sepals that soon fall, no petals, but sometimes a few petal-like staminodes (abortive stamens), numerous fertile stamens, and one to eight stalked or stalkless ovaries or pistils that become the follicles that are the fruits. The roots and rhizomes of black snakeroot have been used medicinally.

Black snakeroot or black cohosh (**C. racemosa**) inhabits woodlands from Massachusetts to Ohio, South Carolina, and Ten-

Cimicifuga racemosa (flowers)

nessee. From 3 to 8 feet tall, this has leaves of three once- or twice-pinnate divisions. The ultimate leaflets, not all heart-shaped at their bases, are coarsely-toothed or lobed, and in outline roundish to oblong. Usually branched, the slender racemes of white flowers are 9 inches to 2½ feet long. The blooms have stamens ⅓ inch long or a little longer and generally one, more rarely two or three, stalked ovaries. The stigmas are broad, flat, and depressed. The seeds, contained in firm-walled follicles, are not chaffy, having rough edges. This kind is pretty in seed. American or mountain bugbane (**C. americana**) much resembles black snakeroot, but does not ordinarily exceed

Cimicifuga americana

6 feet in height. It is distinct in its flowers having three, five, or eight stalked ovaries and almost vestigial stigmas, and the seeds, in papery-walled follicles, being thickly clothed with chaffy scales. This bugbane is endemic in mountain woods from Pennsylvania to North Carolina and Tennessee.

Asian **C. dahurica** resembles black snakeroot, but its leaves have leaflets with heart-shaped bases and its flowers have five stalkless ovaries. Attaining heights of 3 to 5 feet, it has usually twice-divided leaves. The flowers are creamy-white. Beautiful **C. simplex,** the most ornamental of the genus and the latest to bloom, is native to Kamchatka. Its foliage is much like that of *C. americana*. It is 2 to 3 feet tall. Its white flowers, each with two or three stalked ovaries and minute styles, are in dense, usually branchless racemes the ends of which arch gracefully.

Garden and Landscape Uses and Cultivation. Cimicifugas are desirable for thin woodlands and other lightly shaded places, including perennial borders. They associate well with large-leaved ferns. The American kinds are appropriate in native plant gardens. The flowers of all kinds and the seed pods of some are useful for flower

Cimicifuga simplex

arrangements. Cimicifugas need fairly moist soil, and do best in fertile soils that contain generous amounts of organic material. Planting may be done in spring or early fall. Established specimens benefit from an annual spring application of fertilizer and watering deeply at weekly to ten-day intervals during dry weather. So long as they are doing well it is better not to disturb them by lifting and replanting. Propagation is easy by division at planting time and by seeds sown in a cold frame as soon as they are ripe.

CINCHONA (Cin-chòna)—Quinine, Jesuits' Bark. Belonging in the madder family RUBIACEAE, the genus *Cinchona* comprises forty species of opposite-leaved, evergreen trees and shrubs of the Andes. The name commemorates Comtesse de Cinchon, wife of the Spanish Viceroy of Peru, who in 1638 was cured of fever by the use of quinine. Cinchonas have undivided leaves and terminal panicles of small, pink to yellowish-white flowers that are favorites of hummingbirds. The calyx is small and fine-toothed. The corolla has a long tube and five small lobes (petals) with hairy margins. There are five stamens. They do not protrude. The fruits are capsules containing many flat, winged seeds.

For more than 300 years a product of this genus was man's chief defense against and cure for malaria. From about 1630, when Spanish Jesuits in Peru first learned of the virtues of the bark of *C. officinalis*, until the Japanese in World War II captured the Dutch Indonesian plantations of *C. calisaya*, quinine was by far the most effective antimalarial drug. Its use saved millions of lives in India and elsewhere and made living for Europeans possible in West Africa and other fever-stricken regions. During those three centuries the haphazard collecting of Peruvian bark or Jesuits' bark as it was called, from wild trees, resulting in their destruction, was transformed into a highly organized plantation operation with millions of trees of carefully selected, high-

yielding strains being grown under scientific control, and their product carefully harvested, processed and standardized. The debacle wrought by the Japanese forced Americans and Europeans to turn again to the forests of South America as possible sources of supply, and botanists and other experts were sent to reactivate production there. But soon, almost overnight, the need for quinine vanished. Its medicinal use was replaced with synthetic Atabrine and later by related drugs which better served the purpose. Quinine itself was synthesized in 1944, too late to be of commercial significance. And so a product of test tube wizardry usurped this famous drug as others had natural indigo, vanilla, and to a large extent rubber.

The source of most high-yielding strains of cinchona, **C. calisaya** (syn. *C. ledgerana*) of Peru and Bolivia is a tall, white-barked tree. It has oblong leaves up to 7 inches long, and pale pink flowers. Rough, brown bark is characteristic of **C. officinalis** of Colombia and Peru. This tall tree has ovate-lanceolate leaves 3 to 5 inches long, and deep pink flowers. Native from Costa Rica to Venezuela and Bolivia, **C. pubescens** is a tall tree with elliptic leaves up to 1½ feet long and hairy on their undersides, and pink flowers. The sort named **C. succirubra,** probably a variety of *C. pubescens* is a tall tree with ovate to broad-elliptic leaves up to 9 inches long and with pale undersides, and pink flowers with purplish-red calyxes.

Garden and Landscape Uses. Cinchonas are sometimes grown as ornamentals and as items of interest in southern California and other appropriate warm climates, and in greenhouse collections of plants important to man.

Cultivation. These plants thrive in ordinary well-drained soil and are propagated by seeds, and by cuttings planted in a propagating bed provided with a little bottom heat. In greenhouses they succeed if the minimum winter night temperature is about 55°F and the day temperature is five or ten degrees higher. They are watered to keep the soil moderately moist at all times. Light shade is needed in summer.

CINERARIA. Some botanists maintain this as a separate genus, but most include it in *Senecio,* which is the practice followed in this Encyclopedia. The florists' cineraria is dealt with in the next entry. The dusty millers known in gardens as *Cineraria maritima* and *Cineraria maritima candidissima* are, respectively, *Senecio cineraria* and *Senecio vira-vira*.

CINERARIA, FLORISTS'. The plants known to gardeners as cinerarias belong to the botanical genus *Senecio*. There are two distinct groups, the florists' cineraria grown in greenhouses for its abundance of showy

Florists' cinerarias (three figures, above)

flowers in winter and spring and kinds of the dusty miller type with white-felted stems and leaves. These latter are cultivated chiefly as foliage plants for outdoor display in summer. They are dealt with under *Senecio*.

The florists' cineraria, often identified as *S. cruentus*, but in view of its almost certain hybrid origin more correctly and aptly named *S. hybridus*, was developed at the beginning of the nineteenth century in greenhouses in Europe. Surely *S. cruentus* is one of its ancestors, but almost as certainly so probably are several other Canary Island and possibly one Madeiran species.

In short the florists' cineraria is a man-made "species" of complex ancestry, very different from any *Senecio* found in the wild.

Although technically a nonhardy herbaceous perennial, the florists' cineraria (*S. hybridus*) is invariably grown as an annual. Impressive in bloom, it has a short or long, erect stem and large, thinnish, longish-stalked, ovate-heart-shaped to triangular-heart-shaped, angled and toothed leaves, dark green above and white-, purple-, or blue-hairy on their undersides. The daisy-type flower heads, commonly called flowers, are assembled in large to enormous, compact or loose trusses or more or less panicle-like clusters. The individual "daisies," 1 inch to 4 inches in diameter, depending upon the variety or strain of seeds, have a central eye of disk florets encircled with petal-like ray florets. Kinds with double flower heads, that is, heads composed of all or nearly all petal-like ray florets, are occasionally grown. The fruits are seedlike achenes.

Three chief types of florists' cinerarias are recognized. The ones grouped as *S. hybridus grandiflorus* or *S. h. multiflorus* (syns. *Cineraria hybridus grandiflorus, C. h. multiflorus*) are compact and fairly low and carry their blooms just above the mass of foliage. Quite different and less popular than they were earlier are the kinds grouped as *S. h. stellatus* (syn. *Cineraria h. stellata*). Of much looser growth, these have flower heads never as big as the largest of the *S. h. grandiflorus* kinds, borne in much looser panicle-like trusses lifted well above the foliage. Well-developed specimens are 3 to 4 feet tall. Intermediate between these types is the assemblage identified as *S. h. intermedius* (syn. *Cineraria h. intermedius*). All three types are available in a wide range of flower colors from pure white to yellowish-white, brilliant pinks, red, lavenders, purples, blues, and combinations of these.

Garden and Landscape Uses. In California and other places with practically frostless winters and pleasantly warm, dry springs with fairly cool nights, florists' cinerarias are used in outdoor bedding displays. They are also grown for greenhouse and conservatory decoration and for the florists' trade as gift plants and temporary indoor decorations. Under house conditions they remain attractive in fairly cool locations for three or four weeks. After fading they are discarded.

Cultivation. Sow seeds from May to August to have plants that flower from January to April. The earliest sowings produce the earliest flowering specimens. Sow in well-drained pots, pans (shallow pots), or flats in sandy soil containing a fairly high percentage of organic material, or in vermiculite. Cover the containers with a piece of glass and shade them. As soon as

Florists' cinerarias, in a cool sunroom

the young plants appear, remove the glass, but retain light shade. Do not let the soil dry unduly.

When the seedlings are big enough to handle comfortably transplant them 2 inches apart in flats and continue to grow them in an airy, lightly shaded greenhouse or cold frame, preferably one facing north. As growth necessitates, transfer the seedlings to individual small pots, successively to bigger ones. Use rich, coarse soil and make sure the containers are amply drained. Good flowering specimens can be had in pots 5 or 6 inches in diameter, much larger plants in bigger ones. Some growers prefer to use deep (azalea) pans to pots for cinerarias.

Important points to bear in mind are to repot before the roots become too crowded, to keep the soil always moist, to shade sufficiently to prevent the leaves from wilting on sunny days, and to avoid excessively high temperatures. Plants summered in cold frames may remain until there is danger of freezing. Then they must be removed to a greenhouse. From fall to spring let the night temperature be 45 to 50°F, with a daytime increase of not more than from five to ten degrees allowed. On all favorable occasions ventilate the greenhouse as freely as weather permits.

Cinerarias rejoice in cool, humid conditions. To promote such, their pots should be kept standing on a bed of gravel, sand, cinders, or similar material kept constantly moist. They should be generously spaced. If crowded, their lower leaves are likely to turn yellow and drop. When the final pots are beginning to be filled with healthy roots, apply at one- or two-week intervals dilute liquid fertilizer. The chief pests of cinerarias are aphids, red spider mites, and whitefly.

CINNAMOMUM (Cinna-mòmum)—Camphor Tree, Cinnamon Tree, Cassia Bark Tree, Cassia Flower Tree. Aromatic Asian and Australian trees and shrubs of the laurel family LAURACEAE compose *Cinnamo-*

mum. Most of the more than 250 species, including all discussed here, are evergreens. The name is a modification of the ancient Greek one, *kinnamomon.*

Camphor and cinnamon are important products of this genus. The first is obtained by distillation from wood chips and young shoots of *C. camphora,* the latter is bark stripped from shoots of *C. zeylanicum* and sun-dried. As a substitute for and adulterant of cinnamon the less delicately flavored bark of *C. cassia,* known as cassia bark, is employed. Camphor oil and cassia oil, the latter used for flavoring chocolate and similar purposes, are products of *C. camphora* and *C. cassia,* respectively.

Cinnamomums have opposite or rarely alternate leaves, usually with three conspicuous veins from their bases, but in some species only one. The small, usually bisexual, but sometimes single-sexed flowers are in subterminal and axillary panicles. They have perianths with short tubes and six approximately equal segments (petals or, more correctly, tepals). There are up to nine functional stamens in three circles, and one circle of staminodes (nonfunctional stamens). There is one style. The fruits are berries cupped in the persistent bases of the perianths.

The camphor tree (*C. camphora* syn. *Camphora officinalis*) is commonly up to about 40 feet high, but under favorable conditions in the wild may attain 100 feet.

Cinnamomum camphora

Cinnamomum camphora, as a street tree

Cinnamomum camphora, showing branching habit

It has a dense, wide-spreading head and leathery, alternate leaves each with only one main vein. They have ovate to elliptic blades 2 to 5 inches long and one-third to one-half as wide. Their upper sides are lustrous, their undersides glaucous. When crushed, like the twigs, they smell of camphor. The yellow flowers are in short panicles from the leaf axils. The nearly black fruits are ⅜ inch across. This is native to Japan, Taiwan, and China.

The cinnamon tree (*C. zeylanicum* syn. *C. verum*) is up to about 30 feet tall. Native to India and Ceylon, it has opposite, ovate to ovate-lanceolate leaves 4 to 7 inches long by at least one-half as wide, somewhat glaucous on their undersides, and with three main veins from the base. The panicles of yellowish flowers often are longer than the leaves. The pointed fruits are ½ to nearly ¾ inch long. Very closely related to the last, the wild cinnamon (*C. iners*), of Malaya, differs in having leaves three times as long as wide. This attains a height of 60 feet. It is native to China, Japan, and Taiwan.

The cassia bark tree (*C. cassia*) is good-looking and up to 40 feet tall. Native to China, it has opposite, slender-stalked, oblong to almost lanceolate leaves up to 6 inches long, with three well-defined main veins running from their bases to their tips. The flowers are in large, spreading axillary and terminal panicles up to 6 inches long. The spherical fruits are about ¼ inch in diameter. The species grown as *C. cassia* in the United States may be Chinese *C. burmannii.* The latter is smaller, sometimes only a tall shrub, and usually has smaller, dull leaves with the three chief veins less prominent and often not continued to the leaf apex. Also, the flower panicles are smaller.

Native to Japan, the Ryukyu Islands, China, Korea, and Taiwan, *C. japonicum* (syn. *C. pedunculatum*) is a tree with quite long-stalked, alternate, narrowly-ovate or oblong leaves up to 5 or rarely 6 inches

long and 2 inches wide. Their under surfaces are glaucous, and from their base spread three chief veins. The flower clusters, not longer than the leaves, are hairless. The black fruits are less than ½ inch long. Probably native of the Ryukyu Islands and freely planted in Japan, *C. sieboldii* has been called *C. loureiri,* which name properly belongs to another species. Its young shoots have scattered hairs, and the young foliage is white-hairy. The leaves, 3½ to 6 inches long by 1 inch to 2 inches wide, are narrowly-ovate-oblong, more rarely broadly-lanceolate. Their undersides are glaucous. Much shorter than the leaves, the flower clusters are densely-white-hairy. The cassia flower tree (*C. loureiri*) is of medium size. It has alternate or opposite leaves up to 5 inches long.

A native of India and Malaya, *C. glanduliferum* is smaller and more upright than the camphor tree and grows more rapidly. Its glossy leaves are richer green and bigger, and from their bases spread three main veins. Their undersides are glaucous. The flowers are in axillary panicles. Closely related *C. parthenoxylon,* of China and islands of Malaya, has dull leaves and flowers in terminal clusters.

Garden and Landscape Uses. Cinnamomums are admired for their forms and excellent foliage. They make no effective displays of flowers or foliage. None is hardy in the north, but the camphor tree is frequently planted for shade and ornament in the deep south and California, as are to a somewhat lesser extent other species. Because of its aggressive, surface-rooting system and its very wide head the camphor tree is less suitable for small than spacious landscapes. Where it can be displayed advantageously it is a noble specimen. It has the disadvantage of being rather messy when, as new growth begins in spring, it sheds its old leaves. In California *C. glanduliferum* seems to be slightly less hardy than *C. camphora.*

Cultivation. For their best accommodation cinnamomums need deep, well-drained, moistish soil. They do not prosper in wet places. Maintenance consists of any pruning deemed necessary to limit size, periodic fertilizing in soils deficient in nutrients, and in some areas, spraying to control scale insects. Because they resent root disturbance, young plants should be grown in containers until they are planted permanently. Propagation is usually by seed, but cuttings can be rooted under mist and in greenhouse propagating beds.

CINNAMON. This is the common name of the tree *Cinnamomum zeylanicum,* the source of cinnamon. The cinnamon-fern is *Osmunda cinnamomea,* the cinnamon vine *Dioscorea batata.* Wild-cinnamon is *Canella winterana.*

CINQUEFOIL. See Potentilla.

CION. See Scion or Cion.

CIPURA (Cip-ùra). Two species constitute *Cipura*, a South American genus of bulb plants belonging in the iris family IRIDACEAE. Related to *Nemastylis*, and bearing a name of obscure derivation, they have linear leaves and short-tubed white, yellow, or blue flowers in terminal clusters. The blooms have six segments (petals), the inner three very much smaller than the others. There are three stamens. The fruits are capsules. The species sometimes called *C. martinicensis* is *Trimezia martinicensis*.

Native of South America and Trinidad, *C. paludosa* has leaves up to about 1 foot long and white, yellow, or blue flowers 1 inch or more in diameter that open in the morning and remain open for only a few hours.

Garden Uses and Cultivation. The species described, rare in cultivation, adapts to well-drained, sandy peaty soil and multiplies freely by offsets. It thrives in sunny greenhouses where the winter night temperature is about 50°F and that by day five to ten degrees higher. From spring to fall the soil should be kept evenly moist, in winter drier. Summer applications of dilute liquid fertilizer benefit plants in pots well filled with roots. Repotting is done in late winter or spring. Propagation is by offsets and seeds. In warm, essentially frost-free climates cipuras are worth trying outdoors in rock gardens and other suitable locations.

CIRCAEA (Cir-caèa)—Enchanter's-Nightshade. The common name is the most alluring feature of *Circaea* of about a dozen species of the evening-primrose family ONAGRACEAE. Its members are too weedy looking to make much appeal. They are perennial herbaceous woodlanders with opposite, ovate, toothed or angled-margined leaves and tiny, inconsequential, whitish, two-petaled flowers in terminal racemes. These are succeeded by bristly fruits. The name derives from Circe, an enchantress of Greek mythology. Its application is not obvious. The genus is native to north-temperate and arctic regions.

The original enchanter's-nightshade (*C. lutetiana*) is a softly-hairy native of Europe and Asia up to 2 feet tall and branched. The American *C. quadrisulcata* (syn. *C. latifolia*), which has its stems below the flowering portion without hairs, is also sometimes called enchanter's-nightshade. This species is indigenous from New Brunswick to Tennessee and Oklahoma and occurs also in Asia.

Garden Uses and Cultivation. Gardeners occasionally grow enchanter's-nightshade. It thrives with little trouble in shade in moist soil and is propagated readily by seed.

CIRIO is *Idria columnaris*.

CIRRHOPETALUM. See Bulbophyllum.

CIRSIUM (Círs-ium)—Plumed Thistle. Differing only in technical details from the plumeless thistles (*Carduus*), the 150 species of *Cirsium* belong to the daisy family COMPOSITAE. They are of minor horticultural interest. A few kinds may be occasionally planted in naturalistic surroundings to create bold effects. Some, notably the noxious Canada thistle (*C. arvense*), are pests, completely inadmissable to gardens.

Cirsium arvense

Their name is an ancient Greek one derived from *kirsos*, a swollen vein, a condition that thistles were believed to cure. From closely related *Carduus*, plumed thistles differ in the pappus (attached to the ovary and achene) being a tuft or plume of hairs.

Cirsiums are natives of all continents of the northern hemisphere. Mostly medium-large to large, spiny, hardy annuals, biennials, and perennials, they have alternate, pinnately-lobed and toothed, or only toothed leaves. Typical thistles, their flower heads are solitary to many, medium-sized to large. They are composed of numerous disk florets (the type that form the centers of daisies). Ray florets (petal-like ones that encircle the disk in daisy-type flowers) are absent. The involucres (collars of bracts behind the flower heads) are of several rows of overlapping scales, some or all tipped with spines. The fruits are seedlike achenes.

The only plumed thistles worth considering as ornamentals are annuals and biennials. The perennials are likely to become established as pernicious weeds. Among plantable kinds are the bull thistle (*C. vulgare* syn. *C. lanceolatum*), a biennial native of Europe, widely naturalized in waste places and at roadsides in North America. This kind is 2 to 5 feet tall and has pinnately-lobed, more or less hairy leaves, with the lobes again lobed and toothed. The flower heads, in clusters, are 1 inch to 1½ inches long and rosy-purple. Favoring

moist soils from Newfoundland to Saskatchewan, Florida, and Texas, the swamp thistle (*C. muticum*) is 2 to 6 feet in height. It has deeply-pinnately-lobed and toothed leaves and loose clusters of several to many generally long-stalked, rosy-purple flower heads ¾ inch to 1¼ inches long. The fishbone thistle (*C. diacantha*), of Asia Minor, is a biennial about 3 feet tall. It has

Cirsium diacantha

leaves deeply cleft into sharp-spined, lanceolate lobes, with clear white veins, suggestive of the backbone of a fish. The purple flower heads are 1 inch to 1½ inches long.

Garden and Landscape Uses and Cultivation. In semiwild areas and in native plant gardens occasional use may be found for groups of thistles. They need sunny locations. Ordinary soils are satisfactory, the swamp thistle responding best to moist ones. Propagation is by seeds, sown where the plants are to bloom, with the seedlings thinned sufficiently to avoid undue crowding, or sown in a seedbed and the seedlings transplanted while young.

CISSUS (Cís-sus)—Marine-Ivy, Kangaroo Vine, Grape Ivy. About 350 species of chiefly tropical, evergreen vines compose *Cissus* of the grape family VITACEAE. A few are natives of subtropical regions, a few are deciduous, and a few are not vines, a few are succulents. The name, from the Greek *kissos*, ivy, refers to the climbing habit of most kinds.

From closely related *Ampelopsis* and *Rhoicissus* the genus *Cissus* differs in its flowers having sepals, petals, and stamens in fours instead of in fives to sevens. From *Parthenocissus* it differs in its blooms having a fleshy disk or four glands around the ovaries of the flowers and in the absence of disklike tips to the tendrils, the latter usually present in *Parthenocissus*. From *Vitis* the genus *Cissus* differs in having petals that separate and fall individually.

Cissuses have woody or herbaceous, sometimes succulent stems and alternate leaves, undivided or of separate leaflets that spread from the top of the leafstalk in palmate-fashion. Most have tendrils. The little bisexual or unisexual flowers are generally in branched clusters. They are of no significant ornamental value. The fruits, usually blue-black or black and mostly inedible, are berries containing one to four seeds.

Probably the hardiest species cultivated is the marine-ivy (*C. incisa*). This herbaceous-stemmed native from Missouri to Florida, Texas, and Mexico has stems up to 30 feet in length. Its succulent, evergreen leaves are of three leaflets or are deeply divided into three lobes. The leaflets or lobes are ovate to obovate, coarsely-toothed, and about 1 inch long. The flower stalks are branched. The berries are black.

Kangaroo vine (*C. antarctica*), slightly less hardy than the last, is sufficiently cold

Cissus antarctica

resistant to flourish outdoors in the warmer parts of California. It is a vigorous vining native of Australia, with woody stems and undivided, evergreen, ovate or ovate-oblong, smooth-edged or somewhat toothed leaves up to 4 inches long by nearly one-half as wide. The foliage is leathery and lustrous, paler on its under surfaces than above. The leaves are alternate. Opposite each is a stout, forked tendril. At the junctions of the veins on the leaf undersides are often prominent glands. The flowers are few together in axillary clusters. Under favorable conditions the kangaroo vine extends its stems for 60 feet or more. The horticultural variety *C. a. minima* is distinguished by its slower growth, more compact and freer branching habit, and smaller leaves. Australian *C. hypoglauca* is also sufficiently cold resistant for outdoor cultivation in parts of California. It is a handsome, vigorous, tall, woody vine, with leaves of five leaflets each about 3 inches long, shining green above and paler beneath. The middle leaflet is the biggest. The flowers are somewhat decorative. In the bud stage they are bright orange-yellow. The black, bitter fruits are also attractive. Smaller-leaved

and forming a daintier foliage pattern than the above, *C. striata*, of Chile, is also adaptable for growing outdoors in California. Under favorable conditions this evergreen attains a height of 30 feet. Its leaves are of usually five or sometimes three, obovate or spoon-shaped leaflets, toothed at their ends and about 1 inch long. Opposite each is either a three-pronged tendril or a cluster of flowers or berries.

Grape-ivy (*C. rhombifolia*), native of northern South America, more tender to

Cissus rhombifolia

cold than the kinds discussed above, may be grown outdoors in the warmer parts of California. It is a vigorous, tall, woody vine with leaves of three rhombic-ovoid leaflets the veins of which end in conspicuous points. They superficially resemble the leaves of poison-ivy (*Rhus toxicodendron*) and are leathery, glossy, and very persistent. They have brown hairs on the veins beneath. The leaflets are up to 4 inches long. A two-pronged tendril develops opposite each leaf. The fruits are attractive. East African *C. rotundifolia* is a high-climbing, drought-resistant vine, its older stems with four thick, corky wings. The brittle, fleshy, waxy leaves have rather short-hairy stalks and hairless, nearly round blades with heart-shaped bases and round-toothed margins. Yellowish-green, they are 2 to 3, exceptionally 4 inches in diameter. The tendrils and the branched clusters, longer than the leaves, of inconspicuous, green flowers, develop opposite a leaf. The fruits are red berries. South African *C. lanigera* is a vigorous, woody vine with its younger stems and foliage softly-hairy and branchless, or sparingly-hairy and branchless, with sparingly-branched tendrils. Its leaves are of five or less often three coarsely-toothed ovate to obovate leaflets 3 to 5½ inches long. The minute, yellowish flowers are in loose, forked-branched clusters. This sort succeeds in warm frost-free climates.

Tropical vines that succeed in hot, humid environments include these sorts: *C.*

Cissus lanigera

Cissus adenopoda

adenopoda of tropical West Africa has a tuberous rootstock and herbaceous stems. Its green or coppery-green leaves, their uppersides clothed with purple hairs, have deeply-impressed veins and coarsely-toothed margins. Their undersides are wine-red. The pale yellow flowers are in 4-inch-long loose clusters. The fruits are purplish-black. *C. amazonica*, known only in cultivation, is said to be native to Brazil. It has woody stems, and in two ranks, closely-spaced, undivided linear-lanceolate to broadly-ovate-lanceolate, toothless to irregularly-toothed leaves 2 to 6 inches long. Their upper surfaces are metallic-gray-green with silvery-white veins. Their undersides are purplish-red. Flowers and fruits have not been observed. *C. discolor* is a slender-stemmed native of Java. It has oblong-ovate to heart-shaped-ovate, bristly-toothed leaves up to 6 inches long, velvety-green mottled with silvery-white on their upper surfaces, wine-red beneath. The yellowish flowers are in small, dense clusters shorter than the leaves. From the typical species, *C. d. mollis* differs in having leaves with conspicuous white veins and blood-red undersides. *C. erosa*, of Puerto Rico is hairless. It has slender, ribbed or winged stems and thick, shining

Cissus discolor

Cissus erosa

Cissus tuberosa

Cissus juttae, in South Africa

Cissus juttae, young specimen

Florida to tropical America, this has undivided, ovate to oblong, lobeless, fleshy leaves up to 4 inches long and mostly heart-shaped at their bases. Their margins are usually toothed, with each tooth terminated by a tiny bristle. The flowers are pale greenish to purplish. The best growth of aerial roots is attained in warm, highly humid atmospheres. Variety *C. s. albo-nitens* (syn. *C. albo-nitens*) grows rapidly to a great height. It has thin stems and lustrous, metallic silvery-gray, pointed-ovate leaves with green veins and green undersides.

Curious *C. tuberosa* of Mexico is a vigorous vine adapted to arid conditions and so is appropriate for growing with cactuses and other succulent plants. It has thin, rambling, or climbing stems up to several feet long that develop along their lengths irregularly-shaped aerial tubers up to 9 inches long and 3 inches thick. The long-stalked hairless or sparingly hairy leaves have blades 2½ to 4 inches long, deeply palmately-cleft into five lobes that again are lobed or coarsely toothed. A slender tendril sprouts opposite each leaf. The flowers and fruits are unreported.

A few decidedly succulent *Cissus* species are cultivated. Here belongs *C. quadrangula* an interesting vine of distinctive appearance. Its angled stems are thick and fleshy, have four conspicuous wings, and are much constricted at the nodes. Except when quite young, they are usually leafless, but if grown under humid conditions the foliage is retained longer. The leaves are small and kidney- to broadly-heart-shaped. The small greenish flowers are followed by acrid, pea-sized berries. This is a native of tropical and South Africa. Somewhat similar is *C. cactiformis,* also of tropical and South Africa. It differs chiefly in its stems being much thicker, up to 2 inches in diameter, between the nodes.

Remarkable and often grotesque succulent shrubby sorts with much swollen, fleshy stems include those now to be described. By some authorities these and others with similar floral characteristics are segregated as the genus *Cyphostemma.* Native of west tropical Africa, *C. bainesii* (syn. *Cyphostemma bainesii*) is notable because of its much swollen, bottle-shaped trunk, which may be 10 inches in diameter. Up to 4½ feet in height, this species has short-stalked leaves of mostly three ovate or oblong leaflets (the lowermost may consist of a solitary leaflet), unequally serrate and feather-veined. The flowers are in terminal clusters, with glandular stalks. The fruits are coral-red. Astonishing *C. juttae* (syn. *Cyphostemma juttae*), of southwest Africa, has barrel-shaped, gouty stems

leaves with three nearly-stalkless, coarsely-toothed, asymmetrically-ovate to obovate leaflets, paler beneath than on their uppersides. *C. gongylodes* has conspicuously four-angled or four-winged stems that develop at their apex a fleshy tuber that at the end of the growing season falls to the ground to become a new plant. Like *C. sicyoides,* this native of Brazil and Paraguay in humid environments develops a great abundance of very long, slender aerial roots that produce an attractive curtain-like effect. Its leaves have three stalkless, toothed leaflets 4 to 8 inches long, the center one more or less diamond-shaped. The tiny flowers are in crowded, globular clusters. *C. himalayana* (syn. *C. neilgherrensis*) of the Himalayas is a vigorous grower with woody stems that are without tendrils. Its usually glossy leaves are of three, sharply-toothed leaflets 2 to 5 inches long. The flower clusters are about as long as the leaves. The globose fruits are black and about ¼ inch in diameter. *C. sicyoides* is remarkable for the amazing curtainlike effect produced by its abundance of long, slender aerial roots. Native from southern

4 to 6 feet tall that divide above into thick branches. Trunk and branches are green covered with thin, yellowish-green or gray-green skin that on older specimens sheds in papery strips. The stalkless, coarsely-toothed leaves are pointed-ovate, up to 6 inches long by 2½ inches wide, and lustrous waxy-green, with translucent hairs and often reddish veins on their undersides. The fruits are yellow or red. Up to

15 feet tall, **C. crameriana** (syn. *Cyphostemma crameriana*), of southwest Africa, has a corpulent trunk up to 8 feet in girth, interlaced fleshy stems, and thick-stalked, deeply-three-lobed, fleshy leaves, up to 1½ feet long and one-half as wide, more or less covered with gray felt. The leaf lobes are ovate-oblong. The yellowish bark of older plants peels in paper-thin strips. The attractive red fruits are in large clusters. A native of South Africa, **C. oleracea** has a large woody rhizome that annually produces short, succulent stems with undivided, short-stalked, broad-ovate to suborbicular, finely- to coarsely-toothed, fleshy leaves up to 9 inches long by 6 inches wide and glaucous.

Cissus oleracea

Garden and Landscape Uses. Where hardy, the *Cissus* species described are decidedly useful outdoor ornamentals. They are excellent as greenhouse decorations. Some make good window plants. The vines are useful for covering arbors, pergolas, pillars, and other supports and for training against walls. Some, such as *C. hypoglauca* and *C. rhombifolia*, make attractive groundcovers. Several are useful for growing in hanging baskets. Most withstand considerable shade and some, such as *C. discolor*, need it.

The succulents, especially the extraordinary species with bulky, swollen stems, require desert or semidesert conditions and exposure to full sun. They succeed outdoors in southern California and are curiosities well deserving of including in collections of succulent plants outdoors and in greenhouses. Vining *C. quadrangula* and *C. cactiformis* are appropriate as window plants.

Cultivation. The vigorous, vining kinds of *Cissus* succeed in any ordinary garden soil that is well drained and reasonably fertile. When grown in containers they respond to fertilizing regularly from spring through fall with dilute liquid fertilizer, and to being generously supplied with water during that period. In winter they

should be kept drier (but never to the extent that their foliage wilts) and cooler, and not fertilized. The hardier kinds succeed indoors in temperatures of 40 to 50°F at night in winter, but the tropical ones need a 55 to 60°F minimum, and *C. discolor* is best satisfied when the greenhouse temperature never drops below 65 or 70°F. The tropicals thrive best in relatively high humidity, the hardier kinds, especially those with thick leaves, withstand drier conditions. All are very easily increased by cuttings, layers, and seeds.

For succulent sorts sharp drainage assured by coarse, loose, porous soil is essential. One that contains generous amounts of coarse sand, smashed crocks, broken bricks, or other nonorganic material that will expedite the passage of water, but will not decay and break down as does organic material is ideal. A little organic material may be mixed with the soil, but it should not be relied upon alone to provide the necessary porosity. Only when the plants are in growth and leaves are present is moisture needed. Even then the soil should be allowed to become fairly dry between thorough soakings. During the season of dormancy do not water. Indoor temperatures of about 50°F at night suit. By day this may be increased by five to fifteen degrees depending upon the brightness of the weather. Full sun is needed and the atmosphere should be dry. Propagation is easy by cuttings and seeds. The chief pests of cissuses are scales, mealybugs, red spider mites, and aphids.

CISTACEAE—Rock-Rose Family. This family of shrubs, subshrubs, and herbaceous plants includes about 200 species allocated among eight genera. Dicotyledons, they are natives of the northern hemisphere, especially the Mediterranean region, and to a lesser extent, of South America. Many inhabit dry, sandy, or limestone soils and sunny, exposed sites. Frequently they are glandular and more or less aromatic. Rock-roses (*Cistus*) and sun-roses (*Helianthemum*) are among those most often grown as ornamentals.

Plants that belong here frequently have stellate (star-shaped) hairs on shoots, foliage, and sometimes other parts. The leaves are undivided, generally opposite, and often have rolled-back margins. Solitary or in branched clusters, the symmetrical flowers have five sepals, the two outer usually much smaller than the others and bractlike, five, three, or no petals that soon fall, numerous stamens, one style, and three to five stigmas or the stigmas united as one. The fruits are capsules. Cultivated genera include *Cistus*, *Fumana*, *Halimiocistus*, *Halimium*, *Helianthemum*, *Hudsonia*, and *Tuberaria*.

CISTUS (Cís-tus)—Rock-Rose. These typical, Mediterranean region shrubs are

splendidly adapted for cultivation in the warm, dryish climates of many parts of the American west and southwest. Although in sheltered locations one or two kinds can be persuaded to live outdoors in the vicinity of New York City, as a group they are not sufficiently hardy to survive winters in the north. In the wild they occur in dry scrub and open woodland from the Canary Islands through southern Europe and North Africa to Transcaucasia. One kind is sparingly naturalized in California. There are twenty species and many natural and garden-raised hybrids. The name is a modified form of *kistos*, the ancient Greek one for these plants. They are called rock-roses because they often frequent rocky places and have blooms that superficially resemble single-flowered roses. As a classificatory aid this is misleading; rock-roses are not related to true roses, they belong, instead, with sun-roses (*Helianthemum*) in the rock-rose family CISTACEAE. From *Helianthemum* and *Halimium* the genus *Cistus* differs in its seed pods splitting in five or more vertical lines to release the seeds. Also, its blooms are never yellow, although that color may appear in basal blotches on the petals.

Rock-roses are erect, evergreen or partially evergreen, often aromatic, more or less glandular shrubs. Usually hairy, they have opposite, undivided leaves, and large, showy, up-facing blooms that usually last only a few hours and never more than a day. They are produced in such quantities that fine displays are maintained over long periods in early summer. The flowers, rarely solitary, occur more commonly in loose clusters at the ends of the branches and branchlets. They have three or five sepals, five petals, many stamens, and a very short or a long style tipped with a five- or ten-lobed stigma. The fruits are capsules.

Species with white flowers, with or without blotches at the bases of the petals, and three sepals are *C. ladanifer*, *C. palhinhae*, and *C. laurifolius*. The first two have blooms 3 to 4 inches in diameter, those of the last are 2 to 2½ inches across. Native of southwest Europe and North Africa, **C. ladanifer** has the biggest and most beautiful flowers of the genus. Typically they are blotched at the bottoms of their crinkly-edged petals with crimson-purple, but a variant without these blotches, known as *C. l. albiflorus* (syn. *C. l. immaculatus*), occurs. These are vigorous shrubs, 3 to 5 feet or sometimes twice that height, much branched, and with very sticky shoots. Their leaves are sticky, nearly stalkless, linear-lanceolate to lanceolate, 2 to 4 inches long, three-veined, and hairless above, but gray-hairy on their undersides. The flowers, usually solitary, have concave sepals covered with small yellowish scales, hairy, especially along their edges. The style is very short. The seed pod has ten compartments. Not more than 2 feet tall, squat and compact, **C. *palhinhae*** has ovate-spatula-shaped leaves ¾ inch to 2 inches long and white-hairy on their undersides. Its unspotted, solitary blooms, which may be 4 inches wide, are enhanced by a conspicuous group of yellow stamens. This kind is Portuguese. The hardiest species **C. *laurifolius*** under favorable circumstances attains a height of 6 or 7 feet. It has hairy, sticky young shoots, and hairy-stalked, ovate to ovate-lanceolate, wavy-margined, three-veined leaves 1½ to 3 inches long, hairless, sticky and dull dark green on their upper surfaces, and gray-hairy and sticky beneath. The long-stalked blooms, in clusters of three to eight, have yellow basal blotches on their petals. Their styles are very short. The sepals are pointed-ovate, very concave, and hairy. This species is native in the western Mediterranean region.

Species with white flowers having five sepals are *C. monspeliensis, C. psilosepalus, C. salviifolius,* and *C. populifolius.* The first differs from the others in having sepals with rounded instead of heart-shaped bases. Native to the western Mediterranean region, of neat habit, and 2 to 4½ feet in height, **C. *monspeliensis*** has shoots that when young are slightly hairy, and lanceolate to linear-lanceolate stalkless, three-veined, sticky leaves up to 2 inches long by not more than ⅓ inch wide, with inward-rolled edges. Their upper sides are dark green, wrinkled, and sparsely-hairy, their lower ones grayish-pubescent. About 1 inch in diameter, the blooms are in compact clusters of up to eight. The pointed-ovate sepals have many long hairs. The styles are very short. From the last **C. *salviifolius*** differs in having stalked, non-sticky leaves, grayish-green, and wrinkled above, white-hairy on their undersides and downy above, and up to 1¾ inches long and ¼ to 1 inch wide. Its blooms, 1¼ to 2 inches wide, have the bases of their petals usually stained yellow. Their styles are very short. This species is native to the Mediterranean region. Triangular-heart-shaped, long-stalked leaves, 1 inch to 4

inches in length and at least one and one-half times as long as wide and hairless except on their stalks, distinguish **C. *populifolius*,** of southwestern Europe. It has sticky, minutely-hairy young shoots and flowers with very short styles, 1¾ to 2½ inches across, with yellow at the bases of the petals, in clusters of two to six. They have hairy stalks 2 to 3 inches long, and nod in bud. Variety *C. p. major* has wavy leaves usually up to one and one-half times as long as wide. Its sepals are densely covered with white hairs. The plant now called *C. psilosepalus* (syn. *C. hirsutus*) has its home in western Europe. Up to 3 feet in height, it is distinguished by its flat, stalkless, bright green, ovate to lanceolate, three-veined leaves, hairy on both surfaces, and ¾ inch to 2¼ inches long by up to ¾ inch wide. The flowers, in terminal flat-topped clusters of up to five, or occasionally solitary, are 1½ to 2¼ inches in diameter. They are yellow at the bases of their petals and have very short styles. The heart-shaped sepals, covered with shaggy, white hairs, have inrolled margins. The seed capsules are hairy.

Species with purple, rosy-purple, or purplish-red blooms are *C. crispus, C. incanus,* and whitish-hairy *C. albidus.* All have five sepals and styles as long as their stamens. The leaves of the two first have three definite veins, those of the last have one main vein and laterals arranged in pinnate-fashion. Both compact and about 3 feet tall, **C. *crispus*,** of the western Mediterranean region, is distinguishable from *C. incanus* by its usually wavy-edged leaves and nearly stalkless flowers. When young its shoots are clothed with long white hairs. Stalkless, and hairy on both surfaces, the leaves are elliptic-ovate to lanceolate and up to 1¾ inches long and ½ inch, less or very slightly more, wide. In dense terminal clusters, the purplish-red flowers, up to 1¾ inches in diameter, have exceedingly short, hairy stalks and slender styles. Their sepals are pointed-lanceolate. Often called *C. villosus* the species now named **C. *incanus*** inhabits the Mediterra-

Cistus incanus

nean region. Of compact growth and 3 to occasionally 5 feet in height, it has shaggy-hairy young shoots and elliptic to oblong-ovate blunt leaves, up to 2 inches long, with short, flat stalks. They are more or less hairy, especially beneath, and grayish. Occasionally solitary, but usually in lax terminal clusters of up to seven, the purple-pink to rosy-purple flowers, with yellowish bases to the petals, are 1¾ to 2¼ inches in diameter and have grayish-hairy stalks. Their pointed, broadly-ovate sepals are hairy. **C. *albidus*** is a native of southwestern Europe and North Africa. Its name alludes to the whitish down that clothes its young shoots, foliage, and flower stalks and not to its flowers. Bushy, compact, and 5 to 6 feet tall, this species has broad-elliptic to ovate leaves ¾ inch to 2 inches long, up to ¾ inch wide, and conspicuously net-veined on their undersides. Unlike most of the leaves of closely related *C. crispus,* they are without undulating margins. About 2½ inches in diameter, the pale rosy-lilac flowers of *C. albidus* are in crowded terminal clusters of three to eight, have stalks ¾ inch to 1 inch long. The blooms of *C. a. albus* are pure white. A natural hybrid intermediate in characteristics between *C. albidus* and *C. crispus* is named *C. pulverulentus.*

Hybrid rock-roses are numerous and not always easy to identify. Those here described are believed to be the most attractive. As a parent of mixed progeny, *C. ladanifer* has been especially useful. Its offsprings, intermediate between the parents, are often more compact and bloom more freely than *C. ladanifer.* Among the best that it has mothered (the pollen parent in each case being another species) are three with white flowers that have conspicuous, brownish-purple blotches at the bases of the petals, *C. cyprius,* of which *C. laurifolius* is the other parent, *C. loretii,* which has *C. monspeliensis* as a parent, and *C. verguinii* sired by *C. salviifolius.* The result of mating *C. ladanifer* and *C. populifolius* produced **C. *aquilari*,** with large and immaculate white blooms. A *C. ladanifer* hybrid with reddish-purple flowers with

Cistus salviifolius

Cistus crispus

Cistus purpureus

Cistus skanbergii

the bases of the petals marked with dark red is **C. purpureus,** which has as its other parent *C. incanus.* Variety *C.* 'Silver Pink' is of the same parentage. A very good hybrid, one of natural origin between *C. populifolius* and *C. salviifolius,* is **C. corbariensis.** Relatively hardy, it much resembles a small-foliaged edition of *C. populifolius* and sometimes has been called by the invalid name *C. p. minor.* Another natural hybrid, between *C. monspeliensis* and *C. salviifolius* is **C. florentinus.** Intermediate between its parents, it has broader leaves than those of the first-mentioned and lacks its stickiness of the young shoots. Its blooms are larger than those of *C. monspeliensis* and are more like those of *C. salviifolius.* Two natural hybrids that have *C. monspeliensis* as one parent are **C. glaucus** and *C. skanbergii.* The first, which occurs spontaneously in southern France, attains a height of 3 to 4 feet. It has dull green, linear to lanceolate leaves with sticky upper surfaces and hairy undersides. Its white flowers, 1½ to 2 inches in diameter, are in clusters of many. Discovered first on the

Mediterranean island of Lampedusa and later in Greece, **C. skanbergii** is much-branched and up to 3 feet tall. It has bluish-green, linear-lanceolate leaves up to 2 inches long and quite lovely, pale pink flowers 1¼ to 1½ inches wide.

Garden and Landscape Uses. There is little doubt that rock-roses offer great opportunities to Americans who garden in favored climates to exploit them much more extensively than has been done. They revel in dry soils and full sun and are admirably adapted for rock garden plantings, shrub borders fronting, and formal or informal groupings. They are adaptable for seaside locations. Because they resent transplanting (which perhaps has limited their popularity) they should always be grown in containers until they are set in their permanent locations. This should be done before they become over-large or pot-bound.

Cultivation. Provided they are suitably located rock-roses require practically no care. In their early stages they should be encouraged to branch freely by occasionally pinching out the tips of their young shoots. Later, pruning must be done with circumspection. Rock-roses do not respond well to severe cutting back. Ordinarily it is only necessary to remove a small proportion of old, worn-out branches. This should be done as soon as blooming is over. Propagation is by means of seeds sown in sandy, well-drained soil and by terminal cuttings of semimature shoots planted in late summer in a propagating bench in a cool greenhouse or cold frame.

CITHAREXYLUM (Cithar-éxylum)—Fiddle Wood. The vervain family VERBENACEAE includes this genus of 115 species of shrubs and small trees. It has a natural range extending from Florida to Argentina and the West Indies. Its name comes from the Greek *kithara,* a lyre, and *xylon,* wood, and refers to the use of the wood for making musical instruments. The designation fiddle wood derives from the French name

for the strong wood of some species, *bois fidèle.*

The group is characterized by having leaves opposite or rarely in circles of three, small flowers in long, slender terminal or axillary racemes, and berry-like fruits, technically drupes, with a bony stone that divides into a pair of two-seeded nutlets. The flowers are narrowly-bell-shaped and have persistent calyxes, with five tiny lobes and five-lobed, slightly asymmetrical corollas. There are four or five stamens, of which one is usually vestigial or infertile, and a style tipped with a two-lobed stigma.

The Florida fiddle wood (**C. fruticosum**) occurs from central Florida to northern South America and the West Indies. Evergreen or semideciduous, and 10 to 40 feet tall, it blooms and fruits throughout much of the year. Its thick, yellowish-green leaves, 3 to 7 inches long and 1 inch to 2½ inches wide, are mostly elliptic. They are sometimes hairy on the veins beneath and have conspicuously pink or orange stalks ½ to 1 inch long. The fragrant, stalkless or nearly stalkless, white flowers, ¼ inch long or slightly longer, are in terminal and lateral slender racemes 2 inches to 1 foot long. The fruits are globular, glossy, reddish-brown to black, and are up to ⅜ inch in diameter. A closely related species, of the West Indies and South America, **C. spinosum** has thin, hairless, elliptic leaves 3 to 8 inches long and oblongish, glossy, black fruits. It attains a height of 50 feet.

A shrub or tree to 30 feet tall, **C. berlandieri,** of Mexico and adjacent Texas, has four-angled young shoots and oblong to elliptic or ovate leaves 1 inch to 3½ inches long, more or less hairy especially on their undersides, and sometimes toothed toward the apex. The 1- to 3-inch-long spikes of ¼-inch-wide, white or creamy, fragrant

Cistus glaucus

Citharexylum berlandieri

flowers are terminal on short side shoots. The fruits, red and ¼ inch in diameter, blacken when they dry.

Native to Puerto Rico, other West Indian islands, and Central America, **C. caudatum** is an evergreen or partly deciduous, open-crowned tree 15 to 60 feet tall, with slender branches and whiplike branchlets. It is easily distinguished from the species described above by its leaves, when dried, being without a prominent network of secondary veins between the major ones, and by its flowers being definitely stalked. The bluntish, slightly glossy leaves are elliptic and 2 to 5 inches long by ¾ to 2¼ inches wide. The flower racemes, terminal and axillary and 1½ to 3 inches long, are of slightly fragrant blooms about ¼ inch long. The egg- to somewhat pear-shaped fruits, nearly ½ inch long, are orange-brown or black.

Garden and Landscape Uses and Cultivation. In warm, nearly frost-free climates, these graceful small trees and shrubs are popular ornamentals and are well suited for exposed sites. They are adaptable to dry soils and attractive in flower and fruit. Their blooms provide nectar for bees. Propagation is by seeds and cuttings.

CITRANGE. Hybrids between sweet orange (*Citrus sinensis*) and trifoliate-orange (*Poncirus trifoliata*) are called citranges. The

Citrange

hardiest of the citrus group with edible fruits, they succeed 300 to 400 miles north of where oranges can be grown in Florida. The fruits, unsuitable for eating out of hand, are used for preserves and for ades. They are 2 to 3 inches in diameter. Citranges are also employed as understocks for Satsuma oranges. They respond to conditions that suit other citrus fruits. Botanically citranges, of the hybrid genus *Citroncirus*, are identified as *C. webberi*. Varieties are 'Coleman', 'Cunningham', 'Moreton', 'Rusk', 'Rustic', and 'Savage'.

CITRANGEQUAT. The offspring of citranges hybridized with kumquats are called citrangequats. They are hardier than all other edible citrus fruits except citranges.

Unsuitable for eating out of hand, the small fruits are used for preserves and for ades. Citrangequats respond to cultural care that suits other citrus fruits. Varieties are 'Sinton', 'Telfair', and 'Thomasville'.

CITROFORTUNELLA (Citro-fortunélla)—Calamondin. The hybrid genus *Citrofortunella*, its name derived from those of its parents, embraces all hybrids between *Citrus* and *Fortunella* of the rue family RUTACEAE. Its kinds, intermediate between the parents, are nonhardy evergreen trees.

The calamondin (**C. mitis** syn. *Citrus mitis*), old in cultivation, has as one parent *Citrus reticulata* and probably *Fortunella margarita* as the other. One of the hardiest of citrus fruits, this small, dense-headed tree has erect branches and broad-elliptic leaves paler on their undersides than above, with narrowly-winged stalks. Solitary or in pairs, the small, white blooms, at the ends of the branchlets, have usually about twenty stamens. The flattened-spherical, very acid fruits 1 inch to 1½ inches in diameter and of seven to ten segments, have thin, loose, orange-colored skins.

Other citrofortunellas are **C. floridana,** the parents of which are *Citrus aurantifolia* and *Fortunella japonica*, and **C. swinglei**, the parents of which are **C. aurantifolia** and *Fortunella margarita*. Dark green leaves, with paler under surfaces, and white flowers sometimes streaked with pink, characterize **C. floridana.** Its light yellow, ovoid to subspherical fruits contain six to nine segments of pale pulp, usually with one seed in each segment. The horticultural varieties 'Eustis' and 'Lakeland' are cultivated. Its leaves lanceolate, **C. swinglei** has flowers that are pink in the bud stage, but later white. Its obovoid to ovoid or subspherical fruits have pale pulp and light cadmium-yellow skins. Each contains seven or eight segments. For garden and landscape uses and cultivation of these plants see Calamondin.

CITRON. This is the name of a variety of *Citrullus*, dealt with under Citrullus in this

Citrofortunella mitis

Encyclopedia, and of *Citrus medica* of the rue family RUTACEAE. The latter is a close relative of the orange and one of the sorts commonly grouped as citrus fruits. It was the first of these known to Europeans and the first to be cultivated in Europe. It was brought from its native Asia by the armies of Alexander the Great in about 300 B.C. Gradually its cultivation extended through lands bordering the Mediterranean Sea, and it is still grown there in considerable quantities for the thick peels of its fragrant fruits, which are candied and used in confectionary. In citrus-growing regions of the United States citrons are occasionally planted, but not in commercial quantities.

More tender to cold than other citrus fruits, these citrons are suitable only for essentially frost-free areas. Their cultural needs otherwise are those of lemons and oranges. They are propagated by cuttings and by shield budding onto seedling understocks or those of rough lemon. The principal varieties are 'Corsican' and 'Etrog'. The latter is the ethrog, a primitive type the fruits of which are used with religious significance in the Hebrew Feast of the Tabernacles. For further description see Citrus.

CITRONCIRUS (Citron-cìrus)—Citrange. Hybrids between *Citrus* and *Poncirus* of the rue family RUTACEAE are identified as *Citroncirus*. The colloquial name citrange is used for hybrids that have as one parent the sweet orange (*Citrus sinensis*). For more information see Citrange.

CITRONELLA is *Collinsonia canadensis*. Citronella grass is *Cymbopogon nardus*.

CITROPSIS (Citròp-sis)—African-Cherry-Orange. The name of this group indicates its relationship. It derives from that of the genus to which the orange belongs, *Citrus*, and the Greek *opsis*, resembling. Native of Africa, *Citropsis*, of the rue family RUTACEAE, is not widely cultivated. There are ten species. The one described below has been used as an understock for citrus fruit trees and is planted to some extent for ornament.

Citropsises are small trees varying considerably in foliage and flower size. They have solitary or paired spines, or are sometimes spineless, and usually have pinnate leaves of an uneven number (three to seven) leaflets, but in one species the leaves have only one leaflet or leaf blade. The leafstalks of most species are broadly-winged. Much resembling those of the orange, the blooms come from the leaf axils and are generally in clusters or short racemes. They have usually four-, more rarely five-parted calyxes, generally four, sometimes five petals, twice as many stamens as petals, and a short style. The fruits, technically berries, are small, and spherical or subspherical.

A small tree with spines solitary or in

pairs and leaves of three or five leaflets or on fruiting twigs occasionally only one, **C. schweinfurthii** is an attractive ornamental. Its oblong to lanceolate leaflets are toothed or practically toothless, 4 to 6 inches long by one-third or somewhat more wide. The leafstalks are broadly-winged. Many together, the fragrant, white blooms have usually four each sepals and petals and eight stamens with flattened stalks. At their fullest they are about 1 inch in diameter. The subspherical, lemon-yellow, fragrant, sweet fruits are generally in clusters of five or fewer. Their pulp is yellow.

Garden and Landscape Uses and Cultivation. Because of its good-looking foliage, long flowering season, and decorative fruits, *C. schweinfurthii* is worth growing as a greenhouse decorative as well as a landscape ornamental. The last role it can fill only in warm, practically frost-free climates. Its culture and care are essentially those of the lemon and orange.

CITRULLUS (Citrúl-lus)—Watermelon, Bitter-Cucumber or Bitter-Apple. Only the watermelon (*Citrullus lanatus* syn. *C. vulgaris*) and its variety, the citron or preserving-melon (*C. l. citroides*), are commonly grown. For details of their cultivation, see Watermelon. The genus *Citrullus* belongs in the gourd family CUCURBITACEAE. Its name is derived from that of the genus *Citrus* in reference to the orange flesh of the fruits of some kinds. The genus comprises three species of vigorous, frost-tender annual and perennial vines of Africa and probably tropical Asia. Their stems may be up to 40 feet in length and are with or without tendrils. The alternate leaves are deeply-pinnately-lobed. The solitary yellow flowers, from the leaf axils, are unisexual. They have deeply-five-parted, broadly-bell- to wheel-shaped corollas and three each stamens and stigmas. The oblong-cylindrical to nearly spherical, many-seeded fruits are fleshy or dry.

The watermelon (*C. lanatus*) is an annual with rounded or angled, hairy stems, branched tendrils and once- or twice-lobed leaves. Its sulfur-yellow flowers, about 1½

inches across, are five-lobed. The well-known, hard-rinded fruits, often of immense size, are oblong to ellipsoidal. They have white, yellow, greenish, or red flesh in which numerous seeds are embedded. There are many horticultural varieties. The skins of the small, hard, white-fleshed fruits of the citron or preserving melon are candied and used in confectionary.

The bitter-cucumber or bitter-apple (*C. colocynthis*), of India, has purgative medicinal qualities. It is a nonhardy, herbaceous perennial vine, more roughly-hairy than its close relative, the watermelon, and with deeply-twice-divided ovate leaves, up to about 2½ inches in length, and solitary pale yellow blooms. Its round, smooth, variegated green and white fruits, up to 3 inches in diameter, are extremely bitter. This species may be grown under conditions that suit melons and cucumbers as an annual or a perennial. It has been suggested that *C. colocynthis* is the wild ancestor of the cultivated watermelon (*C. lanatus*).

CITRUMQUATS. These are hybrids between the trifoliate-orange (*Poncirus*) and kumquats (*Fortunella*).

CITRUS (Cít-rus)—Orange, Lemon, Shaddock, Grapefruit, Lime, Citron. The fruits of several members of the genus *Citrus* of the rue family RUTACEAE are so familiar that the generic name has become a household word. Even the nonbotanical know them as citrus fruits. Native to southern and southeastern Asia and Indonesia, citruses are naturalized in warm regions elsewhere, including the United States. The name *Citrus* is a classical Latin one originally applied to a quite different tree, but transferred to the citron as early as the first century A.D.

The number of species is controversial. Some botanists have claimed as few as four or five, others as many as one hundred and forty-five. Probably about sixteen is a reasonable estimate. However many species one accepts, their botany is complicated. Several cultivated kinds generally recognized as species no longer exist in the wild, and perhaps never did. It is quite possible that they originated, centuries ago or later, as hybrids of trees cultivated in southeastern Asia and Indonesia. It is well established that species of *Citrus* hybridize with great facility, and that many of the offspring of such matings have the curious ability, following pollination, but without fertilization, of producing seeds that give rise to plants that are exact duplicates of the parent tree. In the twentieth century a large number of hybrids have been raised experimentally in the United States between various kinds of *Citrus* and the related genera *Fortunella* and *Poncirus*.

Citruses are evergreen, usually spiny, small- to moderate-sized trees and shrubs with alternate, undivided, toothed or

toothless, minutely-gland-dotted leaves that, when crushed are aromatic. The leafstalks are generally winged, and except in the citron, are conspicuously jointed at the point of junction with the blade. The usually very fragrant flowers are sometimes solitary, more often in clusters from the leaf axils. Generally they are bisexual. From ¾ inch to 2 inches across, they have four- to five-lobed calyxes, from four to eight, but most commonly five white, pink, or purplish petals, twenty to sixty stamens, and one slender or thick style. The fruits, mostly thick-skinned, are structured in the familiar pattern of the orange and lemon and are known to botanists as hesperidiums. They are yellowish-green, yellow, or orange when ripe.

The fruits of various citruses are esteemed for eating as taken from the trees, for making marmalades, for juicing, for making ades, and for flavoring. The peels of some, notably the citron, orange, and lemon, are candied and used in confectionary and cakes. From a variety of the sour orange, oil of bergamot is obtained, and from the flowers of that species, neroli oil, important in perfumery.

The sweet or common orange (*C. sinensis*) is a dense, conical-headed tree 25 to 40 feet in height. It has usually rather few

Citrus sinensis variety

spines ½ inch to 2 inches long. The glossy leaves, paler beneath than on their upper sides, are 3 to 4 inches in length, sometimes slightly toothed, and have usually thinly-, sometimes more broadly-winged stalks up to 1 inch long. The sweetly fragrant, white blooms, solitary or in clusters of up to six, have four or five sepals and mostly five petals. They are 2 to 2½ inches across and have twenty to twenty-five stamens and a long, slender style. The fruits are the familiar spherical to ovoid or ellipsoid sweet oranges. This kind, probably native of China or Cochin-China, no longer exists as a truly wild species.

Citrullus lanatus variety

Citrus limon variety

Citrus aurantium variety

Mandarin, tangerine, and Satsuma oranges (*C. reticulata* syn. *C. nobilis deliciosa*) are distinguished from sweet oranges by the skins of their fruits fitting very loosely instead of tightly against the interior flesh. Native of China, **C. reticulata** is a small, spiny tree with a dense head of slender branches. Its glossy, lanceolate to ovate-lanceolate leaves, about 1½ inches long, have sometimes slightly-toothed edges, and scarcely-winged stalks. The flowers are smaller than those of the sweet orange. The fruits are flattened-globose and 2 to 3 inches in diameter. This species is hardier than the sweet orange. Hybrids between *C. reticulata* and the grapefruit, itself perhaps of hybrid origin, designated *C. tangelo*, colloquially are called tangelos. Intermediate between the parents, **C. tangelo** has smaller and sweeter fruits than grapefruits, with loose, easily removable skins like mandarin and tangerine oranges. Several horticultural varieties of *C. tangelo* are cultivated.

The sour or Seville orange (*C. aurantium*) differs from sweet oranges and Mandarin oranges in its leaves having very broadly-winged stalks and its fruits being sour. From 20 to 30 feet tall, this native of southeastern Asia and the Philippine Islands develops a dense, rounded head. It is plentifully furnished with long, slender spines, those of older branches stouter and stiffer than younger ones. The leaves are ovate to ovate-oblong, wavy-edged or toothed, and 3 to 4 inches long. Solitary or

Citrus aurantium variety

in clusters, the white flowers, sometimes bigger than those of the sweet orange, are very fragrant. They have twenty to twenty-four stamens. The somewhat flattened, globular, rough-surfaced fruits are about 3 inches in diameter. Hardier than the sweet orange, this is naturalized in Florida. Bergamot (*C. a. bergamia*) is the source of oil of bergamot.

The lemon (*C. limon*), probably of Indian origin, is a rather open-headed, spiny tree 10 to 20 feet tall. Its pointed, elliptic-ovate, toothed leaves, 2 to 4 inches long, have stalks without wings, but may be very narrowly margined. The white flowers, solitary or in pairs and pinkish to purplish in bud and on their outsides, are 1½

to 2 inches in diameter and have reflexed petals. They have twenty to twenty-six stamens. The thin-skinned, acid fruits, usually about 3 inches long, are ovoid to oblong, tapered at both ends, and have a terminal nipple. They are rough- or smooth-surfaced, and light yellow.

The lemandarin, rangpur, or Mandarin lime (*C. limonia*), a hybrid of the lemon and probably the Mandarin orange, is a medium-sized tree with somewhat pendulous branches, few spines, and dull leaves. The white flowers and buds are suffused with purple. The yellowish to reddish-orange, acid, rather loose-skinned fruits have a short nipple and are flattened, spherical, or obovoid. The 'Otaheite' lemon or 'Otaheite' orange is believed to be a dwarf, spineless variant of *C. limonia*. It has small, bland rather than acid, fruits.

The shaddock, pummelo, or pompelmous (**C. maxima** syns. *C. grandis, C. decumana*) is much less hardy than the kinds dealt with above. Native to tropical Asia, it is a compact-crowned tree 15 to 30 feet tall, with usually downy twigs, and with or without spines. Its broadly-ovate, blunt leaves, 4 to 8 inches long, are sparingly-hairy on their undersides. They have somewhat wavy or toothed edges, and stalks with a pair of very broad wings that together form what appears like a secondary heart-shaped blade. Solitary or clustered, the flowers are 2 to 2½ inches wide and have four or sometimes five, thick petals and sixteen to twenty-four stamens. Light lemon-yellow to orange, the fruits are solitary, spherical to somewhat pear-shaped, and 4 to 6 inches in diameter. They have thick, spongy skins, dotted with conspicuous oil glands, and contain bitter or sour pulp and large seeds.

The grapefruit (**C. paradisi**) was given its popular common name because its fruits, unlike those of the shaddock, grow in clusters. Of uncertain origin, it may be a native of southeastern China, or possibly it originated in the West Indies as a seedling sport of the shaddock. Grapefruit trees, from 30 to 50 feet in height, have round to conical heads and are round-topped. Their leaves have round-toothed margins and are pointed- or blunt-ovate. Their stalks are broadly-winged, the pair of wings forming a heart-shaped, small, secondary blade below the main blade, with its apex at the point where the leaf joins the stem. The fragrant flowers are sometimes solitary, more often in clusters of up to twenty. They are white and about 1¾ inches across. The four or five thick petals are slightly reflexed. There are up to twenty-five stamens. The fruits are large, light yellow to orange, and spherical to slightly egg-shaped or somewhat pear-shaped. Their flavor is a mixing of bitterness, acidity, and sweetness.

The lime (*C. aurantifolia*) is an irregularly-branched, sharp-spined shrub or small tree with glossy green, elliptic-ovate to ob-

View of a New York City garden

Citrus aurantiifolia

long-ovate, round-toothed leaves 2 to 3 inches long that have slightly-margined or very narrowly-winged stalks. The flowers are white, small, and in clusters of up to ten. There are twenty to twenty-five stamens in several groups. Nearly spherical to oblongish, often with a nipple at the apex, the acid fruits have thick skins, greenish-yellow to light yellow at full maturity, and greenish flesh. They are 1½ to 2½ inches long. This native of Indonesia and southeastern Asia is naturalized in many tropical and subtropical regions, including Florida.

The citron (this name is also used for a type of watermelon of the genus *Citrullus*) is *C. medica*. This is a large shrub or small tree of irregular growth with leaves, 4 to 7 inches long, that are oblongish and toothed and have stalks without wings. Unlike those of other *Citrus* species discussed here, they are not jointed where they join the blades. The flowers, in clusters from the leaf axils or panicles, are white with purplish outsides. They are 1½ inches wide or wider, and have thirty or more stamens. The fruits have thick, rough,

often lumpy, fragrant skins, are oblong-ovoid to nearly spherical, and when fully ripe are lemon-yellow. They contain only small amounts of acid pulp, and are up to 10 inches long. The citron was brought into cultivation in the Mediterranean region about 300 B.C. by the armies of Alexander the Great, and there the method of candying its peel was discovered. The ethrog, used by Jews in religious ceremonies, is a variety of citron. Its name is also spelled etrog.

Garden and Landscape Uses and Cultivation. These are those generally suitable for Calamondin, Citron, Grapefruit, Orange, Lemon, Lime, and Shaddock as detailed under Encyclopedia entries of those titles.

CITY GARDENING. The urban population of the United States, like those of most other countries, is increasing rapidly in relation to the number of people who live in rural areas. The suburban segment is also increasing. In many regions space and other natural resources important to gardeners are becoming scarce commodities. Yet the urge to grow plants is not less strong among the inhabitants of cities than among dwellers in less artificial surroundings, and the need for the beauty and solace they bring is surely as great. Living plants have a brightening, softening effect on the environment, and in cities especially, their contribution to improving the quality of the air is welcome.

City gardens of necessity are small. They are of two main types, those that occupy ground space, a tiny portion of what the great American horticulturist Liberty Hyde Bailey called the Holy Earth, and those

perched on balconies, terraces, window sills, roofs, and other aerial places. Special information regarding gardening in such locations is given in this Encyclopedia in the entries Backyard Gardens, Roof and Terrace Gardening and Container Gardening. Here we are less concerned with design and methods of cultivation, more with special problems that strictly city environments raise.

The chief restrictions on those who garden in cities are conditioned by polluted air and soil. These may be aggravated by reflected heat from walls and other surfaces, exposure to too constant winds and drafts sweeping through canyon-like spaces between buildings, and too little or more rarely too much light. Coupled with environmental limitations inherent to the locations are the often not inconsiderable difficulties and often formidable costs of obtaining basic supplies, including soil, and, if sizable specimens are employed, plants. Yet despite obstacles plant growing and gardening in cities succeeds wondrously and to the great joy and benefit of those who engage in it, expansively or much more modestly.

Adequate light is necessary for plants to thrive. The intensity of illumination required varies greatly with different kinds, and wise selection helps here. If shade is such that the site is unsuitable for most flowering plants excellent effects can be had by restricting plantings chiefly to evergreen trees, shrubs, and groundcovers. In Japan small gardens so planted are common and charming. But one rarely has to forego flowers entirely. Except in the most extreme shade such hardy bulbs as crocuses, hyacinths, narcissuses, and tulips

View of a New York City garden

to be sure of this it may be well to have a soil test made. Also work into the upper few inches a dressing of a complete garden fertilizer, such as a 5-10-5. If your city garden is of the aerial type you may prefer to substitute an artificial growing mix for soil. See Roof and Terrace Gardening.

A permanently-maintained, 2-inch mulch of loose organic material, such as coarse peat moss, pine bark chips, ground corncobs, or similar materials that may be available, does much to improve rooting conditions for city plants.

Air pollution is something the individual can rarely control to any appreciable extent. One must usually learn to live with it, at least for the short run. The most destructive pollutants are sulfur dioxide and ozone and peroxyacetyl nitrate (PAN) that result from the action of sunlight on smog. For more about these and other pollutants see Air and Air Pollution. Damage by air pollutants usually varies from season to season and time to time, largely in response to the weather. Often it is most severe in spring and fall. As you cannot do a great deal to prevent this, the obvious best procedure, if you garden in an affected area, is to rely to the greatest extent possible upon resistant kinds of plants. Also, avoid using too much nitrogen fertilizer and doing too much watering; both stimulate soft new growth and increase sensitivity to air pollutants. There is no practical chemical treatment for increasing the tolerance of plants to polluted air.

Fences or screens of wood, reeds, plastic, or other appropriate materials, and sturdy wind-resistant hedges or plants set less formally, can be used to protect against damage by wind and reflected heat from walls. Do not use solid screens. Make provision for some air flow through them so that pockets of dead air where little or no circulation is possible are not created. Paling fences with narrow spaces between the slats and louvered screens achieve this and are attractive.

Staking is often more needed in urban gardens than elsewhere, especially where depth of soil is restricted or wind pressure between and around buildings is likely to be high enough at times to topple trees and shrubs inclined to be top-heavy and to cause breakage among other tall plants. Judicious pruning of trees and shrubs serves to reduce these dangers.

Accumulations of grime bedevil city gardeners in some regions. Because of the length of time the leaves are retained these are likely to be heaviest on evergreens. They not only annoy by their appearance, but also reduce the amount of light the foliage receives and so impair the plants' ability to synthesize food. It is undoubtedly helpful to wash off such accumulations, but it is troublesome. Choose carefully what you use. Do not employ laundry soaps, powders, or detergents. Rely upon

will flower gaily in spring. Even though with such conditions it may be necessary to plant new bulbs each fall, not a great many are likely to be needed and the cost is small in comparison to the return. And you may be able to persuade such summer annuals as begonias and torenias to give satisfaction. See Shady Gardens.

Soil in city gardens is often in very poor condition and if in a region of atmospheric pollution is likely to be excessively acid. As a first step inspect it carefully. Dig into it to find what is below the upper inch or two. A discouraging clay or rubble perhaps? Take heart, almost any soil can be greatly improved. About the only condi-

tion that cannot be made better is where the ground is invaded by the roots of some well-established tree, especially if it is a maple, poplar, willow, or some other kind that fills the surface soil with thirsty, hungry roots. In such places settle for paving, gravel, or a wooden deck.

To improve the soil spade in quantities of organic matter. This may be, if you can possibly acquire it, rotted manure or compost or, and these are more generally available to those who garden in cities, peat moss or humus. Almost surely, except where you plan to set such acid-soil plants as azaleas, rhododendrons, and camellias, the ground will be benefited by liming, but

View of a New York City garden

View of a New Orleans city garden

a soap mild enough for a baby's skin. Make a sudsy solution of this in tepid water and with a hand sprayer wet the foliage, but not to the extent that it drips. Allow it to remain five to ten minutes, then with a fine spray of water wash off the solution and, one hopes, much dirt.

Plants for city gardens must be chosen with strict reference to local climate. Many sorts hardy in New Orleans or San Francisco will not survive in Chicago or New York and vice versa. But among kinds climatically adaptable to any particular region are always some better able than others to survive and even prosper in urban environments. Experience indicates that the ones here listed have this ability. If the location is reasonably to their liking most annuals give good accounts of themselves in cities.

Deciduous trees include Cornelian-cherry (*Cornus mas*), crab apples (*Malus*), European white birch (*Betula pendula*), fig (*Ficus carica*), flowering dogwood (*Cornus florida*), fringe tree (*Chionanthus virginicus*), golden-rain tree (*Koelreuteria paniculata*), hawthorns (*Crataegus*), honey-locust (*Gleditsia triacanthos*), Japanese cherries (*Prunus*), Japanese maple (*Acer palmatum*), London plane (*Platanus acerifolia*), *Magnolia soulangeana*, maidenhair tree (*Ginkgo biloba*), ornamental pears (*Pyrus*), pin oak (*Quercus palustris*), pomegranate (*Punica granatum*), redbud (*Cercis canadensis*), Russian-olive (*Elaeagnus angustifolia*), sourwood (*Oxydendrum arboreum*), tree-of-heaven (*Ailanthus altissima*), willow oak (*Quercus phellos*), and willows (*Salix*).

Evergreen trees include acacias, American and English hollies (*Ilex*), Austrian pine (*Pinus nigra*), bull bay (*Magnolia grandiflora*), crape-myrtle (*Lagerstroemia*), *Ficus retusa*, laurel (*Laurus nobilis*), live oaks (*Quercus*), loquat (*Eriobotrya japonica*), pineapple-guava (*Feijoa sellowiana*), and privets (*Ligustrum*).

Deciduous shrubs include *Abelia grandiflora*, azaleas, bush honeysuckles (*Lonicera*), Carolina-allspice (*Calycanthus floridus*), deutzias, *Enkianthus campanulatus*, five-leaf-aralia (*Acanthopanax sieboldianus*), flowering quince (*Chaenomeles*), forsythias, hydrangeas, Japanese barberry (*Berberis thungergii*, privets (*Ligustrum*), rose-of-Sharon (*Hibiscus syriacus*), roses, shadbush (*Amelanchier*), snowberry (*Symphoricarpos albus*), star magnolia (*Magnolia stellata*), sweet pepperbush (*Clethra alnifolia*), viburnums, weigelias, and winged euonymus (*Euonymus alatus*).

Evergreen shrubs include *Aucuba japonica*, azaleas, *Berberis julianae*, boxwood (*Buxus*), camellias, *Fatsia japonica*, firethorn (*Pyracantha*), heavenly-bamboo (*Nandina domestica*), inkberry (*Ilex glabra*), Japanese holly (*Ilex crenata*), leucothoes, mock-orange (*Philadelphus*), oleander (*Nerium*), Oregon-grape (*Mahonia aquifolium*), pittosporums, privets (*Ligustrum*), rhododendrons, viburnums, and yews (*Taxus*).

Deciduous vines include Boston-ivy (*Parthenocissus tricuspidata*), *Clematis paniculata*, Dutchman's pipe (*Aristolochia durior*), grapes (*Vitis*), silver lace vine (*Polygonum aubertii*), Virginia creeper (*Parthenocissus quinquefolia*), and wisterias.

Evergreen vines include Carolina yellow-jessamine (*Gelsemium sempervirens*), English ivy (*Hedera helix*), *Euonymus fortunei*, and *Hedera canariensis* and *H. colchica*.

Evergreen groundcovers include English ivy (*Hedera helix*), lilyturfs (*Liriope* and *Ophiopogon*), myrtle (*Vinca major, V. minor*), purple-leaved wintercreeper (*Euonymus fortunei coloratus*), St.-John's-wort (*Hypericum calycinum*), and sedums.

Herbaceous perennials and biennials include bellflowers (*Campanula*), bleeding hearts (*Dicentra*), bugle weed (*Ajuga*), Christmas-rose (*Helleborus niger*), chrysanthemums, columbines (*Aquilegia*), coral bells (*Heuchera sanguinea*), day-lilies (*Hemerocallis*), evergreen candytuft (*Iberis sempervirens*), forget-me-not (*Myosotis*), foxgloves (*Digitalis*), globe flower (*Trollius*), honesty (*Lunaria annua*), irises, Japanese anemone (*Anemone hybrida*), lily-of-the-valley (*Convallaria majalis*), *Phlox divaricata*, plantain-lilies (*Hosta*), spireas (*Astilbe*), violets (*Viola*), and Virginia bluebell (*Mertensia virginica*).

CLADANTHUS (Clad-ánthus). This genus consists of one or perhaps more species of North Africa and southern Spain. Belonging in the daisy family COMPOSITAE, it has a name derived from the Greek *klados*, a branch, and *anthos*, a flower, in allusion to the blooms being at the ends of the branches.

An attractive, bushy, erect, stiffly-branched annual, **Cladanthus arabicus** has alternate, finely-pinnately-dissected leaves and solitary, bright yellow, daisy-like heads 1 inch to 1½ inches across. Pleasantly scented if bruised or brushed against, and 2 to 3 feet tall, this kind forms mounds of attractive, feathery, pale green foliage among which are displayed the showy flower heads. One flower head ends each stem and from below it arise three to five branches, which in turn terminate in flower heads. This branching habit continues throughout the growing season. The fruits are achenes.

Garden Uses. For summer flower beds and borders and for greenhouse cultivation in pots for late winter and spring display, this kind is very attractive. Its blooms last well in water when cut, but because of the peculiar stiff branching habit do not lend themselves especially well for use in flower arrangements. The plants bloom until killed by fall frosts. In very hot weather their flower heads are fewer and smaller than in cooler periods.

Cultivation. This annual is easily raised from seeds. Outdoors they take three or four weeks to germinate. Sow them in a sunny location in well-drained, fertile soil in early spring and thin the resulting plants so that they stand about 1 foot apart. Alternatively, seeds may be sown indoors about eight weeks before danger from frost has passed and it is safe to transplant the young plants from flats to the open garden. The plants are sturdy enough to stand with little or no staking.

For indoor cultivation, seeds are sown from September to January and the seedlings transplanted as growth demands until they occupy their final pots, which are 5 or 6 inches in diameter. A sunny greenhouse in which the night temperature is 50°F, and the day temperature five to ten or on sunny days fifteen degrees higher, provides ideal growing conditions. When the plants have filled the pots in which they are to bloom with healthy roots they benefit from weekly or semiweekly applications of dilute liquid fertilizer.

CLADOTHAMNUS (Clado-thámnus). One deciduous shrub, little known in cultivation, native from Oregon to Alaska, is the only species of this genus. It belongs in the heath family ERICACEAE, and is closely akin to the eastern North American *Elliottia* and the Japanese *Tripetaleia*. Its solitary flowers distinguish it from both. The name derives from the Greek *klados*, a branch, and *thamnos*, a shrub.

An erect, branching, deciduous shrub up to 9 feet tall, but in cultivation often lower, **Cladothamnus pyrolaeflorus** has angled shoots and alternate or almost opposite, hairless, very short-stalked, obovate-oblong to oblanceolate, toothless leaves 1 inch to 2½ inches long by ¼ to ¾ inches wide. The nodding blooms originate at the ends of the shoots and in the upper leaf axils. They have five narrow, green sepals, and corollas ¾ to 1 inch wide, pink, tipped with yellow. The style is long, curved, and enlarged at the apex. The fruits are small capsules surrounded by the persistent leafy calyxes.

Garden Uses and Cultivation. This shrub is suitable for associating with azaleas, enkianthuses, and other members of the heath family to provide variety and interest. It needs acid soil, never excessively dry, but not saturated, and appreciates a little shade from the hottest sun. It is raised from seeds, sown in sandy peaty soil, and by summer cuttings under mist or in a greenhouse propagating bench.

CLADRASTIS (Cladrás-tis)—Yellow-Wood. One species of *Cladrastis* is a native of eastern North America, the other four of eastern Asia. All are deciduous, smooth-barked trees. They belong to the pea family LEGUMINOSAE. The name comes from the Greek

klados, branch, and *thraustos*, fragile. It refers to the brittleness of the wood.

Cladrastises have alternate, pinnate leaves, with a terminal leaflet and alternately-arranged, short-stalked lateral leaflets. The winter buds, several together and superimposed, are concealed in summer by the enlarged base of the leafstalk and unless the leaf is pulled off are not visible until after leaf fall. The concealed winter buds and the alternate leaflets serve to readily distinguish this genus from closely related *Maackia*. The pea-shaped flowers of *Cladrastis* are in panicles borne in late spring after the trees are in foliage. The fruits are narrow-oblong, compressed pods containing three to six seeds. The yellow-woods are medium-sized, handsome ornamentals deserving of being planted more commonly.

American yellow-wood (**C. lutea**), indigenous in Kentucky, Indiana, Missouri, Tennessee, and North Carolina, usually occupies rich woods and rocky bluffs.

Cladrastis lutea, in bloom

Cladrastis lutea (flowers)

Cladrastis lutea in winter

Cladrastis lutea, roots developed from damaged trunk descending to the ground

Rarely exceeding 50 feet in height, it has a rounded, wide-spreading, densely-foliaged crown. Its leaves have seven to nine, elliptic to ovate, glabrous, bright green leaflets, 3 to 4 inches long. In fall they turn yellow to orange-yellow. The flowers, 1 inch long or longer, white and very fragrant, in loose, pendulous panicles similar to those of wisterias, are borne in early June. American yellow-wood is hardy throughout New England. Its common name refers to the color of its freshly cut lumber.

Chinese yellow-wood (*C. sinensis*) is quite distinct from the American yellow-wood. A rather rare native of western China, it has white or pinkish flowers in erect, branched panicles 5 inches to 1 foot long. It attains a height of 80 feet and has nine to thirteen oblong-lanceolate to oblong, yellowish-green leaves pubescent near their midribs below. The leafstalks and midrib are also hairy. Chinese yellow-wood is hardy in southern New England.

Japanese yellow-wood (*C. platycarpa*) differs from those sorts previously described in that its pods are winged and its leaflets have stipels (tiny appendages) at

their bases. This kind may be 60 feet in height and has leaves with seven to fifteen long-pointed to rounded leaflets slightly pubescent on their veins above as well as on their undersides. The flowers, ½ inch long, or slightly less or more, are in upright panicles. They are white with a yellow spot at the base of the standard petal. This species is about as hardy as *C. sinensis*. It is a native of Japan and China. Another Oriental, *C. wilsonii* grows to a height of about 45 feet and has leaves of seven to nine or rarely more elliptic-ovate to oblong leaflets, nearly glabrous, except for some hairs on the midribs beneath. Its white flowers, about ¾ inch long, are in upright panicles. It is possibly hardy in southern New York.

Garden and Landscape Uses and Cultivation. Yellow-woods are excellent as solitary specimens and avenue trees. They are not particular as to soil, but respond best to those of reasonable fertility. They need full sun. Unless trained to a single stem by pruning when young, they generally branch low down into a number of major massive limbs. This is no disadvantage unless a high-branched shade tree is desired. Because wounds made in spring "bleed" (lose sap) profusely, any needed pruning should be delayed until early summer when root pressure is lower. The most satisfactory method of raising yellow-woods is from seeds sown in fall or spring. They may also be increased by root cuttings and, the Oriental kinds, by grafting in spring onto young plants of *C. lutea*.

CLAM SHELL ORCHID is *Epidendrum cochleatum*.

CLAPPERTONIA (Clapper-tònia). Formerly named *Honckenya*, and still sometimes cultivated as such, *Clappertonia* consists of three species of trees and shrubs, natives of tropical West Africa. They belong to the linden family TILIACEAE. The name honors Hugh Clapperton, a Scottish explorer of Africa, who died in 1827.

Clappertonias are clothed with stellate (starry) hairs. They have lobed or lobeless, alternate leaves and showy, large, blue-violet blooms in terminal racemes. The flowers have three to five each sepals and petals, the latter without glands at their bases, and many stamens. The fruits are capsules.

A much-branched, evergreen shrub with yellowish hairs, *C. ficifolia* (syn. *Honckenya ficifolia*) is about 7 feet tall and as broad. It has purplish or reddish shoots and heart-shaped, roundish or oblong, three- to seven-lobed, toothed leaves of diverse shapes on the same plant. Clustered at the branch ends, the many orchid-like, yellow-centered, mauve or reddish-purple blooms have petals 1 inch to 2 inches long, and numerous stamens of which only a dozen have anthers. The seed capsules, 1 inch to

2 inches long, are clothed with white-tipped, hair-fringed bristles.

Garden and Landscape Uses and Cultivation. The handsome species described is a good general purpose shrub for warm regions. It thrives in Florida, and even if frosted soon recovers. The flowers are displayed in summer and fall. Ordinary garden soils and sites suit this fine shrub. It is propagated by cuttings and by seeds.

CLARKIA (Clárk-ia)—Godetia, Farewell-To-Spring, Red Ribbons, Fairy Fans. Named in honor of Captain William Clark, who accompanied Meriwether Lewis on his explorations of the Rocky Mountains and western North America and who died in 1838, this genus of the evening-primrose family ONAGRACEAE numbers about thirty-five species in western North America and Chile. They are annuals and include several very attractive garden plants adaptable for cultivation outdoors and in greenhouses. Some kinds were previously named *Eucharidium*.

As interpreted by modern botanists, *Clarkia* includes the plants formerly named *Godetia*. They are erect and branched, glabrous or hairy, and have alternate leaves with toothed or toothless margins. The four-petaled blooms (horticultural varieties with double flowers have more petals) are solitary in the leaf axils or in terminal racemes. The petals may be three-lobed. There are four or eight stamens joined to the throat of the calyx. The stigma has four lobes. The predominant flower colors are pink, purple, red, and white. The fruits are capsules.

One of the most important species horticulturally, *C. unguiculata* (syn. *C. elegans*), of California, is cultivated in a num-

Clarkia unguiculata variety

Clarkia unguiculata variety, semidouble-flowered

Clarkia pulchella

Clarkia amoena

Clarkia unguiculata, double-flowered

Clarkia concinna

Clarkia amoena, double-flowered variety

Clarkia amoena, dwarf variety

ber of beautiful garden varieties, the best with double flowers. Except for the calyxes and seed pods this kind is hairless. It is 2 to 4 feet in height, but in greenhouses it may attain 6 feet or more. The stems are reddish and glaucous, the leaves ovate to ovate-lanceolate, the flower buds deflexed. Pink, rosy-purple, or white, with the claws of the petals not toothed at their margins, the blooms have eight stamens. Also commonly cultivated, **C. pulchella**, like *C. unguiculata*, is represented by a number of garden varieties, many with double flowers. Native of northwest America, this kind grows to a height of about 1½ feet, is freely-branched, has narrow, pointed leaves, and, in the wild, generally lilac flowers. The petals are usually strongly-three-lobed and their claws are toothed at their sides. There are eight stamens, four of which are rudimentary. It is probable that some garden varieties of clarkias are

hybrids between *C. unguiculata* and *C. pulchella*. Red ribbons (**C. concinna** syn. *C. grandiflora*), of California, attains a height of 2 feet and has ovate-lanceolate to elliptic leaves and dark pink or lavender-pink flowers with three-lobed, fan-shaped petals, the lobes of which are about of equal width. The flowers have four stamens. Fairy fans (**C. breweri**), of California, about 9 inches tall, has lanceolate leaves and fragrant, dark pink flowers with fan-shaped petals of three lobes, the central one of which is much narrower than its neighbors, and four stamens.

Because it seems likely that for some time to come the species formerly included in *Godetia* will continue to be called godetias by gardeners it is convenient here to treat them together. They all have flowers with eight stamens. The best known horticulturally is the farewell-to-spring (**Clarkia amoena** syn. *Godetia amoena*), an erect or lax-stemmed plant 1 foot to 3 feet tall that in the wild exhibits considerable variation. Its linear to lanceolate leaves are up to 2½ inches long. Often smaller leaves crowd in their axils. The erect flower buds expand into satiny, broad-petaled blooms 2 to 3 inches in diameter and are white, lilac, pink or red, often with a darker patch or

spot at the base of each petal. *C. a. lindleyi*, erect and up to 6 feet tall, has lavender flowers with petals 1 inch to 1½ inches long. *C. a. whitneyi* is distinguished by its young seed capsules having eight instead of four ribs and is usually a lower plant, not over 2 feet in height, with larger flowers. Selected garden varieties of *C. amoena* are offered in seed catalogs. Differing from *C. amoena* in that its older flower buds are deflexed rather than erect, *C. bottiae* (syn. *Godetia bottiae*) is a slender, erect, usually branched native of southern California. From 1 foot to 1½ feet tall, it has looser flower spikes than farewell-to-spring. Its blooms are pinkish-lavender or pink becoming paler toward their centers. They are often flecked with darker markings. The showy-flowered plant previously named *Godetia viminea*, *G. lepida*, and *G. goddardii* is now **Clarkia purpurea viminea.** Erect or lax-stemmed, this has narrow, linear or lanceolate-linear leaves and short, loose spikes of 2-inch-wide lavender to purple-crimson flowers, usually with a dark spot on each petal.

Garden Uses. Clarkias are beautiful for flower beds and borders, are elegant cut flowers, and delightful decorative pot plants.

Cultivation. Clarkias do not withstand hot, humid weather well. Where torrid summers prevail it is important to time the sowing of the seeds so that the plants bloom before adverse weather comes. In most of the United States this means sowing outdoors as early in spring as possible. A delay of even two or three weeks can make the difference between success and failure. In favored areas, such as parts of Florida and California, fall sowing gives good results.

In flower borders seeds may be broadcast in suitably-sized patches, but for cutting it is better to sow in rows spaced 1 foot to 2 feet apart. The seedlings are thinned to 4 to 6 inches. Clarkias need full sun and porous, moderately fertile soil. Excessive amounts of nitrogen encourage lush growth and sparse blooming. A dressing of superphosphate worked into the soil before sowing is usually helpful. Support in the form of twiggy brushwood inserted among the plants, strings stretched between stakes, or some similar device is usually needed.

In greenhouses sow clarkias from September to January. The earlier seedlings give the largest plants. From a sowing made at the beginning of September plants of *C. unguiculata* 6 feet or more in height and one-half as wide, accommodated in 8- or 9-inch pots, can be had in full bloom in April. A January seeding gives plants to be finished in 5- or 6-inch pots that will bloom at approximately the same time, but will, of course, be smaller. For cut flowers, January-sown plants may be set 6 to 8 inches apart in beds or benches.

It is even possible to secure useful bloom for cutting by growing them spaced about 4 inches apart in deep flats.

Indoor sowings are made in pots, pans, or flats. Transplant the seedlings as soon as their second pair of leaves is half grown into flats, spacing them about 2½ inches apart, or set them individually in small pots. The flat method is generally better. From the flats or small pots plant seedlings

in benches or beds or repot them successively into larger pots until the ones in which they are to bloom are achieved. The young plants may be allowed to develop naturally or may be pinched once when about 3 inches high to encourage branching and somewhat shorter growth.

Clarkias revel in a coarse, turfy soil, and it is a joy to the true greenhouse gardener to see their new white roots searching the surface of the soil and advancing into it, as they do when in active growth. Because of this tendency to root at the surface, it is a good plan when potting clarkias into their final pots to keep the soil surface ¾ to 1 inch lower than normal potting calls for and then, when the pots are well filled with roots and hungry new ones can be seen exploring the surface, to add a top dressing of new soil mixed with organic fertilizer.

Full exposure to sun and cool temperatures are absolute necessities. Whenever outdoor conditions permit, night temperatures should be held at 45 to 50°F, day temperatures only a few degrees higher. Plenty of ventilation so that a buoyant atmosphere is maintained is a must. In their early stages clarkias are very sensitive to excessive watering and are likely to damp off or rot near the soil line if this is practiced. When established and well rooted, they need sufficient water to prevent the foliage from wilting and, when in bud, benefit from weekly applications of dilute liquid fertilizer. Neat staking and tying, well before the stems begin to sprawl, are essential.

Diseases and Pests Damping off, stem rot, verticillium wilt, and mildew are the diseases most likely to be troublesome. Aphids and red spider mites are the most common pests.

CLARY. This is the common name of *Salvia sclarea*, a herb much less used now than formerly for flavoring foods and beverages and of importance as a source of oil used in perfumery. In addition to such employ-

Clary

ments, it is ornamental and well worthy of a place in herb gardens and flower borders. It is described botanically under Salvia.

Clary is biennial, flowering in its second season and then dying. For best performance give it a deep, rich soil and sunny location. Sow seeds in a cold frame or outdoors in May or June and transplant the seedlings to a nursery bed, spacing them about 9 inches apart in rows 1 foot apart. In early fall or the following spring transplant to their flowering quarters, allowing about 1½ feet between individuals. Often this species perpetuates itself by self-sown seedlings. To harvest, pick the large leaves separately. Some can be taken each of the two seasons the plant lives. They can be used fresh or dried. To dry, spread newly picked leaves in a shaded, dry, airy place. Crumble the dried leaves by rubbing between the palms of the hands, and store in tightly-stoppered containers.

CLASSIFICATION OF PLANTS. This involves the grouping of different kinds of plants in orderly fashion. Thus they may be distinguished as trees, shrubs, and herbaceous plants, as annuals, biennials, and perennials, as land plants and aquatics, as edible and nonedible plants, as hardy and nonhardy kinds, and in many other useful ways, not of prime importance to botanical classification. Because it is upon their botanical classification that the naming of plants is based it is important for gardeners to become familiar with the principles involved and with at least some of the terms used.

Botanical classification is interpretive. It is based on judgments that of necessity are to a greater or lesser extent subjective. Equally competent botanists studying the same populations of plants may come to different conclusions as to how they should be classified. This does not mean that the choices are unlimited. To be acceptable they must be made within the framework of current botanical thought. If different conclusions are reached, the same plant will be given different names. Such honest differences of opinion are responsible for many name changes that confront and not infrequently confuse gardeners. Such are unavoidable even though sometimes regrettable. That the same plant is known by two names or more does not necessarily mean that one is "right," the others "wrong." It may merely reflect differences of opinion. To illustrate, one botanist as a result of studying pears and apples, concludes that despite some obvious differences the similarities of their flowers and fruits are such that warrant placing them in the same genus. The common pear is then identified as *Pyrus communis*, the common apple as *Pyrus malus*. Another botanist observes the similarities, but reaches the conclusion that the differences between the flowers and fruits of pears and apples are of sufficient importance to justify separating apples as another genus. Under his interpretation the common pear remains *Pyrus communis*, the common apple becomes *Malus sylvestris*. Or at a different level, consider the yellow lady slipper orchid, a widely distributed native of North America, Europe, and Asia. Slight differences exist between the North American and Eurasian plants. These sufficed for some authorities to consider them distinct species, *Cypripedium parviflorum* and *C. calceolus*, respectively. Examination of the available evidence suggested to other botanists that the American population was indeed no more than a variety of the Eurasian, so they applied to it the name *C. calceolus*, variety *pubescens*, or as it is often written, *C. calceolus pubescens*. Even more conservative in outlook, yet other botanists have come to the conclusion that the differences between the American and the Old World plants are not important enough to warrant separate identification and name all *C. calceolus*.

All of the several modern systems of plant classification are based on the premise that the plants that populate the world today have through the millenia evolved from common ancestral stocks and that thus there exists between them various degrees of genetical relationship. Classificatory systems attempt to reflect these relationships by grouping under the same names plants believed to be more closely related and giving different names to those of less intimate relationship. Thus all oaks are placed in the genus *Quercus*, all beeches in the genus *Fagus*. Beeches are considered to be sufficiently distinct from oaks to merit putting them into a different genus. Yet the similarities in the flowers and fruits of oaks and beeches as well as in those of chestnuts (*Castanea*) cause botanists to place all three genera in the family FAGACEAE. Birches (*Betula*), alders (*Alnus*), and some other genera are sufficiently alike among themselves and enough different from members of the FAGACEAE to constitute the family BETULACEAE. Most plant families consist of several or many genera, others of few, some of only one. The ginkgo family GINKGOACEAE, for example, contains only the genus *Ginkgo*, which consists of only one species, *G. biloba*.

Going up the ladder as it were, families are grouped into orders, orders into classes, classes into divisions. These assemblages are not of much importance to gardeners until we arrive at what botanists identify as divisions of the plant kingdom. At this level gardeners generally recognize as discrete ferns and fern allies (the botanists' *Pteridophyta*), which have neither flowers nor seeds, and seed plants (*Spermatophyta*), which produce flowers even though in some kinds primitive, and seeds. They further are generally aware, though some-

times rather vaguely, of the subdivisions of the *Spermatophyta* into conifers (*Coniferae*) and related orders that constitute the gymnosperms (*Gymnospermae*) and the remainder of the seed plants, which are classified as angiosperms (*Angiospermae*). It is of some importance for gardeners to be cognizant of the two classes into which the angiosperms are divided because the success of certain horticultural techniques, grafting, for example, may depend upon whether or not a plant is a monocotyledon (belongs in the *Monocotyledonae*) or a dicotyledon (belongs in the *Dicotyledonae*). Grafting is generally impossible with monocotyledons.

Just as families may comprise one or more genera so a genus may consist of one or more species. Species are the basic units, the building blocks as it were, of plant classification. This does not imply that every individual of a particular species is an exact duplicate of all others. Minor variations occur, but ordinarily not greater than are to be expected from nonhybrid offspring of the same parent. Significant variations within wild populations of species are recognized by botanists and to them are applied such terms as subspecies, variety, and forma, and their status is indicated by adding a third word to the two of the specific name. Some botanical separates of species, as well as other departures from what may be considered the norm, are recognized by gardeners and are frequently of much greater horticultural importance than the typical species. Here belong, in addition to botanically recognized variants, others and hybrids, selected or bred for superior or desirable characteristics and that either breed reasonably true from seeds or are perpetuated by vegetative propagation. These horticulturally significant segments of less than specific rank are identified by names. These may be formed, as are those of botanical subdivisions of species, by adding a word to the end of the specific name, as for example *Cornus florida rubra* for the pink-flowered variant of the normally white-flowered dogwood *Cornus florida*, or by using a name without reference to the parental species as apple 'Mackintosh'. Traditionally, cultivated plants of less than specific rank, whether representative of wild populations or not, have been referred to as varieties. More recently many horticulturists follow the recommendation that the term variety be used only for botanical varieties and that horticultural varieties be called cultivars. There is certain merit in this, but because information necessary for the proper classification of some cultivated plants is incomplete, because the precise application of variety and cultivar as discrete terms involves somewhat complicated conventions for writing the names, which can easily confuse those not botanically oriented, and because to the

vast majority of gardeners it is of small importance whether the plant they grow is a botanical or horticultural variety, the term variety or sometimes horticultural variety is used throughout this Encyclopedia for identified variants of species and hybrids.

CLAUSENA (Claus-èna)—Wampi. Of the thirty species of *Clausena* of the rue family RUTACEAE, one is cultivated to a limited extent in warm parts of the United States. The group, indigenous to tropical Asia, Australia, and Africa, has pinnate leaves, racemes or terminal panicles of small flowers, each with eight to ten stamens, and small, spherical, berry-like fruits. The name commemorates P. Clauson, a Danish botanist, who died in 1632.

The wampi (*C. lansium* syn. *C. punctata*), a native of southern China and Vietnam, is a shrub or tree 15 to 20 feet tall in cultivation in the United States, but up to twice as tall in the wild. Its leaves have an uneven number (five to thirteen) of shiny, resinous-fragrant, ovate-elliptic, leaflets 3 to 5 inches long. Its large, many-flowered panicles, up to 1 foot long or longer, have small white, five-petaled blooms. The nearly spherical to egg-shaped, pale yellow or yellowish-green fruits, about 1 inch in diameter, have an aromatic, slightly acid, white pulp. They may be seedless or have one to five green seeds.

Garden and Landscape Uses and Cultivation. This is an interesting and unusual plant. Its fruits are eaten fresh or used in jellies, pies, and beverages. Individual trees vary considerably in the size and quality of their fruits, and it is desirable to select the best and propagate them by grafting onto seedlings or by taking cuttings of leafy shoots and rooting them under mist; seeds, however, are usually relied upon for the production of new trees. The wampi succeeds in sunny locations in ordinary garden soil. It needs no special care.

CLAVEL DE LA INDIA is *Tabernaemontana divaricata*.

CLAVIJA (Clav-ìja). To the theophrasta family THEOPHRASTACEAE belongs this South American group of more than fifty species of evergreen trees and shrubs. Their name commemorates the sixteenth-century Spanish naturalist Don José de Viera y Clavijo, who died in 1813.

Clavijas are branchless or few-branched evergreen trees and shrubs of curious appearance with, at the tops of their often spiny trunks or stems, a cluster of alternate, rigid, leathery, undivided, sometimes spiny-toothed leaves. The unisexual and bisexual flowers are in axillary, spike-like racemes. They have four- or five-parted calyxes, and white, yellow, orange, or brownish-red corollas, with short, fleshy tubes and four or five lobes (petals). There are four or five stamens and, representing rudimentary stamens, the same number of scales in the throats of the corollas. The style is short. The fruits are berry-like.

Native of tropical South America and Trinidad, *Clavija longifolia* (syn. *C. ornata*), 10 to 12 feet tall, has spiny, toothed, oblanceolate to elliptic, long-stalked leaves up to 2 feet in length and 6 inches wide. In pendulous racemes up to 10 inches long, the orange to brownish-red flowers are borne on portions of the stem below the leaves. The males are fragrant. The fruits are orange.

Garden and Landscape Uses and Cultivation. Because of its distinctive appearance *C. longifolia* and perhaps other kinds are sometimes grown as ornamentals outdoors in the tropics and subtropics, and in greenhouses. Satisfied with ordinary soils and locations, they multiply readily by seeds, and by cuttings set in a greenhouse propagating bed, preferably one with mild bottom heat. In greenhouses a minimum winter night temperature of 60°F is desirable.

CLAW. The sepals and petals of some flowers narrow sharply for some little distance from their bases to form almost stalk-like portions called claws.

CLAW FERN. See Onychium.

CLAY SOIL. See Soils and Their Management.

CLAYTONIA (Clay-tònia)—Spring Beauty. John Clayton, pioneer American botanist, who died in 1773, is commemorated by the name of this genus of the purslane family PORTULACACEAE. It comprises up to twenty species, mostly natives of North America, but also represented by one species each in the native floras of South America, and Australia and New Zealand, and two or three in Asia.

Claytonias, low herbaceous plants and chiefly perennials, often have thick roots, tubers, corms, stolons, or similar underground parts. Their more or less fleshy foliage is hairless. There are one to many basal leaves and usually an opposite or a nearly opposite pair of leaves on the stems. Most commonly in racemes of up to twenty, the flowers are occasionally solitary. Each has two persistent sepals, five almost completely separate, spreading petals, five stamens, and a shortly-three-lobed style. The fruits are spherical to egg-shaped capsules.

Native of damp woods and fields from Nova Scotia to Minnesota, Georgia, Louisiana, and Texas, *Claytonia virginica* is a pretty spring-bloomer that opens its flowers only in bright light. From a small, deep-seated, spherical tuber it sends up one to many stems 3 inches to more than 1 foot tall, with none to many narrow-lin-

Claytonia virginica

Claytonia virginica (flowers)

ear, lanceolate to elliptic basal leaves up to ½ inch wide, and almost or quite as long as the stem. The two leaves on the stem are opposite, 2 to 6 inches long, and without well-defined stalks. The flowers, pink or white with pink veins, ¾ to 1 inch across, are in racemes of up to six. In *C. v. robusta* the plants are up to 1½ feet tall and have leaves up to 1 inch wide. The flowers of *C. v. lutea* are orange-yellow veined with red. Similar to *C. virginica*, but with proportionately broader and distinctly stalked leaves, *C. caroliniana* inhabits cool woodlands from Nova Scotia to Minnesota, and in the mountains to North Carolina and Tennessee.

Western American *C. lanceolata* (syn. *C. multiscapa*) is extremely variable. It occurs throughout most of the western part of the continent from near snowline on high mountains to bogs and deserts. It has a small, nearly spherical corm, and one to several stems 2 to 10 inches tall. There may be no basal leaves, or a few lanceolate to ovate or oblanceolate ones up to 6 inches long, their stalks longer than the blades. The two opposite stem leaves are lanceolate and essentially stalkless. Up to 4 inches long, the racemes are of up to fifteen flowers, pink or white with pink veins, ½ to ¾ inch wide. Variety *C. l. flava* has yellow flowers. Closely related *C. rosea* (syn. *C. lanceolata rosea*), of the Rocky

Mountains, has linear to narrowly-lanceolate basal leaves and rose-pink flowers.

Entirely different *C. megarrhiza*, of high mountains in western North America, has a long, fleshy taproot. Its many fleshy, spatula-shaped to obovate, blunt-ended leaves, up to 6 inches long and over 1 inch wide, have winged stalks. The few to many stems have usually two nearly opposite stem leaves, oblanceolate to linear. In a dense raceme of three to seven, the white to light pink flowers are ½ to 1 inch wide. Variety *C. m. nivalis* has deep rose-pink blooms ¾ inch wide.

Garden and Landscape Uses and Cultivation. Except where cool summers prevail, *C. megarrhiza* is very dificult. It can be recommended only for cautious trial by the keenest rock gardeners, preferably those who have at their disposal an alpine greenhouse. There, it should be tried under conditions as near to its natural mountain rock slides and screes as possible. The other kinds described here are easy to manage in rock and woodland gardens and other lightly shaded places. They delight in soil that contains abundant organic material and is moist at least until the foliage has died down in late spring or early summer. Propagation is very easy by seed.

CLEAN CULTIVATION. This is the practice of keeping ground between crops free of weeds by cultivating the surface shallowly at frequent intervals in contrast to mulching, or in the case of fruit trees, for example, growing them in grass sod.

CLEARING NUT is *Strychnos potatorum*.

CLEISTES (Cleís-tes). Most of the twenty-five species of this genus of New World members of the orchid family ORCHIDACEAE are unknown to gardeners. One, *Cleistes divaricata*, is native from New Jersey to Kentucky, Florida, and Mississippi, the others to Central and South America. The name, calling attention to parts of the flowers being lightly joined to form a tube, is from the Greek *kleistos*, closed.

From clusters of fleshy, often fuzzy roots attached to a short rhizome or rootstock, the leaf-bearing flowering stems of these terrestrial (soil-inhabiting) orchids grow erectly. Solitary or few together in racemes, the blooms have somewhat the appearance of sobralias and are pink, yellow, brown, green, or white. The tropical species are scarcely known in cultivation.

The only species native to the United States *C. divaricata* inhabits swamps and wet woodlands. From 8 inches to 2 feet tall, it has slender stems with a single linear leaf, 2 to 5 inches long above their middles. Below the slightly nodding, usually solitary, fragrant flower is a large leaflike bract. The spreading-ascending sepals are linear, purple-brown, and 1¼ to 2¼ inches long. The petals, oblanceolate with recurved tips, are joined with the slightly longer, purple-veined, greenish or pinkish, beardless lip into a tube. They are pink to rosy-purple, or rarely white, and up to 2 inches long.

Garden Uses and Cultivation. Not easy to tame as a garden plant the species described may be attempted in bog gardens and wet soil areas in rock gardens where it is afforded an environment as similar as possible to those it favors in the wild. Increase is by natural multiplication and careful division. Additional information is given under Orchids.

CLEISTOCACTUS (Cleisto-cáctus). Possibly thirty rather poorly defined species of cactuses, inhabitants of southern South America, constitute *Cleistocactus* of the cactus family CACTACEAE. Their name, based on a characteristic of the blooms, which remain closed rather than having perianth segments (petals) that spread, is from the Greek *kleistos*, closed, and the word cactus.

Cleistocactuses commonly have slender, decumbent, clambering or erect, manyribbed, very spiny stems that often form thickets. The slender, tubular flowers, sprouting from the sides of the stems, curved and sometimes slightly asymmetrical, are clothed with scales, with wool or hairs in their axils. The red, orange, yellowish, or greenish perianth segments (petals) overlap and expand not at all or only very slightly at their ends. The stamens may or may not just protrude. The style does more conspicuously. The fruits are small, spherical, and fleshy.

Commonly cultivated, beautiful *C. strausii* has stems about 4 inches thick that are erect, silvery columns 3 to 6 feet tall, with generally twenty-three to twenty-seven ribs. In clusters of thirty to forty the finehairy and bristle-like spines are mostly white and slightly more than ½ inch in length, but with two to four longer, yellowish centrals. The dark wine-red blooms are about 3½ inches long. Somewhat sim-

ilar *C. jujuyensis* has stems approximately 2 inches in diameter, with about twenty ribs. The spines, in clusters of twenty-five to thirty, are glassy-white except for three centrals, which are deep yellow to brownish and about 1 inch long. The radials are shorter. Approximately 1¾ inches in length, the blooms have bright red perianth tubes and bluish-crimson petals.

Up to 6 feet in height and branching from their bases, the 1- to 1½-inch-thick, twelve- to sixteen-ribbed green stems of *C. baumannii* are thickly covered with clusters of fifteen to twenty pale to dark brown spines, the longest 1 inch to 1½ inches in length, but most much shorter. This species has curiously S-shaped, somewhat asymmetrical, orange-scarlet blooms 2 to 3 inches long. They are succeeded by red fruits approximately ½ inch in diameter. Because of its flowers this kind is sometimes called scarlet bugler.

Other cultivated cleistocactuses include *C. areolatus*, of Bolivia, which has many erect and sprawling stems up to 10 feet long and 2 inches thick. They have thirteen to fifteen low ribs and closely set clusters of seven to nine radial spines less than ½ inch long and one to three centrals about 1 inch long. The flowers, nearly 2 inches long, are carmine. About 3 feet tall, *C. hyalacanthus*, of Argentina, forms stout, twenty-ribbed columns up to 3 feet tall and about 1½ inches thick, clothed with clusters of about twenty-five rather weak, white spines up to ¾ inch long. The dark red flowers are up to 1½ inches in length.

Straight rather than curved red blooms about 2 inches long are borne by *C. smaragdiflorus.* Their petals are margined with emerald-green. This kind has stems ¾ inch to 1½ inches thick, with twelve to fourteen ribs and clusters of twenty-two to twentyeight spines consisting of many radials and several centrals up to ¾ inch long. The

Cleistocactus baumannii

Cleistocactus areolatus

Cleistocactus strausii

Cleistocactus smaragdiflorus

Large-flowered hybrid clematises:
(a) 'Comtesse de Bouchaud'

(b) 'Ernest Markham'

spherical fruits are a little over ½ inch in diameter.

Garden and Landscape Uses and Cultivation. Cleistocactuses are ornamental, free-flowering, generally easy-to-grow components of cactus collections, outdoors in warm desert and semidesert regions, and in greenhouses. Both *C. strausii* and *C. baumannii* serve well as understocks upon which to graft other cactuses. For additional information see Cactuses.

CLEMATIS (Clém-atis) — Virgin's Bower, Traveler's Joy or Old Man's Beard. By some botanists this horticulturally exciting genus has been divided into a number of genera, *Clematis, Atragene, Flammula, Viorna,* and *Viticella.* The conservative and more generally accepted practice of uniting these as *Clematis* is followed here. When this is done there are approximately 250 species and a fairly large number of garden hybrids, many of great beauty. The group belongs in the buttercup family RANUN-CULACEAE and is native in most major parts of the world. Most cultivated kinds are natives of or are hybrids of natives of temperate Asia, Europe, and North America. The name is derived from an ancient Greek one for some climbing plant.

Clematises to most gardeners mean more or less woody, vining plants, mostly of considerable vigor, with often large flowers with wide-spreading petals. This is true of a great many kinds except that in all clematises what are commonly thought of as petals are petal-like sepals. There are clematises, however, that are not vines, but neat, erect, herbaceous perennials of low to medium stature, and those with sepals that instead of spreading to form a wide-open bloom are erect and form a bell- or urn-shaped one. Characteristics of the genus are undivided or pinnate leaves that are opposite and solitary or panicled flowers with generally four or five sepals, no petals, many stamens, and many pistils each of which develops into a one-seeded achene usually with a long, feathery tail that is the persistent style. The clusters of achenes are often showy. In one section of the genus, as botanists who favor dividing it recognize, the outer stamens are developed as petal-like staminodes (modified, nonfunctional stamens). Vining clematises climb by twisting their leafstalks around suitable supports.

The history of *Clematis* in cultivation is a long one in Europe, America, China, Japan, and elsewhere in the Orient. The earliest exotic species brought to England was *C. viticella,* which came from Spain in 1569. We have no certain record of when the first foreigner was introduced into the United States, but certainly considerable interest was being shown in the small-flowered kinds available, such as *C. viticella, C. vitalba,* and native *C. virginiana,* by early in the nineteenth century, and we know *C. florida* was exhibited at Boston in 1838, *C. patens* in 1841, and in an intervening year *C. sieboldii* was exhibited at Philadelphia. The first hybrid clematis of record was raised in England in 1835. The introduction to Europe in 1850 and shortly afterward to America of large-flowered *C. lanuginosa* increased horticultural interest in the genus tremendously. A program of hybridizing, using the new kinds as the chief base, which was carried out mainly in France and England, resulted in the production of a considerable number of stunning varieties some of which are still grown. Other hybrids brought from Japan added to the great wealth of new clematises. A catalog published by the fine nursery firm of Parsons of Flushing, Long Island, New York, toward the close of the nineteenth century, listed no fewer than seventy-three varieties for sale, far more than are available now, more than eighty years later.

Large-flowered hybrid clematises, the most spectacular members of the genus, are among the most magnificent of flowering vines. Their most important parent is *C. lanuginosa.* Others involved are *C. florida, C. patens,* and *C. viticella.* Here is a selection of large-flowered hybrids: 'Belle Nantaise', flowers delicate lavender with six or seven pointed sepals; 'Belle of Woking', flowers double, silvery-mauve; 'Comtesse de Bouchaud', flowers with six pinkish-mauve sepals; 'Crimson Star', flowers garnet-red, sepals six; 'Duchess of Edinburgh', flowers white to greenish-white, double; 'Ernest Markham', flowers magenta-pink, six overlapping sepals; 'Fairy Queen', flowers flesh-pink with a rosy-mauve band down the center of each sepal; 'Gipsy Queen', flowers with six sepals, purple aging to violet; *C. lawsoniana henryi,* flowers with six or seven white sepals, and brown anthers; *C. jackmanii,* flowers with four to six rich blue-purple sepals; 'King Edward VII', flowers flesh-pink with a bright pink band down the center of each of the six or seven sepals; *C. lanuginosa candida,* flowers white, about eight sepals; 'Mme. Baron-Veillard', flowers rosy-lilac, six sepals; 'Mme. Edouard Andre', flowers with six wine-red sepals; 'Mrs. Cholmondeley', flowers light lavender-blue, usually with six sepals; 'Nelly Moser', flowers with eight rosy-mauve sepals banded along their centers with carmine-pink; 'Perle d'Azur', flowers pale blue flushed at their centers with pinkish-mauve, sepals generally six; 'Pink Chiffon', flowers shell-pink, six sepals; 'Prince Phillip', flowers deep blue, six sepals; 'Ramona', flowers with six clear blue sepals; 'The President', flowers light plum-purple with a silvery central band on the undersides of the eight sepals; 'Ville de Lyon', with five or six bright carmine petals and golden yellow stamens.

Ancestors of the large-flowered hybrids include Chinese *C. **lanuginosa*** a species

Clematis lawsoniana henryi

Large-flowered hybrid 'Ville de Lyon'

Clematis jackmanii

Large-flowered hybrid 'Nelly Moser'

Clematis florida sieboldii

probably not now in cultivation in its pure form. A deciduous vine of modest growth, probably not exceeding 6 feet in height, this has thickish leaves of often one, sometimes three, pointed, more or less heart-shaped leaflets up to 5 inches long by 3 inches wide, hairless above, and abundantly gray-woolly on their undersides. From 4 to 6 inches in diameter, and with woolly stalks without bracts, the flowers are borne at the ends of the shoots. They have six to eight, overlapping, wide-spreading, white to light lilac sepals, downy on their outsides. Botanically closely related *C. patens* is also Chinese and probably also native to Japan. In any case it is believed to be commonly grown there. Perhaps no longer in cultivation in its wild form in North America or Europe, *C. patens* is a deciduous vine up to 12 feet tall.

It has leaves of three or five pointed-ovate-lanceolate leaflets 2 to 4 inches long by about two-thirds as wide. Their undersides are downy, their upper surfaces hairless. The solitary blooms, on bractless, downy stalks, and 4 to 6 inches across, have six to eight wide-spreading, long-pointed, not overlapping sepals, believed to be white in the wild. Native to southern Europe and old in cultivation, *C. viticella* is a somewhat woody vine up to about 12 feet tall. Slightly downy at first, its stems are slender and ribbed. Its leaves are pinnately-divided, with primary divisions of each of three leaflets. The latter, often few-lobed, are up to 2½ inches in length, lanceolate to broad-ovate. Rosy-purple, purple, or blue, the 1½-inch-wide flowers on stalks 2 to 4 inches long, are solitary or in clusters of few to several. They have obovate, spreading sepals. The seedlike fruits are without long, tail-like appendages so typical of most clematises. Variety *C. v. albiflora* has white flowers. Dwarf *C. v. nana* is only about 3 feet tall. Double, purplish flowers are borne by *C. v. plena*. Red flow-

ers are characteristic of *C. v. rubra*. A number of more modern variants or more probably small-flowered hybrids of *C. viticella* are in cultivation. Closely related to *C. patens* and by some included in it, **C. florida** is a Chinese species long cultivated in Japan. It differs from *C. patens* in its leaves being commonly of three divisions each of three ovate to lanceolate, toothed or toothless leaflets 1 inch to 2 inches long, and in its flower stalks having at approximately their middles a pair of leaflike, stalkless, lobed bracts. The leaves, lustrous above, have more or less hairy undersides. The flowers have four to six wide-spreading, broad-elliptic to ovate, white or creamy-white sepals with a central stripe of green on their undersides, and spreading, black-purple stamens. In *C. f. sieboldii* the sta-

Clematis florida sieboldii (flower)

Clematis montana rubens

Clematis chrysocoma sericea

mens are replaced by purple, petal-like staminodes (infertile, modified stamens). The blooms of *C. f. plena* are similar, but the staminodes are white. Variety *C. f. fortunei*, with double, creamy-white blooms that change to pink, is by some authorities treated as a variety of *C. patens*.

Beautiful *C. montana* and its varieties are vigorous woody vines. Native to the Himalayas and up to about 25 feet tall, this early-blooming species is not as hardy as its variety *C. m. rubens* and should not be relied upon north of Washington, D. C. or perhaps Philadelphia. The species has 2- to 4-inch-long stalked leaves of three short-stalked, irregularly-toothed leaflets up to

Clematis montana

4 inches long by about one-half as wide. The fragrant flowers, solitary, pure white, and 2 to 2½ inches wide, have stalks 2 to 5 inches in length. The four wide-spreading petals are broad-elliptic. The fruits have long, feathery appendages. Considerably hardier, surviving with sometimes minor killing back in winter as far north as southern New England, *C. m. rubens* is a native of China. Its young shoots and leafstalks are decidedly purplish, and its foliage, more hairy than that of the typical species, is purplish-tinged. The flowers have much the aspect of those of Japanese anemones. Pink and about 2½ inches wide, they have hairy stalks. Other *C. montana* varieties, less frequently grown than *C. m. rubens*, are *C. m.* 'Alexander', with creamy-white flowers; *C. m.* 'Elizabeth', with larger, fragrant, light pink blooms; *C. m. grandiflora*, which has flowers about 3 inches in diameter; *C. m. platyphylla*, which is similar to *C. m. wilsonii*, but blooms earlier and has flowers with broader sepals; *C. m.* 'Tetrarose', a tetraploid variety of great vigor with extra-large, purplish-pink blooms; and *C. m. wilsonii*, which has downy-stalked, white flowers some 3 inches in diameter that come later than other *C. montana* varieties.

Similar to *C. montana*, but less hardy, *C. chrysocoma* (syn. *C. spooneri*), of China, is often erect, bushy, and 8 to 10 feet high, but sometimes climbs and is taller. Its young shoots as well as its leaves and flower stalks are clothed with shaggy, yellow down. The leaves are of three short-stalked, ovate, elliptic, or obovate, lobed or toothed leaflets ¾ inch to 2 inches in length. In axillary or terminal clusters, the pink-tinted, white flowers, 1½ to 2 inches in diameter, are densely-silky-hairy on the outsides of their four spreading sepals. The blooms of *C. c. sericea*, solitary or in pairs, are 3 to 4 inches in diameter and white. Unlike *C. chrysocoma* the variety

never bears supplementary blooms from current season's shoots. A very good hybrid between *C. chrysocoma* and *C. montana*, *C. vedrariensis* is much like the latter, but has slightly larger, pink blooms.

The most popular small-flowered vining clematis, a prolific fall bloomer, has for long been known in gardens as *C. paniculata*. Unfortunately that name rightly belongs to a quite different, evergreen species native to New Zealand. Much more familiar usurpers of the name are Japanese *C. dioscoreifolia* and *C. d. robusta*. A vigorous grower up to 30 feet tall, *C. dioscoreifolia* (syn. *C. maximowcziana*) forms bounteous tangles of slender stems and lustrous foliage. The leaves are of five hairless or nearly hairless, long-stalked leaflets without lobes or teeth, 1 inch to 3 inches in length. The fragrant flowers, 1 inch or so across and with four spreading, white sepals, are in plentiful panicles 3 to 4 inches long. The floral display is followed by clouds of ornamental seed heads with long, feathery, silver-gray appendages. From the typical species *C. d. robusta* is distinguished by its slightly larger flowers and its leaves having sometimes only three leaflets. Closely related to *C. dioscoreifolia*, but less vigorous and not nearly as well known to American gardeners is *C. flammula*, native to southern Europe. Commonly 10 to 12 feet tall, this kind has leaves of three or five toothless, but often few-lobed, hairless leaflets of variable shapes and sizes, usually ½ inch to 1½ inches long, or frequently the primary divisions have three leaflets and the leaves are twice-pinnate. The fragrant, fall-borne, pure white flowers, in panicles up to 1 foot long, are ¾ to 1 inch wide. They are succeeded by decorative white-plumy-appendaged seed heads.

Traveler's joy, or old man's beard as it is known in Great Britain, *C. vitalba* is of the same relationship as the last two dealt with. Native to Europe, North Africa, and temperate Asia, it is a vigorous vine that may attain heights of 40 feet or more. The leaves are pinnate with usually five lan-

Clematis flammula

ceolate to pointed-ovate, stalked, toothed to nearly toothless leaflets up to 4 inches in length. The green-tinged, white, slightly fragrant blooms, ¾ inch across and with four wide-spreading sepals, are succeeded by decorative seed heads of silver-gray plumed fruits.

Native American species with abundant small white flowers with spreading sepals are Virgin's bower (*C. virginiana*) and *C. ligusticifolia*, the first wild from Nova Scotia to Georgia and Kansas, the second from North Dakota to British Columbia, New Mexico, and California. These are less attractive ornamentals than *C. flammula* and *C. vitalba*, from which they differ in having usually unisexual flowers. The blooms, borne in summer and fall, are ¾ to 1 inch across. The leaves of *C. virginiana* are of three ovate to narrow-ovate, toothed leaflets, those of *C. ligusticifolia* of five to seven.

Evergreen *C. armandii*, of China, is an imposing species well adapted to the Pa-

cific Northwest and similar mild-climate areas, but not hardy in the north. It is 20 to 30 feet high and has stalked leaves of three glossy green, hairless, oblong-lanceolate to ovate, stalked leaflets, 3 to 6 inches long and prominently three-veined. Borne in crowded, axillary clusters in spring, usually three on a stalk and 2 to 2½ inches wide, the white to creamy-white flowers of cultivated plants, sometimes distinguished as *C. a. biondiana,* change to rose-pink as they age. They have four to seven narrow-oblong sepals. Wild forms of this species have mostly solitary, smaller flowers with fewer sepals.

Blue or blue-purple or lavender-blue blooms considerably smaller than those of the large-flowered hybrids are borne by a few vining clematises. Here belongs *C. alpina,* native to Europe and temperate Asia. This sort climbs to a height of about 6 feet and has leaves twice- or sometimes thrice-divided into groups of three short-stalked or nearly stalkless, ovate to ovate-lanceolate, toothed leaflets up to 2 inches long. The slender-stalked, violet-blue, bell-shaped flowers have sepals 1¼ to 1½ inches long and a number of petal-like staminodes. The blooms of *C. a. sibirica* are yellowish-white. Allied to *C. alpina,* Chinese and Siberian *C. macropetala* is a vigorous vine up to 15 feet in height. Its leaves are divided in threes and again in threes into nine toothed leaflets. The nodding, 3-inch-wide flowers have four blue-edged, lavender-colored sepals and numerous slender, petal-like staminodes (modified stamens) that result in a double-type bloom. Native from Virginia to Florida and Texas and hardy in New England, *C. crispa* is about 10 feet tall. It has slender stems and leaves of three to seven hairless, ovate to ovate-lanceolate leaflets, sometimes lobed, occasionally replaced by three similar leaflets. Long-stalked, the blooms have sepals, their upper halves spreading, with wavy, paler-colored edges. They are ¾ inch to

1¼ inches long. The fruits are without tails. Native of the Rocky Mountain region, *C. columbiana,* a trailer or climber with stems up to about 10 feet long, has leaves mostly in whorls (circles) of four, each with three pointed-ovate, toothed or toothless leaflets. The blue or purple blooms, nodding at first and solitary, expand to a diameter of about 4 inches. Up to 12 feet tall, *C. jouiniana* is a vining hybrid between *C. heracleifolia* and *C. vitalba.* It has leaves of three or five coarsely-toothed, ovate leaflets up to 4 inches long and great leafy panicles of whitish and grayish-blue to purplish flowers 1 inch to 2 inches in diameter, and with four to six spreading sepals. Variety *C. j. praecox* blooms several weeks earlier than the typical kind. Hybrid *C. eriostemon* has as parents *C. viticella* and *C. integrifolia.* Up to 9 feet tall and a profuse bloomer, it has leaves with up to seven leaflets. Its broadly-bell-shaped, dusky-purple blooms are about 2 inches across.

Small, bright scarlet, urn-shaped blooms characterize *C. texensis,* endemic to Texas. This is root-hardy in southern New England, even though there it is frequently killed to the ground in winter. Even so, in spring it produces new shoots that bloom the same season. About 6 feet tall, this cheerful, semiwoody vine blooms from early summer to killing frost. Its rather leathery, glaucous leaves have four to eight long-stalked, broad-ovate leaflets and in place of a terminal one usually a tendril. The leaflets, sometimes lobed, but not toothed and generally blunt, are up to 3 inches long or even longer. Slender-stalked and solitary, the ¾- to 1¼-inch-long, bell-shaped blooms, much narrowed at their mouths, have slightly spreading tips to their thick sepals. The fruits have long, feathery appendages. The flowers of *C. t. major* are about 1½ inches long. Hybrids between *C. texensis* and large-flowered species and hybrids include a number of very good vines that attain heights of up to about 10 feet and have blooms 1½ to 2 inches long, chiefly in tones of red and pink. Here belong 'Countess of Onslow', flowers with four cerise sepals edged with pink; 'Duchess of Albany,' flowers with four clear pink sepals; 'Etoile Rose', flowers nodding, bell-shaped, with four red-purple sepals edged with silvery-pink; and 'Gravetye Beauty', flowers cherry-red becoming pinker with age. They have four to six sepals.

The best yellow-flowered clematis is *C. tangutica* (syn. *C. orientalis tangutica*). A native of China and hardy in southern New England, this kind is about 10 feet tall. It has slightly-downy stems and, downy when young, pinnate or twice-pinnate, grayish-green leaves, irregularly-toothed and sometimes few-lobed. The slightly-downy flower stalks, 3 to 6 inches long, carry usually solitary, hairless, nodding,

Clematis vitalba

Clematis crispa

Clematis tangutica

bright yellow blooms at first broadly-bell-shaped, later expanding to 3 to 4 inches wide. Their long-pointed, hairless sepals are ovate-lanceolate. The decorative seed heads have feathery appendages. Yellow to greenish-yellow, slightly fragrant blooms 1½ to 2 inches across are displayed by *C. orientalis,* native from Iran to the Himalayas. Solitary or in clusters of few, they have four spreading or recurved, pointed sepals, pubescent on their insides. Not reliably hardy in the north, but naturalized in New Mexico, this hairless vine, up to about 20 feet tall, has pinnate or twice-pinnate leaves of six to nine stalked leaflets or primary divisions. Thin, glaucous, and usually lustrous, the leaflets are ovate to oblong, often lobed or toothed, and sometimes smooth-edged. The seed heads are decorated with feathery appendages. A robust vine up to 25 feet tall, *C. rehderana* has pinnate leaves of seven or nine broad-ovate, coarsely-toothed, prominently-veined leaflets up to 4 inches long. Its nodding, bell-shaped, fragrant flowers, yellowish-white to pale yellow, a little more than ½ inch long, downy on their outsides, are in erect, short panicles presented to produce an almost thalictrum-like effect.

This native of China is hardy about as far north as southern New York.

Nonvining clematises, bushy, herbaceous perennials that die to the ground in winter, like vining kinds can be separated into those with wide-spreading sepals and those with urn- to bell-shaped blooms. Most commonly grown of the first type, *C. recta* is a hardy native of southern Europe. From 2 to 3 feet in height, it has many erect

Clematis recta

stems and pinnate leaves of five to nine short-stalked, pointed-ovate, lobeless, toothless leaflets. The numerous pure white, fragrant flowers, ¾ to 1 inch in diameter, are in abundant, terminal large panicles that overtop the foliage. Flowers up to 2 inches across are characteristic of *C. r. grandiflora,* double flowers of *C. r. plena.* Native to China, *C. heracleifolia* is sometimes slightly woody at its base. It has erect stems 2 to 3 feet tall, leaves of three roundish-ovate, irregularly-shallowly-toothed leaflets with roundish bases, the center one much the biggest, and in short clusters from the ends of the stems and leaf axils tubular, light blue

flowers ¾ to 1 inch long with four reflexed sepals. Individual plants have both unisexual and bisexual blooms. Variety *C. h. davidiana* has leaves with wedge-shaped bases and fragrant, bright blue blooms, males and females on separate plants, in terminal heads of up to fifteen, and fewer in the leaf axils. Native from South Dakota and Idaho to Nebraska and Colorado, *C. scottii* is up to 2 feet tall and has branchless or sparsely-branched stems. Its leaves are stalked and up to 6 inches long, with the upper ones twice- or thrice-pinnate. The leaflets are lanceolate to ovate. Solitary and nodding, the purple flowers are nearly 1 inch long. The seed heads are of plumed achenes.

Clematis scottii

Undivided leaves are a characteristic of southern European *C. integrifolia.* This kind is sprawling to erect and 2 to 2½ feet tall. It has thin, stalkless, ovate-oblong, toothless leaves and solitary, terminal, yellow-stamened, violet-blue or rarely white flowers 1½ to 2 inches long. The seed heads are of plumed achenes. Native of Missouri, Kansas, and Nebraska, *C. fremontii,* called leather flower, has erect, branched or branchless stems 1 foot to 1½ feet tall. Its nearly stalkless, undivided, lanceolate- to ovate-elliptic leaves are netted and leathery. The purple, urn-shaped flowers, with stalks shorter than the leaves, are about 1 inch long. The seed heads are of plumed achenes. Variety *C. f. riehlii,* usually taller, has elliptic leaves. Also called leather flower, *C. ochroleuca* is native from southern New York to Georgia. Up to 2 feet in height, it has branched or branchless stems and nearly stalkless, undivided, toothless, ovate leaves up to 4½ inches long. The nodding, bell- to urn-shaped flowers, solitary at the ends of the stems, have stalks usually longer than the leaves. The blooms, approximately ¾ inch long, are yellowish-white sometimes tinged with purple. The seed heads are of plumed achenes.

Clematis rehderana

Clematis heracleifolia

Clematis integrifolia

Garden and Landscape Uses. Appropriate employments of clematises in gardens and other landscapes relate closely to the growth habits and floral and seed displays of various kinds as well as to their tractability to cultivation in particular areas. Certainly large-flowered hybrids are not for all American gardeners. These superb kinds are choosy as to where they will grow, exhibiting a marked dislike for hot, dry summers and too-acid soils. The latter are more easily corrected than the former. These quite amazing examples of hybridizers' skills are generally well adapted for conditions such as obtain in the Pacific Northwest, but very good examples are also grown in the northeastern and northcentral states. Like all clematises these, most especially these, need cool root runs. They grow best where the ground is shaded by low shrubs or other deep-rooting, not too aggressive plants, peonies perhaps, and where their tops can reach to where they receive good light and free air circulation. But they must not be completely exposed to untempered summer sun. A little dappled shade through the heat of the day is appreciated. If they are planted on the north side of a low wall and their stems trained over its top good results are likely to be had. They associate splendidly with roses, and it is quite practicable to have a climbing rose and large-flowered clematis share the same support. Other vining clematises respond to similar conditions, but some at least are less finicky. All must be afforded support. For the more vigorous, living shrubs of not too rampant or congested growth can supply this if the clematises are planted on their shaded sides. In lieu of living supports, trellises, or for those that die down each year or are pruned hard back, teepees of brushwood serve very well. Wires strained through eyelet bolts driven into walls may be used to carry the stout stems of such strong growers as *C. montana rubens* and *C. dioscoreifolia*. Allot these places in your garden if you possibly can. They are not at all finicky and they will reward you, the first in

spring the other in fall, with bountiful displays of bloom for very little trouble on your part.

Bushy, nonvining clematises that die to the ground each year, most particularly *C. recta, C. heracleaefolia,* and *C. integrifolia,* can be used with telling effect in perennial and mixed flower beds as well as in semi-wild and naturalistic landscapes. Native kinds are useful in wild gardens and other informal landscapes.

Cultivation. For satisfactory results do a thorough job of preparing the ground for clematises. If it is sandy, gravelly, or of other type that dries quickly and is poor in nutrients do your best to correct these disadvantages. Spade deeply and incorporate to a depth of at least one foot and better two, large amounts of decomposed organic matter, rich compost or leaf mold for preference, peat moss if they are not available in sufficient amounts. If you can obtain well-rotted manure mix that in with reasonable abandon. If without access to this precious additive, substitute commercial dried cow manure. It is not as good, but it is the next best thing. A generous sprinkling of bonemeal forked in also makes good sense. It is advantageous to make the planting sites ready a month or more in advance of setting out the plants. This gives the ground time to settle.

Plant only pot-grown specimens, preferably in early spring. Clematises do not transplant well from the open ground. Saturate the root balls an hour or so before planting. Set the plants about an inch deeper than they were in the pots, taking care not to twist, crack, or break the slender stems. This is very likely to happen if the supporting stake becomes loose and disengages from the root ball. Firm the soil well around the roots. Then water thoroughly. A considerable time will elapse before new roots penetrate deeply into the earth or far from the root ball. Make sure that the newly set plants do not suffer from lack of moisture during their first summer. In dry weather drench the ground to a depth of several inches at weekly intervals.

Routine care of established clematises consists of keeping weeds and vigorous neighbor plants from interfering, of maintaining for 4 feet or more around the bases of the plants a mulch of compost, leaf mold, peat moss or other suitable organic material, of watering at regular intervals in dry weather, and of fertilizing in early spring and perhaps once later in the season. Wooden supports may need replacing from time to time, and of course there is pruning.

Pruning must be adjusted to particular kinds of clematises and to the spaces they are to occupy. Do not prune early-bloomers such as *C. alpina, C. armandii, C. macropetala,* and *C. montana* until they have finished blooming. Then without delay do

any cutting you feel is needed to contain them to their allotted space and prevent their stems from becoming hopelessly crowded and tangled. But do no more pruning than that. Refrain also from pruning such early-blooming large-flowering hybrids as 'Belle of Woking', 'Nelly Moser', and 'Mrs. Cholmondeley' in spring. These bear their blooms on short branches from shoots developed the previous year. Cutting out live stems in spring reduces the display of bloom. Kinds, such as 'Comtesse de Bouchard', *C. jackmanii,* 'Mme. Edouard Andre', and 'Mme. Baron-Veillard', that bloom at midsummer or later on shoots of the current year may be cut back in spring to a height of 2½ or 3 feet. This discourages the development of top-heavy specimens with skinny, naked supporting stems. If you do not wish to treat the entire vine as severely as this, prune some of the stems low, others at higher levels.

Propagation of vining clematises is by seed, grafting, cuttings, and layering. The last, although the slowest and least productive, is the surest and easiest method for many amateurs. It is best done in early spring. Seeds in the main are practicable only for multiplying species, not garden varieties and hybrids. Those of some kinds germinate readily, those of others may take a year or longer to sprout. Grafting calls for greenhouse facilities and a fair degree of technical skill. Understocks employed are seedlings or thick pieces of root of almost any vigorous species or variety. Pieces of stem consisting of a single node with one or two buds and a 2- to 4-inch length of stem below it are used as scions. Cuttings of some clematises root more readily than those of others. It is not necessary to make the basal cut beneath a node. Internodal cuttings are as satisfactory if not more so than nodal ones. Plant the cuttings with the node at surface level in a propagating bench in a greenhouse or under mist. Nonvining, clump-forming clematises can be increased by division in early fall or early spring. They can also be transplanted from the open ground with more certainty of success than with most vining clematises.

Pests and Diseases. Insects and other pests that sometimes invade clematises include leaf-eating beetles, crown and root borers, mites, nematodes, scales, and whitefly. The chief diseases are leaf spot and stem rot, some other leaf spots, leaf blight, smut, powdery mildew, and rusts.

CLEOME (Cleò-me)—Spider Flower, Bladderpod. The common spider flower is by far the most familiar of the 200 or so species of this genus of the caper family CAPPARIDACEAE. Mainly, the group is tropical and subtropical, with most of its species native of Africa and the Americas, but outlying representatives extend into temperate parts of the United States and elsewhere. The name is the ancient Latin one of some plant of the mustard family CRU-CIFERAE.

Cleomes are chiefly annuals, but some are shrubs or small trees. Prevailingly ill-scented, they have alternate leaves, sometimes undivided, but more usually of three to seven leaflets that spread in handlike fashion. The purplish, white, yellow, or greenish, asymmetrical flowers are commonly in elongating racemes. They have four sepals, joined or not at their bases, four petals, six or more rarely four, long slender stamens, and a short style or none. The fruits are capsules containing several to many seeds.

The common spider flower (*C. hasslerana*) is often misidentified as *C. spinosa, C. pungens, C. arborea,* and *C. gigantea,* all

pods that spray outward from the stems in all directions.

A good bee plant, **C. serrulata,** native from Indiana to Saskatchewan, Missouri, and Arizona, is 1½ to 3 feet tall. It has branched, nonprickly stems, hairless or with fine hairs when young, and leaves with three narrowly-lanceolate leaflets up to 1½ inches long, hairless at maturity, but sometimes sparsely-pubescent when young. The white, pink, or purplish flowers, with petals under ½ inch in length, the lower parts of which narrow into slender claws, are in dense racemes up to 10 inches long. The long-stalked seed pods are about ¾ inch long. This is an annual.

Yellow-flowered **C. lutea** is similar in habit to the last, but has leaves of five to seven leaflets, and attractive yellow blooms

Cleome isomeris

Cleome hasslerana

names that belong to other species, not all cultivated. Native of tropical America and naturalized in parts of the United States, the spider flower is a coarse, vigorous, strong-scented, clammy-pubescent annual 3 to 5 feet tall, with stout, erect stems usually with short prickles at the bases of the leaves. The latter, all except the uppermost, are long-stalked and have five to seven oblong-lanceolate leaflets. The bracts at the bases of the individual flower stalks are undivided. Very numerous and opening in succession upward over a long season, the flowers, in loose, terminal racemes, are on long, slender, individual stalks. In the typical species they are a rather unattractive purplish-pink, but those of *C. h. alba* and of the similar or perhaps identical 'Helen Campbell' are a good white, variety 'Pink Queen' has blooms of a very pleasing pure soft pink, and there are other named horticultural varieties of distinct hues. In hot, sunny weather the petals, about ¾ inch long, slender below and with broader upper halves, curl by day and open fully only in the evening. The six long, spreading stamens give to the blooms the spidery appearance alluded to in the common name. When the flowers fade they are succeeded by stalked, slender

Cleome lutea

with petals ¼ to ⅓ inch long. Native from Nebraska to Washington, New Mexico, and California, this is an annual.

The bladderpod (**C. isomeris** syn. *Isomeris arborea*) is very different from the other sorts dealt with here. A broadly-rounded, dense or open shrub 2 to 5 or exceptionally 8 feet tall, this variable native of the deserts of California and Baja California frequently inhabits subalkaline soils. It has brittle, widely-spreading branches and ill-scented foliage. Alternate, stalked, and with three toothless, oblong to elliptic-oblanceolate leaflets ½ inch to 1½ inches long and up to ⅓ inch wide, its leaves are dark to grayish-green. In dense axillary racemes up to 6 inches long, the showy flowers, which expand at almost any season, are yellow. They have a four-parted calyx, four oblongish petals ⅓ to a little more than ½ inch long, the two lower spreading more widely than the upper, six long-protruding stamens, and a short style with a round stigma. The fruits are stalked, pendulous, pointed, leathery capsules up to 1¼ inches long by slightly more than ½ inch wide.

Garden and Landscape Uses. Annual cleomes are easily grown. They are useful for flower beds, for filling gaps at the fronts of shrub borders, as cut flowers, and as temporary screens. For satisfactory effects they must be in groups; single specimens make little show. Sunny locations, and ordinary, moderately fertile, well-drained soil are to their liking. They prefer sandy soils to those of a clayey nature.

Cultivation. Cleomes are raised from seed sown in early spring outdoors or from indoor sowings made some eight weeks before it is safe to transplant the young plants to the garden, which may be done as soon as there is no longer danger of frost. If seed is sown outdoors, the seedlings may be thinned to the required distance apart or transplanted. Those from indoor sowings are transplanted 2 to 3 inches apart in flats, or individually into 3- or 4-inch pots, and are accommodated in a sunny greenhouse or approximately similar environment, where the night temperature is about 55°F and that by day is five to ten degrees higher. Outdoors the plants are spaced 1 foot to 1½ feet apart. Little or no staking is needed, and other care is minimal. They bloom for long periods in summer and fall. Cleomes often self-sow. Plants from self-sown seeds of improved varieties of spider flower may revert to the unattractive flower color of the typical species. Such mavericks should be weeded out as soon as their flower color shows.

CLERODENDRUM (Clero-déndrum)—Glory Bower. Formerly spelled *Clerodendron,* this genus of chiefly natives of the tropics and subtropics of the eastern hemisphere consists of 400 evergreen and deciduous vines, shrubs, and trees, many with showy, handsome blooms, some with attractive fruits. It belongs in the vervain family VER-BENACEAE. The name of the genus *Clerodendrum,* derived from the Greek *kleros,* chance, and *dendron,* a tree, is of uncertain application.

Clerodendrums have undivided, smooth-edged, or sometimes lobed or toothed leaves, opposite or in whorls (circles of more than two). The flowers, generally clustered or in panicles, and white, blue, violet, orange, or red, have bell-shaped or less often tubular calyxes with fine shallow teeth or lobes. The corollas have a slender tube and five somewhat unequal, spreading petals. There are four long-protruding, curved stamens and a style that also protrudes conspicuously and is shortly two-branched at its apex. The fruits, technically drupes, are berry-like. Often they are enclosed in the persistent calyxes.

Vining kinds include popular **C. thomsonae** (syn. *C. balfouri*), of West Africa. Most familiar to American gardeners, this

Clerodendrum thomsonae

is a vigorous, evergreen twiner up to about 20 feet in height. It has hairless shoots and opposite, lustrous, pointed, ovate-elliptic, hairless or almost hairless, thin leaves 3 to 6 inches long by up to 3 inches wide. The flowers, in loose, panicle-like clusters from the ends of the branches, have showy, large, creamy-white, strongly-five-angled, baglike, heart-shaped calyxes of erect sepals from which slightly protrude the slender-tubed, approximately 1-inch-long, crimson corollas. The calyxes remain after the corollas fade, gradually becoming pink or violet-pink. The fruits are fleshy and blue. Variety *C. t. variegatum* has leaves edged with light yellow-green, their center portions marbled with light and darker green. West African **C. splendens** is a lower

Clerodendrum splendens

climber than *C. thomsonae*. It has hairless shoots and lustrous, wrinkled, ovate to lanceolate, usually hairless leaves up to 6 inches long by 3½ inches wide. Set fairly closely and many together in terminal panicles about 4 inches across, the flowers, bright red to scarlet or sometimes yellow or white, have corolla tubes ¾ to 1 inch long. After the corollas fall the calyxes become purplish-pink or red.

A semiclimbing, thorny South American shrub, prickly-myrtle (**C. aculeatum**) has hairy shoots, long, arching branches, and slender-stalked, oblong to elliptic-obovate leaves 1 inch to 2 inches in length. In clusters terminating the main stems and side branches, the ½-inch-wide, ¾-inch-long flowers are white with reddish-purple stamens. The fruits are up to ⅓ inch in diameter, grooved and spherical.

Red-flowered, tropical shrubs among clerodendrums include several showy ornamentals. Best known and most popular, **C. speciosissimum** (syns. *C. fallax, C. buchananii fallax*) first blooms when only a foot

Clerodendrum speciosissimum

or so high. Native to Java, this erect, bushy shrub may attain an eventual height of 12 feet. It has bluntly-squarish stems and opposite, pointed, toothed or toothless, broad-heart-shaped, densely-downy leaves, with blades up to 1 foot long and stout, hairy stalks. The showy, brilliant scarlet flowers are in erect terminal panicles, 6 inches to 1½ feet in length. Individual blooms have corollas ¾ inch long by 1½ to 2 inches wide. From the last, handsome, tropical eastern Asian **C. paniculatum** differs in having lobed, hairless leaves and flowers with corolla tubes only ½ inch long. Its fiery-red blooms are in tiered, much-branched, pyramidal, pagoda-like terminal panicles up to 1 foot long and almost as wide. This kind has three- to five-lobed, roundish to ovate, hairless leaves, with blades up to 8 inches long. Of the same relationship and also with bright red flowers are shrubby *C. japonicum* and Chinese

Clerodendrum paniculatum

C. kaempferi. Not infrequently, in cultivation, traveling under the name of the former, **C. kaempferi** differs in its blooms having shorter calyxes. These are shrubs 3 to 6 feet in height with large, heart-shaped, toothed, but rarely lobed leaves, and big terminal panicles of bloom. The fruits are red.

Chinese **C. bungei** is an attractive, distinctive shrub 4 to 6 feet or so tall, naturalized in warm parts of the United States and in Mexico. It has heart-shaped, coarsely-toothed leaves 4 to 8 inches long and nearly as wide. Their under surfaces are hairy. The purple-red blooms, in dense rounded heads 4 to 5 inches in diameter, are displayed in late summer. The foliage when crushed has an objectionable odor. Even if the shoots are killed to the ground in winter, so long as the roots live, new shoots develop that bloom the first year.

White or pink-tinged, sweetly and strongly fragrant double flowers are characteristic of **C. fragrans pleniflorum** (syn. *C. philippinum*). A bushy shrub up to about 8 feet tall, this is a variant of a single-flowered species native to Japan and China.

Clerodendrum bungei

Clerodendrum fragrans pleniflorum

Clerodendrum schweinfurthii bakeri, in fruit

The downy, angled stems carry pointed-roundish-ovate, stout-stalked, shallowly-toothed leaves with blades 3 to 5 inches long. The many-petaled blooms are in densely-crowded terminal heads 3 to 4 inches wide. They have ½-inch-long, purplish to red calyxes. Single-flowered *C. fragrans* (syn. *C. philippinum simplex*) is much less commonly cultivated.

Another clerodendrum with fragrant, white blooms, **C. schweinfurthii bakeri,** sometimes cultivated as *C. bakeri,* differs markedly from all other kinds discussed here in that its 6-inch-wide, crowded clusters of blooms arise from the main stems, sometimes near ground level. A shrub, this native of tropical Africa has downy shoots and oblong to elliptic or slightly-obovate, slender-pointed, coarsely-toothed leaves 3 to 8 inches long. The flowers, approximately 1¼ inches long and wide,

have slender corolla tubes. The attractive fruits, green at first and becoming black at maturity, are seated in creamy-white cups.

Other prevailingly white-flowered tropicals are cultivated. One is *C. colebrookianum* (syn. *C. glandulosum*), an Asian shrub from 5 to 8 feet tall. It has more or less heart-shaped, lobeless and toothless leaves 9 inches to 1½ feet long by almost as wide as their lengths. The terminal and axillary panicles are of numerous, densely-arranged, slender-tubed flowers about 1 inch long. The fruits are blue, with a red, fleshy calyx attached. South African and Madagascan *C. glabrum* is a free-flowering shrub or small tree 4 to 15 feet tall. It has opposite or whorled, essentially hairless, ovate, elliptic, or lanceolate, lustrous leaves ½ inch to 4 inches long by ½ inch to 3 inches wide. The heads of flowers, terminal and at the ends of lateral shoots, are 2

to 2½ inches across, compact, and of many small white blooms. The Turk's head or Turk's turban (*C. indicum* syn. *C. siphonanthus*), of southeastern Asia, is a willowy shrub or small tree. It has opposite or whorled, stout-stalked, pointed, oblong, elliptic, or oblanceolate, toothless, hairless leaves 3½ inches to nearly 1 foot long by up to slightly over 2 inches wide. The flowers, fragrant and opening at night, are in erect panicles up to 1½ feet long or longer by 10 inches wide. White, whitish, or yellow, they have corolla tubes 4 to 6 inches long. The faces of the blooms are almost or quite ¾ inch across, with the petals strongly reflexed. The purplish fruits, ¼ to ½ inch in diameter, have reddish calyxes. A shrub about 10 feet tall, *C. macrostegium* is a native of the Philippine Islands, the Moluccas, and Amboina. It has pointed-ovate, toothed or toothless, velvety leaves up to 10 inches long by 6 inches wide. The clusters of scentless blooms are erect, with conspicuous, whitish-tipped, purplish-blue, leafy bracts. The flowers have whitish calyxes, pure white corollas with purple throats, and white stamens. Native of the Celebes and the Philippine Islands, *C. minahassae* is a shrub or tree up to about 20 feet tall. It has pointed, hairless, oblong-ovate to elliptic leaves with blades 4 to 11 inches long by up to 6½ inches wide. The white or cream, pretty flowers are in terminal panicles of few. Fragrant, they have green calyxes that as they age become red or purple, slender corolla tubes 3 to 3½ inches long, and spreading petals ¾ to 1 inch long. The fruits, about ⅜ inch in diameter, are deep blue. Often grown under the name *C. nutans,* greenish-white-flowered *C. wallichii,* of Fiji and Samoa, has pointed-narrow-ovate leaves 4 to 9 inches long by up to ¾ inch wide, and more or less pendulous, slender, loose panicles of blooms. The flowers are about 1 inch long and 1 inch wide or wider.

Bright blue and violet-blue bicolored flowers are characteristic of **C. ugandense**

Clerodendrum schweinfurthii bakeri

Clerodendrum wallichii

Clerodendrum ugandense

of tropical Africa. A loose, slender-branched shrub 3 to 10 feet tall, this kind has thin, hairless, more or less coarsely-toothed, elliptic to narrowly-obovate leaves 1½ to 4½ inches long by about one-half as wide. In terminal panicles 6 inches or so long and nearly as wide, the daintily-poised blooms suggest a flight of small, brightly-colored butterflies. They have short-tubed corollas with one violet-blue and four light blue petals. The long-protruding, conspicuously-arched stamens are blue. The fruits are black and fleshy. Also with partly blue flowers, but as known in cultivation inferior to the last, African **C. myricoides** is a variable species, an erect or sprawling shrub 3 to 10 feet tall. It has thin, pointed, oblong to ovate leaves, coarsely- or slightly-lobed, 1½ to 3 inches long by up to 2 inches wide. The four upper petals are white, the lower one usually pale blue.

Hardy clerodendrums are not as well known as their merits deserve. Northern gardeners familiar with tropical or subtropical kinds are often unacquainted with *C. trichotomum* and its varieties, which are hardy in southern New York and southern New England. These are deciduous shrubs or small trees that toward the northern limits of their ability to survive may be killed to the ground each winter, but in spring send up abundant new shoots that bloom the same summer.

Clerodendrum myricoides

Clerodendrum trichotomum

Native of Japan and China, **C. trichotomum** is esteemed for its sweetly-fragrant blooms borne in large clusters over a long period in late summer and early fall and for the decorative fruits that succeed them. In more favorable climates this sort becomes a shrub or tree 10 to 20 feet tall, but in the vicinity of New York City it is not likely to exceed 10 to 12 feet. It has pithy stems and branches, downy when young, and stalked leaves, larger on young vigorous specimens than on older ones, with pointed-ovate blades 4 to 10 inches long by about one-half as wide as their lengths. Of soft texture, they are downy on their undersides and stalks, and sometimes are slightly lobed. The loose clusters of flowers, up to 10 inches across, come from the upper leaf axils. From 1 inch to 1½ inches across, they have inflated, reddish calyxes and white, spreading petals. Pea-sized and at first bright blue, a hue that contrasts pleasingly with the persistent calyxes, the fruits turn black as they ripen. Hardier than the species and in fact the most cold-resistant clerodendrum, *C. t. fargesii* (syn. *C. fargesii*), a native of China, differs from the species in not becoming treelike and in its smaller, brighter green leaves being nearly hairless and when they first unfold tinged with reddish-purple. Variety *C. t. ferrugineum* is distinguished by its very dense, almost shaggy hairiness. The species *C. trichotomum* and its varieties spread by suckers that grow from the roots.

Garden and Landscape Uses. Except for *C. trichotomum* and its varieties, which are

Clerodendrum trichotomum fargesii

hardy, and somewhat less hardy, but moderately cold-tolerant *C. bungei*, clerodendrums are generally useful as outdoor ornamentals only in the tropics and subtropics. The great majority need considerable warmth and preferably fairly high humidity for their best performance. A few, notably pretty *C. ugandense*, *C. glabrum*, and *C. bungei*, do well in frostless or nearly frostless Mediterranean-type climates. None is very demanding of the gardener. So long as the climate suits, clerodendrums respond in any reasonably fertile soil. Most prefer just a little shade from the fullest intensity of the sun, but even that they are likely to endure if the ground is not deficient in moisture.

The vines are obvious adornments for pillars, posts, arches, trellises, and like supports that afford opportunity for stems to twine. Shrub and small tree sorts are attractive furnishings for beds and borders and for foundation and screen plantings. The rears of flower beds afford effective sites for *C. ugandense*. Good hedges are fashioned from *C. aculeatum*, a kind planted in some places to check soil erosion.

In greenhouses and conservatories, in ground beds and pots, all kinds can be grown. Most popular are *C. fragrans pleniflorum*, *C. speciossimum*, *C. splendens*, *C. thomsoniae*, *C. ugandense*, and *C. umbellatum speciosum*. Some are occasionally cultivated as window plants.

Cultivation. Seeds, which germinate readily, and cuttings are the usual means of propagation. Cuttings mostly root readily in a warm, humid atmosphere, the more surely if a little bottom heat is supplied. Use firm, but not hard shoots for cuttings. Those of vining kinds are generally a little more reluctant to root than cuttings of shrubby sorts. Best results are from cuttings made of well-ripened, fairly firm wood taken from some distance below the terminations of the shoots. Other methods of increase are by suckers and root cuttings, and by layering.

Soil well-drained and fertile, this last especially important for large-leaved, tropical kinds, suits clerodendrums best. It should be reasonably moist, not wet. In poorish earths regular fertilizing is helpful, as in dry soils are periodic deep soakings when rainfall is inadequate.

Pruning in tropical and subtropical places is usually done, as needed, as soon as flowering is through. Indoors in the north it is better to defer it until late winter or early spring. With vining kinds it consists of thinning out old shoots to reduce crowding, and training in new ones. Shrubby kinds that have reached the sizes at which they are to be maintained are kept compact and shapely by cutting hard back the shoots of the previous year's growth.

Indoors water freely from spring to fall, but in winter keep the soil considerably drier even to the extent that the leaves wilt a little on occasion and some fall. Also, keep them cooler then. A night temperature of 50 to 55°F suits most and for subtropical kinds, such as *C. bungei, C. glabrum,* and *C. ugandense,* 45 to 50°F is adequate. Daytime temperatures may be five to ten degrees higher.

Repot in late winter or early spring, or even as early as mid-January specimens of *C. thomsonae* that are wanted in bloom by late April or May. Prick out from the root balls with a pointed stick as much old soil as can be dislodged without too great damage to the roots, and repot into the same size or bigger containers according to need. Use a loamy mix, coarse rather than finely sifted, and with a potting stick pack it quite firmly, but not rock-hard. Firm potting encourages desirable short-jointed growth.

After potting, put the plants in a cozily-warm, humid place. Water moderately, but not excessively, and spray the tops once, twice, or more times daily with water to stimulate new growth. Repotting into larger containers may be needed in early summer. Do this without disturbing the roots.

Pests and Diseases. These are not usually very troublesome on clerodendrums. The chief pests are mealybugs, scale insects, red spider mites, and occasionally, root nematodes. Leaf spot occasionally affect the foliage and so, infrequently, does a mosaic leaf spot disease.

CLETHRA (Cléth-ra)—White-Alder, Sweet-Pepperbush or Summer Sweet. The only genus of the white-alder family CLETHRA-CEAE is *Clethra.* This comprises thirty to forty, or according to some authorities, more than 100 species of deciduous and evergreen shrubs and small trees, natives of eastern North America, eastern Asia, and Madeira. The name is the ancient Greek one for the alder (*Alnus*), probably applied because of the resemblances of the leaves of some kinds.

Clethras have alternate, stalked, undivided, obovate, ovate, or oblong, sharply-toothed leaves and usually fragrant, white or pinkish flowers, with petals approximately ⅓ inch long, in crowded panicles or racemes at the branch ends. The flowers originate in the axils of densely-hairy bracts and are on stalks that lengthen as the blooms mature. They have a deeply-five-parted, persistent calyx, five separate or nearly separate, erect petals, ten stamens with somewhat flattened stalks and inverted-arrow-shaped anthers, and one three-lobed style. The fruits are many-seeded, spherical or subspherical capsules.

Sweet-pepperbush or summer sweet (*C. alnifolia*) is an extremely hardy, deciduous shrub up to 10 feet tall that forms clumps by suckering from its base. It has short-stalked, hairless or nearly hairless, obovate-oblong, blunt or pointed leaves 2 to 4 inches long, and mostly toothed only along the upper parts of the margins. In erect, densely-short-pubescent racemes or panicles 2 to 8 inches long, the white flowers have stamens with hairless stalks. This native of swamps and moist woodlands is wild from Maine to Florida and Texas. Plants that have their flowers in panicles instead of unbranched racemes are sometimes distinguished as *C. a. paniculata.* They are more ornamental than the others. Occurring sporadically wild, *C. a. rosea* has pink flower buds and pinkish blooms. A deciduous native of the coastal plain from North Carolina to Louisiana, *C. tomentosa* differs from the sweet-pepperbush chiefly in the undersides of its leaves being white-hairy. It is hardy as far north as southern New England.

From other American clethras deciduous *C. acuminata* is distinguished by its short-taper-pointed, broadly-elliptic to oblong leaves, toothed along most of their lengths, and 2 to 6 inches long, and by the stamens of its white flowers, which are in solitary, nodding racemes, being hairy. This endemic of rich mountain woods from Pennsylvania to West Virginia, Kentucky, and Georgia, is less beautiful in bloom than other cultivated clethras, but has the advantage of having attractive, smooth, glossy, cinnamon-brown stems that show to especially good advantage in winter. It attains heights of about 18 feet and is hardy in southern New England.

Among Asian clethras cultivated, *C. barbinervis,* of Japan and Korea, is one of the most admirable. A deciduous shrub or tree, of coarser growth than the other kinds described here, and up to 30 feet in height, it has leaves broadest toward their apexes, and less fragrant blooms than the sweet-pepperbush. The flowers are white and in panicles 4 to 6 inches long held horizontally rather than erectly.

Other deciduous Asian kinds related to *C. barbinervis* are *C. delavayi, C. fargesii,* and *C. monostachya.* Probably none of these is hardy in the north. Except for *C. arborea* the most beautiful of clethras, *C. delavayi* is a shrub or tree up to 10 feet or perhaps more in height, with elliptic-oblong to oblanceolate leaves 2½ to 6 inches long 1 inch to 1½ inches wide, and very downy on their lower surfaces, and panicled, dense racemes, 4 to 8 inches long of white flowers. The leaves of *C. fargesii,* 4 to 6 inches long, are broadest toward their bases, and elliptic- to ovate-lanceolate. The white flowers are in crowded racemes up to 10 inches long. Similar to the last, but its flowers having hairy instead of hairless styles, *C. monostachya,* up to 20 feet tall, has toothed leaves, tapering to both ends, 3 to 5½ inches long by 1 inch to 2 inches wide. Their undersides have small tufts of hair in the axils of the veins. The usually solitary racemes of white flowers are up to 8 inches in length.

Evergreen *C. arborea* is not hardy in the north. Native of Madeira, this beautiful species is 20 to 25 feet high and has nodding, 6-inch-long racemes of small white flowers. The finely-toothed, nearly hairless, oblanceolate to narrowly-elliptic leaves

Clethra alnifolia

are 3 to 6 inches long by 1 inch to 2 inches wide, and glossy on their upper sides. The fragrant, white blooms are in panicles of racemes 3 to 6 inches long. This species stands only a few degrees of frost.

Garden and Landscape Uses. Deciduous clethras are prized because they bear their fragrant blooms in summer, later than those of most flowering shrubs, and for the spicy fragrance of their blooms and the warm yellow to orange of their autumn foliage. The sweet-pepperbush is especially well adapted for planting near the sea. It has the disadvantage, in dryish soil, of being very susceptible to infestations of red spider mites, a pest apparently less fond of Japanese *C. barbinervis,* and not as likely to infest sweet-pepperbushes grown in moist soils. Clethras can be employed effectively for a variety of purposes, in groups, in beds with other trees and shrubs, and singly. They look well at watersides and at the fringes of woodlands. They thrive in full sun and where there is a little shade.

Cultivation. Clethras are easy to grow provided the soil is somewhat acid and does not lack for moisture. They are less well suited for clayey earth than those of a peaty, sandy character. A mulch of organic matter, such as compost or peat moss, maintained about them is helpful. No regular, systematic pruning is needed, but judicious thinning out of older branches at intervals often improves the appearance of these shrubs. Lanky, overgrown specimens can be rejuvenated by cutting them down in late winter to within a few inches of the ground, and following this by fertilizing, and keeping them well watered during dry spells that may occur during the summer. No flowers will be borne for two or three years following such drastic pruning. Because the faded racemes of blooms remain and look untidy, unless seed is to be collected it is good garden housekeeping to pick these off as soon as they cease to be decorative. Clethras are easy to raise from seeds sown in a greenhouse or cold frame in sandy peaty soil kept moist. Cuttings can also be used for propagation. Those of deciduous kinds taken from plants forced into early growth in a greenhouse root most readily. The best cuttings consist of short side shoots with a heel (sliver of the shoot from which they sprang) attached at their bases. Layering, especially appropriate for evergreen species, affords another way of increase. Clump-forming kinds such as the sweet-pepperbush can be divided.

CLETHRACEAE—White-Alder Family. This family of dicotyledons consists of only the genus *Clethra.* For its characteristics see the above entry.

CLEYERA (Clèy-era). In gardens confusion sometimes occurs in distinguishing *Cleyera*

from closely related *Eurya,* both of the tea family THEACEAE. It happens that the specific epithet *japonica* has been applied in both genera so that there are two plants, *Cleyera japonica* and *Eurya japonica.* From *Eurya,* which has unisexual flowers with the sexes on separate plants, *Cleyera* differs in having bisexual blooms with bristly-hairy anthers. The name commemorates Andrew Cleyer, a Dutch physician and botanist, who died in 1688.

Cleyeras are evergreen trees and shrubs. One species inhabits eastern and southern Asia, sixteen are natives from Mexico to South America. They have alternate, stalked, smooth-edged or toothed leaves and, from the leaf axils, solitary or clustered, stalked flowers. The blooms have five sepals of unequal size, five petals, at their bases slightly or not joined, and about twenty-five stamens. The fruits are berries.

Suitable for planting in the south about as far north as Georgia and South Carolina, **C. japonica** (syn. *Eurya ochnacea*) is indigenous in Japan, the Ryukyu Islands, Taiwan, and China. A shrub 4 to 6 feet tall, or in the wild a small tree, it has leathery, short-stalked, narrowly-oblong to ovate-oblong, usually toothless leaves 3 to 4 inches or sometimes more in length, ¾ inch to 1½ inches wide, lustrous dark green on their upper sides, and paler beneath. Solitary or in twos or threes, the fragrant white flowers, produced in late summer, yellow as they age. They are ½ inch to slightly more across and have rather thick petals, stamens with white anthers, and styles with two or three short branches. The fruits are dark red to black, and spherical. Variety *C. j. tricolor* (syn. *C. fortunei*) formerly grown as *Eurya latifolia variegata,* has dark green leaves marked with gray, and along the margins, especially of younger leaves, variegated with creamy-white and red.

Garden and Landscape Uses. This species and its variety are handsome. In cultivation they are slow-growing shrubs chiefly admired for their good growth habits and beautiful foliage, which when it first unfolds is reddish. Their flower and

Cleyera japonica tricolor

fruit displays are less important. The chief uses of these plants are in borders and foundation plantings and in single specimen displays. They succeed in sun or shade in ordinary soils not excessively dry, and endure poor soil drainage remarkably well. They are sometimes cultivated in conservatories and greenhouses.

Cultivation. In subtropical and warm-temperate regions no particular difficulties attend the cultivation of cleyeras. Regular pruning is not required. Any little needed to keep the bushes shapely or restrict them to size is best done in spring. Propagation is by cuttings, 3 to 4 inches long, taken in late summer and planted under mist or in a cold frame or greenhouse propagating bench, and by seeds sown in sandy peaty soil.

In greenhouses cleyeras may be accommodated in large pots or tubs or in ground beds. They grow well in ordinary, fertile, porous soil kept evenly moist, but not constantly saturated. It is beneficial to stand container-grown specimens outdoors in summer. A winter night temperature of about 50°F is satisfactory, with a few degrees more permitted by day. Specimens that have filled their pots with roots benefit from regular applications of dilute liquid fertilizer from spring to fall.

CLIANTHUS (Cli-ánthus)—Parrot's Bill or Parrot Beak or Red Kowhai, Glory-Pea or Sturt's Desert-Pea. One endemic species each of Australia and New Zealand are usually considered to constitute *Clianthus* of the pea family LEGUMINOSAE. Some botanists record a total of three species. They are gorgeous in bloom. The name is from the Greek *kleios,* glory, and *anthos,* a flower. These are frost-tender, evergreen shrubs of more or less lax or trailing habit, with pinnate leaves of an odd number of leaflets. Their extraordinary flowers are in racemes from the axils of the leaves. They have large, upturned standard petals and downturned keels and are much longer than wide. They are structured like the blooms of peas, but are quite different in appearance. They somewhat resemble the flowers of *Erythrina.* Their stamens form a group of nine, and one separate. The fruits are cylindrical, many-seeded pods.

The parrot's bill, parrot beak, or red kowhai (**C. puniceus**), a rare native of New Zealand cultivated by the Maoris before the arrival of Europeans, is one of the few elements in the native flora of its homeland with brilliantly-colored flowers. It is a lax undershrub, 3 to 6 feet tall, with somewhat silky, hairy stems, and hairless or nearly hairless, glossy leaves 3 to 6 inches long, with seventeen to twenty-nine leaflets. The leaflets are ½ to 1 inch long, short-stalked, and linear-oblong and have blunt or slightly notched ends. The flowers, in racemes of six to eight, are 3 inches or more in length. They are crim-

Clianthus puniceus

son, with the standard petal slightly marked toward its center with white. As they age the flower color fades, and tends to become magenta, especially in sun. Variety *C. p. albus* has white blooms.

The glory-pea or Sturt's desert-pea (*C. formosus* syn. *C. dampieri*) is native to the drier parts of Australia. It differs from the last in having much more conspicuously

Clianthus formosus

hairy stems and softly-silky-hairy, sometimes slightly-reddish-tinged, gray foliage. Its blooms, more handsome than those of the parrot's bill, are in racemes of four to six. Its leaves have from fifteen to twenty-one, usually pointed, oblong to elliptic leaflets ½ to 1 inch long. The flowers are glossy, about 3 inches long, and brilliant scarlet with a large purple-black blotch at the base of the erect standard petal. The oblong, hairy, tough seed pods are about 2½ inches long.

Garden and Landscape Uses. The parrot's bill is a satisfactory winter- and spring-blooming outdoor shrub in warm, sunny locations in many parts of California. Even if cut to the ground by frost each year, provided its roots live, it sends forth new shoots that bloom the same season. It

flowers for a long period and does well in partial shade. It may be trained as a vine against a wall or on a trellis or be maintained as a bush. This species can also be grown in cool greenhouses. The glory-pea is less easy to satisfy. Probably it could be grown outdoors in semidesert areas, but usually it is regarded as a greenhouse plant and one that tests the skill of even the most experienced gardeners. It is usually treated as an annual or a biennial.

Cultivation. For the parrot's bill a warm location and thoroughly well-drained soil provide requisite growing conditions. After it is through blooming it is pruned to encourage the production of new shoots that flower the following year. Propagation is by cuttings and by seeds. The glory-pea is raised from seed and is grown in well-drained sandy soil in ground beds, pots, or hanging baskets. British experience is that specimens grown on their own roots are likely to reach a fairly large size and then perish, but this has not happened to plants grown in greenhouses in the vicinity of New York City. Quite possibly the greater intensity of the sunlight they receive, as compared to that of England, is responsible for their persistence. To overcome the problem of the plants dying when partially grown British gardeners recommend grafting while both species are in the earliest seedling stage, the glory-pea onto *Colutea arborescens*. This seems to give them superior vigor and lessen the danger of collapse later. At all times the glory-pea must be grown in full sun and extreme care taken to keep the soil dryish rather than wet. A night temperature of 50°F, with an increase of five to ten degrees by day, is recommended. On all favorable occasions the greenhouse must be freely ventilated. Care must be taken that a heavy, humid atmosphere is not maintained.

CLIFF-BRAKE. See Onychium.

CLIFTONIA (Cliftòn-ia)—Buckwheat Tree or Titi. An endemic of swampy ground from Georgia to Florida and Louisiana, the only species of *Cliftonia* is rarely cultivated. It is an evergreen shrub or tree, occasionally up to 50 feet tall, but usually lower, of the cyrilla family CYRILLACEAE. Its name commemorates Dr. Francis Clifton, who died in 1736.

The buckwheat tree or titi (*C. monophylla* syn. *C. ligustrina*) has alternate, short-stalked, hairless leaves, bluntly-oblong-lanceolate, up to 2 inches long, and with dark green upper surfaces. Produced in early spring, the fragrant, white to pinkish flowers, in terminal racemes, have usually five, rarely up to eight, petals much longer than the sepals. The blooms are about ⅓ inch wide and have ten stamens shorter than the petals. The nearly stalkless stigma is three- to four-lobed. Egg-

shaped, the three- or four-winged, buckwheat-like fruits are about ¼ inch long.

Garden and Landscape Uses and Cultivation. This native American deserves more consideration from gardeners. It is hardy about as far north as Philadelphia and blooms early. For its best comfort it needs fairly moist soil with a reasonable organic content. No special care is needed beyond pruning an occasional misplaced branch to keep the specimen shapely. Propagation is by seeds and by late summer cuttings under mist or in a greenhouse propagating bench. The buckwheat tree blooms as a young specimen when only 3 to 4 feet tall.

CLIMATE. Defined as the composite or general weather conditions of a particular region or place averaged over a number of years, climate has a profound influence in determining which plants can be grown outdoors in prescribed areas and the best methods and timings to achieve satisfactory results with them. Components considered in appraising climates include temperatures, relative humidity, precipitation, evaporation, wind, sunshine, cloudiness, and barometric pressure, the last of lesser importance to plant growth than the others. These factors are not only important singly, but also in relation to each other and in the patterns they display throughout the year. For example, Mediterranean climates are characterized by mild winters, with enough rain to encourage and support plant growth, and hot summers, with low humidity and little or no rainfall. Such conditions are typical of the Mediterranean region, California, and parts of South Africa. This pattern contrasts sharply with the climate of Florida where little rain falls during the mild winter months, much through hot, humid summers. Climates with great extremes of temperature between winter and summer, defined as continental, characterize northeastern and northcentral Europe and parts of Asia. In contrast, the much more equable climate of the British Isles, to a large extent the result of the tempering influence of the Gulf Stream, is oceanic or insular, so called because the climates of many islands are moderated by the large bodies of water that surround them. Desert and semidesert climates result from rainfall insufficient to support other than scanty populations of plants specially adapted to conserve moisture within their tissues and to live through long periods with little or no additional supplies. The annual evaporation in such places, the rate at which moisture is evaporated from an exposed surface of water, exceeds the annual rainfall.

Climate is usually the most important element in determining whether or not specific plants will survive in particular places, but it is not the only one. Others sometimes of equal or greater importance may include the type of soil and the presence or absence of serious pests and diseases. In considering climatic factors it must be remembered that it is often extremes rather than averages that determine what can be successfully grown. This is especially true of temperatures. A drop one winter night to below the critical level for a particular kind of plant will eliminate it no matter what the average winter temperature is. Plants that have withstood average lowest winter temperatures for many years are likely to be killed by a dip in occasional or rare winters to well below that level. Winters of insufficient snowfall in regions where heavy winter coverings are usual may result in the killing of herbaceous perennials that survive under the insulation of heavy blankets of snow. Some climatic factors can be compensated for to greater or lesser extents by such cultural practices as irrigating and watering to make up for insufficient rainfall, mulching and winter covering to insulate against cold and the ill effects of alternate freezing and thawing, and providing shelter belts or other intrusions to temper or deflect wind. In many instances success with vegetables and annual flowers, and some other plants depends upon selecting sowing dates suitable to the climate. The same kinds that do well in New England when sown in spring and that fail miserably if seeded then in Florida or California are likely to prosper in the latter places if sown in fall or early winter. In the north late August or September sowings of lawn grass seed are likely to give very much better results than if made two months earlier or even in the spring, although the latter sowing time is preferable to one made in late spring or early summer.

Gardeners, like farmers, must learn to accommodate to the limits set by their regional climates. Much relevant information about this is obtainable from the publications of the United States Weather Bureau and from State Agricultural Experiment Stations. Information about strictly local manifestations and accommodations should be sought from local gardeners and farmers and can often be deduced to some extent from observing the kinds of plants being grown successfully. In every garden minutely local conditions are responsible for microclimates within the area. These can be important. See Climates and Microclimates.

CLIMATES AND MICROCLIMATES. To a very large extent climate determines the kinds of plants that can be grown outdoors in any particular place. Its most important factors, because they are not subject to appreciable modification by gardeners, are minimum and maximum temperatures and length of the growing season. Palms will not survive winters in the Middle West nor the blue-poppy (*Meconopsis*) of the Himalayas summers near New York City. Watermelons need a longer growing season to develop and ripen their fruits than New England provides.

Certain aspects of climate are more subject to modification than outdoor temperatures and the length of the growing season. Insufficient rainfall and low humidity can be compensated for by irrigating and sprinkling or misting, excessive precipitation to some extent at least by draining the land. Unfortunately there is no practical means of reducing that bug-a-boo of so many plants, excessive humidity accompanying high temperatures. Shade can be arranged to modify summer temperatures, but it also of course reduces light intensity. Removal or reduction of shade accomplishes the reverse. Winter covering may be employed to substitute in part for a thick snow cover. Windbreaks give shelter from drying winds.

Microclimates, small but often significant local variations of climatic factors, exist within almost every garden. It is by taking fullest advantage of these that sensitive gardeners succeed with a variety of plants not generally considered easy in their regions. Ground that slopes to the north is considerably cooler than that with a similar tilt to the south and knowledgeable rock gardeners take advantage of this by giving plants that prefer cool summers, such as many gentians, primulas, and saxifragas, north-facing aspects and reserving sunny, south-inclined locations for such kinds as sedums, sempervivums, and yuccas that respond favorably to high temperatures.

For like reasons trees, shrubs, and vines for espaliering against walls should be selected with thought for the direction in which the wall faces and the tolerance or intolerance of particular plants for reflected heat. Many shrubs, herbaceous perennials, and bulb plants on the borderline of hardiness may prove successful at the base of a wall facing south to southwest, but would fail in less favorable spots.

Shelter from prevailing winds is often important to success and in some places, near the sea and on open plains for instance, may be critical to success. Such shelter may be provided by buildings, walls, fences, hedges, shelter belts, or other plantings or by dunes or higher land. Obviously hilltops are more windswept than locations below their summits.

Air drainage is often highly significant as a microclimate feature. Where adequate, as it usually is partway up slopes and in open flat lands, success may be had with cold-sensitive plants and others, such as many fruit trees that bloom early, that would not prosper at the foots of slopes or in hollows or frost pockets where on nights in winter and early spring cold air collects.

Knowing one's garden and the microclimates it affords and a sensitiveness to the environmental needs of particular plants are basic to the best success with gardening. Such knowledge must be based on careful observation and usually a certain amount of trial and error.

CLIMBER. This, chiefly British usage, is a name for sorts of plants that in North America are usually called vines. It is also applied to some weak-stemmed shrubs with a tendency to climb, such as climbing roses.

CLIMBING. The word climbing forms part of the common names of various plants. Examples include: Australian climbing-lily (*Luzuriaga*), climbing-butcher's-broom (*Semele androgyna*), climbing-dahlia (*Hidalgoa*), climbing-dogbane (*Trachelospermum*), climbing fern (*Lygodium*), climbing-fumitory (*Adlumia fungosa*), climbing-gentian (*Tripterospermum*), climbing hempweed (*Mikania scandens*), climbing hydrangea (*Hydrangea anomala petiolaris*), climbing-hydrangea (*Schizophragma hydrangeoides*), climbing-lily (*Gloriosa*), and climbing-ylang-ylang (*Artabotrys hexapetalus*).

CLINOPODIUM (Clino-pòdium). Belonging to the mint family LABIATAE and consisting of ten species of north temperate regions, *Clinopodium* is closely related to and by some botanists is included in *Satureja*. The name is a Latin one, from the Greek *klinopodion*, basil.

Clinopodiums are herbaceous or subshrubby, aromatic, hardy and nonhardy plants, with square stems, opposite, undivided leaves, and asymmetrical flowers in terminal and axillary clusters (whorls). Their calyxes, more or less two-lipped, are thirteen-veined. The corollas have a straight tube and two lips, the lower three-lobed with the middle lobe the biggest, the upper smooth-edged or notched at its apex. There are four nonprotruding, curved stamens and an unevenly two-branched style. The fruits each consist of four nutlets.

Wild-basil (*C. vulgare* syns. *Satureja vulgaris*, *Calamintha clinopodium*) is a softly-hairy, leafy, hardy, herbaceous perennial. Native of woods, thickets, and waysides throughout Europe, and naturalized in North America, it is 1 foot to 2½ feet tall. It has oblong-ovate, shallowly-round-toothed leaves ¾ inch to 2 inches long by up to 1 inch wide. The rose-purple to paler pink or white flowers have calyxes with long hairs and corollas ½ to ¾ inch long. The lower lip of the corolla is deeply-lobed. The blooms are in rather distantly spaced whorls (tiers), with the terminal whorl larger than the others.

Two American endemics are *C. coccineum* (syn. *Calamintha coccinea*), native of Georgia, Alabama, and Florida, and *C. gla-*

bellum (syns. *Satureja glabella, Calamintha glabella*), which occurs only in limestone soils from Kentucky and Tennessee to Missouri and Arkansas. A hairless or minutely-hairy subshrub 1 foot to 3 feet tall, *C. coccineum* has obovate to spatula-shaped, toothless leaves ½ to ¾ inch long. Solitary or in twos or threes, the showy, scarlet, funnel-shaped blooms are 1 inch to 1½ inches long. They have reflexed petals, the upper one slightly notched. Very different *C. glabellum* has prostrate or spreading stems up to 1¼ feet tall, and narrowly-oblanceolate to elliptic, generally hairless leaves ¾ inch to 1¼ inches long by under ½ inch wide, with two prominent teeth on each side. The pale purple flowers are ⅓ to a little over ½ inch long. Native of the shores of the Great Lakes and elsewhere from Ohio and Illinois to West Virginia, Oklahoma, and Texas, *C. g. angustifolia* differs in having stolons, and stiffly-erect branches 4 to 8 inches tall.

Garden Uses and Cultivation. Clinopodiums are suitable for herb gardens and rock gardens and American species for native plant gardens. Most succeed with ordinary care in sun or light or part-shade and are easily propagated by seeds or cuttings, and some kinds by division. A neutral or somewhat alkaline soil best suits *C. glabellum*.

CLINTONIA (Clin-tònia)—Queen Cup or Bride's Bonnet. The name of this genus of herbaceous perennials of the lily family LILIACEAE commemorates De Witt Clinton, a naturalist and former governor of New York State, who died in 1828. There are half a dozen species, as wildlings equally divided between eastern North America, western North America, and eastern Asia.

Clintonias have creeping, underground rhizomes and a few broad, basal leaves with sheathing stalks. The blooms, solitary or clustered, terminate erect, unbranched stalks. Each has six erect, separate petals (more correctly, tepals), six slender stamens, and a style tipped with a slightly lobed stigma. The fruits are berries.

Clintonia umbellulata

Native in bogs and moist woodlands from Newfoundland to Ontario, New Jersey and Indiana, and in the mountains of North Carolina, **Clintonia borealis** has up to five short-pointed, ovate to obovate, glossy, dark green, hairy-edged leaves that eventually are up to 1 foot long by nearly or quite one-half as wide, but smaller at flowering time. In umbels of three to six, the nodding blooms, on stalks up to 8 inches tall, are greenish-yellow and about ¾ inch in length. The showy, many-seeded fruits, about ⅙ inch in diameter, are blue. Black, few-seeded berries are produced by *C. umbellulata*, which, except for that, and its purplish-brown-spotted flowers being mostly erect, or at most scarcely nodding, is similar to *C. borealis*. This species occurs in mountain woods from New York to Ohio, North Carolina, and Tennessee.

The queen cup or bride's bonnet (*C. uniflora*) inhabits dense mountain woodlands from Montana to British Columbia to California. It generally has two or three obovate to oblanceolate leaves, up to 6 inches long by less than 2¼ inches wide, and more or less hairy. The slender, pubescent, few-bracted or naked flower stalks carry one or two erect, white, pubescent blooms with spreading petals ¾ to nearly 1 inch long. The berries are dark blue. An inhabitant of damp woodlands from Oregon to California, *C. andrewsiana* has several to many deep rose-purple flowers in terminal and often lateral umbels, with stalks 9 inches to 1½ feet tall. They have petals pouched at their bases, up to ½ inch long. The five or six oblanceolate to broadly-elliptic leaves are up to 10 inches long by one-half as wide. The berries are blue.

Garden and Landscape Uses and Cultivation. Suitable for colonizing in woodlands and other shaded places, clintonias are easy to grow in moist soils that contain generous proportions of leaf mold, peat moss, or other agreeable decayed organic matter. They can be raised from seed and

Clintonia borealis

Clintonia uniflora

Clintonia andrewsiana

are easily multiplied by division in spring. A mulch of leaf mold, compost, or similar organic material maintained around them is beneficial.

CLITORIA (Cli-tòria)—Butterfly-Pea. Forty species are included in this genus of the pea family LEGUMINOSAE. Chiefly they are tropical and subtropical, one extends in the wild as far north as New Jersey. The name alludes to a fanciful resemblance of the keel of the inverted flower to a clitoris. The name butterfly-pea is used also for the related genus *Centrosema*.

Clitorias are herbaceous plants and shrubs, usually of vining habit, and trees. Their leaves are pinnate, their flowers pea-like, with calyxes with cylindrical tubes longer than the lobes. The blooms arise from the leaf axils, solitary or in racemes. The fruits are flattened pods. In India *C. ternatea* is sacred to Durga, the wife of the god Shiva. In Malaysia a dye is obtained from its flowers.

The hardiest species *C. mariana* is indigenous from New Jersey to Illinois, Florida, and Texas. In gardens it survives further north. It is hairless and has usually twining stems up to 3 feet long. The leaves are long-stalked and have three ovate to ovate-lanceolate leaflets 1¼ to 2¼ inches long. The flowers, solitary or in pairs, are of two kinds. The first to appear, showy, pale blue, and about 2 inches in length, come from the upper leaf axils. They are followed by smaller blooms that do not open, but give rise to 2-inch-long pods containing sticky seeds.

A warm-climate species, *C. ternatea* is a slender twiner, probably originally Asian, but now widely naturalized in other regions. Its stems are up to 15 feet long, with leaves of five to nine leaflets. The solitary flowers, 1 inch to 2 inches long, are rich blue with white markings, or in variety *C. t. alba*, white. Double-flowered individuals of both the blue- and white-bloomed kinds occur. The pods, 2 to 5 inches long, contain up to a dozen seeds. This pretty vine is not hardy in the north.

Clitoria ternatea

A rare tropical tree, *C. fairchildiana* (syn. *C. racemosa*), of Brazil, has leaves with three slender leaflets. Its fragrant flowers, in racemes of several, are associated with large, ovate, green bracts. The corollas are violet with a red-purple throat. The pods are 6 or 7 inches long. A specimen 25 feet tall grows at Coconut Grove, Florida.

Garden and Landscape Uses. The last-mentioned species is, as known in cultivation, a small tree adapted for southern Florida and other warm, frost-free or nearly frost-free climates. The hardy *C. mariana*, is suitable for native plant gardens and for planting in beds and informal areas as far north as southern New England. Grown as annual or biennial, *C. ternatea* is attractive outdoors in the south and also for providing summer bloom in greenhouses.

Cultivation. Well-drained soil and sun suit these plants. They are propagated by seeds, which germinate more regularly if they are soaked for twenty-four hours in water before sowing. A germinating temperature of 65 to 70°F is satisfactory. In the north plants of *C. ternatea* started indoors in late summer or early in spring and grown in pots in a greenhouse until the following late May, then planted outdoors after the ground has warmed and the weather is settled, give creditable results. This vine must be provided with supports around which its stems can twine.

CLIVIA (Clì-via)—Kafir-Lily. Among the most beautiful, familiar, and easy to grow of the numerous plants of South Africa that so rightly interest gardeners, the genus *Clivia* ranks high. Once named *Imantophyllum*, it belongs to the amaryllis family AMARYLLIDACEAE and consists of four species. Its name compliments an English Duchess of Northumberland, a member of the Clive family, who died in 1866.

Clivias do not have true bulbs, but the expanded fleshy lower parts of their handsome, strap-shaped, evergreen leaves are tightly overlapped to form swollen bases to the fans of foliage. Their roots are thick and fleshy. The flowers, in moderately compact but not densely crowded, often large umbels, at the tops of erect, solid, leafless stalks, are orange-red to nearly scarlet, or rarely yellow. At the base of each umbel are several spathe-valves (bracts). The blooms, funnel-shaped to cylindrical, up-facing or pendulous, have short-tubed perianths with six segments (petals or more correctly tepals), the same number of stamens approximately as long as the perianth, and a three-lobed style. The fruits are shiny, red berries, approximately 1 inch long.

Most familiar to gardeners, and the finest ornamental, *C. miniata* is a robust, stemless species with thick, lustrous, smooth-edged, arching leaves 1½ to 2½ feet in length, 1 inch to 2 inches or more

Clivia miniata

Clivia miniata (flowers)

in width, and slightly narrowed toward their bases. It has upturned, broadly-funnel-shaped blooms, from fifteen to twenty or sometimes more in each spherical umbel. The flowers, on individual stalks 1 inch to 2 inches long, are 2 to 3 inches long and about as wide. They are generally orange-red to flame, usually with yellow centers. In the rare variety *C. m. citrina* (syn. *C. m. flava*) the blooms are clear yellow. In cultivation *C. miniata* has produced truly magnificent variants often referred to as hybrids. Most are probably not of hybrid origin, but are the results of careful breeding and selection of superior types within the species. They are characterized by their larger umbels of more richly colored, bigger, broader-petaled

Clivia miniata citrina

flowers, and their usually broader foliage. A reputed hybrid between *C. miniata* and *C. nobilis*, **C. cyrtanthiflora** shows little or no influence of the supposed first parent and possibly is only a variant of the second. It has narrow-tubed, slightly greenish, pendulous blooms.

Less spectacular than the better forms of *C. miniata*, but as vigorous, stemless **C. nobilis** has two-ranked, rough-edged leaves up to 3 feet in length and umbels of twenty

Clivia cyrtanthiflora

Clivia nobilis

to forty or sometimes more drooping, narrowly-funnel-shaped blooms approximately 1½ inches long, and with the edges of their green-tipped, reddish yellow petals overlapping. From it **C. gardenii** differs in its drooping blooms, 2 to 3 inches long, being fifteen, or fewer, in each umbel.

Distinct from all other clivias in developing as it ages thick, erect stems up to 1½ feet in length, naked below and leafy above, **C. caulescens** has umbels of fifteen to eighteen pendulous, green-tipped, orange-colored flowers approximately equal in size to those of *C. nobilis*.

Garden and Landscape Uses. Clivias are shade-loving plants that in subtropical climates where little or no frost occurs are of inestimable value for planting under trees and elsewhere out of direct sun. They are first-rate greenhouse ornamentals and excellent houseplants, admired for their handsome foliage and spectacular blooms. The latter last well when cut and lend

themselves for use in flower arrangements.

Cultivation. Outdoor cultivation of clivias calls for no special care. They grow best in deep, fertile soil that contains generous amounts of organic matter and is sufficiently loose to preclude the possibility of retaining excessive water, and to permit easy penetration by the thick roots. A spring application of fertilizer, and watering to keep the earth fairly moist from spring until fall and drier during the winter period of partial dormancy, encourage good growth and blooming.

Clivias may remain undisturbed for several years in large pots or tubs. Only when they show signs of beginning to suffer from overcrowding or deterioration of soil quality or drainage is repotting indicated. Specimens in containers under 8 inches in diameter need advancing to larger-sized ones more frequently. They are not plants suitable for permanent retention in small containers. Attend to repotting in spring. In performing the operation take great care not to break or unduly damage the roots, but tease out as much old soil as practicable from between them. Make sure the new pots or tubs are very thoroughly drained and of a size that accommodates the roots comfortably without large amounts of space to spare. Fairly lumpy soil, containing a generous admixture of leaf mold, compost, or peat moss, and enriched with a liberal sprinkling of coarse bonemeal and, if available some dried cow manure, is to the liking of clivias. It is important that it contains enough coarse sand, finely broken brick, or a similar nondecayable material that makes for porosity to ensure it admitting air and permitting free drainage of water for at least a few years. When potting, pack the soil moderately firmly. Immediately afterwards and for the succeeding two or three months keep the plants in a temperature a few degrees higher, and where the atmosphere is more humid, than normal. During this time avoid excessive watering, but do not permit the soil to dry out. Frequent sprinkling of the foliage without saturating the soil is helpful. Routine care of established clivias involves watering generously from late winter or spring until fall, much less generously in winter. Well-rooted specimens benefit from biweekly applications of dilute liquid fertilizer from spring to fall. In summer they may with advantage be stood in a shady place outdoors. If kept in a greenhouse it must be ventilated freely. From fall to the time when new growth begins a night temperature of 50°F, with a few degrees increase by day, is adequate. Later, increases both by night and day of five to ten degrees are in order. Clivias can be forced into early bloom by subjecting them to the higher temperatures and resuming normal summer watering in January or February.

Propagation is most often by division in spring just as new growth begins. It is important to separate the roots so that minimum damage to them results. Seeds afford an easy means of securing new stock, but those taken from superior varieties of *C. miniata* are likely to give plants that show considerable variation, and may be inferior to the parent. Remove the seeds from the berries and sow immediately, without allowing them to dry. In a temperature of 65 to 70°F they germinate in about a month.

Dividing a clivia: (a) Pulling apart an old specimen

(b) A well-rooted division separated from the mother plant

(c) Potting the division separately

CLOCK VINE. See Thunbergia.

CLONE. A clone is the sum total of individuals derived by vegetative increase (asexual propagation) from one original plant. Because no sexual process is involved in the multiplication, members of a clone, except in the unlikely event of a mutation (sport) occurring are genetically identical. They have the same inherent possibilities and respond similarly to identical environments and care. Except for physical separation they are in effect one plant. In this they differ from plant populations raised from seeds or spores, which, except in the unusual circumstance that they have been produced parthenogenetically as a result of the stimulation of female reproductive cells without fertilization by males, vary to a greater or lesser extent in their genetic makeup and are in the fullest sense new individuals.

Clones, not ordinarily identified as such, are of tremendous importance to gardeners. A high proportion of the varieties or cultivars they grow belong to this category. Examples are all exclusively vegetatively reproduced garden varieties such as apple 'Delicious', rose 'Peace', and tulip 'Keizerkroon'. Every plant in existence of each of these is simply a fragment of an original individual selected for desirable qualities and multiplied by grafting, budding, cuttings, division, or other technique of vegetative propagation. Clones may be perpetuated for very long periods, in some cases for centuries. Theoretically there is no limit to how long one can be maintained, but with plants susceptible to virus diseases, invasion of a clonal stock by one or more viruses, coupled with the virtual impossibility of curing infected individuals, results in the viruses being transmitted with each new propagation from an infected plant. This may, over a period, bring about a serious deterioration of the clone. It is the usual reason for the decline in vigor once referred to as the "running out" of a variety. "Breaking" of flowers of tulip clones into distinctive color patterns is also caused in this way. If the original plant of a clone is self-sterile, that is if it is incapable of producing fruits and seeds when fertilized with its own pollen, all individuals of the clone will be sterile to the pollen of all others of the same clone and it is of no avail to plant more than one in the expectation of bringing about fruiting. The remedy lies in planting nearby an individual of a different clone or a seedling that is compatable, that is one that produces pollen that will fertilize the flowers of the self-sterile clone from which fruits are desired.

CLOUDBERRY is Rubus chamaemorus.

CLOVE TREE is Syzygium aromaticum. The clove currant is Ribes odoratum.

CLOVER. The word clover used alone refers to Trifolium. Other plants of which clover forms part of their common names include these: Bokhara-clover (Melilotus alba), bush-clover (Lespedeza), Calvary-clover (Medicago echinus), holy-clover (Onobrychis viciaefolia), Mexican-clover (Richardia), owl's-clover (Orthocarpus purpurascens), prairie-clover (Petalosepalum), sweet-clover (Melilotus), tick-clover (Desmodium), and water-clover (Marsilea).

CLUB. The word forms part of the colloquial names of several plants such as club gourd (Trichosanthes anguina), Devil's club (Oplopanax horridus), golden club (Orontium aquaticum), and Hercules' club (Aralia spinosa and Zanthoxylum clava-herculis).

CLUB ROOT. This prevalent disease of a wide variety of plants of the cabbage relationship, including all vegetables of the brassica group as well as such flower garden plants as candytuft, stocks, sweet alyssum, sweet rocket, and wallflowers, is caused by a soil-inhabiting organism formerly identified as a slime mold, but now believed to be a fungus. The disease, also known as finger and toe, makes its presence known first by the foliage wilting on hot days followed by partial or entire recovery at night. Later, the leaves fail to recover at night and the lower ones become yellow and drop. The roots, distorted into fingered swellings or sometimes into one large misshapen mass, often partially rot.

The spores of club root disease can remain alive in the soil for many years. They do not cause infection if the pH is 7 or above or when the moisture content is less than fifty percent of the soil's water-holding capacity. Control is had by planting only in well-drained soil, by liming to maintain a slightly alkaline condition, by rotating susceptible crops so that they occupy the same ground at intervals of four or five years only, by making sure that young plants set out are free of infection, by not disposing of infected plants by add-

Club-root disease on broccoli

ing them to compost piles or feeding them to animals (the spores pass unharmed through the digestive tract), or in other ways that will enable the wastes to be returned to the soil. As a precaution, about a cupful of a solution of mercuric chloride (a deadly poison), one tablet dissolved in a quart of water, may be poured around each newly planted seedling.

CLUBMOSS. See Lycopodium, and Selaginella.

CLUSIA (Clù-sia)—Copey or Scotch Attorney or Pitch-Apple. Regrettably only one of the nearly 150 species of Clusia, of the garcinia family GUTTIFERAE, seems to be cultivated in the United States. Surely in southern Florida and Hawaii suitable places could be found for more. Strictly tropical and subtropical, the genus occurs in the wild chiefly in the New World, but is represented also in New Caledonia and Madagascar. Its name commemorates Charles de l'Ecluse, his name latinized as Clusius, a distinguished botanist who died in 1609.

These plants usually begin life as epiphytes. They originate from seeds deposited by birds or animals on the branches of trees and germinate and grow there, but without taking nourishment from the tree on which they lodge. They depend for nutrients upon whatever organic debris is available, rain, and air. Their economics in their early lives are those of epiphytic orchids and bromeliads. Unlike orchids and bromeliads, however, clusias are not dependent on other trees for long. Their roots, sent down beside their host's trunk, enter the earth and compete with those of the tree on which they grow for nutrients and moisture. So successful are they that eventually their roots may unite to form a false trunk completely encircling that of the tree, and the guest smothers the host to death in the manner of strangling figs (Ficus). It is not necessary to permit clusias in cultivation to adopt this reprehensible procedure. From their beginnings they grow quite well in soil.

Clusias are evergreen trees and shrubs with usually viscid sap, horizontal branches, four-angled shoots, and opposite, leathery leaves. The blooms, borne at the branch ends, are unisexual or bisexual, with either or both types on individual plants. They have four to six sepals and four to nine petals, usually numerous stamens, and a very short style or none. The fruits are fleshy capsules.

The copey, Scotch attorney, or pitch-apple (C. rosea), of the West Indies, northern South America, southern Mexico, and the Florida Keys, is a tree, possibly up to 60 feet in height, with a dense, wide-spreading crown. Its thick, leathery, obovate leaves, slightly rounded, squarish, or shallowly-notched at the apex, narrow gradually to brief, flattened stalks. They

Clusia rosea

Clusia rosea variegata

Clusia rosea, in fruit

are 3 to 6 inches long by two-thirds as wide, slightly lustrous green above, and dull yellow green on their lower sides. The flowers, white, creamy, or pink-tinged, and with six to eight spreading obovate, fleshy petals, notched at their apexes and about 3 inches across, have somewhat the appearance of magnolia blooms. Male flowers have many fertile stamens joined in a ring, the inner ones in a resinous mass. Females have a ring of sterile stamens and a functional pistil. The blooms are in evidence more or less throughout the year. The fruits of the copey are ball-like and wider than long. They retain the sepals, and at their apex a circle of blackish, flat stigmas. As the fruits ripen they turn from yellowish-green to brown. They are 2 to 3 inches in diameter and contain scarlet pulp and many seeds. Although reputed to be poisonous, they are eaten by bats. The foliage of *C. rosea variegata* is variegated with pale yellow.

The wood of the copey is used in construction, for tool handles and fence posts, for fuel, and for other purposes. It is much subject to infestations of dry-wood termites. The fruits, bark, and other parts contain a resinous latex that has been used for calking the planking of ships, and me-

dicinally. The leaves are easily marked by scratching them with a pin or similar implement, and it is recorded that Spanish conquistadors made playing cards by drawing appropriate designs on leaves of this species. The leaves were also used to some extent as substitutes for writing paper, which, perhaps, accounts for the common name Scotch attorney. An alternative theory holds that this common name reflects, unkindly and undoubtedly slanderously, similarities between the treatment of a host tree by Clusia and a client by a Scottish lawyer.

Garden and Landscape Uses. The copey grows slowly, and in Florida is adapted for the southern part of the state only. There it does not exceed the dimensions of a large shrub. Few plants stand seaside conditions and exposure to salt spray as well as does this useful and handsome ornamental. It flourishes in exposed oceanfront locations and is excellent for foundation plantings and similar uses. It is also very satisfactory for growing in large containers. It has good-looking foliage, but blooms too sporadically to make much floral show at any one time. The copey grows in almost any soil.

Cultivation. In hot, humid climates this plant grows with little care. It stands pruning well, and this may be needed to restrict its size or to shape it. Propagation is by seed, air layering, and cuttings.

CLYTOSTOMA (Clytós-toma). These evergreen, woody vines of the bignonia family BIGNONIACEAE differ from most members of the family in that the tendrils, which in *Clytostoma* usually terminate each leaf of two leaflets (morphologically the tendril represents a third leaflet), are branchless. From *Saritaea*, which also has branchless tendrils, they differ in having elliptic-oblong to ovate-oblong leaflets and narrow calyx lobes. The flowers, in pairs in axillary or terminal panicles, are trumpet- to bell-shaped and have five, long slender sepals and five rounded corolla

lobes (petals). The fruits are prickly capsules containing numerous winged seeds. The name of the genus, which is South American and consists of twelve species, comes from the Greek *klytos*, beautiful, and *stoma*, mouth, in reference to the flowers.

Popular *C. callistegioides* (syn. *Bignonia speciosa*, and in gardens often known as *B. violacea*), native to Brazil and Argentina, is

Clytostoma callistegioides

a beautiful ornamental. Its leaves usually have two wavy-edged, pointed, elliptic-oblong leaflets 2 to 4 inches in length, and a terminal tendril. A few leaves may have in place of the tendril a third leaflet. In dense racemes of six to twelve, the 3-inch-long flowers are 2 to 3 inches across their faces. They have wavy, broadly-oval, spreading petals, the three lower slightly reflexed. They are purplish-mauve, with white throats with two purple lines running to each petal. Blooming is profuse and provides great displays at intervals throughout late spring and summer. The fruits are slender and about 10 inches long. Smaller-flowered *C. binatum* (syn. *Bignonia purpurea*), a native of Uruguay, has ovate-oblong leaflets up to 3 inches in length, and 1-inch-long mauve flowers with white throats.

Garden and Landscape Uses. As ornamentals these beautiful vines are suitable outdoors only for the tropics and warm subtropics and for cultivation in warm greenhouses. They must be afforded supports around which to twine their tendrils. A lattice is satisfactory, or they may be trained to wires stretched across the face of a wall, up posts or pillars, or beneath greenhouse roofs. Vigorous growers, clytostomas require plenty of space

Cultivation. Outdoors, at least one-half a day of sun is needed by these vines. In greenhouses a sunny location with light shade from strong summer sun is appropriate. They grow best in well-drained, fertile soil, with ample moisture. Greenhouse winter night temperatures of 55 to 60°F are needed, with daytime temperatures a few degrees higher. At other seasons more warmth both by night and day is in order. Pruning consists of shortening in late winter or very early spring to within an inch or two of their bases all shoots not needed for further extension of the vines. Propagation is by cuttings made from moderately firm shoots, root cuttings, and seeds sown in sandy peaty soil. Cuttings root most quickly if there is enough bottom heat to keep the temperature of the propagating bed at 70 to 75°F.

CNEORACEAE—Cneorum Family. The characteristics of this family of dicotyledons are those of its only genus, *Cneorum*.

CNEORIDIUM (Cneo-rídium)—Bush-Rue. The name of this one-species genus of the rue family RUTACEAE suggests its resemblance to the Old World genus *Cneorum*, but it belongs in a different family.

Confined in the wild to California and Baja California, the bush-rue (*Cneoridium dumosum*) is a slender-twigged, complexly branched, evergreen shrub, with opposite or clustered, hairless, linear to oblong leaves up to 1 inch long, very narrow, and dotted with tiny glands. From 3 to 6 feet tall and with rather heavily scented foliage, this shrub has fragrant white flowers, pink-tinged on their outsides, in short axillary clusters of two or three, or occasionally solitary. They have four-lobed calyxes and four obovate petals about ¼ inch long. There are eight stamens of two different lengths and a short, flattened style arising from the bottom of the ovary. The greenish to reddish-brown, dotted fruits, fleshy, spherical, and berry-like, contain one or two seeds. They are about ¼ inch in diameter.

Garden and Landscape Uses and Cultivation. Scarcely planted outside its native region or rarely in similar mild dry climates, the bush-rue has limited uses as a general purpose shrub in such regions. It prospers in dry soils in sun. Propagation is by seed.

CNEORUM (Cne-òrum)—Spurge-Olive. This, the only genus of the spurge-olive family CNEORACEAE consists of one species of the Mediterranean region and one, a very rare native of Cuba. They are evergreen shrubs, with alternate, undivided, leathery leaves and axillary flowers with three or four each sepals, petals, and stamens. The fruits have usually three berry-like parts attached to a receptacle (expanded portion of the flower stalk). The name is from *kneoron*, one the ancient Greeks applied to some olive-like shrub.

A hairless shrub of the Mediterranean region that usually grows in limestone soils, *Cneorum tricoccon* is 1 foot to 4 feet tall. Its blunt, oblong, stalkless, glossy leaves are up to 1½ inches long and very narrow. The yellow flowers, borne in early

Cneorum tricoccon

summer, usually in clusters of two or three, are about ⅓ inch across. The brownish, three-segmented fruits have a thin, fleshy covering.

Garden Uses and Cultivation. The species described is not hardy in the north and is suited only to regions of mild winters. It needs plenty of sun and porous well-drained soil. No pruning is required. This shrub is readily propagated by cuttings made from terminal shoots in late summer and inserted in a greenhouse propagating bed or under mist. Because *C. tricoccon* has little ornamental merit its appeal is to the admirer of the unusual rather than to the lover of the spectacular.

CNICUS (Cnìc-us)—Blessed-Thistle or Holy-Thistle or Our-Lady's Thistle. The common names of the only representative of this genus (pronounced nikus) of the daisy family COMPOSITAE refer to an experience recorded of Emperor Charlemagne. At a time when it seemed likely that Charlemagne's army was about to be eliminated as an effective force by bubonic plague he was told by an angel to shoot an arrow into the air and that the plant upon which it dropped would cure the disease. He did so and *Cnicus* was selected. The French Order of the Thistle, founded in the fourteenth century, adopted the blessed-thistle as its emblem. The botanical name, the ancient Latin one of the safflower (*Carthamnus tinctorius*), derives from the Greek *knekos*, a thistle.

Blessed-thistle (*C. benedictus*), a native annual of the Mediterranean region and the Caucasus, branches freely. About 2 feet tall, it has pinnately-lobed, toothed, spiny, oblong leaves and terminal, thistle-like heads of yellow flowers supported on a collar of large, bristly, leafy bracts. The florets are all of the disk kind. The leaves are prettily marbled or blotched with silvery-white. As this plant is native in the Holy Land it is regarded as one of several referred to in the Bible as thistles.

Garden Uses and Cultivation. Of minor importance as an ornamental, this species

is of some interest to the collector of interesting plants and to those who plant collections of Bible plants. It is of the simplest cultivation in sunny places in well-drained soil. Seeds may be sown outdoors in early spring where the plants are to remain, and the seedlings thinned to about 1 foot apart. Once established this plant is likely to reproduce by self-sown seeds.

COACHWHIP is *Fouquieria splendens*.

COACHWOOD is *Ceratopetalum apetalum*.

COAL CINDERS. Less available in most parts of the country than they once were, coal cinders can serve usefully in gardens. They form excellent bases for paths and, when compacted, make paths that may be satisfactory without other surfacing. They are good substitutes for crocks in pots and other plant containers and, sifted to remove all fine ash and pieces that will not pass through a ½-inch mesh, are excellent to mix with clayey soils to improve porosity and drainage. Before adding them to soil they should be leached by exposing them over a period of some weeks or months to several heavy rains or by washing them with water from a hose.

COBAEA (Cobaè-a)—Cup-and-Saucer Vine. The only climbing plants in the phlox family POLEMONIACEAE, belong in *Cobaea*, a tropical and subtropical American genus of possibly eighteen species that by some authorities is segregated as the sole representative of the cobaea family COBAEACEAE. The name commemorates the Jesuit naturalist Father Bernardo Cobo, who lived for many years in America. He died in 1659.

Cobaeas are vigorous woody and herbaceous vines with alternate, pinnate leaves, mostly ending in branched tendrils, and solitary, long-stalked, green or violet flowers in the leaf axils. A conspicuous feature of the flower is the large, leafy, five-parted, inflated calyx. The bell-shaped to cylindrical corolla is five-lobed. There are five protruding stamens and one style. The fruits are capsules containing broadly-winged seeds.

Cup-and-saucer vine (*C. scandens*) of Mexico is the only kind commonly cultivated. Attaining a height of 25 feet or more, this sort is usually grown as an annual although in mild climates and in greenhouses it may be perennial. Its leaves are of two to three pairs of leaflets, the lower ones close to the stem and somewhat eared around it. The flowers have as the "saucer" of the common name a conspicuous calyx. The "cup" of the common name is a greenish-purple to light violet corolla up to 3 inches long and 1½ to 2 inches across. The stamens are strongly curved above their middles. The seed capsules are about 1¾ inches long. Varieties

Cobaea scandens

are *C. s. alba*, with white flowers, and *C. s. purpurea*, with blooms of a deeper purple than the typical species. This vine is of special interest because of the extreme sensitivity of its hooked tendrils, which are modified portions of leaves. Movement becomes apparent if one side of a tendril is rubbed and observed for a few minutes. This rapid response to touch accounts for the great ability of this handsome plant to climb so rapidly and successfully. It blooms for a period of many weeks or even months.

Garden and Landscape Uses. Cup-and-saucer vine gives of its best in full sun in any ordinary garden soil. It is well suited for clothing arbors, pergolas, trellises, porches, and similar supports where screening or verdant drapery is desirable. It needs a fair supply of moisture and will not thrive in areas subject to drought. It is easy to grow in cool greenhouses.

Cultivation. Cup-and-saucer vines can be raised from seeds sown in well-drained soil outdoors where the plants are to remain, or seeds may be sown indoors early and the young plants raised in pots until planting out time, which is about when it is safe to set out tomato plants. Sow the seeds, which usually germinate irregularly, about two months before it is time to set the plants outdoors. Sow them in sandy soil in a temperature of 60 to 70°F. To minimize danger of rotting set each seed with its long edge downward and leave the upper edge flush with the soil surface. It may take up to three weeks for the seeds to germinate. Outdoors, plants may be spaced 1 foot to 2 feet apart. It is, of course, necessary to provide appropriate supports over which the vines can climb. In greenhouses good specimens can be accommodated in pots or tubs 9 inches or more in diameter and be trained to wires or other supports. When grown indoors all side shoots should be pinched back to two leaves or buds, and all very weak shoots should be cut out entirely. The chief pests are red spider mites and aphids. A blight that causes the leaves to curl and dry is sometimes troublesome. No controls have been developed.

COBNUT is *Corylus avellana grandis*.

COBRA PLANT is *Darlingtonia californica*.

COCA or COCAINE PLANT is *Erythroxylum coca*.

COCCINIA (Coccí-nia)—Ivy Gourd. Previously called *Cephalandra*, the genus *Coccinia*, is mostly African but has one species native to the Asian tropics. Belonging in the gourd family CUCURBITACEAE, it consists of thirty species. The name, derived from the Latin *coccineus*, scarlet, alludes to the color of the fruits.

Coccinias are mostly tuberous-rooted, tendril-bearing vines with lobed or angled leaves. Individual plants usually bear blooms of one sex only. The flowers are bell-shaped, the females are solitary, the males either solitary or in clusters and with three stamens. The berry-like fruits are small oblong to roundish.

The ivy gourd (**C. grandis** syn. *C. cordifolia*) of India is prostrate or climbs to a height of 6 feet or more. Its annual stems, which arise from perennial roots, like the leaves are without hairs. The leaves, often shaped like those of English ivy, are lustrous, angled, broadly-triangular-ovate, and up to 4 inches across. The bell-shaped flowers, white and with five sharp corolla lobes, are about 1½ inches long and 2 inches across. The oblong or oval fruits are miniature gourds, scarlet when ripe, and 1 inch to 2 inches long. In India and some other parts of the Old World tropics they are eaten as a vegetable.

Garden Uses and Cultivation. In warm, frost-free regions this kind may be cultivated outdoors as a perennial and it is also an interesting permanent vine to grow in tropical greenhouses. It can be treated as an annual and raised from seeds each year. If the latter plan is followed, in regions of long, warm growing seasons the seeds may be sown directly outdoors in spring, but elsewhere they should be started indoors and grown in a warm greenhouse (night temperature 60°F, day temperature 65 to 70°F) until the outdoor weather is really warm and settled. The seeds are sown indoors eight weeks before it is expected that the young plants will be transferred to the garden. The plants need a sheltered, sunny location and a fertile, loose soil enriched with liberal amounts of compost or other decayed organic matter. They should be supplied with supports around which their tendrils can twine, and throughout the season, with generous amounts of water.

In greenhouses the plants should be favored with a humid atmosphere, adequate supplies of water, and a sunny location with just a little shade from the strongest summer sun. They may be accommodated in soil beds or in pots or tubs 8 or 9 inches in diameter. A coarse, porous soil that contains an abundance of organic matter and that will not become compact as a result of

repeated watering is ideal. After the soil is invaded by healthy roots weekly applications of dilute liquid fertilizer are sustaining and stimulating. Wires or other supports are necessary.

COCCOCYPSELUM (Cocco-cýpselum). This genus of the madder family RUBIACEAE comprises twenty species of tropical American, prostrate, evergreen herbaceous plants. Its name, alluding to the shape of the fruits, comes from the Greek *kokkos*, a berry, and *kypsele*, a vase.

Coccocypselums have opposite, stalked, undivided leaves, and from the leaf axils, stalked heads of little funnel-shaped blooms with four-lobed calyxes, four corolla lobes (petals), four short-stalked stamens, and a two-branched style. The fruits are berries.

Native of northern South America, Central America, and Trinidad, *Coccocypselum guianense* has slender, short-hairy or hairless stems, and hairy or hairless, rather distantly spaced leaves with blades 1 inch to 2½ inches long. Its blooms are purplish. The fruits are china-blue berries ¾ inch long.

Similar to, but more beautiful than the last, largely because of the abundant, long purple or yellow hairs on its stems, leaf-stalks, and veins of the under surfaces of the leaves, *C. hirsutum* inhabits Mexico, the West Indies, and Central and tropical South America. In cultivation it has been misidentified as *C. guianense* and as *C. discolor*.

Garden and Landscape Uses and Cultivation. In the tropics and in tropical greenhouses coccocypselums are useful as ground-covers and for hanging baskets. They grow with little difficulty in ordinary fertile soil and are increased by seed, division, and cuttings. Greenhouses with a humid atmosphere, maintained on winter nights at minimum temperatures of 55 to 60°F, with daytime increases five to fifteen degrees, and at other times of the year warmer, afford satisfactory growing conditions.

COCCOLOBA (Coccó-loba) — Sea-Grape, Pigeon-Plum, Chicory-Grape. Native only of the American tropics and subtropics, *Coccoloba* of the buckwheat family POLYGONACEAE, comprises probably 150 or more species of plants related to buckwheat, to the flower garden annual called prince's feather (*Polygonum orientale*), and to dock weeds (*Rumex*). Its sorts look like none of these. The name, sometimes spelled *Coccolobis*, is derived from the Greek *kokkos*, a berry, and *lobus*, a pod. It alludes to the fruits.

Coccolobas are trees, shrubs and woody vines, mostly evergreens. They have alternate, undivided leaves of various shapes and sizes, and greenish, five-parted, usually small flowers in spikes or racemes. Generally the sexes are on separate plants.

Coccoloba uvifera

The fruits are fleshy and berry-like. Coccolobas are good honey plants, bees are attracted to both the male and female flowers. The wood of the sea-grape has been used for furniture; that of *C. pubescens* is employed in construction and furniture manufacture. It is hard and so heavy that it sinks in water. In all probability the sea-grape was the first New World plant to be seen by Columbus for it is abundant along the coast where he made his first landfall.

The sea-grape (*C. uvifera*) is by far the best known horticulturally. Native of the coasts of southern Florida and the West Indies and those from Mexico to South America, and naturalized in the Hawaiian Islands, this shrub or small tree, rarely more than 25 feet in height, has leathery,

round leaves, heart-shaped at their bases, up to 8 inches in diameter, and with red veins. When young they are reddish-bronze, but as they mature they become green. The white flowers, in dense racemes up to 10 inches long, are succeeded by grapelike bunches of purple fruits each about ¾ inch in diameter. They are used for jelly. There are two variegated-leaved varieties of the sea-grape; in one the variegation is yellow, in the other white.

The pigeon-plum (*C. diversifolia* syns. *C. floridana*, *C. laurifolia*) is a nearly evergreen, round-topped, slow-growing tree with oblong leaves 2 to 4 inches long and short racemes of flowers followed by edible pear-shaped fruits about ½ inch long. It is a native of southern Florida and the West Indies. Another West Indian species,

C. pubescens (syn. *C. grandifolia*) is a sparingly-branched tree up to 80 feet in height, with round leaves 1 inch to 1½ feet across and sometimes, on vigorous sprouts, as much as 3 feet in diameter. They are conspicuously-veined and rusty-pubescent on their undersides. The greenish flowers are in erect terminal racemes about 2 feet long. Another commonly cultivated kind is the chicory-grape (*C. venosa*), a deciduous West Indian tree up to about 30 feet in height, with nearly horizontal branches and drooping, short-stalked, elliptic leaves. The leaves are 3½ to 8 inches long and about one-half as wide; they form two rows along the shoots. The flowers are borne along the stems in dense, slender, erect racemes from the leaf axils. They are succeeded by edible, egg-shaped, white to pinkish fruits up to ¼ inch long.

Garden and Landscape Uses. Coccolobas are tender to frost and can be grown outdoors in the United States only in southern Florida, southern California, and Hawaii. The sea-grape is especially useful in gardens near the sea and because of its distinctive habit of growth, and handsome, large leaves forms a bold and highly ornamental element in the landscape. It tolerates strong sun, wind, and salt, and is one of the first woody plants to colonize sandy shores. It is frequently used for hedges and windbreaks. The pigeon-plum

Coccoloba uvifera, at the seaside with coconut palms

Coccoloba uvifera, in fruit

Coccoloba diversifolia

Coccoloba uvifera, as a hedge

and the chicory-plum are used a lawn trees and street trees and serve these purposes well. In large home grounds, parks and other public places *C. pubescens* can be put to effective use as a specimen tree; it is a slow grower.

Cultivation. For their best development coccolobas require fertile, fairly moist, but well-drained soils; this is true even of the sea-grape despite the fact that natively it often occupies apparently inhospitable sandy strips along shores. They can be transplanted without undue difficulty, but for the best results are grown in containers in their young stages and set out from these. Whatever pruning is needed to develop shapely specimens may be done preferably just before the season of new growth. With the sea-grape this is likely to involve cutting it back fairly severely one or more times to establish a good branch system. Propagation is easy by seeds sown in sandy soil in a temperature of 70°F or more. Layering is also successful and cuttings, treated with a rooting hormone and planted in a sand propagating bed or bench, if possible with a little bottom heat supplied so that the rooting medium is maintained at about 75°F, afford another means of increase. When grown in containers fertile soil that drains readily should be used, and it should be kept moderately moist. In greenhouses they may be grown in sun or partial shade. The atmosphere should be reasonably humid and the minimum temperature 55 to 60°F. When repotting is needed, usually only at intervals of two or three years or longer, it is best done in late winter or spring. Specimens that have filled their containers with roots benefit from occasional applications of dilute liquid fertilizer.

COCCOTHRINAX (Coccothrì-nax)—Seamberry Palm, Silver Palm, Silver-Thatch Palm. This group of about fifty species of palms of the Caribbean region is closely similar to the peaberry palm (*Thrinax*) from which it differs in having flower clusters usually shorter than the leafstalks, flowers with mostly nine to twelve stamens, and fruits that ripen black or nearly black. There are more recondite botanical differences. The name *Coccothrinax* is from the Greek *kokkos*, a berry, and the name of a related genus of palms. It refers to the fruits.

These palms are bisexual, fan-leaved, and mostly small and slender. Their fruits are berry-like and contain a solitary, grooved or lobed seed. They belong in the palm family PALMAE.

Native to southern Florida and the Bahamas, the silver palm (*C. argentata*) attains a height of 15 to 20 feet and has deeply-cleft leaves with many, usually drooping, slender segments that gradually narrow from base to tip and are very silvery on their undersides. Their upper surfaces are glossy green. This kind com-

Coccothrinax argentata, young specimen

Coccothrinax species

monly produces offshoots or suckers around its base. It thrives in sun or shade and is resistant to sea spray. At one time it was confused with *C. argentea*, a very slender native of dryish areas in Hispaniola, that attains heights of 30 feet or more and has a sparse crown of nearly round leaves divided to two-thirds of their depth into thirty or more segments that rarely droop.

Very distinct and handsome *C. crinita*, endemic to Cuba and one of about twenty kinds indigenous there, is up to 30 feet tall. Its trunk from the ground up is densely and thickly covered with a thick thatch of the bases of its old leafstalks interspersed with long strands of fiber produced from the bottoms of the stalks. The handsome, circular leaves, rich, lustrous green above and gray-green beneath, spread in all directions to form a beautiful, rounded crown. The flower clusters, much-branched and at first upright, bend downward under the weight of the very fleshy fruits. The latter, light purple at maturity, are about 1 inch in diameter. In Cuba the hairs of the trunk of this palm are used in pillows, to make brooms, and to secure the mouths of bags of charcoal.

The silver-thatch palm (*C. jamaicensis*), of Jamaica, is related to *C. argentata* from which it differs in its long-stalked flowers and fruits. The flowers are powerfully fra-

grant and whitish. This palm is up to 25 feet in height and has circular leaves up to 5 feet in diameter, divided to beyond their middles into thirty-five to forty lax segments. The undersides of the leaves are bright silvery. They are used in native crafts for making baskets, bags, hats, and other articles.

Garden and Landscape Uses. Seamberry palms are attractive trees suitable for gardens and landscape plantings outdoors in regions of little or no frost and for growing in large pots and tubs in greenhouses. The hardiest is *C. crinita*, which is recorded as surviving temperatures of 27°F at Daytona Beach, Florida when *C. argentata* in the same area was killed. These palms, appropriate for properties of restricted size, are effective as solitary specimens and in groups. They prosper best in full sun, but stand slight shade.

Cultivation. Seamberry palms thrive in ordinary, well-drained garden soil and withstand dryish conditions well. They are propagated by seeds sown while fresh in sandy, peaty soil in a temperature of 75 to 85°F. In greenhouses they can be accommodated in well-drained containers or ground beds in coarse, porous, fertile soil. They need a humid atmosphere, a minimum night winter temperature of 55 to 60°F with an increase of five to ten degrees by day, and a minimum of 60 to 70°F at other seasons, watering to keep the soil moderately moist, but not constantly saturated, and shade from strong summer sun. Applications of dilute liquid fertilizer at biweekly intervals from spring through fall are beneficial to well-rooted specimens. For additional information see Palms.

COCCULUS (Cóccu-lus)—Carolina Moonseed. This genus of about a dozen species takes its name from the Greek *kokkos*, a berry, in allusion to the fruits. It belongs in the moonseed family MENISPERMACEAE and is widely distributed in tropical, subtropical, and temperate regions.

The sorts of *Cocculus* are shrubs and twining vines with alternate, evergreen or deciduous, lobed or lobeless leaves, and small, unisexual flowers in racemes or panicles. The blooms, of no decorative merit, have six each sepals, petals, and stamens and three or six pistils. The fruits are berry-like drupes. In Hawaii the stems of *C. ferrandianus* were used locally for cordage and for making fish traps.

Carolina moonseed (*C. carolinus*), native from Virginia to Missouri, Indiana, Florida, and Texas, is hardy well north of its natural range. An inhabitant of damp woodlands and thickets, it is a vine 6 to 10 feet tall with variously shaped, lobed or lobeless, triangular to ovate, stalked leaves up to 6 inches long and usually hairy on their undersides. The minute male flowers are in slender panicles 4 or 5 inches long. The female flowers, about the same size, are in much shorter panicles. They are succeeded by red fruits ¼ to ⅓ inch long. The flowers are greenish.

Another hardy, deciduous vine, *C. trilobus,* is native from Japan to the Himalayas, and in southeast Asia, Taiwan, and the Philippines. This kind is a pubescent plant with leathery or membranous, ovate, often shallowly-three-lobed, stalked leaves,

Cocculus trilobus

up to 4½ inches in length and nearly as wide. Its tiny, yellowish-white flowers are in narrow, axillary panicles. The black fruits, slightly coated with a waxy bloom, are up to ⅓ inch in diameter.

A handsome lustrous-foliaged shrub up to 15 feet in height, *C. laurifolius* has slightly flattened green shoots and short-stalked, evergreen, oblong-ovate, obovate, or elliptic leaves, with three very distinct longitudinal veins and many parallel transverse secondary ones. The leaves are pointed, not lobed, and up to 4½ inches long by almost one-half as wide. The minute yellow flowers are in axillary and terminal panicles about 2 inches long. The fruits are slightly flattened, black, and about ¼ inch long. This species is native in Japan, China, Taiwan, and southeast Asia. It is not hardy in the north.

Cocculus laurifolius

Garden and Landscape Uses. The Carolina moonseed and *C. trilobus* may be used as foliage vines in partly shaded locations where the soil is not excessively dry. Their fruits make some display in fall. The first is hardy about as far north as southern New York, the other into southern New England. The most useful cultivated member of the genus *C. laurifolius* is unfortunately not hardy in the north, but in the south and other warm regions is a first-rate, general purpose foliage shrub, especially beautiful because of its handsome leaves, and useful for grouping or for planting as a single specimen. It flourishes in part-shade or sun and responds to fertile, reasonably moist soil. This kind also makes an attractive plant for growing in pots or tubs in greenhouses or, in warm climates, outdoors.

Cultivation. These plants grow without trouble and are propagated easily by seeds, cuttings and, the vines, by root cuttings. In pots or tubs *C. laurifolius* thrives in porous, fertile soil kept evenly moist. A greenhouse or similar accommodation, where the winter night temperature is 50 to 55°F, suits it, as does a moderately moist atmosphere. Established plants in containers are benefited by occasional applications of dilute liquid fertilizer from spring through fall. Any pruning and repotting needed is done in late winter or spring.

COCHEMIEA (Coche-mièa). Baja California is the home territory of the five species of small cactuses that belong in this genus. They belong in the cactus family CACTACEAE, are related to *Mammillaria,* and by some botanists are included there. From Mammillarias they differ in having curved, tubular, two-lipped, asymmetrical blooms, smaller, but otherwise rather like those of *Schlumbergera truncata* and *Aporocactus.* The stamens and styles protrude. Rotund to egg-shaped, the fruits, red and smooth, contain black seeds. All species have clustered stems. The name comes from that of a tribe of Indians who once inhabited Baja California.

Not uncommon in cultivation, *C. poselgeri* (syn. *Mammillaria poselgeri*) is of distinctive appearance. Its many prostrate stems, up to 6 feet long and bluish- to grayish-green or in sun tinged reddish, are about 1½ inches in diameter. They are beset with rather widely spaced, somewhat flattened, conical tubercles (projections). From the top of each of these radiate seven to nine comparatively short spines and one spreading, strongly-hooked central 1 inch long or longer. They are dark yellow or red aging to gray. The depressions between the tubercles contain white wool and sometimes bristles. About 1¼ inches long, the handsome shining red blooms are borne near the tops of the stems. A plant in full bloom is spectacular. Very different is *C. halei.* Its stems are clustered, erect,

and 1½ feet long or a little longer by 2 to 3 inches wide. There are woolly hairs in the axils of the short tubercles from the tips of which spread three to six very stiff, straight central spines, 1¼ inches long and not hooked, and ten to twenty stiff radial spines one-half as long. The spines are darker at their extremeties. At first they are reddish-brown, with age they become yellowish or gray. The scarlet blooms, almost or quite 2 inches long, and borne in a circle around the tops of the stems, are succeeded by ½-inch-long, club-shaped fruits.

The erect, dull green stems of *C. setispina* (syn. *Mammillaria setispina*) are up to 1 foot tall and 1½ inches thick. The short tubercles, square at their bases, have white wool between them. From the tip of each extend one to four rigid, hooked central spines, 1 inch to 2 inches long. The nine to twelve spreading radial spines are more or less curved, white tipped with black, and slender. The scarlet flowers, 1¼ inches long, are in a circle around the top of the stem. The fruits are 1¼ inches long, egg-shaped, and red.

Off the western coast of Baja California *C. pondii* (syn. *Mammillaria pondii*) is an island-dweller. Its bright green stems are up to 2 feet tall and up to 2½ inches in diameter. They have short, conical tubercles from which spring six to ten central spines, 1¼ inches long, and fifteen to twenty-five shorter, slender, subsidiary ones in two spreading rows, the outer white, the inner brown. The central spines are dark brown at their extremities. One or two of each group are hooked. There are woolly hairs and bristles between the tubercles. The 2-inch-long flowers are light red. The egg-shaped fruits, ¾ inch long, are pale scarlet. First discovered in 1935, *C. maritima* forms patches up to 3 feet wide of glaucous, blue-green stems, up to about 1½ feet tall, erect, procumbent, or prostrate with their tips upturned. They are covered with conical tubercles with woolly hairs in the depressions between. The clusters of reddish-brown spines are of ten to fifteen radials, under ½ inch long, and four centrals, three ascending and up to ¾ inch long, the lower one hooked and up to 2 inches long. The flowers are scarlet, cylindrical- or trumpet-shaped, and 1¼ inches long. The egg-shaped fruits are about ½ inch long.

Garden and Landscape Uses and Cultivation. These are the same as for mammillarias and other small desert cactuses. Cochemieas come readily from seed. They are often propagated by grafting. For more information see Cactuses.

COCHINEAL PLANT is *Nopalea cochenillifera.*

COCHLEANTHES (Coch-leánthes). By some authorities included in *Chondrorhyncha,* the genus *Cochleanthes,* formerly *Warscewic-*

zella, of the orchid family ORCHIDACEAE is related to better known *Zygopetalum*. Native at high altitudes from Mexico to South America and in the West Indies, it includes about a dozen species. The name, presumably derived from the Greek *kochlos*, a spiral shell, probably alludes to the lips of the flowers.

These interesting orchids are without pseudobulbs and have tufts of loose fans of leaves from the bases of which come solitary, stalked, often very beautiful blooms. The flowers have three sepals and two petals of similar appearance and a lip with two or more lobes.

Native of Cuba, Costa Rica, and Panama, *C. discolor* (syns. *Warscewiczella discolor*, *Chondrorhyncha discolor*) has linear to lanceolate, elliptic, or oblanceolate leaves 1 foot to 2 feet long by about 2 inches wide. Usually horizontal, or more or less drooping, the flower stalks, up to 5 inches long, terminate in a 3-inch-wide, fragrant bloom. The white sepals are sometimes yellowish toward their apexes. The forward-pointing petals are white suffused with purple or violet. The lip, dark violet- to blue-purple, has a conspicuous white to yellowish-white crest. In habit of growth similar to the last, *C. aromatica* (syns. *Warscewiczella aromatica*, *Chondrorhyncha aromatica*) has flowering stalks up to 4 inches long. The fragrant blooms, with yellowish-green to pale green, spreading sepals and petals up to 1 inch

long, have a lip with a lavender disk and violet-blue callus.

Garden Uses and Cultivation. Choice orchids for collectors, the sorts of this genus are usually accounted rather difficult to manage. They are best adapted for growing in hanging baskets in a cool, humid environment. In winter a night temperature of 50 to 55°F, with an increase of five to ten degrees by day, is adequate. A rooting mix of shredded tree fern fiber and sphagnum moss is suggested. This must be kept damp throughout the year. For more information see Orchids.

COCHLEARIA (Coch-leària)—Scurvy-Grass. The twenty-five species of *Cochlearia* are natives of the northern hemisphere, chiefly of sea coasts and mountains. They belong in the mustard family CRUCIFERAE. The name, from the Greek *cochlear*, a spoon, alludes to the shape of the leaves. Scurvy-grass is used as a vernacular name because in times past the plants were esteemed as articles of diet to be taken on sea voyages to ward off dreaded scurvy, since they contain ascorbic acid.

Cochlearias are hairless or hairy annuals, biennials, and herbaceous perennials, with undivided leaves. Their white or purplish flowers, in racemes, have four sepals, four petals that spread to form a cross, and six stamens, two of which are shorter than the others. The fruits are

more or less spherical or ovoid pods, with two rows of seeds in each compartment.

Scurvy-grass (*C. officinalis*), a usually rather fleshy biennial or perennial, has upright stems up to 1½ feet tall, and kidney- to heart-shaped leaves, the basal ones long-stalked. The stem leaves are mostly stalkless, or the lower ones short-stalked. The tiny white flowers are succeeded by lax racemes of small spherical to egg-shaped seed pods. Danish scurvy-grass (*C. danica*) is a slender, not fleshy annual or biennial with both basal and stem leaves stalked, and the lower ones lobed. The flowers are white to purplish. This species chiefly inhabits sea coasts in northern and western Europe.

Garden Uses and Cultivation. Because when bruised scurvy-grass has an unpleasant smell and its flavor is bitter, it is not esteemed as an edible. The only reason for growing these plants is interest in their past uses. They are easily raised from seeds sown where the plants are to remain in lightly shaded or sunny places in moistish soils. They are treated as annuals.

COCHLENIA. This is the name of orchid hybrids the parents of which are *Cochleanthes* and *Stenia*.

COCHLIODA (Cochli-òda). Four or five Andean species of the orchid family ORCHIDACEAE make up *Cochlioda*, the name of which comes from the Greek *kochlion*, a little snail, and refers to the curious shape of the callus on the lip of the flower.

Cochliodas are graceful and beautiful. They have been hybridized with odontoglossums to give *Odontioda*, with oncidiums to give *Oncidioda*, and with miltonias to give *Miltonioda*. Somewhat resembling odontoglossums in aspect, but different in their floral structure, cochliodas have pseudobulbs and one or two oblongish leaves that narrow into stalks at their bases. Their usually bright scarlet or orange-scarlet blooms are in arching racemes that come from the bases of the pseudobulbs. The spreading sepals and petals are nearly alike. The lip has an erect basal claw and a spreading blade with a narrow middle lobe and side ones that often are reflexed.

The three species discussed here, natives of Peru, are much alike. They have grooved, somewhat flattened, egg-shaped pseudobulbs up to about 3 inches long, and more or less linear leaves. Those of *C. densiflora* (syn. *C. noezliana*) are about 10 inches long. Its flowers, almost 2 inches wide, are in rather dense racemes and brilliant scarlet. The tips of their sepals and petals recurve; the callus of the lip has four hairy ridges. This flowers in summer and fall.

Rose-red rather than scarlet or orange-scarlet blooms, in rather loose racemes 1 foot to 1½ feet long, are borne by *C. rosea*

Cochleanthes aromatica

and *C. vulcanica* in fall and winter. The racemes of *C. rosea* tend to be longer than those of *C. vulcanica,* and its 1½-inch-wide blooms somewhat smaller. The leaves of *C. rosea* are generally wider than the leaves of *C. densiflora* and *C. vulcanica.*

Garden Uses and Cultivation. Cochliodas are delightful for orchid collections. They prosper in cool greenhouses under humid atmospheric conditions that suit high-altitude odontoglossums. Shade from strong sun is needed. These orchids flourish when confined to containers rather small for the size of the specimen. No well-marked rest period is required. The plants are watered moderately throughout the year. Potting is done often enough to preclude any danger of the rooting medium, which may be a mixture of osmunda and chopped sphagnum or other material suitable for epiphytic orchids, rotting to the extent that it becomes stale and sour and impedes the passage of air and water. For more information see Orchids.

COCHLIOSTEMA (Cochlio-stéma). Until 1960 only one species of quite remarkable *Cochliostema* of the spiderwort family COMMELINACEAE was known. It had been given two names, *C. odoratissimum* and *C. jacobianum,* but the plants so identified represented a single species and, in accordance with the botanists' rule of priority its name must be *C. odoratissimum.* This kind was discovered about 1847, so well over a century elapsed between its finding and the discovery of the only other known species *C. velutina.* So far as is known, *Cochliostema* is confined in the wild to Central America and tropical South America. The name comes from the Greek *kochlos,* a spiral shell, and *stema,* a stamen. It alludes to the twisted anthers.

This genus consists of stemless or stemmed, succulent, evergreen herbaceous perennials that grow as epiphytes (that is, they perch on trees without extracting nourishment from their hosts, rather than in the ground). Arranged in great rosettes, the succulent leaves overlap at their bases.

Although members of a completely unrelated family, in gross appearance cochliostemas resemble giant bromeliads, at least until they bloom. Then, instead of a flower spike developing from the center of the foliage rosette, as is commonly the case with bromeliads, bracted, panicle-like clusters of blooms of rare beauty are borne from the leaf axils. The flowers each have three sepals and petals and a hooded staminal column, which encloses three spirally-twisted anthers. The stalks of the fertile stamens develop as prominent wings. There are three staminodes (sterile stamens). The fruits are capsules.

Stemless *C. odoratissimum,* of Costa Rica, Ecuador, and Panama, has rosettes of hairless, channeled, broad-strap-shaped

Cochliostema odoratissimum

to oblanceolate leaves up to 3 feet long or sometimes longer. The flower panicles, 1 foot to 2½ feet in length, have pinkish or purple-tinged, whitish stalks. They produce a succession of blooms over a period of several weeks, usually in late summer or fall, but sometimes in late winter and spring. Individual flowers, 2 to 2½ inches across, are fragrant. They have rich blue-violet petals fringed with long hairs and are subtended by large concave bracts. Two of the staminodes are linear and erect, the other shorter and plumose. In National Horticultural Magazine, volume 37, page 43 (1958) Ernest P. Imle states that a specimen given by him in 1952 to L. Maurice Mason, who took it to the Royal Botanic Gardens, Kew, England, was brought to flower there for the first time. This is surely an error for to the certain knowledge of this writer (T. H. Everett) *C. odoratissimum* bloomed well at Kew in 1925 and 1926. It flowered at The New York Botanical Garden in 1941 and 1942.

Semiclimbing or sprawling *C. velutinum,* of Colombia, has stems up to 6 feet in length rooting from the nodes. The spatula-lanceolate leaves are 1¼ to 1½ feet long, green above, maroon on their undersides, and finely-downy on both surfaces. The densely-hairy, crowded panicles have boat-shaped bracts 1¼ inches long among which are crowded flowers with green sepals ¾ inch long and densely-fringed, pale blue-lavender petals of about the same length.

Garden and Landscape Uses and Cultivation. Cochliostemas are striking ornamentals for outdoor cultivation in the moist tropics and for growing in humid, tropical greenhouses where a minimum temperature of 60 to 70°F is maintained. They need good light, with a little shade from strong, direct summer sun. When exposed to much sun the margins of their leaves assume a reddish hue. They grow best in a coarse, loose soil. One that contains an abundance of osmunda fiber or bark chips (of the kind in which orchids are potted), as well as some fibrous turf, peat, or very coarse leaf mold, some coarse sand, and chopped charcoal, is satisfactory. The containers should be thoroughly drained so that water passes through readily. It is important not to permit the roots to become cramped for space, at least not until the plants have attained their full size. At no time should the soil become really dry. In summer abundant supplies of water are needed. Specimens that have filled their containers with healthy roots benefit from weekly applications of dilute liquid fertilizer from spring through fall. Propagation is by division or offsets in early spring and by fresh seeds sown in sandy peaty soil in a temperature of 65 to 75°F. The seeds germinate readily. Flowering plants set seeds freely if the flowers are hand-pollinated when the stigmas are receptive. Seedlings kept growing without check bloom during their first year.

COCHLOSPERMACEAE — Cochlospermum Family. Dicotyledons numbering twenty to twenty-five species distributed in two genera constitute this family of trees and shrubs. They have alternate, usually palmately-lobed leaves and racemes of large, showy, sometimes slightly asymmetrical blooms, with four or five each overlapping or contorted sepals and petals, many stamens, and one style. The fruits are large capsules. The genus *Cochlospermum* is cultivated.

COCHLOSPERMUM (Cochlo-spérmum)—Buttercup-Tree. The most important genus of the cochlospermum family COCHLOSPERMACEAE is *Cochlospermum*. The only other is *Amoreuxia*, which is represented in the natural flora of the southern United States but is not known to be cultivated. There are fifteen to twenty species of *Cochlospermum*, all natives of the tropics and subtropics, many of arid or semiarid regions. The name, derived from the Greek *cochlo*, to twist, and *sperma*, a seed, alludes to a characteristic of the seeds.

Cochlospermums are trees and shrubs, some with short, stout, underground stems, and characteristically deciduous during the dry season. They have alternate, lobed or divided leaves and large, bisexual, symmetrical or slightly asymmetrical, yellow flowers in racemes or panicles. These appear about the end of the dry season before new leaves come and have four or five sepals, the same number of petals, and many stamens. The fruits are capsules.

The buttercup-tree (**C. vitifolium** syn. *Maximilianea vitifolium*) is commonly cultivated in the tropics and subtropics including Hawaii, Florida, and southern California. Native to Central and South America, it is a highly ornamental slender tree 25 to 40 feet in height. In early spring when bright yellow flowers, 3 to 4 inches in diameter, decorate its naked branches, this tree makes a quite gorgeous display. Its flowers are reminiscent of huge single roses. Even as a small, young specimen this tree blooms. The flowers have five sepals, five petals, and many stamens with twisted filaments (stalks). The velvety fruits, about 3 inches long and egg-shaped, contain numerous small, kidney-shaped seeds, with long, silky, white hairs attached. Its long-stalked leaves, 4 inches to 1 foot in diameter, are deeply-five- or occasionally three- or seven-lobed, and toothed. Except that they do not consist of separate leaflets, they have much the aspect of horse-chestnut leaves. A beautiful variety, common in Puerto Rico and the Virgin Islands

Cochlospermum vitifolium, double-flowered

and planted to some extent in Florida, has double flowers with many bright yellow petals 1½ to 2 inches long and numerous orange stamens; there is no pistil, and hence no seeds are formed.

The silk-cotton tree (**C. religiosum**), of India and Burma, is a tree up to 18 feet high. Its deeply-three- to five-lobed leaves, 6 to 8 inches across, are gray-hairy on their undersides. The about 3-inch-wide flowers are bright yellow. This tree is the source of a commercial gum widely used as a substitute for gum tragacanth.

Garden Uses and Cultivation. These are delightful ornamentals for essentially frost-free climates. They need full sun and grow satisfactorily without special care in any ordinary, well-drained garden soil. Propagation is by seeds and by cuttings of leafy shoots. Except for the double-flowered buttercup-tree, which can only be reproduced by cuttings or by grafting onto seedling understocks, seeds provide the preferred means of increase.

COCK'S EGGS is *Salpichroa origanifolia*.

COCK'S FOOT. See *Dactylis glomerata*.

COCKSCOMB. See *Celosia*. Cockscomb-yam is *Rajania pleioneura*.

COCKSPUR CORAL TREE. See *Erythrina crista-galli*. Cockspur thorn is *Crataegus crus-galli*.

COCO-DE-MER or DOUBLE-COCONUT is *Lodoicea maldivica*.

COCO-PLUM is *Chrysobalanus icaco*.

COCOA. See Theobroma.

COCONUT is *Cocos nucifera*. The double-coconut is *Lodoicea maldivica*.

COCOS (Cò-cos)—Coconut Palm. Of the approximately 2,500 species of palms, the coconut is best known to inhabitants of non-palm-growing regions. It has been romanticized in novels, songs, and verse so that to millions of residents of temperate climates who have never seen a palm in its native surroundings the word conjures visions of blue oceans, tropical shores, agreeable natives, monkeys, and coconut palms. Yet few of these millions can know how much they owe the coconut palm for the comforts of their daily lives in lands far removed from tropical shores. It is the most important commercial palm in the world, cultivated so extensively it was estimated that in 1966 plantation grown trees numbered six hundred million, or one for every five persons in a world population of three billion—approximately one coconut palm for every family on earth! And these figures do not include vast numbers planted as neighborhood trees by villagers

in the tropics. But more of the uses of the coconut later, let us consider the tree.

Formerly many species of palm were considered as belonging to *Cocos* of the palm family PALMAE, but modern botanists have transferred all except the coconut to other genera. Thus, the plant once called *C. flexuosa* is *Syagrus flexuosa*, that previously named *C. plumosa* is *Arecastrum romanzoffianum*, and the species formerly known as *C. weddelliana* is *Microcoelum weddellianum*. The name is derived from the Portuguese *coco*, a monkey, and alludes to the fanciful likeness of the coconut fruit to a monkey's head.

The coconut palm (**C. nucifera**) has been cultivated since prehistoric times. It is not now known where it was first native, but

Cocos nucifera, solitary specimen beside the sea

it is presumed to have originated in the islands of the Pacific Ocean. Be that as it may, it has never been found truly wild, every coconut palm is planted by man or derived from such a planting. Despite the fact that the buoyant nuts can float for long distances and for long periods in the ocean and retain their vitality, and are transported by currents, the conditions they meet when stranded are generally so adverse that there are few records of the successful establishment and spontaneous development of washed-up nuts into trees. The chief deterrents to growth are saline water in the soil and crabs that eat the young shoots. To quote the great authority on palms Professor E. J. H. Corner of Cambridge University, England, "There is no island or shore where its presence is not due directly or indirectly to its having been planted by man." The antiquity of the coconut in cultivation is indicated by Sanskrit names for it going back as far as 1000 B.C.

Characteristically the coconut palm has a solitary, roughened, more or less leaning trunk thickened at its base and up to 80 feet tall, topped by a crown of graceful pinnate leaves up to 15 feet long. The long-pointed, narrow leaflets may number a hundred or more and are up to 3 feet long. The leaf bases do not form a crownshaft or cylindrical continuation of the trunk as do those of many palms. Borne among the leaves, the flower clusters are comparatively short and have a stout, grooved, woody bract. The majority of the blooms in each cluster are males. These have six stamens and are very much smaller than the females. The male blooms are on the upper parts of the branches, the females solitary or in pairs at the base of each

Cocos nucifera, a commercial plantation

Cocos nucifera, in fruit

Fruit of coconut with one-half of the outer fibrous husk removed to show the seed or nut inside

branch. Male flowers begin opening about 6 A.M., are fully open two hours later, and drop the same afternoon. Females remain receptive for a few days. The fruits are very large and are covered with a thick fibrous coat. They are one-seeded and have a lining of edible white "meat" and a large central cavity that contains a watery nutrient juice. As they ripen the fruits become green, ochre-yellow, or reddish-orange, and finally dull brown. As is to be expected of a species so widely cultivated for so long, the coconut palm has given rise to many varieties varying in height, fruit characteristics, and so on. One of the more remarkable is the king coconut,

The king coconut

which, contrary to the implication of its name, is a dwarf that begins to fruit when three to four years old and later produces 100 to 400 nuts a year as compared with the 30 or 40 of most common coconuts. The tree is so low that it is often necessary to prop the bunches of developing fruits up so that they do not touch the ground. The king coconut originated in the Adaman Islands, was introduced to Brazil in

1925, and is quite widely distributed, but later it was overlooked and by 1940 only two specimens could be found in Brazil. Then a campaign of propaganda and propagation was instituted and the virtues of the variety, its dwarf size, early bearing, productivity, and sweetness of its fruits, were made widely known and extensive planting was undertaken. In 1962 it was estimated that there were more than ten million king coconut trees in Brazil, all raised in a little more than twenty years as the progeny of two trees.

The coconut is a tropical species, indeed so far as coastal areas are concerned, one might regard its presence as a true indicator of tropical as opposed to subtropical climates. It withstands neither cold nor dryness. In general it is abundant along coasts, but not far inland, wherever there is adequate moisture between latitudes 26 degrees north and south. For millions of native peoples it has supplied nearly all their needs for housing, household goods, implements, food and drink, and fuel. To world commerce its most important contribution is copra, the dried meat of the nuts. From copra is expressed coconut oil, a fat that solidifies at 74°F and is in great demand for margarines and vegetable shortenings. It is used also in cosmetics, shaving creams, shampoos, salves, and soaps. The residue, after extraction of the oil, is excellent cattle feed. In 1966 the annual world harvest of copra was estimated to be more than three million metric tons, with an oil equivalent about fifty percent more than the production of the African oil palm (*Elaeis guineensis*) and more than three times that of the olive (*Olea europaea*). In addition to the uses mentioned, the oil is used as an illuminant, the juice from the cavity of the nuts is a refreshing beverage and that expressed from the meat of the nut is a nutritious liquid, coconut milk. From the young flower clusters sap drunk fresh as palm wine is obtained. It is also converted to vinegar and arrack. The fiber of the outer husk, called coir, is used for mats, ropes, baskets, brushes, and other purposes, the shells of the nuts are made into utensils and other useful and ornamental articles and are converted into the high-grade charcoal used in gas masks. The wood is used for construction and cabinetmaking, and as fuel. The young buds are eaten as palm cabbage or hearts of palm in salads. The meat of the nuts is eaten raw, and shredded and dried as dessicated coconut in confectionary. The leaves are used for thatching and mat- and basketmaking.

Garden and Landscape Uses. The coconut palm is exceedingly graceful and beautiful, one of the most lovely of palms. It is especially well adapted for growing near the sea and will thrive inland provided the climate is uniformly humid and warm tropical. It prospers in Hawaii and southern Florida. It is effective in groups and also as solitary specimens, but care should be taken to plant it only where the falling fruits are unlikely to cause damage, or to remove the fruits from the tree before they are sufficiently ripe to drop. The coconut palm has not proved highly successful as a permanent plant in greenhouses, but young specimens are often grown in greenhouses and conservatories devoted to useful plants.

Cultivation. This palm is at its best in fertile soil that is fairly moist and in full sun. It is propagated exclusively by seeds (nuts), which are planted intact as they fall from the tree in shaded, moist seedbeds outdoors. It is perfectly practicable to plant the nuts where the palms are to grow or they may be started in nursery beds and transplanted once or twice before being set in their permanent locations when three or

Cocos nucifera, seedling

Cocos nucifera, young specimen

four years old. They are planted horizontally with the upper one-third above soil level. In greenhouses fresh seeds germinate readily in a temperature of 80 to 90°F, and the young plants grow in coarse, fertile, well-drained soil kept evenly moist in a humid atmosphere where the minimum night temperature is 70°F and the day temperature 70 to 90°F. For additional information see Palms.

CODIAEUM (Codi-aèum)—Croton. The common name of these spectacular foliage plants is as well known in the tropics and subtropics as are lilac and forsythia in the north. Yet the botanically-minded must be cautious of its application, for there is another genus, of no horticultural significance, of which *Croton* is the correct generic name. The botanical designation of the colorful decoratives called crotons is *Codiaeum*. Its derivation is not clear; it may come from the Greek *kodeia*, a head, and have been given in allusion to the use of the leaves in the homelands of the plants for wreaths used in crowning ceremonies, or perhaps it is derived from a native Malayan name for these plants.

Belonging in the spurge family EUPHORBIACEAE, the genus *Codiaeum* is, almost unbelievably to those not versed in botanical relationships kin to poinsettias, castorbeans, and snow-on-the-mountain, as well as to great organ-cactus-like plants of the African deserts. More acceptable to the casual observer perhaps is their claim to relationship with colorful-foliaged acalyphas and the snow bush (*Breynia disticha*). The similarity between all these plants of such widely diverse aspects lies, of course, in the structure of their flowers.

A half a dozen species are known, which occurs natively in Malaya, Pacific islands, and Australia. None of the wild types is known to be in cultivation. The vast diversity of crotons planted throughout the warm regions of the world are all referable to one botanical variety, *C. variegatum pictum*. Although no one knows for certain where this sort originated, in all likelihood it was in the East Indies. Be that as it may,

Cocos nucifera, as a street tree

iaeum Society for more detailed information about them. The following brief descriptions are of some of the most distinctive and popular sorts: *C. v. amabile,* lanceolate leaves variegated with green, cream, pink, and purple; *C. v. andreanum,* broad, yellow leaves becoming orange-red; *C. v. appendiculatum,* green leaf blades consisting of more than one part held together by bladeless portions of midrib; *C. v. aucubaefolium,* broad leaves with many yellow, sometimes red-tinted spots; *C. v. aureo-maculatum,* small leaves spotted yellow; *C. v. 'B. Comte',* medium-broad leaves, blotched orange-red; *C. v. cronstaedii,* narrow leaves spirally-twisted and variegated with yellow; *C. v. 'Dorothy Reasoner'* medium-broad, orange-red leaves; *C. v. edmontonense,* short, narrow, spiraled, brilliant orange and red leaves; *C. v. 'Ethel Craig',* medium-sized, pointed leaves, mostly yellow and pink; *C. v. 'Fred Sander',* lobed leaves, yellow along mid-vein and sometimes at the centers of the lobes; *C. v. 'General Marshall',* long leaves at first yellow changing to orange and red; *C. v. 'General MacArthur',* somewhat lobed, dark green leaves with orange-scarlet centers. *C. v. 'General Paget',* wavy leaves nearly all bright yellow, with some green; *C. v. interruptum,* purplish-green leaves with red midribs, blade in two parts joined by portion of midrib; *C. v. 'Irene Kingsley',* deeply-lobed, red leaves; *C. v. irregulare,* irregularly-shaped leaves, with yellow and red spots; *C. v. 'Kentucky',* large, shallowly-lobed leaves, bronze, variegated with orange and red; *C. v. 'Madam Fernand Kohl',* large, lobed leaves, veined and mottled with yellow, orange, and pink; *C. v. majesticum,* very long, drooping, narrow leaves, green with yellow ribs that change to crimson; *C. v. 'Marie Dressler',* shallowly-lobed, dark red leaves with pinkish-orange centers; *C. v. 'Miami Beauty',* large, lobed leaves, purplish-green spotted with red and pink; *C. v. 'Monarch',* large, long leaves, green becoming red with carmine and red blotches; *C. v. 'Pitcairn',* broad, recurving leaves veined and irregularly blotched with red; *C. v. 'Polychrome',* slightly-lobed leaves, green and cream shaded with pink; *C. v. 'Rainbow's Starlight',* long, dark green leaves with small rose-pink spots; *C. v. reidii,* large leaves, yellow or pinkish yellow with red veins; *C. v. 'Robert Craig',* leaves brilliant dark red and green; *C. v. stewartii,* leaves with red midribs, veins and margins yellow.

Garden and Landscape Uses. Crotons are among the most useful foliage shrubs for gardens in the humid tropics and subtropics. They are sufficiently densely-foliaged to form good screening and are easy to grow and manage. They succeed in a wide variety of soils, preferring those that are deep, nutritious, reasonably moist, and slightly acid. They prosper in sun or part-shade. For the finest results part-day shade or light dappled shade is better than continuous exposure to blazing sun. They grow surprisingly well near the sea.

A chief attraction of crotons is their brilliant colors, which render them so desirable as accents in foundation plantings and elsewhere, for grouping, and for beds. But, as with all highly colored plants, it is easily possible to misuse them or overdo their use to the extent that the landscape assumes an uneasy feeling of "spottiness" and chromatic discords, instead of harmonies and pleasing contrasts, result. Whereas the value of brilliant-colored crotons properly displayed can scarcely be overstated, it must be remembered that those that shout less loudly, that are arrayed more somberly, chiefly in greens and bronzy greens, have their places too. For hedges and as backgrounds to colorful plants these are often to be preferred to their more flamboyant kin.

As container plants crotons are beautiful and can be used in warm climates to decorate patios, terraces, and similar areas. Crotons are also popular for tropical greenhouses and make quite excellent houseplants. Even in city apartments they succeed much better than many other plants.

Cultivation. Except when the production of new varieties is the objective, crotons are multiplied by cuttings and air layering. Both methods are highly dependable and easy. Cuttings root more quickly if the bed of sand, vermiculite, perlite, or other material in which they are planted is maintained at a temperature of 80 to 85°F. The atmosphere must be highly humid, and shade from direct sun provided. Spring is the most favorable propagating season. Seeds of crotons are as large as small peas. They may be sown ½ to 1 inch apart and covered to their own diameter with soil, in pots or pans (shallow pots) of sandy peaty soil in a temperature of 70 to 80°F. They germinate in one to two weeks.

Young plants, no matter how propagated, are grown in containers until they are of a reasonable size before being planted outdoors. Until newly planted specimens are thoroughly established it is important to keep them well watered. Subsequent care is minimal. In dry weather periodic soaking with water is in order, and light applications of complete fertilizer through the active growing season are helpful. On rich soils a spring and perhaps a fall application are sufficient, but on poor soils once a month is not too often to fertilize. Pruning to shape the plants may be needed from time to time. This is done at the beginning of a new season of growth.

Indoor cultivation of crotons is most practicable in a humid greenhouse with a minimum winter night temperature of 60°F. Day temperatures may rise five to fifteen degrees above that held at night. Good light, with a little shade from strong summer sun, is necessary. The soil should be fertile, loamy, and porous, and the containers well drained. Repotting is done each late winter or spring and, in the case of small plants, about midsummer also. The roots should not be disturbed more than is necessary. Crotons need generous watering, but the soil must not be constantly saturated. Spraying the foliage with water on all days when it will dry before nightfall is beneficial. Plants that have filled their containers with roots are kept in good condition by weekly applications of dilute liquid fertilizer made from spring through fall.

Pests. Crotons are susceptible to infestations with scales, mealybugs, red spider mites, and thrips.

CODONANTHE (Codonán-the). This interesting group of gesneriads (plants of the gesneria family GESNERIACEAE) is native from southern Mexico to southern Brazil, Bolivia, and Peru. Its cultivated members are esteemed by fanciers of its family, particularly for growing in hanging baskets and other suspended containers. There are fifteen species. The name, from the Greek *kodon,* a bell, and *anthe,* a flower, alludes to the blooms.

Codonanthes are epiphytes. This means they grow, as do many orchids and bromeliads, perched on trees without taking nourishment from those they inhabit. They are not parasites. A point of special interest is that almost invariably as wildlings codonanthes grow on the nests of ants and presumably have some special favorable (symbiotic) relationship with the insects. This, however, is not requisite because in cultivation they thrive without such association.

The genus *Codonanthe* consists of low shrubs or stem-rooting, more or less woody vines. Their leaves are opposite, but because of the suppression, near suppression, or early falling of one of each pair may appear to be alternate. When both leaves are evident they may be of equal, subequal, or markedly different sizes. They have short stalks and undivided, lobeless blades, wavy or toothed toward the apexes. Quite frequently there are small red glands on their undersides. The flowers, solitary or in clusters from the leaf axils or other parts of the stems, have a calyx, sometimes two-lipped, deeply-parted into five lobes of unequal size. The white, pink, lilac, or deep purple blooms are funnel- to somewhat bell-shaped, with a prominent to scarcely noticeable spur. They have five rounded corolla lobes (petals) that form a two-lipped face to the bloom. There are four stamens, their anthers joined in pairs or all united, at least until the flowers begin to fade. There is one mouth-shaped or two-lobed stigma. The fruits are fleshy, berry-like capsules, red, pink, yellowish-green, or orange, and contain red, pink, or yellow seeds.

Free-flowering *C. macradenia* has been confused in cultivation and horticultural literature with *C. crassifolia*. Both are trailers with pairs of short-stalked leaves of nearly equal size, hairless to more or less hairy, elliptic broad-ovate, sometimes toothed toward the apex, ¾ inch to about 3 inches long, thick, with only the midvein usually evident, and usually with conspicuous red glands on the under surfaces. Native from Mexico to northern Colombia, *C. macradenia* differs from *C. crassifolia*, the latter native throughout almost the entire range of the genus, in its somewhat bigger blooms being mottled with red along the floor of the inside of the throat and in having there a longitudinal ridge flanked by channels. The flowers are solitary or in pairs from the leaf axils. They are ¾ inch to 1¼ inches long, have densely-rough-hairy, short stalks, except for the mottling inside, and are waxy-white or tinged with pink. They have a short, clearly evident spur, a curved corolla tube, and rounded, spreading, hairy, petals. The fruits are egg-shaped, short-hairy, red-violet to maroon above, paler at the base, and a little more than ½ inch long. The flowers of *C. crassifolia* are solitary or in clusters of up to four from each leaf axil. They are up to 1 inch long or sometimes slightly more longer, with a deeply five-lobed calyx. The corollas, white or sometimes flushed with pink on the outsides of the petals, have a definite short spur. They are yellow inside along the floor of the throat, but neither ridged nor channeled there. The glistening red or seldom pink, ovate fruits, a little more than ½ inch long, contain pink seeds. Light yellow flowers are characteristic of *C. luteola* of Panama. Two distinct forms of this sparingly-hairy species exist, a diploid with toothless leaves up to 1½ inches long and flowers up to 1¾ inches long, and a tetraploid the toothless leaves of which are up to 2 inches long and the flowers of which are 1¾ to 2 inches long. The fruits of this species are undescribed.

Codonanthe crassifolia

Other ornamental species include these: *C. caribaea* (syn. *C. triplinervia*), native of Venezuela, Trinidad, Guadeloupe, and Tobago, has trailing stems and short-stalked, ovate-elliptic to broadly-ovate leaves up to 4 inches long, those of the pairs of nearly equal size. They are hairless to rough-hairy and mostly without prominent glands beneath. One to three from the leaf axils, the 1 to 1¼ inch-long white to yellowish blooms, sometimes with purple spots inside, and each with a prominent basal spur, have their four anthers joined to form a square. The fruits are about ¾ inch long. *C. carnosa* of Brazil is hairless to short-hairy and free-flowering. It has short-stalked elliptic to ovate or obovate leaves up to 3½ inches long, those of each pair of nearly equal size. Four or fewer together from the leaf axils, the flowers are ½ to 1 inch or slightly more long. They are white or sometimes pink-tinged on their outsides, yellow along the floor of the throat. The fruits are glistening red or seldom pink. *C. uleana*, a native of the upper Amazon region, is a shrubby species with thick stems, upright at first, later rigidly pendulous. Its short-stalked, hairless or minutely-hairy, elliptic-ovate, obovate, or oblanceolate leaves 1½ to 3 inches long are prominently coarsely-toothed in their upper halves. The yellow-throated, white or pinkish flowers are 1 to 1½ inches long. The fruits, about ½ inch long, are red or purple.

Garden Uses and Cultivation. Codonanthes are best adapted for humid greenhouses where the night temperature in winter is 56 to 60°F and by day is increased by five to fifteen degrees depending upon the brightness of the weather. They are best accommodated in hanging baskets. The rooting mixture must be porous and consist largely of leaf mold, peat moss, or other agreeable, partially decayed organic material that admits free passage of air and water. A recommended mixture is equal parts of sifted sphagnum moss and peat moss with the admixture of small pieces of charcoal, but other mixes that meet these physical standards are satisfactory. Take care to keep the rooting medium always moist, but not constantly soggy. This calls for carefully timed watering. No resting period with the soil kept dry is necessary or desirable. An application every two weeks of dilute liquid fertilizer benefits well-rooted specimens. Good light is a must but some shade from all direct sun except that of the early morning and early evening is necessary from spring to fall. These plants tend to become straggly after about three years. Then they should be repropagated and new baskets made up. Increase is easily had from cuttings and seeds.

CODONOPSIS (Codonóp-sis). This group of thirty or forty temperate Asian herbaceous perennials is not much known in American gardens. It is closely related to *Campanula* and like it belongs to the bellflower family CAMPANULACEAE. Its name comes from the Greek *kodon*, a bell, and *opsis*, resembling. It refers to the shape of the blooms of some kinds. A difference between *Codonopsis* and *Campanula* is that the seed pods of the former release their seeds from the top, the seeds of the latter escape through small holes in the sides of the capsule.

Codonopsises contain a milky juice. Some species are low and have erect stems, others, vining or semivining, have stems several feet long. Their leaves are short-stalked, alternate or nearly opposite, ovate to elliptic. Often stems and foliage, as well as their whitish, bluish, or yellowish flowers, are unpleasantly musk-scented very much like those of the crown imperial (*Fritillaria imperialis*). There are two sections of *Codonopsis*, one containing kinds with typically bell-shaped blooms, the other kinds with flowers much flatter and more or less saucer-shaped. The blooms have five-lobed calyxes and corollas, five stamens, and one style. The fruits are beaked capsules.

Himalayan *C. ovata*, 6 inches to 1 foot tall, has stems the lower parts of which lie on the ground with the upper parts erect. It is hairy and has leaves up to ¾ inch long. The pale blue, nodding, bell-shaped flowers, 1 inch to 1¼ inches long are on 3- to 6-inch-long stems and are usually solitary. This, the most beautiful species, is rare in gardens. Often the plant cultivated under its name is *C. clematidea*. From western China comes *C. meleagris*, which has long, fleshy roots, stems up to 1 foot tall or slightly taller, and elliptic-oblong to obovate, mostly basal, toothed leaves, glaucous beneath. The nodding, bell-shaped flowers are solitary or in pairs, 1 inch to 1½ inches long, and bluish to cream, veined on their outsides with chocolate-brown. Inside they are violet, sometimes spotted with yellow. Often taller than the foregoing kinds and usually with somewhat vining stems, *C. clematidea*, of central and western Asia, has more globose bell-shaped blooms than those of *C. ovata*. They are grayish-blue, with a purplish blotch at the base or sometimes almost white. Rare *C. convolvulacea* (syn. *C. vin-*

Codonopsis convolvulacea

ciflora) has slender twining stems and narrowly-lanceolate, ovate-lanceolate, or ovate leaves that are alternate and up to 2 inches long. The flowers are solitary, sky-blue, and have five spreading petals. In shape they resemble saucers slit almost to their centers rather than bells. This species is a native of China. Variety *C. c. forrestii* is bigger and more robust.

Native of the Himalayas, trailing or climbing, and somewhat hairy *C. rotundifolia* has terminal, solitary, narrowly-bell-shaped flowers about 1 inch long. They have large, conspicuous, spreading or reflexed, leaflike calyx lobes and yellowish-green corollas veined with purple. The leaves are broad-ovate and round-toothed. Native of northeastern Asia, *C. pilosula* has twining stems and broad-ovate leaves hairy on their undersides. Its flowers are greenish-yellow, bell-shaped, and 1 inch long or longer. Chinese *C. tangshen* is a twiner with, from its leaf axils, long-stalked, bell-shaped, purple-marked, greenish flowers. A climber of eastern Asia, *C. viridiflora* has ovate to lanceolate leaves and long-stalked, bell-shaped flowers terminal on the stems. They are about 1 inch long and have toothed calyx lobes and yellowish-green corollas, with purplish spots at their bases. A beautiful native of the Himalayas, *C. bulleyana* is erect, has nearly leafless stems, and bears solitary, bell-shaped flowers 1¼ to 1½ inches long, slightly constricted toward their mouths and then flared into five spreading lobes (petals), pale grayish-blue or watery-blue, sometimes with dark veinings.

Garden Uses. This is essentially a genus for the lover of rather choice and unusual plants. Most of its kinds are fairly hardy, but need winter protection in cold climates. In the main its flowers are less brilliant and less showy than those of *Campanula*. In some kinds the hues are so muted that color must be regarded as only a secondary attraction. Nevertheless, the finer colored kinds are delightful, and all have an air of quality for which they are esteemed by the plant connoisseur. They are most appropriate for rock gardens and choice collections of herbaceous plants.

Cultivation. Seeds germinate readily if sown in light, well-drained soil. The seedlings develop long, fleshy roots, and it is important not to damage these in transplanting. Because of this propensity for rooting deeply, it is desirable that these plants be set where the earth is congenial to a depth of a foot or more. A light sandy, slightly acid soil is appropriate.

COELIA (Coè-lia). Except one, all species previously named *Coelia* belong in *Bothriochilus*. Lone *C. triptera* (syn. *C. baueriana*) of the orchid family ORCHIDACEAE retains a name given in the erroneous assumption that the insides of the pollen masses were concave. It derives from the Greek *koilos*, hollow.

Native to Mexico, Guatemala, and the West Indies, where it occurs as an evergreen epiphyte, perching on trees or growing on rotting stumps and logs, rocks, or in the ground, *C. triptera* blooms in winter and spring. It has egg-shaped to nearly spherical pseudobulbs, about 2 inches long, that taper into a short stem that supports several arching, longitudinally-ribbed, linear leaves 1 foot to 1¼ feet in length and not more than 1 inch wide. Sprouting from the bottoms of the pseudobulbs, the erect spikes of fragrant flowers have a large brown bract at the base of each short-stalked bloom. The white flowers, about ½ inch long, do not open widely. They have conspicuously winged ovaries.

Garden Uses and Cultivation. These are as for *Bothriochilus*. For more information see Orchids.

COELOGYNE (Coeló-gy-ne). None of the 200 species of these chiefly tree-perching (epiphytic) orchids is well known outside the ranks of orchid fanciers and few are grown even by them. The group belongs in the orchid family ORCHIDACEAE and inhabits warm parts of Asia and lands of the Pacific. Its name, alluding to the hollow stigma, comes from the Greek *koilos*, hollow, and *gyne*, female.

Coelogynes are mostly evergreens. A few kinds are deciduous. They have pseudobulbs thickly clustered along short rhizomes or more widely spaced on longer ones. From one to four, but usually two leaves sprout from the top of each pseudobulb. The flowers are usually in terminal racemes of many, less often in threes or twos, or are solitary. In color and often in pattern of variegation they vary greatly between species. Small to quite large and often fragrant, the blooms have spreading or backward-inclined sepals and petals, three of the former, two of the latter. The sepals are often bigger than the petals. The keeled, usually large lip is three-lobed, with the side lobes erect and enclosing the column, or it may be lobeless. Generally more or less hollowed, it has two or more longitudinal crests. The column, joined to the base of the lip or separate, is erect or arching. There are four pollen masses (pollinia). There are few hybrids in this genus.

Best known, *Coelogyne cristata* when well grown is truly magnificent. Native of the Himalayan region, it forms mats of tightly clustered pseudobulbs. They are yellowish-green, egg-shaped to nearly spherical, and up to about 2½ inches long. Up to 1 foot long by 2 inches wide, the leaves are linear-lanceolate. Each pseudobulb has two leaves. The fragrant flowers are in arching to drooping sprays from the bottoms of the pseudobulbs. Handsome and long-lasting, they are approximately

Coelogyne cristata

Coelogyne cristata (flowers)

4 inches wide. Pure, sparkling white except for five conspicuous golden-yellow ridges on the lip, they have wide-spreading or sometimes somewhat reflexed sepals and petals. Delightful varieties are *C. c. alba* (syn. *C. c. hololeuca*), with pure white blooms, and *C. c. lemoniana* (syn. *C. c. citrina*), its flowers with lips with lemon-yellow ridges.

Other species grown by orchid fanciers include these: *C. asperata,* of Malaya, Sumatra, and New Guinea. This has nearly spherical, clustered pseudobulbs 2 to 4 inches long, each with two oblong-lanceolate, stalked leaves up to 1½ feet long by 2 inches wide. Arching to erect stalks carry a few crowded, 2- to 3-inch-wide, musk-scented blooms. Except for three fringed, brown crests on the three-lobed lip and projections of the same color from its edges they are pure white to creamy-white. *C. corymbosa,* of the Himalayan region, has obscurely-four-sided pseudobulbs about 2 inches high. Each has a pair of short-stalked leaves 6 inches long by 1½ inches wide. The erect flower spikes, from the centers of the new growths, have three to seven loosely-arranged blooms about 2 inches wide, creamy-white, with a white lip relieved by a pair of bright yellow spots margined with brown and with the throat streaked with brownish-yellow. *C. dayana*

Coelogyne corymbosa

Coelogyne dayana

Coelogyne parishii

Coelogyne pandurata

Coelogyne sparsa

dobulbs about 1½ inches in length. The two leaves of each are stalkless, linear-lanceolate to ovate-lanceolate, and up to about 5 inches long by ¾ inch wide. From the tips of the pseudobulbs the blooms come singly or in pairs. Musk-scented, they are about 1½ inches in diameter and have prevailingly greenish-yellow to tan sepals and petals. The yellow lip, its center lobe fringed, is streaked with reddish-brown. *C. flaccida,* native at moderately high altitudes in the Himalayas, has spindle-shaped, two-leaved pseudobulbs about 3 inches long. The leaves are stalked, narrowly-lanceolate, and up to 1 foot long by a little more than 1 inch wide. Somewhat objectionably strongly odorous, the 1½-inch-wide, loosely-arranged flowers are few to many in pendulous to arching racemes up to nearly 1 foot long. They are creamy-white, with a golden-yellow center with three ridges, orange-brown toward their ends, on the lip. *C. pandurata* has straggling rhizomes, with rather distantly spaced, more or less flattened, broadly-ovate to oblongish pseudobulbs up to about 5 inches long. The lustrous leaves are elliptic with tapered bases, up to 1½ feet long by 2½ inches wide. Gracefully arching, the flower stalks may have as many as fifteen blooms. These are fragrant, 3½ to 4 inches wide, and attractive bright green decorated on the rather fiddle-shaped lip with practically black, velvety, longitudinal ridges and markings, a feature responsible for this species sometimes being called the black orchid. This is a native of Borneo, Sumatra, and the Malay Peninsula. *C. parishii* is a small species from Burma. Its angled-cylindrical pseudobulbs, about 4 inches long, have each two elliptic leaves up to 6 inches long. Borne from the tops of the pseudobulbs, two to five together in short loose racemes, the flowers are fragrant and about 3 inches across. They are green with black, and marking on the lip. *C. sparsa,* native to the Philippine Islands, has pointed-ovoid

has longitudinally-pleated, narrowly-conical pseudobulbs up to 10 inches long, each with one or two short-stalked, narrow-elliptic, pointed leaves 1½ to 2 feet long by up to 4 inches wide. The musk-scented, generally two-ranked, long-lasting blooms are in loosely-many-flowered, pendulous racemes 2 to 3 feet long. About 2½ inches wide, the flowers are lemon-yellow to light tan, with the lip crested, and its wavy side lobes lined with chocolate-brown. This is a native of tropical Asia, Borneo, Java, and Sumatra. *C. elata,* of fairly high altitudes in the Himalayas, has narrow-ovoid to nearly cylindrical pseudobulbs 3 to 6 inches long, each with a pair of conspicuously-veined, stalked, lanceolate leaves 1 foot to 1½ feet long by 2 to 3 inches wide. Erect from the tops of the pseudobulbs, the flower stalks carry up to a dozen, long-lived, musk-scented blooms about 2½ inches across. They are of waxy texture and white, except for the three-lobed lip, which has a forked central band of yellow and two crisped, orange-colored ridges. *C. fimbriata,* of China, Vietnam, Thailand, and northern India, has slender rhizomes, with fairly distantly spaced, ellipsoid pseu-

pseudobulbs about 1½ inches long, each with a pair of short-stalked, oblong-lanceolate leaves up to 4 inches long by 1½ inches wide. The flowers are in short arching racemes of up to seven from the bottoms of the pseudobulbs. They are 1 inch to 1½ inches wide, fragrant, and white lightly suffused with green. The brown- or purplish-spotted lip is three-lobed. There is a yellow spot in the throat and three yellow or orange-yellow ridges.

Garden Uses and Cultivation. Coelogynes are choice orchids for inclusion in collections. Mostly they are grown in greenhouses, outdoors, or in lath houses where climates approximate those where they are native. They may be accommodated in pots, pans (shallow pots), hanging baskets, or on rafts or slabs of tree fern trunk. Hanging baskets, suspended pots or pans, or rafts are best for those with pendulous flower spikes. Kinds with long rhizomes are usually best on rafts. Any of the commonly used rooting mediums for orchids may be used for coelogynes, but because they are notoriously impatient of root disturbance, it is well to select one that will not break down and rot after a short time. Osmunda fiber is admirable. When it becomes necessary, repot just before new growth begins. Poor drainage is anathema. The roots of coelogynes need a free circulation of air. Evergreen kinds grow nearly continuously and so must be watered throughout the year to keep the medium in which the roots grow always moderately moist. When leafless, water is withheld from deciduous kinds. The atmosphere must be humid. Sufficient shade to prevent the foliage scorching is necessary. Temperature requirements are related to those of the lands in which the species are native. Coelogynes are conveniently classed as cool (winter night temperature 50°F), intermediate (winter night temperature 55°F), and tropical (winter night temperature 60°F or slightly higher). By day all temperatures may be increased by five to ten or fifteen degrees depending upon the brightness of the weather. During summer and early fall warmer conditions will prevail naturally. Cool climate coelogynes are difficult, and in some cases impossible to grow in torrid regions and where hot summers are experienced unless accommodated in air-cooled greenhouses. Of the kinds described here *C. cristata*, *C. corymbosa*, and *C. elata* need cool temperatures, *C. fimbriata* is satisfied with cool-intermediate temperatures, *C. flaccida* and *C. sparsa* need intermediate temperatures, *C. dayana* will prosper with intermediate-tropical temperatures, and *C. asperata* and *C. parishii* are decidedly tropical in their requirements. For more information see Orchids.

COELOPLATANTHERA. This is the name of orchid hybrids the parents of which are *Coeloglossum* and *Platanthera*.

COFFEA (Cof-fèa)—Coffee. The beverage coffee is prepared from the seeds, called beans, of a few of the forty species of *Coffea* of the madder family RUBIACEAE. Although Brazil is the most important source of plantation coffee, the plants are not natives of South America. The entire genus is endemic to the tropics of Africa and Asia. The generic name is a modification of an Arabic one for the beverage, itself perhaps derived from Caffa, the name of a part of Ethiopia.

Coffeas are evergreen trees and shrubs with mostly opposite, undivided leaves and axillary clusters of small, fragrant, white or creamy, starry flowers that have a five- or less often a four-parted calyx, and a slender, tubular corolla with four to seven spreading corolla lobes (petals). The fruits are berries that contain usually two horny seeds.

The commercial production of coffee beans is a highly skilled agricultural enterprise. The quality and flavor of the product is largely dependent upon the species of

Coffea, the climate and soil in which the trees are grown, and the roasting procedure. Until roasted the beans are flavorless. In general, mild-flavored coffees come from lowland plantations and richer ones from higher elevations. The trees are restricted to manageable size by frequent pruning, and much effort is expended on fertilizing and weed control as well as picking the "cherries," as the ripe fruits are called. Young plantations are shaded, usually by trees of other species. Favorites for this purpose are species of *Erythrina*.

Arabian coffee (*C. arabica*), a native of Ethiopia, is 15 to 40 feet in height and has glossy, thinnish, pointed, oblong-ovate

Coffea arabica (flowers)

Coffea arabica (fruits)

leaves 3 to 6 inches long by about one-third as wide. Its flowers have five or sometimes four petals more than twice as long as they are wide. The fruits are dark red and ½ inch long. There are a number of commercial varieties. They are the chief source of coffee. Congo coffee (*C. canephora* syn. *C. robusta*), of West Africa, has thicker leaves than Arabian coffee and is a more vigorous and bigger tree. Its fruits, reddish when ripe, are about ½ inch in diameter. Liberian coffee (*C. liberica*) differs in having leaves up to 1 foot long and shorter pointed than those of *C. arabica*, and broadest

above their middles, flowers with six or seven petals, and fruits, ¾ inch long, that are black when ripe. It is a native of tropical Africa. Its product is inferior to that of Arabian coffee. Zanzibar coffee (*C. zanguebariae*) rarely is more than 6 feet in height. It has blunt or short-pointed, ovate to obovate leaves 2 to 4 inches long by less than one-half as wide. The flowers have six or seven petals and are succeeded by fruits at first red that ripen black. They are ½ inch long. This sort is native to tropical Africa.

Garden and Landscape Uses. Coffees are glossy-foliaged shrubs and trees that grow rapidly and are attractive ornamentals. Their foliage, flowers, and fruits are all decorative. The fruits ripen in succession over a long period so that at one time the branches are strung with a pleasing combination of green and red berries. Only in frost-free, humid, warm regions will these plants prosper outdoors. In greenhouses they succeed in ground beds, pots, and tubs. They prefer fertile soil with a high organic content. Some effort has been made to popularize them as houseplants, but where winters are cold and much heat is used to keep houses warm they are not likely to succeed because of the dryness of the atmosphere. Fertile, well-drained, moderately moist soil is satisfactory.

Cultivation. As ornamentals coffees are of easy cultivation. They sprout readily from viable seeds sown in a temperature of about 70°F and grow rapidly. In greenhouses they succeed best where the minimum temperature is about 60°F, although they will stand a few degrees lower. The atmosphere must be humid, and they should be afforded as much light as they will stand without the foliage scorching. Good soil drainage is important and so are adequate supplies of moisture. Stagnant water in the soil can be fatal. Coffees are "hungry" plants. If they lack nourishment, the older leaves quickly yellow and drop, and the plants assume a thin, bedraggled appearance. This can happen, too, if they are permitted to become dry from time to time. Especially when they are confined in containers they should be given regular applications of dilute liquid fertilizers. In late winter or spring specimens grown for ornament are pruned to size and shape and are top dressed, repotted, or given whatever other attention they need to promote strong new growth.

Pests. The most likely pests are scale insects, red spider mites, and mealybugs.

COFFEE. The source of coffee is the genus *Coffea*. Some other plants have the word coffee as part of their common names. Here belong coffee berry (*Rhamnus californica*), coffee-tree (*Polyscias guilfoylei*), Kentucky-coffee tree (*Gymnocladus dioica*), and wild-coffee (*Polyscias guilfoylei, Psychotria* *nervosa, P. sulzneri, Royena lucida, Triosteum aurantiacum,* and *T. perfoliatum*).

COGSWELLIA. See Lomatium.

COGWOOD is *Ziziphus chloroxylon.*

COHOSH OR BANEBERRY. See Actaea. Black cohosh is *Cimicifuga racemosa,* blue cohosh *Caulophyllum thalictroides.*

COIX (Cò-ix)—Job's Tears. One attractive species of this genus of five East Indian kinds is cultivated. The group belongs in the grass family GRAMINEAE. Its name is an ancient Greek one for some unidentified reedlike plant. Members of the genus *Coix* are tall, erect, branching, leafy grasses. They have the peculiarity of developing in their flower clusters hard, hollow, beadlike bodies that morphologically are modified leaf sheaths. These form the decorative parts of the clusters and are the "tears" of the only cultivated kind, Job's tears. Male and female flowers are in separate spikelets; the females, except for their styles, are contained within the bead, the male spikelets project beyond the bead tips.

Job's tears (*C. lacryma-jobi*) is an annual, 2 to 6 feet tall, with conspicuously-

Coix lacryma-jobi

jointed smooth stems and long-pointed, sword-shaped leaves ½ inch wide or wider, with prominent midribs. The glossy "beads" or "tears" are solitary at the ends of slender stalks. They are about ½ inch long and slightly less wide. When mature they are pearly-white, lead-colored, or purplish. In *C. l. aurea-zebrina* the leaves are striped with yellow. Several other varieties are cultivated in the Orient. In some places the "beads" are used as food and medicinally. They are commonly made into rosaries, necklaces and other decorative articles. In many parts of the tropics this species is naturalized.

Garden Uses. Job's tears, popular in gardens of yesteryear, is less commonly cultivated than formerly. Modern taste seems to favor more colorful annuals. Yet this old-fashioned favorite should not be neglected. Children find fun in stringing the beads, and sprays of the plant can be cut and dried for indoor bouquets. This is done by taking them just before they reach full maturity, tying them in small bundles, and hanging them upside down in a shaded, dry, airy place until they are thoroughly dry. In days gone by strings of beads of Job's tears were given to teething babies.

Cultivation. Given a reasonably fertile soil not excessively dry and a sunny location, the cultivation of this interesting grass presents no difficulties. Where long growing seasons prevail it may be sown outdoors in early spring and the seedlings thinned to about 1 foot apart. In the north better results are had by sowing indoors some eight weeks before the ground is sufficiently warm and the weather settled enough to transplant the young plants to the garden. This is done at about the time it is safe to make the first sowing of corn. The seeds are sown indoors in a temperature of 60 to 65°F, and the seedlings transplanted to flats or individually to small pots and grown in a sunny greenhouse or window where the night temperature is about 55°F and the day temperature five to ten degrees higher. A sunny location is needed. Before planting out, the plants are hardened for a week or two by standing them in a cold frame or sheltered, sunny place outdoors. A planting distance of 1 foot between individuals is satisfactory. Subsequent care consists of weeding and, in dry weather, watering.

COJOMARIA is *Paramongaia weberbaueri.*

COLA (Cò-la)—Cola, Kola. Like coffee and chocolate the exploited species of *Cola,* of the sterculia family STERCULIACEAE, owe their usefulness to the presence of the stimulating alkaloid caffeine in their seeds. The highly popular cola beverages of the world are prepared from the seeds of *C. acuminata* and *C. nitida,* with additives. Almost all commercial cola nuts are collected in the wild in the forests of tropical West Africa, but some plants are plantation grown in the West Indies and elsewhere. Unlike the seeds of coffee and chocolate, no elaborate preparation of those of cola is needed; they can be chewed fresh as a stimulant and to assuage hunger. The seeds are boiled in water to provide the basis of cola beverages. The name *Cola* is a modification of a native one.

Colas are tropical African trees with alternate, undivided, lobed or lobeless leaves, and flowers that are unisexual on separate trees or are mixed unisexual and bisexual on the same tree. There are 125 species. The blooms have no petals and are in panicles from the leaf axils. The fruits, more or less star-shaped and leathery or woody, consist of several podlike segments each containing eight seeds.

The kinds exploited commercially, *C. acuminata* and *C. nitida,* are similar and may be merely variants of one species. They are evergreen and 40 to 60 feet tall. Their lobeless, pointed, leathery leaves are obovate and up to 10 inches in length. The flowers are yellow, ¾ to 1 inch across, and in panicles. The large fruits contain brown seeds resembling those of chestnuts, and have the odor of roses. When chewed they are at first bitter but leave a sweet taste in the mouth.

Garden and Landscape Uses and Cultivation. Aside from their commercial uses these trees are occasionally grown as items of interest in the tropics and in tropical greenhouses. They prosper in fertile, sandy loamy soil and are propagated by seed. In greenhouses they succeed where a minimum temperature of 60°F and high humidity are maintained.

COLAX (Cò-lax). This horticulturally little favored genus of the orchid family ORCHIDACEAE consists of up to five species, natives of Brazil, related to *Zygopetalum,* but differing in technical details of the pollen masses. The only kind known to be cultivated, *Colax jugosus* with *Zygopetalum* is parent of bigeneric hybrids named *Zygocolax.* The name perhaps alludes to a resemblance of the flowers to butterflies of the same name.

Very beautiful *C. jugosus* has clusters of flattened, long-egg-shaped, smooth pseu-

Colax jugosus

dobulbs 2 to 3 inches long, each topped by a pair of pointed-lanceolate, lustrous, ribbed leaves 6 to 9 inches long and up to 1½ inches wide. In addition, leaves, usually smaller, come from the bottoms of the pseudobulbs. From there also come the erect or arching flower stalks, up to 8 inches tall, sheathed with pale green bracts, and carrying two to four fragrant, fleshy, subglobose, long-lasting blooms 2 to 3 inches wide. The flowers have creamy-white sepals and petals, the latter richly blotched and cross-barred with black-purple, brown-purple or pinkish-purple. The shorter, fleshy, velvety, three-lobed lip, of the same coloring, has many hairy ridges spreading along its center. The bent column is spotted with brown or purple and is hairy at its front.

Garden Uses and Cultivation. The species described here succeeds under conditions favorable to epiphytic (tree-perching) zygopetalums. For more information see Orchids.

COLCHICINE. An alkaloid derived from *Colchicum autumnale*, colchicine has the peculiar ability of influencing normal chromosome formation within animal and plant cells. It is employed by botanists investigating cells and in low concentrations by skilled plant breeders to induce variation by increasing the number of chromosomes. The use of colchicine by amateurs is not practical.

COLCHICUM (Cólchic-um)—Autumn-Crocus, Meadow Saffron. Colchicums are not crocuses. True, superficially they look like them and they are called autumn-crocuses. But closer examination reveals that colchicums have six stamens and a superior ovary. These simple botanical facts without question remove them from the iris family IRIDACEAE, to which true crocuses belong. They are members, instead, of the lily family LILIACEAE. In addition to the six stamens and an ovary above rather than below the petals, they have six petals (or more correctly, tepals) and three long slender styles (in *Crocus* there is one style with three stigmas). The fruits are capsules. The name *Colchicum* comes from Colchis, an ancient name for a country bordering the Black Sea. The genus is confined in the wild to temperate parts of the northern hemisphere of the Old World and consists of sixty-five species. In addition there are some garden hybrids and some double-flowered kinds. The plant treated in this Encyclopedia as *Bulbocodium* is sometimes included in *Colchicum* and so are species of *Merendera*.

Having disposed of the confusion that may arise from "crocus" as part of the popular name autumn-crocus, we may consider the appropriateness of "autumn." Here we are on safer ground. The great majority of colchicums do bloom in fall, although one more tardy (or perhaps more hasty), opens its flowers in spring. To complicate matters further, there are true crocuses that flower in fall as well as the more familiar spring-bloomers. If in doubt as to whether a flower is a *Colchicum* or *Crocus* count their stamens. Another feature of importance to gardeners distinguishes colchicums from crocuses. The latter have slender, grasslike leaves, which, although they sometimes take an unconsciousably long time dying and so delay grass cutting, do not occupy much space. The foliage of colchicums, on the other

The leaves of colchicums, unlike those of crocuses, are broad and lush

hand, is broad, coarse, lush, and requires much space especially when it flops and turns yellow before expiring. Its seasonal exit is exasperatingly long and not pretty. Provided sites are carefully selected, the bad manners of colchicums with respect to their dying foliage are gladly tolerated as are a small inconvenience to pay for the startling beauty of their flowers. Those that present their blooms in fall do so with all the freshness and vitality of the more familiar bulbs of spring. They are in no way upstaged by asters, dahlias, chrysanthemums, and other fall flowers that appear at the same season. Their flowers rise, naked of foliage, like great glowing vases or goblets and withstand the buffets of moderate winds and rains remarkably well considering the somewhat frail appearance of the corolla tubes that serve as stalks to hold the blooms aloft. Being without true stalks, the flowers arise directly from the bulbs.

Compared with some other bulb plants, crocuses, tulips, and hyacinths, for instance, the color range of the flowers of colchicums is limited. It runs strongly to bright lavender-pinks and lilac-pinks, with occasional diversions to deeper purple tones and to white. One exceptional kind, the rare *C. luteum*, has yellow blooms. This species is something of a maverick in another respect; it blooms in spring.

Two of the most popular species are the meadow saffron (*C. autumnale*) and *C. speciosum*. The former is the earlier bloomer, producing its flowers in the neighborhood of New York City about the first week in September. Those of *C. speciosum* do not appear until three weeks or a month later. With usually rose-purple flowers, up to 8 inches tall, and solitary or up to as many as six from a single basal sheath, **C. autumnale** is quite variable in color and includes the excellent white-flowered variety *C. a. album*, which blooms later and has flowers rather smaller than those of the typical species. Even more entrancing, but more difficult to acquire, is the white double-flowered *C. a. album-plenum*. There is also *C. a. flore-pleno*, with double lilac-rose flowers, and *C. a. striatum*, with striped blooms.

Colchicum autumnale

Colchicum speciosum

The leaves of *C. autumnale* and its varieties are lanceolate, 6 to 10 inches long and 1 inch to 1½ inches wide. They develop in spring. Altogether different is the typical lilac-pink-flowered **C. speciosum.** It is highly variable; several varieties are recognized. From its large corms it sends in spring broad-oblong leaves 1½ to 3 inches wide and 1 foot to 1¼ feet long or longer. Three or four blooms are commonly borne from each basal sheath. They have broadly elliptic petals 1½ to 2½ inches long. Varieties are the very lovely white-flowered *C. s. album*, *C. s. bornmuelleri* (*C. bornmuelleri* of gardens), with large, clear, violet-pink blooms that have greenish or white corolla tubes, *C. s. rubrum*, with red flowers, and *C. s. maximum*, with flowers devoid of white in their throats.

Kinds sometimes confused with *C. autumnale* are *C. atropurpureum* (syn. *C. autumnale atropurpureum*) and *C. byzantinum* (syn. *C. autumnale major*). The former is native to Europe, including Great Britain, the latter of Transylvania and the neighborhood of Constantinople. Differing from those of *C. autumnale* the blooms of **C. atropurpureum** have corolla tubes up to 1 inch long that are pink and that deepen to magenta-red as they mature. The flowers

Colchicum byzantinum

of **C. byzantinum** are bigger than those of *C. autumnale* and are in clusters of up to twenty from each basal sheath. Also, its leaves are 3 to 4 inches wide. Related to *C. byzantinum* and not always easy to distinguish from it is **C. cilicicum** (syn. *C. byzantinum cilicicum*), a native of the Taurus Mountains. Its flowers, many from each basal sheath, have keeled petals and are more rounded than those of *C. byzantinum*.

Formerly considered a variety of *C. speciosum*, but now regarded as distinct, **C. giganteum** is, as its name suggests, one of the largest flowered autumn-crocuses. Its blooms do not, however, exceed those of *C. speciosum* in size and are often smaller. They appear later, are deep mauvy-pink, with pale throats, and have the fragrance of honey.

Colchicum cilicicum

Colchicum giganteum

Especially intriguing are autumn-crocuses with blooms checkered like those of the guinea-hen flower (*Fritillaria meleagris*). Here belong *C. agrippinum* (syn. *C. tessellatum*), *C. variegatum* (syn. *C. parkinsonii*), and *C. sibthorpii* (often grown in gardens as *C. latifolium*). The first named is of unknown, possibly hybrid origin; *C. variega-*

Colchicum agrippinum

tum is indigenous from Greece to Asia Minor; *C. sibthorpii* to Greece and Crete. The flowers of **C. agrippinum** come in early fall, two or three from each basal sheath and lilac-purple obscurely checkered with white. A distinguishing feature is that the slightly wavy, glaucous, linear-lanceolate leaves, which appear in spring, are erect. Those of **C. variegatum** are lanceolate and prostrate. The blooms of *C. variegatum* do not appear until late September or October. They are lilac-purple and beautifully and distinctly checkered, although the intensity of the markings varies somewhat in individuals. From the two other checkered-flowered kinds discussed here and from other cultivated autumn-crocuses, **C. sibthorpii** differs in having green pollen. Its lilac flowers are borne one to four from each basal sheath and are 1½ to 2 inches long. The wavy leaves, up to 8 inches long and less than 1 inch wide, spread widely. They appear in spring.

Hybrid autumn-crocuses, chiefly raised in Holland, are freely available and are among the loveliest and most easily grown of fall flowers. They have big substantial blooms of varying shades of lilac-pink, mauve, and lavender-pink and are described in catalogs of dealers in hardy bulbs. They have fancy names and are usually listed under *C. hybridum*. They include 'Autumn Queen', 'Lilac Wonder', 'The Giant', and 'Violet Queen', which have *C. giganteum* and *C. sibthorpii* as parents, and the marvelously beautiful double-flowered 'Waterlily', which is a hybrid between *C. speciosum album* and *C. s. album plenum* and has lilac-mauve blooms.

Garden and Landscape Uses. Colchicums belong not in tidy rock gardens

Colchicum hybridum 'Lilac Wonder'

Colchicum hybridum 'Violet Queen'

where in spring their dying foliage can be so ugly, but in locations sheltered from strong winds at the fronts of perennial beds and shrubbery borders where their inconsiderate spring behavior is less noticeable and less disconcerting. They can be used to advantage in the light shade of widely spaced trees and shrubs. At The New York Botanical Garden effective use is made of them to clothe the ground beneath mollis-type azaleas thus assuring two good seasons of attractive bloom from shared ground. These are agreeable companions, both grateful for cool, moistish, but not wet, deep soil kept fertile by moderate mulching and the occasional application of an organic fertilizer. Under congenial conditions colchicums multiply generously and need lifting, dividing, and replanting every few years. Unlike crocuses, they make little or no appeal to mice or other vegetarian predators, perhaps because of the poisonous qualities they are reputed to possess.

Cultivation. It is important to plant colchicums in August or as early in fall as they can be obtained. Even so their flowers may be pushing from the bulbs when they arrive. No time must be lost in getting them into the earth. To expedite this, it is well to prepare the ground, by deep spading and the incorporation with it of generous amounts of compost, leaf mold, peat moss or other humus-forming material, well before the corms arrive. Shallow planting is needed. Colchicums resent having their corms set deeply. If they are covered with 2 to 3 inches of soil it is enough. The distance between corms may be 6 to 9 inches; they should be spaced not evenly, but in irregular fashion so that an informal, naturalistic effect, as if they had sprung casually from self-sown seedlings, is attained. Care of established plantings calls for little more than the mulching and fertilizing described above, together with watering in dry weather when leaves are in evidence. Under no circumstances should the foliage be cut down until it has died completely.

COLD FRAME. See Garden Frames.

COLEOCEPHALOCEREUS (Coleocephalocèreus). The only species of *Coleocephalocereus* of the cactus family CACTACEAE is an inhabitant of cliffs and rocky places in the region of Rio de Janeiro, Brazil. Botanists of conservative persuasion include it in *Cephalocereus*. Its name derives from the Greek *koleos*, a sheath, and the name of the related genus *Cephalocereus*.

Upright to prostrate and branched from near its base, *C. fluminensis* (syn. *Cephalocereus fluminensis*) has dark green stems 3 to 6 feet long by up to 4 inches thick, with ten to seventeen high, notched ribs. The clusters of pliable pale yellow to grayish spines, sprout from woolly areoles. They are of four to seven radials and usually one central that may be 1¼ inches long. A beardlike pseudocephalium, ¾ inch to 2 inches wide, with white wool and yellow bristles streaks from the top of the stem for up to 3 feet down one side. From this the flowers develop. A little over 2 inches long, they have a few minute scales on the perianth tube. The outer petals are pink, the inner white. The stamens and style are white. The latter is longer than the petals and is tipped with a stigma with about eleven lobes. The smooth, top-shaped fruits, up to ¾ inch long, are purplish.

Garden and Landscape Uses and Cultivation. These are as for *Cephalocereus*, with emphasis on the need for a very well-drained, preferably very sandy, fertile soil. For more information see Cactuses.

COLEONEMA (Coleo-nèma). There are five species of this genus of the rue family RUTACEAE, all South African. They are evergreen shrubs related to *Agathosma* and *Barosma*, from which they differ in their sterile stamens being enfolded in channels in their petals, which characteristic is reason for their name, derived from the Greek *koleos*, a sheath, and *nema*, a thread. From near-kin *Diosma*, the genus *Coleonema* differs in having some sterile stamens. In gardens *C. album* is often misnamed *Diosma reevesii*.

Coleonemas have alternate, slender, gland-dotted, linear or threadlike leaves, and solitary, axillary blooms. The flowers have five-parted calyxes, five-channeled petals, five fertile stamens, and five sterile stamens (staminodes), the latter concealed in the grooves of the petals. The fruits have five sections (carpels).

Most commonly grown is *C. album* (syn. *Diosma alba*). Much-branched and 2 to 3

Coleonema album

feet tall, this rather heathlike species has numerous pointed-linear leaves ¼ to ½ inch long, with tiny marginal teeth, and solitary, white or pale pink flowers from the axils of the upper leaves. The blooms, with spreading, spatula-shaped petals, are about ⅓ inch wide. From 3 to 6 feet tall, *C. pulchrum* has long, slender shoots, with cylindrical, pointed, threadlike leaves ½ to 1½ inches long. More showy than *C. album*, it has flowers almost or quite ¾ inch in diameter, and solitary in the upper leaf axils. They are light pink, darkening toward their centers. Their petals bend backward.

Garden and Landscape Uses. Coleonemas are attractive for outdoor plantings in mild climates, such as that of California, and are useful in many places where low- to medium-sized evergreen flowering shrubs with fine-textured foliage are appropriate. They are also grown in pots in cool greenhouses. They are not particular as to soil, provided it is not alkaline. Sandy peaty earth, neither very wet nor dry, suits them best.

Cultivation. Increase is had from summer cuttings and from seeds sown in spring in sandy peaty soil in a temperature of 55 to 60°F. Young plants are induced to branch by having the tips of their shoots pinched out. Plants in the open appreciate a mulch of peat moss or other suitable organic material. After blooming, they are pruned to keep them shapely and compact. In sunny greenhouses they grow well where the winter night temperature is 45 to 50°F, and the day temperature five to ten degrees higher. These plants do not like a highly humid, stagnant atmosphere and to prevent any suspicion of this the greenhouse must be ventilated freely on all favorable occasions. In summer pot-grown coleonemas benefit from being placed outdoors in full sun with their containers buried to the rims in sand, peat moss, or other material that will keep them at a fairly even temperature and prevent too rapid drying. They are removed to a greenhouse before fall frost. They are watered to keep the soil always moderately moist, but not excessively wet. Well-rooted specimens benefit from mild stimulation with weak liquid fertilizer. In spring, after blooming, the plants are pruned moderately, and two or three weeks later, when new growth begins, they are repotted.

COLEUS (Cò-leus). There can be few gardeners unfamiliar with the common coleus, so highly esteemed for its many beautiful colored-foliaged varieties. Fewer know the green-leaved species that attract by their splendid displays of blue flowers. We shall consider both. Belonging in the mint family LABIATAE, and native of warm parts of the Old World, *Coleus* consists of 150 species. The name, from the Greek *koleos*, a sheath, alludes to the stamens being united into a tube.

Herbaceous perennials, subshrubs, and low shrubs, coleuses have square stems, opposite, toothed, stalked or stalkless leaves, and asymmetrical flowers, generally blue or lilac-blue, in terminal spikelike racemes or panicles. The blooms have five-toothed calyxes, tubular, two-lipped corollas, the upper lip often two-lobed, the lower longer, often three-lobed and frequently containing the two pairs of deflexed stamens and style. The stalks of the stamens are united at their bases into a tube encircling the style. The anthers are joined. The one style has two stigmas. The fruits are of four seedlike nutlets.

The common coleus (*C. hybridus*) consists of a complex of hybrids often incorrectly identified as *C. blumei*, a species that is one, but only one of several probably involved in the parentage of these prevailingly highly colored foliage plants. It is improbable that true *C. blumei*, a native of Java, is cultivated. It is a hairless subshrub, 3 to 6 feet tall, with thin, generally pointed, ovate, round-toothed leaves 3 to 8 inches long and short spikes of small blue-purple flowers. Variety *C. b. verschaffeltii* is more robust and freely-branched and has leaves of various colors. This last is perhaps not in cultivation, plants so named often being *C. hybridus.*

The highly variable hybrids are herbaceous perennials or subshrubs that include trailers as well as erect, freely-branching sorts ordinarily 2 to 3 feet tall, but sometimes attaining twice that maximum. Of many shapes, the 2- to 4-inch-long leaves are prevailingly ovate, with broad or nar-

Coleus hybridus varieties

rowed bases and more or less pointed apexes sometimes extended as long points. With round- or sharp-toothed margins, they may be variously cleft or frilled and are somewhat hairy, especially on their under surfaces. They have a wide range of colors, individual leaves often being patterned with several, including shades of green, purple, red, pink, brown, yellow, buff, creamy-white, and white. The small blue, purple, or whitish flowers are in spikelike thyrses. Another species esteemed for its brightly-colored foliage rather than its spikes of bluish flowers, **C. pumilus** (syn. *C. rehneltianus*), of Ceylon, is a trailer, with smaller leaves than most coleuses. Broad-ovate to roundish and coarsely-round-toothed, they have purplish-brown centers margined with green, and whitish bases. Variety *C. p.* 'Trailing Queen' has slightly larger leaves, pink at their bases. The leaves of *C. p.* 'Red Trailing Queen' have centers of purple-carmine encircled by deep blood-red bordered with green. The carmine-red leaves of *C. p.* 'Lord Falmouth' have rosy-pink centers.

Spires 6 inches to 1 foot long of intensely rich gentian-blue blooms are borne in midwinter by Central African **C. thyrsoideus.** A herbaceous perennial or subshrub 3 to 4 feet tall, this sort has green, clammy-hairy stems and foliage. Its longish-stalked, pointed-heart-shaped, coarsely-double-toothed leaves are up to 6 inches long. The long, narrow panicles are of forking clusters of three to ten flowers each nearly ½ inch long. Also from Central Africa, **C. shi-**

Coleus thyrsoideus

rensis is green-leaved and winter-blooming. Its slender-pointed, broad-ovate, round-toothed, thinnish leaves, 2 to 3½ inches long are hairy, especially on their under surfaces. The flowers, paler blue than those of *C. thyrsoideus,* are in shorter, erect panicles. Yet another native of Central Africa, **C. fredericii,** a perennial, or sometimes a biennial or an annual, is up to 4 feet tall, rough-hairy, and with broad-ovate to heart-shaped, long-stalked, round-toothed leaves, in unequal pairs. The rich purple-blue flowers, with a curved, two-lobed upper lip and a keel-like, ½-inch-long lower one, are in loose, airy panicles. They come in winter.

Other blue-flowered, green-leaved natives of Africa, all perennials, are *C. autranii,* of Ethiopia, and *C. barbatus* and *C. lanuginosus,* of Central Africa. About 3 feet tall, **C. autranii** has stems with long hairs at the nodes (joints). Its long-stalked leaves, up to 5 inches long, are broadly-ovate and heart-shaped to rounded at their bases, short-pointed at their apexes, softly-hairy above, and more bristly-hairy on their undersides. The ¾-inch-long, violet-blue flowers are in handsome spikes about 6 inches long. Wedge-shaped toward their bases, the stalked, ovate to oblong, round-toothed leaves of 2- to 4-feet-tall **C. barbatus** are approximately 1 inch to 5 inches long. Both surfaces are densely-hairy. The bright blue, ½-inch-long flowers are in long, loose panicles. Root tubers are developed by **C. lanuginosus.** Its densely-pubescent, ovate leaves are up to 3 inches long. The flowers are in short clusters in long, loose panicles.

Semisucculent **C. spicatus,** of the East Indies, is a perennial 1 foot to 1½ feet tall with sometimes sprawling or decumbent stems that, like the foliage, are minutely-

Coleus lanuginosus

Coleus lanuginosus (flowers)

Coleus spicatus

clammy-hairy and when bruised emit a pungent odor. Short-stalked and broad-elliptic to nearly circular, the leaves are 1 inch to 2½ inches long and toothed above their middles. The blue to rosy-lavender flowers, ¾ to 1 inch long, each from the axil of an early-deciduous, lavender-tinted bract, are in slender spikes 2 to 4 inches long.

Spanish-thyme or suganda (*C. amboin-*

Coleus spicatus (flowers)

doors in warm frostless countries, in greenhouses, and as a window plant.

Cultivation. Among the easiest of plants to grow, colored-leaved coleuses are simple to raise from seeds, sown preferably in late winter or spring, in a temperature of 65 to 75°F. A small packet ordinarily gives plants with a great diversity of leaf forms and colors. Choice selections and named varieties are regularly increased by cuttings 2 to 4 inches long, which root with facility and quickly at any time in a humid atmosphere, shaded from direct sun, in a temperature of 60 to 70°F. The seedlings or rooted cuttings are set individually in small pots and are successively transferred to bigger ones or eventually planted outdoors or in window boxes or other containers. A minimum night temperature of 55°F, with a daytime increase of five to fifteen de-

(c) Cuttings in propagating bench

icus) in Mexico and the Philippine Islands is called oregano, a name more properly reserved for *Origanum heracleoticum*. It is a stout-stemmed, softly-hairy, aromatic subshrub 1½ to 3 feet tall. Native of India and Indonesia, it has short-stalked, thick, green, coarsely-round-toothed leaves 1 inch to 3 inches long. Its spikes of lavender-pink flowers are not especially showy. The leaves are used as a flavoring herb.

Garden and Landscape Uses. The numerous varieties of colored-leaved coleuses, some bearing such horticultural names as 'Golden Bedder', 'Defiance', 'Frilled Fantasy', and 'Firebrand', others unnamed seedlings, are popular for summer bedding, window boxes, porch boxes and suchlike uses, and as greenhouse and window plants. The spreading and semi-trailing forms are also suitable for hanging baskets. Beautiful blue-flowered coleuses with green foliage are primarily winter-blooming greenhouse plants, although in parts of the tropics they can be used in outdoor flower beds. They are not suitable for house cultivation. Spanish-thyme is cultivated as a novelty flavoring herb, out-

Propagating a coleus by cuttings: (a) Stock plant from which to take cuttings

(d) Rooted cuttings ready for potting

(e) Stock plant from which cuttings were taken may be repotted and grown on

Coleus thyrsoideus, blooming outdoors in the tropics

(b) Preparing a cutting

grees, is suitable for these coleuses. To have shapely, well-branched plants pinch the tips of the shoots out occasionally, especially those of young plants. Large and shapely specimens, 2 to 3 feet tall or taller, trained as pyramids or other forms, can be had in big pots or tubs by repotting each time the containers become fairly filled with roots and by repeatedly pinching out the tips of the shoots each time they attain a length of 5 or 6 inches to stimulate bushiness and shape the specimens. It is even possible to develop tree-form (standard) coleuses. To do this, do not pinch the tip out of the young plant. Instead, train a sin-

Clematis, large-flowered hybrid

Cliftonia monophylla

Clerodendrum bungei

Clerodendrum ugandense

Clivia miniata variety

Clintonia borealis (fruit)

A species of *Clusia*

Clytostoma callistegioides

Codiaeum variegatum variety

gle shoot, removing any side ones that develop when small, by tying it to a stake until it attains a height of 2 feet or so. Then pinch out its tip. Allow a few branches to grow from just below the pinched end. When they are 5 or 6 inches long pinch out their tips. Repeat this with shoots that develop subsequently until a good head is developed. To achieve large specimens it is important never to permit them to suffer from lack of water. If you do, leaves will drop. Also, equal light from all sides is necessary. Once the plants have attained their final pots regular applications of dilute liquid fertilizer will do much to keep them in good health. These colored-leaved coleuses need full sun. Spanish-thyme responds to the same environmental conditions that suit colored-leaved coleuses.

Blue-flowered coleuses grown for their blooms, such as *C. thyrsoideus*, *C. fredericii*, and *C. shirensis*, are easily, and often most conveniently, raised from seeds sown in a temperature of 65 to 75°F. If large specimens are wanted, sow in winter or early spring. Summer sowings give smaller plants. These need fertile, well-drained soil abundantly supplied with compost, leaf mold, or peat moss. Keep it moderately and evenly, but not excessively moist. Grow the plants throughout in a humid greenhouse, shaded lightly in summer. Let the minimum night temperature be 55 to 60°F, and that by day not lower than 60 to 70°F. Always pot carefully. Coleuses dislike serious disturbance of their roots while they are in active growth. During their young stages pinch the tips out of the shoots once or twice to encourage branching, giving the last pinch in July. If the air lacks sufficient humidity, if the soil is allowed to become too dry, if potting is not given attention when needed, or if other environmental needs are neglected, the leaves of flowering coleuses soon turn yellow and die. This results in a shabby appearance and poorer blooms. If it becomes necessary to remove faded leaves, especially those of *C. thyrsoideus*, do so with an upward pull, pulling in a downward direction results in a long strip of skin being taken from the stem. After flowering, keep the plants a little drier and a little cooler for a month to six weeks, then cut them back lightly, top-dress them with rich soil, reintroduce them to a warm temperature, and water more freely. This will result in the production of new shoots suitable for use as cuttings. The chief pests of coleuses are mealybugs, red spider mites, and scale insects.

COLIC ROOT is *Dioscorea villosa*. Yellow colic root is *Aletris aurea*.

COLLARD. Especially popular as a leaf vegetable in the south and much grown where it is too hot for cabbages, collard is a kind of kale that like kale is a develop-

ment of Old World *Brassica oleracea acephala* of the mustard family CRUCIFERAE. From cabbage true collards differ in not forming solid heads. Sometimes cabbages grown for a harvest of greens before they head are called collards.

Soil for collards and its preparation are as for cabbage. For plants to be harvested in fall to get a good start before the onset of hot weather, it is best to sow in early spring in a cold frame or other protection and to transplant as soon as the ground is ready to the garden or field. In regions of very mild winters fall sowing for winter and spring harvesting is practicable. Plant large varieties, such as 'Georgia Blue Stem', in rows 4 feet apart with 3 feet between individuals. Smaller growers, such as 'Green Glaze' and 'Vates', may be set somewhat closer. Subsequent care is that satisfactory for cabbage and kale. Harvest while the leaves are fairly young and tender, before their stalks become tough and stringy.

COLLETIA (Collèt-ia)—Anchor Plant. All temperate and warm-temperate South Americans, the seventeen species of *Colletia* belong in the buckthorn family RHAMNACEAE. The name commemorates the French botanist Philibert Collet, who died in 1718. Colletias are shrubs with often large spines in pairs. Their opposite leaves, except those of young seedlings, soon fall, or may be wanting. Photosynthesis chiefly occurs in the green tissues of the stems and spines, the latter being morphologically modified branches. Solitary or few together at the bases of the spines, the whitish or yellowish, small tubular or urn-shaped flowers have four to six each sepals, petals, and stamens, or the petals may be absent. The stigma is three-lobed. The fruits are dry, leathery, three-lobed capsules. Plants cultivated as *Colletia spinosa* (syn. *C. horrida*) are invariably one of the kinds described below. True *C. spinosa* does not appear to be in cultivation.

The anchor plant (*C. cruciata*) is the most commonly cultivated kind. Its vernacular name alludes to its pairs of very

Colletia cruciata

large, conspicuous spines that provide formidable armament against grazing animals. Except on young seedlings leaves are usually absent, or at most are present for a very brief period on the youngest shoots. They are ovate, toothed, and up to ¼ inch in length. Alternating pairs of spines are at right angles to each other. Mostly the spines are flattened-triangular, ½ to 1½ inches long, and nearly as broad at their bases as they are long. They are stiff and sharply-pointed. Rarely individual shoots have stout, needle-like spines. The yellowish flowers, which appear in fall, are solitary or in twos, fours, or sixes. They are without petals, the showy parts being the five-lobed calyxes. Rigid and much branched, the anchor plant occasionally is 10, but more commonly about 4 feet tall. It is native to Uruguay and southern Brazil.

About as tall, **C. armata**, of Chile, differs in having mostly slender, cylindrical, sharp, grayish-green spines, ½ inch to 1½ inches

Colletia armata

long, very rigid, and usually downy. Quite often there are no leaves; when present they are up to ½ inch long. The fragrant, waxy, white flowers, similar to those of *C. cruciata*, are solitary or in twos or threes, and have protruding stamens. They ornament the branches in fall. Variety *C. a. rosea* has pink-tinged blooms.

Another Chilean, *C. infausta*, is up to 10 feet tall and, like the last, has its younger shoots strongly armed with stout, needle-like spines, but they are completely hairless. The leaves are ⅓ inch long, or wanting. Borne in fall or spring, the tiny blooms are greenish-white, often tinged with red. Their stamens scarcely protrude.

Garden and Landscape Uses. These often grotesque shrubs have an out-of-this-world aspect that appeals to lovers of the unusual. By no means unlovely, they are worth planting in moderation for variety and because of the interest of their forms and flowers. They are not hardy in the north, but are in regions where only light frosts occur. They are plants for hot, sunny locations and dry or dryish soils. In addition to their outdoor uses, the anchor plant and sometimes other kinds are occasionally grown in greenhouses, especially in botanical collections employed to illustrate plant morphology.

Cultivation. Easy to grow and needing little care, colletias thrive in the environments suggested above. No pruning is needed other than any necessary to limit the plants to size or maintain a balanced shape; it is best done in spring. Increase is by cuttings of short side shoots taken with a heel of older wood attached, and inserted in a propagating bench in a cool greenhouse or cold frame in late summer or early fall. In greenhouses a winter minimum temperature of 40 to 50°F is satisfactory. At all times airy, dryish atmospheric conditions indoors are required, and full sun. Watering is done sparingly, especially in winter. The soil must be very well drained, and poorish rather than excessively fertile.

COLLINIA ELEGANS is *Chamaedorea elegans*.

COLLINSIA (Col-línsia)—Chinese Houses or Innocence, Blue Lips, Blue-Eyed Mary. All except two of the seventeen species of *Collinsia* of the figwort family SCROPHULARIACEAE, are natives of western North America; the exceptions inhabit the eastern part of the continent. The group consists entirely of annuals, the cultivated kinds attractive. Their name commemorates a Pennsylvania botanist, Zaccheus Collins, who died in 1831.

Collinsias have undivided, toothless or round-toothed leaves, mostly opposite, but the upper ones often in threes. Those above are stalkless and are gradually reduced in size until the uppermost are only bracts. The flowers come, either solitary or in whorls (circles) of few, in the axils of the upper leaves and bracts. Bell-shaped, pouched at their bases, and strongly two-lipped, they are white, pink, lilac, violet, blue, or bicolored. They have five-parted calyxes and five-lobed corollas, with the upper lip two- and the lower three-lobed. The center lobe of the lower lip is folded longitudinally into a keel or pouch that encloses the four normal stamens, one rudimentary stamen, and the one style. The fruits are capsules containing flattish, more or less winged seeds.

Chinese houses or innocence (*C. heterophylla* syn. *C. bicolor*), one of the most popular kinds, is native to California and Baja California. Up to 2 feet tall and branched or not, it is sometimes somewhat sticky and has more or less pubescent stems. Its green to purplish, toothed, lanceolate to lanceolate-oblong leaves are up to 2½ inches long. The lower ones are short-stalked, those above stalkless. The

Collinsia heterophylla

blooms, in clusters of two to seven, have upper lips usually white or very much paler than the rose-purple or violet lower ones. They are stalkless or nearly so, and up to ¾ inch long or a little longer. Horticultural varieties include one with all white, and one with variegated blooms.

Little or not at all branched, with the upper parts of its stems glandular-pubescent and its leaves hairy on their undersides, *C. tinctoria*, of California, 1 foot to 2 feet tall, has very short-stalked flowers in groups of two to six. They are yellow to greenish-white, dotted or streaked with purple, and ½ to ¾ inch long. Their upper lips are very short. The lower lips are bearded on the upper sides of their lateral lobes and on the outsides of their keels. The toothed or toothless, lanceolate-oblong to ovate leaves are up to 3½ inches long.

Blue lips (*C. grandiflora*), native from California to British Columbia, is up to 1½ feet tall and has the upper parts of its stems hairy and sticky. The oblong to lanceolate, toothed or toothless, mostly stalkless leaves are up to 1¾ inches long. Its blooms, much longer stalked than those of any of the kinds previously discussed, ⅓ to ¾ inch long, are in groups of three to seven. They have violet-blue lower lips, with keels and side lobes of about the same lengths, and shorter, white to purplish upper lips.

Blue-eyed Mary (*C. verna*) is native of rich woodlands from New York to Michigan, Wisconsin, Virginia, Kentucky, and Arkansas. Spring-blooming, and often occurring in large colonies, it is 9 inches to about 1½ feet tall and has stems often prostrate below and finely glandular-pubescent in their upper parts. Except the lowermost, the leaves are stalkless. They are up to 2 inches long and triangular-ovate to oblong-ovate, and may or may not have a few teeth. The stalked flowers, solitary or in groups of up to six, are from a little under to slightly over ½ inch long. They have bright blue lower lips and usu-

ally white, but varying to pale blue or purplish, upper ones.

Garden and Landscape Uses. Collinsias are charming for flower beds and borders, and in regions where native are splendid in wild flower gardens. As cut flowers they last well in water and are delightful. They are pretty as pot plants for blooming in greenhouses in spring. Blue-eyed Mary does best in light shade, the others where they receive a little shade during the brightest and hottest part of the day. None thrives in very hot, humid weather. All appreciate fertile, well-drained, moist, but not wet earth.

Cultivation. Seeds are sown outdoors where the plants are to remain, those of blue-eyed Mary in fall, those of the others, in fall where winters are mild, or in earliest spring. Successional sowings will, where summers are not excessively hot, ensure later bloom. The seedlings are thinned to approximately 6 inches apart. Unobtrusive staking with twiggy brushwood, if available, may be needed to support the somewhat weak stems; otherwise no special care is needed. For spring display in greenhouses seeds are sown in September and the seedlings potted individually, first into small pots, then successively into larger ones until containers 5 or 6 inches in diameter are achieved. The plants are grown throughout in full sun where the night temperature is 45 to 50°F, and that by day is five to fifteen degrees higher depending upon the brightness of the day. On all favorable occasions the greenhouse is ventilated freely. The plants need staking neatly, and after they have filled their final pots with roots, regular applications of dilute liquid fertilizer.

COLLINSONIA (Collin-sònia)—Horse-Balm, Citronella. The eighteenth-century English botanist, Peter Collinson, famous correspondent of the American botanist John Bartram, was extraordinarily interested in American plants. He was instrumental in introducing many to his homeland. It was appropriate, therefore, to honor him by naming a genus endemic to North America after him. The pity is that a more handsome representative of America's flora was not chosen to commemorate this unusual man.

Belonging to the mint family LABIATAE, the genus *Collinsonia* has little to recommend it to gardeners. Its species are coarse and generally unprepossessing, hardy, aromatic, deciduous, herbaceous perennials. One, the citronella (*C. canadensis*), is occasionally accommodated in semiwild places and in gardens of native plants. There are five species, all natives of eastern North America. They have stout, woody rhizomes, erect stems, stalked, toothed, opposite leaves, and terminal panicles of yellowish to yellow, asymmetrical, long-tubed flowers with deeply-fringed center corolla lobes and two or four protruding stamens. The nutlets (commonly called seeds) are solitary.

Citronella (*C. canadensis*) lives in rich woodlands from Quebec to Ontario, Michigan, Florida, Mississippi, and Arkansas. Up to 4 feet in height, its leaves are pointed-ovate or ovate-oblong, with blades 3 to 8 inches long. The leaves diminish in size from the bottom of the stem upward; those toward the top are mostly stalkless. The panicles of lemon-scented, yellowish flowers, 4 inches to 1 foot long, are usually branched. Individual blooms are about ½ inch long and have lower lips almost as long as the corolla tubes. They have two stamens. Commercial oil of citronella is not obtained from this plant but from the citronella grass (*Cymbopogon nardus*).

Garden Uses and Cultivation. Possible garden uses are those indicated at the beginning of this entry. Cultivation poses no problems. Moistish, fertile soil and part-shade afford suitable growing conditions. Propagation is easy by seed and by division in spring or early fall.

COLLOMIA (Col-lòmia). From closely related *Gilia* this genus of mostly western North American and southern South American plants of the phlox family PO-LEMONIACEAE differs in technical details. It comprises fifteen species of erect and prostrate annuals and herbaceous perennials with branchless or branched stems, and alternate, undivided or pinnately-divided leaves. The flowers, solitary and axillary or clustered in heads with leafy bracts, are narrowly-funnel- to trumpet-shaped. They have five-lobed calyxes and five corolla lobes (petals), and may be pinkish-yellow, red, purple, blue, or white. The seeds, of many kinds sticky when wet, give reason for the name *Collomia*, derived from the Greek *kolla*, mucilage. Those described below, the kinds most likely to be cultivated, are annuals.

Variable, 1 foot to 3 feet tall, and indigenous from the Rocky Mountains to California and British Columbia, *C. grandiflora* (syn. *Gilia grandiflora*) has mostly branchless, leafy stems and clusters, usually terminal, and about 3 inches across, of salmon-buff to nearly white flowers. Their corolla tubes, ½ to 1 inch long or slightly longer, are much longer than the calyxes. The stalkless, lanceolate to linear, smooth, shortly-hairy or glandular leaves, up to 2 inches long, gradually diminish in size above until they are represented by leafy bracts. From the last, *C. linearis* (syn. *Gilia linearis*) differs in having pointed, linear to lanceolate leaves, stalkless and up to 2 inches long, and coral-pink to purplish or white flowers, with corolla tubes much shorter in comparison to the calyxes, and from under to scarcely over ½ inch in length. They are in clusters, which are usually terminal. This kind, up to 1½ feet

tall, is very shortly-hairy or glandular and may be branched or not. It is native from Quebec and New York to California, Missouri, and Alaska. South American *C. cavanillesii* (syn. *C. biflora*) is about 2 feet tall and has linear leaves toothed at their apexes and leafy clusters of scarlet blooms.

Garden Uses and Cultivation. For flower borders, edgings, and rock gardens the collomias described are attractive. Their blooms do not last well when cut. They are easy to grow, but dislike torrid weather, being best adapted for regions of cool summers. Even under ideal conditions the blooming period of *C. grandiflora* is very short, but the others provide longer displays. Seeds are sown in spring where the plants are to remain and the seedlings thinned, those of *C. grandiflora* and *C. cavanillesii* to about 6 inches, those of *C. linearis* to 3 to 4 inches apart. Ordinary garden soil and sunny locations suit.

COLOCASIA (Colo-càsia)—Elephant's Ear, Taro, Dasheen, Eddo. Many varieties of one species of the tropical Asian genus *Colocasia* are cultivated, in North America ornamental sorts called elephant's ears, and in the tropics, especially those of the Pacific region, kinds variously called taro, dasheen, and eddo, primarily as food. These last, staples of aboriginal and indigenous peoples, are chiefly esteemed for their tubers, in Hawaii prepared as poi and in other ways, but the leaves are cooked and eaten like spinach and the flower clusters are also used as food.

Being in the arum family ARACEAE, the genus *Colocasia* has characteristics that identify it with its relatives the calla-lily (*Zantedeschia*), jack-in-the-pulpit (*Arisaema*), and skunk-cabbage (*Symplocarpus*). The tissues of these all contain tiny cells filled with numerous minute sharp crystals of calcium oxalate. In contact with the tongue and other mouth and internal parts these behave like tiny slivers of glass and cause intense and lasting pain. Cooking destroys the crystals and it is only after considerable protracted heat that the great majority of varieties of taro can be eaten. From *Alocasia*, which is very closely related, *Colocasia* differs in technical details of the ovaries and their arrangement. The name is an ancient Greek one affiliated with the Arabic *colcas* or *culcas*.

Colocasias are herbaceous, tuberous-rooted perennials, with often large, sometimes immense, shield-shaped leaves with long stalks the bases of which enfold each other. The arrangement of the flowers surely identifies these plants with other members of the arum family. The structure called the bloom consists of numerous tiny flowers and a bract called a spathe. The flowers are unisexual, and crowded in a spikelike spadix (comparable to the yellow slender cylinder in the center of a calla-lily). In *Colocasia* the tip of the spadix, and

for some distance below it, is without flowers, then come male flowers, and below them a small area studded with sterile male blooms, and finally, toward the base of the spadix, female flowers. The spathe (the equivalent of the white trumpet-shaped part in the calla-lily) is longer than the spadix.

Very variable *C. esculenta*, 3 to 7 feet tall, has leaves with satiny blades up to 2 feet long or longer, carried approximately vertically, their tips sometimes reaching almost to the ground. The bases of the leaves are deeply-heart-shaped, and the leafstalk joins the blade some little distance in from its margin. The usually solitary flower stalk is very much shorter than the leafstalks. The light yellow spathe, 6 inches to over 1 foot long, opens widely, and the spadix ends in a flowerless portion that is decidedly shorter than the portion occupied by the male and sterile flowers.

From the typical species, *C. e. antiquorum* (syn. *C. antiquorum*), known as Egyptian taro, differs in the flowerless portion of the

Collomia cavanillesii

spadix being at least as long as the part occupied by the male and sterile flowers and in the spathes being inrolled instead of opening widely. The tubers of this, inferior to those of the species, are used as food in Egypt.

Elephant's ears commonly grown as ornamentals include *C. e. euchlora*, which has leaves with purplish stalks and dark green blades with violet margins, *C. e. fontanesia*, with larger, purple-veined leaves, and *C. e. illustris*, called the black-caladium or imperial taro, the large leaves of which are purplish-black to nearly black between the green veins. A native of tropical southeast Asia, *C. gigantea* is a handsome sort of impressive appearance. It has long-stalked leaves with heart-shaped blades 2 to 2½ feet long by two-thirds as wide and prominently veined.

Garden and Landscape Uses. Elephant's ears are stately plants admired for the bold appearance of their huge foliage. In warm,

Colocasia esculenta antiquorum

Colocasia gigantea

Colocasia esculenta fontanesia

ture of 70 to 75°F, with a fairly humid atmosphere, is satisfactory. To prevent the ground from becoming really dry, liberal applications of water may be necessary. To conserve moisture a mulch of peat moss, bagasse, or any other suitable organic material is helpful. Where winters are cold the tubers are dug before the first frost with as much soil adhering to their roots as possible, the tops are cut off, and the tubers are stored with their roots packed in sand, dry soil, or peat moss in a temperature of 40 to 50°F. For cultivation of dasheen see Dasheen or Taro.

COLONIAL GARDENS. Ample evidence exists that the early American colonists, like those who came later, planted gardens. Undoubtedly the first examples contained only vegetables and a few herbs, culinary, fragrant, and medicinal. But it probably was not long before some green-thumb pioneer housewife set out a few flowers, perhaps a cherished kind brought from the old country, more often native ones that pleased. From such early beginnings derived the simple, utilitarian type of garden common throughout the colonial period and much later in rural New England. Such gardens were functional outgrowths of the way of life and the needs of the people. They were not planned as the word is used today. They were located as close as convenient to the house, barns, and coops where family activities centered. Often some little distance away, where site and soil deemed most favora-

likely to damage its huge leaves. When left in the ground permanently, an annual spring application of fertilizer promotes strong growth. Elsewhere the earth should be deeply spaded and enriched by incorporating with it compost, peat moss, or other decayed organic matter, as well as a liberal dash of a complete fertilizer, before planting. The tubers may be set directly outdoors after the weather is really warm and settled at a spacing of 3 to 6 feet, with their tops covered to a depth of 6 inches or so, but in the north it is better to start them in pots in a greenhouse or other suitable place indoors eight to ten weeks before it is safe to set them outside, which must not be done until settled warm weather has arrived. Indoors a tempera-

Colocasia esculenta illustris

frost-free countries they may be left in the ground permanently; elsewhere they are planted for summer effects and the tubers stored over winter. Except as a food crop in the tropics and for demonstration in collections of plants used by man, taro is rarely grown. Mostly it is cultivated in swampy soil although varieties that stand drier conditions are available.

Cultivation. Elephant's ear succeeds best in deep, moist, fertile soil, in partial shade, and where sheltered from strong winds

Colonial gardens as restored at Colonial Williamsburg, Virginia: (a) Boxwood hedges surround beds of grass, bulb plants, and herbaceous perennials. Crape-myrtle and native holly.

(b) Rectangular beds planted with tulips, pansies, and forget-me-nots are surrounded by low hedges.

(c) Boxwood, as low hedges, outline beds of grass and flowers and, as topiary specimens, define and give character to this garden.

ble, an orchard would be planted and enclosed fields for major crops, such as beans, corn, and pumpkins, established. Nearer home would be one or more fenced plots for growing vegetables needed in lesser amounts, herbs, and flowers. These, like the house and outbuildings, were sited as advantageously as possible in relation to the topography of the land and each other. Their sizes were governed by that of the family. Paths were established for convenience rather than aesthetics. They generally ran directly from house or outbuildings to the garden plots. Within the enclosures they were narrow and sited to best serve cultivation and harvesting. Not unusual with strictly functional arrangements, these procedures often resulted in a special charm, an aesthetic appeal more genuine than contrived effects. The paths were of trodden earth, sometimes surfaced with gravel or crushed clam shells. Inside the garden plots the plants were set without conscious design. They were often in raised, not necessarily symmetrical beds, with sides supported by saplings cut and laid horizontally. Vegetables interspersed with groups of flowers might be in blocks. Often they were planted more irregularly, as fancy pleased at planting time. Tall plants might obscure lower ones.

The tidy-minded Dutch in New York placed greater esteem on symmetry. Their gardens usually had a well-defined walk as a central axis with a series of regularly spaced paths and well-ordered beds at right angles to it, the whole neatly fenced.

Urban gardens, such as those at Colonial Williamsburg, now happily carefully restored, and those of Boston, New York, and Philadelphia merchants were more sophisticated. Often their designs were influenced by contemporary European parterre-type plantings. The few books on horticulture available came from Europe, chiefly from England, and some Americans traveled abroad and corresponded with Europeans. It was usual to set the house on high ground and terrace around it with soil excavated from the cellar. The gardens were near the house and were formal and enclosed by fence, hedge, or more rarely a wall. On either side of a central path, usually on an axis with a door of the house, and ending at an arbor, summer house, steps, sundial, statue, gate, specimen plant, or some other feature of interest, symmetrical patterns of beds and paths at right angles or acute angles to the main path were established. The beds were often outlined with low boxwoods. The design on one side was usually a mirror image of that on the other. Fruit trees, shrubs, flowers, herbs, and if a special plot was not set aside for them elsewhere, vegetables, occupied the beds, planted as the taste and whims of the owner suggested. Because gardening styles in North America

(d) This small garden and sitting area fits appropriately and snugly between buildings. The center beds feature tulips.

(e) A billowy hedge of boxwood surrounds this garden of triangular beds outlined with low hedges of boxwood.

did not change appreciably for some time after 1789, horticulturists and students of landscape architecture usually accept the Colonial period, as it applies to gardens, as being from 1620 to 1840 rather than terminating with the historic establishment of the United States.

Country estates, developed fairly late in the colonial period, were often extensive and included well-planned gardens. Mount Vernon and Monticello are classic examples. These, established by men of wealth and at least some leisure, were to be found in Virginia and near Philadelphia, New York, and Boston. They usually reflected the natural type of landscaping that developed in England in the eighteenth century, and included lawns, lakes, groups of trees, parklike areas sometimes complete with deer, as well as avenues, terraces, and more formal features including parterre-type beds.

The kinds of plants available to colonial gardeners, even late in the period, were much fewer than those we know today. Plant explorers had not yet brought home the plant treasures from Japan, China, or many other parts of the world that we take for granted. Colonial gardeners relied chiefly on European plants augmented by some native ones. Unknown to them were such present day staples as camellias, chrysanthemums, forsythias, Japanese yews, and pachysandra and many hundreds of others. Plants known to have been cultivated in eastern North American gardens prior to 1776 include these, listed in "Arnoldia," a publication of the Arnold Arboretum, in July 1971 together with references to documentation of their authenticity. They are cataloged here under the names used by the colonists, which do not always coincide with names currently used. The modern botanical name is given in parentheses.

HERBS: alkanet (*Pentaglottis sempervirens*), angelica (*Angelica archangelica*), anise (*Pimpinella anisum*), balm (*Melissa officinalis*),

Herbs appropriate for Colonial gardens: (a) Clary

(b) Lavender-cotton

(c) Rosemary

basil (*Ocimum basilicum*), bee flower (*Ophrys apifera*), borage (*Borago officinalis*), burnet (*Sanguisorba officinalis*), camomile (*Chamaemelum nobile*), caraway (*Carum carvi*), catnip (*Nepeta cataria*), chervil (*Anthriscus cerefolium*), chives (*Allium schoenoprasum*), clary (*Salvia sclarea*), comfrey (*Symphytum officinale*), coriander (*Coriandrum sativum*), costmary (*Chrysanthemum balsamita*), cress (*Lepidium sativum*), dill (*Anethum graveolens*), dock (*Rumex patientia*), fennel (*Foeniculum vulgare*), flax (*Linum usitatissimum*), houseleek (*Sempervivum tectorum*), hyssop (*Hyssopus officinalis*), lavender (*Lavandula spica*), lavender-cotton (*Santolina chamaecyparissus*), lovage (*Levisticum officinale*), licorice (*Glycyrrhiza glabra*), madder (*Rubia tinctorum*), marjoram (*Majorana hortensis*), mint (*Mentha arvensis, M. longifolia, M. pulegium, M. piperita, M. spicata*), mustard (*Brassica nigra*), parsley (*Petroselinum crispum*), purslane (*Portulaca oleracea*), rhubarb (*Rheum rhaponticum*), rosemary (*Rosmarinus officinalis*), rue (*Ruta graveolens*), saffron (*Crocus sativus*), sage (*Salvia officinalis*), savory (*Satureja hortensis, S. montana*), skirret (*Sium sisarum*), sorrel (*Rumex acetosa*), southernwood (*Artemisia abrotanum*), sweet cicely (*Myrrhis odorata*), tansy (*Tanacetum vulgare*), tarragon (*Artemisia dracunculus*), thyme (*Thymus serpyllum, T. vulgare*), tobacco

(*Nicotiana rustica, N. tabacum*), woad (*Isatis tinctoria*), and yarrow (*Achillea millefolium*).

VEGETABLE AND FIELD CROPS: artichoke (*Cynara scolymus*), asparagus (*Asparagus officinalis*), barley (*Hordeum vulgare*), bean (*Phaseolus multiflorus, P. vulgaris*), beets (*Beta vulgaris*), broccoli (*Brassica oleracea botrytis*), buckwheat (*Fagopyrum esculentum*), cabbage (*Brassica oleracea capitata*), carrot (*Daucus carota*), cauliflower (*Brassica oleracea botrytis*), celery (*Apium graveolens*), chicory (*Cichorium intybus*), corn (*Zea mays*), cotton (*Gossypium herbaceum*), cucumber (*Cucumis sativus*), dandelion (*Taraxacum officinale*), endive (*Cichorium endivia*), garlic (*Allium sativum*), gourds (*Lagenaria sicereria*), hemp (*Cannabis sativa*), hops (*Humulus lupulus*), Jerusalem-artichoke (*Helianthus tuberosus*), leeks (*Allium porrum*), lentil (*Lens culinaris*), lettuce (*Lactuca sativa*), melon (*Citrullus lanatus, Cucumis melo*), oats (*Avena sativa*), okra (*Abelmoschus esculentus*), onion (*Allium cepa*), parsley (*Petroselinum crispum*), parsnip (*Pastinaca sativa*), peas (*Pisum sativum, Vigna unguiculata*), pepper (*Capsicum frutescens*), pepper-grass (*Lepidium sativum*), potato (*Solanum tuberosum*), pumpkin (*Cucurbita pepo*), radish (*Raphanus sativus*), rampion

Vegetables appropriate for Colonial gardens: (a) Cabbages

(b) Leeks

(c) Pumpkins

(*Campanula rapunculus*), rape (*Brassica napus oleifera*), rye (*Secale cereale*), scurvy-grass (*Cochlearia officinalis*), sorrel (*Rumex acetosa*), spinach (*Spinacia oleracea*), squash (*Cucurbita pepo*), turnip (*Brassica rapa*), vetch (*Vicia sativa*), wheat (*Triticum aestivum*), and yams (*Dioscorea alata*).

FRUITS AND NUTS: almond (*Prunus amygdalus*), apple (*Malus sylvestris*), apricot (*Prunus armeniaca*), barberry (*Berberis vulgaris*), blackberry (*Rubus species*), cherries (*Prunus*

Fruits appropriate for Colonial gardens: (a) Gooseberries

(b) Mulberries

(c) Peaches

Evergreen trees appropriate for Colonial gardens: (a) American holly

cerasus), chestnut (*Castanea dentata*), crab apple (*Malus angustifolia*), cranberry (*Vaccinium macrocarpon*), currant (*Ribes nigrum, R. sativum*), elder (*Sambucus canadensis, S. nigra*), fig (*Ficus carica*), gooseberry (*Ribes grossularia*), grapes (*Vitis vinifera*), hawthorn (*Crataegus monogyna, C. oxyacantha*), hazelnut (*Corylus avellana*), medlar (*Mespilus germanica*), mulberry (*Morus alba, M. nigra, M. rubra*), nectarine (*Prunus nectarina, P. persica*), olive (*Olea europaea*), orange (*Citrus aurantium*), peach (*Prunus persica*), pear (*Pyrus communis*), plum (*Prunus domestica*), pomegranate (*Punica granatum*), quince (*Cydonia oblonga*), raspberry (*Rubus idaeus*), strawberry (*Fragaria virginiana*), and walnut (*Juglans regia*).

TREES, SHRUBS, AND VINES: andromeda (*Eubotrys racemosa*), arbor-vitae (*Thuja occidentalis*), arrow-wood (*Viburnum dentatum*), ash (*Fraxinus americana, F. excelsior*), azalea (*Rhododendron atlanticum, R. canescens, R. indicum, R. periclymenoides, R. prionophyllum, R. viscosum*), bald-cypress (*Taxodium distichum*), bastard-indigo (*Amorpha fruticosa*), bayberry (*Myrica cerifera, M. pensylvanica*), beautyberry (*Callicarpa americana*), beech (*Fagus grandifolia, F. sylvatica*), birch (*Betula lenta, B. nigra*), bittersweet (*Celastrus scandens*), black haw (*Viburnum prunifolium*), bladder nut (*Staphylea pinnata*), box (*Buxus sempervirens*), box-elder (*Acer negundo*), buckeye (*Aesculus octandra*), burning bush (*Euonymus atropurpureus*), butcher's-broom (*Ruscus aculeatus*), butternut (*Juglans cinerea*), button bush (*Cephalanthus occidentalis*), Carolina allspice (*Calycanthus floridus*), Carolina jasmine (*Gelsemium sempervirens*), catalpa (*Catalpa bignonioides*), chaste tree (*Vitex agnus-castus*), cherry (*Prunus virginiana*), cherry-laurel (*Prunus caroliniana*), chinaberry (*Melia azedarach*), chinquapin (*Castanea pumila*), chokeberry (*Aronia arbutifolia*), clematis (*Clematis virginiana*), coralberry (*Symphoricarpos orbiculatus*), cornel (*Cornus alba*), cornelian-cherry (*Cornus mas*), crab apple (*Malus coronaria*), cranberry (*Vaccinium oxycoccus*), crape myrtle (*Lagerstroemia indica*),

(b) *Magnolia grandiflora*

(c) Red-cedar

cross vine (*Bignonia capreolata*), cyrilla (*Cyrilla racemiflora*), devil's walking stick (*Aralia spinosa*), dogwood (*Cornus amomum, C. florida*), elderberry (*Sambucus canadensis*), elm (*Ulmus alata, U. americana*), emerus (*Coronilla emerus*), English ivy (*Hedera helix*), fir (*Picea abies*), flowering almond (*Prunus glandulosa sinensis*), fothergilla (*Fothergilla gardenii*), franklinia (*Franklinia alatamaha*), fringe tree (*Chionanthus virginicus*), golden-rain tree (*Koelreuteria paniculata*), grape (*Vitis rotundifolia*), groundsel tree (*Baccharis halimifolia*), guelder-rose (*Viburnum opulus roseum*), hackberry (*Celtis occidentalis*), hardhack (*Spiraea tomentosa*), hawthorn (*Crataegus crus-galli, C. oxyacantha, C. phaenopyrum*), hemlock (*Tsuga canadensis*), hickory (*Carya laciniosa, C. ovata*),

Deciduous trees appropriate for Colonial gardens: (a) Locust

Evergreen shrubs appropriate for Colonial gardens: (a) Boxwood (*Buxus sempervirens*)

Deciduous shrubs appropriate for Colonial gardens: (a) Carolina allspice

(b) Mimosa

(b) Mountain-laurel

(b) Elderberry

(c) Pagoda tree

(c) Laurel (*Rhododendron maximum*)

(c) Lilac

holly (*Ilex aquifolium, I. decidua, I. opaca*), honey-locust (*Gleditsia triacanthos*), honeysuckle (*Lonicera caprifolium, L. periclymenum, L. sempervirens, L. tatarica*), hoptree (*Ptelea trifoliata*), hornbeam (*Carpinus caroliniana*), horse-chestnut (*Aesculus hippocastanum, A. pavia*), hydrangea (*Hydrangea arborescens*), inkberry (*Ilex glabra*), ironwood (*Ostrya virginiana*), jasmine (*Jasminum officinale*), judas tree (*Cercis siliquastrum*), juniper (*Juniperus chinensis, J. communis*), Kentucky-coffee tree (*Gymnocladus dioica*), laburnum (*Laburnum anagyroides*), lantana (*Lantana camara*), larch (*Larix decidua*), laurel (*Rhododendron maximum*), leatherwood (*Dirca palustris*), leucothoe (*Leucothoe axillaris*), lilac (*Syringa*), linden (*Tilia americana, T. europaea*), lingon (*Vaccinium vitis-idaea*), locust (*Robinia pseudoacacia*), magnolia (*Magnolia grandiflora, M. tripetala, M. vir-*

Vines appropriate for Colonial gardens:
(a) English ivy

(b) Cross vine

(c) Wisteria

phellos, Q. prinus, Q. velutina, Q. virginiana), osier (*Cornus sericea*), pagoda tree (*Sophora japonica*), paper-mulberry (*Broussonetia papyrifera*), pawpaw (*Asimina triloba*), pecan (*Carya illinoensis*), Persian syringa (*Syringa persica*), persimmon (*Diospyros virginiana*), pine (*Pinus strobus, P. taeda, P. virginiana*), plum (*Prunus cerasifera, P. insititia*), poplar (*Populus alba, P. deltoides, P. nigra, P. nigra italica*), potentilla (*Potentilla fruticosa*), privet (*Ligustrum vulgare, L. vulgare italicum*), pyracantha (*Pyracantha*), red-bay (*Persea borbonia*), red-cedar (*Juniperus virginiana*), rose (*Rosa laevigata, R. palustris, R. spinosissima*), rose-acacia locust (*Robinia hispida*), Russian-olive (*Elaeagnus angustifolia*), St.-John's-wort (*Hypericum perforatum*), sassafras (*Sassafras albidum*), savin tree (*Juniperus sabina*), Scotch broom (*Cytisus scoparius*), service tree (*Sorbus torminalis*), shadblow (*Amelanchier canadensis*), Siberian pea-shrub (*Caragana arborescens*), silverbell (*Halesia carolina*), smoke tree (*Cotinus coggygria*), sour gum (*Nyssa sylvatica*), sourwood (*Oxydendrum arboreum*), Spanish-broom (*Spartium junceum*), spice-bush (*Lindera benzoin*), spiraea (*Spiraea salicifolia*), stewartia (*Stewartia malacodendron, S. ovata*), sumac (*Rhus aromatica*), sweet gale (*Myrica gale*), sweet gum (*Liquidambar styraciflua*), sweet spire (*Itea virginica*), sweet-fern (*Comptonia peregrina*), sweet pepper bush (*Clethra alnifolia*), sycamore (*Platanus occidentalis*), thorn (*Crataegus punctata*), trumpet creeper (*Campsis radicans*), tulip tree (*Liriodendron tulipifera*), viburnum (*Viburnum acerifolium*), Virginia creeper (*Parthenocissus quinquefolia*), white-cedar (*Chamaecyparis thyoides*), willow (*Salix alba vitellina, S. babylonica*), winter sweet (*Chimonanthus praecox*), winterberry (*Ilex verticillata*), wisteria (*Wisteria frutescens*), witch-hazel (*Hamamelis virginiana*), withe rod (*Viburnum cassinoides*), yaupon (*Ilex cassine, I. vomitoria*), and yew (*Taxus baccata*).

FLOWERS: aconite (*Aconitum napellus*), asphodel (*Asphodelus albus, Asphodeline lutea*), aster (*Callistephus chinensis, Stokesia laevis*), bachelor's button (*Centaurea cyanus*), balsam (*Impatiens balsamina*), beach-pea (*Lathyrus maritimus*), bearberry (*Arctostaphylos uva-ursi*), bear's ears (*Primula auricula*), bedstraw (*Galium verum*), bee-balm (*Monarda didyma*), bellflower (*Campanula persicifolia, C. pyramidalis, C. rapunculoides, C. trachelium*), bent grass (*Agrostis tenuis*), bergamot (*Monarda fistulosa*), bindweed (*Ipomoea nil*), black-eyed Susan (*Rudbeckia hirta*), bloodroot (*Sanguinaria canadensis*), bouncing bet (*Saponaria officinalis*), candytuft (*Iberis umbellata*), canterbury bells (*Campanula trachelium*), cardinal flower (*Lobelia cardinalis*), carnation (*Dianthus caryophyllus*), catchfly (*Lychnis dioica, L. viscaria*), cat tail (*Typha latifolia*), celandine (*Chelidonium majus*), centaury (*Centaurea centaurium*), Christmas-rose (*Helleborus niger*), cockscomb (*Celosia argentea cristata*), columbine

Flowers appropriate for Colonial gardens:
(a) Asphodel

(b) Bergamot

(c) Cockscomb

(d) Crown imperial

giniana), maple (*Acer platanoides, A. rubrum, A. saccharinum, A. saccharum*), mespilus (*Amelanchier stolonifera*), mezereum (*Daphne mezereum*), mimosa (*Albizia julibrissin*), mockorange (*Philadelphus coronarius*), moosewood (*Acer pensylvanicum*), mountain-laurel (*Kalmia latifolia*), nannyberry (*Viburnum lentago*), New Jersey-tea (*Ceanothus americanus*), oak (*Quercus alba, Q. borealis, Q. coccinea, Q. falcata, Q. marilandica, Q. nigra, Q.*

(e) Daffodil

(f) Marigold

(g) Poppy (*Papaver orientale*)

(h) Primrose

(i) Snapdragon

(*Aquilegia canadensis, A. vulgaris*), coreopsis (*Coreopsis lanceolata*), cornflower (*Centaurea cyanus*), cowslip (*Primula veris*), cranesbill (*Geranium macrorrhizum, G. sanguineum*), creeping Jenny (*Lysimachia nummularia*), crocus (*Crocus vernus*), crowfoot (*Ranunculus aconitifolius, R. acris, R. asiaticus, R. bulbosus, R. gramineus*), crown imperial (*Fritillaria imperialis*), daffodil (*Narcissus biflorus, N. jonquilla, N. odorus, N. poeticus, N. pseudonarcissus, N. tazetta, N. triandrus*), daisy (*Chrysanthemum leucanthemum*), day-lily (*Hemerocallis fulva, H. lilioasphodelus*), deadnettle (*Lamium purpureum*), dittany (*Dictamnus albus*), dogtooth-violet (*Erythronium dens-canis*), elecampane (*Inula helenium*), English daisy (*Bellis perennis*), epimedium (*Epimedium alpinum*), evening-primrose (*Oenothera biennis*), everlasting (*Anaphalis margaritacea*), fall-daffodil (*Sternbergia lutea*), fennel flower (*Nigella damascena*), feverfew (*Chrysanthemum parthenium*), foam flower (*Tiarella cordifolia*), foxglove (*Digitalis purpurea*), galax (*Galax aphylla*), gilliflower (*Dianthus caryophyllus*), gladiolus (*Gladiolus byzantinus, G. communis*), globe-

amaranth (*Gomphrena globosa*), germander (*Teucrium chamaedrys*), golden ragwort (*Senecio aureus*), grape flower (*Muscari botryoides, M. racemosum*), grape-hyacinth (*Muscari comosum*), ground-ivy (*Glechoma hederacea*), guinea hen flower (*Fritillaria meleagris*), harebell (*Endymion nonscriptus*), herb Robert (*Geranium robertianum*), hollyhock (*Alcea rosea*), honesty (*Lunaria annua*), hydrangea (*Hydrangea arborescens*), hyacinth (*Hyacinthus orientalis*), iris (*Iris germanica, I. g. florentina, I. persica, I. pseudacorus, I. pumila, I. sibirica, I. susiana, I.*

variegata), jimson weed (*Datura stramonium*), Joseph's coat (*Amaranthus tricolor*), lark's spur (*Consolida ajacis, C. regalis*), lily (*Lilium canadense, L. candidum, L. martagon*), lily-of-the-valley (*Convallaria majalis*), liverwort (*Hepatica nobilis*), lizard's tail (*Saururus cernuus*), love-lies-bleeding (*Amaranthus caudatus*), lungwort (*Pulmonaria officinalis*), lupine (*Lupinus albus, L. hirsutus, L. perennis*), maidenhair fern (*Adiantum pedatum*), mallow (*Hibiscus moscheutos*), marigold (*Tagetes erecta, T. patula*), marvel of Peru (*Mirabilis jalapa*), meadow-rue (*Thalictrum aquilegifolium, T. flavum*), meadow saffron (*Colchicum autumnale*), mullein (*Verbascum blattaria, V. thapsus*), nasturtium (*Tropaeolum majus*), pansy (*Viola tricolor*), pelletory (*Parietaria officinalis*), peony (*Paeonia officinalis*), periwinkle (*Vinca minor*), phlox (*Phlox carolina, P. maculata, P. paniculata*), pot marigold (*Calendula officinalis*), pinks (*Dianthus plumarius*), poppy (*Papaver orientale, P. rhoeas, P. somniferum*), prickly-poppy (*Argemone mexicana*), primrose (*Primula vulgaris*), red-valerian (*Centranthus ruber*), rocket (*Hesperis matronalis*), rose-campion (*Lychnis chalcedonica*), scabiosa (*Scabiosa atropurpurea*), sea-holly (*Eryngium maritimum*), sensitive plant (*Mimosa pudica*), snapdragon (*Antirrhinum majus*), snowdrop (*Galanthus nivalis, Leucojum aestivum, L. autumnale*), star of Bethlehem (*Ornithogalum umbellatum*), starwort (*Aster amellus, A. tradescantii*), stock (*Matthiola incana*), strawflower (*Helichrysum bracteatum*), sunflower (*Helianthus annuus*), sweetpea (*Lathyrus latifolius, L. odoratus*), sweet william (*Dianthus barbatus*), toadflax (*Linaria vulgaris*), tomato (*Lycopersicon esculentum*), trollius (*Trollius europaeus*), tulip (*Tulipa clusiana, T. gesneriana*), turtlehead (*Chelone glabra, C. obliqua*), valerian (*Valeriana officinalis*), veronica (*Veronica longifolia*), violet (*Viola odorata*), Virginia bluebells (*Mertensia virginica*), wallflower (*Cheiranthus cheiri*), white archangel (*Lamium album*), windflower (*Anemone coronaria, A. hortensis*), winter-aconite (*Eranthis hyemalis*), winterberry (*Physalis alkekengi*), yarrow (*Achillea millefolium*), and yucca (*Yucca gloriosa*).

COLORADOA (Colora-dòa). The only species of *Coloradoa* of the cactus family CACTACEAE, is by conservative botanists included in *Echinocactus, Sclerocactus,* or *Ferocactus*. Its name is an adaptation of that of a state to which it is indigenous.

Found as a native only in a very limited region of Colorado and New Mexico, *C. mesae-verdae* (syn. *Echinocactus mesae-verdae*) is one of the rarest native cactuses of the United States. Its gray-green stems, commonly not more than 3 inches tall and wide, exceptionally attain heights up to 7 inches. They have thirteen to seventeen ribs of tubercles that on young plants are well separated, on older ones they tend to merge. The spine clusters usually consist of eight to eleven yellow to tan radials up to ½ inch long. Very rarely there is one straight or hooked central spine also not more than ½ inch long. Funnel-shaped and ¾ inch to 1¼ inches tall and broad, the fragrant flowers are creamy-white to white, their inner petals somewhat toothed at their tips. Less than ¼ inch in diameter, the cylindrical fruits are first greenish. They change to brown as they ripen.

Garden Uses and Cultivation. This rare cactus is a collectors' item difficult to establish and maintain in cultivation. In view of its rarity and the unlikelihood of collected plants surviving, it should not be taken from the wild, but only attempted from seeds. In the wild *Coloradoa* adapts to extremely alkaline soil and very dry environments. For more information see Cactuses.

COLPOTHRINAX (Colpothrì-nax)—Belly Palm, Bottle Palm, Barrel Palm. A single species of fan-leaved palm, native of Cuba and closely related to *Pritchardia*, constitutes *Colpothrinax* of the palm family PALMAE. Its name derives from the Greek *kolpos*, womb, and *Thrinax*, another genus of palms. It alludes to the swollen trunk.

The belly palm (*C. wrightii*) is endemic to Pinar del Rio and the Isle of Pines, where it grows interspersed with the pine trees after which those localities were named. It attains heights up to 40 feet and has a trunk ridiculously swollen and obese at its middle and tapering below and above in much the fashion of another Cuban palm, the corojo (*Gastrococos crispa*). The more or less circular leaves of the belly palm, 5 feet or more in diameter, are divided into many segments. The fleshy, bisexual flowers are in large, branched clusters originating among the leaves; their stems and branches are enclosed in tubular bracts. The brown, spherical fruits are about ½ inch across.

Where this palm is native the bellied portions of the trunks are hollowed to make casks, water troughs, beehives, and sometimes canoes. The trunks are used as posts and pillars to support huts, the leaves are used for thatching, the fruits to feed hogs.

Garden Uses and Cultivation. This very unusual and interesting species is not known to have been successfully introduced to cultivation. It would, perhaps, respond to the care recommended for *Pritchardia*. In its natural home it grows in sandy lowlands. For additional information see Palms.

COLQUHOUNIA (Colquhoùn-ia). Not hardy in the north, this genus of evergreen or partially evergreen shrubs of the mint family LABIATAE inhabits the Himalayas and southwestern China. Its name commemorates Sir Robert Colquhoun, an early nineteenth-century patron of the botanic garden in Calcutta, India. Differences between its kinds are not easily defined, and the numbers of species are variously regarded as being from two, each with several varieties, to six. They have four-angled, sometimes twining stems, opposite leaves, and tubular flowers with four stamens and a slender style, in terminal spikes or axillary clusters.

Buddleia-like in appearance, **Colquhounia coccinea** comes from the Himalayas. Of loose, somewhat straggly growth and up to 10 or 12 feet tall, it has nontwining, semiwoody, downy stems, and finely-toothed, ovate to ovate-lanceolate leaves 3 to 6 inches in length and mostly one-third to one-half as wide as they are long. Their undersides are more or less clothed with grayish-white down. When crushed the foliage has an apple-like fragrance. In short-stalked clusters of three to five, the scarlet to orange flowers are in whorls (tiers) at the shoot ends and in the axils of the upper leaves, and these compose panicles up to 1 foot in length. They have funnel- to bell-shaped, conspicuously five-lobed, downy calyxes about ½ inch long and more or less hairy, curving corollas 1 inch long or slightly longer, downy, and prominently two-lipped. The three-lobed lower lip, spreading or down-turned, is the larger. The stamens arch under the upper lip. The fruits consist of four seedlike nutlets. Variety *C. c. vestita* (syn. *C. vestita*) which grows in drier places than *C. cocci-*

Colquhounia coccinea vestita

nea, is more hairy. In parts of Bhutan *C. coccinea*, a native of the eastern Himalayas and adjacent territories, is used as hedges around fields.

Garden and Landscape Uses and Cultivation. The colquhounias described are attractive late summer- and fall-blooming shrubs suitable for climates where there is little or no frost. They are grown successfully in California and are conspicuous when in bloom. They grow with little or no trouble in well-drained soils in sunny locations and are easily increased by seeds and summer cuttings. Sometimes plants from self-sown seeds appear in the vicinity of established bushes. Pruning necessary to shape them or contain them to size, and to thin out crowded branches, is done in spring.

COLTSFOOT is *Tussilago farfara.* Sweet coltsfoot is *Petasites fragrans.*

COLUMBINE. See Aquilegia.

COLUMBO, AMERICAN is *Frasera caroliniensis.*

COLUMNEA (Colúm-nea). Highly fashionable among gesneriad fans, columneas are natives of the American humid tropics. They belong to the gesneria family GESNERIACEAE and include 200 species. In addition a fairly large number of handsome horticultural hybrids are cultivated. The name honors Fabius Columna (Fabio Colonna), the author of the first botanical book illustrated with copper plates, who died in 1640.

This genus is closely related to *Alloplectus*, and some of its kinds have been known by that name. Sorts sometimes treated separately as *Trichantha* are included here. Ordinarily columneas are epiphytes. They generally perch on trees rather than grow in the ground, but do not take nourishment from their hosts.

Columneas are subshrubby usually evergreen perennials with erect, spreading, or trailing stems, sometimes rooting from the joints (nodes). The paired leaves are from fairly distantly spaced to so close together that they overlap, and in trailing kinds they form long strands. The leaves of each pair may be similar or very dissimilar in size, or one may be practically absent so that the leaves appear to be alternate. Undivided, they are sometimes toothed. Solitary or several together from the leaf axils, the blooms have five sepals, sometimes toothed and often colored. The corolla has a long or short tube, generally markedly swollen at its base on its back, which erupts erectly or at an oblique angle from the calyx. Commonly the face of the bloom is strongly-two-lipped, the four upper lobes (petals) being joined into a hood that projects in helmet-like fashion, the remaining lobe narrow and usually down-pointing. There are four slender-stalked stamens joined briefly at their bases and with their anthers united to form a square. Usually the ovary is backed by a gland, which is sometimes two-lobed. Less frequently five glands encircle the ovary. The style is long and frequently protruding. The stigma may be mouthlike, club-shaped, or two-lobed. The fruits are many-seeded berries.

Pendulous-stemmed kinds are especially fine for hanging baskets. One of the best known and most popular, **C. gloriosa** in-

Columnea gloriosa

Columnea microphylla

Columnea minor

Columnea microphylla (flowers)

Columnea moorei

deed reflects the meaning of its name. In bloom it is surely impressive, out of bloom quite handsome. Native to Costa Rica, it presents curtains of slender, hairy stems clothed throughout their lengths with pairs of equal or unequal-sized, short-stalked, ovate to oblong-ovate leaves up to 1¼ inches long by one-half as wide. Typically green, and furred with reddish or purplish hairs, their upper surfaces are convex. The flowers are 2½ to 3 inches long. The corollas, rich scarlet clothed with white hairs and with a yellow patch in the throat, have a 1-inch-long tube and a projecting, broad-helmet-like upper lip twice as long as the tube. The lower lip is narrower, bends backward, and is 1 inch long. The fruits are dull white berries broader than long, and about 1¼ inches across. Variety *C. g. superba* (syn. *C. g. purpurea*) has deeper purplish foliage. Also Costa Rican, **C. microphylla** resembles *C. gloriosa* except that its broad-ovate to nearly circular leaves are under ⅜ inch long. Those of each pair are of equal size. A variant has yellow-variegated foliage. Another with slender, hanging stems, **C. allenii,** of Panama, in cultivation has been confused with *C. arguta,* also of Panama. From the latter *C. allenii* differs in its closely-spaced leaves, those of each pair of similar size, being bluntly-oblong-elliptic instead of long-taper-pointed and in its flowers being strongly-two-lipped. It has slender, hanging stems, which are slightly hairy when young. The leaves are about ¾ inch long by ½ inch wide. The flowers are hairy, 2¾ to 3 inches long, red or less commonly orange, with yellow throats. The corolla tube is 1 inch long, the broad, three-lobed, helmet-shaped upper lip nearly or quite twice as long. The latter has down-pointing side lobes. The ribbon-like lower lip is 1 inch long by ⅜ inch wide. Very different **C. arguta** has slender stems, which are red-hairy when young. Its sharp-pointed, lanceolate leaves, those of each pair of equal size, are spaced so closely that they practically overlap. They

are lustrous, have hair-fringed margins, and are often reddish on their undersides. The short-stalked blooms, solitary from the leaf axils, are orange-red with yellow throats. Approximately 2¼ inches long, their corolla tubes account for a little more than one-half their lengths. The upper lip, broader than its ¾-inch length, has large side lobes. The lower lip is elliptic, and ¾ inch long by ½ inch wide.

A handsome native of Ecuador, **C. teuscheri** (syn. *Trichantha teuscheri*) has been mistakenly cultivated as *T. minor,* a name that correctly belongs to a species with mature leaves, unlike those of *T. teuscheri,* devoid of hairs on their upper surfaces except at the margins. A slender-stemmed trailer, *T. teuscheri* is closely related to *T. elegans.* Hairy in all its parts, it has asymmetrical pointed-elliptic leaves with blades 1 inch to 2 inches long and from the leaf axils longish-stalked, narrowly-tubular, 1¼-inch-long blooms, with feathery-bristly sepals and yellow-striped, wine-red corollas and small yellow lower petals. Native of Panama, **C. moorei** (syn. *Trichantha moorei*) has trailing stems up to 1 foot long or

longer, closely furnished with lustrous, fleshy, nearly hairless, broad-elliptic, blunt leaves, approximately ½ inch in length, those of each pair of equal size. Solitary in the leaf axils, the flowers have green calyxes with comb-toothed lobes and conspicuous red hairs. The corollas, 2 inches long or slightly longer, are hairy on their outsides and have red tubes and three clear yellow lobes (petals) and two yellow-margined red ones.

Erect and bushy **C. crassifolia,** of Mexico and Guatemala, is suitable for pots rather than hanging baskets. Introduced to cultivation in the United States in 1954, and about 2½ feet tall, it has linear-lanceolate leaves 2 to 4 inches long by at most scarcely more than ½ inch wide, but usually narrower. Above, they are green and hairless. Their paler undersides are sparsely-hairy. The short-stalked, 3-inch-long, hairy, bright orange-red blooms are solitary in the leaf axils. Their corolla tubes are a little longer than the helmet-like, three-lobed upper lip, the side lobes of which turn upward. The backward-bent lower lip is about

Columnea crassifolia

Columnea illepida

Columnea aureonitens

Columnea lepidocaula

pale yellow corolla densely clothed with silky-white hairs and with an inch-long tube. *C. aureonitens,* of northern South America, erect and little branched, has stems and leaves with white, purple, or golden shaggy hairs. The larger of the pointed-ovate-elliptic to oblanceolate leaves of each pair is up to 9 inches long, the other markedly smaller. In clusters from the leaf axils, the flowers, about 1½ inches long, have scarlet calyxes and yellowish-white to orange corollas clothed outside with golden or red hairs. *C. consanguinea,* of Costa Rica, is a thick-stemmed shrub with pairs of very unequal leaves, the larger up to 1 foot long, the other tiny. They are hairless or almost so, spotted with red on their undersides. The flowers are in clusters, their silky-haired, ¾-inch-long corollas almost hidden among the hairy calyx lobes. *C. erythrophaea,* of southern Mexico, has sprawling stems and asymmetrically elliptic leaves up to 2½ inches long by ⅝ inch wide, those of each pair of nearly equal size. They have green upper surfaces, paler undersides, and are hairy on the veins beneath and along their margins. The solitary flowers on stalks 1½ inches long have a pale orange-green calyx and a strongly two-lipped, 2-inch-long, yellow-throated, orange-red corolla. *C. glabra,* of Costa Rica, has erect stems that

sparsely-hairy leaves, those of each pair markedly different in size, are up to 8 inches long by 4 inches wide. Their glossy, pebbly-textured upper surfaces are bluish-green with a coppery sheen, beneath they are red or green flushed with red. Their margins are round-toothed and recurved. Hidden or partially hidden by the foliage and not readily seen unless the plant is in a hanging basket, the purple-striped flowers are 2 inches long.

Other species cultivated include these: *C. affinis,* of Venezuela, has stout stems with a velvet of red-purple hairs when young. Its purple-hairy leaves, those of each pair very markedly different in size, the biggest up to 9 inches long by 2½ inches wide, are oblanceolate. Almost stalkless, the blooms, in twos or threes from the leaf axils, are 1¼ inches long. Their slender, cylindrical corollas, densely clothed with orange hairs, scarcely expand their petals. There are orange bracts at the bases of the flowers, and the pointed fruits are yellowish. *C. argentea,* of Jamaica, has stout stems and narrow-elliptic, distantly-toothed, gray-hairy leaves up to 5 inches long by 2 inches wide, those of each pair of nearly equal size. The flowers have inch-long stalks and a strongly two-lipped,

½ inch long. About ¾ inch long, the fruits are white tinged with pink or purple. Distinctive *C. lepidocaula* is an erect, bushy species with 3-inch-long, orange, hairy blooms paling to yellow in their throats and on the undersides of the corolla tubes. The corolla tube and helmet-like upper lip are approximately of equal length. The lower lip is scarcely more than ¼ inch long. Native to Costa Rica, up to 1½ feet tall or taller, this sort has rather fleshy, lustrous, elliptic to ovate leaves up to 3½ inches long by approximately one-third as wide.

Somewhat ungainly *C. illepida* (syn. *Trichantha illepida*) is probably Peruvian and has stout, erect, brown-hairy stems and pairs of conspicuously unequal-sized, ovate-lanceolate to oblanceolate, hairy leaves up to 5 inches long by 2 inches wide, on their undersides blotched with red or completely red. The 2- to 3-inch-long blooms, several from each leaf axil, have green calyxes with comb-toothed, but not finely-dissected lobes, and dull yellow corollas tigered with longitudinal stripes of dark brown or maroon.

Notable for its attractive foliage, *C. purpureovittata* (syn. *Tricantha purpureovittata*) of Peru branches from close to its base. Its asymmetrically-lanceolate,

Columnea affinis

Columnea glabra

root from the nodes, short-stalked, oblanceolate leaves about 1½ inches long, hairless above, with somewhat hairy undersides. The scarlet blooms, 2½ inches long or a little longer, hairy on their outsides and pointing toward the tips of the shoots, are solitary in the leaf axils. *C. hirsuta* is Jamaican. A vining sort, this has bright green, slender-pointed, toothed, ovate leaves, the larger of each pair up to 6 inches long. The hairs of their upper surfaces, unlike those of related *C. fawcettii*, stand erect from the surface. Approximately 2 inches long, the flowers are striped red, light yellow, and whitish. *C. hirta*, of Costa Rica, has red-hairy, stiffish

2 inches in length. The longish-stalked flowers, solitary from the leaf axils, are 1¼ inches long by about ⅝ inch wide. They have a finely-hairy, orange-striped, yellow corolla tube and yellow petals. *C. kucyniakii*, of Equador, is an erect shrub up to 6 feet tall or taller, with four-angled stems and pairs of equal clearly-stalked, ovate leaves up to 1 foot long. The lemon-yellow, white-hairy flowers, with two red spots on their lower lobes, nearly 2½ inches long, with stalks as long as the corollas, are in clusters from the leaf axils. *C. linearis*, bushy and up to 1½ feet tall, has spreading stems and, unusual for the genus, bright rose-pink flowers. Its glossy

ing stems upturned at their ends, with brownish hairs. Its closely-set, very fleshy, hairy, short-stalked, broad-elliptic to ovate-elliptic leaves are in pairs, with alternate pairs set at right angles to pairs above and below them. The flowers are solitary from the axils of the upper leaves. They have hairy, brilliant scarlet corollas, a slight spicy fragrance, and are 2½ inches long, with hairy stamens. *C. nicaraguensis*, of Central America, shrubby, is up to 2½ feet high. It has corrugated, lanceolate, satiny leaves, green above, and red on their undersides. Those of each pair vary much in size or sometimes one is not developed, with the result that those that are appear

Columnea hirta

Columnea linearis

Columnea nicaraguensis

rooting stems and pairs of nearly equal, oblongish, densely-hairy leaves ½ inch to 1½ inches long by up to ⅜ inch wide. On short stalks and solitary from the leaf axils, the flowers have hairy, strongly-two-lipped, red corollas 2¼ to 2¾ inches long. Of their total length, the helmet-shaped upper lip accounts for less than one-half. The backward-bent lower lip is slender, about ¾ inch long. *C. jamaicensis*, of Jamaica, has trailing stems and elliptic to ovate, thickish, slightly-toothed, glossy leaves, one of each pair larger than the other. They have ¼-inch-long stalks and blades about

leaves, those of the pairs of equal size, are linear and up to 3½ inches long by ⅜ inch wide. Solitary, hairy, and short-stalked, the blooms are strongly-two-lipped. They are about 1¾ inches long, with the tube and calyx approximately the same length. The upper lip of the corolla, three-lobed and slender, has upturned side lobes. The slender lower lip is approximately ½ inch long. It points downward. This is native to Costa Rica. *C. microcalyx*, of Costa Rica, has short, semipendant stems and lanceolate leaves ½ to 1 inch long, sparsely-hairy on both surfaces. The 2½-inch-long flowers are orange-red, with lemon-yellow at the front of the corolla tube. *C. mortonii*, of Panama, has thickish, trailing or droop-

alternate. The largest are up to 5 inches long by 2 inches wide. Usually in threes from the leaf axils and about 3 inches long, the abundant strongly-two-lipped, hairy flowers are scarlet, with a yellow underside to the over 1-inch-long corolla tube. The upper lip, longer than the tube and three-lobed, is arched. The side lobes are about ¾ inch long. *C. oerstediana*, of Costa Rica, has pendant stems and short-stalked, pointed-elliptic to ovate leaves ½ to ¾ inch long, with hairless upper surfaces and thinly-hairy undersides. The slender, orange-red blooms, 1¾ to 2½ inches long, have calyxes with slightly-toothed or toothless lobes. *C. percrassa*, of Panama, has green, fleshy stems and lustrous, short-stalked, elliptic leaves, dark green above and paler on their undersides. They are up to 1½ inches long. The very slender flowers, from the leaf axils, are scarlet with yellow interiors. They have an arching lip and a shorter, recurved lower one. *C. pilosissima*, of Honduras, is a softly-hairy trailer with slender stems and somewhat glossy, lanceolate-oblong leaves, those of each pair of equal size, up to 1½ inches long by ½ inch wide. About 2¾ inches long, the flowers are reddish-tangerine with yellow stamens and pistil. *C. querceti*, Costa Rican, has semitrailing stems well furnished with pairs of unequal, rather narrow, somewhat waxy pointed-elliptic leaves 2 to 3 inches long and from the leaf axils apri-

Columnea jamaicensis

Columnea mortonii

Columnea pilosissima

cot-yellow flowers almost 1½ inches in length, with urn-shaped calyxes. *C. salmonea*, of Costa Rica, has arching-pendulous stems and pairs of equal, sparingly-hairy, ovate leaves ¾ to 1 inch long, their edges and veins of their undersides finely lined with red. Almost 2 inches long, the handsome wide-mouthed, salmon-pink blooms are hairy on their outsides. *C. sanguinea* (syns. *C. picta, Alloplectus sanguineus*), of northern South America and the West Indies, has rigid, spreading or trail-

Columnea sanguinea

ing stems up to 4½ feet long and pairs of very unequal, nearly stalkless leaves. The biggest are 4 inches to 1 foot long by roughly one-third wide. There is a red patch below their apexes on their undersides. The smaller leaf of each pair is about one-half as wide as its ¾- to 2-inch length. The short-stalked, furry, pale yellow, bellied blooms, ¾ inch long or scarcely longer, are in twos or threes from the leaf axils. Their lobes (petals) are about ⅛ inch long. *C. sanguinolenta*, of Costa Rica, has more or less arching stems and scarcely-toothed oblanceolate leaves; the larger of each pair is up to 4½ inches long. Their upper surfaces are hairless, their undersides, often spotted with red, have hairs lying flat

along their surfaces. The long-stalked flowers, solitary or twinned, have hairy, bellied, scarlet corollas up to 2 inches long. *C. schiedeana* is Mexican. Upright or climbing and less showy than many kinds, it has stems up to 3 feet in length. The leaf pairs are of two similar-shaped individuals markedly different in size. They are asymmetrically-lanceolate to oblong, green with a velvet of white hairs on their upper surfaces, hairy and red-veined beneath. The biggest are 2½ to 5 inches long by ¾ to 1½ inches wide. The flowers, solitary or in pairs from the leaf axils, have two-lipped, maroon-marked, dullish yellow or orange-yellow corollas about 2½ inches long. The helmet-shaped, three-lobed upper lip exceeds the corolla tube in length. The lower lip, slender and bent backward, is almost 1 inch long. *C. translucens*, of Panama, is a small shrub clothed with long hairs intermixed with glands. In unequal pairs, the larger of each up to 2 inches long, the scallop-edged leaves are elliptic-oblong. From 1½ to 2 inches long, the orange flowers have incurved yellow lobes (petals). *C. tulae* is a trailing native of Puerto Rico and Haiti. It has orange or yellow blooms. Yellow-flowered variants are grown as *C. t. flava*. The stems of this species are soft-hairy as are the elliptic to oblong leaves, those of each pair nearly equal, up to 1¾ inches long by up to ¾ inch wide. The solitary, axillary blooms have stalks with bracts. The corollas, strongly-two-lipped, are up to 2 inches long. The slender upper lip has three lobes, the side ones triangular. The under lip is about ¼ inch in length. *C. verecunda*, from Costa Rica, is shrubby and compact, with stems up to 2½ feet long. One leaf of each pair is so rudimentary that the others give the impression of being alternate. They are oblong-lanceolate, up to 4¼ inches long by 1¼ inches wide, green above, wine-red on their undersides, with short, stiff hairs on the veins. On short stalks from the leaf axils are displayed clusters of several yellow or light red blooms 1½ inches long. The upper lip is longer than, the lower lip about as long as, the corolla tube. The fruits are pink. *C. warczewicziana*, of Costa Rica, is a shrub with more or less arching stems and oblanceolate leaves in unequal pairs, the larger individual of each up to 4½ inches long. They are somewhat toothed, hairy on both surfaces, and unlike related *C. sanguinolenta*, never red-spotted beneath. The solitary or paired, long-stalked, scarlet blooms have calyx lobes slightly or not toothed and strongly-bellied corollas 1½ to 2 inches long. *C. zebrina*, of Panama, semierect, has thick stems, much constricted at the nodes, clothed with grayish scales. Each pair of asymmetrical, nearly stalkless, lanceolate leaves consists of one large and one much smaller, the bigger up to 7 inches long by 2 inches wide. The upper surfaces are hair-

Columnea zebrina

less, the lower ones silvery-hairy. In clusters from the leaf axils, the 3-inch-long, tubular flowers are pale yellow, with longitudinal black-purple stripes and white hairs. They have a two-lobed stigma.

Hybrid columneas are fairly numerous, and many are very beautiful. Among them are excellent ones raised in Norway and others at Cornell University. Available kinds are listed and described in specialists' catalogs. Among the best known are these: *C. banksii*, the parents of which are *C. oerstediana* and *C. schiedeana*, has stout-

Columnea banksii

ish, pendulous stems, thickish, lustrous leaves, and scarlet flowers obscurely lined with yellow in their throats. 'Campus Favorite', has arching stems, pairs of leaves of markedly unequal size and bright red flowers with yellow-edged petals. 'Campus Sunset' is a trailer with red-hairy stems, ovate, short-hairy leaves, and canary-yellow and orange blooms 2 to 2¼ inches long. 'Canary', a semitrailer with dark green leaves, those of the pairs unequal in size, wine-red beneath, has brilliant yellow, 2-inch-long blooms. 'Cayugan', has spreading branches, lanceolate leaves, and large crimson blooms with yellow in the throat. 'Cornellian', erect or

Columnea 'Campus Sunset'

Columnea 'Cornellian'

Columnea 'Yellow Dragon', in a basket

Columnea 'Canary'

Columnea 'Ithacan', in a basket

Columnea woodii

Columnea 'Cayugan'

Columnea kewensis variegata

spreading, has lanceolate leaves and flowers in the reddish range and yellow. 'Ithacan', a trailer, has ovate, hairy leaves and 3-inch-long, coral-red blooms with bright yellow undersides. **C. euphora,** its parents *C. gloriosa* and *C. lepidocaula,* has much the appearance characteristics of the first, but its long, pendulous stems are stouter and its ovate leaves bigger, being up to rather more than 2 inches long. About 3 inches in length, the corolla is vermillion suffused with yellow along the top of the tube.

'Othello' is a superior variety of this hybrid. **C. kewensis variegata,** its parents *C. glabra* and *C. schiedeana,* has elliptic leaves variegated with pink and white and bright orange-scarlet blooms. 'Stavanger' is a fine trailer with small, glossy, round leaves and red flowers. 'Yellow Dragon' has large leaves red beneath, and bright yellow, sometimes red-margined flowers. **C. vedrairiensis,** erect and shrubby, has elliptic leaves, large, downy, bright scarlet blooms veined with yellow. **C. woodii,** an inter-

mediate hybrid between *C. crassifolia* and *C. nicaraguensis,* has stems markedly constricted at the nodes, at first erect, later arching downward, but with upcurved tips. The leaves are lanceolate, those of each pair markedly unequal in size, the bigger up to 6 inches long and 2 inches wide. The hairy-tubed flowers, solitary, in twos or threes, are about 3 inches long, and red marked with yellow in the throat.

Garden Uses. Columneas are among the choicest plants for growing, according to the habits of the various kinds, in pots or in suspended containers, such as hanging wire baskets, or attached to slabs of tree fern trunk or cork bark. In the humid tropics they can be similarly accommodated outdoors or be attached as epiphytes on rough-trunked palms or other trees. Where the air is not excessively dry some success may be had with them as window plants and under fluorescent lights. They need adequate warmth and high humidity.

Cultivation. The potting or planting mix must be sufficiently porous to admit air and pass water freely. It should be largely organic and may consist chiefly of leaf mold, coarse peat moss, or tree fern fiber with a little fibrous, loamy topsoil, coarse sand, or perlite and a little crushed charcoal mixed in. The addition of a small amount of crushed limestone or oyster shells benefits some kinds. A rather spongy, open mix is needed. Avoid overpotting. Columneas seem to do best when their roots are just a little crowded. Keep the soil always moderately moist, but not constantly saturated. Specimens that have filled their containers with roots are kept in good condition by biweekly applications of dilute liquid fertilizer. Temperatures of 70 to 80°F are beneficial by day, but from fall to spring 60°F is adequate at night and some sorts, *C. microphylla,* for example, are better at five degrees lower. Too high night levels inhibit blooming. Good light, with sufficient shade to moderate the full intensity of summer sun is needed. Propagation is usually by cuttings, which are easily rooted, and division. Leaf cuttings may also be used but are slower in producing sizable specimens. Plants can also be raised from seed.

COLUTEA (Co-lùtea) — Bladder-Senna. Coluteas are deciduous shrubs, and those described below are hardy about as far north as southern New England. The genus, which in the wild ranges from the Mediterranean region to Ethiopia and the Himalayas, belongs to the pea family LEGUMINOSAE and comprises about twenty-six species. Its name is a modification of the ancient Greek one *koloutea.* The leaves of *Colutea arborescens* possess medicinal properties similar to those of senna with which they are sometimes mixed as an adulterant.

Coluteas have alternate, pinnate leaves, with an uneven number of toothless leaflets. The pea-shaped, yellow, brownish-orange, or light coppery blooms, in long-stalked, few-flowered racemes from the leaf axils, provide a succession for a considerable period in summer. They have five-toothed calyxes, a rounded standard or banner petal with two swellings or folds, and ten stamens of which nine are joined and one is free. The fruits are many-seeded, much inflated, decorative, bladder-like pods, somewhat like those of *Koelreuteria* and *Staphylea.*

Native of southern Europe and North Africa, **C. arborescens** is 12 to 18 feet tall. It is the hardiest kind. A broad, rounded

Colutea arborescens (fruits)

bush, it has downy shoots, and leaves of seven to thirteen elliptic to ovate or obovate leaflets up to 1 inch long, and notched at their ends. The leaves are up to 6 inches long and usually hairy on their undersides. About ¾ inch long, the bright yellow blooms, with beakless keels as long or rarely slightly longer than the wing petals, are in racemes of three to eight. The pods are 2 to 3 inches in length and have at their ends the persistent, greenish style. Variety *C. a. gallica* differs in the ovary being sparsely-hairy instead of smooth. A dwarf, slow-growing variety, *C. a. bullata* has smaller, wrinkled foliage, and smaller blooms. From southern Europe and Asia comes **C. orientalis.** From 6 to 9 feet in height, this is a rounded bush of dense habit. From *C. arborescens* it differs in its 1½-inch-long seed pods curving upward at their ends and in its coppery or reddish-brown blooms having beaked keels. Its shoots are downy. The leaves, up to 4 inches long, are decidedly glaucous, and have seven to nine broad-ovate to roundish leaflets ¼ to a little more than ½ inch long. In racemes of up to five, the flowers are about ½ inch long and wide. The wing petals are shorter than the keel; the standard petal is marked with a yellow spot. Native of eastern Europe, **C. cilicica** resembles *C. arborescens,* but has slightly

Colutea cilicica

larger flowers, with wings longer than the keel and with a spur on the lower edge. From *C. a. gallica,* which it resembles, persistently hairy shoots and conspicuously silvery-hairy ovaries distinguish **C. atlantica,** native to Spain.

A variable hybrid between *C. arborescens* and *C. orientalis* is **C. media.** In appearance it usually more resembles *C. arborescens* than its other parent, but differs in having brownish-red or orange-red blooms, and purplish pods. As a garden plant it is superior to its parents.

Garden and Landscape Uses. Coluteas are too coarse for choice plantings and for most small properties. Their floral and fruit displays do not compare advantageously

with those of many more popular shrubs. Yet they have some advantages and are useful for special locations. They grow well where many better shrubs fail. They are not particular as to soil or situation, provided the ground is not wet and there is not excessive shade.

Cultivation. This is of the simplest. To keep the plants to size and tidy they may be pruned hard in early spring, either annually or at intervals of two or three years. No other care is ordinarily needed. Propagation is by seeds and by summer cuttings, 3 or 4 inches long planted under mist or in a propagating bed in a cold frame or greenhouse.

COLVILLEA (Col-víllea)—Colville's Glory. Sir Charles Colville, an officer of the British army in the Napoleanic war and later Governor of Mauritius, is honored by the name of this deciduous or nearly deciduous tree, the only one of its genus, a native of East Africa and Madagascar. It belongs in the pea family LEGUMINOSAE. It is closely related to, and when not in bloom looks very much like, the royal poinciana (*Delonix regia*), but its leaves are bigger and to a small degree so are the leaflets.

Colville's glory (**Colvillea racemosa**) becomes 50 feet in height and has a spreading head. Its lacy, twice-pinnate leaves, up to 3 feet in length, have numerous leaflets about ½ inch long. The blooms are in dense, cone-shaped or cylindrical, drooping panicles, 8 inches to 1½ feet long, containing some two hundred flowers or more. In bud they look for all the world like great bunches of reddish-orange grapes or, as one distinguished horticulturist said, salmon-pink wisterias. Starting at the tops of the clusters the "grapes" open to show, inside each, a small, central, orange-red petal between two long, narrow dark red ones, and ten protruding bright yellow stamens. The parts surrounding the petals and stamens in the bud stage are sepals. The fruits are cylindrical pods about 6 inches long by 1 inch wide.

Garden and Landscape Uses and Cultivation. Colville's glory is not hardy where winters are more severe than those of southern Florida. It is a warm-climate species. Where it succeeds it is a highly satisfactory ornamental and is useful as a shade tree and a street tree. It grows rather slowly and, in Florida and Hawaii, flowers in fall over a period of two months or more, at a season when, because of its relative scarcity, bloom is especially welcome. Any ordinary soil and a sunny location suit this fine species. It is propagated by seed.

COLVILLE'S GLORY. See Colvillea.

COMANTHOSPHACE. See Leucosceptrum.

COMAROSTAPHYLIS (Comaro-stáphylis)—Summer-Holly. The about twenty species of *Comarostaphylis* of the heath family ER-ICACEAE, closely related to *Arctostaphylos*, are mostly confined in the wild to Mexico and Guatemala. The name, from the Greek *komaros* (*Arbutus*), and *staphyle*, a grape, alludes to the plant's similarity to *Arbutus* and to its clustered fruits. From *Arctostaphylos* and *Xylococcus* this genus differs in its fruits having warty instead of smooth surfaces and more closely resembling those of *Arbutus*. It is distinct from *Arbutus* in having hairy calyxes.

The sorts of this genus are evergreen, erect or spreading shrubs, with alternate, leathery, toothed or toothless leaves often with rolled-back margins. Their few to many small flowers are in solitary or clustered racemes or panicles from the branch ends. They have usually five, rarely four, each sepals and petals, twice as many nonprotruding stamens, and a columnar style tipped with a minute stigma. The fruits are fleshy and drupelike.

Summer-holly (*C. diversifolia* syn. *Arctostaphylos diversifolia*) is a native of southern California and Baja California. Variety *C. d. planifolia* occurs also on islands off the coast. This species is an erect shrub, up to 15 feet in height, with shredding bark and grayish pubescent twigs. Its rounded to elliptic, blunt or pointed, very short-stalked

Comarostaphylis diversifolia planifolia

Comarostaphylis divarsifolia planifolia
(flowers)

leaves are shining green on their upper surfaces and whitish-pubescent beneath. They are 1½ to 3 inches long, closely-toothed, and have strongly rolled-under margins. The white, urn-shaped flowers, with ten or rarely fewer stamens, have short, recurved stalks and are about ¼ inch long. They have tiny corolla lobes. The flowers are in mostly solitary, terminal racemes 1½ to 2½ inches long. Spherical and red, the fruits are about ¼ inch across. Variety *C. d. planifolia* is distinguished by its leaves not being revolute at their margins.

Garden Uses and Cultivation. These are similar to those of the manzanita group of *Arctostaphylos*. The plants are hardy in mild climates only.

COMB FERN is *Ctenitis sloanii.*

COMBRETACEAE — Combretum Family. This family consists of about 600 tropical and subtropical species of dicotyledons allocated among nineteen genera. Most sorts are trees, shrubs, or woody vines, some of which climb by twining, others by hooked spines that are the persistent bases of the leafstalks. One of the best known of the group, the tropical-almond (*Terminalia catappa*) is employed as a shade tree and is grown for its edible nuts in the tropics and subtropics.

Alternate or less often opposite, the leaves of plants of the combretum family are not divided into leaflets. Usually bisexual, rarely unisexual, the symmetrical or sometimes asymmetrical, generally stalkless, small flowers in spikes, racemes, or panicles have calyxes with four, five, or occasionally eight lobes or sepals. The corolla usually has as many petals as there are calyx lobes or sepals, but is sometimes absent. There are two to five or twice as many stamens as calyx lobes or sepals, and a slender style. The dry or drupelike, often winged fruits contain one seed. Genera cultivated include *Bucida, Combretum, Conocarpus, Quisqualis,* and *Terminalia.*

COMBRETUM (Com-brètum). Trees, shrubs, vines, or rarely herbaceous plants of the American and Old World tropics and warm subtropics constitute *Combretum* of the combretum family COMBRETACEAE. They number 350 species. The name is one used by Pliny for some unknown vine.

Combretums have usually opposite, sometimes whorled (in circles of more than two), rarely alternate, undivided, lobeless and toothless leaves. Their generally small, nearly stalkless flowers, in spikes or racemes, are sometimes assembled in panicles. Both bisexual and unisexual flowers are borne. They have bell-shaped calyxes, four or five petals, and long protruding stamens. The fruits are leathery, one-seeded drupes, with four prominent wings.

A vine up to 20 feet tall, *C. coccineum* (syn. *Grislea coccinea*), native to Madagas-

car, has glossy, oblong-lanceolate leaves up to 4 inches long, and in terminal panicles, one-sided spikes of scarlet, ½-inch-long flowers. African *C. grandiflorum* is a vine with short-stalked, oblong-elliptic leaves up to 8 inches long by one-half as wide and scarlet flowers 1 inch to 1¼ inches long in spikes 4 or 5 inches in length, from the leaf axils and terminal. A vine from Trinidad and northern South America, *C. secundum* has short-stalked, pointed, oblong-elliptic leaves up to 6 inches long by approximately one-half as wide. The crowded 2- to 4-inch-long spikes of 1- to 1½-inch-long scarlet blooms come from the axils of the leaves toward the branch ends. Widely spread in its native Central and tropical South America, *C. cacoucia* (syn. *Cacoucia coccinea*) is a magnificent vine or sometimes shrub, with short-stalked, broadly-elliptic leaves up to 6 inches long by 4 inches wide. The handsome spikes of red flowers are 1 inch to 1½ feet in length, the blooms 1 inch to 1½ inches long. The seeds of this species are used as rat poison.

Other sorts cultivated include these: *C. erythrophyllum* is a graceful South African, smooth-barked tree, with opposite or alternate, pointed-lanceolate, hairless, drooping leaves that become yellow and red in winter. The greenish-white flowers have eight stamens. The fruits are four-winged. *C. microphyllum,* of Mozambique, is a straggling shrub, with opposite or alternate, broadly-ovate, 1-inch-long leaves, notched at their apexes. In racemes from the leaf axils shorter than the leaves, the hairy, scarlet, four-petaled flowers are succeeded by four-angled, ellipsoid fruits. *C. paniculatum* is a vigorous, long-stemmed vine, native of tropical Africa. It has hairy branchlets furnished with short spines and alternate, broadly-elliptic to nearly round leaves up to 8 inches long. After the leaf blades fall, the persistent leafstalks are gradually transformed into spines. The four-petaled, 1½-inch-long, coral-red flowers in large terminal panicles are usually displayed when the plant is leafless. The pink to orange fruits are about 1½ inches long.

Garden and Landscape Uses and Cultivation. These attractive vines are suitable for outdoors only in the tropics and warm subtropics. There and in tropical greenhouses, they make good furnishings for pillars, pergolas, and similar supports. In greenhouses they ordinarily do better in ground beds than containers. A mellow soil, well drained, and containing an abundance of leaf mold or peat moss is to their liking. It should be moderately moist. Following blooming, fairly hard pruning is usually advisable. Propagation is by cuttings and seeds.

COMESPERMA (Com-espérma) — Australian Love Creeper. A few more than twenty species constitute this strictly Australian genus of the milkwort family POLYGALA-CEAE. Its name derives from the Greek *kome*, hair, and *sperma*, seed, and alludes to the tufts of hairs usually attached to the seeds. Not hardy in the north, shrubs, subshrubs, and herbaceous, sometimes twining plants constitute *Comesperma*.

Comespermas have alternate leaves, sometimes reduced to scales, and small, asymmetrical blooms in racemes. The flowers have five sepals, the two inner, winglike and larger than the others, are colored. There are three petals, the lower one slightly the shortest. Usually there are six to eight stamens and an incurved, two-lobed style. The fruits are dry capsules.

The Australian love creeper (*C. volubile*) is an attractive, slender-stemmed, often much tangled, woody vine, with rather few small, linear to lanceolate leaves and erect, loose, spikelike racemes of beautiful blue, lilac, or whitish blooms up to ¼ inch in length, in racemes 1 inch to 3 inches long.

Garden and Landscape Uses and Cultivation. As a groundcover for banks and similar places and a vine, in mild, dryish climates the Australian love creeper has merit. It needs well-drained, sandy peaty soil and a sunny location. It is propagated by seeds and by cuttings in a greenhouse propagating bench or similar accomodation.

COMFREY. See Symphytum. Wild-comfrey is *Cynoglossum virginianum*.

COMMELINA (Com-melina)—Day Flower. Day flowers are relatives of and generally resemble tradescantias. From that group they are most easily distinguished by their flowers having one petal smaller, usually markedly so, than the other two. In *Tradescantia* the three are equal. Commelinas number more than 200 Old and New World species of the spiderwort family COMMELINACEAE. The name commemorates three Dutch brothers of the sixteenth and seventeenth centuries. Two, Johann and Kaspar Commelin, became botanists of note, the third did not. Fancifully, the two larger petals of the flower represent

Commelina species

the botanists, the smaller one the other brother. For the plant previously named *Commelina anomala*, see Commelinantia.

Day flowers are deciduous and evergreen, succulent-stemmed annuals and herbaceous perennials. They have jointed, often trailing stems, and alternate, undivided, toothless, linear to ovate-lanceolate, stem-clasping leaves. Their flowers, the individual ones lasting for only a few hours, are clustered, with at the base of each cluster a folded, heart-shaped, leaflike bract. Each bloom has three somewhat unequal-sized sepals, two of which are usually more or less united at their bases. There are three petals the lower of which is markedly smaller than the others, and often of a different color. The stamens and staminodes have hairless stalks and total five or six. One of the stamens curves inward and has a longer anther than the other one or two. The fruits are capsules. All the kinds described below are perennials. None except some of the native North American ones is hardy in the north. The hardiness of the Americans may be deduced from the native ranges. In Hawaii, *Commelina diffusa* and *C. benghalensis* are used as fodder for cattle, and the former, it is reported, sometimes as human food.

Most frequently cultivated is deciduous *C. coelestis*, of Mexico. This kind has fibrous or tuberous roots and erect stems up

Commelina coelestis

to 1½ feet tall. Its leaves are oblong-lanceolate and 3 to 6 inches long. The wavy-edged bracts at the bases of the clusters of ten or fewer flowers are pubescent. The two large petals, nearly ½ inch long, are rich blue in the typical species; in *C. c. alba*, they are white, in *C. c. variegata*, blue and white. Another deciduous Mexican, *C. tuberosa* is a loosely-branched, diffuse plant, with tuberous roots and narrowly-lanceolate leaves often hairy along their edges. The bracts at the bottoms of the six-

Commelina tuberosa

to ten-flowered clusters of flowers may be hairy or almost hairless. The largest two petals are blue and up to ⅜ inch long. Native to tropical Africa and Asia, *C. benghalensis* is an evergreen trailer with broadly-elliptic, fleshy leaves that have wavy, often red-hairy margins. Its flowers are sky-blue. This species has the peculiarity of producing flowers underground that set seeds without being pollinated by other flowers. In *C. b. variegata* the leaves are grayish-green striped with white.

Commelina benghalensis variegata

Kinds native to the United States sometimes planted in gardens include *C. erecta angustifolia*, *C. e. crispa*, *C. diffusa*, and *C. virginica*. The first has upright or decumbent stems 8 inches to 2½ feet long, and linear leaves up to 5 inches in length. Its flowers are blue. This kind is wild from North Carolina to Florida and Texas. Native from Tennessee to Colorado, Indiana, Texas, and Arizona *C. e. crispa* is an attractive upright or decumbent kind, with linear to narrowly-linear-lanceolate leaves up to 3½ inches in length. From the other this differs in having bracts ¾ inch long, or longer and generally abundant white hairs at the bases of the flower clusters. The bracts of *C. e. angustifolia* are up to ¾ inch long, and white hairs are generally absent. Its stems trailing, freely-branched, up to 3 feet in length, and rooting along their lower parts, *C. diffusa* (syn. *C. nudiflora*) inhabits moist woodlands and river banks from Delaware to Missouri, Florida, and Texas. It is also wild in Central and South America and many parts of the Old World. The leaves of this are lanceolate and 2 to 3½ inches long. Its blue flowers are about ½ inch wide. Blue flowers, 1 inch wide, with the lower petal nearly as big as the upper ones are borne by *C. virginica*, an erect species up to 3 feet tall that has long-pointed, linear-lanceolate leaves 4 to 8 inches in length. This sort occurs in moist woodlands from New Jersey to Missouri, Florida, and Texas.

Garden and Landscape Uses and Cultivation. Trailing day flowers serve as groundcovers, those of upright growth for inclusion in flower beds and for planting in informal landscapes. All grow with great ease in porous, fairly moist, reasonably fertile soil and are readily increased by cuttings, seeds, and the tuberous-rooted ones, by tubers. These last can be planted outdoors in spring, lifted in fall, and the roots stored over winter like those of dahlias.

COMMELINACEAE — Spiderwort Family. This family of monocotyledons comprises thirty-eight genera totaling approximately 500 species of chiefly tropical and subtropical, rarely twining, herbaceous perennials sometimes with tuberous roots. The several plants commonly grown as wandering jews belong in the COMMELINACEAE.

Members of the spiderwort family have fleshy, jointed stems and alternate, lobeless and toothless, broad or narrow, more or less succulent, undivided, parallel-veined leaves, with bases that sheathe the stems. The flowers, from the leaf axils or terminal, are solitary or more usually in clusters, which are often umbel-like, or in panicles.

Symmetrical or less often asymmetrical, the blooms have calyxes of three generally separate, less often partly united, sometimes petal-like sepals. The corolla is of three ephemeral petals rarely united below into a tube. There are generally six stamens or three and three staminodes (nonfunctional stamens). Rarely only one stamen is functional. Often the stalks of the stamens and staminodes are thickly clothed with colored hairs. The single style is tipped with a headlike or rarely three-branched stigma. The fruits are capsules. Genera cultivated include *Aploleia*, *Callisia*, *Campelia*, *Cochliostema*, *Commelina*, *Commelinantia*, *Cuthbertia*, *Cyanotis*, *Dichorisandra*, *Geogenanthus*, *Gibasis*, *Hadrodemas*, *Palisota*, *Pollia*, *Rhoeo*, *Setcreasea*, *Siderasis*, *Tinantia*, *Tradescantia*, *Weldenia*, and *Zebrina*.

COMMELINANTIA (Commelin-ántia) — False-Day-Flower or Widow's Tears. This is a genus of one species of the spiderwort family COMMELINACEAE, indigenous to Texas and northern Mexico. Its name is a combined form of those of the allied genera *Commelina* and *Tinantia*.

False-day-flower or widow's tears, *Commelinantia anomala* (syns. *Tinantia anomala*, *Commelina anomala*) is distinguished from *Commelina* by the bracts from which its blooms come being not folded and similar to the upper leaves, but proportionately broader.

A succulent annual up to 2 feet tall, *C. anomala* has tufts of stems, at first erect, but later branched and spreading. Its lower, linear-spatulate leaves, up to 1 foot long, gradually narrow to hair-fringed stalks. The upper leaves, stalkless or short-stalked, are lanceolate to broad-lanceolate and up to 8 inches long. The flowers are in terminal racemes, each with a single leaflike bract at its base. Each bloom has three green sepals, two showy lavender-blue petals about ¾ inch long and broader than long, and a tiny greenish-white petal. There are six stamens, variable in form. The fruits are capsules.

Garden Uses and Cultivation. Not much known outside its home territory, the false-day-flower is sometimes grown in gardens there. In the wild it prefers rich, moist, limestone soils, but adapts to others that are well drained. Shaded or partially shaded locations suit it. It is easily raised from seeds sown in fall or spring and by division and cuttings.

COMPARETTIA (Compa-réttia). Not common in cultivation, *Comparettia* of the orchid family ORCHIDACEAE consists of about seven pretty species, chiefly natives of the Andes of northern South America, with one extending into Mexico and the West Indies. The name commemorates the Italian botanist Andreas Comparetti, who died in 1811.

Comparettias usually grow as epiphytes, which means that they perch on tree branches and trunks without extracting nourishment from them. They have small pseudobulbs, each with one to three comparatively large, leathery leaves. The racemes of flowers are at the tops of arching or pendulous, sometimes branched stalks that come from the bases of the pseudobulbs. One sepal of each bloom is free. The other two are united, and at their bases are prolonged into a prominent tube or horn that encloses the twin spurs of the large, brightly colored lip. The lateral petals converge. The column is separate, and semierect.

Most widely distributed in the wild, *C. falcata* ranges from Mexico to northern South America and the West Indies. It has clustered pseudobulbs about 1 inch long, each with usually one elliptic-lanceolate leaf up to 6 inches long. The slender flower stalks, up to 2 feet or sometimes more in

Comparettia falcata

length, have few to many rich rosy-purple to almost crimson blooms less than 1 inch across, and with a slender spur about as long. Slightly larger, brilliant scarlet flowers, with straighter spurs only about ½ inch long, are borne by otherwise quite similar *C. coccinea*, of Brazil. Each pseudobulb of *C. macroplectron* has two or three leaves 4 to 5 inches long by a little over 1 inch wide. The flowers are about 2 inches across and have a horn as long. In more or less pendulous racemes shorter than those of the other kinds discussed here, they are pale violet or light rose-red sprinkled with spots of deep purple.

Garden Uses and Cultivation. These attractive orchids are suitable for intermediate-temperature and tropical greenhouses, such as suit cattleyas. A humid atmosphere and sufficient shade to prevent the foliage scorching are necessary. Comparettias usually are better in hanging baskets or attached to suspended slabs of tree fern trunk than in pots. If the latter are used they should be very well drained, and rather small for the size of the plant. Osmunda fiber is a satisfactory rooting medium. At no time must the roots be dry, but more abundant supplies of water are needed from spring to fall than in winter. For more information see Orchids.

COMPASS PLANT. See Silphium and Wyethia.

COMPLETE FERTILIZER. Any fertilizer that contains appreciable amounts of the three most commonly deficient plant nutrients, nitrogen, phosphorus, and potassium, is identified as a complete fertilizer. Some soils lack other nutrients, usually referred to as trace elements, not necessarily supplied by a complete fertilizer. These may include boron, manganese, copper and others. See Trace Elements.

COMPOSITAE — Daisy Family. This assemblage of dicotyledons is one of the biggest families of vascular plants, equaled in number of species perhaps only by the orchid family ORCHIDACEAE. Some of its about 900 genera and possibly 20,000 species are known to almost everyone. As wildlings they occur in nearly all parts of the world and include asters, chrysanthemums, dahlias, daisies, dandelions, goldenrods, lettuce, sunflowers, and zinnias, to name but a few of those popular in gardens.

The majority of composites, as members of this family are called, are annuals, biennials, and deciduous or evergreen herbaceous perennials, but included also are a goodly number of subshrubs, shrubs (sometimes vining), and a few trees. Some are twining vines; many are strongly aromatic.

Leaves in this family may be all basal, or alternate or opposite on the stems. They are undivided, with or without lobes and teeth, or variously dissected into few to many leaflets. In some desert sorts they are reduced to scales. The florets or flowers (which must not be confused with the heads in which they are typically crowded and which are usually referred to as flowers), small and very rarely solitary, are commonly in heads of few to many backed or partly surrounded by a collar of bracts called an involucre. The florets, bisexual or unisexual, are of three sorts, the individual heads consisting of one or more. The sorts are, disk, such as form the center of "eyes" of daisies; ray, petal-like ones of the sort that encircle the "eyes" of daisies; and bilabiate, with two lips, the upper with three lobes or teeth, and the lower with two slender, usually recurved lobes. The florets have corollas with four or five lobes or teeth, four or five stamens united by their anthers, and one usually two-branched style. Commonly, but not invariably from their bases sprout a tuft of hairs called a pappus, representing the calyx. The fruits are almost invariably achenes (generally called seeds), usually tipped with a pappus.

Genera cultivated include these: *Achillea, Ageratum, Agoseris, Amberboa, Amellus, Ammobium, Anacyclus, Anaphalis, Andryaia, Antennaria, Anthemis, Aphanostephus, Arctium, Arctotheca, Arctotis, Argyroxiphium, Arnica, Artemisia, Aster, Asteriscus, Athanasia, Baccharis, Baileya, Balsamorhiza, Bellis, Bellium, Berlandiera, Bidens, Boltonia, Brachycome, Brachyglottis, Brickellia, Buphthalmum, Calendula, Callistephus, Calocephalus, Carduncellus, Carduus, Carlina, Carphephorus, Carthamus, Cassinia, Castalis, Catananche, Celmisia, Centaurea, Centratherum, Chaenactis, Chamaemelum, Chaptalia, Charieis, Chrysanthemum, Chrysocoma, Chrysogonum, Chrysopsis, Chrysothamnus, Cichorium, Cirsium, Cladanthus, Cnicus, Coreopsis, Corethrogyne, Cosmos, Cotula, Cremanthodium, Crepis, Crocidium, Crupina, Cynara, Dahlia, Dimorphotheca, Doronicum, Dracopis, Dyssodia, Echinacea, Echinops, Emilia, Encelia, Engelmannia, Erigeron, Eriocephalus, Eriophyllum, Erlangea, Eupatorium, Euryops,*

Evax, Felicia, Gaillardia, Galactites, Gamolepis, Garberia, Gazania, Geraea, Gerbera, Glyptopleura, Gnaphalium, Grindelia, Guizotia, Gutierrezia, Gymnaster, Gynura, Haastia, Haplopappus, Helenium, Helianthella, Helianthus, Helichrysum, Heliopsis, Helipterum, Hertia, Heteropappus, Heterosperma, Heterotheca, Hidalgoa, Hieracium, Homogyne, Hulsea, Humea, Hymenopappus, Hymenoxys, Hypochoeris, Hysterionica, Inula, Jasonia, Jurinea, Krigia, Kuhnia, Lagenifera, Lasthenia, Layia, Leontopodium, Lessingia, Leucheria, Leucogenes, Leuzea, Liatris, Ligularia, Lindheimera, Lonas, Luina, Machaeranthera, Madia, Malacothrix, Marshallia, Matricaria, Melampodium, Microglossa, Microseris, Mikania, Monolopia, Monoptilon, Montanoa, Moscharia, Mutisia, Myriocephalus, Nicolletia, Nothocalais, Oldenburgia, Olearia, Onopordum, Osteospermum, Otanthus, Othonna, Pachystegia, Palafoxia, Parthenium, Perezia, Pericome, Petasites, Phagnalon, Picris, Piqueria, Pityopsis, Pluchea, Podachaenium, Podolepis, Prenanthus, Pulicaria, Raoulia, Ratibida, Reichardia, Rudbeckia, Santolina, Sanvitalia, Saussurea, Scolymus, Scorzonera, Senecio, Serratula, Silphium, Silybum, Solidago, Solidaster, Sonchus, Spilanthes, Stephanomeria, Stevia, Stifftia, Stokesia, Tagetes, Tanacetum, Taraxacum, Telekia, Thelesperma, Tithonia, Tolpis, Townsendia, Tragopogon, Tridax, Trilisa, Tripleurospermum, Tussilago, Urospermum, Ursinia, Venidium, Verbesina, Vernonia, Vittadinia, Wedelia, Wyethia, Xanthisma, Xeranthemum, Zexmenia, and *Zinnia.*

COMPOSITE. As a noun this is a name for any plant of the daisy family COMPOSITAE.

COMPOST and COMPOSTING. In the past it was the practice to call potting soils composts and this is still occasionally done, especially by English writers. More generally the term is reserved for partially decayed organic material prepared in heaps or piles and used, much in the way of manure, for improving the soil. Since automotive traction replaced horse-drawn vehicles dependence upon animal manure as the chief source of organic material for gardens is impractical or impossible for the vast majority of gardeners. Compost is an excellent substitute.

Among the most dedicated organic gardeners the preparation of compost and its use have acquired such a mystique that they have become almost cults, often involving excessive labor and sometimes strange additions and practices (one is sometimes tempted to think, such rites as incantations at midnight by the light of the full moon). There is no need for such nonsense. The preparation of excellent compost is a simple matter.

Save all nonwoody vegetable waste, except any known to contain serious soilborne diseases or pests, for the compost pile. The makings may include leaves, grass clippings, weeds, vegetable garden

Composting: (a) Collecting vegetable wastes for the compost pile

(b) Stacking the "makings" of a compost pile in a wooden-sided bin

(c) A bin with slatted sides affords neat containment of wastes

(d) Or, alternatively, so does a bin with chicken-wire sides waste, skins, peels, decaying fruits and vegetables, hay, straw, cut-down tops of annuals, biennials, and herbaceous perennials, tea leaves, coffee grounds, and indeed practically anything of vegetable origin that can be expected to rot within a reasonable time. Exclude tree and shrub branches, woody prunings, hedge clippings and the like that are much more resistant to decay. It is generally advisable to exclude most animal wastes, except any manures that may be available. Meat and fish scraps for instance may result in disturbing odors and attract animals, including perhaps rats. The addition of egg shells is permissible.

Location of the compost pile needs consideration. Scarcely an object of beauty, it should be screened from casual view, perhaps by a garage or another building, a hedge or a planting of shrubs. Its lack of aesthetic appeal is its only disadvantage. A properly made and managed pile has no unpleasant odor. Do not locate the pile in a hollow where water collects. A partially shaded, sheltered place is better than a sunny, exposed one. If sufficient makings are available it is better to maintain two piles. One, the older, riper, and more decayed, can be drawn from for use as the other is being added to and processed.

Sizes of compost piles may vary within limits. Accumulations of wastes too small to be kept compact and uniformly moist so that microorganisms are stimulated to effect desirable partial decay scarcely qualify as compost piles. Heaps so big that reasonable admission to their interiors of moisture and air are impeded are unsatisfactory because they are slow to decay. Maximum dimensions may be about 12 feet long, 5 feet wide, and after compressing, 5 feet high. As a measure of neatness small piles can be packed within a fence of chicken wire stretched between vertical posts driven into the ground, or built within a bin with openwork sides of wood or brick. With larger piles, the sides of which should slope slightly inward from base to top, it is unnecessary to provide side supports if stacking is done neatly. Keep the top of the pile at all times a little higher around its edges than toward its center so that it will catch rain and can be watered conveniently if it becomes too dry.

To form the pile, spread as a base about 1 foot in depth the material to be composted. If some of an old compost pile can be mixed with this layer it may be advantageous, but is not essential. Compact this layer by treading it firmly. Then, either at the same time or later, as additional material becomes available, add successive layers, firming each in turn. If the makings are dry, but only if this is true, wet them thoroughly as they are stacked. The spreading of a thin cover of soil (low grade topsoil is satisfactory) on top of each 1-foot-thick layer of wastes to be composted is generally beneficial and makes for a better appearance. A generous sprinkling of a phosphatic fertilizer, such as superphosphate, bonemeal, basic slag, or finely ground rock phosphate, on top of each layer results in an enriched product. A light application of ground limestone, or a heavier one if substantial amounts of oak leaves or other acid-forming materials are included, may be spread with each layer. To multiply and prosper, the microorganisms of decay need an adequate supply of nitrogen. If this is in short supply or unavailable the compost making process is delayed or stopped. Larger supplies of nitrogen are needed if a considerable proportion of the wastes contain, as do hay, straw, and semiwoody materials, high proportions of cellulose. Chemical activators containing nitrogen are marketed, but equally good results can be had by sprinkling a complete garden fertilizer such as a 5–10–5 or a 10–10–10 among the wastes as the pile is formed. For a pile that when compacted contains 100 cubic feet use from 50 to 100 pounds of fertilizer. Little or no advantage seems to come from sprinkling piles with cultured inoculants of soil bacteria sold for this purpose and surely there is no gain from the use of brews made from various herbs and so forth that some compost pile cultists advocate.

Turning the pile periodically, at intervals of a few weeks, with a fork or mechanical handler hastens decomposition, but involves considerable labor. If you do this make sure that in the process the outer layers become the inner, and the inner layers the outer of the freshly stacked pile. Although it takes considerably longer, excellent compost can be had without turning. As a good substitute for turning keep the pile covered with a fairly heavy gauge poly ethylene plastic film. This, by retaining moisture and warmth, favors the activities of beneficial microorganisms. Heat is generated as a result of decomposition within compost piles. Certain organisms of decay flourish when the temperature of the pile ranges from 115 to 120°F. At temperatures some twenty to thirty degrees higher these organisms are killed as are weed seeds, insects and their eggs, as well as most disease organisms. At these higher temperatures other organisms that hasten decay take over, and the wastes are converted into clean compost. As this result is achieved the temperature inside the pile gradually decreases until it approximates or equals that of the outside atmosphere. The compost is then ready. It should be used fairly soon. If left standing for a long period it suffers further loss in bulk by continued decay and depletion of nutrients by leaching and evaporation.

COMPTONIA (Comptò-nia)—Sweet-Fern. This, a genus of one species, is not a fern but a close relative of bayberry (*Myrica*). It belongs in the sweet gale family MYRICACEAE. From *Myrica* it differs in having deeply-pinnately-cleft leaves, with appendages at their bases called stipules. Also the plants are bisexual, whereas those of *Myrica* are usually unisexual. Some authorities do not regard the differences sufficiently important to designate *Comptonia* a separate genus and name it *Myrica asplenifolia*. The name sweet-fern refers to the ferny appearance of its foliage and to its fragrance. The generic name commemorates Henry Compton, Bishop of Oxford, who died in 1713.

Sweet-fern (*C. peregrina*), a deciduous bushy shrub, is more or less hairy, and 2

Comptonia peregrina

Comptonia peregrina, catkins

to 4 feet in height. Its foliage when brushed against or bruised is distinctly redolent of bay leaves. The leaves are slender and evenly and deeply incised into many divisions that at a glance might pass for separate leaflets. The male catkins, cylindrical and up to 1 inch long, are clustered and nodding. The females are nearly globose and because of the long, awl-shaped scales that surround them are burlike. The flowers appear in May after the foliage. The natural range of this shrub is from Nova Scotia to Saskatchewan, Minnesota, Kentucky, North Carolina, and Georgia. Chiefly it favors sandy soils. Some botanists distinguish a variety C. p. asplenifolia as being less pubescent and having smaller leaves and catkins than the typical kind.

Garden and Landscape Uses. Sweet-fern is a fairly attractive shrub for dry, acid, peaty soils in full sun. It can be used effectively in wild and naturalistic gardens and in shrub borders, foundation plantings, and sometimes as a single specimen. It is a good groundcover for extensive areas where a plant that can be kept to a height of about 2 feet is not out of scale.

Cultivation. The sweet-fern is easily raised from seeds sown in sandy-peaty soil outdoors or in a cold frame. When young the plants can be transplanted without undue difficulty, but old specimens that have been in one place for a few years are likely to die if moved. Because of this, sweet-fern dug from natural stands often fails. Plants grown in nurseries should be transplanted every year or two to ensure the development of compact root systems. It is always a good plan to prune these plants back severely at transplanting time. The maintenance of established specimens makes no special demands of the gardener. From time to time it may become desirable to prune them back hard to induce new growth from low down and to prevent the bushes becoming too tall for their location. This may be done to any extent desirable in late winter or early spring.

CONANDRON (Con-ándron). Of the three species of *Conandron* recognized by botanists probably not more than one has been cultivated. The group belongs in the gesneria family GESNERIACEAE and is native to Japan and Indochina. Its name alludes to the appendages of the anthers being joined as a cone surrounding the style; it comes from the Greek *konos*, a cone, and *andron*, an anther. From closely related *Streptocarpus* this genus differs in its seed capsules not being twisted. Conandrons are rather tender herbaceous perennials.

From a stout, densely-hairy, fleshy rhizome, **Conandron ramondioides,** in spring, develops one to few, somewhat lustrous, coarsely-toothed leaves, with winged stalks and much-wrinkled surfaces. They attain a maximum size of 1 foot in length by one-half that in width, but those of cultivated specimens are usually smaller. The blooms come in summer. The many flowering stalks are 4 inches to 1 foot tall, usually with up to twelve blooms, but as many as forty on a stalk have been reported. Unlike those of many members of the gesneria family, the flowers are not long-tubular or funnel-shaped. Their five triangular petals spread like the spokes of a wheel. The blooms, which have downy calyxes, are about 1 inch across and purple to lavender with a yellow eye or, in *C. r. leucanthemum,* white. Native to Japan, the Ryukyu Is-

Conandron ramondioides

lands, and Taiwan, in the wild this beautiful relative of the African-violet (*Saintpaulia*) favors humid, shady environments.

Garden Uses and Cultivation. Not hardy where seriously cold winters occur, and averse to hot dry summers, *Conandron* is best adapted for outdoor cultivation in North America in the Pacific Northwest and places with not too dissimilar climates. It is a plant for keen rock gardeners. It may be grown in carefully chosen spots outdoors and in pots in alpine greenhouses and cold frames. It prospers in well-drained soil that contains an abundance of leaf mold or other nourishing decayed organic matter, that is moderately moist, but not wet through the growing season, and that, at least when accommodated in pots in alpine greenhouses or cold frames, is nearly dry in winter. Annual repotting, just as growth begins in spring, is recommended. Shade from strong sun is necessary, as is a humid atmosphere. Propagation is easy by seeds sown in a cool greenhouse or cold frame in pots of sandy peaty soil. The seeds are very fine and scarcely need any covering of soil. Division, and probably leaf cuttings, afford other means of increase.

CONDALIA (Con-dàlia). Natives of the Americas, the eighteen species of *Condalia* belong in the buckthorn family RHAMNACEAE. The name honors A. Condal, an eighteenth century Spanish physician. They are spiny, deciduous shrubs or small trees, with alternate, undivided, generally toothless leaves and umbel-like clusters of small flowers from the leaf axils. The blooms have deeply-five-lobed calyxes, and are without petals, or have five hooded ones that narrow to claws at their bases. Some botanists segregate those with petals in a separate genus, *Condaliopsis*. The styles are notched or shallowy-two- or three-lobed. The berry-like fruits are drupes (fruits structured like plums) and contain a single stone.

Bitter condalia (*C. globosa*) is a rare desert shrub or tree found in the wild in grav-

elly and sandy soils in Arizona and Mexico. At its tallest 15 to 20 feet in height, it develops a dense, broad crown of forking branches, each twig of which is tipped with a rigid sharp thorn. The bark of old specimens is deeply-furrowed and shredded. The few narrowly-oblanceolate to narrowly-spatula-shaped leaves, from ⅙ to nearly ½ inch long, are solitary or in scattered clusters. Singly or in groups of few, the small, petal-less flowers develop in fall. They have five sepals, five petals, and are succeeded by black, nearly spherical, very bitter, juicy fruits up to ⅓ inch in diameter. In the wild this tree lives to a great age. Very similar is C. *warnockii kearneyana* (syn. C. *spathulata*) of the same general region, but its fruits are not bitter. A native of Patagonia, **C. lineata** is a spiny shrub with leathery, spatula-shaped to obovate leaves about ⅓ inch long. It has whitish, petal-less flowers. Its fruits are about ¼ inch long.

Garden and Landscape Uses and Cultivation. In warm, desert and semidesert regions condalias are planted to some extent. They do well in dryish soils in full sun. They are propagated by seeds, which before sowing should be steeped by pouring boiling water over them and leaving them for twenty-four hours. Until they are planted in their final locations they should be grown in cans or other containers.

CONDALIOPSIS. See Condalia.

CONEFLOWER. See Dracopis, Echinacea, Ratibida, and Rudbeckia.

CONFEDERATE. The word confederate forms part of the vernacular names of several plants including Confederate-jasmine (*Trachelospermum jasminoides*), Confederate-rose (*Hibiscus mutabilis*), Confederate vine (*Antigonon leptopus*), and Confederate violet (*Viola priceana*).

CONGEA (Cón-gea). There are eleven species of *Congea*. They belong in the vervain family VERBENACEAE and are natives of tropical Asia. All are vines or climbing shrubs. The name is derived from a native vernacular one.

Congeas have opposite, undivided leaves and large terminal panicles of clustered flowers, with bracts at the bases of the clusters. The funnel-shaped calyx is five-toothed. The corolla, usually not much longer than the calyx, has a slender tube and is two-lipped, with the upper lip of two slender, erect lobes, and the lower of three broader ones. There are four protruding stamens. The fruits are dryish and berry-like.

The kind most commonly cultivated, *C. tomentosa* also is most widely distributed in the wild. It is native from China, Vietnam, and Thailand to India and East Pakistan. A very vigorous, woody vine, it has

Congea tomentosa

usually pointed, elliptic-ovate leaves, 3 to 7 inches long and thickly covered with soft hairs on their undersides.The showiest feature of the very large, loose flower panicles is not the blooms themselves, which are small, white, stalkless, and relatively inconspicuous, but the beautiful bracts that accompany and nearly hide them. The flowers are mostly in groups of seven, more rarely in fives or nines. The bracts, about 1 inch long and woolly-hairy, are in threes and are pink, purplish, or violet, the color changing and intensifying as they age. Normally they begin to color in early fall and remain attractive until spring. This plant may be deciduous or evergreen, depending upon the climate. Its variety, *C. t. nivea,* has more definitely white pubescence. In **C. griffithiana** the bracts are in fours instead of threes. From *C. tomentosa* the quite similar **C. vestita** differs in having stalked flowers.

Garden and Landscape Uses. Only in large conservatories and outdoors in the tropics and warm subtropics can these great vines be satisfactorily accommodated. Where space and appropriate conditions are available they provide beautiful and long-lasting displays. The panicles last well when cut and stood in water, or they may be dried and used as everlastings.

Cultivation. Ordinary soil of moderate moisture content and full sun give satisfactory results with congeas. No special care is needed. Pruning, done in late winter or spring just before new growth begins, consists of cutting out weak and crowded shoots and shortening back, close to their bases, all shoots that have flowered except any needed to extend the vine. Shoots required for extension may have their upper one-thirds removed. Propagation is by seed, layering, and cuttings, although the latter are not always easy to root. When grown in conservatories a minimum winter night temperature of 55 to 60°F is agreeable. Daytime temperatures,

and night temperatures in summer, may be higher. These plants are better in ground beds than in pots or tubs.

CONICOSIA (Coni-còsia). About one dozen South African succulents of the *Mesembryanthemum* relationship are accounted for in this genus. The group belongs in the carpetweed family AIZOACEAE. Its name, from the Greek *konikos,* shaped like a cone, alludes to the fruits.

Conicosias include biennials and perennials. They have upright or sprawling stems and dagger-shaped leaves that technically are opposite, but appear to be arranged in spiraled rosettes from which long flower-bearing shoots develop. The flowers are yellow and daisy-like in aspect, but not in structure. Each is a single bloom, not, like daisies, heads of many florets. A few species do not have their leaves in rosettes. This genus is very similar to *Herrea;* the species formerly known as *Conicosia elongata* is *H. elongata.*

With branchless or few-branched stems 6 inches to 1 foot tall or taller, **C. pugioniformis** has slender, pointed, three-angled, gray-green leaves 6 to 8 inches in length, with deeply-grooved upper sides. They are reddish at their bases. Solitary or in twos or threes, on stalks 4 to 5½ inches long, the lustrous sulfur-yellow flowers are almost or quite 3 inches across. Probably sometimes grown under the name of the last, **C. capensis** attains a height of 6 inches or sometimes more, and has three-angled leaves up to 1 foot long in rosettes at the ends of the stems. Their upper sides, and sometimes their undersides, are slightly grooved. The straw-yellow blooms, 3 inches in diameter, are on stems 4½ to 6 inches long.

Garden Uses and Cultivation. Rare in cultivation, conicosias are beautiful in bloom. They are plants for warm, semi-desert climates and for greenhouses in which conditions favorable to succulents are maintained. Not a great deal of experience with growing them is recorded, but it appears that they may do better if put outdoors in summer. They need very well-drained, sandy soil and full sun. In greenhouses a winter night temperature of 40 to 50°F is sufficient, with a few degrees higher by day permitted. At that season water must be given sparingly because then the plants are semidormant. Propagation is by seeds and cuttings. For more information see Succulents.

CONIFERS. This is a group name for plants of the botanical order or class *Coniferae,* which includes, depending upon interpretation, six to eight families containing 300 to 400 species allotted among forty to fifty genera.

Conifers are gymnosperms. They are trees and shrubs, mostly evergreen, resinous, and usually with a conspicuously developed central trunk and shorter branches often arranged more or less in whorls or tiers. Most usually the leaves are needle-like, as in pines, flat and linear, as in yews, or small, scalelike, and overlapping, as in cypresses. Many conifers have, when young, foliage (identified as juvenile) quite different from that of older specimens. In *Phyllocladus* the tiny leaves are shed early, and the leaflike stems assume the function of photosynthesis. The genera *Glyptostrobus, Larix, Metasequoia, Pseudolarix,* and *Taxodium,* are deciduous.

The flowers of conifers, scarcely recognizable as such by nonbotanists, are always unisexual and usually in cones. Most often both sexes are on the same tree or bush, but in junipers and some other genera males and females may be on separate individuals. The seeds are most commonly in cones, often hard and woody, as are those of firs, pines, and spruces, but in junipers they are berry-like and in yews, plum-yews, and podocarpuses, which by some botanists are not considered to be conifers and which are rarely thought of as such by gardeners, the fruits are berry-like or plumlike.

Conifers, which constitute less than one-quarter of one percent of plant species, have importance disproportionate to that representation. They are dominant as vast forests in many regions, particularly temperate, subtropical, and mountainous ones. They include the tallest (*Sequoia*), most massive (*Sequoiadendron*), and oldest (*Pinus aristata*), living things.

Conifers are of immense commercial importance as sources of lumber, plywood, pulp, naval stores, and other usables. Horticulturally they include numerous splendid and beautiful ornamentals.

Among the most familiar conifers are firs (*Abies*), araucarias (*Araucaria*), incense-cedars (*Calocedrus*), cedars (*Cedrus*), plum-yews (*Cephalotaxus*), false-cypresses (*Chamaecyparis*), cryptomerias (*Cryptomeria*), cypresses (*Cupressus*), junipers (*Juniperus*), larches (*Larix*), dawn-redwood (*Metasequoia*), spruces (*Picea*), pines (*Pinus*), podocarpuses (*Podocarpus*), golden-larch (*Pseudolarix*), Douglas-firs (*Pseudotsuga*), umbrella-pine (*Sciadopitys*), redwood (*Sequoia*), giant-sequoia (*Sequoiadendron*), bald-cypresses (*Taxodium*), yews (*Taxus*), arbor-vitaes (*Thuja*), Hiba-arbor-vitae (*Thujopsis*), torreyas (*Torreya*), and hemlocks (*Tsuga*).

CONIOGRAMME (Coniográm-me)—Bamboo Fern. If the concepts of finely discriminating botanists are accepted, there are about twenty species of this genus; there may, however, be fewer. The group comprises ferns of the pteris family PTERIDACEAE, allied to *Pteris.* It is confined in the wild to the Old World, chiefly of the Malayan region, but ranging from Africa to Japan. The name comes from the Greek *konion,* a cone, and *gramme,* a line, and alludes to arrangement of the clusters of spore capsules.

Coniogrammes have usually short, creeping rhizomes and quite large, generally hairless, pinnate to thrice-pinnate fronds (leaves), with a few large, toothed leaflets.

Bamboo fern (*Coniogramme japonica* syn. *Dictyogramme japonica*), native to Japan, Taiwan, Korea, and China, where it grows in woodlands at low elevations, is the most popular in cultivation. In appearance the bamboo fern much resembles *Pteris cretica*. It is evergreen or nearly so, with the older foliage tending to die about the time the new leaves develop. From 1½ to 2 feet tall, the fronds are once- or twice-pinnate into pointed, toothless leaflets 6 inches to 1 foot long and 1 inch to 2 inches wide; the lowermost leaflets are stalked. The veins merge. A variety with variegated foliage is sometimes cultivated. Native of the Asian tropics and to the Philippines, Fiji, and Hawaii, *C. fraxinea* (syn. *Gymnogramma javanica*) has once- or twice-pinnate, lustrous, pointed leaves 1 foot to 4 feet long. The leaflets are stalkless or nearly so, 3 inches to 1 foot long by ½ inch to 3 inches wide. Their veins do not merge.

Garden and Landscape Uses and Cultivation. Only in frost-free or nearly frost-free regions are these ferns hardy. They may be grown in shade and in lath houses in moist, but not constantly wet, porous soil well supplied with humus. They are also satisfactory for greenhouses, where they prosper in minimum winter night temperatures of 50 to 55°F, with a daytime increase of five to fifteen degrees allowed. They require fairly high humidity and shade from strong sun. Fertile, porous soil containing a considerable proportion of peat moss or other humus material is to their liking. It should be moderately moist at all times. Well-rooted specimens are at their best when supplied regularly with dilute liquid fertilizer. Increase is best obtained by spores; careful division is also successful. For more information see Ferns.

CONIUM (Con-ìum) — Poison-Hemlock. The only place appropriate for the cultivation of these poisonous plants of the carrot family UMBELLIFERAE is in well-ordered botanical collections where they are needed for research or teaching and where their dangerous qualities are well understood and advertised. The genus, the name of which is a modification of the classical Greek name *koneion* for *Conium maculatum*, consists of two species, one a native of Europe and Asia, the other of South Africa. A decoction of the immature fruits of the Eurasian kind, the poison-hemlock, was used by the ancient Greeks to put criminals to death. It is reputed that this was employed by the Athenians to kill Socrates. It is the source of medicines used as sedatives and for other purposes. This spe-

cies is widely naturalized in North America, chiefly in waste places where the soil is moist.

Coniums are not related to hemlock trees (*Tsuga*). They are kin to carrots, parsnips, parsley, and dill. Stout biennials, with thick taproots and freely-branching hollow stems, they have large, finely-pinnately-divided leaves. Their numerous tiny white flowers are in compound flat umbels.

The poison-hemlock (*C. maculatum*) is 1½ to 10 feet tall, with ferny leaves, broadly triangular-ovate in outline, three-

Conium maculatum

or four-times pinnately-divided, and up to 1¼ feet long. The umbels are 1½ to 2 inches in diameter and mostly terminal. A distinguishing characteristic is the purple spots on the hairless stems. All parts of the plant are poisonous, particularly the immature seeds. There are reports of children having been poisoned from using the hollow stems as whistles and of disastrous results overtaking people who mistook the roots for parsnips or the seeds for anise. The poison has a numbing, paralyzing effect in contrast to that of the water-hemlock (*Cicuta*), which causes violent convulsions.

Cultivation. Seeds are sown in early spring in moist fertile soil where the plants are to remain, and the resulting seedlings thinned to 1 foot to 2 feet apart. Every precaution must be taken that children and others do not eat or even put into their mouths any part of this deadly plant.

CONOCARPUS (Cono-càrpus) — Buttonwood or Button Mangrove or Sea-Mulberry. There are two or perhaps only one species of this genus of the combretum family COMBRETACEAE. They are inhabitants of warmer parts of the Americas and tropical Africa and have a name alluding to the form of the fruiting heads, from the Greek *konos*, a cone, and *karpos*, a fruit.

Native of mangrove swamps in tropical and subtropical America, including the West Indies and Florida, button mangrove, buttonwood, or sea-mulberry

Conocarpus erectus

(*Conocarpus erectus*) varies from a prostrate shrub to a tree up to 60 feet tall. Its somewhat fleshy foliage is finely-silky-hairy or hairless. The leaves are alternate, elliptic to broad-elliptic, toothless, and up to 2 inches long. The greenish flowers are in dense heads, from rather less to rather more than ½ inch across, arranged in axillary racemes or terminal panicles. The blooms, ¼ inch or so long, are without petals. They have five each sepals and stamens and one style. The fruits are two-winged drupes arranged in conelike heads. A variant with silver-hairy leaves occurs in Florida. The durable wood makes good fuel and charcoal.

Garden and Landscape Uses and Cultivation. This buttonwood is especially suitable for wet, brackish places and sandy shores, but need not be confined to such. It responds well to more ordinary environments. In warm, frost-free or practically frost-free, humid regions it thrives without special attention in open, sunny locations. It may be raised from seed and is very easy to increase by air layering. Large branches can, by this method, be induced to root.

CONOPHYLLUM (Cono-phýllum). These erect or sometimes looser-growing, succulent subshrubs of low stature belong in the *Mesembryanthemum* section of the carpetweed family AIZOACEAE. Their name derives from the Greek *konos*, a cone, and *phyllon*, a leaf, and alludes to the shape of some pairs of their leaves. There are twenty species, natives of South Africa.

Closely related to *Mitrophyllum*, conophyllums differ in technical details of their fruits (capsules). They have three distinct types of leaves. On nonflowering shoots two pairs develop each season. The lowermost, united only at their bases, with flat upper surfaces, are rounded or keeled below. Mostly they are 1½ to 2 inches long by ¼ to ¾ inch wide and thick. The second two leaves are joined for much or all of their lengths into a cone-shaped body 1¼ to 2 inches long, with only the tip of one leaf free. During the dormant period a new pair of leaves similar to the first forms within the cone, absorbing nourishment from the latter, which is left as a protective, papery sheath. The third type of leaves are bractlike. They develop only on flowering shoots and are separated from the other leaves by a short interval of bare stem. The blooms of *Conophyllum* are similar to those of other members of the *Mesembryanthemum* group in that they suggest daisies, but differ from those familiar flowers in not being heads compounded of numerous florets, but single blooms. White, yellow, pink, or pale violet-red, they have five sepals, many petals in three or four rows, numerous stamens, and five to seven stigmas. They are solitary or clustered.

Much branched and up to 2 feet in height, *C. herrei* has cone leaves about 1 inch long joined for about one-half their lengths, and with their upper parts spreading. The cylindrical, sheath-forming leaves are somewhat longer and about ¼ inch wide. The whitish blooms are about 4 inches across.

Garden Uses and Cultivation. Exceedingly interesting plants for cultivators of choice succulents, conophyllums may be raised from seeds and cuttings and grown in very sandy, thoroughly drained soil in full sun and an arid atmosphere. They need the same general conditions as *Pleiospilos*, *Glottiphyllum*, *Lithops*, and other very fleshy members of the *Mesembryanthemum* tribe. Conophyllums in active growth in summer or fall are kept fairly moist, but throughout their long season of rest the soil must be dry. Full sun is necessary. For further suggestions see Succulents.

CONOPHYTUM (Conó-phytum). Among the most intriguing of an entrancing assemblage of succulent plants, *Conophytum* is one of several genera of stone or pebble plants of southern Africa that greatly appeal to lovers of the unusual and beautiful. They are little plants that amaze by their faithful mimicry of the angular chunks of stone or pebbles, among which they so often grow in the wild, and that are doubly rewarding when they produce their beautiful blooms. The genus *Conophytum* belongs in the carpetweed family AIZOACEAE. It is closely related to *Mitrophyllum* and includes about 400 species restricted in the wild to southwestern Africa, and frequently puzzling to identify to species. The name, from the Greek *konos*, a cone, and *phyton,* a plant, refers to the form of the plant bodies of some kinds.

Conophytums are short-stemmed or stemless. They form clumps of spherical, obconic, or cone-shaped plant bodies each of two fat, fleshy leaves, completely united except for a cleft, notch, or small mouthlike center aperture at the apex indicating a separation between the leaves. From this comes the solitary flower, daisy-like in aspect, but not in structure. Unlike a daisy it is one bloom, not a head of numerous florets. It may be white, pink, magenta, red, or yellow. The blooms are open by day, those of some species are open at night as well. They have tubular calyxes with four to seven lobes (sepals), usually many petals in one or more rows, few to many stamens, and four to seven stigmas. The fruits are capsules.

Representative of kinds cultivated, others are likely to be found in the collections of avid fanciers of succulents, are the following. *C. albescens* is stemless. It forms mats of somewhat laterally-compressed plant bodies that are gray-green clothed with fine, short, white hairs and are 1 inch to 1¼ inches tall by ½ to ¾ inch wide. Their apexes have two rounded, reddish-tipped lobes, with a shallow indentation between. The 1-inch-wide flowers are yellow. *C. bilobum* somewhat resembles the last. Stemless and forming mats of more or less heart-shaped, laterally-compressed, gray-green to whitish-green plant bodies 1¼ to 1¾ inches tall, with two apical red-tipped rounded lobes or ears, this kind has yellow blooms up to 1¼ inches in diameter. *C. flavum* forms cushions of flat-topped or shallowly-concave-topped, obconical plant bodies approximately ¾ inch tall, not lobed at their apexes. Green with darker green dots, and smooth and hairless, they often have reddish sides. Up to slightly more than ½ inch in diameter, the flowers are yellow. *C. frutescens* is a branched shrublet up to 4 inches tall. Its laterally-compressed plant bodies, up to a little more than 1 inch long by nearly as wide, at their deeply-two-lobed apexes are dark green with translucent dots. The rich orange blooms are 1 inch in diameter. *C. gratum* forms loose cushions of pear-shaped plant bodies 1 inch tall and wide, with slightly-saucered apexes. Their smooth, glaucous surfaces are sprinkled with dots, those on the tops dark. The ½-inch-wide blooms are magenta-red. *C. minutum,* as its name suggests, has very small plant bodies. Obconical, they have flat or slightly depressed tops not divided into two lobes,

with the central fissure minutely-hairy. They have smooth, glaucous surfaces and small, yellowish-white blooms. *C. multipunctatum* develops clusters of inverted-pear-shaped plant bodies up to ¾ inch tall by a little more than ½ inch in diameter. Their apexes have a tiny fissure and are freely marked with irregular lines and bordered with numerous minute dots. The white flowers open at night. *C. obconellum,* by some botanists identified as *C. declinatum*, forms mats or cushions of obconical plant bodies up to ¾ inch tall, but often smaller. They have flat or slightly rounded tops, with a small, hairy center fissure and a transverse ridge. They are green, gray-green, or bluish-green, with green to red translucent dots. The flowers, up to a little more than ½ inch wide, are whitish to yellowish. *C. obcordellum* is another with clusters of obconical plant bodies with flat to hollowed, lobeless tops. Up to 1 inch long and wide they generally are bluish-green, with darker dots often in lines. Less often they are gray-green or pink. The sides are mostly pink to dark red. White to creamy-white, the flowers are ¼ to ½ inch in diameter. *C. pearsonii* forms compact cushions of plant bodies resembling those of *Lithops*. Obconical and up to a little more than ½ inch in height, they are flat or shallowly-convex at their apexes, with the edges tending to overhang the sides. Rarely with a few obscure dots and a darker zone encircling the central fissure, the plant bodies are glaucous- to yellowish-green. The violet-pink flowers are up to 1 inch in diameter. *C. simile* forms tight clusters or mounds 2 inches or so high of flattish, heart-shaped, blue-green or gray-green, almost plastic-like plant bodies sprinkled with dots of dark green. From ½ to ¾ inch long, they have a V-notch at the apex. The golden-yellow flowers are ¾ inch in diameter. *C. tischeri,* when old, develops short stems to form mats of broad-heart-shaped, somewhat compressed plant bodies up to about ½ inch long and wide, and with a small fis-

Conophytum minutum

Conophytum simile

Conophytum tischeri

sure at the apex. About ½ inch wide, the flowers are pale lilac. **C. truncatum** (syn. *C. truncatellum*) makes loose cushions that, as they age, develop short stems. Its ob-conical plant bodies, ½ to ¾ inch long, have circular or nearly circular, convex tops about as wide as the body is long, and with a small slit at the center. They are gray-green to bluish-green sprinkled with tiny dark green dots. The ½-inch-wide flowers have slender, straw-yellow to nearly white petals. **C. wettsteinii** forms mats of irregularly circular to broad-elliptic plant bodies of inverted-cone form about ⅓ inch tall by ¾ to 1¼ inches wide. Green to gray-green they are besprinkled with tiny whitish dots. The flowers are violet-purple.

Conophytum truncatum

Conophytum wettsteinii

Garden Uses and Cultivation. These are plants for connoisseur collectors of choice succulents. Not hardy, they are grown in greenhouses under conditions that suit *Lithops* and other small South African stone plants and their close relatives. They can even be grown in windows. They are best accommodated in well-drained pans (shallow pots). Perfectly drained soil, preferably with a moderately high organic content, is essential. Bright light, with slight shade from the fierce sun of summer, is needed. Except during the growing period, which varies somewhat with different species, but is a short period usually in late spring or summer, little or no water is needed, but when the plants are in active growth the soil must be kept moderately moist. A winter night temperature of 50°F is agreeable. By day the temperature may be five to fifteen degrees higher. In favorable weather ventilate the greenhouse freely. Propagation is easy by seeds and cuttings. For more information see Succulents.

CONOSTYLIS (Conó-stylis). Confined in the wild to Australia, this horticulturally little known genus consists of about twenty species of the bloodwort family HAEMO-DORACEAE. Few are sufficiently attractive to be worth cultivating. They are evergreen herbaceous perennials, with leaves in tufts or in two vertical rows from short rhizomes. The flowers, in loose clusters or terminal heads, have woolly perianths with six lobes or segments (petals, or more correctly, tepals). There are six stamens and a cone-shaped style, which gives reason for the name, from the Greek *konos*, a cone, and *stylos*, a pillar. The fruits are capsules.

The leaves of **Conostylis candicans** are rigid, slender, cylindrical, and white-hairy. They are up to 1 foot long, and in tufts. The fairly dense heads of small sulfur-yellow to whitish flowers terminate slender, hairless stalks, 1 foot to 2 feet long, that overtop the foliage and have a short leaf near their centers.

Garden and Landscape Uses and Cultivation. Experience with these plants in North America is not extensive. They may be expected to grow outdoors in California and other places where the climate is warm and for much of the year dry, and to favor well-drained soils supplied with adequate moisture during their periods of active growth. In greenhouses winter temperatures of 50°F at night, with daytime increases of five to ten degrees, are appropriate. The greenhouse should be ventilated freely on all favorable occasions. Propagation is by seed and by division.

CONRADINA (Conrad-ìna). Four or five species of low shrubs of the southeastern United States constitute *Conradina*, of the mint family LABIATAE. Named to honor the American botanist Solomon White Con-

rad, who died in 1831, these plants have clusters of toothless, narrow leaves, with rolled-under margins, and flowers solitary or few together from the leaf axils. The blooms have two-lipped calyxes, the upper lips with three broad, short lobes, the lower with two narrower, longer ones. The asymmetrical, bluish-purple corollas have an erect, curving upper lip and a three-lobed, spreading lower one. There are four stamens, nearly concealed beneath the upper lip or protruding. The fruits consist of four tiny nutlets.

Up to 3 feet in height, **C. grandiflora** has blooms ½ inch long or slightly longer, in stalked clusters of up to a dozen. Because of their rolled margins the narrowly-spatula-shaped leaves appear to be club-shaped. Mostly they are under 1 inch long. Their undersides are white-hairy. This species, native to eastern Florida, blooms most of the year. From it **C. canescens** differs in not being more than 1½ feet tall, in having flowers under ½ inch long, and in its leaves being broader, although, because of their rolled margins, they too appear to be club-shaped, and about ½ inch long. This native of Florida and Alabama blooms in spring.

Garden Uses and Cultivation. Of minor horticultural interest, conradinas may be planted in mild climates and can be employed in naturalistic plantings and native plant gardens. They succeed in sandy soils and may be increased by seeds, by cuttings, and in some cases, by division.

CONSERVATION. Concern about the preservation and wise use of natural resources, including land and landscapes, has developed greatly in the twentieth century as an increasing population has placed additional strains on the environment. The study and practice of methods to achieve these ends constitutes conservation. Many organized horticultural groups, including garden clubs, horticultural societies, and botanic gardens, have been active in promoting conservation and supporting conservation movements, including those designed to obtain favorable legislation.

Large-scale efforts involving the preservation of forests, wetlands, shores, waterways, and soils of necessity are matters for governmental agencies, but these function most effectively only in a climate of understanding, and fullest public support.

Every gardener can and should serve as a conservationist of the land he manages. To this end it is important to prevent soil erosion, to compost vegetable wastes and return them to the Earth, to preserve trees and other wild growth when practicable, but not to hesitate to prune when needed or to remove old declined specimens to afford room for young vigorous ones. No tree lives forever, and under some circumstances cutting down those that have clearly passed their peak or are seriously diseased and the selective cutting of young ones that are crowding others are among the best expressions of conservation. It should be the aim of every gardener and farmer to leave his land and landscape in better condition than he found it.

CONSERVATORY. See Greenhouses and Conservatories.

CONSOLEA (Con-sòlea). Native of the West Indies and one species (*C. corallicola* syn. *Opuntia corallicola*) of the Florida Keys, *Consolea* of the cactus family CACTACEAE is by conservative authorities included in *Opuntia*. Considered separately it comprises a dozen species. Its name honors M. A. Console, a one-timer Curator of the botanical garden at Palermo, Italy.

Consoleas are mostly treelike, with jointless, cylindrical or somewhat flattened, usually spiny trunks and much-branched heads of padlike joints. The abundant, colorful flowers are succeeded by flattish fruits reported to be edible. Some species develop from the ovaries of the flowers, chains of small stem joints that drop to the ground and give rise to new plants.

The most desirable ornamental *C. rubescens* (syn. *Opuntia rubescens*), 9 to 20 feet tall, has reddish stems, their thin, flat ultimate pads oblong to oblong-ovate, up to 10 inches long by one-quarter to one-half as wide. They may be spineless or have clusters of several white spines. Yellow, orange, or red, the flowers are ¾ inch across, the subspherical to obovoid, reddish, spiny or spineless fruits 2 to 3½ inches in diameter. This is native to the West Indies.

Others sometimes cultivated include these: *C. falcata* (syn. *Opuntia falcata*), of Hispaniola, has a trunk up to 5 feet tall and obliquely-lanceolate to curved pads 9 inches to 1¼ feet long by 3 to 4 inches wide. In rather distantly-spaced clusters of two to eight, the spines are up to 1½

inches in length. The small red flowers have an ovary 1 inch to 1½ inches long. *C. moniliformis* is West Indian. From 9 to 15 feet tall, this sort has a flattened trunk thickly covered with spreading spines and a wide-spreading head. Its pads are linear-oblong to obovate, 4 inches to 1 foot long, up to 5 inches wide, and about ½ inch thick. Older ones develop clusters of five to eight yellowish spines and brown glochids. The yellow to orange flowers are 1 inch in diameter. *C. spinosissima*, of Jamaica, up to 15 feet high, has a spiny trunk about 3 inches in diameter and a head with the ultimate pads dull green, narrowly-oblong, and from two to four times as long as wide. They are spineless or have clusters of two or three or solitary, whitish or yellowish spines up to 3 inches in length, and brown glochids. The flowers, yellow at first, change to dull red, are ¾ to 1 inch wide.

Garden and Landscape Uses and Cultivation. These are as for *Opuntia*. For more information see *Cactuses*.

CONSOLIDA (Con-sólida)—Larkspur. This group of about forty species of the buttercup family RANUNCULACEAE was previously included in *Delphinium*. It contains the annual larkspurs of gardens, and differs from *Delphinium* in its members being true annuals, in its flowers having stamens spiraled in five series, in the blooms not having lateral honey leaves (nectaries), and in the two honey leaves they have being joined to form a one-spurred nectary. The name of the genus *Consolida* is an ancient Latin one for an undetermined plant. The genus is native from the Mediterranean region to central Asia.

Consolidas have deeply-pinnately-lobed leaves and spikelike racemes of asymmetrical flowers, with five perianth segments (petals), the upper one spurred. The fruits are podlike follicles.

Rocket larkspurs are horticultural vari-

eties, not hybrids, of *C. ambigua*, and *C. orientalis*. The former, previously named *Delphinium ajacis*, is a native of the Mediterranean region. Hairless or slightly hairy, *C. ambigua*, 1 foot to 3 feet tall, usually branches freely, with the lower parts of the branches spreading nearly horizontally. Its leaves are deeply-cleft, the upper ones into very slender segments. In loose, spikelike racemes or more rarely panicles, the flowers, single or double, are blue, pale pink, or white. The follicles, which narrow gradually at their apexes, are softly-hairy and contain black seeds.

The characteristics given above distinguish *C. ambigua* from *C. orientalis* (syn. *Delphinium orientale*), of southern Europe, which has reddish-brown seeds in hairless follicles that narrow abruptly at their apexes. Its stems are without branches or have erect ones. Up to 3½ feet tall, this species has deeply-lobed leaves, the upper deeply-cleft into slender segments. The flowers, in the wild purplish-violet, in cultivated varieties are bright purple or pink, and single or double.

The branching or forking larkspurs of gardens are derivatives of *C. regalis* (syn. *Delphinium consolida*), of Europe, or more probably of *C. r. paniculata* (syn. *Delphinium paniculatum*), native to southern Europe. They have single or double flowers, white, pink, purplish with white dots, purple, or blue. In the wild the stems of *C. regalis* are up to 2 feet tall and are not much branched, but those of *C. r. paniculata* branch freely, and its flowers are more loosely arranged. In both the stems have down-pointing hairs. The follicles of *C. regalis* are without hairs, those of *C. r. paniculata* may be hairy or hairless.

Garden and Landscape Uses. Larkspurs are among the most satisfactory annuals for outdoors and are admirable for growing in greenhouses in pots, benches, and ground beds. They can be had in a variety of colors and from kinds that do not exceed 1 foot in height to others that attain 3 or 4 feet. The taller ones supply elegant cut flowers. Like many annuals they have a rather short season of bloom and are likely to become bedraggled or fail completely in torrid summers. For these reasons, where summer heat is intense, it is well to rely on larkspurs for early bloom only, and elsewhere to make two or three successional sowings to ensure continuity. Sunny locations sheltered from strong winds best suit larkspurs. They show to prime advantage in front of such dark backgrounds as evergreen shrubs and hedges. As cut flowers larkspurs last well in water. They are cut when the lowermost blooms on the spikes are open and the buds above have yet to expand. If cut before they reach full maturity, tied in bundles, and hung upside down in a shaded, cool, airy place indoors, larkspurs can be dried for use in winter bouquets.

Consolea falcata, at the Royal Botanic Garden, Edinburgh, Scotland

Consolida, horticultural variety

Cultivation. For the finest results outdoors larkspurs must be sown early, as soon as possible in spring, at about the time peas and sweet peas are planted and well before the seeds of less hardy plants. Where winters are not too severe, fall sowing to give plants that bloom early the following summer, or in very mild climates, during winter or spring, is in order. Later sowings may be made to extend the season of bloom. Deep, fertile, well-drained soil is needed; if it be of a limestone character that is no disadvantage. Although seedling larkspurs can be transplanted, they do not take kindly to moving, and it is better to sow where the plants are to stay and pull out surplus seedlings so that the plants that remain are 8 inches to 1 foot apart. Larkspur seeds germinate rather slowly, usually taking two or three weeks. Sowing may be by broadcasting the seed and raking it into the soil surface or by sprinkling it along ½-inch-deep drills (furrows) and covering with soil. For cut flower production the latter is the best method. The drills should be 1 foot to 1½ feet apart. The only attentions needed later are shallow stirring of the soil to admit air and destroy weeds, light staking, if there is danger of storm damage, watering deeply at regular intervals in dry weather, and removing faded flower spikes.

For blooming in greenhouses larkspurs are raised from seeds sown in August or September or in January. The seedlings are transplanted individually to 2½-inch pots and are repotted until they eventually occupy containers 5 or 6 inches in diameter. Alternatively, they are planted in benches or ground beds with a spacing of about 1 foot each way. At all times, but especially in their early stages, great care must be taken with watering. Too wet soil soon causes rotting and death. It is important to keep the surface stirred shallowly. Full sun is needed throughout, except that in spring very light shade from strong sun, given after the flowers are out, prolongs their life. A night temperature of 50°F, increased by five to ten degrees depending upon the brightness of the weather by day, is adequate during the main growing period. The atmosphere must be buoyant, not humid and dank. The greenhouse is ventilated freely whenever outside temperatures permit. Consolidas are long-day plants, producing their flowers when nights are comparatively short. In greenhouses blooming can be hastened by about one month by using artificial lights to increase the day length to about sixteen hours, and by raising the temperature to 58 or 60°F at night and proportionately by day. But this treatment must not begin until the plants have had three months or more in which to make satisfactory stem and foliage growth, otherwise the number and quality of the flowers will be greatly lessened.

Container gardening, each container with one sort of plant: (a) Citrus fruits

CONTAINER GARDENING. There is nothing new about growing plants in containers. It was done by the ancient Romans and earlier peoples, Orientals and Occidentals. Potted plants play a large part in contemporary American life. They are used extensively to ornament homes, offices, and other places and are grown in vast numbers in greenhouses. In addition, plants are cultivated in window boxes, hanging baskets, planters, strawberry jars, urns, barrels, and a wide variety of other receptacles. But container gardening sometimes implies more. It may refer to the employ-

(c) English wallflowers

(b) Lime (*Citrus aurantifolia*)

(d) Polyanthus primroses

C. v. pictum was distributed widely through the islands of the Pacific and some other tropical parts before the arrival of Europeans, and was esteemed for personal adornment, use in various ceremonies, and as a source of medicines. The first croton to arrive in Europe was brought to England in 1804 and named *Croton pictum*. That variety, now correctly identified as *Codiaeum variegatum pictum*, is still grown in Florida, Hawaii, and other places where these plants flourish. The earliest recorded date of crotons being grown in America is 1871.

The genus *Codiaeum* consists of trees and shrubs with undivided, but sometimes lobed, leathery, stalked, hairless leaves having pronounced midribs from which lateral veins spread pinnately. The inconspicuous, unisexual flowers, with usually both sexes on the same plant, are in longish, solitary or paired racemes from the axils of the upper leaves. The calyx has three to six, usually five, lobes. Male flowers have five or six petals and fifteen to thirty or more stamens. The female blooms are without petals. Their styles are separate or are united at their bases. The fruits are spherical capsules.

Very variable **C. variegatum** is a shrub or small tree up to 8 feet, exceptionally 30 feet tall, with leaves of widely varied shapes and colorings. Its numerous variants are by botanists all referred to the botanical variety *C. v. pictum*, but such broad treatment is not satisfactory for gardeners and horticulturists who, of necessity, need a different name for each distinctive kind. Accordingly, horticultural varietal names are used for these. Because specialist breeders are constantly adding to the list of such horticultural varieties it would be futile to attempt to name and describe them all. Consult catalogs of dealers and publications of the American Cod-

A selection of *Codiaeum variegatum* varieties

Codiaeum variegatum (flowers)

(e) Petunias

(f) Mugo pine

Container gardening, with more than one sort of plant in each container: (a) Cinerarias and pansies

(g) Safflower (*Carthamus tinctorius*)

ment of several planted units in such relationship that something approaching a garden-like effect is achieved. Container gardens are more or less mobile gardens. Their components can be rearranged at will. Masonry planters attached to the ground are usually thought of as raised beds rather than containers in the sense that the word is used here.

Container gardening makes most sense where space is limited. It is often the ideal, and sometimes the only practicable way of gardening on roofs, patios, sun decks, and suchlike places. It can be employed to ornament terraces, steps, and similar areas. More and more, plants in ornamental containers are used to bring greenery and flowers to city squares and streets.

Gardening in containers perhaps reached its ultimate development in Hong Kong.

(b) Gloxinias and English ivy

(c) Daffodils, tulips, and grape-hyacinths

(d) Geraniums, ageratums, and *Vinca major variegata*

(e) geraniums and *Vinca major variegata*

(f) Geraniums and *Sedum spectabile*

There, where land is extremely scarce and the subtropical climate favors almost year-around growth, such gardens are more common than what Westerners regard as conventional ones. They are generally extremely well done. Among the great number of kinds of plants successfully used is the beautiful ground orchid *Phaius tankervilliae*. In Hong Kong it is usual to maintain, discreetly screened or some little way from the display area, a nursery garden in which the container plants are grown to near perfection before being moved to the garden proper, and to where permanent plants in containers are moved to recuperate after they are through blooming. In effect the nursery is an outdoor greenhouse. It makes possible in the display garden a year-around showing of unblemished plants at the height of their beauty. This may be a procedure worth adapting and adopting in the warmer parts of America.

Containers may be of wood, natural if of cypress, redwood, or teak that resist decay, or stained, painted, or otherwise treated with a preservative if of other woods. Other possibilities are terra-cotta, glazed earthenware, metal, fiberglass, plastic and concrete. They may include boxes, tubs, barrels, and pots of various

shapes, heights, and lengths, traditional or modern in design. They should not be so vividly colored or otherwise decorated that they compete with the plants for attention. Those used in association should be harmonious. Containers may sit on the floor or ground, but if of wood are better raised on legs or blocks at least an inch or two to allow for air circulation to inhibit rotting. Or planters and other containers can be raised considerably higher for special effects and, incidentally, to eliminate the need for bending while providing routine care.

Soil to remain in containers for several years without replacement need be of somewhat different composition than earth that will be in use for briefer periods, as is for instance that of many pot plants. The trick is to build into it permanent porosity. Organic matter, such as compost, leaf mold, and peat moss, with the addition of such inorganics as sand or perlite, are commonly relied upon to keep potting soils open and aerated. But in time organics decompose, become pasty, and bind the soil particles together. Then, they reduce rather than increase porosity. When mixing soils for long-term use in containers cut down on the amount of organic material. Use no more than ten percent by bulk. Increase proportionately the amount of rather coarse inorganics. This may be sand or better still one of particles that are porous, such as grit-size coal cinders (without fine ash), finely-crushed brick, perlite, or calcined clay of the kind sold for mixing with potting soils. Because its particles are readily compressible vermiculite is not suitable. A mix of this type is of course less nourishing than one with a greater organic content. The discrepency is easily remedied by judicious fertilizing.

Give thought to design in placing the containers. Do not dot them about in higgledy-piggledy fashion. As boundaries you may elect to use long low boxes planted to give something of a low or medium-high hedgelike effect. You can employ sheared hedge plants, such as barberries, privets, yews, Japanese holly, pittosporums, and coprosmas, or perhaps you prefer, or it may be more practicable, to depend upon such annuals as celosias, kochias, marigolds, gloriosa-daisies, salvias, or zinnias, sometimes along with lower kinds, such as lobelias, petunias, portulacas, sweet alyssum, and verbenas that may trail over the edges of the planters. Espaliered trees and shrubs, carefully trained as horizontal cordons are choice boundary items, not beyond the skills of home gardeners to achieve, or vines may be grown on trellises. If walls bound the landscaped area consider ornamenting them with container-grown espaliers or annual or perennial vines. Among suitable subjects for espaliering are fruit trees, yews, pyracanthas, flowering quinces, *Cotoneaster horizontalis*,

and in mild climates, camellias, geraniums, fuschias, plumbagos, and a wealth of other plants. Vines to use on trellises or walls, selected of course with thought for climate, include akebias, allamandas, Boston-ivy, bougainvilleas, clematis, Dutchman's pipe, English ivy, honeysuckle, lace vine, morning glories, passion flowers, plumbago, sweet peas, and Virginia creeper. Climbing roses are effective.

The grouping of containers within the boundaries of the garden may be formal or casual. Often best effects come from informal placement. In well-selected positions arrange clusters of three, five, or more tubs or pots of various heights and sizes and of compatible design. Fill them with plants that complement each other and strive to have the various elements of the garden meld into a harmonious whole. This can be achieved by repeating the use, but not to the extent of becoming monotonous, of the same plants throughout and by giving careful thought to the colors and their compatibility.

Plants suitable for container gardens are many and varied. Of course climate and whether or not you have facilities indoors for winter care of nonhardy kinds impose limitations, but within these give your imagination some rein. Among evergreen shrubs and trees especially adaptable, in addition to those mentioned previously as suitable for hedges, are aucubas, Austrian pines, citrus fruits, fatsias, myrtles, oleanders, photinias, rhododendrons, skimmias, viburnums, and yuccas. Suitable deciduous shrubs and trees include abelias, buddleias, crab apples, crape-myrtles, hydrangeas, hibiscuses, pomegranates, potentillas, privets, rose-of-Sharon, and willows.

Among perennial more or less bulbous plants that do well in containers are agapanthuses, clivias, and strelitzias. For one-season effects spring bulbs, such as hyacinths, daffodils, and tulips, are splendid. These can be effectively combined with such spring bedding plants as English daisies, forget-me-nots, and pansies. Lilies are excellent for summer bloom. Almost all annuals and other summer-blooming bedding plants are good. Especially satisfactory are those that bloom over a long season. These include ageratums, begonias, marigolds, and zinnias and, of course, such excellent kinds as fuchsias, geraniums, and lantanas. The possibilities are pretty near endless.

Herbs lend themselves well to container cultivation. They are especially appropriate for intimate areas near the house, and charming, useful collections can be displayed in small areas. The containers may range from those affording a square foot of planting place to ones three times as big or bigger. A number of sections of large square or round drain pipes fitted with wooden bottoms or, if they are not to be

Colchicum autumnale

Coelogyne cristata

Coleus hybridus, varieties

Colchicum autumnale flore-pleno

Colax jugosus

Collinsia heterophylla

Columnea microphylla

Columnea hybrid 'Yellow Dragon'

Conophytum, unidentified species

moved set directly on paving, are admirable for accommodating herbs and lend themselves to imaginative arrangement. The practical-minded may even essay to try a selection of vegetables. This can be done. Among the most suitable are beets, carrots, cucumbers, lettuce, okra, peppers, radishes, scallions, swiss chard, and tomatoes.

Care of container gardens is not markedly different from that needed by the same plants in the ground. The most obvious additional attention is that of watering regularly to keep the soil desirably moist, but not sodden. Also, permanent plants and often seasonal ones need more frequent fertilizing than their likes in garden beds. But go carefully here. Little and fairly often is the rule rather than massive and infrequent. And do not fertilize so late in the season that trees, shrubs, and perennials are stimulated into late growth that is likely to be subject to winter injury. Quit before the first signs of fall are evident.

Good housekeeping is imperative to keep container gardens in tip-top shape. Not onerous, it requires frequent attention. Here belong such tasks as picking faded blooms, except those of berry-bearing or other ornamental fruiting plants, and removing marred foliage. Make at least a weekly inspection for first signs of pests and diseases and take prompt remedial action if any are found. Attention to staking and tying may be needed as may intelligent persuasive pruning to limit the exuberent and encourage the reluctant. In city gardens accumulated dust and other deposits may make a biweekly or weekly hosing with water, applied with as much force as can be tolerated without damage to the plants, of benefit. In warm weather daily lighter sprayings in early evening, early morning, or both are refreshing.

Maintenance of the soil in an acceptable state calls for special attention where a succession of seasonal plants are set out over a period of a few years without removing and replacing the soil. This last can be a formidable, not infrequently a costly job. To defer it as long as possible do everything practical to keep the soil congenial. It helps greatly to remove the surface 2 or 3 inches annually and replace it with new, mixed with a hand fork as deeply as possible among the old that remains. A light liming once a year is likely to do good. If the containers are to be empty of plants over winter, as soon as the summer occupants are removed fork the soil, leaving its surface as rough as possible.

Winter care of permanently planted containers in regions of severe winters poses problems. Cold damage is not the only hazard. Practicable ways of guarding against that are to rely on plants of unquestioned hardiness, to protect borderline kinds with coverings of one sort or another (see Protection for Plants), or to move the containers indoors. Dehydration, excessive loss of moistures from trunks, branches, and with evergreens foliage, is likely to be more damaging than low temperatures. It occurs when the soil is frozen to the extent that the roots cannot replace moisture transpired by the above-ground parts. Exposure of the latter to sun and wind greatly increases the loss. Obvious offsets are to do all that is possible to prevent the soil freezing solidly for long periods and to wrap the above-ground parts of the plants in burlap or some other shade-giving protection. Freezing may be minimized by standing the containers as closely together as possible in a sheltered spot and piling around them leaves, straw, cut branches of evergreens, or such-like material contained by chicken wire or other restraint. Alternatively, surround each container with a wall of chicken wire or wood standing away from its sides to allow about 1 foot of space to be packed with straw, leaves, Styrofoam, or other insulation, topped off with a polyethylene plastic cover to keep it dry.

CONVALLARIA (Conval-lària)—Lily-of-the-Valley. The only species of this charming genus is a popular, easily grown, deciduous herbaceous perennial. Belonging in the lily family LILIACEAE, lily-of-the-valley has a generic name that stems from the Latin *convallis*, a valley. It alludes to favored habitats of the species. In the wild it occurs throughout much of temperate Europe, and by some botanists is accepted as an original native of mountain woods from Virginia to West Virginia, North Carolina, and Tennessee. Its roots are poisonous if eaten. They are used medicinally.

From horizontal rootstocks, lily-of-the-valley (***Convallaria majalis***) sends each year on very short stems, two or sometimes three erect, pointed, broadly-elliptic, parallel-veined leaves 6 to 9 inches long, and upright, one-sided racemes, 4 to 8

Convallaria majalis

inches tall, of white, sweetly fragrant, nodding flowers ¼ to more than ⅓ inch in length. Each bloom has a globose-bell-shaped perianth, with at its mouth six small, recurved lobes (petals). There are six short-stalked stamens and one style ending in a scarcely-lobed stigma. The fruits are spherical, pea-sized berries, bright red at maturity. They develop only when stocks derived from two or more distinct seedlings are growing in proximity. Varieties *C. m. fortunei* and *C. m.* 'Fortin's Giant' are larger in all their parts than the typical species. Variety *C. m. rosea* has pale, rather dirty pink flowers, on stems triangular in section; variety *C. m. aureo-variegata*, longitudinal stripes of creamy-yellow; and variety *C. m. prolificans* double blooms. For garden and landscape uses and cultivation see Lily-of-the-Valley.

Convallaria majalis aureo-variegata

CONVOLVULACEAE—Morning Glory Family. A number of attractive ornamentals are included in this family of fifty-five genera and some 1,650 species of dicotyledons. Of nearly worldwide distribution, it comprises a few trees, shrubs (often thorny), and woody vines as well as annual and perennial herbaceous plants with or without vining stems. The genus *Dodder* consists of leafless or nearly leafless plants that live as parasites on other plants. Familiar garden kinds that belong here include convolvuluses, dichondra, cypress vine, moonflowers, morning glories, and wood-rose. The sweet potato is also a member of the group. Some members of the family contain medicinal principles.

Plants of the morning glory family frequently have milky sap. Their leaves are alternate, lobeless or lobed, undivided, or pinnate; in parasitic *Dodder* they are reduced to scales. The symmetrical flowers, generally in loose, branched clusters or rarely solitary, have generally five separate, persistent sepals and a lobeless or five-angled or five-lobed corolla that is usually funnel- or bell-shaped or slender-tubular, with wide-spreading lobes. There are five stamens and usually one, two, or three styles. The fruits are usually capsules or more rarely berries. Genera cultivated include *Argyreia, Catystegia, Convolvulus, Cuscuta, Dichondra, Evolvulus, Exogonium, Ipomoea, Jacquemontia, Merremia, Mina, Porana,* and *Stictocardia.*

CONVOLVULUS (Convól-vulus)—Dwarf-Morning-Glory, Bindweed. This genus of the morning glory family CONVOLVULACEAE consists of mostly herbaceous plants, often with trailing or twining stems, and a few low shrubs. There are 250 species widely distributed in the wild in temperate, subtropical, and tropical regions. The name, alluding to the growth habits of many kinds, comes from the Latin *convolvo,* to roll together or entwine. Some species, not cultivated, are troublesome weeds. Here belong the field bindweed *Convolvulus arvensis.* From *C. scammonia,* of Asia Minor, a purgative drug is obtained. The California-rose, previously included in *Convolvulus,* is discussed under *Calystegia.*

Convolvuluses have alternate, stalked, usually lobeless, undivided leaves, characteristically heart- to arrow-shaped. The solitary or loosely-clustered flowers have five-parted, spreading calyxes and funnel- to bell-shaped corollas, with wide-spreading, pleated, slightly five-angled or rarely five-lobed faces. There are five stamens and two oblong or linear stigmas. The fruits are spherical capsules. The pair of bracts beneath the flowers, unlike those of *Calystegia,* which are large and leafy, are small or minute.

An annual or a short-lived perennial ordinarily grown as an annual, dwarf-morning-glory (*C. tricolor* syn. *C. minor*) is native in dryish, open areas in Portugal. From 6 inches to 1 foot in height, and with a spread of up to 2 feet, this kind has trailing and ascending stems. Its stalkless, lanceolate to obovate leaves are pubescent or hairless. Freely produced over a long summer season, and in bright weather remaining open all day, the blooms, in threes, are mostly displayed above the foliage. They have pointed-ovate, hairy sepals. Typically the corollas, 1 inch to 1½ inches wide, are

beautiful azure-blue with yellow throats encircled with a zone of white, but in horticultural variants they are variously spotted, blotched, or striped, or they may be white or red.

Other annuals and kinds grown as annuals sometimes cultivated are these: *C. aureus superbus,* of uncertain botanical standing, is a trailer or twining vine with stems up to 5 feet long. It has heart-shaped leaves and golden-yellow flowers profusely displayed. *C. erubescens,* native to Australia, is a biennial that if started indoors early may succeed as an annual. It has pointed, oblong to spear-shaped leaves and bears over a long period a multitude of small, rose-pink blooms. *C. pentapetaloides* inhabits dry soils in southern Europe. Up to 1 foot tall, it has prostrate stems up to 1 foot long, and stalkless, mostly linear to oblanceolate leaves. Its blue and yellow flowers are under ½ wide. From similar locations in the same general geographic region comes *C. siculus.* This has slender, trailing, rarely twining, non-woody stems 6 inches to 2 feet long. Its leaves are stalked and lanceolate to heart-shaped. Solitary or in pairs from the leaf axils, the up to ½-inch-wide flowers are blue.

A silvery, silky-hairy, evergreen shrub 1 foot to 2 feet tall, *C. cneorum* inhabits limestone cliffs and rocks near the sea. A

Convolvulus cneorum

native of southern Europe, it barely survives outdoors in sheltered places at New York City. For its best comfort it needs a less severe, drier climate. This has linear to oblanceolate leaves up to 2½ inches long by ½ inch wide. The short-stalked flowers, white or tinged with pink and with yellow throats, are in crowded clusters at the branch ends. They are a little more than 1 inch long by 1½ to 2 inches wide and have densely-silky-hairy calyxes. Similar to the last and like it a lime-lover, *C. oleifolius* is also an inhabitant of the Mediterranean region. Its usually pink flowers with more pointed calyx lobes serve to distinguish it from more woody-stemmed *C. cneorum.*

Beautiful *C. mauritanicus* is a trailing, evergreen, herbaceous perennial, slightly

Convolvulus mauritanicus

less hardy than *C. cneorum.* Its shoots and short-stalked, round-ovate leaves, ½ inch to 1½ inches in length, are more or less white-hairy. Blue-mauve to violet-purple flowers with paler throats, ½ to 1 inch in diameter, are plentifully produced over a long season. They are solitary or in groups of up to six.

An erect, pubescent, herbaceous perennial with a somewhat woody base, *C. cantabrica* has stems 4 inches to 1½ feet long. Its leaves are linear to oblanceolate. Solitary or several on a stem, the pink flowers come from the leaf axils and shoot ends. The blooms are ½ to 1 inch wide. This is a native of dry open places from Spain and Portugal to Italy.

A pretty trailer or climber, *C. althaeoides* is a pubescent, herbaceous or subshrubby perennial, with slender stems

Convolvulus althaeoides

about 3 feet long. Variable in shape, its stalked leaves are usually heart- to arrow-shaped, the upper ones at least, deeply-lobed. The flowers, solitary or in groups of up to three or sometimes five, are from the leaf axils. Usually purplish-pink, the blooms are 1 inch to 2 inches in diameter. This kind, which inhabits dry soils in southern Europe, is naturalized in California. Variety *C. a. tenuissimus* (syns. *C. a. pedatus, C. tenuissimus, C. elegantissimus*) is more slender, has hairs flat along the leaf surfaces instead of spreading, relatively deep and

Convolvulus tricolor

narrow leaf lobes, and clearer pink blooms.

Garden and Landscape Uses. The dwarf-morning-glory, highly effective in beds and at the fronts of flower borders, is also suitable for hanging baskets. The other annuals may be used similarly. Of the other kinds dealt with here *C. althaeoides* and its variety, *C. cneorum,* and *C. mauritanicus* are suitable for rock gardens and, the last two, for clothing banks and as groundcovers in mild, dryish climates. Convolvuluses are sun-lovers that need well-drained, dryish rather than too moist, only moderately fertile soil.

Cultivation. Annual *C. tricolor* is raised from seed sown outdoors in spring where the plants are to bloom or started indoors six to eight weeks earlier in a temperature of 60 to 65°F and the young plants carried in pots until danger from frost is over, and it is safe to plant them outside. Seedlings from outdoor sowings are thinned out early to prevent overcrowding. A distance of about 1 foot between individuals is satisfactory. In mild climates outdoor sowings in fall give plants that bloom early. Other annual kinds respond to the same conditions and care. The single-flowered type of the California-rose can be raised from seed or by division. The last is the method used for increasing the double-flowered variety. Other perennial kinds may be raised from seed, by division, or from summer cuttings rooted under mist or in a cold frame or cool greenhouse propagating bench. No special care is needed with established specimens except that if the perennials are planted in climates on the borderline of their winter hardiness it is well to give them the protection of a winter covering of branches of pine or other evergreen or some suitable substitute.

COOLABAH TREE is *Eucalyptus microtheca.*

COON TAIL is *Ceratophyllum demersum.*

COONTIE. This is a common name for some sorts of *Zamia.*

COOPERATIVE EXTENSION AGENT. See next entry.

COOPERATIVE EXTENSION SERVICE. This extraordinarily valuable service, operated as a partnership of federal, state, and county governments, is of inestimable value to gardeners as well as farmers, orchardists, and others interested in growing plants. Its agents are based in offices in 3,150 county seats in the United States, one in nearly every county. These Extension Agents have access to the resources of the land grant colleges of their respective states and of the United States Department of Agriculture. They stand ready to offer helpful advice to all who ask. The telephone numbers of their offices are listed in phone books under County Government.

Among the many useful services Cooperative Extension Agents, formerly known as County Agricultural Agents, perform are distributing federal and state bulletins and other publications on gardening, landscaping, and pest and disease control measures, answering questions, giving advice, and lecturing on these subjects, and often, usually for a small fee, making soil tests. In some regions "tip-a-phone" services that permit gardeners to dial for current "what to do" advice, with the programs changed weekly or oftener, are offered. In addition, some Agents conduct newspaper columns and radio and television shows directed toward gardeners. The Cooperative Extension Service is an aid and boon to gardeners. Use it.

COOPERIA. See Zephyranthes.

COPA DE ORA is *Solandra maxima.*

COPAIFERA (Copaíf-era). Evergreen trees of tropical America and Africa, some of which are commercial sources of oleo-resins called balsams and of the lumber called greenheart, constitute *Copaifera.* There are twenty-five species, which are chiefly South American. They belong in the pea family LEGUMINOSAE. The name is derived from a Brazilian one for one of the balsams obtained from them, and the Latin *fero,* to bear. Copaiferas have leathery, pinnate leaves, and panicles of small white flowers. The fruits are pods. Of little horticultural importance, they may sometimes be included in collections of plants useful to man. They thrive under conditions appropriate for trees of the humid tropics and are propagated by seeds and by cuttings. Native to tropical America, **C. officinalis** has leaves with two to five pairs of asymmetrical leaflets besprinkled with tiny translucent dots.

COPERNICIA (Coperní-cia) — Carnauba Palm or Wax Palm. The genus *Copernicia* of the palm family PALMAE consists of thirty species of New World palms. More than one-half are natives of Cuba. Its name honors the Polish scholar, scientist, physician, and theologian, Copernicus (Nicolaus Koppernigk), who died in 1543.

Copernicias are fan-leaved trees with trunks covered with the persistent bases of old leaves. On young specimens the leaves are often undivided, but old ones are palmately-cleft. The flower clusters are axillary and much-branched. Each globose or ovoid fruit contains one seed. The carnauba or wax palm is the source of the most important vegetable wax, carnauba wax, which has a high melting point and is esteemed for shoe, furniture, and other polishing pastes, high luster varnishes, and candles. At one time it was much used in phonograph records. The wax, obtained from both wild and cultivated trees, occurs as an exudate on the leaves and is recovered by drying and scraping the young leaves and depositing the scrapings in boiling water. The wax melts and floats. After cooling it is collected and formed into cakes.

The carnauba palm (**C. prunifera** syn. *C. cerifera*) has a trunk 30 to 45 feet tall topped by a dense, rounded crown of leaves that because of their waxy coating are grayish. The leaf blades are almost round and 3 to 4 feet wide. They are cleft to or below their middles into as many as sixty narrow, bifid segments. The leafstalks have scattered spines. The erect or spreading flower clusters are up to 6 feet in length. The fruits, ovoid to nearly spherical, are 1 inch long. This is native in Brazil.

Other species include **C. alba** (syn. *C. australis*), native from Brazil to Argentina, usually in wet soils. This attains heights of 70 feet or more. Its rounded leaf blades are 2½ feet across, whitish and with many rusty dots beneath. They are cleft into nearly fifty segments and have spiny stalks. The trunks are used for construction and as telephone poles, the leaves for thatching. Cuban **C. hospita** is 10 to 18 feet tall and has grayish, waxy leaves almost 4 feet

Copernicia hospita

long. One of the handsomest of Cuban native palms, **C. vespertilionum** is a hybrid that Cubans call *jata de los murcielagos.* Thus both scientific and vernacular names refer to the bats that are fond of nesting among the dead foliage that hangs from the crown of this kind. This palm is 40 to 50 feet tall.

The petticoat palm (**C. macroglossa** syn. *C. torreana*) is also Cuban. Usually 10 to 12 feet tall, it has leaves 4 to 5 feet in diameter. These hang on for many years after they are dead and form a skirt or petticoat around the trunk. From the leaves of *C. baileyana* Cubans make baskets and hats. The foliage of this is covered with wax much like that of the carnauba wax palm. The wax is used locally.

Copernicia macroglossa

Garden and Landscape Uses and Cultivation. Copernicias are not much grown in the United States. They are occasionally planted in Hawaii, southern California, and southern Florida, mostly in palm collections. The genus is especially well represented at the Fairchild Tropical Garden, Miami, Florida. The hardiest species, *C. alba*, has withstood a temperature of 22°F at Daytona Beach, Florida without vital injury, although the foliage was seriously damaged. Copernicias succeed in ordinary garden soils that are not excessively dry, but they are often slow-growing, especially in their young stages.

The carnauba wax palm is of interest in conservatories where plants of use to man are displayed. It thrives in a ground bed or a large container in a coarse, fertile, porous soil with good underdrainage. The greenhouse atmosphere should be fairly humid and the soil watered freely from spring through fall, and more moderately in winter. Shade from strong sun is needed and a minimum winter night temperature of 60°F. At other seasons the night temperature should be at least a few degrees higher and always day temperatures may exceed night ones by five or ten degrees. Applications at biweekly intervals from spring to fall of dilute liquid fertilizer are of benefit to well-established specimens. Propagation is by fresh seed sown in sandy, peaty soil in a temperature of 75 to 85°F. For additional information see Palms.

COPEY is *Clusia rosea*.

COPIAPOA (Copia-pòa). Recent evaluations of *Copiapoa* of the cactus family CACTACEAE place the number of its species at more than thirty. All are natives of Chile. The generic name is a modification of Copiapo, a province of Chile.

These are solitary, or rarely clustered, globular, conical, or columnar cactuses, heavily felted with woolly hairs at their tops, from whence the flowers emerge.

The blooms are yellow, or yellow with reddish outer petals. They are broadly-funnel-to bell-shaped. The stigmas are yellow and lobed. Their small fruits, smooth and shiny with green scales at their tops, contain large black seeds.

Not many kinds are cultivated. One sometimes grown, *C. cinerascens* (syn. *Echinocactus cinerascens*), is spherical to cylindrical, 3½ to 4½ inches thick, and has about twenty sharp ribs that broaden at the areoles (from which the spines arise). Each areole produces one central spine about ½ inch long and seven to nine slightly shorter ones. At first brown, they become black and, finally, gray. The flowers are yellow, their inner perianth segments (petals) toothed. Although in the wild *C. coquimbana* (syn. *Echinocactus coquimbanus*) forms clusters up to 3 feet wide, in cultivation its plant bodies are often solitary. Spherical or conical, they are up to 2 feet tall, about 4½ inches wide, and have ten to seventeen ribs that bulge conspicuously at the areoles. From each of the latter comes one or two central spines ¾ to 1 inch long, and eight to ten shorter radial ones. The spines at first black are later gray. The predominantly clear yellow, bell-shaped flowers have greenish or reddish-green outer petals and yellow stamens, styles, and stigmas. The blooms are 1¼ inches long by about as wide. Sometimes more than 3 feet tall and with cylindrical stems 4½ inches in diameter, *C. cinerea* (syn. *Echinocactus cinereus*) is a handsome kind, stone-gray to chalky-white, with about eighteen rounded ribs with crowded areoles. From each areole sprouts five or six spines, but most of these fall leaving only one stout, erect, black spine. The spines of each cluster vary in length, the biggest being about ¾ inch long. The yellow flowers are 1 inch wide and 1 inch long or a little longer.

Cactus fanciers' collections may also include *C. gigantea,* which grows to be 3 feet tall and has clusters of stems, globular when young but later columnar, about 8 inches in diameter, and at their tops very spiny. Their tops have brownish-yellow to reddish wool, and there are fourteen to twenty-two ribs that bulge at the areoles. The dark-tipped, yellowish spines, of approximately equal length, are ⅓ to ⅔ inch long and are in clusters of eight or nine. The flowers are yellow.

Garden and Landscape Uses and Cultivation. These are the same as for most desert cactuses and are discussed under Cactuses. It is important that copiapoas be rested by keeping the soil fairly dry through the winter. This promotes blooming. Also, they should be grown in full sun and be given little or no fertilizer.

COPPER LEAF is *Acalypha wilkesiana.*

COPPERTIP. See Crocosmia.

COPROSMA (Cop-rósma)—Mirror Shrub. Approximately one-half of the ninety species of botanically bewildering *Coprosma* are endemic to New Zealand, the others are wild in Australia, Tasmania, Borneo, Java, Chile, and in Hawaii and other Pacific islands. They are evergreen shrubs or trees of the madder family RUBIACEAE. None is hardy in the north. The name, alluding to the unpleasant odor of the foliage of some kinds, is from the Greek *kopros,* dung, and *osme,* a smell. The botanical complexity of this genus largely results from the frequency with which species in the wild hybridize. The kinds described here are all natives of New Zealand.

Coprosmas have undivided, opposite leaves, sometimes grouped, and small flowers solitary or in clusters. The small greenish or white blooms come from cups formed by the joining of a pair of bracts. Males and females are most commonly on separate plants, but it is not unusual for a few of the opposite sex to be intermingled with those of the sex that predominates. They have calyxes with four or five usually minute teeth, or these may be lacking in male flowers. The tubular, narrowly-funnel- to somewhat bell-shaped corollas have four or five lobes (petals). There are four or five stamens and usually two styles. The fruits are fleshy drupes (of plumlike structure).

Mirror shrub (*C. repens* syn. *C. baueri*), of New Zealand, and its varieties are by far the most familiar to American gardeners. Variable, this species ranges from shrub dimensions to a tree 25 feet tall. Its highly glossy, broad-oblong to ovate-oblong, short-stalked, fleshy leaves, 2½ to 3½ inches long by about two-thirds as wide, have blunt or notched apexes, recurved margins, and undersides paler than the upper. The flowers are in branched clusters. The orange-yellow to orange-red, ovoid fruits are ⅓ inch long. Variety *C. r. variegata* has foliage blotched with yellowish-green. The leaves of *C. r. marginata* are edged with yellow, the smaller leaves of *C. r. argentea* are silver-variegated.

Coprosma repens variegata

Coprosma repens marginata

Sometimes called creeping coprosma, *C. kirkii*, believed to be an intermediate natural hybrid between *C. repens* and *C. acerosa*, has linear to linear-oblong leaves about 1 inch long. Up to 6 feet tall and wide, but often very much lower, *C. acerosa* has flexible, prostate or sprawling stems and paired or clustered, leathery, bluntly-linear leaves, yellowish-green, and up to ½ inch long by up to ⅒ inch wide. The solitary flowers are at the ends of short branchlets. The delicate blue fruits, often flecked a darker blue, are spherical and ¼ inch in diameter.

Related to *C. acerosa*, wiry-stemmed *C. propinqua* is a spreading shrub up to 10 feet tall or sometimes taller, with interlacing branches and opposite or clustered leaves. The latter are linear to broad-oblong, mostly up to ½ inch long, and paler on their undersides than above. The flowers are solitary or in small clusters. The translucent fruits are pale blue sometimes flecked with lighter and darker blue. New Zealand botanists recognize several varieties of this variable species. A natural hybrid between *C. propinqua* and *C. robusta* is *C. cunninghamii*. This up-to-15-feet-tall shrub has leathery, linear or linear-lanceolate leaves about 2 inches long. Its pale translucent fruits are about ¼ inch long.

A variable shrub or tree, mostly not over 10 feet, but sometimes up to 20 feet in height, *C. lucida* has short-stalked, thick, narrow-ovate to broad-elliptic leaves, glossy dark green above and paler beneath, and 4½ to 7 inches long. The male flowers are in dense, sometimes branched clusters, the females in clusters of three or four. The orange-red fruits are ⅓ inch long or a little longer. Variety *C. l. angustifolia* has narrower leaves than the species. A low kind that forms cushions or mats occasionally up to 6 feet across, *C. petriei* has leathery, nearly stalkless, toothless leaves, usually clustered on short branchlets, and up to ⅓ inch long. They are dark green above, paler on their under surfaces, and narrowly-elliptic to oblong or obovate. The solitary, greenish flowers are terminal. The spherical fruits, ¼ inch or slightly more in diameter, are purplish-red to light blue.

Garden and Landscape Uses. Coprosmas are excellent outdoors in mild, nearly frostless regions, and *C. repens* and its varieties as indoor pot plants. They are esteemed for their foliage and ornamental fruits. Low, trailing kinds can be used with good effect in rock gardens. The hardiest seem to be *C. petriei* and *C. acerosa*. Coprosmas do well in coastal areas. Because it stands shearing well, *C. repens* and its varieties are good hedge plants.

Cultivation. Ordinary soils, well drained but not excessively dry, and sunny locations, or at most part-day shade, are to the liking of coprosmas. No special care is needed beyond whatever pruning is necessary to shape the specimens or limit their size. This may be done at any time, preferably just before new growth begins. Propagation is easy by cuttings of firm, but not hard shoots planted in a greenhouse propagating bench, a cold frame, or under mist, in sand, sand and peat moss, vermiculite, or perlite. Layering also is an easy means of increase. Seed may also be used.

As greenhouse and window plants *C. repens* and its varieties succeed in fertile soil kept evenly moist. They dislike excessively high temperatures. In winter 50°F at night is appropriate, and by day an increase of five to fifteen degrees. Pruning to shape and repotting is done in late winter or spring. Occasional pinching out of the tips of the shoots of young specimens to secure a desirable shape may be necessary.

COPTIS (Cóp-tis)—Goldthread. This delightful genus of dainty woodlanders inhabits cool parts of the northern hemisphere around the globe. There are about ten species of *Coptis* of the buttercup family RANUNCULACEAE. The name comes from the Greek *koptein*, to cut, in allusion to the divided foliage.

Goldthreads are lowly perennial herbaceous plants, with slender, spreading underground rhizomes and evergreen, ferny foliage. Their blooms are small white buttercups carried singly or few together on slender stems to a height somewhat exceeding the foliage, which means from 2 to 6 inches above the ground. Bisexual or unisexual, with plants often having flowers of one sex only, the blooms have five to seven petal-like sepals and five or six smaller, linear, hooded, club-shaped organs, which some botanists believe to be staminodes (petal-like stamens), and others to be petals. The sepals are the showiest parts of the flower. There are numerous normal stamens and three to seven pistils, which develop into stalked dry fruits (follicles). In their northern and mountain homes these plants bloom in May and June, but in lowland gardens not too far north they are precocious. At New York City cultivated specimens are in flower before March is gone, they scarcely await the arrival of spring.

The most commonly cultivated goldthread is native in Greenland and in eastern North America as far south as North Carolina, Indiana, and Iowa. Botanists differ as to whether this sort should be considered the same species as a very similar kind that occupies Alaska and eastern Asia. The differences are minor, the chief being that the Alaskan-Asian plant has larger fruits. Those who consider the two to be one species use the name *C. trifolia*. Botanists who accept them as two species reserve *C. trifolia* for the Oriental and Alaskan kind and call the plant that grows in Greenland and eastern North America *C. groenlandica*. Fence straddlers regard the latter a variety of the former, which they designate *C. trifolia*, and name the kind from Greenland and eastern North America *C. trifolia groenlandica*. Although this may be a bit confusing to nonbotanists, it is of small import to the gardener whose chief concern is with growing them. These variants, be they one species or two, have bright yellow, slender rhizomes and shining leaves of three nearly stalkless, shallowly-lobed or toothed, obovate leaflets. The flowers are white, about ½ inch across, with spreading sepals.

Other species that may appear in cultivation are *C. asplenifolia*, native from British Columbia to Alaska, with leaves of five long-stalked, pinnately-divided leaflets, and flowers with narrow, whitish sepals; *C. laciniata*, of California to Washington, with three-lobed leaflets pinnately-three-to five-cleft, and whitish blooms; and *C. occidentalis*, of Washington to Oregon and Montana, with leaves of three pinnately-divided leaflets and greenish or yellowish sepals. Asian species include *C. quinquefolia* and *C. japonica*, natives of Japan. With leaves of five irregularly-toothed leaflets, *C. quinquefolia* has usually solitary white flowers on stems up to 6 inches tall. The leaves of *C. japonica* are also divided into five leaflets; its white flowers are solitary, or two or three together, on stems up to 10 inches long.

Garden Uses. The goldthreads are choice and precious things for inclusion in woodland gardens, rock gardens, native plant gardens, and bog gardens. They are carpeters, not too fast spreading; in fact they usually extend themselves more slowly than the gardener wishes. They are at home in moist, acid soils and shade, in association with small ferns, partridge-berry, shortias, and other nonaggressive lovers of cool woodland conditions.

Cultivation. The chief deterrents to success with these plants are alkaline soils, dryness, and hot summers. A cool spot should always be found for them. A north-facing slope or location behind an old tree stump or rock is often acceptable in gardens where otherwise they fail to thrive. Happily located they give no particular trouble. A shallow mulch of peat moss maintained about them is helpful, and in winter a light covering of branches of evergreens or forest litter protects the foliage from damage by sun and wind. Seeds, sown in moist sandy peaty soil as soon as they are ripe, provide a ready means of increase, and the young seedlings transplant without difficulty provided ample soil is retained about their roots. Careful division in spring is also a satisfactory method of increase.

CORAL. As part of the common names of plants the word coral is used in these instances: coral-bean (*Sophora secundiflora*), coral bells (*Heuchera sanguinea*), coral blow or coral plant (*Russelia equisetiformis*), coral drops (*Bessera elegans*), coral plant (*Berberidopsis corallina*, *Jatropha multifida*, and *Russelia equisetiformis*), coral root (*Corallorhiza*), coral shower (*Cassia grandis*), coral tree (*Erythrina*), coral vine (*Antigonon leptopus* and *Kennedia coccinea*), and scarlet coral-pea (*Kennedia prostrata*).

CORALBERRY is *Symphoricarpos orbiculatus*.

CORALLITA is *Antigonon leptopus*. White-corallita is *Porana paniculata*.

CORALLODISCUS (Corallo-díscus). About eighteen Asian species of the gesneria family GESNERIACEAE constitute *Corallodiscus*, a close ally of *Chirita* and *Didymocarpus*. The name derives from the Greek *korallion*, coral-like, and *diskos*, a disk or plate.

These plants have a rhizome that produces a basal rosette of leaves, with conspicuously indented veins. Their flowers are stalked and have a five-parted calyx and a tubular corolla with an upper lip of two lobes and a lower lip of three larger lobes. There are two pairs of stamens and a short style tipped with a two-lobed stigma. The fruits are capsules.

Himalayan *C. lanuginosus* (syn. *Didymocarpus lanuginosus*) has obovate leaves up to about 2½ inches long and approximately one-half as wide. The 3- to 4-inch-long flowering stalk carries few to several about ½-inch-long light blue or purplish blooms.

Garden Uses and Cultivation. These are as for *Chirita*.

CORALLORHIZA (Corallo-rhìza)—Coral Root. The majority of the fifteen species of this genus of the orchid family ORCHIDACEAE occur wild in North America, the others elsewhere in the northern hemisphere. They are saprophytes (plants devoid of or essentially without chlorophyll that derive their nourishment from dead organic matter). The name, alluding to the appearance of the roots, comes from the Greek *korallion*, coral, and *rhiza*, a root.

From clustered masses of coral-like roots these orchids produce erect racemes with pinkish, brown, yellow, or purplish succulent stems that toward their bases have a few sheathing scales representing leaves. There is no other foliage. The small blooms, in terminal racemes, are not showy. The narrow sepals and petals are similar. They spread, or are more or less erect and project forward over the slightly incurved column, which is shorter than the sepals and petals. The oblongish to roundish, decurved lip often has two upturned, ridged lobes or may be lobeless.

An inhabitant of rich woodlands from Newfoundland to British Columbia, North Carolina, Indiana, and California, ***Corallorhiza maculata*** has pinkish-purple stems 9 inches to 1½ feet tall. The upper 2 to 6 inches is a raceme of up to forty prominently spurred, brownish-purple flowers, with white, purple-spotted or purple-suffused lips that have a large middle lobe and two smaller side ones. From the last *C. striata* differs in its flowers having sepals, petals, and a lobeless lip marked with longitudinal purple lines. It is native from Quebec to British Columbia, New Mexico, and California.

Garden Uses and Cultivation. Like many saprophytes coral roots are nearly impossible to tame with any degree of reliability. Taken from their natural surroundings and transplanted to gardens or pots they may linger for a year or two, but invariably prove transient. It is wisest to leave these orchids undisturbed where they grow naturally. For more information see Orchids.

CORCHO is *Microcycas calocoma*.

CORCHOROPSIS (Corchor-ópsis). A few eastern Asian species of the linden family TILIACEAE comprise this genus of annuals, the name of which is derived from that of the allied genus *Corchorus*, and the Greek *opsis*, similar to. They have stalked, undivided, toothed leaves, with three or five chief veins and small, solitary, yellow flowers from the leaf axils. The blooms have five persistent sepals slightly shorter than the five obovate petals, ten to fifteen functional stamens, and five longer nonfunctional ones (staminodia). There is one style. The fruits are cylindrical capsules.

Scarcely decorative enough to warrant much attention from gardeners, ***Corchoropsis tomentosa*** (syn. *C. crenata*) is erect, loosely-branched, and 1 foot to 2 feet tall or perhaps sometimes taller. Its alternate, ovate, bluntly-toothed leaves, up to 3½ inches long and nearly 2 inches wide, are densely covered with stellate (starlike) hairs. With slender stalks ¾ inch to 1¼ inches long, the blooms are ½ to ¾ inch across. They come in summer and fall. The seed capsules are about 1 inch long.

Garden Uses and Cultivation. Only in specialized botanical collections is this sort worth growing. It succeeds in ordinary soil

in sunny locations and is raised from seed sown outdoors in early spring.

CORCHORUS (Cór-chorus)—Jute, Jews'-Mallow. The plant at one time named *Corchorus japonicus* is *Kerria japonica*. The genus *Corchorus*, of the linden family TILIACEAE, has no horticultural significance except that as the source of jute it may be grown in educational displays of plants useful to man. Its name is derived from the Greek *koreo*, to purge, and *kore*, the pupil of the eye, in allusion to supposed medicinal virtues of the plants for treating eyes. There are about 100 species, natives of warm regions. Mostly tall annuals, but including some subshurbs, corchoruses have alternate, undivided, toothed leaves and tiny yellow blooms, with five each sepals and petals, and many stamens. The fruits are many-seeded capsules.

Jute (*C. capsularis*), native of India, is an annual up to 15 feet tall, with oblong leaves some 4 inches long. It has spherical, wrinkled seed pods. From it the jews'-mallow (*C. olitorius*) differs in its seed pods being elongated and beaked. Jute fiber is obtained from both species, which are cultivated in Asia and parts of Africa for this product. By retting (soaking in water until the soft tissues are rotted sufficiently to allow the tougher fibers to be separated from them) jute fibers are prepared for use. In warm countries the young shoots of the kinds described above are used as potherbs.

Garden Uses and Cultivation. The limited horticultural possibilities of these plants are given at the beginning of this entry. Jute and jews'-mallow grow rapidly in warm, humid climates from seed sown in spring where the plants are to remain. They succeed in ordinary soils in sunny locations.

CORDIA (Cór-dia)—Geiger Tree. The borage family BORAGINACEAE is chiefly known to gardeners from a number of attractive herbaceous and subshrubby plants it contains, such as anchusas, forget-me-nots, mertensias, and heliotropes. In the tropics it runs to shrub and tree species of less familiar genera. Only a few of these are cultivated, among them *Cordia*, a genus of 250 species of trees, shrubs, and woody or more or less herbaceous climbers of warm parts of the Old and the New World. The name honors the German botanist Valerius Cordus, who died in 1544. Some botanists segregate *Cordia* and a few other genera that traditionally have been included in the borage family as the ehretia family EHRETIACEAE.

The leaves of cordias, usually alternate, are undivided and toothed or toothless. Bisexual or unisexual, the blooms are densely clustered in heads or are two-ranked in looser spraylike arrangements called scorpioid cymes. They have tubular or bell-shaped calyxes, with three to five teeth, and corollas that are funnel-shaped or that have slender tubes and four or more spreading lobes (petals). There are as many stamens as corolla lobes. The style is four-cleft or four-lobed. The fruits, those of some species edible, are technically drupes. They are small, and to a greater or lesser degree are enclosed by the persistent calyxes.

The geiger tree (*C. sebestena*) is an evergreen shrub or tree up to about 30 feet tall, native from the Florida Keys and the West Indies southward. It has ovate, scratchy-hairy leaves, sometimes with wavy margins, 4 to 9 inches long. The stalked blooms, with five to twelve burnt-orange to scarlet, crinkled petals, are 1 inch to 2 inches long and commonly 1 inch or more in diameter. In loose, terminal heads, they are displayed almost throughout the year. The edible, fragrant, white fruits, enclosed in husklike calyxes, are ¾ to 1 inch in length.

Called wild-olive in its native Texas (it also extends into Mexico), *C. boissieri* is an evergreen tree up to about 25 feet tall. Its clusters of crepy, white, yellow-centered, 1½-inch-wide flowers, except in color, much resemble those of the geiger tree. The velvety-hairy, wavy leaves are ovate to ovate-oblong, and up to 5 inches long.

Less well-known species include *C. alliodora* (syn. *Cerdania alliodora*), an evergreen tree up to 90 feet tall, but often much shorter, that in the wild ranges from the West Indies and Mexico to South America. Its pointed, elliptic, oblong, to obovate leaves, 4 to 8 inches long, are onion-scented when crushed. The small white blooms, in large showy clusters, are produced in abundance. In bloom rather like the geiger tree, but its bright orange flowers somewhat larger and with twelve to seventeen corolla lobes (petals) *C. dodecandra*, of Mexico, is up to 100 feet tall. It has very rough, gray-green, 6-inch-long, broad-oblong to very broad-elliptic leaves. A Cuban species with burnt-orange blooms, smaller, but otherwise much like those of the last, *C. angiocarpa* is a slow-growing evergreen with 4-inch-long, rigid, gray-green leaves. A somewhat straggly, freely-branched shrub, *C. lutea*, of South America, has broad-elliptic to nearly round leaves, 3 to 4 inches long, and clusters of trumpet-shaped, bright yellow, ¾-inch-long blooms, succeeded by small, white, reportedly edible fruits. Less showy in bloom, but an attractive shade tree, *C. obliqua*, of Cochin-China, has broad-elliptic leaves, and white flowers that are not especially showy. Another cordia of secondary interest so far as its flowers are concerned the cereza (*C. nitida*), of Puerto Rico, has red, cherry-like fruits. This attains heights of up to 70 feet, but is often much smaller. Its leaves are elliptic to obovate and 3 to 5 inches long. The tiny, yellowish to whitish flowers are many together in clusters.

Deciduous and about 40 feet tall, *C. myxa* (syn. *C. dichotoma*) has stiffish, hairless, sometimes wavy-toothed, broad-ovate leaves up to 5 inches long. Its white, bisexual and unisexual flowers, up to ½ inch long, are in large clusters. The tan fruits, about ¾ inch in diameter, were formerly used medicinally. This kind is native from India to Australia.

Garden and Landscape Uses and Cultivation. Cordias are good-looking, useful, in some cases slow-growing, general-purpose landscape adornments for warm regions. The hardier ones, including *C. boissieri*, *C. myxa*, and *C. obliqua*, withstand a few degrees of frost. All are satisfied with ordinary garden soils, and succeed in open locations. No systematic pruning is needed. Propagation is usually by seed. Cuttings and air layering are alternatives.

CORDON. A tree or bush, usually of a kind grown for its edible fruit trained to a single stem by repeatedly cutting back all branches to induce them to develop as fruiting spurs, is called a cordon. A double cordon has two stems trained in this way. Among fruits that lend themselves to growing as cordons are apples, pears, European (vinifera) varieties of grapes, red and white currants, and gooseberries. See Espalier.

CORDYLINE (Cordy-line)—Ti Plant, Cabbage Tree. Although the members of this group are often cultivated under the name dracaena they are distinct from the genus *Dracaena* in their botanical characteristics. The most easily observable differences are that cordylines have creeping rhizomes, whereas dracaenas do not, and that in *Cordyline* there are three bracts at the base of the flower stalks, which are not present in *Dracaena*. A more recondite reason for their separation is that cordylines have six or more ovules in each cell of the ovary, and dracaenas one. The genus *Cordyline* numbers fifteen species, all of tropical, subtropical, and warm-temperate regions. It belongs to a group of the lily family LILIACEAE that some botanists separate as a distinct family, the AGAVACEAE, but that others retain as the subfamily AGAVOIDAE. This subfamily includes, among other genera, *Agave*, *Dracaena*, *Phormium*, *Sansevieria*, and *Yucca*. The name *Cordyline*, which is pronounced as four syllables, is derived from the Greek *kordyle*, a club, and alludes to the thickened roots. Except for the differences noted, cordylines are very like dracaenas. They are mostly natives from New Zealand and Australia to Malaya, but one species is indigenous in South America.

Cordylines have branched or branchless stems, either solitary or clustered, and leaves that are usually crowded toward their tops, but sometimes are more scattered. The flowers are white or tinted yellowish, pinkish, or lavender. The corolla has six spreading petals (more correctly, tepals). There are six stamens and a three-lobed style. The fruits are fleshy, spherical, three-seeded berries.

The ti plant (*C. terminalis*) is one of the most familiar. A shrubby kind, up to 12 feet in height, it is indigenous from continental Asia to Australia and Hawaii. In the latter island it has always been important to the inhabitants. Its leaves were used to make hula skirts, raiment, and sandals, to wrap fish and other foods, as plates, and for thatching. They also served as fodder for cattle and horses. The thick roots provided human food in times of scarcity, and from them an intoxicating beverage was prepared. It was believed that a hedge of ti planted around a dwelling kept away evil spirits. In other parts of its native range the ti was put to similar uses. This species usually has broader leaves than other cultivated cordylines. They are 1 foot

Cordyline terminalis

Cordyline terminalis variety

Cordyline terminalis variety, in bloom

to 2½ feet in length, up to 6 inches in width, and arranged in close spirals toward the tops of the stems. Typically, they are narrow-oblong to paddle-shaped, but there are numerous varieties and their leaves vary considerably in size, shape, and color. They are shiny and flexible and have deeply-channeled stalks 2 to 6 inches long. The color of the foliage varies according to variety from deep green to rich deep red, with the young leaves often much more brilliantly colored than the old. The flowers of the ti plant, in 1-foot-long, slender-branched panicles, are delicate lavender or sometimes red. The fruits are red. For available varieties of the ti plant the catalogs of specialists in tropical plants should be consulted. These are especially worthy: *C. t. amabilis*, with wide, green to bronze leaves variegated with pink and cream; *C. t. angusta*, with long, narrow, coppery-green leaves edged with red and with purplish undersides; *C. t. 'Baby Ti'*, a small-growing variety with slender metallic-green leaves suffused with coppery-red and margined with red; *C. t. baptistii*, with broad, deep green to bronzy leaves striped with creamy yellow and pink, *C. t. 'Madame Eugene Andre'*, with red-margined green to bronzy older leaves and brilliant pink young ones; and *C. t. tricolor*, with broad leaves variegated over a base of green with cream, pink, and red.

From New Zealand come other popular kinds, including one commonly called *Dracaena indivisa* by gardeners, but correctly identified as *Cordyline australis*, another correctly named *C. indivisa*, and *C. banksii*. In New Zealand the first named is known as the cabbage tree and attains a huge size for a plant of the lily family. One reliable botanist described a specimen with a hollow trunk so large that a native had fitted it with a door and was using it as a storeroom. This species is remarkably tenacious to life; cut trunks deposited on a sea beach and rolled about for several months by the tide budded vigorously after they were washed ashore by particularly high water,

and it is recorded that hollowed trunks used as a chimney for a fire for several months produced an abundance of new shoots after the fire was no longer used.

The New Zealand cabbage tree (**C. australis**) has a large rhizome-like base and a branched or branchless trunk up to 40 feet high. The ends of the trunk and branches are crowned with dense clusters of gracefully arching leaves 1 foot to 3 feet long by 1 inch to 2½ inches wide. The numerous creamy-white, fragrant flowers are in much-branched, erect panicles 2 to 4 feet long and one-half as wide. The fruits are white or bluish and about ¼ inch in diameter. There are a number of horticultural varieties of *C. australis*, the best known of which are *C. a. atropurpurea*, with the lower parts of the leaves and their midribs below purplish; *C. a. aureo-striata*, with its leaves banded longitudinally with creamy-yellow lines; *C. a. cuprea*, with narrow, coppery-red-brown leaves; and *C. a. doucetii*, with

Cordyline australis

Cordyline australis, old specimen, in Ireland

Cordyline australis, in bloom

Cordyline australis, young specimen

Cordyline australis atropurpurea, in containers

Cordyline australis doucetti

dark olive-green, narrow leaves variegated with creamy-white and with pink-tinted edges.

From *C. australis* and its varieties *C. indivisa* differs in being smaller, often not more than 10 feet in height, and in having drooping panicles of white blooms tinged purple on their outsides. Its thick leaves have yellow or reddish midribs and are leathery and pliable. From them the Maoris extracted a very fine fiber they used for making clothing. Like those of *C. indivisa*, the flower clusters of *C. banksii* hang downward. This kind is distinguished by its occasionally branched slender stems, 4 to 10 feet tall, its leaves, 3 to 6 feet long and 1½ to 3½ inches wide, and its pure white flowers. The Australian *C. stricta* has a slender stem or stems, sometimes 12

Cordyline stricta

feet tall, but usually shorter. They are sometimes branched. Its closely set leaves are 1 foot to 2 feet long and up to 1 inch wide. They may be green or bronzy-purplish. Its flowers are lilac. Variety *C. s. discolor* is distinguished by its especially dark bronzy-purple foliage.

Garden and Landscape Uses. Decidedly tropical, *C. terminalis* and its varieties are adapted for outdoor cultivation in the continental United States only in southern Florida, but they are excellent and colorful garden plants for Hawaii and other warm, humid places. They are also splendid for growing in tropical greenhouses and for use as indoor ornamentals, although their need for a highly humid atmosphere usually prevents them from prospering as permanent houseplants. The other kinds discussed above are cooler climate plants that will even stand light frosts. Nevertheless, they can be grown outdoors only in the milder, nearly frost-free parts of the United States. Their need for a humid atmosphere is less pronounced than that of *C. terminalis*. These warm-temperate cordylines are striking landscape features, resembling somewhat the giant yuccas of the United States and Mexico and usable in much the

same way as accents. They succeed well as pot or tub plants and as small specimens are more adaptable as houseplants than *C. terminalis*.

Cultivation. In appropriate climates no difficulty attends the cultivation of cordylines, nor are they difficult to handle as pot plants, except that *C. terminalis* must have a very humid atmosphere. All need well-drained soil of average or better than average fertility that is not excessively compact. That in which *C. terminalis* is grown must be constantly moist for the best results, but not saturated. The other kinds withstand somewhat drier conditions. Specimens cultivated in containers should be repotted or top dressed in spring according to their needs. Throughout the spring-to-fall growing season, well-rooted plants respond to weekly or biweekly applications of dilute liquid fertilizer. Some protection from intense sun is appreciated by *C. terminalis* and its varieties, but the other kinds revel in full sun. Propagation of cordylines is easily done by seed sown in sandy peaty soil in a temperature of about 75°F for *C. terminalis* and 60 to 70°F for the others and also by terminal and sectional stem-cuttings, by air layering, and by separating fat sections of the roots, sometimes called toes, and planting them in a sandy peaty soil in a warm propagating bed. Under greenhouse cultivation a minimum winter night temperature of 60 to 65°F is satisfactory for *C. terminalis*, for the others 50 to 55°F is adequate. The chief pests of cordylines are mealybugs, scales, and red spider mites.

Cordyline terminalis: (a) Removing "toes" for propagation

(b) "Toes" ready for planting

COREMA (Cor-èma)—Broom-Crowberry. The most familiar species of this genus of the crowberry family EMPETRACEAE inhabits northern North America from Newfoundland to New Jersey, chiefly in coastal areas. The only other species occurs in the Canary Islands, the Azores, and Europe. From the crowberry (*Empetrum*) these differ in their flowers being in terminal heads and without petals and in having five or fewer stigmas. The name comes from the Greek *korema*, a broom, and alludes to the twiggy growth of the plants.

Coremas, low, heathlike, much-branched shrubs of stiff habit, have small, narrow, undivided leaves with rolled-under margins scattered along the stems or almost in whorls (circles of more than two). Male and female blooms are on separate plants. They have three or four somewhat petal-like sepals and the same number of stamens. The slender styles have two to five stigmas. The usually three-seeded fruits are small, berry-like drupes.

The North American broom-crowberry (*Corema conradii*) forms broad mats up to 6 feet in diameter and 1½ feet in height. Its purplish flowers are few together. Its leaves are up to ¼ inch long. The fruits are dryish and very small. From 1½ to 2 feet tall, *C. album,* of the Azores, Canary Islands, Spain, and Portugal, is erect. It has markedly downy shoots and dark green, narrowly-linear leaves ¼ inch long or a little longer, with very strongly-rolled-under edges. They are mostly in circles of three. The stalkless flowers make little display. Following pollination, those of female plants develop three-seeded, white berries. This species is not hardy in the north.

Garden and Landscape Uses and Cultivation. These are suitable for sandy, acid soils and appropriate for planting in association with heathers and heaths. They respond to the same conditions as heathers and heaths and like them are propagated by seed and by summer cuttings. The American species is the more commonly cultivated. It may be employed as a groundcover, especially in rock gardens and native plant gardens. For more information see Heath or Heather Garden, and Heather.

COREOPSIS (Coreóp-sis)—Tickseed. This genus of 120 species inhabits North America, the Hawaiian Islands, and tropical Africa and belongs in the daisy family COMPOSITAE. It includes plants previously known as *Calliopsis* and *Leptosyne*. They are annuals, biennials, herbaceous perennials, or rarely shrubs, several greatly esteemed for their showy blooms and ease of culture. The name *Coreopsis* is derived from the Greek, *koris*, a bug, and *opsis*, resemblance, and refers to the appearance of the fruits, as also does the colloquial name tickseed (this name is also sometimes applied to *Bidens*).

Coreopsises have usually opposite, lobeless, lobed, or variously dissected leaves and daisy-type flower heads on loosely-branched or branchless stalks. The involucre (collar of leafy bracts just beneath the flower head) in cultivated kinds consists of two rows of eight bracts each, united at their bases. The ray florets, yellow or more rarely pink or bicolored, are in a single row except in horticultural varieties with semi-double or double blooms. The fruits are seedlike achenes.

Three cultivated perennial species are distinguished by having leaves divided like the fingers of a hand (palmately), the other perennial kinds grown in gardens have leaves either undivided or divided in pinnate (feather-like) fashion. The annual species have lobeless or pinnately-lobed leaves. The best-known palmately-leaved kind is the thread-leaf tickseed (**C. verticillata**), a charming native from Maryland to Alabama and Florida, that grows 1 foot

Coreopsis verticillata

to 3 feet tall and has stalkless leaves, 2 to 3 inches long, dissected into slender, almost hairlike, segments. Its flower heads, few together in clusters, are 1 inch to 2 inches across, and each has about eight bright yellow ray florets. The disk florets are yellow. The larkspur tickseed (**C. delphinifolia**) is a leafy species, 2 to 5 feet in height, with stalkless leaves divided into linear segments and clusters of slender-stalked flower heads 1 inch to 2 inches across, with yellow rays and purple-brown disk florets. The larkspur tickseed is native from Virginia to Georgia and Alabama. The third kind with palmately-divided leaves is **C. tripteris,** which occurs as a native from Ontario to Georgia and Louisiana, and is 3 to 10 feet tall and anise-scented. It has stalked leaves 2 to 5 inches long and divided into three oblong-lanceolate segments. Often its upper leaves are undivided. The flower heads, 1½ to 2 inches across, are in clusters. Each has seven or eight yellow rays and yellow disk florets that become brown or purplish with age.

The best-known perennial kinds with undivided or pinnately-divided leaves are *C. lanceolata* and *C. grandiflora* both of which are usually somewhat hairy and 1 foot to 2 feet in height. In **C. lanceolata** the stems are leafy only toward their bases and the leaves are mostly undivided. The leaves of **C. grandiflora** are mostly divided, and the stems bear leaves throughout their length. The leaves of *C. lanceolata* are lobeless or have one or two small lateral lobes,

Coreopsis grandiflora

whereas the upper stem leaves of *C. grandiflora* are usually divided into three or five lanceolate or linear lobes. Both species have flower heads 1½ to 2½ inches across with about eight toothed, yellow florets and yellow disks. Those of *C. l. florepleno*, with many more florets, are in effect double flowers. The native range of *C. lanceolata* is from Michigan to Florida and New Mexico, of *C. grandiflora* from Missouri to Florida and New Mexico.

Less robust **C. auriculata** is a very pretty native from Virginia to Florida and Louisiana. It attains a height of 1½ feet, is leafy toward its base, and has many erect flower

Coreopsis auriculata

stalks bearing solitary flower heads 1½ to 2 inches in diameter. Each has about eight three-toothed, yellow ray florets and a yellow disk. The long-stalked leaves of this are ovate to round, often with one or two small lobes near the bases of the blades. Rosy-pink ray florets distinguish **C. rosea** from other cultivated tickseeds. Native from Nova Scotia to Georgia, this sort is 1 foot to 2 feet in height. It has linear, undivided leaves or the lower ones twice- or

Coreopsis rosea

thrice-pinnately-cleft into narrow lobes. The short-stalked flower heads are ¾ to 1 inch across and have about eight toothed or lobed rays and a yellow disk. Smaller *C. r. nana* has lilac-pink ray florets.

Two handsome tender perennials from the Pacific Coast are *C. maritima* (syn. *Leptosyne maritima*) and *C. gigantea* (syn. *Leptosyne gigantea*). A stout, fleshy, few-branched shrub that inhabits dunes and rocky cliffs along the shores of California, Baja California, and offshore islands, **C. gigantea** ranges from 5 to 10 feet in height and is nonhairy. It has a main trunk up to 4½ inches in diameter, and alternate, pinnately-finely-divided leaves in tufts near the branch ends. The yellow flower heads, each with ten to sixteen ray florets, are 1½ to 3 inches in diameter. The sea-dahlia (**C. maritima**) is 1 foot to 3 feet in height and has fleshy, hairless stems and leaves, the latter twice-pinnately-divided into narrow-linear segments. The solitary flower heads, 2½ to 3½ inches in diameter, have stalks up to 1 foot long. Both ray and disk florets are yellow. This species, indigenous to the coast of southern California, although perennial, is often cultivated as an annual.

Annual tickseeds often cultivated include the golden wave (*C. basalis* syns. *C. drummondii*, *Calliopsis drummondii*), *C. tinctoria* (syn. *Calliopsis bicolor*), *C. atkinsoniana* (syn. *Calliopsis atkinsoniana*), and *C. bigelovii*. The golden wave (**C. basalis**) is a hairy or hairless native of Texas, 1 foot to 2 feet

Coreopsis maritima

in height, with pinnately-divided, stalked leaves that mostly have oblong to ovate divisions except that those of the upper leaves may be linear. The long-stalked flower heads, 1 inch to 2 inches in diameter, have about eight toothed or lobed ray florets and a dark purple disk. The rays are yellow, with brownish-purple bases. Differing from the golden wave in having the segments of its leaves linear to linear-lanceolate and in having the outer involucral bracts much shorter rather than about as long as the inner ones, *C. tinctoria* is a nonhairy native from Minnesota to British Columbia, Louisiana, and California, 1 foot to 3 feet in height, branching above and with its leaves twice-pinnately-divided and the lower ones long-stalked. The flower heads, ¾ inch to 2 inches across, are on slender, branched stems. They have seven to eight yellow, toothed or lobed ray florets and a purplish-brown disk. In variety *C. t. atropurpurea* the rays are uniformly cinnamon-red. Variety *C. t. flore-pleno* has double flowers, and *C. t. nana* is more dwarf than the typical species. The species *C. atkinsoniana* resembles *C. tinctoria*, differing only in technical differences in the flowers. It is a native of the Pacific Northwest. Endemic to California, *C. bigelovii* is commonly misidentified in gardens as *C. stillmanii* (syn. *Leptosyne*

Coreopsis tinctoria

stillmanii), a species that seems not to be cultivated. Slender, hairless, and 6 inches to 2 feet tall or sometimes taller, it has stems leafy in their lower parts. The mostly basal leaves are pinnately-divided into three to seven narrow segments. The solitary, long-stalked flower heads, ¾ inch to 1½ inches in diameter, have each about eight coarsely-toothed, yellow ray florets and a yellow disk. A double-flowered variety is sometimes grown.

Other kinds less commonly cultivated include these: *C. californica* (syn. *Leptosyne californica*), a native of southern California often misidentified in gardens as *C. doug-*

Coreopsis bigelovii

lasii, is an annual up to 1½ feet tall. It has mostly basal, linear-threadlike leaves, sometimes few-lobed and 4 to 6 inches long. Its flower heads are 1 inch to 1½ inches across and bright yellow, with often paler tips to the ray florets. *C. calliopsidea* (syn. *Leptosyne calliopsidea*) is a native annual of California up to 2 feet in height with leafy stems. The upper leaves are pinnately-divided. The flower heads, 3 inches in diameter, are bright yellow. *C. cardaminefolia* is a hairless, branching annual, 6 inches to 2 feet in height, with once- or twice-pinnate leaves and flower heads about 1 inch in diameter that have dark purple disks and yellow ray florets with brown-purple bases. It is native from Kansas to Arizona. *C. douglasii* (syn. *Leptosyne douglasii*), about 1 foot tall, has narrow-linear or finely-pinnately-divided leaves, all basal. The flower heads, 1 inch to 1¼ inches in diameter, are bright yellow. It is a native of southern California and Arizona. *C. nuecensis* (syn. *C. coronata*) is a hairless or sparsely-hairy annual, a native of Texas, that is about 2 feet tall and has obovate or spatula-shaped, oblong leaves, the lower

Coreopsis douglasii

Coreopsis pubescens

ones sometimes three-parted. The 2-inch-wide flower heads have dark-lined, yellow ray florets. *C. palmata*, the finger tickseed, is indigenous from Indiana to Nebraska and Louisiana. About 3 feet tall, it has palmately-cleft (handlike) leaves of three deep lobes. Its flower heads, about 2½ inches across, have bright yellow, toothed ray florets and yellow disks. *C. pubescens* is often mistakenly cultivated as *C. auriculata*, a name that correctly belongs to another species. A native from Virginia to Louisiana and Florida, it is up to 4 feet in height and has oval leaves, undivided or with small side lobes. Its flower heads, with yellow, lobed rays and yellow disks, are about 2½ inches in diameter.

Garden and Landscape Uses. Coreopsises are excellent decoratives for flower gardens and cut flowers. Some, especially *C. lanceolata*, and in the west *C. maritima* and *C. gigantea*, adapt well to naturalizing in semiwild places. They mostly stand summer heat well and are appropriate for sunny locations and ordinary, well-drained garden soils. They are less satisfactory in wet, clayey earths, but *C. rosea* favors moist soil. With the exception of the sea-dahlia (*C. maritima*) and *C. gigantea*, all the perennial kinds discussed are hardy in the north. The exceptions winter satisfactorily outdoors in frost-free or nearly frost-free climates only. In addition to their values for summer displays outdoors, the annual kinds, and the sea-dahlia cultivated as an annual, are well worth growing in greenhouses for late winter and early spring flowering.

Cultivation. Selected varieties of perennial tickseeds are increased by careful division of the plants in early spring or early fall. The perennial species can also be propagated in this manner, but they are easily raised from seed sown in May in cold frames or outdoor seed beds. The seedlings, as soon as they are big enough to handle, are transplanted to nursery beds in rows about 1 foot apart, with about 6 inches allowed between the plants in the rows. By fall or the following spring they will be of a size suitable for transplanting to their flowering locations and they will bloom the following season. Spacing of 9 inches to 1½ feet, according to vigor of the kind, will prove satisfactory. Routine care of the perennial kinds is simple. An annual spring application of a complete fertilizer promotes good growth and so does regular shallow cultivation or mulching between the plants. Faded flower heads should be removed promptly, and in very dry weather periodic watering is helpful. Some kinds may need staking. In regions of cold winters protection in the form of a light covering of salt-hay or of branches of evergreens checks harm likely to result from alternate freezing and thawing.

Annual tickseeds are raised by sowing seeds in early spring, or in mild climates in fall, outdoors where the plants are to bloom. If needed for cut flowers seeding may be done in rows 1 foot to 1½ feet apart; if the plants are to form decorative patches in flower gardens the seeds are broadcast. In either case they are covered with soil to a depth of about ¼ inch. If in patches the seedlings are thinned to about 6 inches, if in rows to about 4 inches apart. Alternatively, seeds can be started indoors in a temperature of 55 to 60°F six or seven weeks before the young plants are to be set in the garden, an operation that may take place as soon as all danger of frost is passed. Until planting out time, the seedlings are grown, spaced 2 inches apart, in flats in a sunny greenhouse with a night temperature of 50°F and daytime temperatures five or ten degrees higher.

For greenhouse flowering in late winter and spring, seeds of the annuals and of the sea-dahlia are sown from September to January, and the seedlings are transferred individually to small pots and repotted into larger containers as growth demands. At their final potting they may be accommodated one in a 5-inch pot or three in a pot 6 or 7 inches in diameter. When the final pots are filled with healthy roots weekly applications of dilute liquid fertilizer are beneficial. Environmental conditions suitable for raising young plants for outdoor gardens are satisfactory for those to be bloomed indoors. Early flowering can be induced by using artificial light to lengthen the day. The sea-dahlia and *C. bigelovii* are very satisfactory as cut flowers when grown spaced 9 inches to 1 foot apart in soil beds or benches in greenhouses. They produce over a long period.

Diseases and Pests. Tickseeds are subject to leaf spots, rust, stem and root rots, and botrytis disease as well as to the virus diseases called yellows and curly top. The chief insect pests are aphids, four-lined plant bug, and the cucumber beetle.

CORETHROGYNE (Corethrogỳ-ne). Endemic to California, this group of the daisy family COMPOSITAE consists of three species of aster-like, herbaceous perennials sometimes woody toward their bases. In allusion to the brushlike appendages to their styles, the genus is named from the Greek, *korethron,* a brush made of twigs, and *gyne,* a style. Its young stems and foliage are clothed with white cottony hairs, which often fall away later. From *Aster* these plants differ in details of the hairs accompanying the ovaries and fruits (seeds).

Species of *Corethrogyne* (pronounced with five syllables) are very variable, and several varieties of two species, including the one described below, are recognized by botanists. They have alternate leaves and solitary or clustered flower heads, each with many florets. The central, disk florets are yellow, the petal-like ray florets are violet, lilac, or lilac-pink.

With stems woody and generally prostrate or semiprostrate at their bases, then becoming erect, and from 6 inches to 1½ feet tall, **C. californica** blooms in late spring or summer. Its stems may branch above. Their lower parts are thickly clothed with stalked, linear to spatula-shaped or obovate leaves, sometimes toothed at their apexes, and mostly ¾ inch to 2 inches long, but the uppermost often reduced to bracts. The dense cottony hairiness of the stems and foliage is generally retained by the lower parts of the plant, but on the upper parts it is deciduous. Solitary from the ends of the branches, the stalked flower heads are from slightly under 1 inch to 1½ inches wide. They have thirty to forty spreading, violet-purple to lilac rays, and a yellow eye.

Garden Uses and Cultivation. Adapted for sunny sites and ordinary, well-drained soils, the plant described is chiefly suited for wild gardens and naturalistic plantings in California and places with climates not too dissimilar. It is unlikely to prosper in the east, but little data are recorded regarding its hardiness. Propagation is by seed and by division.

CORIANDER. See Coriandrum.

CORIANDRUM (Corián-drum)—Coriander. The unpleasant association by the ancient Greeks of the scent of this plant with

Coriandrum sativum

bedbugs is apparently reflected in their name. It is a modification of *koriannon,* the Greek name for coriander and presumably derived from *koris,* the bedbug. There are two species, natives of the Mediterranean region. They belong in the great family of umbellifers, the carrot family UMBELLIFERAE. One kind, the coriander (*Coriandrum sativum*) is naturalized in North America, especially in the south. This is the species cultivated for its fruits (commonly called seeds), which when ripe and dry have a pleasant odor and taste in contrast to the fetid properties of the immature fruits. They are used for flavoring and seasoning and as an ingredient of curry powder.

Corianders are slender, unpleasantly strong-smelling, hairless annuals, with pinnately-finely-divided leaves, the lower ones of which are broader and toothed, and small, compound few-flowered umbels of tiny white, pink, or purplish blooms. The outer flowers of the umbels exceed the inner ones in size. The nearly spherical ribbed seeds are under ⅕ inch long.

The cultivated coriander (**C. sativum**) is 1 foot to 3 feet tall and has its upper foliage finely dissected into threadlike segments. It somewhat resembles anise (*Pimpinella*) from which it is more readily distinguished by the stink of its foliage than by its form. It blooms in summer. Its cultivation is simple. Any ordinary garden soil, not excessively rich with nitrogen, and a sunny location suit it. In soils too rich its flavor and scent are less strongly developed. Seeds are sown in spring or in mild climates in fall, in rows 1 foot to 1½ feet apart and the young plants are thinned to 3 or 4 inches in the rows. The stalks are cut after the flowers become quite brown, but before the fruits (seeds) have scattered. They are dried and the seeds, after cleaning, are stored in tightly-stoppered jars.

CORIARIA (Cori-ària). The genus *Coriaria* is the only one of the coriaria family CORIARIACEAE. It includes fifteen or more species and is native from the Mediterranean region to Japan, Central and South America, and New Zealand. Its botanical relationship uncertain, it has been considered to be allied to the ANACARDIACEAE, to which family *Rhus* belongs, and to the SIMAROUBACEAE, the family of *Ailanthus*. Its members are mostly shrubs, sometimes small trees, and rarely subshrubby herbaceous plants. The name, from the Latin *corium*, a hide or leather, alludes to some species being used for tanning.

The fruits of some coriarias are said to be eaten in their native countries but those of certain kinds are surely poisonous. In view of the scanty and conflicting evidence, it seems wise to avoid eating any parts of these plants.

Coriarias have undivided, parallel-veined leaves, chiefly opposite or whorled (in circles of more than two), but toward the ends of the shoots sometimes alternate. Generally they are disposed along the branches in a manner that gives the impression that they are the leaflets of pinnate leaves. Their inconspicuous, small, greenish flowers are in racemes from the shoot ends or from shoots of the previous season. They have five each sepals and petals, ten stamens, and a slender style. In some species male and female blooms are in separate racemes. The fruits are berrylike, the succulent part composed of the petals, which after the flowers are polli-

Coriaria japonica, in bloom

Coriaria japonica, in fruit

nated enlarge, become fleshy, and surround the five or rarely up to ten, seedlike nutlets.

The hardiest coriarias are *C. japonica*, of Japan, and *C. terminalis*, from western China. The first is a shrub, the other a herbaceous perennial, woody at its base. In sheltered places these survive outdoors with some winter protection as far north as southern New England, but generally should not be relied upon in climates much colder than that of Philadelphia. From 2 to 6 feet or sometimes taller, **C. japonica** has four-angled, pithy shoots and nearly stalkless, taper-pointed, ovate to ovate-lanceolate leaves, with three principal veins. Those on lateral branches are 1½ to 3½ inches long, but the leaves of vigorous sucker growths from the bases of the plants may be 5 inches long and proportionately broad. The unisexual racemes of flowers come along the shoots of the previous year. They are 1½ to 2½ inches long, in twos or threes. Those of male blooms are smaller than female racemes. As the fruits ripen they change from green to bright red to black.

Herbaceous **C. terminalis** spreads by underground rhizomes, and from a woody base sends up annual, four-angled shoots to a height of 2 to 4 feet. Broadly-ovate to ovate-lanceolate, and with five to nine main veins, the nearly stalkless leaves, rough-hairy on their veins beneath, are 1 inch to 3 inches long. Those of the primary shoots are much bigger than the leaves on side branches. The flowers are in terminal

Coriaria terminalis xanthocarpa

racemes 6 to 9 inches long. The black fruits, almost ½ inch in diameter, are remindful of those of pokeweed (*Phytolacca americana*). Especially lovely *C. t. xanthocarpa* has beautiful, translucent yellow fruits.

Dangerously poisonous if its fruits, foliage, and probably other parts are eaten (even goats have been poisoned by it), **C. myrtifolia**, a bushy shrub 4 to 6 feet in height, is a common inhabitant of southern Europe and other parts of the Mediterranean region. Its very short-stalked,

hairless, pointed-ovate, three-veined leaves, 1 inch to 2½ inches long, are opposite or in threes. The 1-inch-long racemes of blooms come from shoots of the previous year or from the ends of young shoots. The shining, black fruits are up to ¼ inch in diameter. The leaves of *C. myrtifolia* are used for tanning and for making ink, and its fruits, mashed in water, serve well as fly poison. Reports vary about the possible poisonous properties of other species.

Native from Mexico to Peru, **C. microphylla** (syn. *C. thymifolia*) is a pleasing shrub that because of the arrangement of its numerous small leaves has a distinctly ferny aspect. From 3 to 12 feet tall, it has lanceolate-oblong to oblong-ovate leaves up to ¾ inch long. The slender racemes, axillary and from the ends of the young shoots, are up to 5 inches long. The purplish fruits contain five to eight seeds. This, a poisonous species, has been used in Mexico to kill animals. It is also employed for making ink. A nearly related New Zealand kind is **C. kingiana** (syn. *C. thymifolia undulata*). New Zealand **C. arborea** is similar to and perhaps identical with South American *C. ruscifolia*. A shrub or small tree, it has broad-ovate leaves 1½ to 3 inches long and pendulous racemes 6 inches to 1 foot long. The fruits are dark purple.

Garden and Landscape Uses and Cultivation. As foliage and fruiting plants, the cultivated coriarias are quite handsome and add interest and variety to landscapes and shrub plantings. They prosper in full sun or part-day shade and prefer deep, loamy, moderately moist soils of reasonable fertility. Seeds afford the most satisfactory means of propagation. Summer cuttings can be rooted under mist or in a greenhouse or cold frame propagating bed. Some species can be increased by division.

CORIARIACEAE—Coriaria Family. The characteristics of this family of dicotyledons are those of its only genus, *Coriaria*.

CORK BARK. The outer bark of the cork oak (*Quercus suber*) is the cork of commerce. Highly resistant to rot, even under moist conditions, slabs of it are employed for practical and ornamental purposes. Suspended by chains or wires they may be used as baskets or rafts in or on which to grow such epiphytic plants as bromeliads and orchids. Skillfully employed to cover frameworks of metal pipe or rot-resistant woods, such as black locust or cedar, they are used in the fabrication of artificial "trees" upon which epiphytes are grown in naturalistic fashion. Window and porch boxes are sometimes faced with cork bark.

CORK TREE. See Phellodendron. The commercial source of cork is the cork oak (*Quercus suber*).

CORKSCREW FLOWER is *Vigna caracalla*.

CORKSCREW RUSH is *Juncus effusus spiralis.*

CORKWOOD. This is a common name of *Entelea arborescens* and *Leitneria floridana.*

CORM. Corms are bulblike organs that differ from bulbs in having food-storing parts that are short, stocky, solid, underground stems rather than concentric or overlapping, fleshy scales (modified leaves). They are usually enclosed in one or more membranous layers of dead leaf bases that store no food. Some sorts renew annually, the old ones dying and being replaced by new ones, and multiply by producing tiny corms called cormels. Familiar examples of corms are those of crocuses, freesias, and gladioluses.

CORMEL. Small, young corms such as develop freely around the corms of gladioluses are called cormels.

CORN. This name is commonly applied to the principal grain crop of a country or region. In North America corn means *Zea mays* and is dealt with in the next entry. In England wheat is called corn. The word is also employed in the names of some other plants. Among these are: broom-corn, chicken-corn, durra-corn or Kafir-corn (*Sorghum bicolor*), broom-corn-millet (*Panicum miliaceum*), corn cockle (*Agrostemma githago*), corn-lily (*Ixia* and *Veratrum californicum*), corn-marigold (*Chrysanthemum segetum*), dealt with in this Encyclopedia under 'Chrysanthemums, Annual', corn-plant (*Dracaena fragrans*), and squirrel-corn (*Dicentra canadensis*).

CORN. Of all vegetables sweet corn is probably America's favorite. Although better suited for largish and big plots than small ones, scarcely anyone who has a vegetable garden is willing to forego attempting to raise at least a few ears of corn. Given a sunny location and fertile, well-drained, reasonably moist soil this is easy to do in practically all parts of the United States, but in the south drought, excessive

heat, and the ravages of the corn earworm make it difficult to obtain worthwhile results in mid-summer.

Sweet corn, like field corn, is botanically *Zea mays,* of the grass family GRAMINEAE. The species is not known in the wild. The sorts now grown are descendants of kinds cultivated by Indians when Europeans discovered the New World and for untold ages before. Corn is America's greatest contribution to the food basket of the world.

Good corn is grown in a wide variety of soils, but most to its liking is a deep, clay-loam. Prepare it for planting by spreading a generous layer of manure, compost, or other partly decayed organic matter and a liberal dressing of superphosphate, and by spading, rototilling, or plowing these in to a depth of 8 or 9 inches. It is better not to add fertilizer to the drills or hills before planting, since it seems to retard germination.

Seeding must be delayed until the ground is fairly warm and the weather settled. Make the first sowing about the time apples are in bloom, in the vicinity of New York City in early May. There is no advantage in sowing prematurely. In ground too cold and wet the seeds are likely to rot. Prior to sowing, treat the seed with a commercial crow-repellent to discourage crows and starlings from taking the seeds from the ground. To assure a continuous supply of ripe ears make successional sowings of one variety every two weeks or sow an early and a midseason variety every three or four weeks. Let the last sowing be of an early variety and time it so that the ears will mature at least ten days before killing frost. Near New York City this means the end of June or a little later if an early variety is used for the last sowing. Late sowings can conveniently follow crops of such early short-season ones as spinach, lettuce, beets, peas, and early potatoes. The number of sowings can be reduced by seeding at one time several varieties carefully chosen for the various lengths of time they take to come into bearing. Early varieties are listed as taking sixty to sixty-

eight days, mid-season varieties seventy-five to eighty days, late varieties eighty-five to ninety days, but varying weather conditions can cause crops to take up to two weeks longer to reach maturity than the catalogs estimate.

Sow corn in blocks of at least three or four rows rather than in single rows. In this way cross-pollination, necessary for well-filled ears and accomplished by the

Block sowing of corn in a home garden

wind, is best promoted. The rows need not be long. For early, low varieties 2 feet between them will suffice. Taller, late varieties need 3 feet. Most gardeners sow in drills 1 inch to 2 inches deep instead of in hills. One-quarter of a pound of seed is sufficient for 1,000 feet of row. Space the seeds 3 to 4 inches apart. If sowing in hills is preferred, space them 3 feet apart and drop five or six seeds in each. Later thin the young plants to three to each hill.

When the plants are about 6 inches high pull out all surplus, leaving those that remain 10 inches to 1 foot apart. At this time apply a side dressing of a high nitrogen fertilizer and stir the surface soil shallowly with a cultivator or hoe. As the heights of the plants increase hill soil to a height of 4 to 5 inches about their bases.

Routine care consists of frequent cultivating to keep the surface soil loose and weed-free and dressing lightly every three weeks, until the ears begin to show silk, with a complete fertilizer. There is no gain and probably some disadvantage in the old practice of removing suckers from the bases of corn plants. Left, the leaves of these synthesize foodstuffs that benefit the developing of plants. Harvest about the time the silks of the ears turn brown, when the kernels are in the full milk stage, and before their skins toughen. In hot weather the ears soon pass the edible stage. For best eating the ears should be in the pot within 30 minutes or so of picking. If they must be kept longer wrap them, unshucked, in a damp cloth and keep them in a cool place.

Varieties of sweet corn are of two types,

Corn

Rows of corn sown to provide succession, the later sowing is at the right

Ornamental corn, with colored kernels

Variegated-leaved corn

open-pollinated and hybrids. The former include 'Country Gentleman', 'Early Evergreen', 'Golden Bantam', 'Golden Midget', and 'Stowell Evergreen'. Hybrid sweet corns include 'Carmel Cross', 'Evergreen', 'Golden Cross Bantam', 'Iochief', 'Span-cross', and 'Marcross'. Self-saved seeds of hybrids will not give plants true to type. Variegated-leaved corn and varieties with cobs with colored kernels are grown for ornament. For these see Zea.

CORN EARWORM. Also known as tomato fruitworm and tobacco budworm, this very destructive insect, the larva of a moth, is the most serious pest of corn, and does much damage to geraniums, globe artichokes, strawberries, sunflowers, tomatoes, and some other plants. When fully grown about 2 inches long, these larvae are yellowish, green, or brown, with lengthwise paler and darker stripes. For recommended controls consult a Cooperative Extension Agent or State Agricultural Experiment Station.

CORN SALAD or LAMB'S LETTUCE. By proper management and with little trouble, this comparatively little known, but excellent salad can be harvested in many parts of North America throughout or nearly throughout the year. It is extremely hardy. In the vicinity of New York City harvesting can be done up to the end of December from plants protected only with polyethylene plastic. In frames or green-

Corn Salad

houses kept at about 40°F at night, somewhat higher by day, cutting may continue all winter.

Corn salad, botanically *Valerianella locusta*, and native to Europe, is naturalized in parts of North America. It succeeds in any ordinary garden soil and yields succulent makings for salads in weather too hot for lettuce to prosper. For summer use sow in early spring, for fall and winter use in August. Several varieties, some with green others with bronze or reddish leaves, are obtainable from European seed firms. Fewer are available in the United States.

Prepare the soil as for most vegetable crops, and sow in shallow drills about 1 foot apart. Thin the seedlings to 6 to 8 inches apart in the rows. Beyond keeping down weeds by surface cultivation or mulching no other attention is required. In the north before killing frost cover the rows of plants in the open garden with long, low tents of polyethylene plastic film supported by simple wooden frameworks. Leave the ends of the tents open to permit free air circulation and to prevent the temperature inside from becoming too high on sunny days.

CORNACEAE—Dogwood Family. Here belong a dozen genera totaling about 100 species of deciduous and evergreen dicotyledons. Most are shrubs or trees, or more rarely, woody vines. A minority are herbaceous perennials. As a family they are most abundant in temperate and warm-temperate regions, a few sorts inhabit mountains in the tropics.

Members of the dogwood family have alternate or more frequently opposite, undivided, usually stalked leaves. Bisexual or unisexual, the flowers are in crowded clusters, umbels, panicles, or heads, those of the flowering dogwoods (sorts of *Cornus*) adorned with showy, petal-like bracts so that the whole inflorescence has a flower-like aspect. The true flowers are small and have calyxes with four or five teeth or lobes or they may be without calyxes. There are four or five petals or none and four or five stamens. The single style ends in a lobed stigma. The fruits are one- or two-seeded drupes or berries, rarely united to form a compound fruit. Cultivated genera are *Aucuba*, *Cornus*, *Corokia*, *Griselinia*, and *Helwingia*.

CORNEL. See Cornus.

CORNELIAN-CHERRY is *Cornus mas*.

CORNFLOWER is *Centaurea cyanus*. The perennial cornflower is *Centaurea montana*.

CORNISH-MONEYWORT is *Sibthorpia europaea*.

CORNUS (Cór-nus)—Dogwood, Cornel, Cornelian-Cherry. Constituting *Cornus* of the dogwood family CORNACEAE, the about forty species of this genus are, except for two herbaceous perennial kinds, deciduous or rarely evergreen trees and shrubs, nearly all natives of the northern hemisphere. Most inhabit temperate regions, a few extend the natural range of the genus into the subtropics. The name comes from the Latin one for *C. mas*. Some authorities split the group into several genera. The name dogwood is without canine significance. It is a corruption of the old name, dagwood or dagger wood applied because daggers for skewering meat were made from the wood of some sorts.

Dogwoods have generally opposite, rarely alternate, undivided, lobeless leaves, with few pairs of conspicuous lateral veins that curve inward at their tips. Those of many species have hairs that lie flat on the leaf surface, joined to it only at their centers. The flowers, in terminal clusters, with or without an involucre (collar of bracts beneath a cluster or head of flowers), are tiny. This is true even of the kinds called flowering dogwoods (a somewhat ridiculously named group because all dogwoods flower). The showy floral parts in flowering dogwoods are petal-like involucral bracts. The real, tiny flowers are in a cluster at their bases. These true flowers, as in all dogwoods, have four each sepals, petals, and stamens and one style. The fruits are small drupes (fruits constructed like plums). They are favorites of birds.

These sorts, trees with flower clusters surrounded by large, spreading, conspicuous, petal-like bracts, are of great decorative merit. They fall into two groups, those in which the berry-like fruits are in clusters, but quite separate from each other, and those in which the fruits of each cluster are united into a fleshy, strawberry-like compound fruit. Two species native to the United States belong to the first group.

Eastern flowering dogwood (*C. florida*) is one of the most handsome and beloved trees of North America. Native from Massachusetts to Florida, Ontario, Texas, and Mexico, and usually 15 to 20 feet tall, it occasionally attains nearly twice that maximum. This sort has elliptic to ovate, conspicuously-veined, sparsely-hairy or hair-

Cornus florida

less, pointed leaves 3 to 6 inches long, glaucous on their undersides. In fall they assume attractive, rather somber tones of reddish-purple. The little greenish-white to yellowish flowers are in clusters encircled with white to pinkish, or in horticultural varieties, white, pink, or red, broadly-obovate, petal-like bracts indented at their apexes and up to 2 inches long. The ellipsoid fruits, in close clusters, are berry-like and bright red. Varieties are *C. f. pendula*, with rigid weeping or pendulous branches; *C. f. pluribracteata*, the flower clusters of which are encircled by more than the usual four bracts so that they look like double flowers; *C. f. rubra*, with red or pink bracts, its flower buds less hardy to cold than those of the typical species; *C. f. welchii*, its foliage beautifully variegated with green, creamy-white, and pink, but a sparse bloomer; and *C. f. xanthocarpa*, which has yellow fruits. More recent horticultural selections are 'Apple Blossom', with soft pink and white bracts, 'Cherokee Chief', its bracts rich ruby-red; 'Cloud Nine' which blooms when young and in southern California tolerates lack of winter cold and hot summers better than other varieties; 'New Hampshire', reported to have flower buds that withstand lower

temperatures than usual; 'Spring Song', with rose-red bracts; 'Sweetwater Red', with red bracts and reddish foliage; and 'White Cloud', which is exceptionally free flowering.

Western flowering dogwood (*C. nuttallii*) is one of the floral splendors of North America. Unfortunately it does not prosper in the East, although a specimen in the garden of Mr. Harold Epstein, Larchmont, New York has lived for many years and sometimes blooms. Similar in gross aspect to the Eastern flowering dogwood, *C. nuttallii* differs in several well-marked details. Up to 75 feet tall and with branches less obviously tiered than those of its counterpart, this sort occurs in the wild from British Columbia to southern California. It has short-stalked, elliptic to ovate or obovate, briefly-pointed leaves 3½ to 4½ inches long. They have five or six pairs of veins and are somewhat hairy, more so when young. Their undersides are slightly glaucous. In fall they color beautifully with tones of yellow and red. The clusters of little flowers are encircled with usually six, less often five or four, generally pointed, spreading, obovate to oblong white or pink-tinged, petal-like bracts 2 to 3 inches long or longer. They do not enclose the

Cornus florida, in bloom

Cornus florida, showing central clusters of tiny flowers

Cornus florida welchii

Cornus florida, showing clusters of flowers encircled by petal-like bracts

Cornus florida (fruits)

Cornus florida welchii (foliage)

Cornus nuttallii

Cornus kousa, showing clusters of flowers surrounded by petal-like bracts

Cornus mas

flower buds through the winter. The fruits, those of each cluster separate, are bright red or orange, a little over ⅓ inch long. The inflorescences (flower heads) of *C. n.* 'Colrigo Giant' are exceptionally large. Variety *C. n. eddiei* has white-edged leaves. A hybrid between the Eastern and Western flowering dogwoods named 'Eddie's White Wonder' offers considerable promise and is apparently much hardier than *C. nuttallii*, but less so than *C. florida*.

Kousa dogwood (*C. kousa*), of Japan and Korea, unlike American flowering dogwoods blooms in summer after its foliage is well out. It differs also in that the fruits of each cluster are united into pinkish, fleshy aggregates. Hardy into southern New England, this beautiful, somewhat variable species attains heights up to slightly in excess of 20 feet. Elliptic-ovate and pointed, its leaves are up to 4½ inches long. Their upper sides are dark green, the lower have somewhat hairy surfaces and tufts of brown hairs in the axils of the main veins. In fall they turn brilliant red. The flower heads are much like those of American flowering dogwoods. The little blooms, closely grouped in stalked, spherical heads, are encircled by four pointed, creamy-white or sometimes slightly pink-tinged elliptic-ovate to oblong-ovate bracts 1 inch to 2 inches long. Variety *C. k. chinensis*, described from material collected in China, does not appear to be botanically very dis-

Cornus kousa (fruits)

Cornus mas (flowers)

tinct from the typical species. It has flower heads with bracts 2 to 3½ inches long. Similar variants occur in Japan as well as China. Variety 'Milky Way' is said to bloom unusually freely.

An evergreen or partially evergreen flowering dogwood, *C. capitata* is native from China to the Himalayas. Not hardy in the north, where it grows satisfactorily it makes a magnificent specimen up to 40 feet tall and often much wider than tall. This sort has leathery, elliptic-oblong leaves 2 to 4 inches in length with hairs lying parallel with the leaf surfaces. The clusters of small flowers have each four to six ovate, creamy-white to sulfur-yellow petal-like bracts 1½ to 2 inches long. The red fruits coalesce into strawberry-like heads 1 inch to 1½ inches in diameter.

Cornelian-cherry (*C. mas*), of central and southern Europe and western Asia, and very similar *C. officinalis*, of Japan and Korea, are admired for their attractive flowers borne in abundance in late winter or earliest spring. They are deciduous shrubs or trees up to about 25 feet high, or the Japanese sometimes rather taller. Un-

Cornus mas (fruits)

Cornus kousa

Cornus officinalis

Cornus officinalis (flowers)

Cornus macrophylla

Cornus officinalis, flower cluster

like those of flowering dogwoods, the short-stalked flower clusters have very small, not showy, yellowish bracts that drop as the flowers mature. The chief differences between the Cornelian-cherry and its Japanese twin are that the leaves of the former are without the conspicuous tufts of hairs in the axils of the veins on the undersides of the leaves that are characteristic of C. officinalis and that its individual flowers have stalks not appreciably longer instead of twice as long as the bracts of the involucre. Both kinds have short-stalked ovate to elliptic, pointed leaves up to 4 inches long, with three to five pairs of lateral veins and hairs on both surfaces. The flowers, which appear in abundance well before the foliage, are bright yellow. (The pleasantly-flavored scarlet fruits from a little under to slightly over ½ inch long are ellipsoid.) Variety C. m. alba has white fruits, C. m. aurea has yellowish foliage, C. m. flava has yellow fruits, C. m. nana is a dwarf variety, and C. m. variegata has leaves edged with white.

Quite different C. macrophylla, of China and the Himalayas, is a tree up to about 50 feet tall, with flowers and fruits more nearly like those of the shrubby kinds than those of the other tree kinds described

here. Hardy perhaps as far north as New York City, it has long-pointed, elliptic-ovate to elliptic-oblong leaves 4 to 6 inches in length, glaucous and slightly hairy beneath, with nearly hairless upper surfaces. In conspicuous panicle-like clusters 3½ to nearly 6 inches across, the yellowish-white flowers are succeeded by spherical, bluish-black fruits.

The only dogwoods with alternate leaves are **C. alternifolia,** of North America, and similar, fast-growing **C. controversa,** of Japan and China. The first is 10 to 25 feet tall, the other, under favorable conditions a rapid grower, attains heights of 40 to 60 feet. Both have branches that spread horizontally in distinct tiers. Mostly toward the ends of the shoots the slender-stalked leaves of the American species are elliptic-ovate, 2¼ to 4½ inches long, have five or six pairs of lateral veins. Their undersides, somewhat glaucous, have irregularly-diverging hairs, parallel with the surface. Of small decorative appeal, the white flowers, in clusters 2 to 2½ inches wide, are succeeded by glaucous, blue-black, red-stalked, berry-like fruits about ¼ inch in diameter. The leaves of C. controversa, 3 to 6 inches long, are similar except that they have six to nine pairs of lateral veins and the hairs of the undersides of the leaves are straight and parallel. The clusters of white flowers are 3 to 6 inches wide. Bluish-black, the ¼-inch-wide fruits are appreciated by birds. Variety C. c. variegata has white-edged leaves.

Colored-stemmed shrubs include a number of worthwhile dogwoods. Especially noteworthy are C. alba and C. sericea and their varieties. Native from Siberia to Manchuria and Korea and extremely hardy, **C. alba** is erect and up to 10 feet tall and characteristically has blood-red stems. Whitish on their undersides, its ovate to elliptic leaves are 1½ to 3½ inches long. They have five or six pairs of lateral veins. They turn red in fall. The yellowish-white flowers, in clusters 1½ to 2 inches across, are succeeded by sometimes slightly blue-tinged, white fruits containing pointed-

ended stones decidedly longer than broad. Siberian dogwood, brilliant coral-red barked C. a. sibirica, is one of the very best colored-stemmed hardy shrubs. Similar, if not identical is variety 'Coral Beauty'. Black-purple stems distinguish C. a. kesselringii. Varieties with variegated foliage are C. a. argenteo-marginata, a splendid shrub with white-edged leaves, C. a. spaethii, the leaves of which are margined with yellow; and C. a. gouchaultii, similar to the last but with leaves variegated yellowish-white and pink. Red osier dogwood (**C. sericea** syn. C. stolonifera), native from Newfoundland to Manitoba, Nebraska, Virginia, and Kentucky, does not differ greatly from C. alba, but it spreads more vigorously by suckers. Identifying characteristics are that the stones in its fruits are not longer than broad and have rounded bases. Its leaves, similar to those of C. alba, become reddish in fall. Typically the younger stems are brilliant red. Those of C. s. flaviramea are bright yellow, those of C. s. nitida an attractive green. Rarely more than 2 feet tall, C. s. kelseyi is compact and freely-branched.

Blood-twig dogwood (**C. sanguinea**) has less brilliant red stems than those of C. alba and C. sericea. A native of Europe, up to about 12 feet tall, it has broad-elliptic to ovate leaves, hairy on both surfaces, with three to five pairs of veins, and up to 3½

Cornus sericea

Cornus sericea flaviramea

Cornus sericea kelseyi

Cornus racemosa

Cornus canadensis

Cornus racemosa flowers and foliage

Cornus canadensis (fruits)

inches long. In fall they turn rich red. In clusters 1½ to 2 inches wide, the dullish white flowers are followed by purple-black fruits. The foliage of *C. s. variegata* is marked with yellowish-white, the stems of *C. s. atrosanguinea* are deep red, and those of *C. s. viridissima* are green, as are its fruits.

Other shrub dogwoods include *C. amomum,* a 3- to 10-foot-tall native of moist and wet soils from southern Canada to Illinois, South Carolina, and Alabama. This much resembles *C. alba* and *C. sericea,* but its red stems are less brilliant. Ovate to broad-elliptic and 2 to 4 inches long, the leaves, with four to seven pairs of veins, are reddish-hairy on the veins beneath. In fall they turn red. The yellowish-white flowers, in clusters 1½ to 2½ inches wide, are succeeded by blue to light blue or grayish-blue fruits. Gray dogwood (*C. racemosa*), of eastern North America, is without brightly colored stems. Up to about 15 feet in height and dense and bushy, this has lanceolate to elliptic leaves not more than one-half as wide as long, with three or four pairs of veins. Its flat clusters of creamy-white blooms are 1½ to 2½ inches wide. Red-stalked, the fruits change as they ripen from a leaden hue to white. Its provenance uncertain, *C. hessei,* probably a native of northeast Asia, rarely is very

compact, or may be more than 1½ feet tall. It has crowded, slender-pointed, narrowly-elliptic to lanceolate leaves 1½ to 3 inches long and about one-third as wide. Its flowers and bluish-white fruits are in clusters ¾ to little more than 1 inch across. This slow-growing, dark-foliaged kind is hardy in New England. Chinese *C. paucinervis,* up to 5 to 10 feet tall, has narrow-elliptic leaves up to about 4 inches long, and where winters are not too severe, it is partially evergreen. The white flowers, in clusters 2½ to 3½ inches wide, are succeeded by black fruits. This kind is hardy in southern New England.

Herbaceous perennial dogwoods, of which there are only two, are forest floor plants, their heights measured in inches rather than feet. Despite their lowly statures and the fact that their stems die to the ground each winter, the most casual observer is likely to recognize them as close relatives of the tree-type flowering dogwoods. Their tight clusters of tiny flowers are decorated by four spreading, white, petal-like bracts. Their spherical fruits, separate but clustered, are bright red. Best known is the bunchberry (*C. canadensis*), a native of moist, acid woodlands and bogs from Greenland to Alaska, New Jersey, in the mountains to West Virginia, to Minnesota and California, and in eastern Asia. Bunchberry forms mats of spreading rhizomes from which arise erect stems 4 to 8

inches tall bearing distant pairs of miniature leaves and at their top an apparent whorl (circle of more than two) of four or six pointed, lanceolate to oblanceolate or ovate leaves 1½ to 3 inches long, with two or three pairs of lateral veins starting from the base to almost the middle of the leaf. From the center of the leaf cluster is borne on a slender, erect stalk a solitary flower cluster the bracts of which are up to ¾ inch long. Very like bunchberry, but not exceeding 6 inches in height and with a few pairs of well-developed leaves below the terminal cluster, and often with branches coming from the axils of these, *C. suecica* inhabits wet woodlands and rocky places of northern Europe, northern Asia, Greenland, and North America from eastern Canada to Alaska. From the bunchberry this differs in the usually two pairs of lateral veins of the leaves arising from very close to the base of the leaf. The white bracts surrounding its flower clusters are under ½ inch long.

Garden and Landscape Uses. According to kind, dogwoods serve splendidly many different garden and landscape employments. The bunchberry and its herbaceous perennial relative belong in carefully cho-

Cornus amomum

sen sites in native plant gardens and rock gardens. Tree flowering dogwoods are magnificent as lawn specimens, at the fronts of woodlands, and for sprinkling through open woodlands, but always remember that much shade seriously diminishes the abundance of bloom. The Cornelian-cherry and its Asian counterpart are splendid reminders that spring is at hand. They are seen at their best against backgrounds of evergreens and can be usefully placed as lawn specimens or as landscape groups. They do well under city conditions. In winter, especially when the landscape is blanketed with snow, the colored-stemmed dogwoods show to splendid advantage. They are especially suitable for stream banks, the shores of lakes, and the margins of ponds. They hold the soil and in favorable light reflect beautifully in the water. They can also be used to good effect in mixed shrub plantings where the more or less suckering growth of some kinds is not objectionable. Other dogwoods can be usefully used similarly and in general landscaping.

Choice and often tantalizing to grow, bunchberry and frequently less-accommodating *C. suecica* are most likely to succeed in strongly acid, moist, peaty soil in partial shade. They can be raised from seeds freed from surrounding pulp, sown as soon as possible after they ripen in a cold frame or protected bed outdoors. They can also be increased by cutting and transplanting large clumps or sods from well-established colonies. Needless to say these should not be taken from wild stands where the plants are not abundant.

Cultivation. The majority of dogwoods are not troublesome. The tree types ask no more than a fairly good soil not excessively dry, and with reasonable organic content. They flourish in sun or where they get just a little shade. No pruning is ordinarily needed. Lawn specimens benefit from an annual late winter or early spring application of a complete garden fertilizer and from deep watering at seven- to ten-day intervals in dry weather. It is advisable to keep a ring a foot or two wide free of grass around the base of the trunk.

Shrub dogwoods respond to average soils not excessively dry. They are best in full sun, but tolerate part-shade. Those admired for their colored stems need regular pruning to encourage strong new shoots, which are the most brilliantly colored. The procedure is to cut out one-half to two-thirds of the older stems in early fall. Old, rankly overgrown specimens can be restored by cutting them in spring to almost ground level and applying a generous dressing of fertilizer. If the ground tends to be infertile, a spring application of fertilizer will help develop an abundance of vigorous shoots.

Pests and Diseases. The eastern flowering dogwood is subject to invasion by borers. Newly planted specimens and trees in poor health are especially susceptible. Because of this, the trunks of newly transplanted specimens should be wrapped (see Tree Wrapping). Other pests that affect dogwoods include several kinds of borers and a minute midge that causes club-shaped galls or swellings on the twigs. Pruning out in summer and burning infested twigs is recommended to control the midge. Scale insects infest dogwoods and so does a species of leaf miner. The chief diseases of these trees and shrubs are crown canker, fungus flower and leaf and twig blights, leaf spots, and powdery mildew.

CORNUTIA (Cor-nùtia). Honoring Jacques Cornutus, French physician and traveler, *Cornutia* of the vervain family VERBENACEAE consists of perhaps one dozen species of evergreen shrubs and trees of tropical America and the West Indies. They have usually four-angled branches and opposite leaves. Their flowers are in terminal clusters. Each has a sometimes slightly curved, cylindrical corolla tube and four spreading corolla lobes. There are two functional stamens and two imperfect ones called staminodes. The fruits are small berries.

Native to Mexico and Central America, *C. grandifolia* is a shrub or tree up to about 25 feet tall with densely-hairy twigs and pointed, elliptic to elliptic-ovate, hairy leaves that sometimes exceed 1 foot in length, and are often toothed above their middles. The blue or purple ½-inch-wide flowers are in pyramidal panicles 6 inches to 1½ feet long. West Indian *C. pyramidata,* a shrub commonly 3 to 12 feet tall, or sometimes taller, in its homelands frequently inhabits limestone soils. It has broadly-ovate, pointed, toothed leaves 3 to 4 inches long that are softly-hairy above and more heavily hairy on their under surfaces. The slightly two-lipped flowers are crowded in showy panicles up to 1 foot, but mostly less, long. They are violet-blue with yellow eyes. Each flower is about ½ inch across.

Garden and Landscape Uses and Cultivation. These are attractive ornamentals for warm, frost free-climates and for cultivating in greenhouses and conservatories where a minimum winter night temperature of 55°F is maintained. A moderately humid atmosphere and fertile, well-drained, soil agree with these plants. Any needed pruning should be done in spring. Container-grown specimens benefit from dilute fertilizer applied from spring to fall. Cuttings and seeds provide ready means of propagation.

COROKIA (Cor-òkia). Botany sometimes brings together strange bedfellows. This is certainly true of the dogwood family CORNACEAE, which contains not only the well-known American and Oriental flowering dogwoods and other members of the genus *Cornus,* but also about six species of curious evergreen shrubs and trees, of New Zealand, the Chatham Islands, and the South Pacific island of Rapa, of the genus *Corokia.* The name is the Maori one.

Corokias have alternate leaves and usually clustered, small, starry flowers, with five-lobed calyxes, five petals, and five stamens. These are succeeded by small fruits, one-seeded and with the persistent calyx clinging to the apex. Cultivated kinds, all except *C. macrocarpa* natives of New Zealand, have the undersides of their leaves clothed with white hairs.

The most familiar sort is *C. cotoneaster,* which forms a dense bush, up to 10 feet tall, of numerous rigid, twisted, tortuously-interlaced, black branches and alternate or clustered, roundish-ovate to obovate leaves up to 1 inch long, dark green above, and with flattened stalks. Its flowers, silky-hairy on the outsides of the petals and about ½ inch wide, are solitary or in clusters toward the ends of short twigs. The red fruits are about ⅓ inch long. Differing from the last in having linear-lanceolate leaves, lustrous above, and 1½ to 6 inches long, and in slender, terminal panicles 1 inch to 2 inches long, ½-

Cornutia pyramidata

Corokia cotoneaster

inch-wide yellow flowers, with petals downy on their outsides, *C. buddleioides* is up to 12 feet tall. It has gray-felted shoots and orange-red fruits about ⅓ inch long. Native to the Chatham Islands, *C. macrocarpa* grows up to 20 feet tall. It has narrowly-ovate to oblong-lanceolate leaves, tapered at both ends, 2 to 4 inches long and up to 1 inch wide. The flowers up to ½ inch wide are in axillary clusters of three to eight. The fruits are egg-shaped, orange-yellow, and about ¼ inch long.

A natural hybrid between *C. buddleioides* and *C. cotoneaster*, beautiful *C. virgata* is 6 to 10 feet tall or sometimes taller. Much branched, it has slightly zigzagged, but not tortuously-intertwined shoots, when young clothed with white down. Its spatula-shaped leaves are ¼ inch to 1¾ inches long, have dark, glossy green upper surfaces, and are covered with white down beneath. The yellow flowers, borne with great freedom in clusters of three near the ends of the shoots and almost ½ inch wide, are succeeded by egg-shaped, ¼-inch-long, orange-yellow fruits. The kind named *C. cheesemanii* is also a natural hybrid between *C. buddleioides* and *C. cotoneaster*. It is 6 to 12 feet tall, and has white-felted shoots, oblanceolate, obovate, or elliptic-oblong leaves ¾ to 2 inches long, with flattened stalks, and terminal and axillary clusters of two to four blooms. The fruits are oblongish, orange-red, and somewhat under ½ inch long. The branches are not twisted or interlaced.

Garden and Landscape Uses and Cultivation. Corokias do not survive low temperatures. They are cultivated outdoors in the milder parts of Great Britain, in California, and elsewhere where winters are not severe and in pots in greenhouses where winter night temperatures are 40 to 50°F. They need well-drained, fertile soil and sunny locations and are increased by summer cuttings and seeds. Any pruning needed to keep them shapely or to restrict their size should be done immediately after flowering.

COROLLA. The corolla is the part of the flower consisting of the petals. These are generally the second set of floral organs from the outside, or the first, in rare instances when the calyx is lacking. Commonly white or brightly colored, they frequently serve to attract insects and birds useful in effecting pollination. The petals may be separate or joined for a greater or lesser part of their lengths into a corolla tube. In a few instances, as in *Pectinaria*, they are united at their apexes. The corolla and calyx together form the perianth. When the perianth parts (the sepals and petals) are similar they are correctly identified as tepals, but if they are petal-like in appearance, as they are in tulips and lilies, for example, they are commonly, but somewhat inaccurately called petals.

CORONILLA (Coron-ílla)—Crown-Vetch. Annual and perennial herbaceous plants and low shrubs are the constituents of this genus of the pea family LEGUMINOSAE. Few of the approximately twenty species are cultivated. The group is native to Europe, the Mediterranean region, western Asia, and the Canary Islands. Its name is a diminutive of the Latin *corona*, a crown. It alludes to the flower clusters.

Coronillas have usually pinnate leaves with an uneven number of toothless leaflets; rarely they have three leaflets or one. The long-stalked, headlike clusters of yellow, purplish, or white, pealike flowers arise from the leaf axils. The blooms have bell-shaped, usually slightly two-lipped calyxes and ten stamens, nine of which are united and one separate. The fruits are cylindrical or longitudinally-angled or ridged pods without constrictions between the seeds.

Crown-vetch (*Coronilla varia*) is a coarse, vigorous, herbaceous perennial 9 inches to 2 feet tall or taller, with more or less prostrate or ascending stems and leaves with up to twelve pairs and a terminal one of oblong to obovate leaflets. On stalks considerably longer than the leaves the heads of pink to purple or white flowers, the individual blooms approximately ½ inch long, are borne in summer. The four-angled pods are about 2 inches long. Native to Europe and Asia, and naturalized

Coronilla varia

Coronilla varia (flowers)

in North America, this species is cultivated as fodder. It is hardy at least as far north as Maine.

Another herbaceous perennial, *C. cappadocica,* of Asia Minor, is considerably more refined in appearance. Its minutely-white-hairy, erect stems are about 1 foot tall. The foliage is glaucous. Each leaf has nine to eleven wedge-shaped-ovate leaflets ½ to ¾ inch in length. The bright yellow blooms, about ½ inch long, in heads of six to nine atop long stalks, are succeeded by down-pointing, four-angled pods, beaked at their ends. This sort is probably much less hardy than crown-vetch.

Coronilla cappadocica

Annual *C. cretica* attains a maximum height of about 3 feet. Its leaves have seven to seventeen obovate-oblong leaflets ¼ to ¾ inch long. The flower heads, of three to nine white or pink, ¼-inch-long blooms are followed by slender, beaked, four-angled pods 1¼ to 3 inches long. This species inhabits southern Europe from Italy eastward.

Scorpion-senna (*C. emerus*), a dense shrub up to 7 feet in height, is hardy about as far north as southern New England. In mild climates it is evergreen; toward the northern limits of its tolerance it loses its foliage in winter. The leaves are glaucous and have five to nine obovate leaflets ½ to ¾

Coronilla emerus

inch long. On stalks about as long as the leaves, the pale yellow flowers, tipped with red, in clusters of up to five, are borne in spring and summer. The obtusely-angled, cylindrical seed pods are 2 to 4 inches long. This native of Europe occurs as far north as southern Norway. Variety *C. e. emeroides* (syn. *C. emeroides*) has flower stalks longer than the leaves, and up to eight blooms to each head. It is less hardy than *C. emerus.*

Variable *C. valentina,* of the Mediterranean region, a hairless shrub not reliably hardy in the north, is 2 to 4 feet tall. Its leaves have seven to fifteen obovate leaflets, notched at their ends, and up to ¾ inch long. The heads of up to twelve fragrant, yellow blooms, each flower almost ½ inch long, are on stalks generally longer than the leaves. The somewhat flattened seed pods have two blunt angles. In *C. v. glauca* (syn. *C. glauca*) the leaves have five to seven ovate or lanceolate, thinnish leaflets. A variant has leaves prettily variegated with yellow.

Very different, *C. juncea,* a shrub of the western Mediterranean region, is 1 foot to 3 feet in height. It has gray-green, rushlike stems that bear leaves with three to seven fleshy, linear to oblong leaflets up to 1 inch long. The leaves soon fall. The heads of

Coronilla valentina

Coronilla valentina variegata

blooms consist of five to twelve flowers ¼ to ½ inch long. The bluntly-four-angled seed pods are up to 2 inches long. It is not reliably hardy in the north.

A shrub usually under 1 foot in height, *C. minima* inhabits dry, open places in western Europe. Procumbent, it has leaves with five to thirteen elliptic to obovate or nearly round leaflets up to ½ inch in length. The yellow flowers, about ⅓ inch long, are in heads of up to ten or rarely fifteen. Less than 1½ inches long, the seed pods are four-angled. This sort is not believed to be hardy in the north.

Garden and Landscape Uses. The crown-vetch (*C. varia*) has limited uses for the fronts of flower borders, rough rock gardens, and as a groundcover for poor soils. It also serves to control erosion on banks and slopes. It has been much advertised as a solution to problems connected with such sites and has been planted along highways for soil stabilization. It has the disadvantages of not being evergreen and of being too coarse for refined garden areas. For the fronts of flower borders and in rock gardens, *C. cappadocica* is suitable. Annual *C. cretica* can be employed to add variety to flower borders. All the above kinds need full sun and prosper in poorish, well-drained, ordinary soils including those of a limestone nature. Shrubby coronillas are attractive and are suitable for shrub borders and similar plantings, and the lower ones for rock gardens. They grow readily in sunny locations in well-drained soils.

Cultivation. The crown-vetch and other herbaceous perennial coronillas are usually raised from seed, and so, of course, is annual *C. cretica.* Propagation of shrubby kinds is by seeds and cuttings. The latter may be made from soft new shoots and planted in a greenhouse propagating bed with bottom heat, or from firmer shoots taken in late summer and rooted under mist or in a propagating bed in a cold frame or greenhouse.

Routine care of coronillas is minimal. In fall tidiness is promoted if the old stems of herbaceous perennials are cut at about ground level. Shrubby kinds are pruned to shape and the shoots thinned out, if needed to prevent excessive crowding, in late winter.

COROZO (Corò-zo)—American Oil Palm. As a vernacular name corozo is used for a number of different palms including *Acrocomia media* and *Gastrococos crispa,* but the botanical genus *Corozo* consists of only one species, the American oil palm. This native of tropical Central and South America is a member of the palm family PALMAE. Common in Panama, it always occupies swampy, shady places. The generic name is the native one of the plant.

The American oil palm is similar in general appearance to the African oil palm (*Elaeis guineensis*), but is much lower and

smaller and differs in botanical details. As a producer of commercial oil it is of little importance. Even in the American tropics the palm planted for oil is the African kind. Nevertheless, the fruits of the American species contain an oil very similar to that of the African oil palm. In early colonial days it was used by Europeans to make candles. Indian women in some parts of its native range use it to dress their hair.

The American oil palm (*C. oleifera*) has a trunk with its ascending portion up to 6 feet tall and usually its lower end reclining like a log on the ground, emitting roots and often curved. Its pinnate leaves, up to 24 feet long, consist of numerous slender, drooping leaflets. Their stalks are spiny and spread at all angles like great plumes to form a fountain-like crown. The flower clusters, dense and crowded, are borne close to the crown of the tree. The red fruits are about 1 inch long.

Garden and Landscape Uses and Cultivation. The American oil palm is an attractive ornamental, but is little known in cultivation. It has been successfully grown in southern Florida and would undoubtedly succeed in humid tropical climates in ordinary garden soil, in part-shade, and in tropical conservatories. Its seeds germinate if sown while fresh in sandy, peaty soil in a temperature of 80 to 90°F. For additional information see Palms.

CORPUSCULARIA. See Delosperma.

CORREA (Cor-rèa)—Australian-Fuchsia. As now interpreted, seven to twelve species of Australian and Tasmanian nonhardy evergreen shrubs of the rue family RUTACEAE constitute *Correa,* the name of which commemorates the Portuguese botanist Jose Francesco Correa da Serra, who died in 1823. But the genus is poorly understood botanically, and further study may lead to the acceptance of more or fewer species.

Correas have opposite, undivided, toothless leaves with small translucent dots, and like the stems commonly furnished with stellate (star-shaped) hairs. The usually pendulous flowers have generally four-lobed or toothed calyxes, four petals that in some species are united into a tube, in others separate to their bases, eight stamens in two groups of four, and a style with four short lobes.

Commonly cultivated are *C. reflexa* (syn. *C. speciosa*), its varieties and hybrids. From 3 to 8 feet tall, and furnished with rusty hairs, this species has slender branches and short-stalked, or almost stalkless ovate to ovate-oblong, leathery leaves, ½ inch to 1¾ inch long by up to ¾ inch wide, dark green above, and densely covered underneath with whitish hairs. The flowers, solitary or in twos or threes, at the branchlet ends, are about 1 inch long and ¼ inch wide. Their stamens protrude. They vary in color from red through yellowish-green to white. A hybrid between *C. reflexa* and *C. pulchella,* named **C. harrisii,** has tubular, 1-inch-long scarlet flowers with protruding yellow stamens. Another hybrid or varietal selection, **C. magnifica,** is remarkable for its chartreuse flowers. Other horticultural varieties, including one named 'Silver Bells', are cultivated.

Correa harrisii

Previously identified as a variety of *C. reflexa,* Tasmanian **C. backhousiana** (syns. *C. ferruginea, C. splendida*) has been given status as a distinct species. From 3 to 10 feet tall, it has slender stems, and ovate to broad-elliptic leaves ½ to 1½ inches long, whitish-hairy on their undersides. The 1-inch-long flowers, solitary or in clusters of usually not more than three, have cup-shaped, scarcely-toothed calyxes, and pale citron-green corollas shading to tawny russet toward their tips. The anthers are cream. From the above, **C. pulchella** (syn. *C. neglecta*), considered by some botanists to be a variety of *C. reflexa,* differs in having rosy-red blooms. Native of Australia and Tasmania, **C. alba** is rather rigid about 4 feet tall and has its shoots and the undersides of its foliage clothed with soft hairs. The leaves are ovate to obovate and up to 1½ inches long. The flowers, ¾ to

Correa backhousiana

1 inch long, and from one to three at the branchlet ends, are white to pink.

Garden and Landscape Uses and Cultivation. Correas are elegant and charming shrubs for general purpose use in California-type climates and are attractive as greenhouse plants. They grow willingly in fertile, sandy peaty soil that is well drained, but not excessively dry.

As greenhouse plants they succeed under conditions that suit acacias and many other woody plants of Australia. Excessive warmth from fall to spring is harmful. Then, a night temperature of 45 to 50°F is adequate, with five to ten degrees increase by day. Whenever the weather is favorable the greenhouse should be ventilated freely. Highly humid atmospheres are not to the liking of correas. During the summer they are best stood outdoors with their containers buried nearly to their rims in a bed of sand or similar material. Pruning to shape, and any repotting needed, is done after flowering. Well-rooted specimens benefit from applications of dilute liquid fertilizer during the summer growing season. Propagation is easy by summer cuttings in a greenhouse propagating bed or under mist. When procurable, seed can be used for propagation.

CORRYOCACTUS (Corryo-cáctus). Twenty-one or according to some interpretations considerably fewer species of cactuses of the cactus family CACTACEAE, belong here. Native to Peru, Bolivia, and Chile, *Corryocactus* has a name based on that of a Mr. T. A. Corry, chief engineer of a Peruvian railroad, who facilitated botanical explorations of his region, and the word cactus.

Corryocactuses have strongly-ribbed, thick and columnar to slender, cylindrical stems, usually branched mostly from their bases, upright or procumbent, and very spiny. Bell-shaped, and with short calyx tubes that have scales with felt and spines in their axils, the yellowish or reddish blooms have wide-spreading perianth segments (petals). The fleshy fruits are spherical and have clusters of deciduous spines.

Freely-branching *C. brachypetalus* (syn. *Cereus brachypetalus*) and *C. brevistylus* (syn. *Cereus brevistylus*) are natives of Peru. Attaining heights of 6 to 12 feet, **C. brachypetalus** has numerous massive, rigidly erect stems with usually seven or eight ribs and clusters of about twenty spines of markedly unequal size, the biggest up to 6 inches long, most much smaller, spaced about ¾ inch apart. The deep orange, broadly-funnel-shaped flowers are 1½ to 2½ inches in diameter. Very prominently six- or seven-ribbed, the stiffly erect, pale green stems of **C. brevistylus** are 6 to 10 feet tall and less massive than those of *C. brachypetalus.* They have clusters of fifteen spines of varied lengths, the biggest up to 10 inches long, spaced 1½ inches apart. The flowers are bright yellow and 4 inches

Corryocactus melanotrichus

wide. The stems of **C. melanotrichus,** of Bolivia, are up to 1½ inches in diameter and have seven or eight shallow ribs with clusters, ¾ inch apart, of twelve spines. Each cluster consists of one to three central spines up to 2 inches long and others approximately ½ inch long. The pink blooms are 2¾ inches in diameter.

Garden and Landscape Uses and Cultivation. Not common in cultivation, these cactuses are suitable for collections and for growing outdoors in warm, semidesert regions and in greenhouses. Their needs are those of most cactuses of their growth habits. The first discussed above are reported to be good understocks upon which to graft other cactuses. For additional information see Cactuses.

CORSICAN MINT is *Mentha requienii.*

CORSICAN SANDWORT is *Arenaria balearica.*

CORTADERIA (Corta-dèria) — Pampas Grass. It is a pity these noble grasses are not hardy in the north. In milder climates, such as those of California and the southwest, they provide elegant ornamentation for gardens and their great flower plumes, fresh or dried, are handsome for indoor decorations. Dried, they are offered for sale by florists, usually after they have been dyed a variety of bright analine hues.

Belonging in the grass family GRAMINEAE, the genus *Cortaderia,* of South America and New Zealand, consists of two dozen species of perennials. They form massive clumps, and individual plants are usually male or female, occasionally bisexual. At one time they were included in the related genus *Gynerium* and in gardens are still sometimes called by that name. From *Gynerium* our present genus differs in that its leaves are crowded at and near the bases of the flower stalks, whereas those of *Gynerium* are spaced along them. The name *Cortaderia,* a native Argentinian one, means cutting, and refers to the rough-edged leaves.

The pampas grass (*C. selloana,* syns. *C. argentea, Gynerium argenteum*), a native of Argentina, forms great clumps of long, narrow, recurved, grayish leaves, with rough margins, and tufts of hairs in the throats of their sheaths. The flower stalks spray skyward to form a magnificent fountain of large fluffy flower plumes, those of female plants with white or pinkish-silky hairs, the males without hairs. They are up to 3 feet long and on especially vigorous plants may attain a height of 20 feet; more usually, in bloom they are 6 to 12 feet tall. This species is naturalized in parts of Cal-

Cortaderia selloana

Cortaderia selloana, plumes of bloom

Cortaderia selloana pumila

ifornia. Variety *C. s. pumila* is more compact.

Other sorts cultivated include *C. jubata,* often wrongly named *C. quila,* of South America, which is represented in cultivation by female plants only. Much like *C. selloana,* it is distinguishable by its looser, purplish or yellowish flower panicles of rather smaller spikelets. The plant grown as *C. rudiuscula* is very like *C. jubata.* The species of Chile and Peru to which the name *C. rudiuscula* rightly belongs differs in details of its flowering parts, notably in the flowers being narrower. New Zealand *C. richardii* (syn. *C. conspicua*) has branchless stems up to 10 feet tall.

Garden and Landscape Uses. These grasses are seen to best advantage as specimen clumps on lawns or in other locations where the surrounding herbage is low and where they are not crowded by competitors. If evergreens provide a backdrop so much the better. A location sheltered from strong winds is advantageous. Female plants produce the showiest and most durable flower heads. Deep, well-drained, fertile soil that never lacks moisture during the season of active growth develops the finest specimens.

Cultivation. Where climate, soil, and site are right, cultivation presents no problems. Propagation is by seed or by division in spring. When preparing the soil it should be made physically agreeable and fertile to a depth of 2 or 3 feet; usually it is desirable to spade in generous amounts of decayed organic material, compost, rotted manure, or peat moss, and bonemeal. Once planted, these grasses may remain undisturbed for a great many years, and no further manipulation of the under soil is practicable. Care of established clumps consists of removing old flower heads and dead leaves, supplying their roots with an application of a complete garden fertilizer in spring, and watering copiously in dry weather.

CORTUSA (Cort-ùsa)—Alpine Bells. Some botanists recognize ten species of this genus of primrose relatives, others fewer. As wildlings they occur from central Europe to Japan and Sakhalin Island. They belong in the primrose family PRIMULACEAE. Their name commemorates the Italian botanist Jacobi Antonii Cortusi, who died in 1593. From primroses (*Primula*) cortusas differ in having their stamens attached at the bottom of the corolla tube.

Cortusas are herbaceous perennials, usually hairy, with all basal, long-stalked, heart- to kidney-shaped or roundish, lobed, and toothed leaves exceeded in length by the flower stalks, which are topped by umbel-like clusters of pink, purplish, or yellow blooms. The flowers have five-lobed, persistent calyxes and bell-shaped, short-tubed corollas, with five spreading lobes (petals). There are five stamens with very short filaments (stalks) located at the base

of the corolla tube and a branchless, slender style. The fruits are plump, roundish capsules containing many small seeds.

Alpine bells (*C. matthiolii*) occurs in several varieties from central Europe to northern China, Korea, Sakhalin, and Japan. It is a hardy, summer-flowering mountain plant that favors wet rocks and moist glades. Typically it has hairy, long-stalked leaves up to 4 inches across, broadly-kidney-shaped to roundish, shallowly seven- to nine-lobed, and with rounded teeth. The flower stalks, up to somewhat over 1 foot long, terminate in four to twelve nodding pink blooms, somewhat frayed at their petal margins, that turn violet as they age. Each is about ⅓ inch long and on a long individual stalk.

Garden Uses and Cultivation. Alpine bells is charming for open woodlands and other shaded places where the root run is cool and moist and contains abundant humus. It is as easily raised from seeds, sown in a cold frame or cool greenhouse, as common primulas, and needs the same care. It can also be increased quite readily by division, in spring or early fall, and by root cuttings. In late summer the foliage dies, and the plant remains dormant until spring. Like its primrose relatives, this plant flourishes with a mulch of leaf mold, peat moss, or rich compost.

CORYANTHES (Cory-ánthes)—Bucket Orchid. The extremely complicated structure of the blooms of *Coryanthes,* described as probably the most complex and fascinating of the entire orchid family ORCHIDACEAE, has been responsible for much wonderment and speculation among botanists. Every detail of their structure seems designed to attract pollinating insects and by a series of ingenious contrivances to assure cross-fertilization.

Endemic to the American tropics, *Coryanthes* comprises seventeen species, all epiphytic (tree perching, but not taking nourishment from the host tree). Most often they grow in association with other plants in the nests of fierce ants. If the plant or nest is disturbed the ants rush out and attack the intruder. Collecting these orchids in the wild can be decidedly uncomfortable. The name *Coryanthes,* from the Greek *korys,* a helmet, and *anthos,* a flower, alludes to the form of the lip of the bloom.

In many ways *Coryanthes* resembles *Stanhopea,* but differs in its flowers having lips with helmet-shaped (bucket-like) terminal lobes and in the upper sepal and two petals being much smaller than the two backward-pointing lateral sepals. The clustered pseudobulbs are often proportionately longer than those of stanhopeas, and commonly are deeply longitudinally grooved and usually more or less hidden by aerial roots. The flowers, in racemes of few or solitary, are on thick, often sharply pendulous stalks from the bases of the pseudobulbs. They are fleshy and mostly curiously and strongly fragrant. They last for a few days only. The prevailing flower colors are cream-white to pale green, frequently spotted with vinous-purple or red. Considerable variation is likely to occur even in the same species.

The remarkable feature of the bloom is the bucket of the lip. This faces upward and has an incurved rim. From two horns that project from the base of the column a watery fluid drips into the bucket, keeping it filled to the level of an overflow conduit. Positioned above the bucket is a dome of fleshy tissue attractive to bees. Crowding on this, the insects jostle each other and one occasionally falls into the bucket. Its wings wetted, it cannot fly out, and the incurved lip prevents it from crawling out of the top of the bucket. The only means of escape is by squeezing through the narrow overflow conduit. In doing this, the bee brushes first against the stigma where pollen it may previously have gathered is deposited, and then against the anther where it picks up a new charge of pollen to take to the next bloom it visits. Thus pollination, and as a result cross-fertilization, of *Coryanthes* is effected.

A native of Venezuela, *C. bungerothii* has ovoid-oblong pseudobulbs 2 to 3 inches long. The flowers, solitary atop stalks up to 1½ feet long, have reddish-purple-spot-

Coryanthes bungerothii

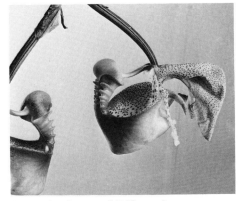
Coryanthes bungerothii (flower)

ted, whitish-green sepals and petals, the lateral sepals up to 6 inches long and 2 inches wide, the upper sepal 2 to 2½ inches long. The petals are about 3 inches long and very narrow. The lip, spotted with brown internally, has a yellow center lobe. Ovoid to conical, strongly-ribbed, clustered pseudobulbs, up to 6 inches long by 1½ inches wide, are characteristic of *C. maculata* (syn. *C. hunterana*). Each has from its base a pair of pleated, leathery, pointed-lanceolate leaves 1 foot to 2 feet long by up to 4 inches wide. The flowers,

Coryanthes maculata

usually two or three to each raceme, are about 4 inches long and wide. They generally have more or less purple-spotted, yellow sepals and petals and a fleshy lip of similar coloration. This kind is native from Panama to Brazil. Rather stouter pseudobulbs, each with a pair of usually narrower leaves from the base, are typical of *C. speciosa* (syn. *Gongora speciosa*). Indigenous from Central America to Brazil and Trinidad, it has flowers about 2½ inches wide in racemes of two to five. They vary considerably in color, being most commonly yellowish-brown with a tawny-yellow bucket. The usually strongly-ribbed pseudobulbs of *C. macrantha* (syn. *Gongora macrantha*) are subconical and up to 6 inches long. They have a pair of pointed, lanceolate-elliptic leaves about 1 foot in length. The blooms are 5 or 6 inches wide. Their sepals, the lateral ones twisted, and undulate petals are yellow to flesh-pink spotted with purplish-red. The red-spotted lip is brownish-yellow. A variety *C. m. alba* is cultivated. The species is native to Guyana, Venezuela, and Trinidad.

Garden Uses and Cultivation. Collectors' orchids, rare in cultivation, these need the same care and conditions as stanhopeas, but are generally more difficult to satisfy. Suspended baskets or similar containers that allow the hanging flower spikes to develop without hindrance are needed by the kinds with descending spikes of bloom. For more information see Orchids.

CORYDALIS (Corý-dalis). Several charming species belong in this genus of more than 300, but only a few are at all common in cultivation. They are ferny-foliaged an-

nuals or sometimes biennials or are herbaceous perennials, with flowers that generally resemble those of bleeding hearts (*Dicentra*), but differ in having only one spur. The genus *Corydalis* belongs in the fumitory family FUMARIACEAE and in the wild is distributed throughout the northern hemisphere, with in addition, one species in the mountains of East Africa. The name is an adaptation of the Greek *korydallis*, the lark. It was applied because the flower spur suggests the spur of the bird. Most kinds are hardy.

Corydalises are erect or less commonly trailing or climbing. Some of the perennials have rhizomes or tubers. The alternate, opposite, or apparently whorled (in circles of more than two) leaves, both basal and on the stems, are pinnately-divided. The very asymmetrical flowers are in racemes. They have two small, scalelike, deciduous sepals or none, four petals of which one has a basal spur, six stamens, a slender style, and two stigmas. The fruits are capsules.

A pretty annual or biennial, *C. sempervirens* is native from Nova Scotia to the midwest, and in Alaska. Its freely-branched, slender stems are 1 foot to 2 feet high. Like the foliage they are glaucous. The short-stalked to nearly stalkless, twice- or thrice-pinnate leaves 1 inch to 1½ inches in length, have lobed or lobeless segments. Freely disposed and abundantly produced in few-flowered racemes, the dainty blooms, ½ to ¾ inch long, are bright pink to purplish-pink decorated with yellow at the tips of the petals and spur, or less commonly are white with yellow decorations. They are produced over a long summer period.

Two pleasing perennials that differ only slightly except for the color of their flowers are *C. lutea* and *C. ochroleuca*. Both are native to southern Europe. The first has golden-yellow, the second yellowish-white blooms. These have short rhizomes and fibrous roots. From 6 inches to 1 foot tall or a little taller, they have leafy stems. The leaves are triangular, two- or three-times

Corydalis lutea

Corydalis ochroleuca

divided, and 3 to 4 inches long. Those of *C. ochroleuca* have flattened stalks and are more glaucous than those of *C. lutea*. The long-stalked racemes are densely flowered. The short-spurred blooms are ¾ inch long.

Especially lovely, **C. nobilis** of central Asia in aspect suggests a very robust *C. lutea*. A perennial, it has fleshy roots and rhizomes. From 1 to 2 feet tall, its stout stems each have two or three once- or twice-pinnate leaves with their ultimate segments wedge-shaped and deeply-lobed. Fragrant and ¾ inch or a little longer, the golden to paler yellow flowers, each tipped with reddish-brown, are crowded in showy heads up to 2½ inches long.

Chinese *C. cheilanthifolia* and *C. saxicola* (syn. *C. thalictrifolia*) are worthy perennials. Not infrequently, inferior *C. ophiocarpa*, a Chinese and Himalayan species, masquerades in gardens under their names. Nearly stemless and forming a basal mass of twice- or thrice-pinnate leaves up to about 9 inches long, with stalks one-third to one-half as long as the blades, *C. cheilanthifolia* has flower stalks that bear sometimes branched, terminal racemes of ½-inch-long, bright yellow blooms, about as long as the leaves. The flowers have

Corydalis saxicola

straight, ascending spurs and upturned styles. Native on limestone cliffs, *C. saxicola* attains a height of about 1½ feet. Its spreading, long-stalked leaves 1 inch to 3 inches long are less divided into broader segments than those of *C. cheilanthifolia*. They have five to seven wedge-shaped lobes. The yellow blooms, ½ to 1 inch long, have short, blunt spurs and outer petals with short, high crests. Also Chinese, *C. tomentella* is a perennial 6 to 9 inches tall with a woody, spreading rhizome and finely-divided long-stalked, gray-green leaves sometimes tinged with pink. Its green-tipped, bright yellow flowers are in erect racemes 4 to 8 inches long.

Corydalis tomentella

The perennial sorts now to be considered have tubers and bloom in spring. They have foliage that dies down completely with the approach of summer. Their flowers are in tones of purplish-pink to rosy-purple or red, or rarely are white. Most commonly cultivated, variable European **C. bulbosa** (syns. *C. halleri*, *C. solida*) has ovate-spherical, solid tubers about 1 inch in diameter and flowering stems 6 to 8 inches tall with a conspicuous bractlike scale near their bases. Each stem bears two or three long-stalked pinnate leaves, with the secondary divisions cleft, but not deeply enough to form separate leaflets. The light-

Corydalis bulbosa

purplish-pink to reddish or occasionally white flowers ¾ to 1 inch long are without sepals and have a usually slightly curved spur. They are in erect terminal racemes of mostly ten to twenty. The floral bracts are palmately-lobed. Variety *C. b. densiflora* has the secondary divisions of its leaves divided into separate leaflets and floral bracts much more dissected than those of the typical species.

In aspect similar to *C. bulbosa* and often confused with it, **C. cava**, of central Europe, differs in having tubers up to 2 inches in diameter characteristically hollow or very markedly depressed, in the lower parts of the stems being without bracts, in the floral bracts not being lobed, and in the flowers having small sepals and a more markedly down-turned spur. With age the tubers split into several small parts each of which becomes a new plant. The flowers of *C. c. albiflora* are white.

Lovely **C. cashmeriana**, of the Himalayas, is less hardy than other tuberous sorts discussed here and is perhaps naturally rather short-lived. From 3 to 9 inches tall and of neat habit, it has clusters of small spindle- to narrowly-egg-shaped tubers, branchless stems, and bright green, three-times-divided leaves. The clear blue flowers, with partly white side petals, are ¾ to 1 inch long and in short raceme-like clusters of up to eight at the tops of the stems.

Garden and Landscape Uses and Cultivation. Primarily rock garden and wall garden plants, although the taller kinds may be used at the fronts of flower borders, at the bases of walls, and in similar locations, corydalises are in the main accommodating and easily grown. Some rarer high mountain species, such as *C. cashmeriana*, are more difficult and are best adapted for growing in alpine greenhouses under the care of a specialist. Nearly all prefer light shade and moistish, but not wet, sandy soil that contains abundant organic matter and drains freely. As long as they are doing well, it is inadvisable to transplant them unless there is a good reason for doing so. Early fall and early spring are the best seasons for moving most kinds, but the best time to transplant the tuberous kinds is immediately after the foliage dies down. Established plants benefit from annual mulching with leaf mold, good compost, peat moss, or similar material.

The annuals are propagated from seeds. Self-sown seedlings of *C. sempervirens* often spring up abundantly in areas where the species grew the previous year. The germination of the seeds of perennial kinds is much more unpredictable. Many fail after being stored dry. The usual method is to sow in a cold frame or in a cool greenhouse as soon as the seeds ripen on the plants, but even then up to a year may elapse before seedlings appear. Storing the seeds in slightly damp peat moss in polyethylene plastic bags in a temperature of about 40°F for several weeks before sowing might give good results. It offers a field for experiment.

CORYLOPSIS (Cory-lópsis) — Winter-Hazel. Belonging to the witch-hazel family HAMAMELIDACEAE, this genus is indigenous from Japan, Korea, and China to the Himalayas. Its name is derived from that of the genus *Corylus*, and the Greek *opsis*, similar to.

The sorts of *Corylopsis* are delightful early-flowering, deciduous shrubs that charm with profusions of pendulous tassels of yellow or slightly greenish-yellow, mildly fragrant flowers well before their new foliage arrives. More graceful and less stridently colored than the better-known forsythias, they bloom at about the same time and are as worthy as ornamentals. Of the twenty known species about one-half are in cultivation, and of these seven are hardy about as far north as New York City, but two are slightly hardier. Even near New York City their blooms are sometimes nipped by late frosts and the twig ends of the more sensitive may be killed by winter cold. These are chances worth taking.

Winter-hazels are densely-branched, twiggy shrubs of widely-spreading or broadly-pyramidal shape. Their smaller branches, those from which the flowers hang so pleasingly, are slender. The hazel-like leaves are alternate, conspicuously pinnately-veined, and usually broadly-ovate. Their margins are furnished with bristly teeth terminating the lateral veins. The flowers, mostly from ¼ to ⅓ inch in length, are few to as many as twenty in drooping, catkin-like spikes or racemes with large bracts at their bases. An unusual feature is that, after the flowers fade, regular foliage leaves grow from the axils of the lower, flowerless bracts and are full sized when the fruits are developed. From the blooms of their relatives the witch-hazels (*Hamamelis*) those of *Corylopsis* differ, not only in being in racemes rather than clusters, but in having five instead of four each sepals, petals, and fertile stamens. Alternating with the stamens are five staminodes (sterile stamens) that in some kinds are forked. There are two styles. The fruits are two-beaked, woody capsules containing two glossy black seeds.

Three wide-spreading shrubs from 6 to 8 feet in height are C. *pauciflora*, C. *spicata*, and a hybrid between them. Native to Japan, **C. pauciflora** has, as its name suggests, the shortest, fewest-flowered racemes of any winter-hazel, but they are produced in such great quantities that its floral display rivals those of other kinds. Its young shoots are hairless and its pointed leaves asymmetrically ovate to broad-ovate. On their undersides the leaves are slightly glaucous and thinly-hairy along the veins. They are 1½ to 3 inches long. The racemes, each of two or three widely expanded, soft

Corylopsis pauciflora

Corylopsis sinensis

yellow flowers nearly ½ inch long, are up to ¾ inch long. The fruits are not pubescent. Also native to Japan, **C. spicata** has silky-hairy shoots and racemes, 1 inch to 2 inches long, of six to twelve bright yellow blooms with hairy calyxes studded on pubescent stalks. Their stamens, about as long as the petals, have purple anthers. The non-fertile stamens are forked. The bracts are hairless on their outsides. Round-ovate to round-obovate and with asymmetrical bases, the leaves, up to 4 inches long, are glaucous and pubescent on their undersides and have densely-hairy stalks up to an inch in length. The fruits are pubescent. The hybrid between C. *pauciflora* and C. *spicata* is intermediate in its characteristics.

Attaining 15 feet in height, and with proportionately narrower leaves than C. *spicata*, the Chinese **C. sinensis** differs also in having racemes of twelve to eighteen flowers with bracts hairy on their outsides. The racemes have hairy stalks and are 1 inch to 2 inches long. The pale yellow blooms have stamens shorter than the petals. Their calyxes are hairy. The young shoots of this species are downy. Its asymmetrically obovate leaves, slightly glaucous and pubescent on their undersides, especially on the veins, are 2 to 4½ inches long. The fruits are pubescent.

Corylopsis spicata

Free-blooming and charming, primrose-yellow-flowered **C. veitchiana** grows to a height of about 6 feet and is of erect habit. Its young shoots are hairless. When they unfold, the elliptic to ovate leaves, which grow to be 3 to 4½ inches long, have a purplish tinge and are sparsely-silky-hairy on their undersides, but both conditions soon disappear. The many-flowered racemes, 1 inch to 2 inches long, have pubescent stalks. The bracts with flowers in their axils are silky-hairy on their outsides, but the outer surfaces of the flowerless basal ones are hairless. The calyxes are pubescent. The stamens have reddish-brown anthers and are as long as or a little longer than the petals. The sterile stamens (staminodes) are forked, the calyxes hairy, and the fruits pubescent. This species is Chinese.

Sometimes a small tree, but more usually a tall shrub, Chinese **C. wilsonii** has hairy young shoots and ovate to elliptic, pointed leaves 2½ to 4½ inches long, slightly glaucous and sparsely-hairy on their undersides, with the hairs most abundant on the veins. The stalks of the leaves are densely-hairy. From 2 to 3 inches long, the racemes of narrow-petaled blooms have hairless calyxes and stamens slightly shorter than the petals. A noticeable feature of this species is that all bracts, including the basal ones, are densely-hairy on both sides. The fruits are without hairs.

The young shoots of **C. willmottiae** are hairless. This Chinese species, up to 12 feet tall, has ovate to obovate leaves 1½ to 3½ inches in length. Their undersides are slightly glaucous and hairy, chiefly on the veins. Their stalks are sometimes slightly pubescent. The crowded, many-flowered racemes of pale yellow blooms are 2 to 3 inches long and have pubescent stalks. Like the fruits, the calyxes are hairless. The stamens, the sterile ones forked, are much shorter than the petals.

One of the showiest winter-hazels is **C. platypetala**, this because its flowers have especially broad petals and are many together in racemes 1½ to 2¼ inches long.

Corylopsis platypetala

Corylopsis platypetala (flowers)

Corylopsis platypetala (foliage and fruits)

The young shoots are hairless, but they have small glands, as have the leafstalks. A native of China, this species is about 12 feet tall. Its short-pointed leaves, hairless except when they first unfold, are ovate to elliptic and 2 to 4 inches long. The pale yellow flowers, with broad, hatchet-shaped petals and hairless calyxes, are on stalks without hairs. Their stamens, the sterile ones forked or notched, are shorter than the petals. The fruits are hairless. Variety

C. p. laevis is distinguished by its young shoots and its leafstalks being without glands.

The two hardiest, but unfortunately least showy species are *C. glabrescens* and *C. gotoana*, both native to Japan. From 15 to 20 feet tall, *C. glabrescens* has very slender younger branches and shoots without hairs. Its leaves are pointed-ovate, 1½ to 4 inches long, hairless on their upper surfaces, and sparingly-hairy beneath. Their margins have prominent, short, bristly teeth. The pale yellow, ½-inch-long flowers, in hairless racemes ¾ inch to 1½ inches long, have stamens one-half as long as the petals. The sterile stamens are deeply-forked. The fruits are hairless. Very similar *C. gotoana* is distinguished from the former by having obovate leaves and slightly smaller flowers, with stamens almost as long as the petals.

Kinds tender in the north and much less commonly cultivated are *C. griffithii*, of the Himalayas, and the Chinese *C. yunnanensis*. Related to *C. spicata*, *C. griffithii* differs in having larger leaves and longer racemes with bracts hairy on both sides. Also, its stamens are much shorter than the petals. Related to *C. platypetala*, the species *C. yunnanensis* differs in its leaves being sparsely-pubescent beneath, its bracts being somewhat hairy on their outsides, and in having larger flowers with pubescent calyxes.

Garden and Landscape Uses. Winter-hazels flaunt their blooms less blatantly than such brilliant spring-flowering shrubs as azaleas and forsythias. Theirs is a quiet, refined beauty that appeals to lovers of the delicate and graceful. Because of this, they compete less noisily with companion plants and can be used with tasteful effect to enhance the loveliness of such bulbs as squills, glory-of-the-snows, crocuses, and early daffodils, as well as other spring flowers. In addition to their floral display these shrubs have quite attractive bluish-green foliage that turns yellow in fall. Their fruits are not ornamental. Toward the northern limits of their hardiness they must be afforded shelter. This can be done, as at Princeton University, by espaliering them against walls or, better still because of the rich green foil provided for their blooms, by locating them in front of evergreens sufficiently tall to break cold winds. Indeed, an evergreen background anywhere sets these shrubs off to fine advantage. Deep, rather sandy soil, of a peaty character that does not dry excessively, is best for winter-hazels, although they are fairly accommodating in this respect. They grow well in part-day or lightly dappled shade where the sun reaches them in spring.

Cultivation. Little care is needed. No regular pruning is necessary unless they are espaliered, except that in the north the removal of branch ends that have been

winterkilled may be required. Espaliered specimens are handled by thinning and cutting back the flowering shoots as soon as the blooms fade. The maintenance of an organic mulch is good practice, and in dry weather watering may be desirable. Propagation is by seeds sown in winter or spring in a temperature of about 60°F in sandy peaty soil, by cuttings made of firm, but not hard shoots in summer and inserted preferably under mist, and by layers that root readily if the soil is kept moist.

CORYLUS (Córy-lus)—Hazel, Hazelnut, Filbert, Cobnut. Native to North America, Europe, and Asia, the fifteen species of hazels or hazelnuts and filberts are deciduous shrubs or small trees of the genus *Corylus* of the birch family BETULACEAE, or according to those who prefer the segregation, to the hazel family CORYLACEAE. The name is an ancient Greek one. Some coryluses are cultivated for their edible nuts, some as ornamentals.

Hazels and filberts have alternate, undivided, stalked, ovate to oblong, toothed leaves. Their inconspicuous flowers are unisexual with both sexes on the same plant. They appear before the foliage. Male blooms are solitary in the axils of the upper bracts of slender, pendulous catkins. Without calyxes or corollas, they have four to eight two-branched stamens. Female blooms, only the two red stigmas of their styles protruding and scarcely discernible as individuals, are enclosed in small, leafy buds. Usually clustered at the ends of short branches, the fruits consist of an egg-shaped to pointed-cylindrical nut partly or completely encased in an involucre (husk) of leafy bracts.

The American hazel (*C. americana*) is native in light woodlands and thickets from Maine to Saskatchewan, Georgia, Missouri, and Oklahoma. A shrub up to 10 feet tall, this kind has glandular-pubescent branchlets and pointed, ovate to broad-elliptic leaves 2½ to 5 inches in length, finely-pubescent on their undersides, slightly so above. Usually paired, sometimes solitary or in groups of up to six, the fruits have husks consisting of a pair of

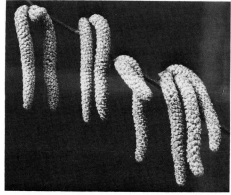

Corylus americana (catkins)

irregularly-lobed bracts united at their bases, about twice as long as the roundish nuts. From this species the beaked hazel (*C. cornuta*) differs in having branchlets without glands and in the husks of its nuts being tubular, abruptly narrowed above, and projecting as a long beak beyond the ovoid nut. Native in rich woodland clearings, borders of woodlands, and thickets from Quebec to Saskatchewan, Georgia, and Missouri, this about 10-foot-tall shrub has pointed, ovate to ovate-oblong, double-toothed leaves 1½ to 4 inches long, softly-hairy on their undersides, nearly hairless above. Differing from *C. cornuta* chiefly in its leaves having more heart-shaped bases and more densely-hairy under surfaces, *C. c. californica* (syn. *C. californica*) is native from British Columbia to California.

Common or European filbert (*C. avellana*), a shrub up to 15 feet tall, has glandular-pubescent branchlets. Its roundish-elliptic to obovate leaves, pubescent on the veins on their undersides, are pointed, 2 to 4 inches long, and slightly heart-shaped at their bases. They are toothed and sometimes shallowly-lobed. The fruits, solitary or in clusters of up to four, have husks of two irregularly-lobed bracts exceeded in length by the roundish-egg-shaped nut. Of this sort there are many varieties including the cobnut (*C. a. grandis*), a large-fruited kind cultivated particularly for its nuts, and others esteemed as ornamentals. One of the most striking decoratives, *C. a. contorta* attains up to one-half the maximum height of the species. It has branches and branchlets so grotesquely twisted and curled that it is known as Harry Lauder's walking stick. Variety *C. a. pendula*, with strongly pendulous branches, when grafted high on an understock of the typical species makes a striking weeping specimen. The foliage of *C. a. aurea* is soft yellow. That of *C. a. fusco-rubra* (syn. *C. a. purpurea*) is a less attractive purple than that of *C. maxima purpurea*.

Richly-colored purple foliage distinguishes *C. maxima purpurea*, a native of the Balkan Peninsula. This is an ornamental variety of the giant filbert (*C. maxima*), a native of southern Europe and a shrub or

Corylus avellana contorta

Corylus avellana contorta, young specimen

broad-headed tree up to 30 feet tall. Its somewhat larger leaves resemble those of *C. avellana*, from which it differs in the husks of the fruits being tubular and slightly narrowed above the nut. Many varieties of the filbert and of hybrids of it and *C. avellana* are cultivated for their nuts, which are longer and bigger than those of *C. avellana* and are hidden within the involucres.

Other cultivated hazels include the Turkish hazel (*C. colurna*) and the Japanese hazel (*C. sieboldiana*). A symmetrical, pyramidal tree up to 75 feet tall, *C. colurna*

has glandular-pubescent branchlets and sometimes somewhat lobed, broadly-ovate to obovate leaves, 3½ to 4½ inches long with hair on their veins beneath. The catkins of the male flowers are 2 to 3 inches long. The clustered fruits have husks deeply cleft into recurved, narrow, glandular lobes. This kind is hardy through most of New England. A shrub up to about 15 feet in height, Japanese *C. sieboldiana* is hardy in southern New England. It has elliptic to oblong or obovate, slightly-lobed leaves 2 to 4 inches long, nearly hairless except on the veins of their undersides. Young leaves often have a central patch of red. The nuts, in tubular, bristly husks much narrowed beyond the end of the nut, are conical. In *C. s. mandshurica*, the more decidedly lobed leaves are on average somewhat longer and have usually heart-shaped bases.

Other Asian species include the Chinese hazel *C. chinensis*, a native of China that there attains heights up to 120 feet, and *C. heterophylla*, of Japan and northeast Asia, a shrub or tree up to about 20 feet in height. Both are hardy in southern New England. Broad-headed *C. chinensis* has pointed ovate to ovate-oblong leaves up to 7 inches long and hairy on the veins beneath. The clustered fruits have husks much contracted above the subspherical nuts and divided into linear lobes. The leaves of *C. heterophylla* are nearly round to obovate, abruptly pointed, 2 to 4 inches long, and usually lobed near their apexes. They are hairy on the veins on their undersides. Solitary or in clusters of usually not more than three, the nuts have husks of bracts cut into triangular, sometimes few-toothed lobes usually longer than the subspherical nuts.

Garden and Landscape Uses. Except in the Pacific Northwest, hazels are little grown for their nuts, but those of native species, inferior to those of good varieties of the European hazel, are collected and eaten. Kinds most commonly cultivated for ornament are those with colored foliage and those with curiously crooked or pendulous branches. Handsome *C. maxima purpurea* and *C. avellana aurea* contrast pleasingly as neighbors. The Turkish hazel makes a quite handsome specimen. Hazels succeed best in deep, loamy, well-drained, slightly acid to neutral soils in sun or part-day shade. To color most effectively, purple- and yellow-leaved varieties need full sun. When cultivated for their nuts, heavier cropping is likely when several varieties are grown together.

Cultivation. Hazels as ornamentals present few problems to the gardener. They are easily propagated, the varieties by grafting onto understocks of European or Turkish hazel, by suckers, by layering, and by summer cuttings of firm, but not hard shoots planted under mist or in a greenhouse propagating bench. Species may be increased by these methods and by

Corylus avellana (catkins)

Corylus colurna

seeds sown as soon as they are ripe in a cold frame protected from animals or by mixing them with slightly damp peat moss or vermiculite and storing them at 40°F for three months, and then sowing them in sandy, peaty soil in a greenhouse. Insects that may infest hazels, especially if they are not in thrifty condition, are aphids, lacebugs, mites, and scale insects. For information about growing hazels for their nuts see Filbert.

CORYNABUTILON (Coryn-abùtilon). Endemic to Chile, *Corynabutilon* of the mallow family MALVACEAE consists of three species of shrubs and small trees. From *Abutilon*, which otherwise it closely resembles, *Corynabutilon* differs in technical details of the styles of its flowers. Its name comes from the Greek *koryne*, a club, in allusion to the form of the style branches, and the name of the genus *Abutilon*.

Perhaps the only sort cultivated, **C. vitifolium** (syn. *Abutilon vitifolium*) is a handsome tall shrub or small tree that withstands a few degrees of frost. Under favorable conditions it may become 30 feet tall, but more often is closer to one-half that height. Its downy, long-stalked leaves, 3 to 6 inches long, have three, five, or rarely seven long, narrow, coarsely-toothed lobes. Pale-blue to lavender-blue best describes the color of the 2½- to 3½-inch-wide, wheel- or shallowly-saucer-shaped flowers that look much like those of a mallow or hibiscus. Those of variety *C. v. album* are pure white.

Garden and Landscape Uses and Cultivation. The species described is a handsome ornamental for outdoor planting in warm frost-free and nearly frost-free climates. It succeeds in well-drained soil of moderate fertility in sun or part-day shade. Propagation is easy by seeds and by cuttings. A large majority of plants raised from seeds of *C. v. album* have flowers like

Corynabutilon vitifolium

Corynabutilon vitifolium (flowers)

those of the parent plant, those of a minority are pale blue.

CORYNOCARPACEAE — Corynocarpus Family. The characteristics of this family of dicotyledons are those of its only genus, *Corynocarpus*.

CORYNOCARPUS (Coryno-cárpus)—New-Zealand-Laurel or Karaka. A genus of four or five species, this is the only one of the corynocarpus family CORYNOCARPACEAE. So far as is known it is confined in the wild to New Zealand, the New Hebrides, and New Caledonia. The species described below is naturalized in Hawaii. The name, from the Greek *koryne*, a club, and *karpos*, a fruit, alludes to the shape of the fruits.

Corynocarpuses are trees, related to hollies (*Ilex*), with alternate, undivided, lobeless and toothless leaves. Borne in panicles at the ends of the branches, the bisexual flowers have five sepals, five petals joined to the bases of the sepals, and five stamens that alternate with an equal number of more or less petal-like staminodes (nonfunctional stamens). There are one or two styles. The fruits are drupes, that is, they are structured like plums.

The New-Zealand-laurel or karaka (*Corynocarpus laevigata*) is a good-looking native of New Zealand that attains a height of up to 50 feet and has a spreading crown of stout branches and lustrous, evergreen foliage. Its thick, blunt, short-stalked, obovate-oblong leaves are mostly 4 to 6 inches in length, sometimes rather shorter or longer, and 2 to 3 inches in width. They are dark green. The greenish-yellow flowers are in stout, rigid panicles up to about 8 inches long. Their minutely-toothed petals are up to ⅕ inch in length. The staminodes, also minutely-toothed, are spatulashaped. The very short style is capped with a roundish stigma. Egg- to football-shaped, the orange fruits are 1 inch to 1½ inches long. Their flesh is edible, but the seeds are bitter and poisonous, unless subjected, as they were by the Maoris, to long cooking. Variety *C. l. variegata* has yellow-margined leaves.

Garden and Landscape Uses and Cultivation. The New-Zealand-laurel is an attractive ornamental for general landscaping in the tropics, subtropics, and warm-temperate regions. It is not hardy in the north. It grows well in ordinary garden soil and is propagated by seed.

CORYPHA (Corỳ-pha)—Talipot Palm. Most famous of this group of eight massive palms of Indomalaysia, southeast Asia, and Ceylon is the talipot palm (*Corypha umbraculifera*), which has rightly been described by the American botanist W. H. Hodge as Nature's biggest bouquet. Dr. Hodge refers to the enormous flower cluster, or inflorescence, which consists of tens of thousands of small blooms massed in a gigantic panicle 20 to 30 feet high and two to three times as massive and big as the inflorescence of the krubi (*Amorphophallus titanum*), a plant frequently cited as the "world's largest flower." But it should be pointed out that neither the talipot palm nor the krubi bear the largest known blooms. Their inflorescences are aggregations of numerous small blooms, with attendant stalks, bracts or spathes, and other floral parts. The largest single flower is that of the astonishing parasitic plant *Rafflesia arnoldii*, of Malaya. The genus *Corypha* is monocarpic, that is to say its members bloom and fruit once, usually when between 20 and 80 years of age, and then die. It belongs in the palm family PALMAE. The name is from the Greek *koryphe*, summit, and refers to the leaves being in a terminal crown.

Coryphas have tall, straight, massive trunks topped by a head of immense, rigid, orbicular, fan-shaped leaves deeply divided into numerous segments and with long spiny stalks. The immense inflorescences, terminal, pyramidal, erect, and much branched, tower to great heights above the foliage. The small, tubular flowers are bisexual and have three petals and six stamens. The globose fruits are about as big or somewhat bigger than cherries.

The talipot palm (*C. umbraculifera*), which is the national floral emblem of Sri Lanka (Ceylon) and a native of that country and southern India, attains a height of 60 to 80 or occasionally 100 feet and has a trunk sometimes 3 feet in diameter ringed with scars left by fallen leaves. Its leaves have stalks up to 10 feet long with rather short spines, often in pairs, and blades from 10 to 16 feet wide and cleft to their middles into from eighty to a hundred segments. The cream flowers, with the scent of sour milk, are in broad panicles of such gigantic size that their development and

Corypha umbraculifera

Corypha umbraculifera, in fruit at Chapman Field, Miami, Florida

that of the fruits that follow completely exhausts the starchy food supply that throughout the life of the plant has been stored in the trunk in anticipation of this event. About a year passes from the opening of the first flowers to the ripening of fruit and the dying of the palm. The fruits, 1½ inches in diameter, are white.

In its homelands the talipot is put to many uses. From the starchy material in its trunk a kind of sago is prepared. The leaves are employed for thatching, basket- and matmaking, and for handcrafting into sunshades and umbrellas. The horny white seeds are used for making buttons, beads, and trinkets, and a fish poison is prepared from the fruits. In ancient times the leaves were prepared to serve as writing paper, the writing being impressed on them with a metal stylus.

Other kinds include these: *C. elata*, of Burma and Bengal, up to 70 feet tall, has leaves 8 to 10 feet across with strongly-spined stalks. Its olive-colored fruits are about 1 inch in diameter. *C. taliera*, of Burma, is up to about 30 feet in height and has leaves more deeply cleft than those of the talipot palm. The flower clusters are 20 feet tall or taller. *C. utan*, of the East Indies, is up to 70 feet tall and has leaves about 10 feet wide. Its flower clusters are about 15 feet high.

Garden and Landscape Uses and Cultivation. In the United States, these huge tropical palms can only be expected to succeed outdoors in Hawaii and southern Florida, whereas only the loftiest of large conservatories could possibly accommodate and provide for them indoors. Outdoors or in, they need plenty of space when they attain their mature size. They surely are not plants for small gardens or sites where space is limited. These palms succeed in any deep, well-drained soil not excessively dry. They flourish in sun and need plenty of humidity and heat.

Propagation is by fresh seed sown in sandy, peaty soil in a temperature of 80 to 90°F. For other information see Palms.

CORYPHANTHA (Cory-phántha). The genus *Coryphantha*, of the cactus family CACTACEAE, is native from southern Alberta, Canada, to central Mexico. According to a conservative estimate there are twenty to thirty species. Botanists who base species on lesser differences estimate the number to be sixty-five. Some authorities include the group in nearly related *Mammillaria*. The name is derived from the Greek *koryphe*, a summit, and *anthos*, a flower. It refers to the position of the blooms.

Differences between *Coryphantha*, *Mammillaria*, and *Escobaria* are admittedly rather weak and not always constant. Generally coryphanthas are distinguished from mammillarias by the possession of a groove along the top of each tubercle and by their flowers being near the centers of the tops of the stems instead of closer to the rims. They differ from escobarias by the outer segments (petals) of the flowers usually not being fringed with hairs and by the seeds not being dotted with minute pits.

Coryphanthas have spherical, ovoid, or cylindrical, solitary or clustered stems 1 inch to 6 inches in length and up to 3 inches in diameter, solitary or those of some species forming broad mounds of 200 or more. They are covered with lumpy protrusions (tubercles) that do not coalesce to form ribs and that sometimes with age fall from the lower parts of the stems. Except in very young specimens, these are grooved along their upper sides and in the groove are one or more glands. Each tubercle is tipped with an areole bearing a cluster of spines of generally two types, an outer group of usually straight ones radiating or being arranged in comblike fashion, and fewer inner spines, more or less erect, and straight, curved, hooked, or twisted. Coming from near the apex of the stem, the flowers are bell-shaped and erect. The perianths have short tubes with few or no scales and spreading petals. Ovoid to oblong, the fleshy, green or yellowish fruits are without scales.

The pincushion cactus (*C. vivipara*), as now understood, includes a number of varieties previously listed as separate species and widely distributed in deserts in the United States and Mexico. Its very spiny stems are depressed-spherical to egg-shaped or cylindrical, from 1½ to 6 inches long by 1½ to 3 inches wide. Some variants form mounds up to 1 foot tall consisting of up to 200 or more closely packed stems. Each areole has twelve to forty usually white radial spines and three to ten 1-inch-long, straight central ones white below, and pink, red, or black in their upper parts. The flowers, 1 inch to 2 inches in diameter, have outer petals fringed with hairs. Open for only a few hours, they are pink, lavender-pink, red, or yellow-green. The ellipsoid fruits, ½ to 1 inch long, are green. Varieties include *C. v. arizonica* (syn. *C. arizonica*), which has solitary or clustered, ovoid stems 2 to 4 inches long by 2 to 2½ inches wide, *C. v. deserti* (syn. *C. deserti*), which has usually solitary stems 3 to 6 inches long by 3 to 3½ inches wide.

Other kinds cultivated are fairly numerous. Among them are those now to be described. Specialist dealers in cactuses list additional kinds in their catalogs. *C. bergeriana*, of Mexico, has club-shaped stems up to about 5 inches tall by one-half as wide. The tubercles have glandular grooves. The spine clusters are of eighteen to twenty gray radials up to a little more than ½ inch long and four often recurved, yellowish centrals ½ to ¾ inch long. The yellow-centered, white flowers are about 1½ inches long. *C. clava* is Mexican. It has a cylindrical to club-shaped stem up to 1 foot tall and 4 inches wide. The tubercles, with one or two red glands at the base of the groove, are arranged rather openly. The starry spine clusters are of seven to ten radials about ½ inch long, and one ¾-inch-long central. The pale yellow blooms, tinged red toward their outsides, are 3 inches or sometimes more in diameter. *C. echinus*, endemic to Texas, is spherical to somewhat egg-shaped or conical, 1 inch to 2 inches in diameter, and up to 3 inches in height. Usually solitary, it occasionally forms small clumps. The spine clusters are of sixteen to thirty radials, when young yellowish to whitish, later grayish, usually with dark tips, and three or four much stouter, brown centrals ½ to ¾ inch long. Clear sulfur-yellow, with brownish-green outer petals, the flowers are 3 inches wide or rather wider. The plant sometimes called *C. pectinata* appears to be an unstable variant of *C. echinus*, in which the spine clusters are without centrals. *C. elephantidens* is a Mexican with a flattish-spherical stem up to about 5½ inches tall and 7 inches wide, with large, closely-packed tubercles.

Coryphantha elephantidens

The spine clusters are of six to eight curved, stout radials about ¾ inch long, yellow at first becoming brown later, and with dark tips. There are no central spines. The blooms, rosy-red to carmine with darker stripes, are 3 to 4 inches wide. *C. erecta*, of Mexico, has clustered, cylindrical, yellowish-green stems up to 1 foot tall. The clusters of needle-shaped spines are of eight to fourteen radials ½ inch or less long and two to four centrals. The yellow flowers have very slender petals. *C. gladiispina* has solitary, ovoid to ellipsoid stems up to 4 inches in height and 2½ inches in width, usually less spiny toward their bases than above. In each spine cluster there are seventeen to twenty radials, the several upper ones brushlike, and four 1-inch-long centrals. The 2½-inch-wide flowers are yellow. This sort is native of Mexico. *C. macromeris* of Mexico forms clumps of prominently tubercled, more or less cylindrical stems up to 8 inches tall. The spine clusters are of many white radials and some black centrals. The purple flowers are up to 3 inches across. *C. minima* (syn. *C. nelieae*), of Texas, has egg-shaped to cylindrical stems, mostly solitary and under 1 inch long by ¾ inch wide. The spine clusters are of thirteen to fifteen club-shaped radials up to ⅛ inch long and three much stouter spines, one shorter than the other, that some authorities refer to as radials; the others are centrals. The spines, pinkish at first, become yellowish. The rose-purple blooms are from a little more than ½ to 1 inch in diameter. *C. ottonis* is a Mexican with almost spherical stems up to 3 inches wide or wider and up to 5 inches high. The spine clusters are of eight to twelve radials about ⅓ inch long and three or four centrals ¾ inch long. The white flowers are about 1½ inches in length. *C. pallida* has single or clustered, spherical stems up to 5 inches wide. The clusters of twenty or more white spines usually include three centrals. The 2½-inch-wide blooms are lemon-yellow, with the outer parts greenish. This is native to Mexico. *C. palmeri*, of Mexico, has nearly spherical stems about 4 inches tall, somewhat under 4 inches wide, with closely set tubercles. Its spine clusters are of eleven to eighteen black-tipped, yellowish radials and one ¾-inch-long, hooked central. Pale yellow to nearly white, with broad, brownish stripes on their outsides, the blooms are about 1¼ inches long. *C. poselgeriana* is a Mexican with a high-centered, nearly spherical stem and large tubercles with glandular, woolly grooves. In each spine cluster there are five to seven bulbous-based radials up to 2 inches long and one central of approximately the same length. Salmon-pink or less commonly yellow, the flowers are up to 2 inches long. *C. pseudechinus* has a stem up to 3½ inches tall with clusters of eighteen to twenty-five rigid, grayish radials, about ½ inch in

length, and a single bulbous-based central, approximately twice as long. The flowers, ¾ inch long by 1¼ inches wide, are violet-pink, sometimes with greenish throats. This is a native of Mexico. *C. salm-dyckiana* is a Mexican with solitary or clustered stems, spherical to club-shaped and up to 6 inches tall. The spine clusters have mostly ten to fifteen whitish radials up to a little more than ½ inch long and one to four bulbous-based black or nearly black centrals up to almost twice as long. The flowers are light yellow tinged with red. They are up to 1½ inches long. *C. werdermannii*, at first spherical, but later more elongated, is a Mexican with clusters of twenty-five to thirty radial spines, and four brownish-white centrals. The flowers are golden-yellow and up to 3 inches in diameter.

Garden and Landscape Uses and Cultivation. Coryphanthas are attractive for rock gardens and similar locations in warm desert and semidesert regions and for inclusion in greenhouse collections of succulents. They succeed under conditions and general care that suit mammillarias and many other desert cactuses. For more information see Cactuses.

CORYTHOLOMA. See Sinningia.

COSMOS (Cós-mos). Although in the wild some species of *Cosmos* are perennial herbaceous plants, those cultivated are either annuals or are treated as such. They are attractive summer- and fall-bloomers. The genus, of the daisy family COMPOSITAE, has approximately twenty-five species, all natives of the warmer parts of the Americas. The name, alluding to their decorative merits, is from the Greek *kosmos*, an ornament.

Cosmoses have opposite, usually much pinnately-divided, but sometimes simply lobed or lobeless leaves. Solitary or in loose panicles, the daisy-form flower heads have central disks of tubular, bisexual florets and showy, spreading, petal-like, sterile, white, pink, red, yellow, or orange rays. Horticultural varieties, in which some or all of the flower heads are "double flowers," are common. The involucres (collars of bracts at the backs of the flower heads) are of two rows, with the bracts united at their bases. Cultivated species are all natives of Mexico.

Source of most garden varieties, *C. bipinnatus* is a stout, freely-branching, hairless or more or less hairy annual 4 to 10 feet tall, with leaves twice-divided into linear, almost threadlike segments. The long-stalked flower heads, in wild plants up to about 3 inches across, but often larger in cultivated kinds, have white, pink, or crimson ray florets with blunt, toothed ends and yellow disk florets. Horticultural varieties, commonly listed in seedmen's catalogs, include early and late flowering ones, and anemone- and double-flowered

Cosmos bipinnatus, variety

varieties. In anemone-flowered varieties, the florets of the central disk are transformed into crowded cushions of short, petal-like parts colored like the ray florets, in double-flowered varieties they are petal-like, as long as the ray florets, and of the same color.

Yellow cosmos (**C. sulphureus**) is a showy, satisfactory garden annual with light to deep, rich yellow or orange-yellow blooms in late summer and fall. It is 3 to 6 feet in height and has pubescent leaves two- or three-times-pinnately-divided into lanceolate lobes. The flower heads, 1¾ to 3 inches across, have strongly-toothed ray florets.

About 1½ feet in height and in the wild a tuberous-rooted perennial, **C. diversifolius** has leaves undivided or pinnately-cut into five or seven leaflets. Its long-stalked flower heads, about 2 inches in diameter, have yellow disks and lilac-pink to rose-pink ray florets. Very similar, but somewhat taller and with smaller flower heads with red disks and rich, dark red, velvety, rays is the black cosmos (**C. atrosanguineus**).

Garden and Landscape Uses and Cultivation. Cosmoses are gay plants for beds

Cosmos sulphureus

and borders and as cut flowers. Tall varieties serve usefully as temporary screens. They are easily raised from seeds sown outdoors where the plants are to remain or, more commonly in the north, are sown indoors in a temperature of about 60°F six to eight weeks before planting out time and the young plants set in the garden from small pots or flats (shallow boxes) after there is no danger from frost. Until planting out time they are grown in a sunny greenhouse or under similar conditions where the night temperature is 50°F and that by day from five to ten degrees higher. The soil must not be rich; if it is, growth is too exuberant, blooming is delayed, and flowers are fewer. Well-drained, sandy, poorish earth gives best results. Because they are easily damaged by high winds sheltered locations are best for these plants. They need full sun. Varieties of *C. bipinnatus* may be spaced 2 to 3 feet apart. Plants of *C. sulphureus* and *C. atrosanguineus* may be allowed 1½ to 2 feet and those of *C. diversifolius* about 1 foot between individuals. Tall kinds need staking. Other than the removal of spent flower heads and thorough soakings with water during protracted dry weather, no regular attention is needed.

COSSONIA. See Raffenalida.

COSTMARY is *Chrysanthemum balsamita.*

COSTUS (Cós-tus)—Crape Ginger or Malay-Ginger, Spiral Flag. Few of the 150 species of this genus of the Old and the New World tropics are familiar to gardeners, but those are decidedly ornamental. The group belongs to the ginger family ZINGIBERACEAE and consists of sturdy, low to tall herbaceous perennials. The name is an ancient one, thought to be derived from the Arabic.

Costuses generally have thick, fleshy roots and almost always leafy stems, with the undivided leaves arranged spirally instead of, as is usual in the ginger family, in two ranks. The showy, white, yellow, or red flowers are in conelike, broad-bracted heads that terminate the leafy stems or separate leafless stems. They have tubular, cleft corollas with one enlarged, bell-shaped, petal-like staminode (sterile stamen) that forms the lip and decorative portion of the bloom. The only other genus in the family known to be in cultivation that has spirally arranged leaves is *Tapeinocheilos,* but its flowers are very different from those of *Costus.*

Central American *C. malortieanus* (syn. *C. zebrinus*) forms suckering clumps 2 to 3 feet tall of erect stems, with scattered along them, recurved, fleshy, bright emerald-green, elliptic to obovate leaves up to about 1¼ feet long and approximately one-half as wide. They have longitudinal stripes of darker green curving outward from the

Costus malortieanus

Costus malortieanus (flowers)

midribs. The reddish-streaked, yellow flowers are in heads about 2½ inches long. Attaining a maximum height of about 10 feet, *C. stenophyllus,* a native of Costa Rica, forms dense bamboo-like clumps of arching stems marked with pale, horizontal cross bands, having along much of their upper parts long, drooping, narrow leaves. The flowers, which come from the bases of the plants, are yellowish. From 1 foot to 2 feet tall, *C. igneus,* of Brazil, has long to elliptic-lanceolate leaves 3 to 6 inches in length, reddish on their undersides, rich green above, and crowded toward the tops of the stems. The nearly circular, orange-red flowers, about 2½ inches across, are few together among inconspicuous bracts. Another South American, *C. pulverulentus* (syn. *C. sanguineus*) has coppery stems, along which are spiraled, curving, obliquely-pointed-elliptic, fleshy leaves with short, purplish-red stalks and velvety bluish-green blades with a central clear silvery, longitudinal band, which has slender lines of gray angling outward from it. The undersides of the leaves are rich blood-red. Terminating leafy stems, the spindle-shaped heads of bloom, up to 3 inches long by 2 inches wide, have orange-red bracts and red flowers up to 1½ inches long and sometimes with a yellow lip.

West Indian *C. macrostrobilus* attains heights of 6 feet or more and has obovate, hairy leaves up to 1 foot long by one-third as wide. The flower heads have green-tipped, pink bracts and pink- and yellow-tinged white blooms 4 inches in length, with three-lobed lips. From the last, *C. villosissimus,* also West Indian, differs in its proportionately broader leaves being more densely clothed with darker reddish hairs. The yellow flowers, about as big as those of *C. macrostrobilus,* rest in a nest of green bracts.

Crape-ginger or Malay-ginger (*C. speciosus*) has short-lived, white blooms with yellow centers. They are 3 to 4 inches across and in heads or spikes up to 5 inches long and with blunt bracts. This is a native of tropical Asia and Indonesia. It has branching, reedlike stems, somewhat woody at their bases, and 4 to 10 feet tall. The leaves, with silky undersides, are up to 1 foot long by 3 inches wide. They are spiraled along the stems rather than clustered at their tops.

Sometimes called African spiral flag, *C. lucanusianus* in many respects is similar to the crape-ginger, but it has shorter topmost leaves forming nests in which the flower heads sit. The flower heads have pointed, greenish bracts. The leaves are silky-hairy on their undersides. The red-margined lip of the flower is about 2 inches across and yellow at its center. The stamen is petal-like and tipped with red. This is native to tropical West Africa.

Stemless *C. spectabilis* of the Sudan has a spiral rosette of usually four overlap-

Costus spectabilis

ping, obovate leaves up to 1 foot long by 8½ inches wide. Its few short-stalked flowers, from the center of the rosette, have lanceolate petals 1½ to 2 inches long and an obovate, pale yellow lip 3 to 3½ inches long by 1½ to 2 inches wide.

Garden and Landscape Uses and Cultivation. Costuses are satisfactory only where high temperatures and humidities prevail. They belong in the humid tropics and in

greenhouses that simulate such environments. Rich, well-drained earth, fat with decayed organic matter, such as compost, leaf mold, manure, or peat moss, and always moist, but not waterlogged, gives the best results. Well-rooted specimens in pots or other containers respond to frequent applications of dilute liquid fertilizer. Propagation is most commonly by division, and by cuttings made by cutting the stems into pieces 1 inch to 3 inches long.

COTINUS (Cótinus)—Smoke Bush or Smoke Tree. At one time included with sumacs in the genus *Rhus,* the two or three species now separated as *Cotinus* differ in having panicles of flowers and fruits with many sterile flower stalks clothed with silky hairs. These form the "smoke" responsible for their common names. The group is indigenous to the northern hemisphere. *Cotinus* belongs in the cashew family ANACARDIACEAE. The name is an ancient Greek one of uncertain meaning. The name smoke tree is also applied to *Dalea spinosa.*

Cotinuses are deciduous shrubs or small trees with alternate, undivided, slender-stalked, toothless leaves. They are hardy in southern New England. Their flowers are tiny, yellowish or greenish, and in sizable, loose, feathery, terminal panicles. They have five-parted calyxes one-half as long as the petals, of which there are five, five stamens, shorter than the petals, and three short styles. Individuals may be unisexual or have unisexual and bisexual flowers on the same plant. The tiny berry-like fruits are few, flattened, and beaked on one side.

The European and Asian smoke bush or smoke tree (*C. coggygria* syn. *Rhus cotinus*) is much the handsomer species in flower and fruit, provided care is taken to secure a form or variety of known "smoke"-producing propensity. Wholly male trees develop no fruiting panicles, and specimens raised from seeds are likely to vary greatly in the quantity and quality of their displays. Up to 15 feet tall, this native of eastern Europe to the Himalayas and China is broad, dense, and shrubby. Its leaves, 1½

Cotinus coggygria

Cotinus coggygria purpureus

to 3½ inches in length, are broad-elliptic to obovate. They are without hairs. In fall they turn yellow or orange-yellow. Up to 8 or 9 inches long, the grayish to pinkish flower and fruit panicles are heavily plumose. As they pass into the fruit the stalks of the numerous sterile flowers are elongate. In variety *C. c. purpureus* the leaves and fruiting panicles are purplish. Improved selections of this kind have been given horticultural names. Here belong such richly colored kinds as 'Notcutt's Variety', 'Flame', and 'Royal Purple'. The exact shade of color is likely to vary with different soils and environmental conditions; ordinarily most of the purple fades by late summer.

The American smoke tree (**C. obovatus** syns. *C. americanus, Rhus cotinoides*) is indigenous from Tennessee to Alabama and Texas. A shrub or small tree up to 35 feet in height, it has mostly obovate leaves, 2½ to 5 inches long and when young silky-hairy on their undersides, and in fall becoming brilliant orange to red. The flower and fruit panicles, the sterile flower stalks with brownish or purplish hairs, make little show. They are 4 to 6 inches long.

Garden and Landscape Uses. The European and Asian smoke bush, an old-time inhabitant of gardens, is deservedly well thought of. Its rich purple-leaved varieties are especially attractive. It makes a good lawn specimen and is fine for planting among other shrubbery where it has plenty of room to spread without being crowded by its neighbors. Its display of flower and fruit panicles lasts for a long time and rarely fails to attract favorable comment. The American smoke tree is much less decorative in flower and fruit, but it does make a rich display of brilliantly colored foliage in fall. Both kinds need full sun and give of their best in dryish, poorish soils. In fertile earth they grow too lushly and tend to flower sparsely. Toward the northern limits of their hardiness they should be given sheltered locations.

Cultivation. Once established, these plants need no particular care. They can be propagated by seed, by layering, and by root cuttings.

COTONEASTER (Coton-e-áster). Among the fifty species of this northern hemisphere genus of the rose family ROSACEAE are many attractive shrubs. Not all are hardy in the north, and regrettably, the group is subject to fire blight disease and infestations of lacebugs and red spider mites. Despite these pessimisms, cotoneasters deserve consideration for landscape planting. They are closely related to hawthorns (*Crataegus*), from which they are readily distinguished by being thornless and by their leaves being without lobes or teeth. The name is derived from the Latin *cotoneum*, the quince, and *aster*, a suffix implying inferior.

Cotoneasters are low to tall shrubs or rarely small trees. They include evergreen and deciduous kinds and some that behave one way or the other depending upon climate and other circumstances. Although they vary greatly in habit of growth, they are remarkably similar in their flowers, and except for color, their fruits. The blooms, white or delicate pink, but often darker in the bud stage, are generally in clusters of few to many terminating short branches from the leaf axils. Less often they are solitary. Each is ¼ to ⅓ inch wide and has a five-lobed, top-shaped, persistent calyx, five spreading or erect petals, about twenty stamens, and two to five styles. The berry-like, red, black, or rarely yellow fruits are little pomes (fruits constructed like apples) containing two to five seedlike nutlets. The remains of the calyx is attached at the top of the ¼- to ⅓-inch-wide fruits. Because cotoneasters have the ability to set fertile seeds without the flowers being pollinated, a procedure known to botanists as apomixis, (in effect a kind of parthenogenesis) many minor variants are perpetuated as populations within some species. This tends to confuse the demarcation of species, and makes identification difficult.

Probably the most popular cotoneaster in America is **Cotoneaster horizontalis** (syn. *C. davidiana*). Sometimes called rock spray, this native of western China is hardy in all except the coldest parts of New England. A most distinct species, 2 to 3 feet tall, it has horizontal branches from which side branches spread in a flat plane to give a fishbone-like or vaguely fernlike effect. Semievergreen or deciduous, it has broad-elliptic to nearly round leaves ⅓ to ½ inch long. Before they fall they turn crimson. The pinkish flowers, solitary or in pairs, have upright petals. The fruits, generally with three nutlets, are bright red and not more than ¼ inch in diameter. Variety *C. h. perpusillus* has leaves up to ¼ inch long. In *C. h. variegatus* the leaves are edged with white. An especially vigorous variant is *C. h. robustus*.

Cotoneaster horizontalis, in flower

Cotoneaster horizontalis, in fruit

Very slow-growing **C. adpressus,** of western China, does not exceed 1 foot in height. It differs from *C. horizontalis* in not branching in stylized fishbone pattern and in its leaves usually having wavy margins. Its flowers, pinkish, and solitary or in pairs and with erect petals, are succeeded by bright red fruits with usually two nutlets. This is about as hardy as *C. horizontalis*. Variety *C. a. praecox* is more robust and ripens its fruits earlier. Nearly as hardy, **C. dammeri,** of central China, has long, trail-

Cotoneaster horizontalis

Cotoneaster horizontalis perpusillus, in fruit

Cotoneaster adpressus, in flower

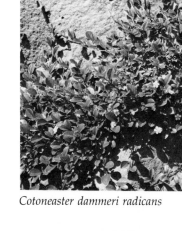

Cotoneaster dammeri radicans

Cotoneaster adpressus praecox, in fruit

Cotoneaster dammeri radicans (flowers)

ing stems that often root into the ground. Its lustrous, elliptic-oblong, evergreen leaves are ¾ to 1 inch long and hairy only when young. Usually solitary, the flowers have wide-spreading petals. The fruits are bright red. Variety *C. d. radicans* is distinguished by its smaller, more wrinkled foliage, the leaves blunt and often notched at their ends. Its flowers are solitary or paired. Often misidentified as the last variety, *C. d. major* is more vigorous than the species, and has leaves 1 inch to 1½ inches long. Another vigorous variety, or possibly hybrid, is *C. d.* 'Skogholm'. It attains a height of 1½ feet, but does not fruit freely. Similar to *C. horizontalis* is **C. apiculatus,** of western China. Sometimes called cranberry cotoneaster, this has almost round leaves with a minute sharp point at the apex. Its pink flowers have erect petals. They are succeeded by bright red berries larger than those of *C. horizontalis.*

A group of related evergreen species usually about 3 feet in height that have flowers with widely spreading petals are *C. microphyllus, C. congestus,* and *C. conspicuus.* The first is hardy in southern New England, the others are somewhat more tender. All are native to the Himalayas, with *C. microphyllus* extending into China. With spreading branches, **C. microphyllus** has obovate to obovate-oblong leaves up to ⅓ inch long, lustrous above, and densely-grayish-hairy on their undersides. Usually solitary, rarely in twos or threes, the white flowers are succeeded by conspicuous red fruits. An excellent variety, *C. m. cochleatus* is quite prostrate. The undersides of its leaves have long, white hairs. The smallest-leaved cotoneaster, *C. m. thymifolius* is prostrate and low, although with age it may attain heights of 1 foot to 2 feet. It has narrowly oblong-ovate leaves up to ⅜ inch long, with their margins rolled under. Flowers and fruits are smaller than those of *C. microphyllus.* By some regarded as a variety of *C. microphyllus* but usually considered to be a distinct species, **C. congestus** has obovate to broad-elliptic leaves sparingly pubescent at first, not densely-woolly on their undersides, and soon hairless. Its habit is its chief distinctive feature. Instead of spreading widely, its branches are congested and often more or less contorted and downswept. It forms a compact, rounded mound. The pinkish, solitary flowers have spreading petals. The berries are bright red. Typically low, but with variants that have been segregated as *C. permutatus* up to 8 feet tall, **C. conspicuus** has arching branches, and ovate to linear-obovate evergreen leaves up to ¼ inch long. Their upper sides are lustrous and hairless, their undersides gray-hairy. The solitary, white flowers have spreading petals, and red-purple anthers. The fruits are

bright red and comparatively large. This is a native of Tibet. Variety *C. c. decorus* is prostrate.

Cotoneasters normally about 6 feet tall or somewhat taller, are numerous. Among the best of those hardy in southern New England, the last being the hardiest, are *C. bullatus floribundus, C. b. macrophyllus, C. dielsianus, C. divaricatus, C. zabelii,* and *C. hupehensis.* All are deciduous. The first is a very good variety of an inferior Chinese species. Of open habit, it has long, arching branches, and ovate to oblongish leaves about one-half as wide as their 1½ to 3½ inch lengths. Their undersides are covered with a felt of grayish or yellowish hairs. Their upper sides are puckered between the veins and sparsely-hairy. The pinkish flowers, their petals erect, are in clusters of ten to thirty at the ends of short branchlets. The chief attraction of this kind is its 2-inch-wide clusters of brilliant red, spherical to pear-shaped fruits about ⅓ inch long, that face upward and decorate the branches in two rows, one along each side of the stem. A variety of *C. bullatus* probably less hardy than *C. b. floribundus* is *C. b. macrophyllus.* This is distinguished by its larger leaves, up to 6 inches long. Native to central China, **C. dielsianus** has graceful, arching, slender branches and ovate leaves ½ inch to 1½ inches long by about two-thirds as wide, that are orange and red in fall. They are clothed on their undersides with a felt of white hairs that becomes brownish as the leaves age. The pinkish flowers have erect petals and are in clusters of up to seven at the ends of short side shoots. Round to pear-shaped, the scarlet fruits, about ¼ inch long, contain three or four nutlets. Variety *C. d. elegans* (syn. *C. elegans*) has thinner, more persistent leaves, and orange-red fruits. A broad shrub from western China, and in fruit one of the most beautiful cotoneasters, **C. divaricatus** has roundish elliptic to ovate or obovate leaves up to 1 inch long by a little more than ½ inch wide, but often smaller, with a sharp point at their apexes. The lustrous short-stalked leaves,

Cotoneaster dielsianus, in fruit

Cotoneaster divaricatus

sparsely-hairy on their undersides have three or four pairs of veins. They turn orange and red in fall. Mostly in threes, but occasionally solitary, the pink flowers have erect petals and are succeeded by ⅓-inch-long, egg-shaped, red fruits with usually two nutlets. The red, broadly pear-shaped fruits of **C. zabelii,** of central China, also contain two nutlets. They are spherical to somewhat pear-shaped, downy, and about ⅓ inch long. The shrub, which sometimes attains a height of ten feet, has dullish, mostly broad-elliptic to ovate, blunt or sometimes pointed leaves ½ inch to 1½ inches long by up to two-thirds as wide. Their upper sides are loosely-hairy. Beneath they are covered with a felt of yellowish hairs. In fall they turn yellow. Small, and in clusters of up to ten, the pink flowers have erect petals. Variety *C. z. min-*

iatus is more compact, and has smaller, orange-colored fruits. From central and western China, **C. hupehensis,** up to 6 feet tall, has long arching branches. Its leaves, up to 1½ inches long, are broad-elliptic to ovate. Their undersides are thinly-gray-hairy. In clusters of up to one dozen, the flowers have spreading white sepals. The fruits, comparatively large, are scarlet and contain two nutlets. Hardy in much of New England, its leaves turn yellow in fall.

Kinds normally exceeding 6 feet in height, may be considered as those hardy at least as far north as southern New England and those generally more tender or hardy only in quite sheltered locations there. Among the hardiest are **C. racemiflorus soongoricus** (syn. *C. soongoricus*), *C. multiflorus calocarpus,* and *C. simonsii.* Native of central Asia, the first is a graceful variety of a less well-known, more tender species that occurs in the wild from North Africa to Turkestan. It differs from the species in having less hairy leaves, and generally more profuse displays of pink fruits, although at the Arnold Arboretum at Boston, Massachusetts these are not borne abundantly every year. It adapts to dry soils better than most kinds. A wide shrub to about 8 feet tall, *C. r. soongoricus* has distinctly gray-green foliage, with leaves up to 1½ inches long. Its flowers have spreading petals. The fruits are ⅓ inch in length. Very like the last, but scarcely as hardy, **C. multiflorus calocarpus** differs most obviously in its fruits being brilliant red. A Chinese representative of a species that has a wide range in Asia and also occurs in Spain, it differs

from typical *C. multiflorus* in having longer, narrower leaves and larger fruits. Its flowers have spreading petals. From 10 to 12 feet tall and of stiffish, erect form **C. simonsii** is a sometimes semievergreen native of the Himalayas. It has very short-stalked, broad-elliptic leaves ¾ inch to 1¼ inches long in two opposite rows, paler on their undersides than above, and hairy there. The ½-inch-wide white blooms, with erect petals, are in clusters of up to four. The scarlet fruits are ⅓ inch long or slightly longer.

Black fruits are characteristic of *C. foveolatus* and *C. lucidus;* before they are fully ripe, however, they are red. A native of central China, and up to about 10 feet tall, **C. foveolatus** has slender-pointed, broad-elliptic to ovate leaves 1 inch to 3½ inches long by approximately one-half as wide. Sparsely-hairy on their undersides and hairy along their margins, in fall they turn bright orange or scarlet. In clusters of up to seven, the erect-petaled blooms are pinkish. The fruits contain three or four nutlets. This species is hardy in southern New England. Native of northern Asia, including Siberia, **C. lucidus** not surprisingly is one of the hardiest cotoneasters. About 10 feet tall, and of dense habit, it has shiny leaves, hairless except for a sparse distribution on their undersides. They are ovate to broadly-elliptic and ¾ inch to 2 inches long. The pinkish flowers are in short-stalked clusters of up to ten. They have erect petals. The fruits contain three or four nutlets.

Somewhat more tender cotoneasters up to 10 feet tall or taller are *C. franchetii, C. pannosus, C. rotundifolius,* and *C. salicifolius floccosus.* These survive outdoors in not-too-exposed locations near New York City, but cannot be depended upon much further north. Semievergreen or evergreen and with orange-scarlet fruits, **C. franchetii** of China is up to 10 feet tall. It has slender, arching branches, woolly when young. Its oval to elliptic leaves, with stalks not more than ¼ inch long, are ¾ to a little over 1 inch in length, and about

Cotoneaster zabelii in fruit

Cotoneaster franchetti, in fruit

one-half as wide. Their upper surfaces, hairy at first, are lustrous. Their undersides are clothed with white or brownish felt. In clusters of up to fifteen, the pink-tinged blooms at the ends of short side shoots have felted calyxes and erect petals. The oblong fruits have usually three nutlets. Variety *C. f. sternianus* may be slightly hardier than the typical species. Its fruits are nearly spherical to obovoid instead of oblong. In *C. f. cinerascens*, the leaves are larger and clothed with a felt of gray hairs.

Rather like *C. franchetii* in appearance, and as hardy and also a native of China, the silver-leaf cotoneaster (*C. pannosus*) is evergreen or semievergreen. Lower than *C. franchetii*, it has arching, slender branches, and tapering, dullish, elliptic leaves up to 1 inch long by about one-half as wide. They are conspicuously white-felted on their lower surfaces and have stalks up to ¼ inch long. In rounded clusters of fifteen or twenty, the flowers have spreading white petals and woolly calyxes. The fruits are dull red.

Evergreen, partially evergreen, or sometimes deciduous, handsome **C. rotundifolius** (syn. *C. prostratus*) of the Himalayas is one of the finest species. Up to 12 feet tall,

Cotoneaster rotundifolius

but often somewhat lower, it has long, arching branches and broad-elliptic to nearly round leaves ½ to ¾ inch in length, with wedge-shaped bases. Their undersides are sparsely-hairy. Above they are lustrous, and usually hairless. White, with spreading petals and purple anthers, the flowers are in twos or threes or solitary. The red fruits, large for a cotoneaster, contain two nutlets. The berries of this kind remain attractive far into the winter. They apparently do not appeal to birds. Variety *C. r. lanatus* has leaves with densely-hairy undersides, and flowers in clusters of three to five or occasionally more.

The Chinese species **C. salicifolius** is little cultivated, but its hardier variety *C. s. floccosus* is popular. Evergreen or semievergreen, it is up to 15 feet tall, and sur-

Cotoneaster salicifolius floccosus

Cotoneaster salicifolius floccosus (flowers)

Cotoneaster salicifolius floccosus (fruits)

vives outdoors at New York City and in sheltered locations somewhat further north. Its gracefully arching branches bear pointed, lanceolate or narrowly-ovate, short-stalked leaves up to 2½ inches long by ¾ inch wide. They become purplish in fall. Their upper surfaces are hairless and wrinkled, their undersides clothed with gray-white hairs. The flowers, in clusters of nine to fifteen at the ends of short lateral branchlets, have spreading petals. The bright red fruits contain usually three nutlets. A vig-

orous variety with somewhat larger leaves and fruits, the latter with usually three nutlets, is distinguished as *C. s. rugosus*. Other horticultural varieties, some low or spreading, are cultivated. Here belong *C. s. repens* and *C. s.* 'Herbstfeuer' ("Autumn Fire").

Not hardy in the north, but suited for milder climates such as those of the Pacific Coast are *C. frigidus*, *C. henryanus*, *C. glaucophyllus*, and *C. lacteus*. One of the tallest cotoneasters, Himalayan **C. frigidus,** under

Cotoneaster frigidus (fruits)

favorable circumstances becomes a shrub or small tree 20 feet in height. It has deciduous, dull bluish-green, elliptic to obovate leaves 3 to 6 inches long and 1 inch to 2 inches wide. When young, their undersides are densely-woolly. The flattish, many-flowered clusters of bloom are at the ends of short, lateral twigs. The flowers have spreading petals. They are succeeded by showy, bright red fruits that contain two nutlets. Several hybrids between *C. frigidus* and other species are known in Europe. One, *C.* 'Cornubia', is cultivated in California. A broad shrub up to 20 feet or more tall, it has pointed, ovate-lanceolate leaves 4 to 5 inches long, slightly downy on their undersides. It is one of the finest cotoneasters in fruit. The fruits are brilliant red and abundantly produced.

Semievergreen to evergreen **C. henryanus** is up to about 12 feet tall. Native of China, it has arching, pubescent branches and pointed, oblong to oblong-lanceolate leaves, at first grayish-hairy on their lower surfaces and sparingly-pubescent above. They are 2 to 4½ inches long, and approximately one-third as wide. The white flowers have spreading petals and purple anthers. The brownish-red to deep crimson fruits contain two or three nutlets.

Perhaps not in cultivation, **C. glaucophyllus** is a deciduous native of China. Its variety, *C. g. serotinus*, is a handsome evergreen that under exceptionally favorable circumstances may attain a height of 30 feet. It has shoots downy at first, later hairless, and pointed, elliptic to oval leaves 1 inch to 3 inches long by almost or quite one-half as wide. When young, they are downy on their undersides. The flowers, with white, spreading petals, are in clusters at the ends of short side shoots. Bright red, the fruits have two or three nutlets. They are attractive late in the year.

Closely related to *C. glaucophyllus,* and an evergreen up to 12 feet tall, **C. lacteus** of China differs from that species in its more prominently-veined leaves having more persistently hairy undersides. From its allies *C. salicifolius* and *C. henryanus,* it differs in its proportionately wider leaves, up to 2¼ inches long by 1¼ inches wide. The white flowers, in clusters, are succeeded by red fruits. This cotoneaster is the one often cultivated under the name *C. parneyi.*

Garden and Landscape Uses. Cotoneasters can serve usefully many garden purposes. True, the majority make no very great display of bloom, although a few, such as *C. multiflorus, C. racemiflorus,* and *C. salicifolius floccosus,* are by no means without floral merit. Their greatest appeals are their attractive habits, neat foliage, and handsome displays of berry-like fruits. Low and prostrate kinds, such as *C. dammeri, C. horizontalis, C. apiculatus,* and their varieties, and *C. microphyllus thymifolius,* are admirable for rock gardens, banks, and similar places. For espalier training against walls *C. horizontalis* is particularly well

A hedge of *Cotoneaster lacteus*

A bank planted with *Cotoneaster apiculatus*

Cotoneaster horizontalis, espaliered on wall

suited, and other kinds can also be used effectively in this way. Many lend themselves for use in beds, borders, foundation plantings, the fringes of woodlands, and similar places. Some make good hedges, but because fire blight, if it appears, is likely to severely damage a planting of that kind, it is probably unwise to depend upon them for that purpose. Another interesting way of growing and displaying cotoneasters is as standards (tree-form specimens). These are had by grafting onto understocks that form the trunk of the combined specimen. Rather commonly mountain-ash (*Sorbus*) has been used for this purpose, but experience in New Jersey suggests that a better understock, at least for American conditions, is the Washington thorn (*Crataegus phaenopyrum*). The grafting is done in late winter in a greenhouse. Most kinds are highly attractive when grown as standards. Especially fine are *C. multiflorus, C. salicifolius floccosus,* and *C. horizontalis,* but others can be just as lovely.

Cultivation. Cotoneasters adapt to a wide variety of soils, but not too wet or poorly drained ones. Nor can they be expected to give of their best in extraordinarily dry soil. They need full sun, or at most tolerate part-day shade. Systematic pruning is not needed, but some to shape or to contain the plants to size may be desirable. It is advisable to do this before the bushes become too large and grow out of

hand. Severe cutting back is less commendable. Propagation is by seeds, freed from surrounding pulp, and sown as soon as ripe in a cold frame, or stratified in damp sand or peat moss contained in a polyethylene bag and then sown in pans (shallow pots) or flats of sandy soil in a cool greenhouse. Summer cuttings root readily under mist or in a cold frame or greenhouse propagating bed.

COTTON. See Gossypium. Other plants with common names that include the word cotton are cotton-grass (*Eriophorum*), cotton gum (*Nyssa aquatica*), cotton-rose (*Hibiscus mutabilis*), lavender-cotton (*Santolina chamaecyparissus*), red-silk-cotton tree (*Bombax malabaricum*), silk-cotton tree (*Ceiba pentandra*), and wild-cotton (*Hibiscus moscheutos*).

COTTONSEED MEAL. This organic fertilizer is the residue of cotton seeds from which oil has been expressed for commercial use. See Fertilizers.

COTTONWEED is *Otanthus maritimus.*

COTTONWOOD is *Populus deltoides.*

COTULA (Cótu-la)—Brass Buttons. Widely distributed as natives of temperate and warm-temperate parts of the southern hemisphere, the seventy-five species of

Cotula belong to the daisy family COMPOS-ITAE. The name, from the Greek *kotyle*, a small cup, has reference to the little basins formed by the stem-clasping bases of the leaves of some kinds.

Cotulas include annuals and herbaceous perennials, the latter, with the exception of *C. squalida*, which survives outdoors at least as far north as New York City, probably not hardy in the north. They are often more or less prostrate, tufted or creeping plants, with alternate, usually pinnate, pinnately-lobed, or rarely lobeless or nearly lobeless leaves. The yellow flower heads are composed of disk florets (the kind that form the eyes of daisies) without encircling ray florets, but in some kinds the outer corolla lobes of some of the florets are so enlarged that they have somewhat the appearance of rays. The heads may be short- or long-stalked. They have involucres (collars) at their backs of one or more rows of bracts. The fruits are seedlike achenes.

Brass buttons (**C. coronopifolia**), native of South Africa and naturalized in California, is a slightly fleshy, hairless perennial 4 inches to 1 foot tall and much branched from the base. It has stems up to about 1 foot long, the lower parts more or less prostrate and rooting into the ground, the upper parts ascending. The linear-lanceolate to oblong-lanceolate leaves, up to 2 inches in length, may be lobeless, but more commonly are irregularly cut into rather broad lobes. The button-like, bright golden-yellow flower heads, produced in considerable profusion on stalks of varied lengths from the leaf axils and branch ends, are about ½ inch in diameter.

Very different **C. squalida** (syn. *Leptinella squalida*), of New Zealand, forms a dense mat 1 inch or 2 inches high of slender, wiry, rooting stems and crowded foliage. Its fernlike leaves, usually hairy, but sometimes nearly hairless, are obovate and 1 inch to 2 inches long by under ½ inch wide. They are pinnate toward their bases and pinnately-lobed above, with the segments deeply toothed, usually along their upper edges only. Male and female florets are in separate flower heads, those of females ⅓ inch wide and with the silky-hairy, purplish bracts of the involucre curving over and covering them. Heads of male florets are smaller. This sort is perennial. Another New Zealander, **C. atrata** is a creeping perennial 4 to 6 inches tall with rather fleshy, glandular-hairy, narrow-oblong to narrow-obovate, deeply-pinnately cleft or pinnate leaves up to 1½ inches long. The dark purple to nearly black flower heads on stalks 2 to 2½ inches long are up to ¾ inch in diameter. Those of *C. a. dendyi* are pale yellow to brown; those of *C. a. luteola* are yellow.

Annual **C. barbata,** a native of South Africa, is tufted and silky-hairy. Its leaves are mostly basal or nearly so. Above their middles they are cleft into a few, some-

Cotula atrata luteola

times forked, bristle-pointed, narrow lobes. The long-stalked, globose flower heads about ¼ inch in diameter are borne in such numbers that they make a very pleasing show.

Garden and Landscape Uses and Cultivation. In mild, dryish climates where little or no frost occurs brass buttons is attractive for flower beds, rock gardens, and other places where it can be displayed to advantage. Although it makes no appreciable floral display, *C. squalida* is a pleasing foliage plant for rock gardens, where it may be used to cover deciduous small bulbs, and for planting in chinks between paving stones. Both species need damp soils. Brass buttons needs a sunny location, the other succeeds in sun or partial shade. Propagation is easy by division and by seed. A pretty plant for the fronts of flower beds and similar locations, *C. barbata* is easily raised from seeds sown in spring where the plants are to remain, and the seedlings thinned out to forestall overcrowding. Or seeds may be sown indoors earlier and the young plants transferred to the garden when there is no longer a threat of frost. As is true of many South African annuals, this kind does not stand up well in hot, humid weather. In regions of mild winters and fairly cool summers *C. atrata* and its varieties are pleasing rock garden plants.

COTYLEDON. As a botanical term cotyledon is used for the one, two, or rarely more leaves of the embryonic plant contained in a seed. When the seed germinates these, the cotyledon leaves or seed leaves, are the first to develop. Often they are markedly different in shape, thickness, and size from all succeeding leaves or true leaves as they are sometimes called. If you take a bean seed and remove its outer coat you find a pair of fleshy bodies each with one rounded, one flattened side. They are the cotyledons. Typical of nearly all plants belonging to the great group called dicotyledons, the seeds of beans have two cotyledons. If you take the seed of an on-

ion or a palm tree, or if you take a grain of wheat and uncover its inside, you will find only one cotyledon. These plants belong to the group called monocotyledons.

COTYLEDON (Coty-lèdon). The genus *Cotyledon* of the orpine family CRASSULACEAE, comprises some forty species, all except *C. barbeyi* of East Africa and southern Arabia, natives of South Africa. The name, alluding to the deeply-cupped leaves of *Umbilicus rupestris* formerly *Cotyledon umbilicus*, derives from the Greek *kotyle*, a cavity. Many other plants once named *Cotyledon* are contained in other genera including *Adromischus, Chiastophyllum, Dudleya, Echeveria, Orostachys, Pachyphytum,* and *Umbilicus*. In gardens some of these are still occasionally grown under the name *Cotyledon*.

Cotyledons are mostly low, evergreen or deciduous, succulent shrubs and subshrubs. They have opposite or alternate, undivided, fleshy leaves and terminal, umbel-like clusters of mostly nodding, yellow-red or greenish blooms that have the tips of their petals reflexed. The flowers have a five-parted calyx and a corolla with a cylindrical, sometimes angled perianth tube cleft above into five segments (petals). The ten stamens are mostly contained within the corolla. There are five separate stigmas, with a slender scale at the base of each. The fruits are podlike follicles.

A tall, freely-branched shrub, **C. barbeyi** has opposite, stalkless, gray-green leaves, obovate or obovate-wedge-shaped, 2½ to 5 inches long, 1 inch to nearly 3 inches wide, and flat. The long flowering stalk carries many orange to reddish blooms, minutely-glandular-hairy on their outsides. Erect to nearly prostrate and about 2 feet tall, **C. decussata** has opposite or nearly opposite, more or less red-tipped, linear to lanceolate or wedge-shaped leaves 1½ to 4 inches long, cylindrical to almost flat and pale to dark green, with a waxy coating. The red-striped, yellowish blooms are numerous.

A subshrub that forms small clumps of procumbent shoots upturned at their tips and up to 9 inches long, **C. jacobseniana** is attractive. Its crowded, dull green or sometimes waxy-coated leaves, nearly cylindrical or somewhat flattened on their upper and lower sides, are ¾ to 1 inch long. They taper to both ends. Charming **C. ladysmithiensis** is up to 8 inches in height. Its stems become distinctly woody as they age. Chiefly at the shoot ends, the stalkless or nearly stalkless, fleshy, obovate leaves, 1¼ to 2 inches long by ½ to ¾ inch wide, have a few coarse, blunt teeth at their apexes. They are clothed with coarse white hairs. Up to about 6 inches tall, the flowering stalks carry a cluster of brownish-red blooms with paler interiors.

Variable and vigorous **C. orbiculata** (syn. *C. ausana*) includes several varieties, some

Cotyledon ladysmithiensis

Cotyledon orbiculata oophylla

C. orbiculata. Displayed in tall, loose panicles, the long-stalked, nearly 1-inch-long, reddish-tipped, orange-yellow flowers are pendant.

The erect, short, thick stems of **C. reticulata** branch repeatedly from near their bases to form a rounded bushlet 6 inches to 1 foot tall. The nearly subcylindrical leaves, fleshy, hairless, and tipped with a short fine point, in groups of four to eight, are approximately ¾ inch long. The small, greenish-yellow flowers are on freely-branched stalks 2½ to 3½ inches long. After the flowers fade the stalks become woody and spiny and persist. This species is reported to be poisonous. Another poisonous species, **C. wallichii** has fleshy and usually branched stems 1 to 1½ feet tall by 1 to 1½ inches in diameter. In evidence only during a short growing period each year, the leaves of this sort are clustered at the apexes of the stem and its branches. They are subcylindrical, slender, hairless, and 2 to 4 inches long. The branched flowering stalk, 1 foot long or a little longer, displays blooms that are predominantly green and ½ to ¾ inch long.

Garden and Landscape Uses and Cultivation. Cotyledons are for the most part among the easiest succulents to cultivate.

Cotyledon orbiculata

Cotyledon paniculata

Cotyledon reticulata

with leaves resembling those of pachyphytums. Common in cultivation, this species typically is a freely-branched shrublet or shrub 1½ to 5 feet in height with upright or procumbent branches and usually narrowly-red-edged, intensely waxy-white or waxy-light-gray foliage. The stalkless or nearly stalkless, bluntish, broad-obovate to spatula-shaped or more rarely linear leaves narrow downward. They are 1½ to 5 inches long by up to 2 inches wide or slightly wider. Characteristically their under surfaces are rounded, their top sides flattish. The yellowish-red blooms are in long-stalked, chandelier-like clusters. The thick, sausage-shaped leaves of C. o. *higginsiae* are about 1½ inches long and mealy-white edged with brownish-red. Pretty C. o. *oophylla* has short, thick stems and up-pointing, densely-waxy-white, pointed-egg-shaped leaves with a dark moon-shaped mark at the apex. The flowers are orange-red. Closely related to C. *orbiculata* and sometimes treated as a variety of it, **C. dinteri** has thick stems and deciduous, cylindrical leaves 1 inch to 4 inches long. The flowers are pendant and yellow-green.

Distinctive **C. paniculata** may be 5 feet tall or sometimes taller. It has thick, gouty stems covered with yellow-brown, papery skin. The leaves are deciduous, alternate,

gray-green, and 2 to 3 inches long by 1 inch to 2 inches wide. They are crowded in rosettes at the tops of the stems. The flowers are red with green stripes. Popular **C. undulata** is especially beautiful. From 1 foot to 1½ feet in height and with its younger parts thickly dusted with a floury coating of white wax, this has opposite, rhomboidal-ovate, usually upward-angled to erect leaves, 3 to 4½ inches long by 2½ inches wide, conspicuously waved along their apical margins. This last feature distinguishes this species from closely related

Cotyledon undulata

Cotyledon wallichii

In warm, semidesert climates they are admirable for planting outdoors under conditions that suit other dry-country plants. They also do well in greenhouses and as window plants. The soil should be reasonably nourishing, well aerated, freely-drained, and moderately moist. For deciduous sorts it should be kept dry during their leafless periods. Do not practice overhead watering with kinds, such as *C. undulata*, that are heavily coated with waxy meal. To do so removes this coating and reduces the beauty of the foliage. Well-rooted samples in active growth benefit from an occasional application of dilute liquid fertilizer, but fertilization must not be overdone. Propagation is easy by cuttings, leaf cuttings, and sometimes by division. Seeds germinate readily, but if taken from plants in mixed collections they may give rise to hybrid offspring. For more information see Succulents.

COUNTY AGRICULTURAL AGENT. See Cooperative Extension Service.

COURANTIA. See Echeveria.

COUROUPITA (Cour-oupìta)—Cannon Ball Tree. Visitors to the tropics are frequently intrigued by the remarkable cannon ball tree. In flower and fruit it is a decided curiosity, since its large and strangely shaped blooms, to the botanically uninitiated vaguely suggestive of those of orchids, spring directly from its trunk and larger branches, and its huge fruits resemble nothing so much as rusty cannon balls. The genus *Couroupita*, which belongs in the lecythis family LECYTHIDACEAE, numbers about twenty species of trees, natives of tropical America and the West Indies. Its name is a modification of a native one.

The only species cultivated is the cannon ball tree (**C. guianensis**), a deciduous native of northern South America and Trinidad, and planted elsewhere in the tropics and warm subtropics, including southern

Couroupita guianensis

Couroupita guianensis (flower)

Couroupita guianensis (fruits)

Florida, as a curiosity. Attaining a height of about 50 feet, it has alternate, toothed or toothless, pointed, oblong leaves 8 inches to 1 foot long and mostly clustered toward the ends of the branches. On its lower trunk are masses of tangled, interlacing, leafless branches 1 foot to 4 feet long that carry the flowers and fruits. The flowers are many together in clusters of mostly from fifty to one hundred, but commonly only one or two of each cluster produce fruits and apparently only then if they are fertilized with pollen from another tree. Individual blooms are curiously structured and 4 to 5 inches wide. They have six waxy petals that curve forward and are yellow with a suffusion of crimson on their outsides and wine-red inside. The numerous lilac-tinted stamens are in two series on a broad, fleshy, petal-like column bent almost double like the shell of a slightly gaping clam. Into a central hole in the disk formed by the lower set of stamens, the blunt pistil fits and the curve of the staminal column brings the upper set of stamens over it to form a canopy. The flowers are fragrant. In breezy weather the globular fruits, 6 to 8 inches in diameter, swing against each other and the trunk and make a clattering noise. Hard shelled and dark brown, they contain many seeds and a malodorous pulp and are heavy enough to cause serious harm if they fall on a person or an animal. They take about eighteen months to ripen. Uncollected fallen fruits rot and emit a rather foul odor.

Garden Uses and Cultivation. The cannon ball tree is primarily a curiosity for planting where its falling fruits are unlikely to cause damage. It grows in ordinary soil and is propagated by seed.

COUTAREA (Cout-àrea). The name of this genus of the madder family RUBIACEAE is an adaptation of a native Guinean one. Entirely American, *Coutarea* occurs from southern Mexico to Argentina and comprises seven species of evergreen trees and shrubs. Its leaves are opposite, undivided, short-stalked, and pointed-ovate. Showy and fragrant, the whitish, greenish, pinkish, or purplish, funnel- to bell-shaped, symmetrical or asymmetrical blooms are solitary or few together in clusters at the branchlet ends. Unusual for the family, in which the flower parts are nearly always in fours, they have five- to eight-lobed calyxes and corollas. The fruits are capsules.

A hairless or sparsely-hairy shrub or small tree, but in Peru reported to sometimes exceed 100 feet in height, **C. hexandra** (syn. *Portlandia hexandra*) has short-pointed, ovate to broad-elliptic leaves 3 to 5 inches long and showy, white to purplish, asymmetrical flowers 2 inches long. The blooms, mostly in twos or threes, are grouped to form larger terminal clusters that have much the appearance of those of some deciduous azaleas. They have six-lobed calyxes, six corolla lobes (petals), and six protruding stamens. Variety *C. h. pubescens* differs in having densely-pubescent foliage and blooms. The hard, heavy wood of *C. hexandra* is used locally for construction and furniture. Its bitter bark has been employed as a substitute for quinine.

Garden and Landscape Uses and Cultivation. In the tropics and warm subtropics the species described above is a very worthwhile ornamental. It grows satisfac-

Coutarea hexandra

torily in ordinary, fertile soil in sun or part-day shade and is useful as a decorative shrub or small tree. Propagation is by seeds and by cuttings.

COVER CROPS. This term applies to plants grown on a short-term basis to protect the soil from the harmful effects of leaching and erosion and to smother weeds. When such crops are raised for the primary purpose of improving the soil by adding organic matter they are called green manures. There is no sharp line of distinction between the applications of these terms. Cover crops are especially helpful for temporarily stabilizing sloping land and banks and for checking the leaching of nutrients. This last they do by absorbing and converting them into organic compounds, which they store in their tissues. After the

Spading in a cover crop

cover crop is plowed, rototilled, or spaded into the ground it gradually decays, forming humus and releasing its nutrients in forms available to living plants. In the southern and central United States, winter cover crops play important parts in vegetable garden routine. In northern regions where the ground is frozen for much of the winter they are less commonly used. Plants useful as cover crops are discussed under Green Manuring.

COW. As part of the common names of plants cow forms part of the following: cow herb (*Saponaria vaccaria* and *Vaccaria pyramidata*), cow-lily (*Nuphar*), cow-parsnip (*Heracleum*), cow-pea (*Vigna unguiculata*), and cow tree (*Brosimum galactodendron*).

COW VETCH is *Vicia cracca*.

COWAGE is *Mucuna pruriens*.

COWANIA (Cowán-ia). Four or five species of evergreen shrubs or small trees of the rose family ROSACEAE are the only rep-resentatives of *Cowania*, a genus restricted in the wild to the southwestern United States and Mexico. Its name commemorates an English merchant and amateur botanist, James Cowan, who traveled in Mexico and Peru, and from there sent plants to England.

Cowanias have pinnately-lobed or -toothed, alternate or clustered, gland-dotted and more or less sticky, leathery, stalked leaves and white to purple blooms, the unisexual and bisexual ones on the same plant. The flowers have five-parted, tubular or bell-shaped calyxes, five spreading petals, numerous stamens, and one to twelve pistils. The seedlike fruits are achenes with, attached to them, the persistent, elongated, feathery styles. From related *Fallugia* the present genus differs in having no bracts at the bottom of the calyx and in the pistils being fewer.

Most likely to be cultivated **C. mexicana stansburiana** (syn. *C. stansburiana*) is a variety of a Mexican species, from which it differs in its grayish rather than brownish

Cowania mexicana stansburiana

bark and its leaves having toothed lobes. Usually not more than 10 feet tall and often lower, this aromatic shrub has three- or five-lobed leaves up to ½ inch long, white-hairy on their undersides, and rolled-under margins. The short-stalked, white to pale yellow blooms, ½ to ¾ inch wide, have usually two achenes. Native from Colorado to California, New Mexico, and Mexico, *C. m. stansburiana* inhabits dry slopes and canyons.

Garden and Landscape Uses and Cultivation. Perhaps hardy in southern New England, the shrub discussed above is much better adapted to drier climates. In regions where it is native, and in places where similar climates prevail, it is planted to some extent as a general-purpose ornamental. Sharply drained soil and sunny locations are to its liking. Propagation is usually by seed, but cuttings may also be used.

COWBERRY is *Vaccinium vitis-idaea*. Mountain cowberry is *V. v. minus*.

COWITCH. This is the common name of *Mucuna pruriens* and *Urera baccifera*. For cowitch tree see *Lagunaria*.

COWSLIP. The cowslip of Great Britain is *Primula veris*. In New England *Caltha palustris* is called cowslip. The American-cowslip is *Dodecatheon*, the Cape-cowslip *Lachenalia*, the Virginia-cowslip *Mertensia virginica*.

COYAL. See Acrocomia.

COYOTE BRUSH is *Baccharis pilularis consanguinea*.

COYOTE-MINT is *Monardella villosa*.

CRAB APPLE. See Malus.

CRAB GRASS. Among the chief weed pests of lawns the crab grasses *Digitaria ischaemum* and *D. sanguinalis* rank high. They are annuals, the seeds of the previous year of which germinate in late spring or early summer and some of the current year's crop in succession later. Crab grasses grow vigorously in hot weather; with the coming of frost they turn brown and die. Because of their short periods of greenness, their coarse appearances, and the unsightly color and texture patterns they give to lawns they are ill-regarded.

Control and elimination of crab grasses is most surely achieved by creating environments more favorable to desirable lawn grasses than to them. There is no gain in killing a crop of the pest if conditions afterward remain better suited to the growth of it or other weeds than to the sorts of grass you would like to have.

The establishment of deep, fertile seed beds for new lawns and timely attention to advantageously scheduled and performed maintenance of established ones do much to solve the crab grass problem by encouraging desirable rival grasses. The important points to keep in mind are to encourage growth by fertilizing and watering in spring and fall when blue grass and other wanted lawn sorts can benefit, and with lawns invaded by crab grass not to fertilize or water more frequently than absolutely necessary during the summer.

Another very helpful way of discouraging crab grass is to avoid mowing too closely. A height of 1½ to 2 inches in summer should be minimal. Higher cutting, so the permanent grasses will tend to shade and prevent the germination of crab grass seed, may also be practiced as a temporary measure.

Chemical crab grass killers, selective and if carefully used as manufacturers direct not dangerously harmful to desirable grasses, are sold by dealers in garden supplies. There are two types, preemergent that prevent the seed from germinating and postemergent that kill crab grass plants. No matter how effective these are in performing their appointed tasks, they will not prevent future invasions of the weed pest. Only adequate improvement of the lawn environment by good cultural practices can do that.

CRAB'S EYE is *Abrus precatorius.*

CRACCA. See Tephrosia.

CRAMBE (Crám-be)—Seakale. Not widely known in gardens, *Crambe,* of the mustard family CRUCIFERAE, contains some species of interest as ornamentals and one, seakale, esteemed as a vegetable in Europe, although less well known in America. The name is an ancient Greek one.

Ranging in the wild from the Canary Islands to western Asia, *Crambe* comprises twenty-five species of mostly glaucous, fleshy, annuals, biennials, and herbaceous perennials. They commonly have large, pinnately-lobed leaves of bold appearance. The white or yellowish flowers, abundantly borne in branched racemes, and typical of the family, have four sepals, four petals that spread in the form of a cross, six stamens of which four are longer than the others, a slender style, and a slightly lobed stigma. The spherical to ovoid podlike fruits contain a single seed.

Seakale (*C. maritima*), an inhabitant of the seacoasts of Europe, is a perennial with

Crambe maritima

thick, branched stems, and somewhat fleshy, hairless, ovate-oblong, rather cabbage-like leaves, more or less lobed, and up to 2 feet long. On stalks up to 3 feet tall, the numerous pure white, ½-inch-wide flowers, are crowded in fine display. The nearly globular, pea-like seed pods are ¼ inch wide or somewhat wider.

Ornamental *C. cordifolia,* a native of the Caucasus, is a perennial that differs from the last in being somewhat hairy, especially the leafstalks, and in having seed

pods not more than ¼ inch in diameter. This kind attains heights of 3 to 6 feet or more and has impressive foliage. The leaves, the lower ones heart-shaped at the bases, are lobed and toothed. The leafless flower stalks carry the numerous flowers well above the foliage to form broad and showy sheaves of slightly ill-scented blooms.

Garden and Landscape Uses and Cultivation. For the cultivation of seakale as a vegetable see Seakale. As ornamentals the species discussed here are useful for flower beds and for planting informally. They are hardy at least as far north as New York City, and possibly in areas with colder climates. Deep, fertile, well-drained, not acid soil and sunny locations are to their liking. They are easy to raise from seeds and begin blooming well in their third year. Root cuttings, which are an alternative means of increase, give plants that bloom at least one year earlier. Crambes dislike root disturbance, thus they should not be transplanted oftener than necessary. In cold climates the protection of a winter mulch is advisable. Distance between plants in the garden may be 2 to 3 feet.

CRANBERRY. Three species of *Vaccinium* are known as cranberries. The name, a modification of crane berry, alludes to the fanciful resemblance of the unopened flower buds to the neck, head, and beak of a crane. The plants so named are the American or large cranberry (*V. macrocarpon*), the European or small cranberry (*V. oxycoccus*), and *V. vitis-idaea* and its variety *V. v. minus,* the latter called mountain cranberry. Entirely different and botanically unrelated *Viburnum opulus* and *V. trilobum* are called cranberry-bush, and the last, highbush-cranberry. The Chilean-cranberry is *Ugni molinae.*

The only cranberries cultivated as crop plants, and the ones dealt with here, are *V. macrocarpon* and its varieties. Because of their highly specialized requirements these are not ordinarily grown in home gardens, and their commercial cultivation is essentially restricted to Massachusetts, New Jersey, Wisconsin, and to much smaller acreages in Washington, Oregon, and Canada. In addition, a certain amount of fruit is picked from wild stands. Cranberries were first cultivated on Cape Cod in 1818, but it was not until the middle of the nineteenth century that they attained substantial importance as a crop.

For their successful cultivation cranberries need moderately cool summers. Soil requirements are exacting. They grow satisfactorily only in bogs of acid mucks, peats, or what in cranberry-growing areas is called gray sand bottoms or savannahs. Although some bogs, called dry bogs, are managed without winter flooding, the usual and often necessary practice is to arrange for them to be covered with water from

late fall until spring and to flood them for shorter periods at other times. Flooding with acid water to the extent that the plants are completely submersed, is accomplished by gravity flow or by pumping. It is as important to be able to drain the bogs rapidly as it is to be able to flood them. Flooding is done to protect the plants from winter injury by heaving out of the ground, from desiccation, from damage to the flower buds by late frosts, and sometimes to control insects.

Cranberries are propagated by cuttings planted in spring in a 3- to 4-inch layer of sand spread over the bog. Every three or four years an additional ¼ to 1 inch of sand is spread. The crop is harvested with hand scoops or rakes or sometimes by machine. A well-managed cranberry bog may remain productive for sixty years or more. Anyone contemplating the commercial production of cranberries, which involves a very considerable initial investment, should seek advice from an Agricultural Experiment Station in a state where cranberries are an important crop.

CRANESBILL. See Geranium.

CRAPE. This word is employed as part of the common names of the following plants: crape-ginger (*Costus speciosus*) crape-jasmine (*Tabernaemontana divaricata*), and crape-myrtle (*Lagerstroemia*).

CRASPEDIA (Cras-pèdia). Perennial herbaceous plants and annuals of the daisy family COMPOSITAE, craspedias are confined as wildlings to New Zealand, Australia, and Tasmania. There are possibly seven species, but their relationships and geographical distributions are not always clear. The name is from the Greek *kraspedon,* an edging, and alludes to the feathery hairs that accompany the fruits (achenes).

Craspedias have undivided leaves, in basal rosettes or alternate, and branchless flower stalks terminating in dense, spherical clusters composed of many small heads of three to eight florets accompanied by leafy bracts. There are no ray florets (the equivalents of the petal-like ones of daisies). All are of the disk type (like those of the eyes of daisies). The flowers are usually white or yellow, rarely pink or purplish.

Occasionally cultivated *Craspedia uniflora,* of Australia, Tasmania, and New Zealand, is usually an annual. It has rosettes of broad-ovate, mostly long-stalked basal leaves 2 to 6 inches long and up to ½ inch wide and smaller, stalkless stem leaves. The leaves are generally margined with tangles of white, cottony hairs and have their surfaces more or less clothed with shorter hairs, with a few longer ones intermixed. New Zealand botanists recognize a number of subspecies, some with more or less sticky foliage and one in

Craspedia uniflora

which the marginal hairs are inconspicuous. The flower stalks 6 inches to 1½ feet in length are terminated by white or yellow, globular, flower clusters ¾ inch to 1¼ inches wide. In Australia this species is called billy buttons.

Two other New Zealanders are sometimes cultivated. A perennial up to 1 foot high, **C. alpina** (syn. *C. uniflora lanata*), thickly clothed with grayish-white hairs, has a rosette of obovate leaves up to 4 inches long that as they age lose much of the hair from their upper surfaces. The yellow or white flowers are in heads ¾ inch across. As tall as the last, perennial **C. incana** is densely clothed with a wool of snow-white hairs. It has obovate-spatula-shaped leaves, the basal ones up to 4 inches long, the others successively smaller up the stems. The yellow flower heads are up to 1¼ inches wide.

Garden Uses and Cultivation. The species described, sometimes including *C. uniflora*, survive where winters are nearly frostless. It is often better, however, to treat *C. uniflora*, and perhaps the others, as an annual by sowing seeds in a greenhouse or similar accommodation about eight weeks before the last frost is expected and growing the seedlings in pots or flats until planting out time. Alternatively, seeds may be sown in early spring outdoors, and the seedlings thinned sufficiently to prevent harmful crowding. A porous soil and sunny location are needed. These plants are suitable for rock gardens.

CRASSULA (Crássu-la). Jade Plant. This, the type genus of the orpine family CRASSULACEAE, is widely dispersed in the wild, but by far the greater number of its approximately 300 species, including all described here, are natives of South Africa. The name, a diminutive of the Latin *crassus*, thick, alludes to the succulent foliage and stems of most species. The plant sometimes called *Crassula coccinea* is *Rochea coccinea*.

Members of the genus *Crassula* are non-hardy, evergreen, herbaceous perennials and subshrubs some with tuberous roots and a few, not known to be cultivated, annuals. Crassulas have opposite, generally stalkless leaves, undivided, lobeless, and rarely toothed, and frequently tapering toward their bases and often with the bases of those of each pair united. In many kinds the leaves are crowded in rosettes. The flowers, in loosely-branched clusters or more compact heads, are often showy. Most commonly they are pink, red, or white, less often yellow or greenish. Each has usually five, sometimes four sepals, petals, stamens, and pistils. The fruits are little podlike follicles.

The jade plant (**C. argentea** syn. *C. portulacea*) is commonly cultivated in greenhouses, as a house plant, and in warm,

Crassula argentea

Crassula argentea (flowers)

dry climates outdoors. It is a dense, much-branched evergreen shrub up to about 10 feet tall. It has thick, succulent stems with symmetrically disposed branches and more or less pointed, glossy green, thick, fleshy, obovate to spatula-shaped leaves often margined with red. The plant cultivated as **C. obliqua**, but not the one to which that name indeed belongs, is very like the jade plant and is probably a variety of it. A var-

iant with white-streaked leaves is grown as *C. obliqua folius-variegatis*. Usually not blooming until it is quite large and old, those plants have clusters of small, starry, white or pinkish flowers at the shoot ends. Similar to the jade plant, but its blunter leaves gray, finely-red-dotted and with red edges, **C. arborescens** is also a shrub. Like the jade plant, unless big and old, it rarely blooms. Believed to be a hybrid between *C. argentea* and *C. arborescens*, the crassula named 'Blauwe Vogel' ('Blue Bird') much resembles *C. arborescens*. Its foliage is bluish-glaucous, its ¾-inch-wide flowers are white faintly tinged with lavender. This kind seems to bloom at a much earlier age and with greater regularity than *C. arborescens*.

Very distinct **C. falcata** (syn. *Rochea falcata*) has stiffly-erect, mostly branchless stems 1 foot to 3 feet tall hidden through-

Crassula falcata

out their lengths by the united bases of its rigid, strongly obliquely-sickle-shaped, fleshy, gray, 3- to 4-inch long, down-curved leaves, the blades of which are held with their edges pointing more or less upward and downward. The flattish, up-facing, crowded clusters of small, crimson to orange-red or sometimes paler blooms are terminal. Hybrids of *C. falcata* include **C. pulverulenta,** the other parent of which is *C. schmidtii*. Erect and often exceeding 1 foot in height, it has spreading, lanceolate leaves in pairs joined at their bases, with alternate pairs set at right angles to the pairs immediately below and above them. The leaves are 2 to 3½ inches long, and about ¾ inch wide, their upper surfaces channeled, their lower sides rounded. They are sprinkled with minute pustules. The small red flowers are in branched, flat-topped, clusters. Another *C. falcata* hybrid, this with *C. mesembrianthemopsis* as its other

Crassula 'Morgan's Beauty'

Crassula lactea

parent, is charming C. 'Morgan's Beauty' (syn. 'Morgan's Pink'). Choice and dwarf, this has rosettes of light gray-green, shortly-hairy, blunt, broad-obovate, spreading leaves and nearly stalkless clusters of beautiful pink blooms. There are also hybrids between C. *falcata* and C. *perfoliata*.

A dainty winter-bloomer, *C. multicava,* usually under 1 foot tall, has round-ended, moderately fleshy, broad-obovate, hairless leaves 2 to 3 inches long and with the bases of the stalks of each pair joined. The little, pink-centered, ¼-inch-wide, starry flowers are abundantly produced in loose, airy panicles. Each has four spreading, pointed petals. As the blooms fade numerous little plantlets develop on the branches of the panicles. This does well in sun or part-shade. Another attractive winter- or spring-bloomer, bushy *C. lactea,* is showier in bloom than the last. Its fragrant flowers are in dainty, longish-stalked, denser panicles than those of C. *multicava*. A subshrub 6 inches to 1 foot tall that does well in part-shade, it has obovate, flat, fleshy leaves marked near their margins with minute white dots. They are 2 to 3 inches long.

Other kinds fairly commonly cultivated include these: *C. alstonii* has stems, sparingly branched from below, up to 4 inches tall, and thickly clothed with four rows of closely overlapped, roundish, fleshy, gray-green leaves some ¾ inch wide. The small white flowers are in branched, loose clusters carried well above the foliage on erect stalks that come from near the stem apexes. *C. argyrophylla* has clustered, eventually prostrate stems and up-pointing to spreading, obovate, stalkless, minutely-hairy, dull green leaves 1¼ to 2 inches long by one-half or more as wide, the young ones brown-edged. The 3- to 4-inch-long flowering stems, with one pair of small bracts, bear branched, flat-topped clusters, ¾ inch to 1½ inches wide, of white blooms. *C. barbata* has rosettes of soft, fleshy, green leaves, their margins bearded with long

hairs. The slender, spirelike flower stalk rises to a height of nearly 1 foot and bears in the axils of its bracts small, short-stalked clusters of white flowers. The flowering stems die, but new ones come from their bases. *C. columnaris* is a choice kind with clusters of erect stems from 1½ to 4 inches tall, densely clothed with overlapping, nearly circular, fleshy leaves cupped inward at their apexes and forming a plaited pattern along the stems. The yellowish-orange to white, scented flowers are in terminal hemispherical heads about 1¾ inches in diameter. *C. comptonii* has tiny, nearly or quite stemless rosettes of minutely-gray-hairy, obovate to nearly ovate leaves and heads of pale yellow flowers. *C. congesta* (syn. *C. pachyphylla*), in flower up to 4 inches tall, is usually single-stemmed, but sometimes has clusters of shoots with thick, fleshy, pointed, broadly-ovate, hairless gray-green leaves about 1 inch long, flattened above, with rounded undersides. Alternate pairs are at right angles to each other. The bases of each pair are united. The white flowers are in dense, hemispherical heads about 2 inches in diameter. *C. cooperi* has lanceolate-spatula-shaped leaves up to ½ inch long, rounded on their undersides, and with hair-fringed margins. They are light green to reddish, and finely pitted on their upper sides. The light pink flowers in groups of three to seven are in short-stalked clusters. They are freely-borne. *C. corymbulosa,* little-branched from the base, grows erectly. Alternate pairs of its up to 1½-inch-long, tapered-triangular, gray-green, darker-dotted leaves are set at right angles to those below and above. The white flowers are many together in pyramidal clusters at the ends of stems up to about 1 foot tall. *C. deceptrix,* charming and generally not more than about 2 inches tall, forms groups of erect stems sparingly-branched from their bases. They are completely hidden by four rows of fat, roundish-triangular, keeled, whitish-gray leaves patterned with a network of minute bumps. The rather insignificant white blooms are

Crassula multicava

Crassula deceptrix

in slender-stemmed, branched clusters from near the apexes of the stems. **C. fusca,** 9 inches to 1 foot in height, has hairless stems and tapering, narrowly-white-edged, rusty-red to green leaves up to 1½ inches long by up to ½ inch wide, slightly channeled on their upper sides, and those of each pair joined at their bases. The rose-pink to white flowers are in fairly dense panicles 3 to 4 inches in length. **C. gillii** forms clusters of flat rosettes 2 to 3 inches wide of broadly-ovate, green leaves, their upper sides slightly rounded, and hairless except along the margins. The white flowers are in close heads about 1¼ inches wide atop slender stems about 4 inches long. This species does not send out runners. **C. hemisphaerica** forms mats of low rosettes, each almost circular and com-

Crassula justi-corderoyi

Crassula marnierana

Crassula hemisphaerica

posed of a brief stem carrying up to about ten closely superimposed pairs of wide-spreading, nearly round, flat, or slightly down-curved, hair-fringed leaves, the bases of each pair united. The little blooms are white. **C. hirta,** out of bloom about 6 inches tall, has narrow, lanceolate-awl-shaped, erect, spreading, and recurved leaves up to 4 inches long by ⅓ inch wide or slightly wider, the bases of each pair united. They are sprinkled with minute tufts of rough white hairs. The white flowers are in small heads at the branch ends of the up to 1-foot-tall stalks. **C. justi-corderoyi** has clusters of erect, rather loosely-foliaged stems and pink to reddish blooms. At flowering time it is up to 4½ inches tall. The leaves angle outward and upward. Pointed-ovate-lanceolate and fleshy, those of each pair are united at their bases. They are 1 inch to 1½ inches long by ⅓ inch wide or a little wider. They have flat or slightly channeled upper surfaces, rounded undersides, and are sprinkled and margined with longitudinal rows of minute white hairs and tiny dark dots. Its white flowers are in branched clusters. **C. lanu-**

ginosa, barely 8 inches tall, is a sparingly-branched subshrub with rather widely spaced pairs of thick, more or less pointed, softly-hairy ovate leaves and flowers in small branched clusters. **C. lettyae** has clustered, loose rosettes approximately 4 inches wide of blunt to somewhat pointed, flat, fleshy, bright green leaves about 2 inches long by approximately one-half as wide. Their margins, most noticeably toward their apexes, are fringed with white hairs. The slender, spirelike flowering stems attain heights of 8 or 9 inches. Somewhat distantly spaced along them in the axils of small bracts are little, essentially stalkless clusters of white flowers. **C. lycopodioides,** a very distinctive kind, as its name indicates has much the aspect of a *Lycopodium.* It exists in many varieties, differing in height, thickness of stem, amount of branching, and other details. Typically it is a subshrub up to 1 foot tall. It has slender, erect stems that branch sporadically. Throughout their lengths they are densely clothed with little, triangular, tightly overlapped yellowish-green leaves. Minute yellowish blooms develop in the leaf axils. **C. macowaniana,** 1 foot tall or taller, has branching, thick stems and thick, hairless, yellowish-green leaves up to a little more than 2 inches long by somewhat under 1 inch wide. The light pink flowers are in shortish-stalked, hemispherical, branched clusters approximately 3 inches wide. **C. marnierana** is a white-flowered species allied to *C. rupestris,* with branching stems 2 to 4 inches tall and densely furnished throughout their lengths with tightly overlapping, short, thick leaves. **C. mesembrianthemopsis** is a beautiful little mimic that, in the wild, like stone plants of the *Mesembryanthemum* relationship, closely resembles the pale grayish limestone among which it often grows. It makes clumps of neat rosettes, the flattened, triangular apexes of its leaves being all that are visible of plants that in their natural habitats grow in sand. The little white flowers are in practically stalkless clusters 1 inch wide or a little wider nestled tightly in the centers

of the rosettes. **C. nealeana** has many leafy stems up to 8 inches in length, all except the young ones trailing, and in contact with the ground, rooting. They are thickly strung with pairs of hairless, reddish-edged, glaucous-blue, pointed, broadly-ovate leaves, slightly over ¼ inch long, and with the bases of each pair joined. Alternate pairs are at right angles to each other. The upper surfaces of the leaves are concave, dotted red toward their tips, the undersides rounded, and keeled toward their apexes. Tiny white flowers are in branched, interrupted panicles up to 2 inches long. **C. orbicularis** has rosettes approximately 3 inches in diameter of spreading, pointed-obovate, 1-inch-wide, hair-fringed leaves. From the rosettes long slender stolons push out and produce new rosettes at their ends. From 6 to 10 inches tall, the slender, leafless flower stalks, loosely-branched above, carry many little clusters of white flowers. **C. perfoliata,** 2 to 3 feet tall, has erect stems and grayish-glaucous, pointed-lanceolate, spreading leaves 4 to 6 inches long by up to 1½ inches wide at their bases. Alternate pairs are at right angles. The upper sides of the leaves are more or less channeled. The undersides are rounded and somewhat keeled. Its scarlet flowers are in branched, dense clusters, up to 8 inches in diameter, that top long, leafless

Crassula perfoliata

stalks. Variety *C. p. albiflora* has white flowers. **C. perforata,** up to 2 feet tall, has long, slender, almost branchless stems, with, spaced along them, pairs of pointed-broad-ovate, spreading leaves, the bases of each pair united and completely encircling the stem. The leaves, light gray-green with tiny red dots, have bristle-haired margins. The tiny flowers are in slender spikes. **C. picturata** much resembles *C. cooperi* and is

Crassula perforata

sometimes known in gardens as *C. c. major*. It forms mats of stemless rosettes with pointed-ovate leaves ½ to 1 inch long, arranged in four rows. They have upper surfaces with tiny pits and red undersides. Their margins are fringed with hairs. The many little white blooms are in branched clusters on white-haired stalks up to 2 inches high. **C. punctulata** is bushy, clump-forming, and 6 to 10 inches tall. It has slender, branched stems and pointed-cylindrical, green leaves, powdery-glaucous toward their apexes, up to ⅓ inch long. The narrowly-tubular, up-facing white flowers are in small terminal clusters. **C. pyramidalis** has stiffly-erect stems up to about 3 inches tall very densely and completely covered with four vertical rows of broadly-triangular-ovate leaves that overlap as tightly as roof shingles and produce a strongly four-angled effect to the leafy growths. The tight, stalkless, flattish clusters of little white flowers sit attractively on the tops of the stems. **C. quadrangularis** is a larger edition of *C. socialis*. Its squarish

Crassula pyramidalis

rosettes of green, broadly-ovate leaves are 1½ to 2 inches wide. They are hairless except for minute hairs along the margins. About 6 inches tall, the flower stalks, branched at their tops, carry little clusters of small, pure white flowers. **C. radicans** is a more or less spreading subshrub up to more than 1 foot in height, its flat, usually hairless leaves from somewhat under to a little over ½ inch long. Its small white flowers are in crowded clusters. **C. reversisetosa** is a densely-matting, dwarf species, with rosettes up to ¾ inch wide of fleshy, green, broadly-ovate to nearly round leaves with minute red spots. Stems and leaf margins are bristly-hairy, the hairs on the stems directed backward. The trailing flowering stems, 4 to 6 inches in length, have pairs of leaves joined at their bases. They produce new rosettes from the leaf axils which, in contact with the ground, root. The white blooms are in terminal clusters. **C. rosularis** is somewhat variable. It has basal rosettes of rather loosely-arranged, oblong-lanceolate, lustrous green leaves, hairless except along the margins, 2 to 3 inches long and ½ to ¾ inch wide. The little white flowers are clustered at the branch ends of erect panicles 8 to 10 inches tall that develop from the centers of the rosettes. This kind appreciates light shade. **C. rupestris** (syn. *C. perfossa*) is much like *C. perforata*, its pairs of light-gray-green leaves appearing as though threaded at regular intervals along the thin, lax stems. They are broadly-ovate, about 1 inch long, and much-rounded on their undersides. The united bases of each pair completely encircle the stem. The little flowers are yellowish. **C. schmidtii** forms mats, in bloom about 4 inches high. It has hairy stems and spreading, pointed, linear-lanceolate, green leaves with flat or slightly concave upper surfaces approximately 2 inches long by ¼ inch wide. Reddish beneath, they are dotted with tiny pits and have minutely-white-hairy margins. The rose-red to deep carmine-red flowers are in loose, branched clusters. It appreciates light shade. **C. s. alba** is a white-flowered variant. **C. sericea,** 6 to 8 inches tall, has spreading to upright, softly-silky-hairy, elliptic to obovate, bluntish, green leaves, with flattish upper surfaces and rounded lower ones. They are up to 1 inch long by ¾ inch wide or a little wider, the younger ones often brownish-edged. Those of each pair are joined at their bases. Alternate pairs are at right angles to each other. The branchless or forked-branched flowering stems carry few flowered clusters of white blooms. **C. socialis** is a neat, low matting kind with alternate pairs of pointed-broad-ovate, hairless, green leaves, each pair at right angles to the pairs above and below, forming distinctly four-sided rosettes. The bases of each pair of leaves are united. The leaf edges have minute-horny teeth. The largest leaves are not over ⅓ inch long. Plentifully pro-

Crassula socialis

duced, the small terminal clusters of white flowers are held aloft on erect stalks 1 inch to 1½ inches long. **C. spathulata** thrives in semi-shade. A robust trailer, this has slender, branching, squarish stems and stalked, pointed-heart-shaped, minutely-round-toothed, hairless, glossy green leaves ½ to 1½ inches long, the pairs fairly distantly spaced. The starry, pink flowers are in loose, branched clusters. This is suitable for hanging baskets. **C. susannae** forms carpets of closely-packed, 1-inch-wide rosettes of flattened leaves. The leaf tips have a silvery, frosted appearance so that the rosettes have much the aspect of those of *Titanopsis*. The tiny flowers, in clusters at the apexes of 1-inch-high stalks, are white and fragrant. **C. tecta** forms low clumps of loose rosettes of grayish-green, egg-shaped leaves, semicircular in cross section and thickly clothed with brief, thick, white hairs. They are 1 inch to 1¼ inches long by nearly ½ inch wide. The white flowers are in dense, rounded heads about ¾ inch wide that top usually branchless, erect, nearly leafless stalks 3 to 4 inches tall. **C. teres** is a little gem, well known to succulent fanciers, and somewhat like *C. columnaris*. Of neat appearance, it forms a cluster of erect stems approximately 2 inches tall symmetrically clothed with broad, brownish-green, translucent-margined, tightly overlapping leaves, cupped inward at their apexes. These plaited-patterned, leafy columns are in season capped with flattish, crowded

Crassula teres

clusters about 1½ inches wide, of fragrant, white flowers. **C. tetragona,** up to 3 feet in height, is an upright subshrub. Its slender stems are clothed with somewhat distantly spaced, up-curving, nearly cylindrical, tapered, green leaves 1 inch long or a little longer by approximately ¼ inch wide. The flowers are small and white. C. 'Tom Thumb' much resembles a smaller C. *rupestris,* one of its parents, the other being C. *brevifolia.* Compact and in bloom up to 4 inches tall or a little taller, the hybrid has brownish-edged, fleshy, broadly-ovate leaves scarcely ¼ inch long strung along its slender stems. The bases of the leaf pairs are united. The tips of young leaves are reddish. The flowers are white, in short-stalked, flattish heads at the terminations of the stems. **C. trachysantha** (syn. C. *mesembryanthoides*) is a bushy, bristly-hairy subshrub up to 1 foot tall. Its tapering, slim, cylindrical leaves, incurved toward their tips, are thickly clothed with backward-pointed white hairs. They are ½ to ¾ inch long. Yellowish-white, the flowers are in dense, small clusters at the terminations of erect stems that have a few pairs of distinctly spaced small leaves below the flower.

Garden and Landscape Uses. Crassulas are of prime interest to collectors of succulents. The species described here and others less well known are treasured by fanciers of such desert plants. None is hardy. All thrive with little care in greenhouses and adapt well as window plants. According to habit they are suitable for pots, pans (shallow pots), and hanging baskets. In warm, dry regions they are appropriate for rock gardens and the taller ones for other uses, for instance, as container specimens to decorate patios and terraces.

Cultivation. Like the vast majority of succulent plants cultivated crassulas demand thoroughly drained soil. They will not tolerate wet feet. For preference the earth in which they grow should be coarsely textured. Potting soils can be brought to this state by the inclusion of liberal amounts of broken crock, crushed brick, crushed oyster shells, gritty cinders, fine grit, or perlite. The earth, however, should not be poor in nutrients. These plants respond best to a reasonable level of fertility. Although desert plants, not all crassulas appreciate brilliant sun. Some grow natively in the shade of shrubs. In cultivation in sunny regions, such as California, these are at their best in partial shade. In places with less intense light they may stand, and indeed be better with, full exposure. If the foliage shows signs of scorching (the most exposed paling or browning) provide some shade. Indoors, winter night temperatures of about 50°F suit, and by day, five to fifteen degrees higher, depending upon the brightness of the weather. The atmosphere should be buoyant and airy, not heavily humid. Moderate watering from spring through fall, with much less frequent applications in winter, is needed. It is always well to let the soil become nearly dry before soaking it. Very well-rooted specimens may be given occasional, say once a month, applications of dilute liquid fertilizer in summer, but this must not be overdone. It is better to err by not fertilizing than to overuse fertilizers. Potting is not usually necessary every year, but generally is every second or third year, except for quite large specimens. It is best done in spring. In years when potting is not done, pick away as much of the surface earth as possible and replace it with new. Crassulas are very easily increased by cuttings, leaf cuttings, division, offsets, and seeds. For more information see Succulents.

CRASSULACEAE—Orpine Family. The majority of the 1,500 species of the thirty-five genera of this group of dicotyledons are natives of South Africa, the Mediterranean region, southcentral Asia, and Mexico. But the orpine family, by no means exclusive to those lands, is nearly cosmopolitan in its natural distribution, although scarce in Australia and islands of the Pacific. Among its most familiar sorts are kalanchoes, sedums, and sempervivums.

Chiefly natives of dry or seasonally dry habitats and practically all succulents or semisucculents, these plants are mostly evergreen herbaceous perennials and subshrubs. A few are annuals. Their fleshy, generally stalkless leaves, chiefly undivided, occasionally pinnate, are alternate, opposite, or in whorls (circles of more than two). The flowers, rarely unisexual with the sexes on separate plants, are prevailingly bisexual. Symmetrical and in branched clusters, heads, spikelike arrangements, or racemes, they seldom are solitary. The calyx is of three to thirty, but most often four or five sepals, commonly separate, less often partly joined at their bases. There are as many petals as sepals, usually free, rarely united. Generally there are as many or twice as many stamens as petals and usually as many pistils, each with a scale-like gland at its base. The fruits are follicles. Genera in cultivation include *Adromischus, Aeonium, Aichryson, Chiastophyllum, Cotyledon, Crassula, Cremneria, Cremnophila, Dudleya, Echeveria, Graptopetalum, Graptoveria, Greenovia, Kalanchoe, Kalorochea, Lenophyllum, Monanthes, Orostachys, Pachyphytum, Pachyveria, Rochea, Rosularia, Sedadia, Sedum, Sempervivella, Sempervivum, Tacitus, Thompsonella, Umbilicus,* and *Villadia.*

CRATAEGOMESPILUS (Crataego-méspilus)—Haw-Medlar. This is the name of graft-hybrids or chimeras of English hawthorn (Crataegus *monogyna*) and the medlar (Mespilus *germanica*), which gives reason for the name. They belong in the rose family ROSACEAE. Sexual or true hybrids between these genera are named *Crataemespilus.* As chimeras the tissues of *Crataegomespilus* contain cells of both parents living in intimate association, but each cell carries only the chromosomes and genes of one parent. Crataegomespiluses are not true hybrids.

The origin of these chimeras is interesting. About 1895 an English hawthorn in the garden of M. Dardar, at Bronvaux near Metz in France, developed two branches from close to the union of its understock and scion that showed characteristics intermediate between those of hawthorn and medlar. The branches differed from each other. They were propagated by grafting and the progeny were thought to be hybrids; later they were named graft-hybrids, and still later chimeras.

Resembling a medlar more than a hawthorn, **C. dardari** is a deciduous tree up to 20 feet tall with rather pendulous, more or less thorny branches and ovate, oblong, or obovate leaves 1½ to 4 inches long and less than one-half as wide. They are sometimes finely-toothed, but are lobeless and are pubescent on both surfaces. The white blooms, in clusters of up to twelve, are 1½ inches wide. The fruits resemble those of medlars, but are smaller and clustered. In addition to this characteristic growth this tree produces occasional branches with leaves,

Crataegomespilus dardari

Crataegomespilus dardari

flowers, and fruits identical with those of *C. d.* 'Jules d'Asnieres', others having leaves, flowers, and larger, solitary fruits typical of the medlar, and occasionally some that are pure English hawthorns.

The chimera *C. d.* 'Jules d'Asnieres' (syn. *C. asnieresii*) is a quite lovely flowering tree. Fifteen to 20 feet in height, it has downy shoots sometimes with a few thorns, and obovate to broadly-ovate leaves, some deeply lobed like those of the English hawthorn, others without lobes. When young they are scurfy, and their undersides are softly-hairy. They are 1½ to 3 inches long by about two-thirds as wide. The flowers, in hairy-stalked clusters, are white fading to pinkish and distinctly larger than those of English hawthorn. The fruits are about as big as those of that tree, but are brown and downy.

Garden and Landscape Uses. Because of their interesting histories, as well as for their not inconsiderable decorative appeal, these chimeras are worth planting where small, flowering trees can be used advantageously. They are good lawn trees, thriving under conditions that suit apples and crab apples, that is to say in fertile, well-

Crataegomespilus dardari 'Jules d'Asnieres'

Crataegomespilus dardari 'Jules d'Asnieres'

drained soil in sun. They are hardy in southern New England.

Cultivation. This is the same as for crabapples (see Malus). Ordinarily no regular care is needed beyond the removal of any branches from the interior of the trees that tend to become crowded. Propagation traditionally has been by grafting or budding onto English hawthorn understocks, but it is possible that cuttings would root under mist. When taking scions of the chimeras for grafting those intermediate between hawthorn and medlar must be selected. If the occasional shoots of pure hawthorn or pure medlar are used as scions the trees so propagated will be of those species.

CRATAEGUS (Crat-aègus)—Hawthorn, Thorn, Thornapple, Haw. No one really knows how many species of *Crataegus* there are. More than one thousand have been described as inhabiting North America, and there are others in Europe, North Africa, and Asia. Most authorities agree that many of these so-called species should not be considered species, that they are only variants of more definable species or are hybrids. The situation is complex and confusing, and much research must be done before it is clarified, if it ever is. The matter is not of prime importance to gardeners because, although there are a number of kinds useful to them, the group as a whole is not of prime horticultural importance. Hawthorns certainly do not rate with their close allies the crab-apples or with flowering cherries as garden ornamentals.

A realistic estimate of the number of species of *Crataegus* is, perhaps, in the vicinity of 200. They are most numerous in North America where there are certainly over 100 species. The genus takes its name from the Greek *kratos*, strong, and alludes to the wood. It consists of deciduous trees and shrubs, usually thorny, of the rose family ROSACEAE. They have alternate, undivided, but usually lobed or toothed leaves, often varying in size, shape, and other characteristics on the same plant and bisexual flowers, rarely solitary, in twos or threes, but more often in many-flowered flattish clusters. Prevailingly they are white or cream, less often pink or red. There are five sepals, five petals (more in double-flowered varieties), five to twenty-five stamens, spaced alternately in one to five circles, and with stalks that broaden at their bases, and one to five styles. The fruits are structured like miniature apples and, like apples, are what botanists call pomes. They are spherical to egg- or pear-shaped, mostly red, but in a few species and varieties yellow, blue, or black. Each contains one to five bony nutlets usually called seeds.

Because native species of hawthorn are so abundant in North America and in many places are readily collectable from

the wild, the number that may be brought into occasional cultivation locally is considerable. Many of these are similar and not superior to those here described. In addition to native kinds, two European species and varieties of them are commonly planted, as are a few Asians.

English hawthorn and may are common names applied without discrimination to two fragrant-flowered species, *C. laevigata* (syn. *C. oxyacantha*) and *C. monogyna*, both popular in Great Britain where they are much used for hedges surrounding fields as well as for ornament. They are also planted, especially in their pink- and red-flowered varieties, in North America. From a practical point of view there are no significant differences between these species; botanically they are distinguished by *C. laevigata* having flowers with two or three styles and fruits with two seeds and *C. monogyna* having blooms with one, or rarely two styles and one-seeded fruits. These species hybridize. Unlike many hawthorns they make no show of fall foliage color. Up to 15 feet in height, **C. laevigata** has spreading branches and strong thorns about 1 inch long. Its broadly-ovate to obovate, hairless leaves have wedge-shaped bases and usually three to five toothed lobes. They are up to 2 inches long. In its typical form the flowers are white with red anthers and in clusters of five to twelve, each bloom with twenty stamens. The red, ellipsoid fruits are from

Crataegus laevigata

Crataegus laevigata (flowers)

Crataegus monogyna

Crataegus phaenopyrum (foliage and fruits)

slightly under to slightly over ½ inch long. This species is indigenous to Europe and North Africa. The very similar **C. monogyna** may attain a height of 30 feet and inhabits Europe, North Africa, and western Asia. Both species are hardy through most of New England.

Varieties of English hawthorn include variants of both *C. laevigata* and *C. monogyna*. Notable among the former are *C. l. aurea*, with yellow fruits; *C. l. paulii*, with bright red, double flowers; *C. l. plena*, with double white blooms; *C. l. punicea*, with single carmine flowers with white centers; and *C. l. rosea*, with single red flowers with white centers. Varieties of *C. monogyna* are *C. m. compacta*, a compact, thornless kind; *C. m. pteridifolia*, with deeply-lobed leaves; *C. m. rosea*, with pink flowers; *C. m. stricta*, a slender, dense, col-

umnar tree, and *C. m. pendula*, with drooping branches, and the Glastonbury thorn.

The Glastonbury thorn (*C. m. biflora*) is of particular interest because of the legend attached to it. Its peculiarity is that it bears white blooms in fall (at New York City about November, in England around Christmas) as well as in spring. The story told of this variety is that after the crucifiction of Christ, Joseph of Arimathaea came to England to introduce Christianity and preached at Glastonbury on Christmas Day. Failing to impress his hearers, he thrust his hawthorn staff into the ground and prayed for a miracle. His prayer was answered by it bursting immediately into bloom. The annual midwinter blooming of the Glastonbury thorn is said to commemorate this Christmas Day miracle.

Two outstanding native Americans are the Washington thorn and the cockspur thorn, both much favored for landscaping. Undoubtedly the first (**C. phaenopyrum**) is

one of the finest of its genus. Broadly columnar and about 30 feet in height, it is densely-branched and has slender thorns and glossy foliage that becomes brilliant orange-red to scarlet in fall. The broad-ovate to triangular leaves, 1¼ to 2¾ inches long, are three- to five-lobed and sharply-toothed. In many-flowered clusters, the white flowers are ½ inch wide and have twenty stamens with yellow anthers. The fruits, ⅓ inch long and nearly spherical, are glossy scarlet and remain attractive long into the winter. Handsome in bloom and fruit and of good habit, the Washington thorn is native from Virginia to Alabama and Missouri and is naturalized further north. Its variety *C. p. fastigiata* is of upright, slender habit. The cockspur thorn (**C. crus-galli**) is a tall shrub or rounded or flat-topped tree, 25 to 35 feet in height, with a dense crown of horizontal branches and many long, slender thorns. It is native from Quebec to North Carolina and Kansas. From ¾ inch to 3 inches long, its lustrous leaves are obovate to oblong-obovate, with wedge-shaped bases and usually rounded apexes. They are not lobed, but their upper margins are sharply-toothed. In fall they become orange to scarlet. The white flowers, many in a cluster, exceed ½ inch in diameter and are showy. They have ten to thirteen stamens with pink anthers. The subspherical red fruits, ⅜ inch in diameter, remain attractive through much of the winter. Variety *C. c. splendens* has very glossy foliage.

One of the finest hawthorns, **C. prunifolia** is thought to be a hybrid between the eastern North American species *C. crusgalli* and *C. succulenta macracantha*. Up to about 20 feet tall, it forms a rounded, densely-foliaged head, commonly wider than high, with its branches often reaching the ground and having sharp thorns 1½ to 3 inches long. Its roundish to elliptic or obovate, lobeless leaves with glossy, hairless upper surfaces and dull, sometimes slightly downy undersides, are 1½ to 3½ inches long and finely-toothed almost to their bases. In fall they turn glowing crim-

Crataegus laevigata paulii

Crataegus monogyna biflora

Crataegus crus-galli (foliage and fruits)

Crataegus prunifolia (flowers)

son. Borne in rounded clusters, the ¾ inch-wide flowers each have ten to fifteen stamens with pink anthers. The fruits are nearly spherical, ⅝ inch long, and rich red. Broader leaves, hairy flower stalks, and fruits that drop early distinguish this hawthorn from *C. crus-galli*.

Long-lasting fruits are also a feature of *C. nitida,* a round-headed or sometimes flattish-topped tree up to 30 feet in height, indigenous from Illinois to Arkansas. It has spreading branches and usually is thorny. The lustrous, more or less lobed leaves are elliptic to oblong-ovate, pointed, coarsely-toothed, and ¾ inch to 3 inches long. They turn brilliant orange and red in fall. In many-flowered clusters, the white flowers, each with fifteen to twenty stamens tipped with light yellow anthers, are ¾ inch in diameter. The dull red fruits are subspherical, about ⅜ inch long, and contain two to five seeds. Other hawthorns with fruits that last well into the winter are *C. viridis*, of central and southeastern United States, and the hybrid *C. lavallei* (syn. *C. carrierei*), which has as parents *C. crus-galli* and *C. pubescens*. Both are hardy through most of New England. About 20 feet tall, *C. lavallei* has pubescent young shoots and is sparsely furnished with stout thorns. Its pointed, ovate-oblong to elliptic leaves are irregularly toothed along the upper parts of their margins and pubescent be-

neath and to a lesser degree on their upper sides. In fall they become an attractive bronzy-red. The good-looking, ¾-inch-wide white flowers have red center disks and fifteen to twenty stamens with red or yellow anthers. The stalks of the many-flowered clusters are hairy. Orange-red to brick-red, the fruits are ellipsoid, about ⅝ inch long, and decorative. A spreading tree up to about 35 feet in height, *C. viridis* is sparingly furnished with slender thorns and has oblong-ovate to elliptic, toothed leaves ¾ inch to 2¼ inches long, glossy on their upper sides and paler beneath. At first with hairs, they lose these except on the veins beneath as they become older. The white flowers, many together in clusters, are about ½ inch wide. They have twenty stamens with pale yellow or sometimes red anthers. The spherical fruits, up to ¼ inch in diameter and bright red to orange-red, usually have five seeds.

Early to fruit and early to drop its fruit (they ripen in August and fall in September), *C. arnoldiana* is native from Massachusetts to Connecticut and New York. Up to about 35 feet tall, it has zigzagged branchlets and formidable thorns 2 to 3 inches in length. The coarsely-double-toothed, shallowly-lobed leaves are broad-ovate and 2½ to 4 inches long. When young pubescent, at maturity they become shining and hairless above, remaining slightly hairy on their undersides. The white, ¾-inch-wide flowers have ten stamens with pale yellow anthers. About ¾ inch long, the subspherical, pale-dotted, red fruits are pubescent at their ends. They have three to four seeds.

Large edible fruits, for which the trees are cultivated abroad, characterize the azarole (*C. azarolus*), of the Mediterranean region and western Asia, and *C. pinnatifida major*, a Chinese kind. The azarole, little known in America and probably not hardy north of southern New York, is a slightly thorny or thornless tree, up to 30 feet in height, similar to the English hawthorns, but differing in its fruits having three to five seeds without cavities on their inner surfaces. The deeply-lobed, ovate to obovate leaves, up to 2¾ inches long, at first pubescent on both sides, with age become almost hairless above. The ½-inch-wide, white flowers have twenty stamens with purple anthers. The fruits of the azarole are whitish, yellow, or red, and roundish and apple-flavored. In China the fruits of *C. pinnatifida* as well as those of its variety *C. p. major*, are candied, strung on sticks, and offered for sale in the markets; they resemble miniature candied red apples. Those of *C. p. major* are 1 inch in diameter, those of the species smaller. Their stamens number twenty. These trees are nearly thornless, about 20 feet tall, and have leaves divided almost to their midribs into toothed lobes. Both are hardy in southern New England.

Exceptionally large fruits are also borne by *C. mollis* and *C. punctata*. Indigenous from Ontario to Virginia, South Dakota, and Kansas, *C. mollis* has a broad, rounded crown, attains a height of about 35 feet, and has thorns 1 inch to 2 inches long. Its broad-ovate, shallowly-lobed leaves are 2½ to 4 inches long and sharply-double-toothed. When young they are hairy on their undersides, but at maturity the pubescence persists only on the veins. The 1 inch-wide, white flowers, with red central disks, have twenty stamens. The stalks of the flower clusters are densely-hairy. Pubescent and usually pear-shaped, the bright red fruits with pale dots are about 1 inch long and have four or five seeds. They ripen when the foliage is still green and show to good effect against it. Picturesque and with widely-spreading horizontal branches, *C. punctata* is native from Quebec to Georgia and Illinois. It is either thornless or has stout short thorns. Its young shoots are pubescent. The leaves are obovate, 2 to 4 inches long, sometimes lobed, always sharply toothed toward their apexes, and hairy beneath. The flowers, many together in hairy-stalked clusters, are white, ½ to ¾ inch across, and have twenty stamens with pale yellow to pink anthers. The fruits, which give reason for the botanical *punctata*, are spotted, dull red, roundish to pear-shaped, and ¾ to 1 inch long. In variety *C. p. aurea* they are yellow.

The hardiest hawthorns of merit are *C. succulenta* and a hybrid between it and *C. laevigata paulii* named *C.* 'Toba'. These survive in southern Canada. Native from Quebec to Massachusetts and Illinois, *C. succulenta* is from 10 to 20 feet in height and has stout branches and thorns up to 2 inches long. The leaves are broadly-elliptic to obovate, 2 to 3¼ inches long, and coarsely-double-toothed. Glossy and hairless above, they have hairs on the veins of their lower surfaces. Many together in hairy-stalked clusters, the white flowers, ¾ inch across, have fifteen to twenty pink-anthered stamens. The spherical, bright red fruits, ⅝ inch in diameter, contain two to four seeds. With glossy foliage, fragrant, double, deep rose-pink blooms, and red fruits ½ inch in diameter, *C.* 'Toba' resembles English hawthorn in habit, but is considerably hardier.

Other red-fruited kinds of decorative value include *C. coccinioides* and *C. pruinosa*, both hardy in southern New England. The first is native to the central United States, the latter from Ontario to Virginia and Illinois. Round-topped and dense-headed *C. coccinioides* is about 20 feet tall and has glossy, dark purple spines 1½ to 2 inches long. Its rounded, broadly-ovate, double toothed, lobed leaves, at first hairy on the veins beneath, become nearly hairless at maturity. They are 2 to 3½ inches long and dull green turning to orange and scarlet in fall. Five to seven together, the white flowers, about 1 inch in diameter, have twenty pink-anthered stamens. The glossy, orange to red, roundish fruits, ¾ inch in diameter, have usually five seeds. About the same height as the last, **C. pruinosa** has firm-textured foliage, reddish when very young, later bluish-green above and paler beneath, and white flowers with pinkish anthers. It has a spreading crown and many strong thorns. Its pointed, elliptic to ovate leaves, up to 2 inches long, are triangularly-lobed and irregularly-, often doubly-, toothed. In clusters of usually five to ten, the flowers are ¾ to 1 inch in diameter and have twenty stamens with usually pink anthers. The fruits change from apple-green to red and finally dark purple. They are up to ½ inch in diameter and contain sweet, yellow flesh and three to five seeds.

Sometimes called thicket hawthorn, **C. intricata** is 6 to 10 feet tall and has curved thorns 1 inch to 1½ inches long and

Crataegus intricata

shallowly-irregularly-toothed, broad-ovate leaves 1 inch to 3 inches long. The white flowers, more than ½ inch wide, are succeeded by dull, reddish-brown, spherical fruits nearly or quite ½ inch in diameter. This native from Massachusetts to Michigan, North Carolina and Indiana, makes effective thorny thickets and hedges.

Yellow to orange-yellow fruits are the outstanding feature of the central Asian *C. wattiana*, a species hardy as far north as southern New England. It is probably the handsomest yellow-fruited hawthorn (others are *C. laevigata aurea* and *C. punctata aurea*, already referred to under their respective species). Up to about 20 feet tall, **C. wattiana** has short spines, or none, and bright green, ovate, hairless leaves, 2 to 3¼ inches long, with sharply-toothed lobes. Its flower clusters are 2 to 3 inches in diameter. The white blooms have fifteen to twenty stamens with cream anthers. The roundish bracts are about ⅓ inch in diameter.

With blue fruits ⅓ inch in diameter the pomette bleue (**C. brachyacantha**) is the only hawthorn with fruits of its color. A stout tree up to 50 feet in height, it is indigenous to the southern United States. It is not hardy in the north. It has spreading branches and many short, curved spines. The leaves, obovate-oblong to elliptic, are up to 2 inches long. They are slightly lobed and have rounded teeth. When young they are pubescent, but become smooth later. The white flowers, ⅓ inch across and with fifteen to twenty stamens, fade to orange and are followed by bright blue, slightly waxy-coated fruits.

Garden and Landscape Uses. As ornamentals the better hawthorns have many good qualities to commend them to prospective planters, but alas, they have off-setting bad ones that must be taken into consideration. Not the least of the latter is their distressing susceptibility to a number of quite serious pests and diseases. To maintain hawthorns in really attractive condition is likely to involve annual spraying as well as other details of care. On the credit side they are shapely and often with age become picturesque with rugged trunks and dense irregular crowns of wide-spreading horizontal branches. Because of their moderate size they are suitable for small properties and limited spaces, and they tolerate environments unfavorable to many trees and shrubs, adapting themselves well, for instance, to city and seaside conditions and withstanding exposure to wind. They are not fussy about soil. They succeed in those of an alkaline character as well as in neutral or acid ones. Even in poor, comparatively infertile soil, they do surprisingly well, although they make faster growth in more nutritious soils. The thorny kinds are formidable barriers, and many can be sheared to form tall, strong hedges. Some, such as the English hawthorns and the Washington thorn, make good low- or medium-height hedges. The foliage of hawthorns is dense and generally pleasing in texture, pattern, and color. That of some kinds colors well in fall. The blooms, although less showy than those of crab-apples and flowering cherries, are quite decorative and are usually succeeded by attractive fruits, which in some species remain long into the winter and provide welcome color, particularly when seen against a background of evergreens or snow.

Disadvantages possessed by hawthorns, in addition to their susceptibility to disfigurement or worse from pests and diseases, are that they do not transplant especially well, particularly in large sizes, that in some locations the thorns can be a nuisance or even a danger, and that because of these most kinds are difficult to handle and to prune.

Cultivation. When planting hawthorns it is better to set out young specimens than

large ones. The latter are usually slow or reluctant to reestablish themselves. They should be dug with large root systems and be relocated before these have any opportunity to dry. If the soil is poor it pays to improve it in the vicinity of the roots prior to planting by mixing in organic matter and fertilizer. Early fall or early spring are appropriate planting seasons. Pruning, involving reducing the tops of the plants by one-third to one-half, either by judicious thinning and shortening of the shoots or by cutting the plants back will do much to ensure their survival. Immediately following planting the ground should be soaked and during dry weather in the first summer weekly irrigations promote success. Routine care of hawthorns, aside from any necessary spraying to control pests and diseases, includes giving attention to whatever pruning is needed. This consists mostly of thinning overcrowded branches, especially from the interior of the tree, cutting out unwanted sucker shoots, and encouraging the leading shoots of young trees by shortening any others that show a tendency to outgrow them.

Seeds afford the most satisfactory way of raising hawthorns, except double-flowered, pink- and red-flowered, and other horticultural varieties. [These last are grafted or budded onto the species to which they belong or onto other kinds of hawthorns. The cockspur thorn (*C. crus-galli*) is a satisfactory understock for most kinds.] Seeds do not usually germinate well until the second, or sometimes third, season after sowing, although if sown fresh in the fall they sometimes will sprout the following spring. The seeds may be sown in cold frames, in outdoor beds protected from rodents and other causes of disturbance, or in pots or pans (shallow pots) sunk to their rims in the ground. A shaded location where the soil can be kept uniformly moist, but not constantly saturated is needed for the seed bed. A 2-inch layer of peat moss or other mulch spread over the bed helps to retain moisture. It should be removed at the end of a year, when sprouting is expected. An alternative method is to sow in flats, stack these on top of each other in a shaded place, and cover them with black polyethylene film. There they remain until the spring after sowing when they are spaced out in anticipation of germination.

Hedges of hawthorn need shearing once or more times a year. Should they become thin and bare at their bases they can be thickened and restored to good condition by a procedure known in Great Britain as laying. This consists of chopping or cutting part way through, close to ground level, the trunks and branches, bending them over in the line of the hedge at an angle of 30 to 45 degrees, and securing them to stakes driven into the ground at intervals or to occasional branches of the hedge left erect. When the operation is completed the hedge is sheared tightly. In response to this treatment new erect shoots develop and interlace with those laid at an angle to form a dense impenetrable barrier.

Pests and Diseases. The most common afflictions are aphids, borers, caterpillars, leaf miners, red spider mites, and scale insects, as well as cedar-apple rust, fireblight, fungus leaf blight, powdery mildew, and twig blight.

CRATAEMESPILUS (Cratae-méspilus). This, the correct name of sexual bigeneric hybrids between *Crataegus* and *Mespilus* of the rose family ROSACEAE, must not be confused with the so-called graft-hybrids or chimeras of those genera named *Crataegomespilus*. The latter are not true hybrids.

Old in cultivation and of uncertain origin, but probably a natural hybrid, *Crataemespilus grandiflora* (syn. *Crataegomespilus grandiflora*) is a round-headed tree up to about 30 feet tall, with drooping lower branches. Its broad-elliptic to obovate leaves, 2 to 3½ inches long and downy on

Crataemespilus grandiflora (flowers)

both surfaces, are finely-toothed and often have angular lobes near their apexes. Generally in twos or threes, the 1-inch-wide flowers are pure white and look much like those of the medlar. The yellowish-brown spherical fruits, approximately ¾ inch in diameter, contain mealy flesh and two small stones. From the last *C. gillotii* (syn. *Crataegomespilus gillotii*) differs in having toothless, lobed leaves and smaller flowers with two styles. Discovered in France in 1875, this is believed to be a natural hybrid between the medlar and *Crataegus monogyna*.

Garden and Landscape Uses and Cultivation. These are handsome small trees well suited for use as lawn specimens. They respond to conditions and care that suit their parents. They are propagated by grafting onto English hawthorn. These trees are hardy in southern New England.

CRATEROSTIGMA (Cratero-stígma). Twenty species of Africa and Madagascar compose *Craterostigma*, of the figwort family SCROPHULARIACEAE. The name, from the

Greek *krateros*, stout, and *stigma*, alludes to the conspicuous stigmas of the flowers.

This is a genus of nonhardy herbaceous perennials and annuals with all basal, plantain-like leaves and flowers in spikes, racemes, or less often solitary. The blooms have a tubular, five-ribbed calyx, and a tubular, two-lipped corolla with a concave upper lip, a three-lobed, spreading lower one. There are four stamens and one style. The fruits are capsules.

Native of Ethiopia, *C. pumilum* (syn. *Torenia auriculaefolium*) has a rosette of broadly-ovate, hair-fringed leaves and erect, slim,

Craterostigma pumilum

flowering stalks scarcely exceeding the foliage, each with one or two pale lilac, ¾-inch-wide blooms blotched with lavender, and with white and yellow in their throats.

Garden Uses and Cultivation. A rare and interesting plant for inclusion in collections of tropical plants and for planting in rock gardens and similar locations in the humid tropics, the species described is easily raised from seed and responds to environments that suit begonias.

CRATOXYLUM (Crató-xylum)—Mempat. Indigenous from China to Borneo and the Philippine Islands, *Cratoxylum*, of the St.-John's-wort family HYPERICACEAE, comprises a dozen species. Its name, sometimes spelled *Cratoxylon*, and of obvious application is from the Greek *krateros*, strong, and *xylon*, wood.

Cratoxylums are trees and shrubs with opposite, undivided, toothless, thin leaves besprinkled with tiny, semitransparent dots visible when the leaf is held to the light. The flowers, one to five on stalks from the leaf axils or in terminal panicles, are yellow, white, or pink. They have five each sepals and petals, many stamens, and three to five styles. The fruits are capsules, differing from those of related *Hypericum* in splitting along the partitions instead of midway between them, and in containing winged seeds.

A handsome, aromatic, deciduous shrub or tree with flattened branches and short-

stalked, narrowly- to broadly-elliptic leaves usually 1½ to 3 inches, but sometimes up to 4 inches long, that on their undersides are glaucous, the mempat (*C. polyanthum* syns. *C. ligustrinum, Hypericum pulchellum*), is native from China to the Philippine Islands. Its slightly fragrant, pink flowers, solitary or in clusters of two or three from the upper leaf axils, are ⅓ to ½ inch wide. Their stamens are in three bundles.

Garden and Landscape Uses and Cultivation. In warm climates the mempat is cultivated as an ornamental. It succeeds in ordinary soil in sun or part-day shade and is easily raised from seeds and cuttings.

CRAWFURDIA. See Tripterospermum.

CREAM. The word cream appears in the common names of these plants: cream bush (*Holodiscus discolor*), cream cups (*Platystemon californicus*), and cream sacs (*Orthocarpus lithospermoides*).

CREEPER. As part of their common names the word creeper belongs in these: Australian bluebell-creeper (*Sollya heterophylla*), Australian love creeper (*Comesperma volubile*), evergreen trumpet creeper (*Phaedranthus buccinatorius*), railroad creeper (*Ipomoea cairica*), Rangoon creeper (*Quisqualis indica*), snow creeper (*Porana*), trumpet creeper (*Campsis*), and Virginia creeper (*Parthenocissus quinquefolia, P. inserta*).

CREEPERS. Plants that spread by trailing above ground or underground rooting stems are sometimes called creepers. The Virginia creeper (*Parthenocissus quinquefolia*) is a good example. In British usage the word creeper is often used for tight-clinging vines, such as *Ficus repens*.

CREEPING. This word is employed as part of the common names of various plants including these: creeping bent (*Agrostis*), creeping Charlie (*Lysimachia nummularia* and *Pilea nummulariaefolia*), creeping-cucumber (*Melothria pendula*), creeping devil (*Machaerocereus eruca*), creeping-forget-me-not (*Omphalodes verna*), creeping Jennie (*Lysimachia nummularia*), creeping-myrtle (*Vinca minor*), creeping pearlberry or creeping-snowberry (*Gaultheria hispidula*), creeping sailor (*Saxifraga sarmentosa*), and creeping-zinnia (*Sanvitalia procumbens*).

CREMANTHODIUM (Creman-thòdium). In reference to the more or less drooping flower heads, the name *Cremanthodium* derives from the Greek *kremao*, hanging, and *anthodium*, flower head or capitulum. It is that of a genus of fifty-five species of the daisy family COMPOSITAE, natives of the Himalayan region and China.

Cremanthodiums are related to arnicas. They have undivided leaves and solitary, large, daisy-type flower heads with bisexual disk florets and a single row of yellow or pinkish, petal-like ray florets. The fruits are seedlike achenes.

Not common in cultivation, *C. arnicoides* is possibly not reliably hardy in the north. About 3½ feet tall, it has broad-ovate, mostly basal leaves and flower heads 2½ inches across that resemble small sunflowers with bright yellow ray florets and darker disks.

Garden Uses and Cultivation. These are as for *Arnica*.

Cremanthodium arnicoides

CREMNERIA (Crem-nèria). Hybrids between *Cremnophilia* and *Echeveria* of the orpine family CRASSULACEAE are correctly identified as *Cremneria*, a name formed by combining parts of those of the parent genera. Until 1975, when the name *Cremneria* was first applied, such plants were included in *Echeveria* and as such are likely to be found in gardens.

In appearance cremnerias are characteristically intermediates between their parents. They include *C. expatriata* (syn. *Echeveria expatriata*), the parents of which are *Cremnophylla linguifolia* and probably *Echeveria microcalyx*, *C. mutabilis* (syn. *Echeveria mutabilis*), the parents of which are *Cremnophila linguifolia* and probably *Echeveria scheeri* or perhaps *E. carnicolor*, and *C. scaphylla* (syn. *Echeveria scaphylla*), the parents of which are *Echeveria agavoides* and *Cremnophylla linguifolia*.

Garden Uses and Cultivation. These are as for *Echeveria*.

CREMNOPHILA (Cremnó-phila). Closely related to *Echeveria*, the two species of *Cremnophila*, of the orpine family CRASSU-

LACEAE, have by some authorities been included in that genus. From it they differ, as they also do from *Pachyphytum*, which they otherwise much resemble, in technical details of the flowers. The name, alluding to the natural habitats of these plants, comes from the Greek *cremno*, a cliff, and *phylla*, a leaf.

Cremnophilas are succulent perennials with thick, spreading or erect stems and alternate, fleshy leaves in rosettes or more or less scattered along the stems. Originating in the leaf axils, the flowering stalks, their lower parts furnished with much smaller leaves than those of the stems, terminate in narrow panicles of flowers with five-lobed calyxes, five petals, ten stamens united at their bases, and five erect carpels. The fruits are capsules.

The first species to be named, *C. nutans* (syn. *Sedum nutans*), of Mexico, has stems 3 to 5 inches long and spoon-shaped, obovate, long-persistent leaves 1 inch to 3 inches in length by up to 1½ inches wide by ½ inch thick and pointed at their apexes. The curved, bent, or drooping flowering stalks, 4 to 8 inches long, end in about 4-inch-long panicles of many blooms with green sepals and slightly longer, bright yellow petals. Well known in gardens, *c. linguifolia* (syn. *Echeveria linguaefolia*), of Mexico, has branched, often decumbent stems up to 1 foot long. The

Cremnophila linguifolia

broad-spatula-shaped to obovate, thick, dark green leaves, flat on their upper sides, and round beneath, are in rosettes at the branch ends. They are 1 inch to 3½ inches long and up to ¾ inch wide. From 1 foot to 1½ feet long, the flowering stalks end in raceme-like panicles of many whitish blooms.

Garden Uses and Cultivation. These are as for *Echeveria*.

CRENATE. Meaning shallowly-round-toothed, this term is frequently used to describe the margins of leaves and other plant parts.

CREOSOTE. One of the oldest and most effective wood preservatives, creosote must be used with considerable caution in gardens. It is deadly to plant tissues and even when considerably diluted with water can work great harm. Creosote fumes also damage plants. Never treat wood to be used in the construction of cold frames, or stakes, trellises or other supports for plants with creosote and do not paint wounds made by cutting limbs or branches with it.

CREOSOTE BUSH is *Larrea divaricata*.

CREPIS (Crè-pis)—Hawk's Beard. Of the approximately 200 species of this genus of the daisy family COMPOSITAE, twelve are natives of the Americas, the remainder of Europe, Asia, and North Africa. The New World kinds, all North American, are reinforced there by the naturalization of several Old World species. The name, used by Pliny for some unidentified plant, is derived from the Greek *krepis*, a sandal or foot, and as applied to *Crepis* has no obvious significance.

Milky-juiced annuals, biennials, and herbaceous perennials, crepises have lobeless and toothless or once- or twice-pinnately-lobed leaves, mostly basal, with a few much reduced in size and alternate on the stems. As usual with plants of the daisy family, the structures ordinarily called flowers are flower heads consisting of many florets, which are the true flowers. In hawk's beard all the florets are strap-shaped like those of dandelions, and all are bisexual. The plant sometimes called *C. barbata* is *Tolpis barbata*. The seedlike fruits are achenes.

Very few hawk's beards are of horticultural interest. The most commonly cultivated, *C. rubra*, is a showy annual, freely-branching and 1 foot to 2 feet tall. Native from Italy to Greece, and nearly hairless, it has several leafy stems. The leaves are lobed, the basal ones stalked, and those above stalkless. The solitary, long-stalked, bright red to deep pink flower heads are up to 1½ inches wide. In variety *C. r. rosea*

Crepis rubra

the flower heads are pink. Those of *C. r. alba* are white or whitish.

A free-blooming, yellow-flowered perennial, **C. blattarioides** is a native of the mountains of Europe. From 1 foot to 1½ feet tall, it has a stout taproot and stalked, oblanceolate, toothed basal leaves up to 6 inches long. The stem leaves are smaller, lanceolate, and stalkless. The flower heads are in clusters of up to five.

Crepis blattarioides

A European native of mountain meadows, **C. aurea** is a hardy, showy-flowered perennial with generally solitary, orange or reddish-orange flower heads ¾ inch to 1¼ inches wide topping stalks up to 10 inches tall. All the foliage is basal. The oblanceolate, hairless, shiny leaves are shallowly-lobed.

A yellow-flowered hardy perennial, native in dry, open places from Alberta, South Dakota and New Mexico to British Columbia and California, **C. occidentalis** is up to 1½ feet tall and has gray-hairy stems and foliage that may become nearly hairless with age. The lower leaves, up to 1 foot long, are oblong-lanceolate and pinnately-lobed or -toothed, the upper ones stalkless and stem-clasping. Up to 1 inch wide, the flower heads are in clusters of up to twenty.

Garden Uses and Cultivation. Hawk's beards present no problems of cultivation. They grow vigorously in almost any well-drained soil in full sun and may be planted in flower beds, and *C. aurea* in rock gardens. In very hot, humid weather the annual kind suffers and does not bloom as freely as under less torrid, drier conditions. Seeds of this are sown in early spring, or in mild climates in fall, where the plants are to remain, and the seedlings thinned to 6 inches apart. Perennial hawk's beards can be raised from seeds sown outdoors or indoors and the young plants grown in pots or in an outdoor nursery bed until big enough to set out in their permanent locations.

CRESCENTIA (Crescén-tia)—Calabash Tree. Five tropical American species of ever-

green trees of the bigonia family BIGNONIACEAE form this genus. Only one is common in cultivation. Native of dryish regions of Mexico and Central America, the calabash tree is grown in most hot climates. In many places distant from its original habitat, it occurs spontaneously as an escape from domestication. The name *Crescentia* commemorates the Italian writer on agricultural topics Pietro Crescenzi, who died in 1321.

Crescentias are hairless. Their undivided leaves are alternate, and fairly widely spaced or clustered at the nodes. The flowers are large and yellowish, with reddish or purplish veins. They have a two-parted or deeply-five-cleft calyx and a tubular corolla with five fluted lobes (petals). There are four usually nonprotruding stamens in two pairs, and a slender style tipped with a two-lobed stigma. The fruits, far different in appearance to what the term implies to the nonbotanist, are technically berries. They have hard skins and contain numerous seeds surrounded by pulp. The shells of the fruits of the calabash tree are put to many uses in the tropics. They are cleaned and cured, usually by slow smoking, and made into hula rattles and maracas as well as cups, bowls, vessels, ornamental boxes, and similar articles. They take a high polish and often are decorated by carving intricate patterns into their surfaces and by painting. By skillful tying of the fruits when they are young selected parts are constricted and mature specimens can be had in a variety of shapes. In Mexico the poisonous pulp of the fruits is used medicinally and the seeds, after roasting, are eaten. In the West Indies the young fruits are pickled and eaten. The wood of the calabash tree is used as fuel, and for constructing boats.

The calabash tree (**C. cujete**), occasionally 40 feet tall, more commonly does not exceed one-half that height. Its branches spread widely and horizontally. They are

Crescentia cujete

furnished with irregularly-spaced, knob-like protuberances bearing tufts of leaves. Within the tufts the leaves vary considerably in size, the largest being about 7 inches long. They are dark glossy green and broadly-lanceolate. The pendulous blooms are solitary and evil smelling, the odor intensifying as they mature. They open at night and undoubtedly attract carrion-eating insects that effect pollination. Short-stalked, and funnel-shaped, they are about 2 inches long and yellowish-purple or greenish-yellow, with dull purple or brown veins. They are borne on the trunk and larger branches and are succeeded by the quite amazing, spherical fruits up to 20 inches in diameter. A calabash tree with a heavy crop of fruits hanging from its trunk and branches is a strange sight to those unacquainted with the species. The tree sometimes named *C. alata* is *Parmentiera alata*.

Crescentia cujete (fruits)

Garden and Landscape Uses. Although scarcely a handsome species because of its gawky, branching habit, the calabash tree is of sufficient interest as a curiosity to find a place in many tropical gardens. When in fruit it never fails to attract the attention of visitors from temperate climates who are unaccustomed to trees that bear fruits from their trunks and major branches or that have fruits as large as those of *C. cujete*. A special use for this species in humid warm climates is as a host upon which to grow epiphytic orchids, bromeliads, and other tree-perching plants. Such kinds seem to thrive particularly well when attached to the bark of calabash trees. In the United States this species can be grown outdoors in southern Florida, southern California, and Hawaii. It is sometimes cultivated as a curiosity in large greenhouses. The tree grows rather slowly, in any ordinary garden soil.

Cultivation. No particular care is needed to cultivate the calabash tree. It may need pruning occasionally to keep it within bounds. In greenhouses it may be accommodated in a ground bed or in a large tub in rich, coarse soil kept always evenly moist. Well-rooted specimens benefit from regular applications of dilute liquid fertilizer from spring through fall. The atmosphere should be humid. A minimum winter night temperature of 55 or 60°F is recommended, with the day temperature five to ten degrees warmer. At other seasons considerably higher temperatures are in order. Propagation is by seed and by air layering.

CRESS. This word forms parts of the common names of several plants, most, but not all members of the mustard family CRUCIFERAE. For a discussion of cresses grown as salads see the following entry. The common names of these plants include the word cress: bitter-cress (*Cardamine*), blister-cress (*Erysimum*), Brazil- or Para-cress (*Spilanthes oleracea*), Indian-cress (*Tropaeolum*), lake-cress (*Armoracia aquatica*), penny-cress (*Thlaspi*), rock-cress (*Arabis*), stone-cress (*Aethionema*), violet-cress (*Ionopsidium acaule*), and winter or upland cress (*Barbarea vulgaris*).

CRESS. This name is used for more than one kind of plant of the mustard family CRUCIFERAE esteemed for their piquant leaves used as salads, seasonings, and garnishes. For watercress see Watercress. Common garden cress (*Lepidium sativum*) is grown as an annual. Sow seeds in the open as early in spring as the ground can be worked and make successional sowings at two- or three-week intervals. Choose a sunny spot where the soil is fertile and moist, or can be kept so, and sow fairly thickly in shallow drills spaced about 1 foot apart. Leaves will be ready for cutting in one month to six weeks, and a second crop will grow if the first harvesting is not too severe. In hot summer weather this cress does not do well and soon runs to seed, but in the north make sowings again in August for a fall crop of salad greens. If you garden in mild climates you may sow at intervals through fall and winter.

Early or Belle Isle cress (*Barbarea verna*) is grown as a biennial. Sow seeds under conditions advised above for common cress in very early spring. By summer the plants will have attained fair size, but usually they are not as much used for salads then as through fall, winter (in mild climates), and early spring. In spring the plants soon go to seed.

Traveling Americans are sometimes puzzled by the little green seedlings that in Great Britain and some other parts of Europe are included in salads and sprinkled on sandwiches. They are baby plants of common cress and of white mustard (*Sinapis alba*) or black mustard (*Brassica nigra*). They are raised by sowing the seeds thickly in shallow flats of soil and keeping them moist in full light in a temperature of 50 to 55°F at night and slightly warmer during the day. The seeds are merely pressed into the surface, not covered with soil. Children enjoy raising these crops by sowing them on pieces of flannel laid across plates or other suitable containers that can hold just enough water to keep the flannel moist. When mustard and cress (as companions the two are considered superior to either eaten alone) are to be harvested together, the cress is sown, in a separate container, about three days before the mustard.

CRETAN BEAR'S TAIL is *Celsia arcturus*.

CRETAN-MULLEIN is *Celsia cretica*.

CRICKETS. Familiar insects related to and resembling grasshoppers, crickets are well known for the chirping sound males make by rubbing together modified parts of the front wings. Some sorts live in trees and are called tree crickets, some that, in addition to flying, burrow in the ground are called mole crickets. Several species damage plants including berry fruits, cucumbers, grapes, tomatoes, watermelons, and many others. The chief control is by poison baits.

CRIMSON FLAG. See Schizostylis.

CRINDONNA is *Amarcrinum*.

CRINITARIA VULGARIS is *Aster linosyris*.

CRINODENDRON (Crinodén-dron). Two natives of temperate South America comprise *Crinodendron*, of the elaeocarpus family ELAEOCARPACEAE. The name is from the Greek *krinon*, lily, and *dendron*, a tree, it alludes to the shape and texture of the flowers. In gardens these plants are sometimes known by the synonym *Tricuspidaria*.

Crinodendrons are evergreen shrubs or small trees with usually alternate, toothed leaves and solitary, nodding, thick-stalked flowers from the leaf axils. The blooms have a bell-shaped, obscurely five-toothed, deciduous calyx, five toothed or lobed petals, and numerous stamens. The fruits are leathery capsules.

Attaining a maximum height of 30 feet, but in cultivation usually lower, *C. hookeranum* is of stiff habit. It has pointed, oblong-lanceolate leaves 1½ to 5 inches long by up to one-third as wide, toothed except toward their bases, dark glossy green above, paler beneath, and downy on the midribs above and on the midribs and veins beneath. The numerous urn-shaped flowers, fleshy, rich coral-red, and 1 inch to 1½ inches long, hang from shoots of the previous year. Their petals end in three small teeth. Up to about 30 feet in height, *C. pa-*

Crinodendron hookeranum

tagua has oval leaves up to 3 inches long by one-half as wide, shallowly- and rather coarsely-toothed, and dull dark green above and markedly paler beneath. The bell-shaped white flowers, about ¾ inch long, have petals three-toothed at their ends.

Garden Uses. Crinodendrons can be grown outdoors and in lath houses in California and similar mild climates, but they will not tolerate much frost. They are plants for cool, well-drained, moist, acid soils containing an abundance of organic matter. They do not succeed in exposed places, where summers are very hot, or in windy, dry, and sunny locations. Appropriately sited, they are handsome and choice evergreens that add quality to garden plantings.

Cultivation. Propagation can be by cuttings of firm, half-ripened shoots in late summer or fall. The young plants should be grown in pots in a fertile, peaty soil until large enough to be planted in their permanent locations. At no time should the soil dry out. If drainage is good, as it should be, water generously. Frequent spraying of the foliage with water is highly beneficial. Scale insects are likely to infest these plants.

Crinodendron patagua

CRINODONNA is *Amarcrinum*.

CRINUM (Crì-num)—Crinum-Lily, Milk-and-Wine-Lily, Poison Bulb. Crinums, or crinum-lilies as they are often called, are handsome, large bulb plants of the amaryllis family AMARYLLIDACEAE. As natives they inhabit warm and tropical parts of the Old and the New World. The name *Crinum* is derived from the Greek *krinon*, a lily. There are 100 species or more.

Crinums include evergreen and deciduous species. Their large, sometimes immense bulbs often taper gradually upward to form thick, stemlike necks. The stalkless, thick leaves are strap- to sword-shaped. The flowers are in umbels of few to many atop leafless, solid stalks. At the base of each umbel is a pair of large spathe valves (bracts). Individual blooms are large and short-stalked or stalkless. Their perianths have long, slender, cylindrical, curved or straight tubes and six similar or nearly similar lobes, generally called petals, but more correctly tepals. These spread

Crinum, unidentified species

widely or recurve from the top of the slender perianth tube or point forward to form funnel- or bell-shaped blooms of lily-like aspect. They are white or nearly white, often suffused or striped with wine or purplish-red. There are six usually down-pointing stamens and a long, slender style terminating in a small, headlike stigma. The fruits are capsules with large, green seeds.

Species native to the Americas include the Southern swamp crinum (**C. americanum** syn. *C. roozenianum*), endemic to wet soils and swamps from Georgia to Florida and Texas. The short-necked bulbs of this are up to 4 inches in diameter. The narrowly-strap-shaped leaves are sparingly-toothed, 1 foot to 2 feet long by 1½ to 2 inches wide. Fragrant, the white, individually stalkless or nearly stalkless flowers, up to six in each umbel, have

slim, green to purplish, straight perianth tubes 4 to 5 inches in length, and wide-spreading petals up to ½ inch in width, as long or shorter than the perianth tube. The spreading, conspicuous stamens have pink or red stalks. The stalks supporting the umbels are 1 foot to 2 feet tall.

Other New World natives are West Indian *C. caribaeum* (syn. *C. floridanum*) and, from tropical America, *C. erubescens* and *C. kunthianum*. With strap-shaped leaves, at their middles 3 to 4 inches wide, **C. caribaeum** has umbels of three or four blooms each with a straight perianth tube 3 to 4 inches long, and nearly as long as the tube, spreading, linear petals. Short-necked bulbs 3 to 4 inches in diameter are characteristic of **C. erubescens.** Its many thin, strap-shaped leaves, with slightly rough edges, are 2 to 3 feet long by up to 3 inches wide. In umbels of usually up to six, but sometimes twice as many, the fragrant blooms are white inside and purplish-pink outside. They are stalkless or have exceedingly short stalks, narrow, straight, erect perianth tubes 5 to 6 inches long, and recurved lanceolate petals about one-half as long. The stamens, which do not protrude, have burgandy-red stalks. Its short-necked bulbs up to 3 inches in diameter, **C. kunthianum** has many more or less wavy, smooth-edged, strap-shaped leaves, 2 to 3 feet long by 2 to 3 inches wide. The umbels top stalks about 1 foot tall. They have few blooms, white or white with purplish outsides, without or nearly without individual stalks. The slim perianth tubes are 6 to 8 inches long. The wide-spreading lanceolate petals are not more than 3 inches long.

Frequent in southern gardens, South African **C. bulbispermum** (syns. *C. longifolium*, *C. capense*) is probably the most commonly cultivated crinum in North America. It has long-necked, flask-shaped bulbs and several to many roughish-margined, long-pointed, deciduous leaves 2 to 3 feet long by 3 to 4 inches wide, which arch upward and then downward. On stalks 1½ to 3 feet long, the umbels are of eight to fifteen trumpet- to bell-shaped flowers with individual stalks 1 inch to 2 inches in

Crinum bulbisperum

length. The blooms are white to pink with a rose-pink streak down the center of each petal. The outsides of the petals are stained reddish-purple. The oblong petals, up to 1 inch wide, about equal in length the 3- to 4-inch-long, curved perianth tubes. The stamens are nearly as long as the petals. Native of woodlands in South Africa, beautiful **C. moorei** differs from *C. bulbispermum* in its deciduous, short-pointed leaves having smooth edges. They are 2 to 3 feet long and 3 to 4 inches wide. The

Crinum moorei

blooms resemble those of the belladonna-lily (*Amaryllis belladonna*), but appear with the foliage rather than when the bulb is leafless. The umbels of bloom terminate stalks 2 to 4 feet long. Very fragrant, the flowers, on individual stalks 1½ to 3 inches long, are pale to deep pink or sometimes white. They have a curved perianth tube 3 to 4 inches long equaled in length by the oblong, 1- to 1½-inch-wide petals. The pink-stalked stamens are much shorter than the petals. The bulbs of *C. moorei*, 6 to 8 inches in diameter, have thick necks up to 1 foot long. An intermediate hybrid between the last two species, **C. powellii** is a very lovely pink-flowered crinum. White flowers are borne by *C. p. album*.

Other South African species, both deciduous, are *C. campanulatum* (syn. *C. aquaticum*) and *C. macowanii*. As its synonym

Crinum powellii

Crinum powellii (flowers)

suggests, **C. campanulatum** is a lover of wet soils, but requires drier conditions during its dormant period. It has comparatively small bulbs and narrowly-strap-shaped, channeled, square-ended leaves 3 to 4 feet long. In umbels of up to fifteen or more, its flowers are up to 5 inches long. They are pink, with a deeper stripe down the back of each petal, and have curved perianth tubes from which the petals expand to form a broad bell. The free parts of the petals are pointed and recurved. The bulbs of **C. macowanii** are spherical and 8 inches or more in diameter and taper to a very long neck. The bright green, spreading, often wavy leaves are 2 to 3 feet long. The stalks supporting the umbels of sweetly scented, nodding, trumpet- to bell-shaped blooms are about the same length. The flowers are white or pink, with a rose-pink stripe down the center of each petal. Their curved perianth tubes are about 3 inches long and greenish toward their bases. The petals are shorter than the tube and curve outward at their apexes.

Several tropical African crinums are cultivated. A native of Angola, **C. fimbriatulum** has narrow leaves 2 to 3 feet long, with hairs along their margins. The umbels are of few funnel-shaped greenish-white blooms with corollas with a narrow tube 4 to 5 inches long, and broad petals striped red down their centers. Its homeland Zanzibar, **C. kirkii** has very pointed, strap-shaped, finely-toothed leaves 3 to 4 feet long by 4 inches wide or wider. The 1- to 1½-foot-long stalks carry umbels of one dozen or more short-stalked or stalkless, trumpet-shaped flowers, white with a wide red band down the outside of each petal. They have greenish perianth tubes about 4 inches in length and petals about as long, exceeding 1 inch in width, and longer than the stamens. An inhabitant of shallow waters in West Africa, **C. natans** has submersed bulbs and floating, strap-shaped, wavy leaves up to 3 feet long and 1½ inches wide. The umbels of flowers, with scarcely evident individual stalks, top a common stalk that lifts them above the

water. White, tinged with purple, they have a perianth tube 5 to 6 inches long and petals about one-half as long. Native of bogs and shallow waters in West Africa, **C. purpurascens** has stolons, and strap-shaped leaves up to 2 feet long and 1 inch wide. Borne on stalks not more than 1 foot long, the umbels are of up to ten purple-tinged, white flowers, with a perianth tube 5 to 6 inches in length, petals about one-half as long, and red stamens. About 1 foot tall, evergreen **C. podophyllum,** of tropical Africa, has a few oblanceolate, toothed leaves up to 1 foot long by 2 inches wide. Its white, funnel-shaped flowers, without or with very short individual stalks, have perianth tubes 5 or 6 inches long.

Showy **C. scabrum** has a large bulb and several firm, strap-shaped, rough-edged leaves 2 to 3 feet long by not more than 2 inches wide. Its handsome, funnel-shaped blooms, with or without very short individual stalks, are white with crimson keels to the petals. They have greenish, curved perianth tubes 3 to 5 inches long and pointed, oblong petals about 3 inches long and not surpassed by the stamens. Toothed- or crinkle-margined leaves up to 2 feet long by 2 inches wide are characteristic of **C. sanderanum.** Native of tropical Africa, this kind has umbels of few trumpet-shaped white flowers, with red keels along the outsides of the petals, and without or with very short individual stalks. The perianth tubes are 5 or 6 inches long. The petals are 3 to 4 inches long by under 1 inch wide. This is sometimes called milk-and-wine-lily.

The popular milk-and-wine-lily is a variety of **C. latifolium,** of tropical Africa and tropical Asia. The species has a long, thick-necked bulb up to 6 inches in diameter. Its many leaves are strap-shaped, 2 to 3 feet long, 3 to 4 inches wide, and rough-edged. In umbels of ten to twenty, and with very short individual stalks, the funnel-shaped flowers have a curved, green perianth tube 3 to 4 inches long. The petals, white with red keels, are 3 to 4 inches long by about 1 inch wide. The milk-and-wine-lily (**C. l. zeylanicum** syn. **C. zeylanicum**), of tropical Asia, has larger bulbs and fewer, shorter, wavy leaves with somewhat rough margins. Atop tall, purple-tinged stalks, the umbels are of up to twelve individually very short-stalked, trumpet-shaped, red-striped, white, fragrant blooms, each of which opens for only one evening and night. The flowers have curved perianth tubes 3 to 6 inches long and pointed, oblong-lanceolate petals 3 to 4 inches long by 1 inch wide. Shorter than the petals, the stamens point downward.

Poison bulb (**C. asiaticum**) is robust and very variable. Its ominous vernacular name alludes to a variety of it, the bulb of which is reported to be poisonous and is used medicinally in Asia. Native to tropical Asia, where it commonly grows along sea-

shores, this is hardy in the deep south. It forms clumps of tapering, cylindrical bulbs 1 foot long or longer by nearly one-half as thick, with columnar, stemlike necks. It has twenty to thirty crowded, arching, pale bluish-green, tapering leaves up to 3 feet long by 3 to 5 inches wide. Held not much above the foliage, the spidery, pure white or sometimes pink, fragrant blooms are in clusters of twenty or more atop two-edged stalks. They have individual stalks under 1 inch long. The perianth tube is 3 to 4 inches long. The linear petals, about 3 inches long by up to ½ inch wide, droop or curve downward. The red-stalked stamens are about 2 inches long. St.-John's-lily (**C. a. sinicum**) has broader leaves and larger flowers. Another native of tropical Asia and Indonesia, **C. amabile** is very large. Much like *C. asiaticum*, it has comparatively small bulbs with necks up to 1 foot long or longer. Its numerous strap-shaped, smooth-edged, tapering, green leaves are 3 to 4 feet long by 3 to 5 inches wide. The more or less two-edged stalks supporting the umbels of twenty or more blooms are 2 to 4 feet tall. The very fragrant flowers with individual stalks 1 inch to 2 inches long have narrow, straight or slightly curved, crimson perianth tubes 3 to 4 inches long. The wide-spreading petals, 3½ to 6 inches long or sometimes longer by not more than ½ inch wide, spread widely. Longer than the stamens, they are curved, deep crimson on their outsides and white flushed with red inside. Similar **C. augustum,** of Mauritius and the Seychelles Islands, is larger than *C. amabile*. Its bulbs attain a diameter of 6 inches, and its leaves are less tapered. Its less heavily scented blooms, deeper red on their outsides, have petals up to 7 inches long and usually somewhat over ½ inch wide.

Australian **C. pedunculatum** (syn. *C. australe*) is similar to *C. asiaticum*, but has flowers with green perianth tubes and white petals. The linear petals are considerably shorter than the 4-inch-long perianth tubes. The stamens have red stalks. This kind, up to 5 feet tall, in the wild often grows along seashores. Another Australian, **C. flaccidum** has bulbs 3 to 4 inches in diameter and long, flat leaves up to a little more than 1 inch wide. The umbels of up to fifteen strongly fragrant, white flowers terminate stalks 1 foot to 2 feet long. The perianth tubes are 2 to 4½ inches long, the wide-spreading petals up to 3 inches long and ¾ inch wide. The stamens are nearly as long as the petals.

Garden and Landscape Uses. Crinums are magnificent, free-flowering ornamentals, where hardy splendid for beds and other plantings outdoors and for containers to decorate patios, terraces, and similar places. They are especially effective at watersides. They prosper in greenhouses, in ground beds, and in large pots or tubs, but because of their considerable size are generally unsuitable for small structures. Most crinums are not hardy in the north, but *C. bulbispermum*, *C. moorei*, and *C. powellii* prospered outdoors for many decades at the base of a south-facing wall at The New York Botanical Garden in New York City.

Mild frostless or nearly frostless climates are necessary for the successful outdoor cultivation of most crinums. Many, including the only native species *C. americanum*, succeed in Florida and along the Gulf Coast, and some further north. There are others that need more tropical conditions.

Crinums can be displayed with good effect as single specimens or grouped informally. For their best success they need deep, fertile, moist soil. Some kinds do well in wet soil. Light shade from the strongest sun is appreciated. In cold climates the few that survive outdoors are benefited by the protection of a winter covering of a thick layer of salt hay, leaves, branches of evergreens, or similar material.

Cultivation. Plant the bulbs of crinums 2 to 3 feet or more apart according to the size individual kinds are expected to attain. Outdoors set the bulbs with their tips just beneath the ground surface. In pots or tubs the long necks stand above soil level. Except for the removal of faded blooms and dead leaves crinums need no special care. They are benefited by yearly, or in poor soils, twice- or thrice-yearly applications of a complete garden fertilizer. Container specimens must be watered freely from spring through fall, but kept much drier in winter. Well-rooted specimens benefit from weekly applications from spring to fall of dilute liquid fertilizer. Indoor temperatures must be adjusted according to the preferences of individual kinds. Those native of the southern United States and South Africa succeed with winter night temperatures of about 50°F, with daytime increases of five to fifteen degrees. Kinds from more tropical areas need a minimum winter night temperature of 60°F, with corresponding increases by day. All appreciate a humid atmosphere. In the north container specimens of cool climate kinds may be put in a lightly shaded place outdoors during the summer and brought indoors before fall frost. Propagation by fresh seeds is very easy, in fact the seeds sometimes begin to germinate before they are shed from the capsules. Sow the seeds, covering them to about their own depth, in very sandy soil, in a temperature of 60 to 70°F.

CRISTARIA (Cris-tària). Forty species of the mallow family MALVACEAE constitute *Cristaria*. The name, of uncertain application, presumably derives from the Latin *crista*, a tuft, plume, or comb.

Natives of temperate South America, cristarias are mallow-like herbaceous plants

with lobed leaves and usually rosy-violet flowers, solitary or in racemes or panicles. They have five-parted calyxes, five petals, many stamens united in a tube, and as many styles as carpels. The fruits are ripe carpels arranged like those of *Malva*, in a ring.

Sometimes cultivated, *C. glaucophylla*, of Chile, is 1 foot or so high and is clothed throughout with soft, stellate (star-shaped) hairs. Its long-stalked leaves have blades, ovate in outline, or three leaflets, conspicuously lobed or cleft. The center leaflet much exceeds the others in length. The flowers, in panicles from the tops of the stems, are ¾ inch long.

Garden Uses and Cultivation. In regions of fairly mild winters and not excessively hot and humid summers the species described is useful for flower gardens. It succeeds in ordinary well-drained soil in sunny locations and is easily raised from seed.

CRITHMUM (Críth-mum)—Samphire. To the carrot family UMBELLIFERAE belongs samphire (*Crithmum maritimum*). As its specific epithet suggests it is an inhabitant of sea coasts from western Europe to the Crimea. It favors crevices in rocks and cliffs, often those bathed with salt spray. More rarely it occurs on beaches. The only species, it is sometimes cultivated in herb gardens, more rarely for ornament. Its leaves are pickled and eaten. The common name is derived from an old French one, a corruption of Saint Pierre, the patron saint of fisherman. The botanical name comes from the Greek *krithe*, barley, and alludes to the resemblance of the fruits to grains of barley.

A fleshy, hairless, densely-branched, hardy perennial herbaceous plant 1 foot to 2 feet in height and often with a somewhat woody base, *C. maritimum* has fleshy, solid, longitudinally-ribbed, pliant stems, and narrow, glaucous, twice- or thrice-pinnate leaves, with thick, juicy, awl-shaped leaflets about ½ inch long, the bases of which encircle the stems. Lasting but for a short time, the white or yellowish flowers, up to ¹⁄₁₀ inch across, are in umbels 1¼ to 2½ inches wide of eight to twenty smaller umbels. The five-petaled blooms have no calyxes. The green to deep purple fruits (seeds) are ovoid and slightly winged.

Garden Uses and Cultivation. Away from the sea, an environment it seems to need for its best development, samphire is often troublesome to grow. It needs full sun and a very thoroughly drained, dryish soil or a saline moistish one. It is raised from seed.

CROCANTHEMUM. See Helianthemum.

CROCIDIUM (Cro-cídum). The daisy family COMPOSITAE contains the one species of *Crocidium*. It is a spring-blooming annual, native from California to British Columbia.

Its name is a diminutive of the Greek *kroke*, thread, and alludes to the persistent woolly hairs in the leaf axils.

Up to 6 inches tall, *C. multicaule* forms tufts of erect stems terminating in solitary golden-yellow flower heads about ¾ inch across, with ray florets (those that suggest petals) under ¹⁄₁₀ inch long. When young the foliage is woolly-hairy, but soon the hairs fall. The leaves are confined to the lower one-half or one-third of the stems. Obovate to spatula-shaped, they are about ½ inch in length. Sometimes they have a few marginal teeth. The basal leaves are fleshy. The fruits are achenes.

Garden Uses and Cultivation. This is an interesting plant for rock gardens, the fronts of flower borders, and similar locations. It is satisfied with ordinary, well-drained garden soil. In mild climates the seeds are sown in fall, in harsher climates in spring, where the plants are to remain, and the seedlings are thinned so that they do not crowd each other.

CROCKING. This is a gardening term for the placing over the drainage holes in the bottoms of pots and other plant containers crocks (shards) or substitutes for crocks to promote the escape of surplus water. When crocking, first position over each hole, hol-

Crocking: (a) Place a large crock concave side down

(b) Add smaller crocks

(c) Cover with coarse leaves or similar material

low side down, a fairly large crock. If the container is 5 inches in diameter or larger set on top of this a layer of smaller crocks, also with their concave sides down. Add sufficient smaller crocks to level the surface, and cover with a layer of coarse dry leaves, hay, moss, or similar material to prevent soil washing into and clogging the drainage. With containers under 5 inches in diameter, the secondary crocks may be dropped in without positioning, but without disturbing the main crock placed over the drainage hole. The amount of crocks to use varies with the size of the container, character of the soil, kind of plant, and watering practices. Ordinarily a 1-inch layer is adequate in a 5-inch pot, a thicker layer in larger ones.

CROCKS. Pieces of broken clay flower pots (shards) are called crocks. They are chiefly used in the bottoms of pots and other plant containers to cover drainage holes and promote the escape of surplus water. Substitutes for crocks include coal cinders, broken oyster shells, pieces of charcoal, and devices made of wire or plastic. Crocks smashed or crushed to sizes that pass through a ¼-inch mesh are good for mixing with potting soils to promote porosity and drainage.

CROCOSMIA (Crocós-mia)—Coppertip. Some authorities admit only *Crocosmia aurea* to this genus, referring the remaining species that others include to nearly related *Tritonia*. Here, *Crocosmia* is accepted as comprising six species. It belongs in the iris family IRIDACEAE and has a name, derived from the Greek *krokos*, saffron, and *osme*, smell, given because when soaked in water the dried flowers have a saffron-like odor. In the wild the genus is restricted to South Africa.

Crocosmias are deciduous, herbaceous perennials with underground, bulblike organs called corms. They have erect, wiry, usually branched stems with, toward their bases, linear to sword-shaped, parallel-

veined, stem-sheathing leaves. The flowers are in spikes or in panicles of spikes. They have tubular perianths with six lobes (petals) of nearly equal size, which are oblong to somewhat spoon-shaped. There are three stamens and a three-branched style. The fruits are capsules.

An old garden favorite that in the wild favors partially shaded places where the soil is moist, **C. aurea** has fans of light green leaves up to 1½ feet long by under ½ inch wide. Each has a prominent midvein. In panicles about twice as tall as the foliage are many slightly down-facing blooms. These are about 2¾ inches across and have golden-orange faces and deeper orange outsides to their petals. In variety *C. a. maculata* three of the petals are marked with chocolate bands.

Also long in cultivation, **C. pottsii** (syn. *Tritonia pottsii*) is 3 to 4 feet tall. It has erect, two-ranked, bluish-green leaves, the longest of which approach in height that of the flower stems and are up to ¾ inch wide. The montbretia-like blooms, orange-yellow flushed with red, have curved, broadly-funnel-shaped perianth tubes about 1½ inches long and petals about one-half that length that do not spread widely. The stamens, tipped with red anthers, are about one-half as long as the perianth.

Most spectacular of crocosmia species is **C. masoniorum** (syn. *Tritonia masoniorum*). This magnificent kind carries its glowing orange or orange-scarlet blooms in branchless, gracefully arching to nearly horizontal, one-sided spikes. They are up to 2½ feet tall and just top the foliage. The petals of the 1½-inch-wide, up-facing blooms spread widely. The stamens protrude conspicuously. The leaves, about 2 inches wide and tapering to both ends, have several prominent veins.

A swarm of varieties that resulted from hybridizing *C. aurea* and *C. pottsii* are grouped together as **C. crocosmaeflora.** In gardens these are called montbretias. For more information see Montbretia.

Garden and Landscape Uses and Cultivation. Crocosmias are among the few cultivated summer-flowering bulb plants of South Africa. Most bulb plants native there

Crocosmia masoniorum

Crocosmia crocosmaeflora

bloom in spring. In regions where severe winter freezing does not occur, crocosmias may be grown permanently outdoors. They are splendid for grouping in flower beds and at the bases of walls and similar places, and supply excellent cut blooms. In regions where they are not hardy, they may be planted in spring, lifted in fall, and stored over winter. Their cultivation is dealt with in more detail under Montbretia.

CROCUS (Crò-cus). It is necessary to make a clear distinction between crocuses and autumn-crocuses. True crocuses (some of which bloom in fall and are called autumn-flowering crocuses) belong to the genus *Crocus* of the iris family IRIDACEAE, autumn-crocuses belong to *Colchicum* of the lily family LILIACEAE. One species of *Colchicum* blooms in spring. The two genera are easily distinguished. The flowers of crocuses have three stamens and one style with three stigmas, those of colchicums have six stamens and three long, slender styles. For Chilean-Crocus see Tecophilaea.

As wild plants crocuses are found from the Mediterranean region to Afghanistan. There are seventy-five species. One, the source of saffron, has been cultivated in Europe and Asia since the times of the ancient Greeks and Hebrews. In the Bible saffron is referred to, together with spikenard, calamus, and cinnamon. A beautiful yellow dye and spice, it is readily soluble in water and much used for coloring and flavoring curries and other foods and in liqueurs and medicines. It is obtained from the stigmas and ends of the styles of the flowers. Those from about 4,000 blooms are required to yield an ounce of the dye. The name *crocus* is the ancient Greek one for saffron.

Crocuses do not have bulbs. The organs that look like them are technically corms. Unlike bulbs, which are composed of concentric layers of scales as are onions, or of scales arranged like shingles on a roof as are lily bulbs, corms are solid. The corms of crocuses live for one season only. The planted corm shrivels and is replaced by the time the flowers and foliage fade by one or more new corms. This is repeated yearly. Crocuses have grasslike, linear leaves that come with or after the blooms; if with them, they continue to lengthen after the flowers fade. There is no above-ground stem; the flowers rise directly from the earth. The apparent stem is part of the slender corolla tube. The flowers, white and almost all colors except red, have six petals (correctly tepals). The ovary is below the petals. The fruits are small, usually oblong, capsules, hidden among the leaves either at or just below the ground surface.

Although spring-flowering crocuses are best known, not all kinds bloom then. In favorable climates, but not where winters are so severe that the ground freezes for long periods, a parade of species may be had in flower from late July to April. In harsh climates the coldest winter months are omitted, but even in the vicinity of New York City crocuses have been known to bloom in all the months mentioned, although to have flowers in January is decidedly unusual and infrequent. For convenience crocuses are divided into two groups, those that bloom in winter or spring and those that bloom in late summer or fall. The latter, the autumn-flowering crocuses are, as explained earlier, entirely different from autumn-crocuses or colchicums.

Spring-blooming crocuses of quite gorgeous garden varieties, mostly developed by industrious Hollanders and consequently often called Dutch crocuses, are among the best known of garden plants. Together with tulips, daffodils, and hyacinths, they parade across the pages of every catalog of spring bulbs. Many American gardens sport at least a few, and with them the interest of most gardeners in the genus *Crocus* ends. Yet to know only these is to have but the slightest of nodding acquaintance with a most delightful race that, in addition to the wild species, includes a goodly number of varieties, most of which are hardy in the north, and nearly all are easy to grow. In the catalogs of bulb specialists who list more than the run-of-the-mill kinds for which there is considerable popular demand, there are offered at least some of the two or three dozen species of crocuses that are available to gardeners. They are worth seeking.

Dutch crocuses, fat and opulent achievements of the horticulturists' art, are available in a wealth of named varieties. They are hybrids and selections of *C. vernus,* a species of central and southern Europe that itself exhibits considerable variation. In the wild its blooms are lilac or white, often striped with purple. The stamens have white stalks (filaments) and lemon-yellow anthers, and the branches of the style (upper part of the pistil) are orange-red. But in size of bloom and range of

Dutch crocuses, three varieties

flower color the Dutch crocuses go far beyond the relative modesty of *C. vernus* in its native alpine haunts. They may be had in pure whites, yellows, oranges, lavenders, blues, and royal purples, and various combinations including strikingly striped and feathered blooms. Unlike tulips and daffodils, they do not run to double-flowered kinds. So fat are the flowers of some that even when closed they appear to be bursting, perhaps with just pride at their development in the hands of their Dutch breeders. It would be futile to describe the numerous varieties here. That is adequately done in catalogs of spring bulbs, easily available to all. A point to remember is that the yellow-flowered varieties bloom earlier than those of other colors.

Scarcely less prolific of varieties than *C. vernus,* ancestral parent of the Dutch crocuses, is *C. chrysanthus.* It and its progeny are smaller flowered than Dutch crocuses. They bloom earlier, near New York City in

Dutch crocus 'Pickwick', white with lilac stripes

March. Undoubtedly crocuses of the *C. chrysanthus* complex include hybrids as well as selected varieties. It is impossible to sort out which is which and, as they are offered in catalogs as varieties of *C. chrysanthus*, they are so treated here. They are a delightful lot, as lovely as the Dutch crocuses, and exhibiting almost as wide a range of flower colors. Native of southeastern Europe and Asia Minor, even in the wild *C. chrysanthus* exhibits considerable variation. It has cup-shaped blooms 1 inch long or longer and ranging in color from white through lilac and sulfur-yellow to the bright yellow most typical of this kind, and variously penciled and feathered on the outsides of the blooms with shades of brown and red-purple. The anthers are orange, usually tipped with black. The styles, with undivided branches, are red to orange. One natural variety, *C. c. fuscotinctus*, is sometimes cultivated. It is a dwarf with golden-yellow flowers tinged outside with brown.

Garden varieties (including some of hybrid origin) are numerous and are listed and described in the catalogs of specialists. They include pure white, golden-throated 'Snow Bunting', soft yellow 'Cream Beauty', delicate blue 'Blue Pearl' and 'Blue Peter', as well as rich yellow 'E. A. Bowles' (one of the most beautiful and satisfactory of all crocuses), orange-yellow 'Canary Bird', and clear blue 'Princess Beatrix'. Plants raised from seeds of varieties of *C. chrysanthus*, provide a delightful kaleidoscope of color when in bloom.

Many other crocuses that bloom in spring or in winter are well worth growing. Most of those discussed are fairly available, others may turn up as the result of diligent search through specialist catalogs and other sources of promise. Two of the earliest to bloom are *C. korolkowii*, native from Turkestan to Afghanistan, and *C. sieberi*, which hails from Greece to Crete. At The New York Botanical Garden these on occasion have expanded their blooms in January,

Crocus chrysanthus 'E. A. Bowles'

Crocus chrysanthus 'Canary Bird'

but February more commonly greets their first opening. The star-shaped flowers of *C. korolkowii* are glistening golden-yellow with dark bronze centers and bronze markings on the outsides of the petals. They have orange anthers and stigmas. The blooms of *C. sieberi* are typically delicate lilac-blue with orange-yellow throats, orange anthers, and orange-red stigmas. There are named varieties such as 'Firefly', with its outer petals almost white; 'Hubert Edelsten', which has deep purple blooms blotched white at their tips; and 'Violet Queen', with globular violet-blue flowers. These species and varieties are reliably permanent perennials when conditions are at all to their liking. Even so, they do not make themselves as much at home in most American gardens as do *C. tomasinianus* and *C. imperati*, which are not only permanent, but multiply by self-sown seeds. This is particularly true of *C. tomasinianus*, a native of the Balkan Peninsula, allied to *C. vernus*, but with a beard of hairs

Crocus tomasinianus

in the throats of the flowers. It has slender-tubed flowers easily toppled by rain or wind and looks belying unpromising in the bud stage. The whole flower in bud and its lower parts later are whitish tinged with silvery-lilac. They open to reveal satisfying and lovely deeper tones of lavender. The anthers are light orange, the branches of the stigma bright orange. Variety *C. t.* 'Whitewell Purple' has reddish purple flowers, those of *C. t.* 'Ruby Giant' are

deep violet-purple, and *C. t.* 'Barr's Purple' has lilac flowers of comparatively large size.

Noble, as its name implies, the Italian *C. imperati* is of considerably more substantial appearance than *C. tomasinianus*. In bud this very early bloomer is buff, striped with dark purple. The inner surfaces of the outer petals and both sides of the inner ones rosy-purple, the exact shade varies with individuals. The anthers are yellow, the branchless, but fringed stigmas brilliant scarlet. There is a rare white-flowered variety, *C. i. albus*.

Limestone soils in Asia Minor support *C. fleischeri*. Its glistening white flowers, with yellow throats, are of moderate size and have deep orange anthers and stigmas. This kind blooms very early.

Early, but not the earliest blooming kinds, include the outstanding cloth of gold crocus *C. angustifolius* (syn. *C. susianus*), which as its name indicates, has bright yellow flowers. One of the most prolific bloomers, this native of the Caucasus and southwest Russia is so easily contented that in many gardens it multiplies from self-sown seeds. On the outsides its outer petals are feathered with brown, but on sunny days the flowers open into stars of pure, brilliant gold. It has orange anthers and orange-red, long-branched styles. A charming miniature is *C. a. minor*, which blooms somewhat later than the typical species and has flowers of deeper hue and leaves narrower and more erect. Also yellow-flowered and one of the finest and most satisfactory of mid-season spring crocuses, vigorous *C. flavus* (syn. *C. aureus*) hails from southeast Europe and Asia Minor. It comes in many variations, all beautiful. Mostly, their blooms are ample, and when open cup-shaped. They are bright golden-yellow with pale yellow anthers and stigmas. In *C. f. sulphureus concolor*, the blooms are pale yellow. This variety blooms later than the deeper-yellow-flowered type and is much less vigorous. It self-sows.

The Scotch crocus (just how a native from Italy to western Asia acquired this name is not clear) also blooms between the flowering of very earliest and latest spring-blooming kinds. It is *C. biflorus*. Of modest, but charming mein, this species produces a succession of rather small blooms, white on their insides and with their yellowish outsides feathered with bluish-purple. The anthers are orange, the undivided branches of the stigma brilliant orange-red. Its variety, *C. b. alexandri*, has blooms of pure white with a lustrous purplish exterior, those of *C. b. argenteus* are narrower petaled, yellower on their outsides, and on their insides are white to silvery-lilac and more or less lined with lavender or purple on the outer petals. Variety *C. b. weldenii* has blooms of pure white, tinged blue-gray on their outsides. The blooms of *C. b. w.*

Crocus biflorus

'Fairy' are pure white tinged gray on their outsides, and *C. b. pusillus* is similar to *C. b. argenteus,* but has smaller flowers. The Scotch crocus in many gardens increases quite rapidly by multiplication of its corms, but as it is sterile it produces no seeds.

Other mid-season bloomers are *C. ancyrensis, C. dalmaticus, C. minimus, C. olivieri, C. suterianus,* and *C. versicolor.* Asia Minor is the native land of *C. ancyrensis,* which has more or less globular flowers of bright golden-yellow flushed with purple on their outsides. The anthers are light orange, the undivided branches of the stigma orange-red.

Crocus ancyrensis

Having lilac or lavender flowers about the same color as those of *C. sieberi* and with yellow throats, *C. dalmaticus* has blooms of fairly large size with yellow anthers and stigmas. It is a native of coastal parts of the eastern Adriatic.

An appealing little species is *C. olivieri,* of Greece to Rumania. It has rather small, bright orange-yellow blooms with brilliant orange anthers and orange, deeply-two-cleft branches to its stigma. Closely related to *C. olivieri,* but with narrower leaves, and much more addicted to multiplying rapidly by offsets, *C. suterianus* has its home in Asia Minor. Its flowers, of fairly large size,

Crocus olivieri

are orange-yellow, the petals somewhat over 1 inch long. They have orange anthers and orange, two-cleft stigma branches. Resembling a miniature *C. imperati* and indigenous to Corsica and Sardinia, *C. minimus* blooms much later than that species. The name is somewhat misleading; it is by no means the smallest crocus.

Common in the south of France, *C. versicolor* is an old-time garden favorite, which at one time, was cultivated in many varieties, most of which seem to have gone out of existence. One remains; *C. v.* 'Cloth of Silver' has slightly lilac-tinted, white blooms with a purple blotch on each outer petal. In typical *C. versicolor* the flowers are plain pale to dark purple or have their inner petals conspicuously feathered with a hue darker than the background, a pattern unique among crocuses. It has yellow anthers and orange-red stigmas. This species and its varieties are very vigorous.

Imposing *C. etruscus,* native to Italy where it is found only in the Etruscan

Crocus etruscus

mountains, flowers after *C. imperati,* to which it is closely related, and later than most crocuses of spring. As cultivated in gardens it has robust flowers with lavender-gray undersides to the outer petals (in this it differs from botanical descriptions of it which call for buff or buff-green coloring on the outsides of the blooms). The anthers are orange as are the undivided

branches of the stigmas. Variety Zwanenburg', of *C. etruscus,* has clear gray-lilac blooms; it flowers even more freely than the typical species. A yellow-bloomed second-early species, *C. balansae* has very much smaller flowers than the last, yet they make a good display. Sweetly fragrant, they are bright orange-yellow flushed on the outsides of the outer petals with reddish-brown. The anthers are orange, the stigmas bright red. This crocus has its home in Asia Minor. Indigenous to Corsica, *C. corsicus* has petals with buff outsides feathered with purple veins. The insides of the petals are lilac, the anthers orange, and the stigmas bright red. It is not always easy to establish in gardens.

One of the latest spring-blooming crocuses is *C. candidus,* which usually makes its bow in late March or April. It has medium-sized flowers ranging in color from white to orange-yellow, veined on their outside with brown and purple. The anthers are orange, the much-divided stigmas creamy-white. This species is native to Asia Minor. More commonly cultivated *C. c. subflavus* has broad-petaled, light yellow flowers, often with the outer petals flecked with bronzy-purple. The anthers and stigmas are yellow.

Crocuses that bloom in fall are perhaps less thrilling, but not less beautiful than those of late winter and early spring. A greater variety of other flowers competes for attention in autumn than spring, and human hunger for color in the garden is less acute than earlier in the year. Consequently, autumn-flowering crocuses are less well known than vernal ones. Only occasionally are they grown in American gardens. They deserve greater popularity. The saffron crocus (*C. sativus*) belongs here. Less important as the source of coloring for foods and medicines than it was formerly, this is not known in the wild, but is believed to have originated in Asia Minor. It has narrowly-linear leaves, eventually 1 foot to 1½ feet in length, and fragrant, reddish-purple, lavender or white flowers 1½ to 2 inches long. They have styles with long, blood-red branches. The anthers are yellow. It does not produce fruiting capsules.

That great American gardener and writer about gardens, the perceptive and knowledgeable Louise Beebe Wilder, wrote "If half a dozen autumnal crocuses are to be chosen, they might include *kotschyanus, speciosus* 'Pollux', *pulchellus, longiflorus, medius,* and *cancellatus.*" It would be hard to improve on that list except, perhaps, to add other varieties of *C. speciosus.*

First to bloom and perhaps finest of autumn-flowering crocuses is *C. speciosus,* a kind widely spread throughout southeastern Europe, Asia Minor, and Iran. Satisfyingly permanent in gardens, it has large and brilliant blue-lilac flowers, more nearly true blue than those of any other species.

Convolvulus cneorum

Consolida, a garden variety

Convolvulus tricolor

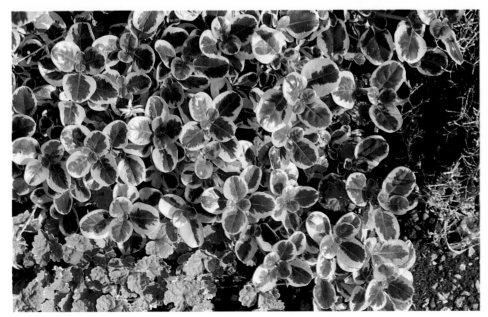

Consolida, a garden variety

Coprosma repens marginata

Corallorhiza maculata

Cornus alba 'Coral Beauty'

Cornus officinalis

Cornus florida, in bloom

Cornus florida, in fruit

Crocus speciosus

The petals are finely veined with blue on their insides and dotted with dark blue on their outsides. The anthers are large and chrome-yellow. The much-branched stigmas are brilliant red. This species multiplies rapidly by offsets and by seeds. There are several varieties, all easy to grow. Among the most notable are these: *C. s. aitchisonii*, later flowering than the typical species and with blue-veined, pale lavender petals almost 3 inches long is the largest flowered wild crocus; *C. s. albus* has large pure white blooms; *C. s.* 'Artabir' has big pale blue flowers, darker on their insides than outsides and with dark featherings on their outsides, which open a few days earlier than those of the typical species; *C. s.* 'Cassiope', the latest flowering *C. speciosus* variety, has flowers of analineblue with faint yellowish bases; *C. s.* 'Oxonian' has rich dark violet-blue blooms; *C. s.* 'Pollux', slightly later blooming than the typical species, has very large flowers, the inner petals of which are almost white, the outer ones blue-lavender of an entrancing shade, with darker veinings marking the petals.

Soon after *C. speciosus* the blooms of *C. cancellatus* and *C. pulchellus* expand. In late September, well before the first frost blackens dahlias, their flowers are fully open at The New York Botanical Garden. Unattended by leaves when first they present themselves, they remain in bloom a month or more, and the developing foliage is evident before the last flowers fade, although it does not reach full size until spring. The globose flowers of *C. cancellatus* are variable in color, the ones most commonly cultivated being delicate lavender feathered with darker markings. This is the common coloration of the wild plants in Asia Minor; those from European locations usually have white blooms and represent the variety *C. c. albus*. The flowers of the species have yellow anthers and bright red, much-divided stigmas. Quite distinct is *C. c. cilicicus*, a September-blooming variety with long, narrow-oblong petals that reflex when expanded and are rosy-lilac with darker

veinings. The anthers are white instead of yellow as in the typical species. Ranging as a wildling from Greece to Asia Minor, *C. pulchellus* is of medium size and broadpetaled. It has satiny, bright lilac, fragrant flowers, rich yellow in their throats. The anthers are white, the much-divided stigmas orange-yellow.

Next to bloom of Mrs. Wilder's recommended six are *C. medius* and *C. kotschyanus*. They follow closely on the heels of *C. speciosus*, at New York opening their first blooms in early October, before their leaves appear. Limited in its natural distribution to the mountains from Mentone, France, to Genoa, Italy, pretty *C. medius* accommodates readily to garden condi-

Crocus medius

tions and multiplies quite astonishingly. One of the showiest species, its flowers are rich, deep lavender-purple with conspicuous dark purple stars in their throats. Pale orange anthers and brilliant scarlet, muchdivided stigmas are characteristic of *C. medius*. A rare white-flowered variant, exhibiting the same purple star in its throat, occasionally appears. More fragile in appearance than many autumn-flowering crocuses, but actually sturdy, *C. kotschyanus* (syn. *C. zonatus*) comes from Mount Lebanon, Mount Herman, and the mountains of Cilicia. It is a grateful kind, accommodating easily to garden environments and under reasonably favorable circumstances increasing quite rapidly. Its flowers are rosy-lilac with a clearly marked ring of orange-yellow dots in their throats. The petals, up to 2 inches long, have a tracery of slightly darker veins on their insides. The anthers are cream, the stigmas yellow. The latest of Mrs. Wilder's favorites to bloom is *C. longiflorus*, which she selects as, next to *C. pulchellus*, her favorite autumn-flowering crocus. It is quite lovely, and the more appreciated because of the lateness of its advent. There is clearly more than one form of this species, for both in Mrs. Wilder's garden and at The New York Botanical Garden different plantings bloomed in late October and late November. The foliage of *C. longiflorus* develops with the flowers. The latter are wonderfully scented and pinky-lilac with orange

throats. The anthers are orange. The stigmas, often divided, are scarlet.

Other autumn-flowering crocuses are worth considering. Perhaps best known under its synonym *C. iridiflorus*, the very distinct *C. byzantinus* has its home in southeastern Europe. Because its deep purple outer petals are about twice as long as the delicate lilac inner ones the blooms do indeed, as its botanical synonym indicates, have a strong superficial likeness to those of irises. The anthers are orange, the much-divided stigmas lavender. When in bloom this species is naked of foliage; the leaves do not appear until spring. Unfortunately *C. byzantinus* is less predictable than some autumn-flowering crocuses. It does not thrive in all gardens, even though conditions seem to be suitable. It should be tried in peaty soil in light shade. Native of the Asturias Mountains of Spain, *C. asturicus* has mauve to lilac or occasionally white flowers that appear with or after the foliage. Hairy in their throats and with pointed petals about 1½ inches long, they have white-stalked stamens and a bright red, much-divided style. The blooms of *C. a. atropurpureus* are dark mauve.

The latest crocus to bloom, at least at The New York Botanical Garden, where, if it were any later, it would be the first to flower in the new year, is *C. laevigatus fontenayi*. In favorable seasons it lasts well into December, opening its first blooms very early in that month. This variety differs from its parent species in having larger flowers that are rosy-lilac, tinted on their outsides with buff, instead of white, feathered on their outsides with crimson-purple, and with orange throats. The species *C. laevigatus*, a native of Greece, blooms somewhat earlier than *C. l. fontenayi*. The blooms of both are deliciously fragrant. Those of the species have petals about 1 inch long. The anthers are white, the feathery stigmas bright yellow.

Garden and Landscape Uses. On the whole crocuses are easy to grow and dependable, but some qualifications are necessary. Although most are thoroughly hardy in the north, a few will not survive winters more severe than those of Washington, D.C. With a small minority a somewhat reverse situation exists. Unless they are subjected to fairly long and decently cold periods of dormancy they pine and disappear. For them climates more salubrious than are typical of, say, Virginia, are unfavorable. That such variations in the requirements of crocuses exist is not surprising when we consider that the genus disports itself naturally under conditions as diverse as those of the low lands immediately bordering the Mediterranean, of the high Alps, Pyrenees, and Caucasus, and the mountains of Asia Minor. The wonder is that more species are not finicky as to extremes of winter cold and summer heat. Those who garden where winters

Crocus asturicus

A sheltered spot at the foot of a south-facing wall induces early bloom

are, to humans, either rugged or extremely mild, had better try unknown crocuses in small quantities before investing more heavily.

A more serious inhibiting cloud broods over the horizon of would-be planters of crocuses. In many areas mice, squirrels, chipmunks, rabbits, birds, and other foraging creatures regard plantings as their private kitchen gardens, sources of delectable, nourishing root vegetables and tender, juicy greens. Gnawing mouse, nibbling rabbit, and pecking bird are equally destructive. The first concentrates on the bulbs, rabbits prefer the foliage, and a few kinds of birds appear to take fiendish delight in tearing apart the buds and flowers. These feathered friends also pull out and consume small corms. An oft-blamed creature, but one probably innocent of devouring crocuses is the mole; it serves, however, as accessory before the fact by constructing tunnels that provide mice ready access to the choice tidbits planted by the gardener. Suggested remedies are not lacking, nor is any completely effective. Where the law permits dedicated gardeners sometimes resort to shooting. Traps of various kinds are undoubtedly helpful. Poison baits may be used with discretion. To discourage birds, stretching a network of black thread above the plants before they come into bud is recommended, but how effective this is is open to doubt. Another suggested check on the depredations of birds (based on the theory that they attack crocus blooms to satisfy their thirst) is to place saucers of water on the ground.

A slightly lean and ambitious cat patrolling the area can afford relief by discouraging or destroying warm-blooded enemies of crocuses. As a last resort the corms, but not the foliage and flowers, may be protected by planting them inside cages of small-mesh chicken wire.

The placement in gardens of crocuses calls for some thought. Dutch crocuses can be effective in groups at the fronts of shrub borders and flower beds and so can some of the strong-growing species, such as C.

susianus, and its varieties, and C. tomasinianus, but for the most part the species, and most surely the frailer kinds, give the best results when tucked away in agreeable spots in rock gardens and other well-cared-for areas. For porch boxes and roof gardens Dutch crocuses have much to commend them. True, under such conditions they may need to be renewed frequently, perhaps even yearly, but they are inexpensive and the reward is great. All spring-blooming crocuses may be planted in containers and forced into early bloom with great ease. Pans and bowls of them in full flower when window panes are iced and snow covers the outdoors provide a great lift for the spirit.

Spring-blooming crocuses, especially the Dutch kinds, are frequently recommended for planting in lawns and other grassy areas. In such places they certainly can create lovely and satisfying pictures, but such use presents certain problems. First, if the turf is thick they are unlikely to persist for many years. They are much more likely to

Dutch crocus 'Pickwick', at front of flower border

Crocuses, easily forced, make delightful pot plants (two figures above)

Dutch crocuses in a grassy area

be permanent and to multiply in rather sparse grass, yet where the soil is sufficiently deep and nourishing for their own well-being, a combination not too common, grass is likely to flourish. The difficulty can, of course, be overcome by planting new crocus corms from time to time to replace those that disappear. Certainly when used to decorate lawns crocuses can be lovely, especially when scattered with

some abandon beneath high-branched, deep-rooting trees, where, when they besprinkle the grass with their blooms, they receive enough sun to assure that their cups will open, and the garden is gayer because of the surety that the arrival of spring is imminent.

Another problem is created by the need to allow the foliage of the crocuses to die completely before it is cut. This means that grass mowing must be delayed until very late spring or early summer and lawn areas in which the plants grow may look unacceptably unkempt for a long period, and after the first cutting appear scalped and yellow for a while.

Covers other than grass are often needed for areas where crocuses are planted. Shallow-rooted perennial groundcovers may be satisfactory, but the best plan with the spring-blooming types is to oversow with shallow-rooting annuals, such as sweetalyssum, California-poppies, or portulacas or, in rock gardens, with such dainty kinds as *Gypsophila muralis* or *Ionopsidium acaule*.

Cultivation. Gritty, well-drained soil, in good heart and fertile to a depth of at least 6 inches is best for crocuses. They find extremely unfavorable sticky, clayey earth or that which lies wet for long periods. Nor do they favor manure in any abundance. A little, old and rotted, is acceptable, but generally a liberal admixture of compost, humus, or peat moss should be depended upon to raise the organic content of the soil. Full sun during the season when the foliage is in evidence is best for crocuses.

The corms are planted as early as they can be obtained from dealers. This is especially important with fall-blooming kinds. They are set with their bases 3 to 4 inches beneath the surface. Unless the plantings become obviously congested they may remain undisturbed for years, being encouraged by an annual spring application of a complete garden fertilizer. If transplanting is undertaken it should be done immediately when the foliage has died, which may not be until July. The corms are cleaned and sorted into sizes. Autumnflowering kinds are replanted immediately. This may be done, too, with springblooming kinds, or they may be stored in a cool, dry place and planted in early fall.

For early flowering indoors, crocuses are planted in pots or pans (shallow pots), with a distance of about one-half the diameter of the corms between them, in fall. They are set just beneath the surface in porous soil. The planted containers are buried under 6 inches or so of sand, peat moss, or ashes, outdoors or in cold frames, or are placed in a cool, dark cellar or similar place and left until after the end of the year. From January on, in successive batches, they are brought into a greenhouse, or placed in a window in a cool room, kept well watered, and there come into bloom. A forcing temperature of 50 to

60°F is adequate. After flowering, the plants are kept watered and growing as long as the foliage remains green. After it has died the soil is allowed to dry, and later the bulbs may be planted in the outdoor garden. Propagation of crocuses is by natural increase of the corms. Natural species are also easy to raise from seeds sown thinly as soon as ripe or in early spring in a cold frame, in flats, or in a bed of porous soil. Allow the young plants to remain undisturbed for two complete growing seasons, then transplant them about 2 inches apart in a nursery bed. Most will bloom in their third or fourth year.

CROSNES. See Artichoke, Chinese or Japanese.

CROSS or CROSSBRED. Plants the parents of which are distinct varieties of the same species or are the progeny of the results of such are known as crosses or crossbreds. The terms are also sometimes applied to hybrids between different species.

CROSS FERTILIZATION. When as a result of cross pollination a sperm or male reproductive cell of one plant unites or fuses with an ovule (egg) or female reproductive cell of another, cross fertilization is achieved.

CROSS POLLINATION. This consists of the transfer of pollen from one plant to the stigma of the flower of another of a different species or variety.

CROSS VINE is *Bignonia capreolata*

CROSSANDRA (Cross-ándra). This is one of many genera of the acanthus family ACANTHACEAE that provide good ornamentals. It consists of fifty species of herbaceous plants and shrubs of the Old World tropics. The name is from the Greek *krossos*, a fringe, and *aner*, male, and refers to the fringed anthers.

The toothed or toothless leaves of crossandras are opposite or in whorls (circles of more than two). Their yellow, orange, orange-red, or white flowers, in dense, four-sided, terminal and axillary spikes with overlapping bracts, have five-lobed calyxes and long, slender-tubed corollas, with very asymmetrical faces of five spreading lobes (petals). There are four stamens in two pairs and a shortly twolobed style. The fruits are four-angled capsules.

A shrub or subshrub 1 foot to 3 feet tall, *Crossandra infundibuliformis* (syn. *C. undulaefolia*) is a native of India. Pubescent in its upper parts, it has lustrous, pointed, narrowly-ovate or lanceolate, more or less wavy leaves, 3 to 5 inches long, and hairless or with very short hairs. Its showy salmon-orange flowers are in dense, downy spikes about 4 inches long. They have ¾-

Crossandra infundibuliformis

inch-long corolla tubes and a spreading three- to five-lobed lip 1 inch wide or wider. Variety C. i. 'Mona Wallhed', up to 1½ feet tall, has glossy, blackish-green foliage and salmon-pink blooms. Both the species and its variety bloom while quite small.

Quite different from the last and native to East Africa, C. pungens is a bushy subshrub up to 2 feet tall. It has oblongish-elliptic, olive-green leaves very clearly veined with creamy-white, and light orange flowers in spikes with hair-fringed, ovate bracts.

A rarely branched species not more than 6 inches tall, C. guineensis, of tropical West Africa, has red, hairy stems each with two to four pairs of wavy-margined, elliptic to obovate leaves 3 to 5 inches long that are deep green with pink reticulations. The slender-tubed, pale lilac flowers are a little more than ½ inch across and not as wide.

Garden and Landscape Uses and Cultivation. Only in the moist tropics and subtropics are crossandras suitable for growing outdoors, elsewhere the protection of a greenhouse is necessary. One, C. infun-

Crossandra pungens

Crossandra guineensis

dibuliformis, is often recommended as a window plant, but rarely does very well under such conditions because of lack of adequate humidity. For their best satisfaction soil that contains an abundance of organic matter and is moist, but not wet is needed, as is shade from strong sun. Indoors, the minimum winter night temperature should be 60°F, with a rise of five to fifteen degrees by day. Young specimens of all except C. guineensis are pinched occasionally to encourage branching. Those that have filled their pots with roots are given regular applications of dilute liquid fertilizer. Repotting is done in spring. Propagation is easy by cuttings and seeds.

CROTALARIA (Crotal-ària)—Rattle Box, Sunn-Hemp. Not many of this extensive genus of the pea family LEGUMINOSAE find favor as ornamentals. A few are grown as cover crops in warm regions and the sunn-hemp is cultivated for its fiber used for ropes, nets, canvas, packing material, and paper. There are more than 500 species of Crotalaria, the majority Asian, but the group is widely distributed elsewhere particularly in the tropics and subtropics of both the Eastern and Western Hemisphere. The name, from the Greek krotalon, a rattle, alludes to the fruits.

Crotalarias are shrubs, subshrubs, herbaceous perennials, and annuals. Their leaves are undivided or separated into three, five, or seven leaflets that spread finger-like from the tip of the leafstalk. The pea-shaped flowers, generally yellow to brownish-yellow, less commonly purplish or blue-purple, are solitary or in racemes. Of their five petals, the standard or banner is usually largest. The keel is curved and beaked. There are ten united stamens and a conspicuously curved, sometimes hairy style. The fruits are spherical or oblong inflated pods containing many seeds that rattle when the mature pods are shaken.

South African C. capensis is an evergreen shrub 4 to 10 feet tall. It has shoots when young silky-hairy and minutely-hairy or hairless, stalked leaves with three broadly-ovate leaflets ¾ inch to 2 inches long, blunt, except for a brief, sharp apex.

The fragrant, conspicuously veined, bright yellow flowers, over 1 inch in length, are striped with red-brown. They are in loose terminal and axillary racemes up to 6 inches long. The pubescent seed pods are 1½ inches long.

Canary bird bush (C. agatiflora), of East Africa, is a loose shrub up to 12 feet in height, if untrimmed, as broad. Its gray-green leaves are of three ovate leaflets up to about 3 inches long. Terminal racemes up to 1¼ feet in length display the attractive, curiously birdlike flowers. They are greenish-yellow and have standard or banner petals 1½ inches long. The tips of the keels are stained with brownish-purple. In California this species blooms through much of the year.

Sunn-hemp (C. juncea) is an annual 4 feet tall or taller with downy stems and scarcely-stalked, undivided, linear-oblong leaves 1½ to 4 inches long. The flowers, in terminal racemes up to 1 foot long, and with a reflexed standard or banner petal 1½ inches across and wings and keel one-half as long, are rich yellow. Sunn-hemp, the earliest fiber plant to which reference is made in Sanskrit, has been cultivated in Asia for thousands of years. It does not occur as a wild plant.

Garden and Landscape Uses. The shrub crotalarias described here are attractive for outdoors in mild, dryish, frostless or nearly frostless regions. The South African kind is sometimes grown in greenhouses. They succeed in ordinary, well-drained soils in sunny locations or part-day shade. A rank, somewhat untidy shrub if left untended, C. agatiflora needs pruning two or three times a year to restrain its exuberance and keep it in bounds and acceptable. Given this attention, it blooms attractively over a long season. Even if injured by frost it soon recovers. In greenhouses these crotalarias respond to conditions and care that suit acacias. Sunn-hemp is suitable for including in collections of plants useful to man.

Cultivation. The shrub species described here succeed with little care. The chief attention needed is pruning to keep them shapely. They are propagated by seeds and by cuttings of firm, but not hard shoots planted under mist or in a propagating bed in a greenhouse or cold frame. Sunn-hemp can be raised from seeds sown outdoors in fertile, well-drained soil in a sunny location at about the time corn is sown, or plants may be raised indoors from earlier sowings, be grown in pots, and planted outdoors when the weather is warm and settled.

CROTON. This is the common name for Codiaeum. It is also the botanical name of another genus of the spurge family EUPHORBIACEAE, one without horticultural importance, but the source of the powerful purgative croton oil.

CROW POISON. See Nothoscordum.

CROWBERRY is *Empetrum nigrum*. The broom-crowberry is *Corema*.

CROWEA (Cròw-ea)—Wax Flower. Four Australian species are the only representatives of *Crowea*, of the rue family RUTACEAE. The name commemorates the British botanist Dr. James Crowe, who died in 1807. By some botanists *Crowea* is included in closely related *Eriostemon*.

Croweas are evergreen shrubs with alternate, undivided, glandular-dotted leaves and solitary or less commonly paired flowers from the leaf axils or ends of the stems. They have a five-lobed calyx, five petals, and ten woolly stamens. The fruits are of five carpels.

In its native land called wax flower, **C. saligna** (syn. *Eriostemon crowei*) is 1 foot to 1½ feet tall and hairless, with narrow-elliptic to lanceolate leaves up to 2½ inches

Crowea saligna

long and ½ inch wide, or occasionally bigger. The beautiful rose-pink to red, starlike blooms, solitary from the leaf axils, have spreading petals and yellow anthers. They are 1¼ to 1½ inches in diameter. From the last **C. exalata** differs in having slimmer

Crowea exaltata

branches, mostly blunt, narrow-linear leaves often less than 1 inch long or rarely attaining 1½ inches, and pink, red, or rarely green flowers about ¾ inch wide terminating the branches, or if from the leaf axils, having stalks with leafy bracts.

Garden and Landscape Uses and Cultivation. Croweas are hardy only in warm, frost-free or essentially frost-free climates of the Mediterranean type. The species described is also suitable for cool, airy greenhouses. It prospers in well-drained soil of a peaty character kept evenly moist, but not saturated. In greenhouses a winter night temperature of 40 to 50°F is adequate, with that by day rising by ten to fifteen degrees. Prune to shape and repot in early spring. Seed affords the best method of propagation. Increase can also be had by cuttings and by grafting onto understocks of *C. alba* or *Eriostemon buxifolia*.

CROWFOOT. See Ranunculus.

CROWN. This word is employed in the common names of some plants including crown beard (*Verbesina*), crown flower (*Calotropis*), crown imperial (*Fritillaria imperialis*), crown-of-gold tree (*Cassia spectabilis*), crown-of-thorns (*Euphorbia milii splendens*), giant crown-of-thorns (*Euphorbia keysii*), and crown-vetch (*Coronilla varia*). Crown is also used in other ways. The head of a tree, the branched part, is termed the crown, and so is the cluster of leaves that terminates the trunk of a palm, cycad, tree fern, or other plant of similar habit of growth. The word is also used for the crowded cluster of buds and connected parts at the top of the root systems of many herbaceous perennials, such as for example, chrysanthemums, delphiniums, and peonies. The corona, a circular appendage or circle of appendages at the centers of some flowers, for instance narcissuses, is also called a crown.

CROWN GALL. This serious bacterial plant disease, its causal agent *Erwinia tumefaciens* (syn. *Agrobacterium tumefaciens*), affects many different kinds of plants including trees and shrubs. It is common on roses, fruit trees and bushes, and grape vines. The characteristic symptom is the development, chiefly on the roots, but sometimes on stems, and with grafted plants frequently at the graft union, of irregular, rough-surfaced, rounded swellings or galls, from pea- to sometimes baseball-sized. At first soft and whitish, with age they harden and become brown.

The bacteria gain admittance to the tissues only through wounds. The best preventive is the rigid exclusion of all newly acquired plants suspected of infection (with especial concern about roses and fruit trees and bushes with any suspicious bumps on the roots or close to the bases of the stems

or trunk). Discard affected plants, and do not plant susceptible ones in the same locations for two years. Some success has been had by dipping the roots in a solution of streptomycin. There is some evidence that the disease is less prevalent among plants in acid soils as compared with the same kinds in neutral or alkaline ones.

CROWNSHAFT. This is the name of the apparent trunklike extension of the true trunk in some palms. It is formed of the erect, much-broadened bases of the overlapping leafstalks wrapped tightly around each other.

CRUCIANELLA (Crucian-élla). Some thirty species or more of the madder family RUBIACEAE constitute *Crucianella*, the name of which is derived from a diminutive of the Latin *crux*, a cross, and alludes to the positioning of the leaves of some species.

Crucianellas include annuals, herbaceous perennials, and subshrubs of dry, open habitats. As natives they are indigenous from the Mediterranean region to central Asia. They have four-angled stems and leaves in whorls (circles) of four to eight, or the upper ones alternate. The small, white, pink, or blue flowers are in bracted spikes or clusters. Funnel-shaped, they have a much reduced calyx or none, a four- or five-lobed corolla, five stamens, and a branched style. The fruits are capsules.

Most familiar, annual **C. stylosa** (syn.

Crucianella stylosa (two figures above)

Phuopsis stylosa) is native to the Caucasus. From 6 inches to 1 foot tall, it has slender stems with whorls of six to eight slender-pointed, narrowly-lanceolate leaves fringed with fine, bristly hairs. Its tiny pink to deep pink flowers are in rounded heads about ½ inch across.

Other sorts cultivated, all annuals, include these: *C. angustifolia,* of southern Europe, has stems up to 1½ feet long and, mostly in whorls of six or eight, linear-lanceolate to linear, rough-hairy leaves with recurved margins. The minute white to yellowish flowers are stalkless. *C. latifolia* is a southern European species with more or less lax stems up to 1½ feet long and obovate to linear-lanceolate leaves in whorls of four to six. The little whitish flowers are in slender spikes. Each has a tubular, lobeless calyx and a narrowly-funnel-shaped corolla with five spreading lobes (petals). There are five stamens and a long-protruding style.

Garden Uses and Cultivation. Crucianellas are useful for the fronts of flower beds and in rock gardens. They grow satisfactorily in ordinary well-drained soil, in sunny or lightly shaded locations, and come readily from seeds sown where the plants are to remain. The seedlings should be thinned sufficiently to prevent undue crowding.

CRUCIFERAE—Mustard Family. Crucifers is an embracing common name for the members of this extremely important dicotyledonous family that includes such well-known vegetables and salads as broccoli, brussels sprouts, cabbage, cauliflower, collards, cress, horseradish, kale, kohlrabi, mustard, radishes, rutabaga, and watercress, as well as a wide variety of ornamentals and many common weeds.

The vast majority of the about 3,200 species are annuals, biennials, or herbaceous perennials; a few subshrubs. They are apportioned among 375 genera.

Species of crucifers are most numerous in temperate and cold regions throughout the world. None is poisonous and before it was discovered that lime juice was a preventative of scurvy British naval officers were taught to recognize members of the family and to gather them in lands they touched as additions to the crew's diet.

Crucifers are generally easily recognizable by their flowers and seed pods. They have mostly alternate leaves, undivided and lobeless and toothless or variously lobed, toothed, or sometimes apparently pinnately-divided. The flowers have four sepals and four petals (rarely the petals are wanting) that spread in the form of a cross (which gives reason for the names crucifer and *Cruciferae*). There are six stamens, of which two are shorter than the others, and one style. The fruits are dry, long and slender or short and broad, cylindrical or flat pods.

Genera belonging in the mustard family include *Aethionema, Alyssoides, Alyssum, Anastatica, Arabis, Armoracia, Aubrieta, Aurinia, Barbarea, Berteroa, Biscutella, Brassica, Bunias, Cardamine, Caulanthus, Cheiranthus, Cochlearia, Crambe, Dentaria, Diplotaxis, Draba, Eruca, Erysimum, Fibigia, Goldbachia, Heliophila, Hesperis, Hutchinsia, Iberis, Ionopsidium, Isatis, Kernera, Lepidium, Lesquerella, Lobularia, Lunaria, Malcolmia, Matthiola, Moricanda, Morisia, Nasturtium, Notothlaspi, Orychophragmus, Peltaria, Petrocallis, Phoenicaulis, Physaria, Raffenaldia, Raphanus, Ricotia, Rorippa, Schivereckia, Schizopetalon, Smelowskia, Stanleya, Stenodraba, Thelypodium,* and *Thlaspi.*

CRUEL PLANT is *Cynanchum ascyrifolium.*

CRUPINA (Cru-pìna). One species of this group of four annuals of southern Europe and western Asia is occasionally cultivated in flower gardens. The name is derived from a Belgian or Dutch vernacular name. The genus *Crupina* belongs in the daisy family COMPOSITAE and is allied to *Centaurea.*

Crupinas have pinnately-divided or pinnately-lobed leaves and long-stalked flower heads with purple florets and an involucre (basal collar) of many long, slender bracts. Their seedlike fruits are achenes. Most likely to be cultivated is *C. vulgaris* (syn. *Centaurea crupina*), which is 1 foot to 2 feet in height and of rather loose habit. Its stems are slender, erect, and branched. Except for its broad-elliptic cotyledon leaves and the pair next above, the leaves are pinnately-deeply-dissected into slender, finely-toothed segments. The flower heads are about 1 inch long and wide.

Garden Uses and Cultivation. Of use to add interest and variety to flower gardens, *C. vulgaris* also may be used as a source of cut flowers. It is easily raised from seeds sown outdoors in early spring in a sunny place in well-drained, porous soil. Seed is sown where the plants are to remain, and the seedlings are thinned to about 9 inches apart. No special care is needed.

CRYOPHYTUM. See Mesembryanthemum.

CRYOSOPHILA (Cryo-sophíla). Nine species of Mexico and Central America belong in *Cryosophila* (previously named *Acanthorrhiza*) of the palm family PALMAE. They are little known horticulturally. The name is derived from the Greek *kryos,* cold, and *phileo,* to love, but does not seem apt. Cryosophilas have trunks clothed with numerous aerial roots the lower ones developing into down-pointing, stiff, sometimes branched spines. Their leaves are fan-shaped and deeply divided into many narrow segments. The creamy-white or pinkish bisexual flowers are succeeded by small globose to oblong fruits.

Perhaps best known is Mexican *C. nana* (syn. *Acanthorrhiza aculeata*), which attains a height of up to 40 feet and has round leaves up to 4 feet in diameter with stalks 3 to 4 feet long. On young specimens they are deeply divided into many segments, but on older trees they are only three- or four-cleft. The leaves are dark green above and whitish beneath, the segments lax and drooping. The flowers are in dense, spike-like clusters.

Garden and Landscape Uses and Cultivation. These attractive and unusual palms are primarily plants for collectors. They need humid, tropical conditions and thrive in any ordinary garden soil in full sun. They may be grown in greenhouses in large containers or in ground beds. They need the same general cultivation as *Acanthophoenix.* For additional information see Palms.

CRYPSINUS (Cryp-sìnus). The genus *Crypsinus* of the polypody family POLYPODIACEAE comprises some forty species of Asian ferns. It previously was included in *Polypodium.* The name, alluding to the clusters of spore capsules being in tiny pits, comes from the Greek *kryptos,* hidden, and the Latin *sinus,* a pocket.

Crypsinuses are small- to medium-sized ferns that perch on trees as epiphytes. They have creeping rhizomes shaggy with black fibers, and spaced rather distantly along them, lobeless or pinnately-lobed, more or less leathery fronds, which are usually of two sorts, fertile and sterile. Sometimes the fertile (spore-bearing) fronds are narrowly-linear, usually with a single row of spore capsule clusters on either side of the midrib.

Very beautiful *C. glaucus* (syn. *Polypodium glauco-pruinatum*), of the Philippine Islands, has sterile, stalked, lobeless leaves with paddle-shaped blades 2 to 3 inches long by ½ to ¾ inch wide and longish-stalked fertile ones of an overall length of 1½ feet and a width up to 9 inches. These last are deeply cleft into variously-sized, narrow, pointed lobes. They are dull green above and richly glaucous-blue on their undersides. The large, circular clusters of spore capsules are in single rows on either side of the midribs of the upper leaflets.

Garden Uses and Cultivation. These are as for the hare's foot fern (*Phlebodium aureum*) and *Drynaria.* For more information see Ferns.

CRYPTANTHA (Crypt-ántha). Mostly natives of western North America, but with some in southern South America, about 100 species belong here. The group belongs in the borage family BORAGINACEAE. The name, from the Greek *kryptos,* hidden, and *anthos,* a flower, alludes to the minute flowers of the species first discovered.

Cryptanthas are annuals, biennials, herbaceous perennials, and subshrubs, usually bristly-hairy, and with their first leaves opposite and later ones alternate. The white or rarely yellow flowers are in one-sided spikes or racemes, in the bud stage coiled. Generally the calyx is five parted. The corolla characteristically has a brief tube closed at its throat with scales and five spreading lobes (petals). The fruits are of one to four seedlike nutlets. Closely related to *Eritrichium*, this genus, of which sixty-five species are recorded as being indigenous to California alone, is of small horticultural importance.

Native of California and Baja California, **Cryptantha intermedia** (syn. *Eritrichium intermedium*) is a rigid, erect, branching annual, very densely clothed with bristly white hairs, and 6 inches to nearly 2 feet tall. Its leaves, lanceolate to linear, are up to 2 inches long. In twos to fives, the bractless spikes, 2 to 6 inches long, are composed of white flowers up to ¼ inch wide. A more definitely desert species, **C. barbigera** (syn. *Eritrichium barbigerum*) is a usually lower plant with inconspicuous white flowers up to ¹⁄₁₀ inch across. It ranges from California to Utah, New Mexico, and Mexico. Native in dry open places from the Great Plains states to British Columbia, Washington, and Oregon, **C. celosioides** (syn. *Oreocarya glomerata*) is a branched or branchless biennial or short-lived perennial up to about 1½ feet tall. It has spatula-shaped to linear leaves up to 2½ inches long and white flowers almost ½ inch wide.

Garden Uses and Cultivation. The first sort described is probably sometimes cultivated under the name of the second. It and others are in regions of its natural occurrence occasionally grown in native plant gardens. Sowing seeds in fall where the plants are to remain, on well-drained, sunny sites, and thinning out the seedlings to avoid crowding are recommended procedures.

CRYPTANTHUS (Crypt-ánthus)—Earth Star. An especially attractive genus of the pineapple family BROMELIACEAE, named *Cryptanthus*, consists of twenty-two species endemic to Brazil. To its members is applied the colloquial name earth star. A few of its sorts are likely to be familiar even to city dwellers. They are frequent components of dish gardens displayed in florists' windows. Some are natural species, others highly attractive varieties or hybrids. The name, from the Greek *krypto*, to hide, and *anthos*, a flower, is not entirely apt, the unisexual blooms, although nestled among the foliage, are clearly visible.

Cryptanthuses are mostly small, nonhardy perennials. In the main they are terrestrial, which means they grow in the ground, not perched on trees like most bromeliads. They have stolons from which grow rosettes of usually stalkless, spreading, evergreen, rigid leaves with finely- or more coarsely-toothed edges. The flowers, in dense heads, are white and have three sepals, the bases of which form a tube, three wide-spreading, longer petals slightly joined at their bases, six slender-stalked stamens, and a slender style. The very small seeds are contained in dry, berry-like fruits.

Popular **C. acaulis** forms flat rosettes of mostly stalkless, gray-scurfy, lanceolate to pointed-triangular leaves with wavy, weakly-spiny margins and undersides thickly covered with white scales. Variety *C. a. ruber* has purplish-bronze centers and margins to its leaves, which are covered with light tan scurf. Unlike most species of this genus, **C. bahianus** has its rosettes of foliage lifted on stems 4 to 6 inches above the soil surface. It has rigid, spreading to recurved, long-lanceolate, spiny-toothed leaves, apple-green with bronzy-red wavy margins. Their undersides are silvery. Spreading, stalked, broad-elliptic to spatula-shaped, pointed leaves of thin texture form the loose, somewhat asymmetrical rosettes of **C. beuckeri**. They have spiny margins, scaly undersides, and above are conspicuously marbled with darker green

Cryptanthus acaulis

Cryptanthus beuckeri

on a lighter green background. Variable **C. bivittatus** typically has spreading, rather wavy, triangular-lanceolate leaves up to about 9 inches long by 1 inch to 1½ inches wide with sharp-toothed edges. Their upper surfaces are green with a pair of broad, buff or pale greenish, longitudinal bands with dull purplish-red bases. Their undersides are brown. Smaller *C. b. minor* (syn. *C. roseo-pictus*) has proportionately shorter, olive-green leaves with two pale bands. They have a tint of salmon-pink and in strong light the foliage assumes coppery tones. The olive-green to bronzy leaves of **C. bromelioides** spread in all directions to give a more erect plant than most cryptanthuses. Sword-shaped and taper-pointed, they have wavy, toothless margins. Especially attractive *C. b. tricolor* has leaves striped lengthwise with ivory-white and at their bases and margins suffused with rose-pink. Quite large rosettes of wavy-margined, broadly-ovate to oblanceolate or spatula-shaped, bronzy-purple leaves with a slight frosting of silvery scales are characteristic of **C. diversifolius**. In habit **C. lacerdae** much resembles C. acaulis. Its bright green leaves are decorated with silvery stripes down their centers and margins.

Handsome **C. zonatus** has rosettes of wavy, strap-shaped-lanceolate, finely-toothed leaves up to about 9 inches long

Cryptanthus bivittatus

Cryptanthus bromelioides tricolor

Cryptanthus zonatus

Cryptanthus osyanus

and 1½ inches wide. Their upper surfaces are cross-banded in zebra fashion alternately with brownish-green and ashy-gray or light tan. Their undersides are more or less gray- or white-scaly. Especially striking *C. z. zebrinus* has more intense variegation. The leaves are purplish-bronze banded with silvery-beige. Other varieties are *C. z. fuscus*, its leaves reddish-brown, especially in winter, with silver-gray cross bands and *C. z. viridis*, the undersides of the leaves of which are green and without scales. Very like *C. zonatus* is **C. fosteranus.** The chief differences are its larger size (its collector reports a specimen in Brazil more than 2½ feet across) and thicker, more rigid leaves. Spreading widely, they are coppery-green to purplish-brown with cross-bands of tan. A hybrid between *C.*

beuckeri and *C. lacerdae* is named *C. osyanus.* Hybrids between *Cryptanthus* and *Billbergia* are named *Cryptbergia.*

Garden and Landscape Uses and Cultivation. Among the most delightful of bromeliads, earth stars or cryptanthuses are suitable for outdoor rock gardens and similar uses in warm, humid climates. As indoor plants they grow readily in greenhouses and terrariums. They do surprisingly well as room plants and are popular for use in dish gardens. They succeed in well-drained, moist, but not wet soil that is fairly loose and contains an abundance of leaf mold, rich compost, peat moss, or similar humus-forming material. Light shade from strong sun is needed. Earth stars prosper under artificial light. For more information see Bromeliads or Bromels.

CRYPTBERGIA (Crypt-bérgia). This name, constructed from those of the parent genera, is used for hybrids between *Cryptanthus* and *Billbergia* of the pineapple family

BROMELIACEAE. Such hybrids are more or less intermediate in characteristics between their parents, have the same uses, and respond to the same conditions and care. The offspring of *Cryptanthus beuckeri* and *Billbergia nutans*, named *Cryptbergia* 'Mead', has 1-foot-long, narrow, erect leaves mottled green and pink. It is grown as an ornamental foliage plant. Another, *C. rubra*, has a broad rosette of bronzy-red foliage and small white flowers in a central cluster. The parents of this are *Cryptanthus bahianus* and *Billbergia nutans.*

CRYPTOCARYA (Crypto-càrya). This genus of between 200 and 250 trees and shrubs of the tropics and subtropics, excluding central Africa, belongs to the laurel family LAURACEAE. The fruits of *Cryptocarya moschata*, called Brazilian nutmegs, are used as a commercial spice. The name comes from the Greek *kryptos*, hidden, and *karyon*, a nut. It alludes to a characteristic of the fruits. The tree previously named *C. miersii* is *Beilschmiedia miersii.*

Cryptocaryas have alternate, subopposite or opposite, more or less leathery leaves. The flowers are small and in panicles. They have a long-tubed, six-parted perianth, nine fertile stamens, three staminodes (aborted stamens), and one style. The fruit is completely or almost completely enclosed in the usually woody perianth tube.

Native to southern Chile, densely-foliaged *C. rubra* (syn. *C. peumis*) is up to 50 feet tall. Its short-stalked leaves, alternate, subopposite or opposite, are broadly-ovate to nearly round. They attain maximum dimensions of about 3½ inches long by nearly 2 inches wide, but often are considerably smaller. Their upper surfaces are lustrous green, their undersides glaucous.

Cryptanthus zonatus zebrinus

Cryptanthus fosteranus

Cryptocarya myrtifolia

The midrib is prominent. The minute, yellowish blooms are in many-flowered, axillary panicles. South African *C. myrtifolia* is a slender-branched tree with pointed, broadly-elliptic-lanceolate to elliptic, leathery leaves 1 to 1½ inches long, at first finely-hairy, hairless later. The tiny flowers are in panicles 1 to 1½ inches long from the leaf axils.

Garden and Landscape Uses and Cultivation. In southern California and elsewhere in warm climates, cryptocaryas are occasionally planted as ornamentals, chiefly for their attractive evergreen foliage. Little is reported about their cultural needs. It is to be expected that they will grow under ordinary garden conditions. Propagation can be by seeds and probably by cuttings.

CRYPTOCEREUS (Crypto-cèreus). This genus of two species of Mexico and Central America of the cactus family CACTACEAE is related to and by some botanists included in *Selenicereus*. Its name, derived from the Greek *krypto*, to hide, and the name of the genus *Cereus*, refers to it having eluded detection by botanists until 1946 when it was discovered in Chiapas, Mexico.

The species first described, and cultivated by cactus enthusiasts is **Cryptocereus anthonyanus**. An *Epiphyllum*-like cactus, this has climbing stems 3 feet long or longer and up to 6 inches wide, with at

Cryptocereus anthonyanus

intervals clusters of flat, leaflike, deeply-lobed branches. The lobes are 1 inch to 2 inches long by up to slightly more than ½ inch wide. They have areoles, each with three tiny spines. The flowers, approximately 6 inches long by nearly or quite as wide, have many narrow petals, the outer ones reddish-purple, the others cream. They open at night.

Garden Uses and Cultivation. These are as for *Epiphyllum*. For additional information see Cactuses.

CRYPTOCORYNE (Crypto-còryne). Several choice species of this tropical Asian

genus of the arum family ARACEAE are treasured by fanciers of aquarium plants. The group consists of aquatic and swamp plants and numbers about fifty species. A great deal of confusion exists as to their correct identification. This is especially true of cultivated kinds. The problem of proper naming is especially difficult because many cryptocorynes bloom rarely or not at all in aquariums, and the species vary much in appearance according to light, water depth, soil, and other environmental factors. The name of the genus is from the Greek *krypto*, hidden, and *koryne*, a club, and alludes to the spadix being completely concealed within the tube of the spathe.

Cryptocorynes are perennials with creeping, branching rhizomes and strongly-ribbed, linear, lanceolate, elliptic, or heart-shaped leaves. The submersed leaves may differ in shape from the above-water ones of the same species. Like other members of the arum family, their flowers are in structures analogous to those of the calla-lily. They are minute and form a central column called a spadix (the equivalent of the yellow cylindrical feature of the calla-lily) from the base of which arises a bract called a spathe (the equivalent of the white petal-like organ that surrounds the spadix of the calla-lily). The fruits are follicles.

In *Cryptocoryne* the lower part of the tube formed by the spathe is separated by a narrow neck from the upper part. Only a small opening connects the two. At the bottom of this lower chamber are the ovaries surrounded by a circle of glands that produce an odor, inviting to pollinating insects, which enter the chamber through the neck. The stamens occupy the upper part of the lower chamber just under the neck and are surrounded by a membrane that prevents the immediate escape of the insects. Later, the membrane bends back making it possible for the insects to depart and carry the pollen to other blooms.

Cryptocorynes in cultivation include strong-growing **C. affinis**, a native of the Malay Peninsula, with lanceolate or oblong leaves widest at their bases, dark green above, often more or less claret beneath and with paler midribs and lateral veins. The spathes are up to 1¼ feet long with the inner surface above velvety black-purple. This is one of the best species for aquariums. The beautiful **C. balansae**, of Thailand and adjacent territories, has corrugated, narrowly-lanceolate, bright green leaves with blades up to 1½ feet long. The spathe is about 4 inches long. In **C. beckettii**, of Ceylon, another easy-to-grow kind, the leaves are oblong-lanceolate, olive-green to reddish-brown above, usually purplish or pinkish beneath. They often have wavy margins. The blades are up to 6 inches long. The leafstalks are purplish-green to violet. The spathe, 4 to 5 inches long, is yellow or greenish-yellow with a dark pur-

Cryptocoryne balansae

Cryptocoryne beckettii

ple throat. A native of Thailand, **C. blassii** has elliptic to egg-shaped leaves with blades up to 3 inches long, marked with wine-red or brownish-red and with long, slender, purplish leafstalks. The spathes are yellow and horizontally wrinkled. Having a wide natural range in southeastern Asia and New Guinea, **C. ciliata** has oblong-lanceolate leaf blades 6 inches to 1½ feet long, 2 to 4 inches wide, and light green on both sides. They have a prominent mid-vein and, branching from it, six to ten curving lateral veins. Its spathes, 10 inches to 1½ feet long, are fringed at their margins. They are yellow marked with purple inside and yellow and purplish-red outside. In bloom this is one of the most beautiful kinds.

Native to the Indo-Malayan Archipelago, **C. griffithii** has egg-shaped leaves, flat or slightly wavy, velvety green above and paler on their undersides. Sometimes they are mottled or veined with purple. The blades are up to 4 inches long by one-half as wide. The spathe is 4½ to 9 inches long, purple in its upper part, warted or deeply ribbed, and yellow or orange in the throat. This is one of the most free-blooming kinds. The name of a closely similar species, a rare sort in cultivation, **C. cor-**

Cryptocoryne griffithii

Cryptocoryne cordata

data, is sometimes misapplied to *C. griffithii.* The slender tail, up to 6 inches long, that terminates the spathe is a distinguishing feature of *C. johorensis,* native to the Malay Peninsula and Singapore. Its long-stalked, heart-shaped leaves have blades 2 inches long, with crenated margins. Including the tail, the spathe is up to 1 foot in length, corrugated above, and purple. This species is attractive in bloom. Similar to the last in its spathe ending in a long tail, shade-loving *C. longicauda,* of Borneo, has less-broadly-heart-shaped leaves that when young are usually distinctly crinkled. Its spathes are pinkish-purple outside and purplish-red inside. They are up to 10 inches long including the tail. A native of Ceylon, *C. lutea* has ovate to ovate-elliptic, very dark green leaves with veins purple on the undersides, and blades about 4 inches long and often corrugated. Its yellowish spathes are 4 to 5 inches long. Also from Ceylon comes *C. nevillii,* a very satisfactory low aquarium plant, with narrowly-ovate or lanceolate submersed leaves and shorter and broader above-water leaves. The former may have blades 3 inches long by 1 inch wide. The spathe is 2 to 3 inches in length, warty, and purple inside. Indian and Malayan *C. retrospiralis* has a spathe with a long tail or projection very characteristically coiled in a tight spiral. The spathe, without being uncoiled, is 4 or 5

inches long and red or purple inside. The leaves are narrowly-lanceolate with blades 6 inches to 1 foot long and a little over ½ inch wide. They are deep green and corrugated. From New Guinea comes *C. versteegii* with heart-shaped to nearly triangular, deep green or bluish-green leaves that have blades up to 3 inches long. The spathe is 3 to 4 inches in length, warty, and purple inside with a yellow throat. The leaves of *C. willisii* of Sri Lanka (Ceylon) are lanceolate with blades up to 5 inches long or occasionally longer. Their upper sides are olive-green, beneath they are paler. Generally dull purple suffuses parts of all of the leaves, which have wavy margins. Above-water leaves are darker colored and larger than submersed ones. The spathes are 4½ inches in length and greenish-yellow, with pale green throats. Very similar to if not identical with this is *C. undulata.* A native of Thailand and southwest Asia, *C. wendtii* has brownish to reddish, or, when grown in shade, green, lanceolate leaves with blades up to 4 inches long and 1½ inches wide, with crisped margins. The spathes are about 3 inches long and purplish-green, with a white-spotted purple throat.

Garden Uses. Cryptocorynes need warm or hot, highly humid environments such as can be provided in aquariums and tropical greenhouses. Most can be kept submersed, many can be grown either completely submersed or partly so in shallow water where their leaves reach into the air. A few, notably *C. ciliata* and *C. versteegii,* seem to prefer to be in mud with their leaves entirely in the air. Cryptocorynes are among the most lovely and popular aquarium plants.

Cultivation. For the best results, a bed at least 4 inches in depth of sandy soil or unwashed river sand with a fairly high clay and organic content should be provided. Soft, slightly acid water is generally most desirable. For *C. ciliata* and *C. lutea* slightly brackish water is suitable. The water temperature should be 68 to 85°F. It is advantageous if the floor of the aquarium is warmed. Cryptocorynes resent transplanting. It usually takes a few months for them to fully recover after being moved. Flowering is encouraged with some species, including *C. beckettii, C. lutea, C. retrospiralis,* and *C. willisii* by lowering the water level so that they are only partly immersed. Good top light is important, but strong sunlight is detrimental. It is often advantageous to grow floating plants such as *Azolla* over cryptocorynes to give slight shade. Propagation is chiefly by runners and by division.

CRYPTOGAM. Botanically plants are separated into cryptogams and phanerogams. Phanerogams include all those that, if they reproduce sexually, do so from seeds.

They are plants that ordinarily have flowers. Cryptogams produce neither flowers nor seeds. Their sexual reproduction is by spores. Among the most familiar sorts are ferns and mosses. Others of horticultural interest include certain equisetums, lycopodiums, marsileas, salvinias, and selaginellas. The name cryptogam, from the Greek *kryptos,* hidden, and *gam,* marriage, alludes to the details of sexual reproduction being less obvious than with phanerogams.

CRYPTOGRAMMA (Crypto-grámma) — Rock-Brake, Parsley Fern. These ferns, of the pteris family PTERIDACEAE, are generally considered as being confined in the wild to northern North America, Europe, and Asia. There are two to four species in these regions, depending upon the interpretations of different authorities; some botanists recognize a Chilean fern as belonging in this genus. The name, which alludes to the rows of clusters of spore capsules being concealed by the turned-under leaf margins, is from the Greek *kryptos,* hidden, and *gramme,* a line.

Cryptogrammas are evergreen or deciduous and have fertile fronds (leaves) different from the shorter, spreading barren ones. Both have stalks that, except for their bases, are green.

Short rhizomes and fronds in dense clusters are characteristics of the parsley fern (*Cryptogramma crispa*) and its American variety *C. c. acrostichoides* (syn. *C. acrostichoides*), the latter a native of bleak habitats from the northern United States and southern Canada to the Arctic, and distinguishable from European and Asian populations by having thicker leaves, and in minor technical details. This fern has hairless fronds, the sterile ones 2 to 5 inches long, with stalks accounting for about one-half their length. The triangular-ovoid blades are twice-pinnate or twice-pinnately-lobed, with the major divisions alternate or nearly so. The ultimate segments are approximately elliptic and toothed. Erect and up to 8 inches tall, the fertile fronds have triangular-ovate blades considerably shorter than their stalks and cut into much finer, linear divisions than the barren fronds.

Slender rock-brake (*C. stelleri*) has creeping rhizomes and scattered instead of clustered fronds. It inhabits limestone rocks from Newfoundland to Quebec, Alaska, New Jersey, West Virginia, Illinois, Idaho, and Washington, and also occurs in Asia. Its fronds are translucent and flimsy in comparison with the opaque, thicker ones of *C. crispa.* The weak-stalked sterile ones are 4 to 6 inches long and up to 1½ inches wide. They have ovate, twice-pinnate blades with five or six pairs of opposite or alternate major divisions. The ultimate segments are chiefly obovate and under ½ inch long. Longer than the barren fronds, the fertile ones are twice- or thrice-pinnate into final segments linear-lanceolate and up to ¾ inch long.

Garden and Landscape Uses and Cultivation. In general these are ferns for collectors and others who delight in the somewhat rare and unusual. They are adapted for rock gardens and similar situations where cool, damp root runs and shade assure conditions favorable to growth. For the slender rock-brake, crushed limestone mixed with the soil is advantageous. Increase is by spores and by division. For further information see Ferns.

CRYPTOLEPIS (Cryptó-lepis). The one dozen species of *Cryptolepis* belong in the milkweed family ASCLEPIADACEAE, or according to those who accept the division, in the periploca family PERIPLOCACEAE. Natives of warm parts of the Old World, they have a name derived from the Greek *kryptos*, hidden, and *lepis*, a scale. It alludes to the scales inside the corolla tube.

These are twining vining, or sometimes erect shrubs. They have opposite leaves, and flowers few together, loosely arranged in umbel-like, forked clusters, from the leaf axils or terminal. The blooms have deeply-five-cleft calyxes with five scales at their bases, a corolla with a cylindrical or bell-shaped tube, and five spreading or reflexed, twisted, linear lobes (petals). There is a corona or crown of five scales attached near the middle of the inside of the corolla tube. The fruits are spreading, paired, hairless, podlike follicles.

Erect and about 3 feet tall, *C. longiflora* has long-pointed, willow-like leaves 3 to 5 inches long, often tinged with red, clustered toward the ends of its stems. The slender, white flowers, 1 inch or more in length, are among or just above the foliage. The nativity of this species is unknown.

Garden and Landscape Uses and Cultivation. The species described is useful in beds and for the fronts of shrub plantings. It prospers in ordinary well-drained soil outdoors in warm, essentially frostless regions. It may be increased by seeds and by cuttings.

CRYPTOMERIA (Crypt-omèria)—Japanese-Cedar. The most important and one of the noblest conifers native in Japan *Cryptomeria* does not develop as splendidly in America or Europe as in its native land. Nevertheless it can be quite handsome and is especially pleasing when comparatively young. The only species, *C. japonica* is represented in southern China by variety *C. j. sinensis*.

A member of the taxodium family TAXODIACEAE, this genus is closely related to the big-tree or giant sequoia (*Sequoiadendron*) and the redwood (*Sequoia*), from both of which it differs in the construction of its cones. The name is from the Greek *krypto*, hidden, and *meris*, a part, but its significance is obscure.

In Japan this is a favorite for avenues, locations near shrines and temples, and gardens. One of the most famous avenues is at Nikko, leading to the tomb of the Shogun Ieyasu Tokugawa founder of the Tokugawa dynasty. The establishment of this planting is of historical interest. After Ieyasu died and was buried at Nikko his successor commanded every Diamyo in the land to send a stone or bronze lantern to be installed in the temple grounds at Nikko. All except one, too poor to meet the expense, did as instructed. He asked that, instead, he be permitted to plant cryptomeria trees to flank the road along which pilgrims would pass to the tomb of Ieyasu, to provide them with shade. His request was granted, the long rows of trees were planted at the beginning of the seventeenth century, and they have been maintained ever since. Whenever a tree dies or is destroyed by storms it is replaced. The great British plant explorer "Chinese" Wilson (E. H. Wilson) reported seeing a mile-long avenue of cryptomerias ranging in height from 120 to 180 feet in Japan, the trees approximately 650 years old. It is believed that *Cryptomeria* forms nearly one-third of the total forest resources of Japan. Its fragrant, coarse-grained wood is strong and durable and is widely employed for construction and building purposes, interior trim, furniture, boxes, and for many other uses. It is repellent to insects. Slabs of the bark are used for roofing.

Sometimes called Japanese-cedar, *C. japonica* at times exceeds 150 feet in height and occasionally has a trunk diameter of 8 feet. Usually it is smaller, but heights of

Cryptomeria japonica

Cryptomeria japonica, near Boston, Massachusetts

Cryptomeria japonica (male cones)

100 to 125 feet are not uncommon. Pyramidal, with a straight, tapered trunk stoutly buttressed and fibrous at its base, it has reddish-brown bark that sheds in strips. Its branches, commonly in whorls (tiers), are drooping or horizontal with the branchlets spreading or pendulous. Its flattened, blunt-ended, awl-shaped leaves, ½ to 1 inch long, are arranged spirally and radiate in all directions, but point forward and curve inward. Their bases clasp the shoots. The male cones are clustered in the leaf axils near the ends of the shoots, the females are at the tips of short branchlets. Individual trees bear both sexes. The fruiting cones, which mature at the end of the first season, are subglobose and consist of twenty to thirty woody, wedge-shaped scales each with two to five spinelike processes at their tips. Each scale bears two to five seeds. Many varieties of *Cryptomeria* have been described, especially in Japan, but few are cultivated in America. The one to which reference is most commonly made, *C. j. lobbii*, is said to have lighter green foliage and to be more compact and slightly

Cryptomeria japonica lobbii

Cryptomeria japonica spiralis

hardier than the typical species, but it is doubtfully distinct. The Chinese variety, *C. j. sinensis*, is looser in habit than the Japanese type and less ornamental. Its branchlets are more slender and pendulous, and its cones usually have no more than twenty scales with only two seeds to each. Variety *C. j. araucarioides* has short leaves on slender, drooping, widely-spaced branchlets, *C. j. cristata* has many of its branchlets joined to form cockscomb-like terminations and is a slow-grower, *C. j. elegans* is dense and bushy with leaves that spread widely or even backward from the shoots, are longer and softer than those of the typical kind, and turn from bluish-green to bronzy-red in winter, *C. j. nana* (syn. *C. j. lobbii nana*) is a low-growing, round-topped kind with dark foliage, and *C. j. spiralis* has slender branchlets and sickle-shaped leaves that twist curiously around the shoots.

Garden and Landscape Uses. Crytomer-

Cryptomeria japonica araucarioides (female cones)

ias are hardy in sheltered places in the vicinity of New York City and even in parts of southern New England, but they thrive better in climates not harsher than that of Washington, D.C. They are beautiful and have a great deal of character. Their elegant foliage has the general appearance of that of the big-tree (*Sequoia*) of California. Cryptomerias are splendid for planting along avenues and are very effective in groups with the individuals spaced closely enough so that they eventually nearly touch, but do not crowd, allowing their individual tops to rise like spires or minarets from a common mass of basal greenery. As accent trees cryptomerias are pleasing. They possess a slight somberness of habit and appearance that lends dignity to planned landscapes and subtlety complements architecture. They are, of course, quite ideal for use in Japanese gardens, and the dwarf varieties are appropriate for rock gardens. Cryptomerias thrive in ordinary well-drained garden soils. For their well-being they must have a deep fertile soil that supplies abundant moisture, but not one in which water gathers and stagnates. Good subsurface drainage is as essential as continuous access of the roots to moisture. They need shelter; cryptomerias are not subjects for exposed wind-swept locations. They grow well in sun, but stand a little shade.

Cultivation. Cryptomerias can be transplanted with minimum risk even when quite large. Their routine culture calls for no special care. Pruning is not normally needed except that in regions a little too cold for best growth a certain amount of winterkilling may occur, and then it is necessary to prune out dead branchlets or branches in spring. Propagation is by seeds sown in sandy peaty soil in a cold frame or a cool greenhouse in spring, or by cuttings, 3 to 4 inches long, inserted under mist or in a propagating bench in a very humid greenhouse or cold frame in summer or very early fall.

Diseases. Leaf-blight and leaf-spots sometimes cause browning of the foliage. Cutting out affected twigs usually gives sufficient control. Spraying with a copper fungicide is also recommended.

CRYPTOSTEGIA (Crypto-stègia)—Rubber Vine. Two or three species of milky-juiced, opposite-leaved, woody vines, of Africa and Madagascar, constitute *Cryptostegia*, of the milkweed family ASCLEPIADACEAE, or according to botanists who favor splitting that assemblage, of the periploca family PERIPLOCACEAE. Their sap contains rubber, but the plants are no longer commercially important as a rubber source. The name comes from the Greek *krypto*, hidden, and *stego*, a cover, and alludes to the crown of scales inside the corolla. The name purple-allamanda is sometimes used for *C. grandiflora*.

Cryptostegias have opposite, undivided leaves, and thrice-forked clusters of allamanda-like flowers. The blooms have leafy calyxes of five sepals with many glands at the base and short-tubed, funnel-shaped corollas with five lobes (petals) and a crown of five, sometimes two-parted scales, not exposed to view. There are five stamens united around the stigma and with their anthers adherent to it. The fruits are paired, podlike follicles with the parts spreading horizontally. These plants are poisonous to livestock.

African **C. grandiflora** is a vigorous vine or scrambling shrub with short-stalked, lustrous, hairless, broad-elliptic to oblongish leaves 3 to 4 inches long, about one-half as wide, and with short-pointed apexes. The showy, lilac-purple or lilac-pink blooms are darker when young. They are 2 to 3 inches across and have the scales of their crowns deeply cleft into two slender lobes. The hard, strongly-angled seed pods are up to 4 inches long. This species is naturalized in the West Indies and other tropical places. From the last, **C. madagascariensis**, of Madagascar, differs in having reddish-purple blooms with calyxes only about ¼ inch long (those of *C. grandiflora* are twice that length) and in the scales of the crowns of its flowers not being cleft.

Garden and Landscape Uses and Cultivation. The species described are fairly commonly cultivated in tropical gardens as strong-growing ornamental vines. They thrive in ordinary soils and locations and need little attention other than occasional pruning to limit them to alloted space. Propagation is by seed, cuttings, and layering.

CRYPTOSTEMMA. See Arctotheca.

CRYSTALWORT is *Riccia fluitans*.

CTENANTHE (Ctenán-the). Belonging to the maranta family MARANTACEAE, the tropical South American genus *Ctenanthe* is closely related to *Stromanthe*. It differs in having green instead of colored floral bracts and in its flowers having two petal-like staminodes. There are about fifteen species. The name derives from the Greek *kteis*, a comb, and *anthos*, a flower. It alludes to the arrangement of the bracts.

Ctenanthes are *Calathea*-like tropical herbaceous plants admired for their beautiful evergreen foliage. They have branched stems and undivided, lobeless, toothless, basal and stem leaves. The flowers are crowded in terminal, bracted spikes or racemes. Each has three separate, somewhat parchment-like sepals, a short-tubed corolla with three nearly equal lobes (petals) hooded at their apexes, a short staminal tube, one fertile stamen, and two outer petal-like staminodes (abortive stamens).

Often misnamed *Bamburanta*, which is a synonym for *Hybophrynium*, attractive *C. compressa* is a 2 to 3 feet tall. It has wiry stems and angled from them unequal-sided, paddle-shaped leaves with blades about 1¼ feet long by about one-third as wide. They have waxy-green upper surfaces, gray-green undersides. Some naked reedlike stems topped by tufts of two to four plantlets are also produced. More slender *C. c. luschnathiana*, of daintier appearance, has reedlike stems with at their tops narrower, oblong, light green leaves with obscure darker featherings. Miscalled giant bamburanta, 6 feet-tall *C. oppenheimiana* is dense and bushy. Its leathery, pointed-elliptic, downy-stalked leaves, wine-red

Ctenanthe oppenheimiana

on their undersides, are above dark green relieved by broad silvery bands curving outward and upward from the midrib. Highly variegated *C. o. tricolor* is lower. Its foliage is green with silvery-gray bands angling outward from the midrib, irregularly patched with cream-white and sometimes with some of the wine-red color of the under surfaces showing through.

Other kinds cultivated include *C. glabra*, which has short-pointed, broad-paddle-

shaped, somewhat obscurely dark-veined, yellowish-green leaves, lustrous green on their undersides and with flattened stalks with long, reddish sheaths. The short-pointed, spreading, oblong leaves of *C. humilis* have quilted, bright green upper surfaces and grayish-green lower ones. Tufts of short, leafy branches are borne at the ends of erect, naked stems. About 2½ feet tall and with stolons, *C. kummerana* has pointed-elliptic to lanceolate leaves very distinctly variegated in young speci-

Ctenanthe kummerana

mens, less markedly so or not or at all in older ones. When most intense the pattern is of alternate silvery and dark green bands curving outward and upward from the midrib. The undersides of the leaves are purple. The short-pointed, oblong leaves of *C. lubbersiana* are green mottled and dappled with yellow. Their undersides are paler. Reedlike, downy stems and pointed-oblong-elliptic, yellowish-silvery leaves with narrow green bands running outward and upward from midrib to margin are characteristic of *C. setosa*. The under surfaces are purple.

Garden and Landscape Uses and Cultivation. These are as for *Calathea*.

Ctenanthe setosa

CTENITIS (Cte-nìtis). Some 150 species of this genus of ferns of the aspidium family ASPIDIACEAE are distributed in tropical, subtropical and warm-temperate regions. Most nearly related to *Dryopteris*, they

have usually short, ascending or erect, rarely creeping rhizomes or stems and two- or three-times-pinnate, broad-based leaves (fronds) with midribs hairy on their upper sides. The round to kidney-shaped clusters of spore capsules are on the veins on the undersides of the fronds. The name, derived from the Greek *kteis*, a comb, has no obvious application.

Native to New Zealand, Australia, Tasmania, and Norfolk Island, *Ctenitis decomposita* (syns. *C. pentangularis*, *Nephrodium pentangularum*) has creeping, branching, scaly rhizomes with, spaced along them, fronds up to 3 feet tall with stiff stalks and triangular to triangular-ovate, palish green, two- or three-times-finely-pinnately-divided blades from 4½ inches to 1 foot long or longer and nearly as broad as long, with the ultimate segments ⅓ to ½ inch long and toothed. In the wild this species occupies wet soils and margins of woodlands.

A beautiful Floridian, *C. sloanii* is sometimes called comb fern and Florida tree fern. The last designation seems a little ambitious because its erect stems are not over about 1½ feet tall. This has graceful, twice-pinnate fronds that are responsible for the plant's impressive spread of up to 12 feet.

Garden and Landscape Use and Cultivation. The dense clumps of *C. decomposita* are attractive for embellishing lightly shaded locations in such mild climate areas as California and the Pacific Northwest. This species withstands five or six degrees of frost and succeeds in moderately dry as well as moister soils. The Florida species behaves best in moist soil and a humid atmosphere where it is not subjected to frost. The best soil for ctenitises is a fairly well-drained one that contains an abundance of leaf mold, peat moss, or other decayed vegetable matter and is always moist. An annual top dressing with similar organic material is helpful. In greenhouses a winter night temperature of about 50°F, with a daytime rise of up to fifteen degrees, is satisfactory. At other seasons warmer environments are in order. High humidity and shade from strong sun are necessary. Propagation is easy by division which is best done in spring just as new growth begins. New plants can also be had from spores. For more information see Ferns.

CUACHILOTE is *Parmentiera edulis*.

CUBAN. This word forms part of the common names of these plants: Cuban bast (*Hibiscus elatus*), Cuban-hemp (*Furcraea hexapetala*), and Cuban-lily (*Scilla peruviana*).

CUCKOO FLOWER. This name is applied to *Cardamine pratensis* and *Lychnis flos-cuculi*.

CUCKOO PINT is *Arum maculatum*.

CUCUMBER. In addition to its employment as the name of varieties of *Cucumis sativus*, dealt with in the next entry, the word cucumber forms part of the common name of various plants including these: African-horned-cucumber (*Cucumis metuliferus*), bitter-cucumber (*Citrullus colocynthis*), bur- or star-cucumber (*Sicyos angulatus*), creeping-cucumber (*Melothria pendula*), cucumber tree (*Magnolia acuminata* and *Averrhoa bilimbi*), Indian cucumber-root (*Medeola virginiana*), large-leaved cucumber tree (*Magnolia macrophylla*), mock- or wild-cucumber (*Echinocystis lobata*), squirting-cucumber (*Ecballium elaterium*), wild-cucumber (*Echinocystis lobata* and *Marah*), and yellow cucumber tree (*Magnolia cordata*).

CUCUMBER. Cucumbers, varieties of *Cucumis sativus*, of the gourd family CUCURBITACEAE, have been cultivated for at least 2,000 years. Their ancestral home is not surely known, but is believed to be tropical Asia or Africa. Certainly it is Old World. Cucumbers are grown outdoors nearly everywhere in warm climates and, where summers are hot enough, in temperate ones. They are also raised in greenhouses. As a summer outdoor crop they can be had in most parts of the United States, as a

Young cucumber plants in hills

Cucumber (foliage and male flowers)

Cucumber allowed to trail on the ground

Cucumber tied to a wire support

Cucumber ready for harvesting

winter one only in the very far south and Hawaii. Their fruits are favorites for salads and pickling.

Soil for this crop must be decidedly fertile and well supplied with decomposed organic matter, such as half-rotted manure or compost. The objective is to stimulate rapid, continuous growth. As a precaution against transmitting disease be sure the compost does not contain the remains of plants of the gourd family, such as cucumbers, gourds, melons, pumpkins, or squashes. Prepare the ground by spading, rototilling, or plowing. In home gardens it

is usual to sow in hills spaced 4 to 6 feet apart each way, but satisfactory results can also be had by sowing in drills 5 to 7 feet apart with the plants 2 to 3 feet apart in the rows. The hills, usually raised a few inches above the level of the surrounding soil, are made by forking three or four good shovelfuls of manure or compost into a circular area about 2 feet in diameter and heaping a flat mound of soil over this. Mix well into each hill one-half pound of 5-10-5 fertilizer or enough of a fertilizer of other formulation to supply an equivalent amount of nitrogen. If you are sowing in rows, allow one pound of 5-10-5 to each 10 feet and mix it, and some additional organic matter, along each row to a depth of 8 inches or more.

Defer sowing until the soil temperature is at least 55°F. To permit earlier planting, this may be reached, by spreading, two or three weeks before the expected sowing date, over the locations where the seeds are to be sown, a 1-inch layer of leaf mold or other dark colored organic material that will absorb more sun heat than a lighter colored surface. Sow six seeds in each hill. Cover them to a depth of about ½ inch. Later thin the seedlings to three in each hill. In rows, sprinkle three or four seeds to each running foot and thin the seedlings to 8 to 10 inches apart. As an alternative to sowing outdoors, start seeds in pots or other containers or in pieces of turf turned upside down, in a greenhouse or other place indoors where a temperature of 60 to 70°F can be maintained. Do not do this until three to four weeks before the expected date for setting the plants outdoors. This should not be until the ground temperature is 55°F or higher, which is not until late May or early June in the vicinity of New York City.

It is helpful to protect newly sown seeds and newly set out plants with parchment paper tents of the Hotcap type or with plastic domes or similar devices. These serve as little greenhouses, affording shelter from inclement weather and giving some protection against insects that appear early. Be sure to have an opening in each tent to forestall the possibility of the inside temperature rising to unacceptable levels when the sun shines.

Care through the growing season includes keeping the soil around the plants shallowly cultivated and, should the weather be dry, watering copiously and often enough to prevent wilting. The length of the cropping season and the size of the crop is increased by harvesting all fruits before their seeds become hard. Be vigilant to circumvent cucumber beetles, which not only feed on the leaves, but also transmit very serious diseases.

Cold frames and heated frames or hotbeds make possible earlier crops than can be had outdoors. They can also be used to grow cucumbers in climates too cool for their satisfactory development without such aid. Seeds can be sown or young plants started indoors can be set in cold frames two or three weeks earlier than in the open garden. In heated frames and hotbeds even earlier beginnings are practicable. Plant in hills, mounded or on the flat, three plants to each. Allow 16 to 20 square feet for the vines of each hill. Thus, one hill to each standard 6 by 3 feet frame sash is sufficient. Whenever temperatures at night are below 60°F or those by day are below 70°F keep the sash closed or closed except for an opening judged big enough to give needed ventilation and prevent the inside temperature from rising above 85 or 90°F. When nights are cool close the sash early enough in the afternoon for the temperature inside to rise to 90°F, or even a little higher, thus trapping enough heat to prevent it from dropping too low during the night. Maintain a humid atmosphere inside the frame by spraying the foliage with water in bright weather.

As a greenhouse crop cucumbers are not difficult to manage. English forcing varieties are best for this purpose. They need a minimum night temperature of 60°F and a humid atmosphere. By day the temperature may with advantage rise ten to fifteen degrees or more over that kept at night. Shade from strong sun, just enough to prevent scorching the foliage, is needed. The usual practice is to plant in mounds of soil placed at intervals of 1½ to 2 feet along the benches. Make sure the soil is fertile, loose, and contains a large proportion of coarse leafmold, compost, or half-rotted manure. Do not sift it. At first each mound may consist of about one-half a pailful or so of the mix. Later, whenever searching, young white roots reach the surface add to the mounds by spreading up to an additional inch of the mix over them. This may be necessary every week or so. It is far better to do this top dressing a little at a time than to add a great deal of soil in one operation.

Supports must be furnished. These may be wires stretched horizontally about 1 foot beneath the greenhouse roof and spaced about 9 inches apart. A stake from the base of the plant to the first wire affords early support for the young plant and serves as a lead. Training consists of allowing the main stem to grow unpinched until it reaches the top wire, a distance of perhaps 4 to 6 feet. Its tip is then pinched out. Remove any side shoots that develop below the first wire. Pinch out the tips of all others just beyond their second leaf and of secondary shoots that come as a result of that pinch just beyond the first leaf. Keep male flowers, those with three stamens and without an obvious ovary behind the corolla, picked off the vines.

Varieties of cucumbers are of three chief types, slicing, pickling, and English forcing. Slicing varieties include 'Ashley',

'Surecrop Hybrid', 'Straight Eight', and 'White Spine'. Good pickling sorts are 'Chicago Pickling', 'National Pickling', 'Ohio M.R.12', and 'Ohio M.R. 17'. The last two are strongly resistant to mosaic disease. The basic English forcing kind is 'Telegraph'.

Pests and Diseases. Success with cucumbers depends largely upon pest and disease control. The chief insect depredators are the striped cucumber beetle and the spotted cucumber beetle. Both feed on the undersides of the foliage and transmit serious diseases among cucumbers and other plants of the gourd family. They are controlled by regular and frequent applications of contact insecticides. Aphids and red spider mites also infest cucumbers. Bacterial wilt disease causes sudden wilting and collapse of the plants. It cannot be cured. Mosaic disease, which causes a yellow mottling of the foliage, is transmitted by aphids and other insects. Prevention of these diseases, for which there are no cures, is based on controlling insects and using resistant varieties. Other diseases include anthracnose, leaf spots, powdery mildew, and scab.

CUCUMIS (Cucùm-is)—Cucumber, Melon, West Indian Gherkin. The most familiar members of *Cucumis*, of the gourd family CUCURBITACEAE, are melons, cucumbers, and West Indian gherkins, grown for their edible fruits. A few others are occasionally cultivated as ornamentals or as interesting curiosities. The about twenty-three wild species are natives of Africa and southern Asia. The geographical origins of the ancestral stocks of cultivated cucumbers, melons, and West Indian gherkins are not surely known, but it seems certain that they were of the same provenence. This is almost certainly true of the West Indian gherkin. Once believed to be a native of the New World, it is now fairly well established that it was brought to the Americas more than 300 years ago incidental to the slave trade. Fairly strong, but by no means conclusive evidence is offered by some investigators that the cucumber is a development of a species native to India, and rather convincing arguments are advanced that melons have a similar origin, but they may be African. The name *cucumis* is an ancient Latin one.

These frost-tender, vining, herbaceous plants are annuals or are commonly grown as such. They have alternate, usually angled, lobed, or divided leaves and branchless tendrils. The flowers are unisexual with both sexes usually on the same plant. Solitary, or the males sometimes in clusters, they come from the leaf axils. They have a five-lobed calyx and a flat-bell- to saucer-shaped, deeply-five-parted, yellow or whitish corolla. The males have three stamens, the females three to five stigmas. The fleshy, hairless, pubescent, or prickly

fruits are many-seeded and covered with a distinct rind. Morphologically they are berry-like, technically they are pepos.

The cucumber (*C. sativus*) has angled, roughly-hairy stems and more or less shallowly-five-lobed, irregularly-toothed leaves with the terminal lobe the biggest. The flowers are 1 inch to 1½ inches in diameter, the males mostly in threes, the females usually solitary. The fruits are slightly angled and usually prickly, at least when young. They are long-cylindrical to nearly globular. Variety *C. s. sikkimensis* has orange-brown to reddish-brown fruits about 1 foot long.

Cucumis sativus sikkimensis

Melons, muskmelons, and cantaloupes are included in highly variable *C. melo*, which is represented in cultivation by several distinct races. It has softly-hairy to nearly hairless, slightly-ribbed or angled stems and usually more or less five-angled, sometimes shallowly-three- to seven-lobed leaves, rounded, ovate, or kidney-shaped in general outline. The flowers, up to 1 inch in diameter, are usually unisexual, but occasionally are bisexual. The males are in clusters, the females solitary. Tremendous variation exists in size and other characteristics of the fruits, including the color of their flesh and the appearance of their outsides, which may be heavily netted, smooth, or rough. For horticultural purposes *C. melo* may be divided into the following groups: *C. m. cantalupensis*, the true cantaloupe, little known in America, has rough, warty, or scaly, medium-sized fruits without netting. *C. m. chito*, the mango melon or garden-lemon, has small, acid, smooth, mottled, but not netted fruits. *C. m. conomon*, the Oriental pickling melon, has small, smooth, oblong, acid-flavored, white-fleshed fruits, mottled, but not netted, that are mushy when ripe. *C. m. dudaim*, the pomegranate melon or Queen Anne's pocket melon, has musky-scented, pubescent fruits up to 2 inches in diameter, on very pubescent vines; it is naturalized in parts of the southern United States. *C. m. flexuosus,* the snake melon or serpent melon, has straight or coiled, smooth, slender, cucumber-like fruits 1 foot to 3½ feet long. *C. m. inodorus,* the winter melon or white-skinned melon, has smooth or corrugated fruits, without netting, that ripen late and can be stored for a few weeks. *C. m. reticulatus,* the netted muskmelon or nutmeg muskmelon, com-

Cucumis melo reticulatus

monly but incorrectly known in North America as cantaloupe, has medium-sized, strongly-netted fruits with deep salmon-orange to green flesh.

The West Indian gherkin, bur gherkin, or gooseberry gourd (*C. anguria*) has slender, trailing, rough-hairy stems and leaves, the latter deeply-palmately-three- or five-lobed. The flowers, up to ½ inch in diameter, are succeeded by prickly, greenish-fleshed, egg-shaped to oblong fruits 1½ to 2 inches long, yellow and often green-striped when ripe. The male blooms are clustered, the females solitary.

Cucumis anguria

Ornamental species include the hedgehog or teasel gourd (*C. dipsaceus*) and the African-horned cucumber (*C. metuliferus*). These are sometimes grown as decoratives and as curiosities. Resembling the melon (*C. melo*), *C. dipsaceus* has oblong to nearly spherical fruits, 1 inch to 2 inches in

length, thickly covered with long spines that become dry and hard. The fruits of *C. metuliferus* are about 4 inches long and red and spiny. Its heart-shaped leaves are somewhat three-lobed.

Garden Uses and Cultivation. For further information about cucumbers and melons consult the entries Cucumber, and Melons and Muskmelons, in this Encyclopedia. The West Indian gherkin responds to the same cultural care. The ornamental species as well as the mango melon (*C. melo chito*) are grown as annual gourd vines. For information about them see Gourds, Ornamental.

CUCURBITA (Cu-cùrbita) — Pumpkin, Squash, Yellow-Flowered Gourd. Pumpkins, and some kinds of squashes and gourds, are familiar examples of *Cucurbita,* of the gourd family CUCURBITACEAE. The genus consists of fifteen species of frost-tender annuals and herbaceous perennials (usually grown as annuals), natives of the New World, chiefly of the warmer parts of North America. The name is the classical Latin one for gourds.

Cucurbitas are vining or bushy. They have succulent, furrowed or angled, often more or less prickly stems. Their alternate, undivided leaves are shallowly- to deeply-palmately (in hand fashion) -lobed. The large, creamy-white to deep orange-yellow, unisexual blooms, both sexes on the same plant, are solitary in the leaf axils. They have bell-shaped calyxes and corollas. Male flowers are long-stalked and have three stamens, their anthers are usually joined. The short-stalked female blooms have a thick style with three two-lobed stigmas. The fruits, strange as it may seem when one contemplates a squash or pumpkin, morphologically resemble berries. Technically they are pepos. Small, large, or immense, they have fleshy or fibrous interiors, many seeds, and soft or hard rinds.

Four species of *Cucurbita* have varieties commonly cultivated for their edible fruits, but the names pumpkin, squash, and marrow applied to most of these do not correlate with the species. Rather they reflect the culinary uses to which their fruits are put. So it is that varieties called pumpkins and others known as winter squashes are to be found in all four species, and marrows in two. Summer squashes and the yellow-flowered ornamental gourds of this genus belong to *C. pepo,* the cushaw squashes to *C. mixta.*

Most variable of cultivated cucurbitas, *C. pepo* is vining or bushy and prickly and harsh to the touch. It has rigid, erect, conspicuously-lobed leaves, broadly-triangular to ovate-triangular in outline, and 6 inches to 1 foot wide. The flower stalks, scarcely or not swollen where they join the fruits, are five-angled. The corolla lobes are erect or spreading. Varieties of this

species include field and pie pumpkins, and scallop, patty pan, crook-neck, straight-neck, cocozelle, zucchini, acorn, and marrow squashes.

The ornamental, yellow-flowered gourds, varieties of *C. pepo,* are sometimes distinguished as *C. p. ovifera.* They are long-stemmed vines with many tendrils and broad-ovate, angled or lobed leaves not over 8 inches in diameter. Their small, inedible, hard-shelled fruits are of various shapes and are uniform in color or striped. These plants are closely related to *C. texana,* a species native to Texas.

Variable, but less so than *C. pepo* is *C. maxima.* It includes varieties of pumpkins and winter, turban, and marrow squashes. Mostly trailing vines, or rarely bushy, the plants are softer and not as harsh and prickly to the touch as *C. pepo.* Rounded in outline, their leaves are scarcely or not lobed. The flowers are pale to deep yellow and have calyxes with short, narrow lobes. The lobes of the corolla are usually reflexed. The soft- or hard-shelled fruits are spherical, top-shaped, or cylindrical. The varieties 'Banana', 'Boston Marrow', 'Hubbard', and 'Mammoth Chili' belong here.

Other pumpkins and winter squashes are varieties of *C. moschata,* a softly-velvety-hairy, vining species with shallowly-lobed, roundish or slightly pointed leaves often marked with whitish spots along the veins. The male blooms have tubeless or short-tubed calyxes, often with leafy lobes. The corolla lobes are wide-spreading or reflexed. Generally big, and spherical, cylindrical, or flattened, the fruits are on stalks that thicken considerably at the point of attachment. Identified with this species are varieties 'Butternut' and 'Kentucky Field'.

The varieties 'Cushaw', 'Japanese Pie', and some other squashes, as well as some varieties of pumpkins, belong in *C. mixta.* Vining plants, not harsh to the touch, these have large leaves usually with white blotches and more or less lobed. The flowers are yellow to orange-yellow or green. Hard- or soft-shelled and of dull appearance, the fruits vary considerably in shape and size.

Ornamental gourds belonging in *Cucurbita,* other than those described above as varieties of *C. pepo,* include the Malabar gourd (*C. ficifolia*) and the calabazilla (*C. foetidissima*). Archeological discoveries prove that the first, a perennial at one time thought to be a native of Asia, has existed in America for at least 5,000 years. A stout, rather woody stemmed vine, *C. ficifolia* has five-lobed, roundish to ovate, sometimes marbled leaves and flowers with bell-shaped, short calyxes. The hard-shelled, white-fleshed, roundish egg-shaped, white-striped, green fruits, up to 1 foot long, have black seeds. Also perennial, *C. foetidissima* is native to desert regions of the southwestern United States and Mexico. Not very prickly, it has long, trailing stems and shallowly-scalloped, more or less triangular, gray-hairy leaves. The flowers resemble those of *C. pepo*. Nearly spherical, the yellow-variegated, smooth, green fruits are about as big as oranges. They are not edible.

Garden and Landscape Uses and Cultivation. For these see Pumpkin, Squash, and Gourds, Ornamental.

CUCURBITACEAE—Gourd Family. Important horticulturally, the gourd family comprises 640 species of dicotyledons grouped in 110 genera. Among its members are such popular vegetables, salads, and fruits as chayotes, cucumbers, gourds, melons, pumpkins, squashes, and watermelons.

Chiefly natives of warm regions, cucurbits, as members of this family are called, include annual and perennial herbaceous plants and woody-stemmed vines and shrubs. Of the cultivated genera only *Xerosicyos* has woody stems. Other sorts grown in gardens are herbaceous and except certain bush varieties of squash vines, all except the squirting-cucumber (*Ecballium*) have tendrils.

Cucurbits have alternate, usually undivided, palmately-lobed leaves and generally branched or branchless tendrils. The flowers are symmetrical and unisexual, with commonly both sexes on the same plant, but sometimes segregated on separate plants. They have five-lobed calyxes and a corolla with five petals or five lobes. There are three stamens (two of which represent pairs of stamens that are united) and usually three carpels, each with one stigma. The fruits, technically pepos, often large and fleshy, are more rarely dry and papery.

Among cultivated genera are *Benincasa*, *Bryonia*, *Citrullus*, *Coccinea*, *Cucumis*, *Cucurbita*, *Cyclanthera*, *Diplocyclos*, *Ecballium*, *Echinocystis*, *Gerrardanthus*, *Gurania*, *Ibervillea*, *Kedrostis*, *Lagenaria*, *Luffa*, *Marah*, *Melothria*, *Momordica*, *Sechium*, *Sicana*, *Sicyos*, *Thladiantha*, *Trichosanthes*, and *Xerosicyos*.

CUDRANIA (Cud-rània). This Old World group of four species of deciduous or evergreen trees and sometimes climbing shrubs is so closely related to the American osage-orange (*Maclura pomifera*) that a hybrid between the latter and *Cudrania tricuspidata* exists. Intermediate between its parents, this is *Macludrania hybrida*. Native from eastern Asia to New Caledonia and Australia, *Cudrania* belongs in the mulberry family MORACEAE. The name is a modification of its native Malayan one, cudrang.

Cudranias often have spiny branches. Their undivided leaves are alternate, short-stalked, and sometimes three-lobed, but not toothed. The unisexual, minute blooms are crowded in short-stalked, spherical heads from the leaf axils. They are without petals. Each has a four-parted calyx. The

males have four stamens, the females an ovary enclosed by the perianth, an undivided or two-parted style, and a slender stigma. The compound, rind-covered fruits have fleshy insides. In the Orient the foliage of *C. tricuspidata* is fed to silkworms.

Native to Korea and China, and planted to a limited extent in Japan, *C. tricuspidata* is a deciduous shrub or tree rarely up to 50 feet tall, and much more commonly up to one-half that height. Its branches, especially the lower ones, have stout spines ¼ inch to 1¼ inches long. The leaves, ovate to broad-elliptic or obovate and sometimes three-lobed, are up to 4 inches long. The midrib above and the veins on the undersides have very short hairs. The clusters of green or yellow flowers are solitary or paired. Edible, the red shiny fruits, about 1 inch in diameter, have a hard, rough rind.

Garden and Landscape Uses and Cultivation. Not hardy in the north, in regions of milder winters *C. tricuspidata* is planted for interest and ornament. It can be used as a hedge plant. Its flowers make no ornamental display. It adapts to ordinary, well-drained soil in sun or light part-day shade and is propagated by seed and by summer cuttings under mist or in a greenhouse propagating bench.

CUDWEED. See Gnaphalium.

CULCASIA (Cul-cásia). About twenty tropical African species of nonhardy, evergreen herbaceous perennials, most of which are stem-rooting vines, constitute *Culcasia* of the arum family ARACEAE. The name is derived from *kulkas*, the Arabic one for *Colocasia antiquorum*.

Culcasias have lanceolate, elliptic, or paddle-shaped leaves with long-sheathing stalks and a long-stalked inflorescence consisting of a club-shaped spadix with tiny flowers without perianths, enveloped by a bract called a spathe. The flowers on the lower one-third of the spadix are females, those on its upper one-half, males. The fruits are berries.

About 2 feet tall and native of the Cameroons, *C. mannii* has leaves with conspicuously veined, pointed-elliptic to ellip-

Culcasia mannii

tic-oblong blades 8 to 10 inches long. The inflorescences, on stalks 2 to 4 inches long, have a spathe, white on its inside, about the same length as the stalk and approximately one-half as wide. The spadix is shorter. The male flowers are white, the ovaries of the females, and the fruits, red.

Garden Uses and Cultivation. The species described is desirable in warm, humid climates and in tropical greenhouses as a decorative foliage plant. It prospers in shade and in soil containing an abundance of organic matter and kept evenly moist. Propagation is easy by cuttings and by seeds.

CULTIVAR. This is a partial replacement for the term variety as previously universally applied to particular plant groups in English language references and as employed in this Encyclopedia. In classifying plants it is used to identify discrete populations maintained in cultivation. It is a substitute for the older terms horticultural variety and garden variety and as such is a useful concept adopted in the International Code of Nomenclature for Cultivated Plants and rapidly gaining acceptance among sophisticated horticulturists. The purpose of introducing the term cultivar was to make it possible to distinguish by the manner in which the names of certain plants are presented whether or not they apply to botanical varieties, that is, to populations of less than specific rank recognized by botanists and that perpetuate themselves in the wild, or whether they identify populations of plants selected or bred by man that mostly are maintained only in cultivation and presumably and probably would be lost were man's interest in them to cease. The chief importance to gardeners as distinct from simply an interest they may have in the subject of distinguishing between botanical varieties and horticultural varieties or cultivars is that whereas the former, of which comparatively few are grown, come true from seeds, the latter are unlikely to. But correct understanding and application of the terms variety (or more properly *varietas*) and cultivar as spelled out in the rules of the International Code calls for more technical knowledge about horticultural and botanical nomenclature than the majority of gardeners possess and misapplications lead to confusion. This is especially true of conventions established for writing names. To illustrate, if a plant is a botanical variety its name is to be written entirely in italic in one of these ways, *Rhododendron catawbiense variety album* or *Rhododendron catawbiense var. album*. Not infrequently in horticultural writings, although this is not in strict accordance with the botanical code, the form *Rhododendron catawbiense album* is used. Cultivar names under the code rules are to be presented in one of these fashions, *Rhododendron carolinianum* cultivar

'Album', *Rhododendron carolinianum* cv. 'Album', or *Rhododendron carolinianum* 'Album', part of the name in italic, part in roman letters. Cultivars named after 1959 cannot, according to the Code, be given Latin-form designations; they must have modern language names, for instance 'White Beauty', 'Godfrey's White', 'Mary Smith', or other "fancy" name instead of 'Album'. This is excellent. Less satisfactory is that Latin-form cultivar names in use before 1959 are retained. Although changing them would create considerable confusion, it is improbable that this would be greater than the confusion that arises from the common practice of changing plant names to conform with the rule of botanical priority.

The decision usually not to differentiate between natural varieties and cultivars (horticultural varieties) in this Encyclopedia will be deplored by some. It has been made in the belief that in the vast majority of cases it is of little or no importance to gardeners to know to which of these groups the plants they grow belong. Because in speaking of plants gardeners are unlikely to make such distinction and because writing their names in accordance with the Code's slightly complicated system is, in ordinary usage, likely to lead to confusion and error and in any case cannot be done accurately without knowing the category to which each particular plant belongs, it was thought better not to make the distinction here. And so, without criticism of those who wish to follow the Code, Latin-form names of all categories in this work are in italic, "fancy" names in roman type enclosed in single quotation marks.

CULTIVATION. This term is used in more than one sense. It may allude to all the procedures employed in raising and caring for particular plants, including initial preparation of the soil, propagation, planting, potting, pruning, watering, fertilizing, staking, and disease, pest, and weed control. Thus, one refers to the cultivation of apples, peas, roses, and strawberries. Similarly, the cultivation of a garden or section of a garden, such as a perennial border, implies giving attention to all such matters related to the needs of the plants in the designated area.

In a more restricted sense the term cultivation alludes to stirring the upper inch or two of soil between growing plants with a hoe or a tool or implement called a cultivator. This is done to eliminate weeds, facilitate the admission of air and water, and conserve moisture. Its very great effectiveness in checking moisture loss is based on the creation of a loose surface layer that soon dries into a soil or dust mulch. This is a strong barrier to water that rises by capillarity from deep in the soil and escapes as vapor into the atmosphere. If the loosened soil is compacted by rain or

walking upon, the capillary system to the surface is restored and maximum water loss begins again. Therefore wise gardeners cultivate from spring to fall after every soaking rain, but not until the ground has dried sufficiently so that it does not stick to shoes and tools, and as far as possible they avoid walking on cultivated ground. In long periods of dry weather cultivating between rainfalls is beneficial. Take care not to stir so deeply that roots are injured, 1 inch to 2 inches is adequate. Pulverize the loosened earth finely. Do not leave it in lumps or clods.

CULTIVATORS. Cultivators are tools or implements for stirring (cultivating) the surface soil between growing crops. To accomplish this they have usually prongs that are pulled or pushed through the ground, but some types have instead rotating blades. Power cultivators, usually used only for large areas, are available. For home gardens hand cultivators with long or short broom-handle type shafts and hand-pushed wheel cultivators are generally adequate. The latter are best suited for use between plants in rows as in vegetable gardens and nurseries. Wheel hoes come with one or two wheels. For home gardens one wheel is usually preferred. The prongs are commonly detachable and replaceable with hoe type and plow type blades. The

A wheel cultivator in use

A hand cultivator in use

latter are useful for hilling (mounding) soil about the bases of corn and other plants. Hoes, really cultivators in which blades substitute for prongs, are discussed under Hoes and Hoeing.

CULVER'S ROOT or CULVER'S PHYSIC is *Veronicastrum virginicum*.

CUMIN. See Cuminum. Black-cumin is *Nigella sativa*.

CUMINUM (Cùmin-um)—Cumin. One insignificant annual of the Mediterranean region constitutes this genus. Its only horticultural recommendation is as a subject for herb gardens. Its fruits (usually called seeds) are employed for flavoring and in curry powder. They have a bitterish, balsamic flavor. Cumin (**Cuminum cyminum**) is an umbellifer, a member of the vast carrot family UMBELLIFERAE. A slender annual, all of its parts except its seeds are hairless. About 6 inches tall and branched above, it has leaves cut into threadlike segments ½ inch to 2 inches long. The tiny white or pink flowers are in compound, few-flowered umbels. The slightly flattened fruits are narrowly-oblong. The name *Cuminum* is of classical origin.

Cultivation. Because it requires a comparatively long growing season, in eastern North America cumin gives greater satisfaction south of Philadelphia than north of that city. It succeeds in any ordinary soil in full sun. Seeds are sown in spring and the seedlings thinned to 2 or 3 inches apart. Subsequent care consists of keeping weeds pulled. The crop is harvested in fall after the seeds turn brown, but before they are scattered. The plants are pulled up and dried thoroughly in an airy place. The seeds when released are stored in tightly stoppered containers.

CUNILA (Cu-nìla)—Maryland-Dittany or Stone-Mint. One of the about fifteen species of perennial herbaceous plants and low shrubs that constitute this genus, which ranges in the wild from the United States to Uruguay, is cultivated. It is the Maryland-dittany or stone-mint, which inhabits dry and rocky woodlands, in acid soil, from New York to Indiana, Illinois, Missouri, South Carolina, Tennessee, Arkansas, and Oklahoma. Cunilas belong in the mint family LABIATAE. The name is a Latin one for some species of the family.

Maryland-dittany (**Cunila origanoides** syn. *C. mariana*) is a strongly aromatic plant with freely-branched, square stems 9 inches to 1¼ feet tall, and with more or less woody bases. Its leaves are opposite, nearly stalkless, ovate to triangular-ovate, usually few-toothed, and without hairs. The nearly symmetrical flowers, crowded in the leaf axils and in larger terminal clusters, are mostly shorter than the leaves against which they nestle. They have tu-

Cunila origanoides

Cunninghamia lanceolata

bular, rose-purple to white corollas ¼ inch long or a little longer, each with a spreading, three-lobed lower lip, an erect two-lobed upper one, and two long-projecting stamens. The leaves are distantly spaced so that the blooms are in tiers. The fruits consist of four seedlike nutlets.

Garden Uses and Cultivation. This species, grown in flower borders, informal areas and in gardens of native plants, is also suitable for rock gardens. It succeeds in part-shade or sun in any ordinary garden soil and is easily raised from seed sown in spring or fall and by division in spring.

CUNNINGHAMIA (Cunninghám-ia) — China-Fir. Two or three species of conifers of the taxodium family TAXODIACEAE constitute this genus of evergreen trees. They are natives of eastern Asia nearly related to *Cryptomeria,* but differing in having flat, lanceolate leaves that, although arranged spirally on the shoots, are except on leading shoots twisted at their bases so that they spread in two ranks. In gross appearance cunninghamias somewhat resemble certain araucarias, especially *Araucaria angustifolia,* but differ in that their cones have three seeds with each scale instead of one. Their name commemorates James Cunninghame, a botanical collector who discovered the genus in China in 1702.

Cunninghamias are beautiful trees with widely-spreading branches, which are pendulous at their extremities, and arranged in tiers. Their leaves, finely-toothed along their margins, live for five years or more and, after they die and turn brown, remain attached to the branches for several more years. The primitive flowers are formed in fall and open in spring. The male catkins are in clusters at the shoot ends, the females, which develop into

fruiting cones, appear one to three together, near the ends of the shoots. The cones are subglobose and remain on the trees after their seeds are dispersed.

Commonest in cultivation, **Cunninghamia lanceolata** (syn. *C. sinensis*) of southern and central China attains a maximum height of 150 feet and under forest conditions has a straight trunk without branches for up to one-half its length. When growing in the open its lower branches are retained. Its brown bark shreds and exposes the newer, reddish bark beneath. The leaves of this kind curve backward and are 1 inch to 2¾ inches long and up to ³⁄₁₆ inch wide. Its egg-shaped or spherical cones are 1 inch to 1½ inches in diameter. The variety *C. l. glauca* is distinguished by the bluish-green color of its foliage. The fragrant wood of this species is greatly appreciated in China and is used to a considerable extent for coffins as well as many other purposes such as construction, interior trim, boat building, and boxes. The bark is used as shingles for roofing houses. An endemic native of Taiwan, **C.**

Cunninghamia lanceolata (foliage)

konishii as a young specimen very much resembles the Chinese species, but as it matures differences become apparent. The most important of these are that its leaves are smaller and not curved, and its cones are not so large. This species grows to a maximum height of 150 feet and has brownish, scaly bark and slender leaves on young specimens, up to 1¼ inches long, but on older trees usually up to ¾ inch. The cones are almost globose and up to 1 inch across.

Garden and Landscape Uses. China-firs are beautiful ornamental evergreens adapted for planting in mild climates. They thrive best in part-shade and in any ordinary well-drained, moderately fertile soil that never is excessively dry. They may be used effectively singly or in groups, but ample room should be left between individuals to allow for their development. The hardiest is *C. lanceolata,* which in very sheltered places survives in southern New York, but does not prosper there. Probably not hardy north of Washington, D.C. is *C. konishii.*

Cultivation. Cunninghamias need no spherical care in the matter of cultivation. They thrive without difficulty if the environmental needs suggested above are met. Keeping the ground around them mulched with compost, peat moss, or other suitable organic material is helpful, as are regular waterings during long periods of dryness. In climates just a little too cold for their best performance, parts of branches are likely to be killed back in winter and these should be removed promptly. The best means of propagation is by seed sown in sandy peaty soil, but cuttings taken in summer, of leading, upright-growing shoots, may be rooted under mist or in a propagating bench in a greenhouse or a cold frame. China-firs that are killed back or cut back produce new shoots freely from the stumps and are regenerated by this means in the forests of their homelands. For further information see Conifers.

CUNONIA (Cu-nònia)—Red-Alder. The natural distribution of *Cunonia* is highly interesting. One species, the red-alder of South Africa, is endemic to that land, the others are natives of faraway New Caledonia, an island near Australia. They are not related to the true alders (*Alnus*), of the Northern Hemisphere, but belong in the cunonia family CUNONIACEAE, a group akin to the saxifrage family SAXIFRAGACEAE. The name honors John Christian Cuno, who gardened in Amsterdam, Holland and who in the eighteenth century published a catalog of the plants he grew.

Cunonias are trees and shrubs of six species with opposite, stalked leaves that consist of three leaflets or of more arranged pinnately. They have dense cylindrical spikes of tiny white or creamy-white flowers, each with a five-parted calyx, five petals, and ten stamens. The fruits are beaked capsules.

The only kind likely to be cultivated is the red-alder (*C. capensis*). This handsome evergreen tree attains a height of 60 feet and a trunk diameter of 6 feet, but on poor soils is sometimes no larger than a good-sized shrub. It inhabits moist mountain forests. It is hairless and has dark green

Cunonia capensis

Cunonia capensis (flowers)

leaves of five to seven oblanceolate, toothed leaflets 3 to 4 inches long. Conspicuous are the large, spoonlike leaf appendages (stipules) that enclose the buds, but soon fall. Candle-like in appearance, the racemes of creamy-white blooms develop from the leaf axils. They are 4 or 5 inches long. The stamens protrude conspicuously from the tiny flowers.

Garden and Landscape Uses and Cultivation. The red-alder blooms while small and is suitable for growing in large pots or tubs or planted in ground beds in greenhouses. It is attractive as a container-grown specimen for decorating patios, terraces, steps, and other architecturally dominated areas. It blooms in summer. In frost-free or essentially frost-free climates it is good for outdoor planting. The red-alder thrives in any ordinary, fertile, well-drained soil that does not lack for moisture, in light shade or sun. It requires no special care. Any pruning needed to keep it shapely or restrict its size is done in early spring. At that time, too, necessary repotting of container-grown specimens should receive attention. Specimens in pots and tubs benefit from weekly applications of dilute liquid fertilizer from spring to fall. In greenhouses they accommodate well to a minimum winter night temperature of 50°F and slightly warmer conditions during the day. Propagation is easy by cuttings taken in summer and by seeds sown in sandy peaty soil in a temperature of 55 to 60°F. The seeds should not be covered, but merely sown on the surface of soil and kept evenly moist.

CUNONIACEAE—Cunonia Family. Trees and shrubs totaling 250 species contained in twenty-six genera constitute this family of dicotyledons. Chiefly natives of the Southern Hemisphere, but extending to Mexico, the West Indies, and the Philippine Islands, the group has leathery leaves often of three leaflets, or pinnate, opposite or in whorls (circles of three or more). The small, usually bisexual flowers, in spikes or panicles, have four or five each sepals and usually smaller petals, or petals may be lacking, eight, ten or many stamens, and one to five pistils. Usually capsules, the fruits are less commonly drupes or nuts. Genera cultivated are *Ackama, Callicoma, Ceratopetalum, Cunonia,* and *Weinmannia.*

CUP or CUPS. One or the other of these words form part of the common names of these plants: cream cups (*Platystemon californicus*), cup-and-saucer vine (*Cobaea scandens*), cup fern (*Dennstaedtia*), cup flower (*Nierembergia*), cup-of-gold (*Solandra maxima*), cup plant (*Silphium perfoliatum*), fringe cups (*Tellima grandiflora*), painted cup (*Castilleja*), and queen cup (*Clintonia uniflora*).

CUPANIA. See Cupaniopsis.

CUPANIOPSIS (Cupani-ópsis) — Carrot Wood. Closely related to the American genus *Cupania*, from which, with the Greek suffix *opsis*, similar to, its name is derived, *Cupaniopsis* comprises sixty species of trees and shrubs of Australia and Polynesia. It belongs to the soapberry family SAPINDACEAE.

The sorts of *Cupaniopsis* have alternate, pinnate leaves and in panicles or panicle-like assemblages from the leaf axils, small unisexual and bisexual flowers with five each sepals and petals, usually six to eight stamens, and one style. The fruits are three-lobed capsules.

The Australian carrot wood (*C. anacardioides* syns. *Cupania anacardioides, C. sapida*) is often misidentified as *Blighia sapida.* An attractive, shapely, deep-rooting, evergreen tree, with much the aspect of the

Cupaniopsis anacardiopsis

carob (*Ceratonia siliqua*), it ordinarily attains heights of approximately 30 feet and is about two-thirds as wide as tall, but in the wild is sometimes 50 feet tall. When young open-headed, with age it becomes more densely-branched and heavier-foliaged. Its leathery leaves have three to five pairs of blunt, toothless, broad-oblongish leaflets about 4 inches long and with a prominent mid-vein. Its flowers are cream tinged with green. The leathery, three-lobed fruits are almost ¾ inch in diameter. Variety *C. a. parvifolia,* with very much smaller leaves, has been described.

Garden and Landscape Uses. A clean and beautiful shade tree for use on streets, lawns, patios, and in other locations in California and elsewhere where warm, dryish climates are enjoyed, the Australian carrot wood grows rather slowly. When young it is more tender to cold than at maturity when it will endure, for short periods, temperatures down to 22°F. It is splendid for seaside locations and withstands arid atmospheres and poorly drained soils.

Cultivation. No special care is needed. Pruning to establish satisfactory branch structure and to thin out branches that tend to crowd may be done. Propagation is by seed.

CUPHEA (Cù-phea)—Cigar Flower. Chiefly natives of the American tropics and subtropics, the 200 to 250 species of *Cuphea,* belong in the loosestrife family LYTHRACEAE. They are annuals, herbaceous perennials, sometimes somewhat woody, and shrubs. The name, from the Greek *kyphos,* a hump, alludes to the swollen or spurred calyxes.

Usually more or less clothed with clammy or sticky hairs, cupheas generally have opposite leaves, the alternate pairs set at right angles to each other. Sometimes they are in whorls (circles of three or more). Rarely they are alternate. They are stalked or stalkless, undivided, ovate to lanceolate, elliptic, or linear. The flowers, one to three together from the leaf axils or from the stems between the leaves, are asymmetrical. Their tubular calyxes, often colorful and showy, have twelve longitudinal veins and are distinctly pouched or spurred at their bases. Their apexes are six-lobed, with the upper lobe the biggest. There are most often six, sometimes two or no, usually early, deciduous petals. There are generally eleven, sometimes fewer and sometimes twelve, stamens of two lengths and one style tipped with a headlike or rarely two-lobed stigma. The stamens and style may protrude from or be included in the calyx tube. The fruits are capsules.

The cigar flower (**C. ignea** syn. *C. platycentra*) is most familiar to gardeners. Native to Mexico, this is a bushy subshrub with wide-spreading, slender branches, up to about 1 foot tall. Nearly or quite hairless, it has sharply-pointed, lanceolate to ovate-lanceolate, stalked leaves 1 inch to 2½ inches long. The solitary, slender-stalked flowers, from the leaf axils or the stems near the leaf axils, are about ¾ inch long. The showy parts are their narrow, six-toothed calyxes. There are no petals. The calyxes are red, tipped with a darker band and with a white mouth. The coloring suggesting the ashy, burnt end of a tiny cigar. Another Mexican, this with

flowers with six minute petals shorter than the twelve teeth of the calyx, is **C. micropetala.** A hairy to nearly hairy shrub 1 foot to 2 feet tall, it has pointed, short-stalked, lanceolate to oblong-lanceolate leaves 2 to 4½ inches long. In long, terminal racemes, its 1-inch-long blooms, the showy parts of which are the hairy calyxes and protruding stamens, are yellow with the bottom of the calyx tube and the stamens red. The calyx tube is narrowed below its oblique mouth.

A neat, densely-leafy shrub, **C. hyssopifolia** is popular in mild climates and is sometimes grown in greenhouses and windows. Much-branched, 1 foot to 2 feet tall, and hairy, this kind has narrow, scarcely-stalked, linear to lanceolate leaves sometimes 1 inch long, often considerably smaller. The very many ¼-inch-long, lilac to violet or white, stalked blooms, swollen

Cuphea micropetala

Cuphea hyssopifolia

above at their bases, have six petals of nearly equal size. The stamens do not protrude. This species is native to Mexico and Guatemala.

Shrubby or subshrubby kinds with quite large flowers with showy petals include Mexican **C. llavea** and *C. l. miniata,* the latter distinguished by its hairy stems. Erect and few-branched, these sorts are 2 to 2½ feet tall. They have short-stalked, pointed-ovate leaves, 1 inch to 3 inches long and hairy, particularly on their undersides. Mostly they are opposite, the upper ones nearly so. Solitary and nearly stalkless, the 1-inch-long blooms have twelve-toothed calyx tubes pouched at their bases and a pair of brilliant red or rarely white, flaring,

Cuphea llavea

wavy petals. They are few together in terminal, leafy racemes. From the last, **C. hookerana,** native of Mexico and Central America, differs in having long-pointed, linear-lanceolate leaves 3 to 4 inches long. The flowers are in loose, sometimes branched racemes. They have reddish or green and red calyxes, a pair of large violet-purple to red petals, and often up to six smaller ones.

Hybrids between *C. llavea* and *C. procumbens,* correctly identified as **C. purpurea,** are grown under many horticultural

Cuphea ignea

Cuphea hyssopifolia (flowers and foliage)

Cuphea purpurea

names such as 'Avalon' and 'Firefly'. They are often misnamed *C. miniata*, a designation properly a synonym of *C. llavea miniata*. More or less intermediate between their parents, these hybrids are up to 1½ feet tall, and have ovate-lanceolate leaves and quite large and showy rose-pink to pinkish-violet flowers.

Annual cupheas are sometimes grown in gardens. Here belongs *C. lanceolata,* a sticky Mexican that attains 4 feet in height and has lanceolate leaves up to 3 inches

Cuphea lanceolata

long. Its solitary blooms have purplish calyx tubes and six petals, two of which are large and purple, the others tiny. Also sticky, *C. viscosissima,* native of eastern North America and about 1½ feet high, has ovate-lanceolate leaves up to 2 inches long. The blooms have two dark purple petals and four smaller, whitish ones. As its name indicates, the stems of sticky *C. procumbens* sprawl. They are clothed with purplish hairs. The flowers have purplish calyx tubes tipped with green and two petals larger than the other four. This species is native to Mexico.

Garden and Landscape Uses. The cigar flower may be grown permanently in outdoor flower beds in climates with little or no frost. Elsewhere it is excellent for summer bedding. It also is a good pot plant for greenhouses and window gardens. For these last uses *C. hyssopifolia* also serves well and in mild climates can be used in rock gardens, for edgings, and at the fronts of flower beds. The other kinds described above are useful, according to their permanence, as annuals or longer-lasting inhabitants of flower gardens. The flowers of many are useful for cutting. They last well in water. Only *C. viscosissima* is hardy in the north.

Cultivation. Cupheas are not fussy. They succeed in ordinary, reasonably fertile soils in sun or part-day shade. The shrubby and herbaceous perennial kinds are easily increased by cuttings, and some by division. They can also be grown from seeds. If started from seeds early indoors many of the herbaceous perennials bloom the first

year. Shrubby kinds are pruned to shape in late winter or spring, and all of them, like all the perennials, benefit from an application of a complete garden fertilizer then. Seeds of annual cupheas, and of herbaceous perennial kinds treated as annuals, are sown outdoors, where the plants are to bloom, at about the time the first corn is sown or are given a head start by sowing eight to ten weeks earlier in a greenhouse or some approximately similar place. A temperature of 70 to 75°F is suitable indoors, but after the young plants are well up this should be reduced by about ten degrees. As soon as they are large enough to handle, transplant individually to small pots and pinch out the tips of the shoots just above the third or fourth pair of leaves. This induces branching. Grow the young plants in full sun, and two or three weeks before they are to be planted outdoors harden them by putting them in a cold frame or sheltered place outside. When planting allow 6 inches to 2 feet apart according to the expected growth of the species or variety.

CUPID'S DART is *Catananche caerulea.*

CUPRESSACEAE—Cypress Family. An important family of gymnosperms, the CUPRESSACEAE comprises nineteen genera and 130 species, some represented in gardens by many horticultural varieties. The family has a very wide natural distribution. Many of its sorts are sources of valuable lumber. The fruits of junipers are used to flavor gin. Cedar oil is obtained from *Thuja occidentalis.*

Evergreen, resinous trees and shrubs with primitive unisexual flowers, the sexes on the same or different plants, constitute this family. Usually scalelike, less commonly linear, the leaves are opposite or in whorls (circles of more than two) of three or four. The flowers are in small strobiles (cones). Male cones develop from the leaf axils or are terminal at the short ends. Female cones are terminal or on short side branches. The fruits are woody, leathery, or those of junipers, fleshy, berry-like cones.

Genera cultivated include *Actinostrobus, Austrocedrus, Callitris, Calocedrus, Chamaecyparis, Cupressocyparis, Cupressus, Fitzroya Juniperus, Libocedrus, Microbiota, Pilgerodendron, Platylcadus, Tetraclinis, Thuja, Thujopsis,* and *Widdringtonia.*

CUPRESSOCYPARIS (Cupresso-cýparis)—Leyland-Cypress. This is a bigeneric hybrid between the Monterey cypress (*Cupressus macrocarpa*) and the Nootka false-cypress (*Chamaecyparis nootkatensis*), belonging in the cypress family CUPRESSACEAE. Its name is derived from those of its parent genera. It was raised at Leighton Hall in Wales in 1888. The original seedlings had the Nootka false-cypress as their

female parent. In 1911 two seedlings of the reverse cross were raised at Leighton Hall. The hybrid was named *Cupressocyparis leylandii* in 1926. Slight differences in the seedlings, evidenced in growth habit, color, and degree of ease with which they can be propagated, are perpetuated in their vegetatively propagated progeny, some of the most important of which have been given horticultural varietal names such as 'Green Spire', 'Haggerston Gray', 'Leighton Green', and 'Naylor's Blue'.

The leyland-cypress (*C. leylandii*) is columnar and more closely resembles the Nootka false-cypress than its other parent, but has the compact branching habit of the

Cupresscyparis leylandii

Monterey cypress. Its branchlets are flattened and its leaves keeled. The latter do not smell as strongly as those of Nootka false-cypress when bruised. The cones, intermediate between those of its parents, are about ¾ inch long and usually have five seeds to each scale. This kind is a vigorous, rapid grower. In England a specimen was 100 feet tall in 1964, and average growth there is reported to be 60 feet in 30 years. In California plants from cuttings were 15 to 20 feet tall in five years.

Garden and Landscape Uses. Leyland-cypress is a splendid evergreen for use as a specimen and for planting in groups. It is especially useful for screens and windbreaks and can be pruned or sheared to form excellent tall hedges. It thrives in a wide variety of soils, even some not fertile enough for many other evergreens. It is probably hardy as far north as New York.

Cultivation. This fine hybrid is easily propagated by cuttings inserted under mist in late summer or early fall. The young plants grow rapidly under nursery conditions and transplant easily, as do larger specimens. When used as a hedge, pruning or shearing should be done just before new growth begins in spring.

Hedge of *Cupressocyparis leylandii*

CUPRESSUS (Cuprés-sus)—Cypress. It is important not to confuse this group of evergreen conifers with other plants called cypress, some of which are conifers and some of no close botanical kinship. Especially they must be distinguished from the false-cypresses (*Chamaecyparis*), with which they were once united, and from the swamp-cypresses (*Taxodium*), both of which are conifers. Except as to the application of the name they are not likely to be confused with the standing-cypress (*Gilia*), summer-cypress (*Kochia*), or cypress-vine (*Quamoclit*), all of which are annuals.

The true cypresses (*Cupressus*) include about twenty species of beautiful trees, or rarely shrubs, of North America, the Mediterranean region, the Himalayas, and eastern Asia. Many are extraordinarily graceful and have beautiful feathery foliage, others are as starkly defined and rigid as an exclamation point. Among the best known are the Monterey cypress (*C. macrocarpa*) and the Italian cypress (*C. sempervirens*). The name of the genus *Cupressus* is from the Greek *kus*, to produce, and *parisos*, equal, and alludes to the symmetrical growth of *C. sempervirens*. The genus belongs in the cypress family CUPRESSACEAE. The mourning-cypress formerly *C. funebris*, is *Chamaecyparis funebris*.

As a group cypresses are more tender than false-cypresses (*Chamaecyparis*), only one or two tolerate fairly cold winters and even they do not have nearly the resistance of such low temperatures as the hardiest false-cypresses. The chief differences between cypresses and false-cypresses are that the former have six to twenty seeds to each cone scale and the latter five or less (usually two) and that the cones of false-cypresses, with the exception of *Chamaecyparis nootkatensis*, ripen their seeds in one season. Also, the branchlets of cypresses are usually quadrangular, those of false-cypresses, with few exceptions, are conspicuously flattened, a condition rare in *Cupressus*.

Like false-cypresses and other members of the family CUPRESSACEAE, cypresses produce two distinct kinds of foliage. In their early life seedlings have needle-like juvenile leaves that radiate from all around the shoots and somewhat resemble those of certain junipers, but with the passing of time these are gradually replaced in transitional stages by adult foliage consisting of opposite pairs of small scalelike leaves pressed closely against the shoots. In a few horticultural varieties the production of juvenile foliage persists through the life of the plant, but this phenomenon is not nearly as common among cypresses as it is with false-cypresses and arbor-vitaes (*Thuja*). The fruiting cones of cypresses, which mature in their second year and usually exceed ½ inch in diameter, are globular and have three to seven pairs of woody scales. They remain attached to the branchlets for an indefinite period after the seeds are dispersed.

Some cypresses are important timber trees. Their lumber is aromatic or sometimes rather unpleasantly odorous and tends to discourage insects. It is used for a wide variety of purposes including buildings, interior trim, furniture, telephone poles, and fencing. In Europe chests for storing clothes are often made from the wood of the Mediterranean cypress, which species supplied the lumber for the gates of Constantinople and the great doors of St. Peter's Church in Rome, which were sound after being in use for eleven centuries. In the Orient temples are constructed and religious statues fashioned from the wood of *C. torulosa* and it is burnt as incense.

The Monterey cypress (*C. macrocarpa*) is native in a very limited area along the coast of central California. Old specimens may be 70 to 90 feet tall and as picturesque as ancient cedars of Lebanon with their stout branches spreading widely to form broad crowns carrying dense masses of not strongly fragrant foliage. The leaves are slightly swollen at their ends. The trunks are stout and covered with thick bark, reddish at first, but becoming whitish with age. As younger trees Monterey cypresses are pyramidal and highly ornamental. They have thick branchlets and bright to dark green leaves that are obscurely or not at all glandular. Its cones are 1 inch to 1½ inches

Cupressus macrocarpa, native specimen by the sea, Monterey, California

Cupressus macrocarpa, as a roadside tree, Monterey, California

Cupressus macrocarpa, in southern California

Cupressus, unidentified species in Scotland

Cupressus macrocarpa 'Golden Pillar', young specimen

long. Varieties C. m. 'Golden Pillar' and C. m. lutea have yellow young shoots and leaves. Variety C. m. stricta forms a narrow column.

Other native American species include the inaptly-named Portuguese cypress (C. lusitanica), which first came to the attention of botanists in a monastery garden in Portugal in the seventeenth century. It had presumably been brought from its native Mexico, Guatemala, or Honduras by some plant-loving traveler. This sort has also been called the cedar-of-Goa, apparently in the mistaken belief that it was a native of India. Its home is definitely in the New World. It has only immigrant status in Portugal. A wide-spreading, graceful tree with pendulous branchlets not in flat sprays and glaucous-green foliage usually without glands, C. lusitanica attains a maximum height of about 100 feet and has reddish-brown bark. Its cones, ½ inch or somewhat more in diameter, especially during their first year are very glaucous. The variety C. l. benthamii has sprays of branchlets flattened in one plane in frond-like fashion. Very similar is C. l. knightiana,

Cupressus lusitanica

Cupressus lusitanica (cones)

the only differences being that its branchlets are arranged even more regularly, and the leaves are usually more glaucous. With long pendulous, whiplike branchlets, twice-pinnately-divided at their ends, C. l. flagillifera is a curious, rare kind.

The most easterly North American sort is the Arizona cypress (**C. arizonica**), a tree occasionally 70 feet tall, that inhabits Arizona, New Mexico, and Mexico. It has a stout trunk with reddish-brown or dark bark and usually glaucous-green or gray-green leaves with glands, sometimes inconspicuous, on their backs. Two varieties are recognized, C. a. glomerata with cones slightly over ½ inch across and short branchlets, and C. a. minor with cones up to 1 inch in diameter and longer branchlets.

The hardiest cypress in fairly common cultivation is C. macnabiana, but C. bakeri may prove to be equally as cold resistant. A native of California and Oregon, **C. macnabiana** lives outdoors in parts of New England, but for its best development needs a somewhat milder climate. A tree 30 to 40 feet in height, or sometimes shrubby, it has grayish, fibrous bark and wide-spreading branches. Its slender branchlets are not in one plane. The pale green or glaucous foliage is pleasantly fragrant. There is a conspicuous gland on the back of each leaf. The globose cones are ¾ to 1 inch in diameter.

The Italian cypress, the very formal, very narrow, strictly erect tree that is such an admired feature of Italian gardens and cemeteries and is prized for landscape planting wherever it will grow, presents something of a problem nomenclaturally. It actually is not a good species, but is a form or variety of a wild kind that spreads its branches much more widely than the tree we have in mind. This broader type, the Mediterranean cypress, is a natural species that has been known as C. sempervirens, and the Italian cypress has been referred to as C. s. stricta, a perfectly appropriate arrangement that seemed to reflect the relationship. Now, someone has discovered that the original plant described by Linnaeus was in fact the slender, fasti-

giate variety, the Italian cypress, and so it becomes C. sempervirens, the 'stricta' is discarded, and the naturally wild, wide-spreading tree must properly be called C. sempervirens horizontalis, which gives a completely false picture of the relationship between the two entities, but satisfies the rules of botanical nomenclature. One has a suspicion, and perhaps a sneaking hope, that for a long time to come gardeners will call the Italian cypress C. sempervirens stricta, even though we shall follow the new arrangement here.

The origin of the Italian cypress (**C. sempervirens**) is not known, but it is an ancient one. Up to 80 feet tall, it has upright branches that are responsible for its characteristic outline. When raised from seed a proportion of the seedlings retain or ap-

Cupressus sempervirens

proximate the fastigiate habit, but for the best and surest results it should be propagated vegetatively from the most desirable forms.

The wild prototype, the Mediterranean cypress (C. sempervirens horizontalis) inhabits southern Europe and western Asia and is somewhat variable. It is the cypress of ancients, to which frequent reference is made in the classics. Often it lives to a great age. The Mediterranean cypress attains a height of 100 to 150 feet and has foliage and comparatively large cones like those of the Monterey cypress to which this species is most closely related. From the Monterey cypress the Mediterranean cypress differs in having more slender branchlets and leaves not thickened at their tips.

Asian cypresses include some of the most beautiful conifers, but are rare in cultivation in America. They number four. The Himalayan cypress (**C. torulosa**) inhabits moist limestone soils in mountainous regions in the drier parts of western China, where it is up to 150 feet in height. As a young specimen it is pyramidal, but has a wide-spreading head at maturity. It has brown, peeling bark and horizontal or

Cupressus sempervirens horizontalis

Cupressus cashmeriana at the Royal Botanic Gardens, Kew, England

Cupressus forbessii

Cupressus sempervirens horizontalis (cones)

Cupressus cashmeriana (foliage)

Cupressus goveniana

up-pointing branches with flattened sprays of branchlets, pendulous at their tips. The backs of the leaves usually bear glands. The spherical or ellipsoid cones are about ½ inch long. Variety *C. t. corneyana* has more conspicuously drooping branchlets that are not in flattened sprays.

Beautiful *C. cashmeriana* has never been found in the wild. Its name suggests Kashmir as its native land and there are reports that it is a native of Tibet, but no reliable record supports this and it is possible that it originated in cultivation. One of the finest, if not the finest example of *C. cashmeriana*, a tree about 70 feet tall, is at Isola Madre, Lake Maggiore, Italy, and there is another attractive, but smaller tree cultivated indoors in the Temperate House in Kew Gardens, England. This species has foliage intermediate between typically juvenile and typically adult. The leaves are not scalelike, yet are not more than ¹⁄₁₆ inch long and they hug the shoots except for their spreading tips. Narrow and pyramidal *C. cashmeriana* has ascending branches and very long pendulous flat sprays of compressed branchlets with glaucous, gray-green foliage. The cones, about ½ inch in

diameter, have ten scales with about ten seeds each.

Additional kinds include these: *C. abramsiana,* of California, is up to 40 feet in height, densely-branched, and has light green foliage. *C. bakeri,* the Modoc cypress of California, is a slender-trunked, pyramidal tree 30 to 90 feet tall with grayish foliage. *C. b. matthewsii,* of California and Oregon, is similar, but is up to 100 feet tall and has longer branchlets. *C. duclouxiana,* of China, is an elegant tree up to 150 feet tall closely related to *C. sempervirens horizontalis,* but with slenderer branchlets and smaller cones. *C. forbesii,* the Tecate cypress, a tree of California up to 30 feet tall, has bright green to dull green foliage. *C. glabra,* of Arizona, up to 50 feet tall, is similar to *C. arizonica,* but has smoother bark, more glaucous foliage, and larger cones. *C. goveniana,* of California, is up to 50 feet tall or taller and has inconspicuously glandular, fragrant, light green or yellowish-green leaves. *C. g. californica* is a shrubby variety with long, drooping branchlets and spreading leaves that are intermediate between typical juvenile and typical adult ones. *C. guadalupensis,* en-

demic to Guadalupe Island off the coast of Baja California, is up to 65 feet tall and has cherry-like bark that peels to reveal red under-bark. *C. montana,* a native of Baja California, is up to 70 feet tall and has grayish-brown bark, which peels on young branches only and shows grayish-red beneath, and thickish branchlets. *C. nevadensis,* of California, is up to 30 feet tall and has bark that rarely peels and grayish-green foliage. *C. pygmaea,* the Mendocino cypress of California, is closely related to *C. goveniana.* Despite its specific name, under good conditions it becomes 150 feet in height, but is usually lower, sometimes only a few feet tall. Its foliage is dark, dull green. **C. sargentii,** of California, is a slender or bushy tree up to 45 feet tall, sometimes taller. It has gray to nearly black bark and dull green or gray-green, fragrant foliage. **C. stephensonii,** the Cuyamaca cypress, which is up to 50 feet in height, has thin, cherry-red bark and blue-green or gray-green foliage. It is endemic to San Diego County, California.

Garden and Landscape Uses. Among the true cypresses are some of the loveliest of evergreen trees and shrubs. Unfortunately, with one or two exceptions, they are hardy only in mild climates such as those of California, the Gulf States, and Hawaii. For colder regions their near relatives, the false-cypresses (*Chamaecyparis*), may be substituted. The hardiest cypresses are C. macnabiana, C. arizonica, and C. bakeri. As solitary lawn specimens and as accent plants, cypresses rank highly, and many kinds can also be grouped with pleasing results. Some, including the Monterey cypress, do well close to the sea. All stand shearing well and are excellent for screens and hedges. The Monterey cypress is very commonly employed in that way. Cypresses withstand dry atmospheric conditions better than false-cypresses and flourish in a wide variety of soils.

Formally sheared specimen of *Cupressus macrocarpa*

Cultivation. For the best results under cultivation cypresses need a deep, reasonably fertile and moist, well-drained soil; the Monterey cypress grows well in limestone soils. They need full sun and a reasonably pure atmosphere. They are not well suited to the polluted air of cities. If a good, unbroken root ball is taken with them most cypresses transplant without difficulty. To ensure having such balls it is advisable to transplant young specimens in nurseries every two or three years. The Monterey cypress is rather sensitive to root disturbance and is best set out in its permanent location when young. The best time for shearing or other pruning is just before new growth begins in spring or immediately after summer growth is completed. Keeping the ground mulched with compost, peat moss, or other organic mulch is beneficial to these trees, and in dry

weather periodic deep soaking of the soil with water promotes vigor and health. Cypresses can be increased by seeds sown in sandy peaty soil or by summer cuttings under mist or in a propagating bench in a very humid greenhouse or cold frame.

Pests and Diseases. Cypresses are subject to cankers, crown gall, and juniper blight and may be infested with aphids, mealybugs, caterpillars, and scale insects. Control of cankers consists chiefly in cutting out and burning affected parts and completely removing badly infected trees. Spraying with copper fungicide may also be helpful.

CURARE. See Strychnos.

CURCULIGO (Curcù-ligo). The *Curculigo* familiar to generations of gardeners and botanists as *C. capitulata* or *C. recurvata* is sometimes named *Molineria recurvata*. Its foliage is very like that of a young palm, but curculigos and palms are not related. Curculigos belong in the amaryllis family AMARYLLIDACEAE. The name is from the Latin *curculio*, a weevil, and alludes to the beaked ovary. There are fifteen species or more in the tropics and subtropics of the Old and the New World.

These are stemless evergreens with short or somewhat tuber-like rhizomes from which sprout tufts of foliage. The leaves are leathery and pleated lengthwise. Often they are deeply cleft or split into two parts or lobes. In dense heads or spikes close to the ground, the flowers are small and not very conspicuous. Each has six spreading petals (or more correctly, tepals), six short-stalked stamens, a brief style, and three stigmas. The fruits are fleshy capsules.

Native to tropical Asia and Australia, **C. capitulata** (syn. *C. recurvata*) has short, stout rhizomes from which arise many erect, firm-textured leaves with gracefully recurving, broad-elliptic blades. They have toothless margins and long, grooved stalks. Often they divide at the apex into two lobes. The leaves are 2 to 3 feet long or occasionally longer and 2 to 6 inches wide. The yellow flowers, ¾ inch across, are in crowded heads atop short, brown-hairy

Curculigo orchioides

Curculigo orchioides (flowers)

stalks that bend downward at their ends. Varieties with variegated foliage are *C. c. striata,* which has leaves with a central white band, and *C. c. variegata,* the leaves of which are longitudinally striped with white. Much less common than *C. capitulata* is **C. orchioides,** of the East Indies, Taiwan, southern China, and Japan. It has lanceolate leaves not over ¾ inch wide and yellow flowers about ¾ inch across. In the Orient its rhizomes are candied and eaten.

Garden and Landscape Uses and Cultivation. In Hawaii, southern Florida, and other places with warm, moist climates, curculigos are splendid foliage plants for shaded places where the ground is moist, but sufficiently well drained to admit air freely. They withstand a touch of frost. In regions where winters make their permanent cultivation outdoors impracticable, these plants may be set out in summer and be wintered in a greenhouse or other light place in a temperature of 50°F or more. Curculigos are also fine for permanent cultivation in greenhouses and conservatories either in containers or in ground beds. When cultivated in pots or tubs they need ample drainage and coarse, rich, porous soil. For their best development high humidity and a minimum temperature of 60°F are required. Shade from strong sun is essential. From spring through fall watering should be copious; less water is given in winter, but at no time should the soil be allowed to become really dry. Regular applications of dilute liquid fertilizer from spring through fall do much to encourage vigorous growth and good foliage color. Repotting, normally necessary only at intervals of a few years, is best done in late winter or early spring. Then, too, stock can be increased by division, the common means of propagation.

CURCULIOS. These are a group of mostly small weevils that as larvae and adults feed on various plants and plant parts. Several garden plants are eaten by one or more kinds. Among the most common are those that infest apples, seedling cabbages, cauliflowers, turnips and related plants, grapes, plums, quinces, rhubarb, and roses. Up-to-date information on control measures can be had from County Cooperative Extension Agents and State Agricultural Experiment Stations. For additional information see Weevils.

CURCUMA (Cúr-cuma)—Turmeric. This group of more than fifty species has a natural distribution from India and Malaya to Australia. It belongs in the ginger family ZINGIBERACEAE and has an Arabic name. Its members have thick, fleshy, branching, lumpy rhizomes and roots often with tubers. From the rhizomes come clusters of few leaves with more or less paddle-shaped blades, surrounded at their bases by bladeless sheaths that form a false stem. The inner leaves of the clusters have fairly long, channeled stalks, the outer ones shorter stalks or none. The proportions of width to length and the sizes of the leaf blades vary in each cluster. The margin of the midribs is often suffused with purple. Conelike or broadly spindle-shaped, the flower heads are on separate scaly shoots or terminate the leafy ones. They have large, broad bracts joined to neighboring bracts for halfway up their edges to form distinct pouches, each containing two to several blooms that last for only a few hours, and are white, yellow, pink, or purplish. The short two- or three-toothed, tubular calyxes, are split partway down one side. The corolla tube and stamen tube join to form a tube that expands above in cuplike manner. There are two conspicuous petal-like staminodes (abortive stamens) and one fertile stamen. The staminodes have their inner edges folded under the hooded upper petal. The lip points forward or is reflexed and is flanked by two up-curved lobes. There is one broad fertile stamen. The fruits are capsules.

Some species of *Curcuma* are esteemed for their starchy, edible roots. East Indian arrowroot is obtained from *C. angustifolia.* The midribs of some kinds supply useful fiber. Some are used as perfumes, and several are employed in native medicines and in religious ceremonies. The most important species, tumeric, is an ingredient of curry powder and is employed to color rice and other foods. It has also been used for dying cloth, but the results are not fast. Paper impregnated with tumeric is used as a chemical indicator of acidity and alkalinity.

Turmeric (**C. domestica**) for long was wrongly identified as *C. longa.* Commonly grown in Hawaii and extensively elsewhere in the tropics, its deciduous leaf clusters are 1½ to 2 feet tall. They come from rhizomes that are yellow or orange on their insides. The leaves have thin blades 8 inches long by 3 inches wide or sometimes the leaves are larger. The cylindrical flower heads, about 5 inches in length, terminate the leafy shoots. The uppermost bracts are pink and without flowers, those below are green and protect two

or more pale yellow blooms. The native habitat of this species is not known. Very similar, **C. australasica,** a native of northern Australia, differs in having only slightly aromatic, internally white rhizomes. The upper bracts of its flower heads are rose-pink. It is planted as an ornamental. An Indian species, the aromatic rhizomes of which are used medicinally and for flavoring, **C. pallida** (syn. *C. zedoaria*) has leaves up to about 2 feet long by 6 inches wide. On short stalks separate from the leafy ones, the flower heads, about 4 inches long by 3 inches wide, have red or purple sterile bracts and those enclosing the flowers green and red. The flowers are yellow.

A handsome kind, **C. petiolata** is sometimes called queen-lily. Native of India, it has leaves with blades up to 10 inches long by 6 inches wide and, separately, flower heads 5 or 6 inches in length on stalks nearly as long. The flowerless upper bracts are purplish-brown, the others green. The blooms are yellow. Flower heads, on short stalks separate from the leaves, about 8 inches long by 3 inches wide and with yellow-orange bracts are borne by **C. roscoeana,** native to Malaya.

Larger plants with leaves up to 4 feet long by 1 foot wide are *C. elata* and *C. latifolia,* both yellow-flowered natives of India. On stalks 8 or 9 inches long the flower heads, about 8 inches long by 3 inches wide are borne. In **C. elata** the bracts with flowers are greenish and those without violet, in **C. latifolia** the flower-producing bracts are striped with red, the flowerless ones are red with white bases.

Garden and Landscape Uses. Curcumas are suitable for shaded locations in the tropics and subtropics and make good tropical greenhouse plants. Their foliage is interesting and the flower heads reasonably attractive. Turmeric and some other kinds are appropriate for including in collections of plants of economic importance to man.

Cultivation. The chief needs of these plants are deep, moist, but not saturated, fertile soil, high temperatures, high humidity, and shade from strong sun. Propagation is chiefly by division of the rhizomes, at the beginning of the growing season, and by seed. In greenhouses a minimum winter temperature of 55 to 60°F is needed and considerably higher temperatures at other seasons. Water is given freely from spring to fall. In winter when the plants are leafless, only enough is given to prevent the soil from becoming completely arid. Summer applications of dilute liquid fertilizer benefit well-rooted plants.

CURLY-GRASS FERN is *Schizaea pusilla.*

CURMERIA. See Homalomena.

CURRANT and FLOWERING CURRANT. See Ribes. The Indian-currant is *Symphor-*

icarpos orbiculatus, the currant tomato *Lycopersicon pimpinellifolium.*

CURRANTS. Currants are bush fruits less popular in North America than in Europe, but easily managed and well worth growing in home gardens. They are practicable only where summers are fairly cool and humid, generally north of Washington, D.C., westward to the Mississippi River, and in the Pacific Northwest. They are especially esteemed for jams, jellies, and juices.

Of the three chief types, black currants (varieties of *Ribes nigrum*) may not be planted in most parts of the United States. This because of their great propensity for serving as alternate hosts to the white pine blister rust disease. Other currants, less active as transmitters of the disease, are forbidden in certain regions where white pine is important as a timber tree. Before planting check with your Cooperative Extension Agent or State Agricultural Experiment Station. In Canada currants are generally allowed and, being extremely hardy, are suitable for planting in most regions adapted to agriculture.

Red currants and white currants, the latter less well known than the red, are varieties of *Ribes sanguineum,* which are bushes some 5 or 6 feet tall. In addition, a variety

Black currants

Red currants

of the buffalo currant (*R. odoratum*) named 'Crandall' is grown to some extent in the Midwest.

Currants grow readily in any reasonably fertile garden soil that is well drained, preferring those that are clayish to sandy ones. This is especially true of black currants. They appreciate cool rooting conditions; therefore, land sloping to the north affords advantage. Although they prefer full sun, they will succeed where they are in shade for part of each day, which makes is possible to cultivate them on the northwest sides of buildings and fences and to interplant apples with them in home orchards. Do not, however, expect them to compete successfully with the roots of nearby shade or forest trees. Be sure that the site had good air drainage. The flowers come early and in low lying frost pockets are likely to be injured. Preparation of the soil for planting is as for apples.

Planting is best done in fall, alternatively in early spring. Allow 5 to 6 feet between rows, 3 to 4 feet between plants in the rows. Greater distances between the rows are needed in commercial plantations where wide machines are used for cultivating.

Set the young plants somewhat deeper

Pruning currants

than they were previously. This encourages the development of branches from at and below ground level. Prune them to a height of 8 to 10 inches, and if planted in fall, mound soil around them to reduce the likelihood of them being heaved out of the ground by alternate freezing and thawing.

Prune annually in late winter before the leaf buds expand. Red and white currants fruit chiefly on spurs on two- and three-year-old and older wood, black currants chiefly on one-, two-, and three-year-old branches. Renewal pruning based on cutting out all branches at the beginning of the fourth year, as well as any younger ones that are weak or ill placed, is followed for black currants. Prune so that the bushes are left with three or four branches that are three years old, three or four that two years old, and three or four that are one year old. Red and white currants may be

pruned in the same way, or branches may be retained for more than three years and be spur pruned by cutting all side shoots back to within 1 inch or less of their bases. By following this method of pruning, these currants can be trained as espaliers and cordons. Whichever pruning plan you follow eliminate any canes infested with borers. Do not shorten retained branches.

Fertilize in early spring. Manure is excellent, but commercial fertilizers may be used instead. The chief element needed is nitrogen. Apply a fertilizer, such as nitrate of soda at 500 pounds, to supply total nitrogen at 80 pounds to the acre. Approximately 3 ounces of nitrate of soda or an amount of other fertilizer that will supply the same total nitrogen is sufficient for each mature bush. Spread the fertilizer evenly over the entire root area, well beyond the spread of the branches. Summer management chiefly involves controlling weeds by shallow cultivation or mulching and picking the fruit. Propagation of currants is easy by hardwood cuttings. The chief varieties are 'Red Lake', 'Wilder', and 'White Grape' with white fruits. The chief pests are aphids, borers, San Jose scale insects, and the leaf-eating larvae, known as the imported currant worm.

CURRY LEAF is *Murraya koenigii.*

CURRY PLANT is *Helichrysum angustifolium.*

CURTONUS (Cur-tònus). Long known in gardens as *Antholyza paniculata,* the only species of this South African genus of the iris family IRIDACEAE is now separated as *Curtonus paniculatus.* Its name is from the Greek *kurtos,* bent, and *onus,* an axis. It alludes to the zigzagged flower stalks.

From *Antholyza* this differs in its stems not being continued beyond the uppermost blooms as extensions bearing bracts without flowers in their axils. Like *Gladiolus* and several other members of the iris family, **C. paniculatus** has corms (underground food storage organs that resemble bulbs, but are solid throughout instead of being composed of concentric scales as are onions, for example). The corms are large and roundish. From them develop stems bearing thin, rigid, sword-shaped, parallel-veined leaves up to 2 feet long by 3 inches wide. Above, the stem branches form a flexible-stalked open panicle, each branch of which is a many-flowered dense spike of rich orange-red and yellow, tubular, narrowly-funnel-shaped blooms. The curved perianth tube, about 1½ inches long, is abruptly narrowed below its center and divides above into six linear-oblong perianth lobes (petals), the upper one exceeding ½ inch in length and conspicuously longer than the others. There are three stamens. The style is three-branched. In bloom this plant is 3 to 4 feet tall.

Garden and Landscape Uses. As an early summer-bloomer for grouping in

flower beds and borders and for cutting for floral decorations this species is deserving of more attention than it receives. Its airy grace, cheerful color, and ease of culture commend it as worth-while. In contrast with many other flower garden plants, the vertical lines of its erect foliage produce pleasant effects. As a cool greenhouse plant it is very satisfactory for winter and early spring blooming.

Cultivation. This is similar to that of gladioluses. The corms are planted outdoors in spring, after the ground has warmed a little, at a depth and spacing of 3 to 4 inches. They need a fertile, agreeable soil, of a type that suits most vegetables and flower garden plants, and a sunny location. No staking is ordinarily required. When the flowers are cut a fair amount of foliage should be left to produce food to build good corms for the following year. Where frost does not penetrate the ground to the extent that the corms are frozen they can remain in the ground overwinter. Elsewhere, after the foliage has died naturally and completely or after it has been damaged by fall frosts, the corms are lifted, cleaned off, and stored dry in a temperature of 40 to 50°F until the next planting season. Increase is by natural multiplication of the corms and by seed.

In greenhouses corms may be planted in September or October, about eight in an 8-inch pot, in well-drained, porous, fertile soil. They are grown throughout in a night temperature of 45 to 50°F, with a daytime rise of five or ten or, on very sunny days, fifteen degrees. Until the newly planted corms have made generous root growth watering is done with care so the soil is not constantly sodden. When the containers are filled with roots more generous watering is in order, and then weekly soaking with dilute liquid fertilizer is beneficial. When blooming is through every effort should be made to retain the foliage as long as possible and to this end watering is continued. Only when the leaves begin to die naturally should watering be tapered off (by reducing the frequency of applications) and finally stopped. Then the corms are removed from the soil and stored in a dry, cool, shaded place until the next planting time.

CURUBA is *Sicana odorifera.*

CUSCUTA (Cus-cùta)—Dodder or Love Vine. The plants of this genus are not cultivated, at least not intentionally, but they frequently invade cultivated areas and are pests of the first order. They are parasites, serious weeds that harm many different kinds of plants. Because their life history and mode of growth is so different from that of most plants, and so interesting, it is worth-while to explain it. Dodders belong to the morning glory family CONVOLVULACEAE. The name *Cuscuta* is of unknown derivation. There are many species, most of very similar appearance, which are difficult to identify as to kind except by experienced botanists. Their seeds can germinate after they have lain in the ground for many years, possibly ten or twenty. Those from any one season's crop, although they are shed and reach the ground at one time, do not sprout together. They give rise to new plants over a period of several years.

Dodder or love vine seedlings have neither roots nor leaves. Each consists of a threadlike stem. When this emerges from the soil it rotates slowly and, if it comes into contact with any slender object, wraps itself around it. Should this, fortunately for the dodder, be the stem of a suitable host plant it sends into it rootlike suckers and begins to absorb moisture and nourishment, a parasitic procedure it will follow throughout its life. Then the lower part of the dodder stem withers and its upper

Cuscuta parasitizing English ivy

parts lose connection with the ground. Henceforward the dodder is completely dependent upon its host. If the young dodder fails to find an acceptable plant to which it can attach itself it soon dies. Following its successful establishment the dodder grows rapidly, branches and rebranches, and forms tangled networks of leafless, slender, pale, greenish, yellow, or reddish twining stems that relentlessly and continuously exact their toll. The stems may bridge gaps and infest nearby plants. Some tropical species clamber over shrubs and small trees to a height of many feet. In the north they are less vigorous, but no less harmful. From late spring to late summer, minute white, yellowish, or pinkish flowers in tight clusters are borne along the stems. These produce many seeds thus assuring abundant future crops of the weed.

Dodder is extremely difficult to control. Under garden conditions the best practice is to cut out and burn every scrap, or to burn it *in situ* with a weed flame gun, as soon as infestation appears. This means sacrificing stems, branches, and tops of

cultivated plants, but that cannot be avoided if the weed is to be eliminated. Because of the ability of dodder seeds to lie dormant so long, this treatment must probably be continued for several years. One must be ruthless. Dodder infestations usually result from the introductions of seeds in topsoil, earth around the roots of plants, and manure. They can also be carried on shoes and implements and be passed in viable condition through the digestive systems of animals and birds. Dodders are practically cosmopolitan in their natural distribution. There are 170 species.

CUSH-CUSH is *Dioscorea trifida.*

CUSHION-PINK is *Silene acaulis.*

CUSSONIA (Cus-sònia). Endemic to Africa and nearby islands, *Cussonia,* of the aralia family ARALIACEAE, comprises twenty-five species of thick-stemmed, small evergreen trees and shrubs. Its name honors Dr. Peter Cusson, French physician and professor of botany, who died in 1785.

The leaves of *Cussonia* are alternate, long-stalked, and have usually five to nine, but sometimes more leaflets radiating from the tops of the leafstalks. The small, bisexual blooms, in racemes, dense spikes, or in panicles of spikes or umbels, have minute toothed or toothless calyxes, five petals, and the same number of stamens. There are two or three styles. The fruits are small berries.

From 10 to 15 feet tall, **C. paniculata,** of

Cussonia paniculata

South Africa, is a round-topped tree with leathery leaves of eight to twelve leaflets 6 inches to 1 foot long and 1½ to 4 inches wide, smooth-edged or deeply- and coarsely-pinnately-lobed, with the lobes spine-tipped. The little white or yellowish, stalkless flowers are crowded in spikes 2 to 3 inches long, disposed in panicles up to 1 foot long. From the last, **C. spicata,** native from South Africa to Zambia, differs in its leaves having six to eight obovate leaflets 1 inch to 2 inches long, on young specimens undivided, at a later stage of

the tree's growth three-parted, or on old trees pinnate. The stalkless spikes of small yellow flowers are in umbel-like clusters of eight to twelve. A shrub or tree 8 to 12 feet tall, South African **C. thyrsiflora** has leaves of six to eight mostly undivided, obovate leaflets 1 inch to 2 inches long. The flowers are white, stalked, and in racemes 3 to 4 inches long, arranged in umbel-like clusters.

Garden and Landscape Uses and Cultivation. The handsome, bold-foliaged species described are well adapted for outdoor landscaping in warm, dryish, frostless or nearly frostless regions, such as California, and as indoor decorative pot and tub specimens. Succeeding in ordinary, well-drained fertile soil in sun or part-shade, they are propagated by seeds and air layering. In greenhouses a winter night temperature of 45 to 50°F is appropriate, with an increase of up to fifteen degrees by day. On all favorable occasions ventilation should be given freely. Avoid keeping the soil and atmosphere excessively moist. Well-rooted specimens benefit from occasional applications of dilute liquid fertilizer.

CUSTARD-APPLE is *Annona reticulata.*

CUTHBERTIA (Cuth-bértia)—Roselings. This genus of three species is endemic to the southeastern United States. Some authorities classify it as *Phyodina,* some retain it in *Tradescantia.* Related to the last, it belongs to the spiderwort family COMMELINACEAE. The name probably commemorates Alfred Cuthbert, an amateur botanist of Augusta, Georgia. He died in 1932.

Cuthbertias are nonhardy herbaceous perennials with cordlike roots, basal foliage, and tufted or solitary stems each with three or four leaves. The leaves, half-rounded or flat, are slender and pointed. The flower clusters terminate longish stalks. Unlike those of *Tradescantia,* they are without one to three leaflike bracts at their bases, but have instead small or minute bracts very different from the foliage leaves. The short-lived blooms have three pink-tinged, green sepals, three spreading, pink petals, six stamens, their stalks with purple hairs, and one style. The fruits are capsules.

In dense, grassy tufts up to 8 inches tall or sometimes taller, **Cuthbertia graminea** has many slender stems. Its half-rounded leaves are erect, slender, and pointed. The pretty, bright pink blooms, ¾ to 1 inch wide, are not held higher than the leaves. From the last, **C. ornata** differs in its stouter stems being solitary or few, not in dense tufts, and in its pink blooms, 1¼ to 1¾ inches across, carried well above the foliage, having petals with conspicuously round-toothed edges. From the previous two, **C. rosea** differs in its stems often sprawling with age and its narrowly-linear leaves being flat. The ¾- to 1-inch-wide pink blooms are carried above the foliage.

Garden and Landscape Uses and Cultivation. Chiefly adaptable for wild gardens, rock gardens, and in their home territories native plant gardens, cuthbertias are easy to grow. They prefer sandy, well-drained soil and will stand light shade. Propagation is by seed and by division.

CUTTING GARDEN. Whenever practicable it is advisable to set aside a part of the home plot as a cutting garden, an area devoted to raising flowers for indoor use. This makes much more sense than harvesting blooms from display beds and borders, first because to do so reduces the decorative splendor of the garden, and second because flowers in tidy rows in cutting gardens can be grown with less effort than when planted in more complicated patterns.

A cutting garden can be large or small depending upon the quantities of flowers needed and the availability of space and labor to plant and care for them, but even a small plot, one 20 by 10 feet say, will produce many flowers if carefully planned and managed. The trick is to rely upon kinds that come into bloom at times when you are likely to need flowers and not to have them in much greater quantities than you can use. If you vacation in August, say, omit plants that bloom chiefly then.

Locate the garden in an open, sunny area, not too far from the house and for preference screened from the rest of the garden. An excellent arrangement is to combine flowers for cutting and vegetables. Both are most conveniently grown in rows and need essentially the same soil preparation and care. As with vegetables, flowers ordinarily grown for cutting respond to any reasonably good earth improved by deep spading or rototilling and the application of a dressing of a complete fertilizer, and if too acid, some lime to bring it to only a slightly acid or neutral condition.

In planning a cutting garden take into consideration the length of time the different kinds of plants will occupy space. Probably a larger proportion of the area will be given to hardy perennials than is true of vegetable gardens. Even so, you will most likely want some annuals, perhaps biennials, and probably such nonhardy perennials as dahlias, gladioluses, and tuberoses. Keep the hardy perennials together in one part of the plot, kinds that occupy the ground for shorter periods in another. This makes for greater ease in preparing the ground for the temporary crops.

Quantities to plant depend upon need and available space. There is no point in having many more flowers than you can conveniently use or give to friends. Remember that many annuals, such as baby's breath, larkspur, and love-in-a-mist, have short seasons of bloom and all of one sowing come into flower at the same time. Short blooming seasons are characteristic too of many bulb plants, for example, daffodils, gladioluses, tulips, and of such perennials as irises and peonies. The seasons of annuals and bulb plants can often be extended by making successional sowings or plantings. Those of some perennials and bulb plants, irises, lilies, and narcissuses, among them, can be prolonged by selecting early, midseason, and late varieties. Yet other plants suitable for cutting gar-

These special gardens provide flowers for cutting: (a) Tulips

(b) Irises

(c) Delphiniums

(d) Acidantheras

(e) Lilies

(f) Feverfew

dens are of the cut-and-come-again class that produce blooms over much longer periods. Dahlias, marigolds, and tithonias belong here. Flowers suitable for cutting gardens are very numerous. Individual preferences must govern choice of kinds and varieties. Roses will surely be included wherever climatic conditions make their cultivation possible. Here are lists of other flowers worth considering.

Hardy perennials include achilleas, anchusas, asters, astilbes, baby's breath (*Gypsophila paniculata*), balloon flower (*Platycodon*), bee-balm (*Monarda*), bellflow-

(g) Gladioluses

ers (*Campanula*), blanket flowers (*Gaillardia*), blazing stars (*Liatris*), bleeding hearts (*Dicentra*), boltonia, butterfly weed (*Asclepias tuberosa*), Christmas-rose (*Helleborus niger*), chrysanthemums, cimicifugas, columbines (*Aquilegia*), coneflowers (*Rudbeckia*), coreopsises, day-lilies (*Hemerocallis*), delphiniums, doronicums, eupatoriums, false-dragonhead (*Physostegia*), false-indigos (*Baptisia*), fleabanes (*Erigeron*), foxtail-lilies (*Eremurus*), garden-heliotrope (*Valeriana*), geraniums (*Geranium*), geums, globe flowers (*Trollius*), globe-thistles (*Echinops*), goat's beard (*Aruncus*), heleniums, heliopsises, irises, Japanese anemones, lily-of-the-valleys (*Convallaria*), loosestrife (*Lysimachia*), lupines (*Lupinus*), lychnis, matricarias, meadow-rues (*Thalictrum*), monkshoods (*Aconitum*), mulleins (*Verbascum*), mullein-pink (*Lychnis coronaria*), nepetas, painted-daisies or pyrethrums (*Chrysanthemum coccineum*), penstemons, peonies, perennial-pea (*Lathyrus latifolius*), pinks, poker plant (*Kniphofia*), poppies (*Papaver orientale*), potentillas, primroses (*Primula*), purple-loosestrife (*Lythrum*), ranunculuses, red-valerian (*Centranthus ruber*), salvias, Shasta-daisies (*Chrysanthemum maximum*), statices (*Limonium*), Stokes'-aster (*Stokesia*), sunflowers (*Helianthus*), sweet rocket (*Hesperis*), Thermopsis, thrifts (*Armeria*), veronicas, violets (*Viola*), and Virginia bluebells (*Mertensia*).

Annuals worth considering include anchusas, arctotises, baby's breath (*Gypsophila elegans*), balsams (*Impatiens*), blanket flowers (*Gaillardia*), blue lace flower (*Trachymene*), browallias, calendulas, California poppies (*Eschscholzia*), candytufts (*Iberis*), carnations (*Dianthus*), celosias, China-asters (*Callistephus*), chrysanthemums, clarkias, collinsias, coneflowers (*Rudbeckia*), coreopsises, cosmoses, cynoglossums, dimorphothecas, everlasting flowers (*Ammobium, Helichrysum, Helipterum, Xeranthemum*), gilias, globe-amaranths (*Gomphrena*), godetias (*Clarkia*), heliotropes, jacobaeas (*Senecio elegans*),

larkspurs (*Consolida*), lavateras, leptosynes (*Coreopsis*), linarias, love-in-a-mist (*Nigella*), love-lies-bleeding (*Amaranthus*), lupines (*Lupinus*), Madagascar-periwinkle (*Catharanthus*), marigolds (*Tagetes*), Mexican tulip-poppy (*Hunnemannia*), mignonette (*Reseda*), monkey flowers (*Mimulus*), nasturtiums (*Tropaeolum*), nicotianas, patience plants (*Impatiens*), petunias, phloxes, salpiglossis, salvias, snapdragons (*Antirrhinum*), snow-on-the-mountain (*Euphorbia marginata*), statices (*Limonium*), stocks (*Matthiola*), sunflowers (*Helianthus*), sweetpeas (*Lathyrus*), sweet sultans (*Centaurea*), tithonias, ursinias, venidiums, wax begonias (*Begonia semperflorens-cultorum*), wishbone flower (*Torenia*), and zinnias.

Biennials of merit as cut flowers include English wallflowers (*Cheiranthus*), forget-me-nots (*Myosotis*), foxgloves (*Digitalis*), honesty (*Lunaria*), pansies (*Viola*), Siberian wallflowers (*Erysimum*), and sweet wiliams (*Dianthus*).

Bulb plants, kinds with bulbs, tubers, or other bulblike organs, include calla-lilies (*Zantedeschia*), calochortuses, dahlias, English bluebells and Spanish bluebells (*Endymion*), flowering onions (*Allium*), gladioluses, grape-hyacinths (*Muscari*), hyacinths, English irises, Dutch irises, and Spanish irises, lilies, montbretias (*Crocosmia*), narcissuses, including daffodils, Peruvian daffodils or ismenes (*Hymenocallis*), snowdrops (*Galanthus*), spider-lilies (*Hymenocallis*), squills (*Scilla*), tiger flowers (*Tigridia*), tuberoses (*Polianthes*), tulbaghias, and tulips.

CUTTINGS, PROPAGATION BY.

Multiplying plants by cuttings is one of the most popular and efficient means of vegetative propagation. Professionals and amateurs alike employ it extensively. As compared with raising plants from seeds, it has the advantage of the resulting individuals being genetically identical with the original plant and thus normally showing no visible differences. This is by no means always true of plants from seeds. Seedlings commonly exhibit minor, and hybrids and horticultural varieties, often major, variations from their parents. In gardeners' parlance they may not "breed true."

Cuttings are pieces of plants that placed in favorable environments may be expected to generate new individuals. They may be rootless parts of stems, leaves, rhizomes, tubers, or bulbs or portions of roots without stems or leaves. Most familiar are stem cuttings, known to generations of home gardeners as "slips." Theoretically perhaps any plant can be propagated by cuttings of one part or another, but in practice this by no means true. There are many that even the most skilled propagators with all the advantages of the most sophisticated equipment have failed to reproduce in this way. There are numerous others that, for practical reasons, are more

efficiently multiplied by seeds, divisions, grafts, or layers or in other ways. The facility with which particular plants reproduce from cuttings is determinable only by experience and experiment. In some cases it varies between varieties and perhaps individuals of the same species. Guide lines and suggestions as to the practicability of the propagation of specific sorts by cuttings are given throughout this Encyclopedia.

The chief concerns of gardeners intent on growing plants from cuttings are selecting the most suitable plant parts, determining the best time to take them, and providing environments conducive to their growth. A common denominator applicable to all plants is to be sure the cuttings are taken from disease-free, normally vigorous stock. Be very certain that plants infected with virus diseases are not used as sources of cuttings. Another general rule is to make all cuts as cleanly as possible with a sharp knife or razor blade. Although it is true this may not be of prime importance with some easy-to-root items, such as willows and privets, crushed tissues do not heal as quickly as those that are cleanly cut with a sharp tool, and frequently result in failure. Under suitable conditions the wounds heal, and by cell multiplication develop a covering of callus tissue. New roots develop from the base of the cutting or sometimes along its length, and shoots or leaf buds from near its apex and often lower.

Root-inducing substances, usually commerical preparations containing indolebutyric acid and commonly called hormones, are extensively employed by commercial propagators and can be used with good effects by others. They accelerate the rate of rooting of cuttings of many kinds of plants, although with some the quicker

Dipping the ends of cuttings in a root-inducing powder

rooting is followed by a slower rate of growth later so that cuttings rooted without this aid give plants that soon catch up. Root-inducing substances in powder form are the easiest to apply. Follow the manufacturer's directions carefully. Too much can harm the cuttings. The best way of application is to spread a little of the powder on a sheet of plastic or wax paper, dip the cut bases only of the cuttings in this, shake off any surplus, and plant immediately. An important point to remember is that the use of root inducers does not relieve the gardener of the necessity of providing as ideal an environment as possible for the cuttings or of giving them meticulous care. Use these substances as aids, not crutches.

The surroundings provided for cuttings must meet certain standards. Often it must be rather special. A primary requirement is that it prevent the cuttings from drying, from losing more moisture by transpiration than, until they have developed new roots with root hairs capable of absorption, they can replace. The atmosphere must be humid, yet not to such excess that fungus rots and other disease organisms develop. Drafts, because they are drying, must be avoided.

Light is essential for cuttings with leafy, or in the cases of cactuses and some other plants with leaflike, stems. Compromise is often essential. The brighter the illumination within the limits the sort of plant normally finds agreeable, the faster rooting takes place, but increased light speeds water loss by transpiration and if this amounts to more than can be replaced the cuttings wilt, dry out, and eventually die. Furthermore, exposure to direct sun, if the cuttings are within an enclosure, is likely to raise the temperature to fatal levels. Cuttings, therefore, except those planted outdoors under constant mist, and those of very fleshy cactuses and other succulents, must be shaded from direct sun. Temperatures favorable to growth, and these vary considerably with different kinds of plants, must be maintained. Frequently the optimal ones are a little higher than those that suffice or are best for established

plants of the same kinds. Often there is an advantage in having the rooting medium warmed by heating cables or other means to a few degrees higher than it would be without such aid. Levels of 70 to 80°F are generally appropriate. This is called giving bottom heat.

The rooting medium is the other important part of the cuttings' environment. Many materials are or can be used. Among the most popular are sand, vermiculite, perlite, and mixtures of these and peat moss. Others include sphagnum moss, sphagnum moss mixed with sand, sandy soil, and water. The last is most often employed by amateurs, and with many kinds of plants is successful. It has the disadvantage of inducing roots that are best adapted to the medium in which they have been developed, and unless they are very carefully managed during the transition period

Rooting cuttings in water: (a) Stem cutting of *Plectranthus*

the young plants are likely to go into shock and suffer a decided setback when transferred to soil. This likelihood can be reduced by, at the first planting, using a very sandy soil and for a week or two afterward keeping them in a decidedly humid atmosphere.

Whatever the rooting medium, it must provide the cuttings with adequate supplies of moisture and air, be as free as practicable of rot-causing fungi and bacteria, be favorable to root growth, and be fairly inexpensive. For most plants a slightly acid reaction seems advantageous, and with acid-soil sorts such as azaleas and rhododendrons more acidity is desirable. Rarely is one rooting substance significantly better than another. The chief function of the medium is to assure adequate supplies of moisture and air. A wide variety of materials that do this can be utilized for rooting cuttings. With mediums other than water, it is important that they be well drained and porous. To assure sharp drainage make sure the cutting bed is underlain with a layer of coarse gravel or other material of similar character and that receptacles in which cuttings are planted are similarly adequately drained.

To ensure adequate drainage, a layer of coarse gravel is covered with fine nylon net as a base for the rooting medium

In most cases it is advisable to pack the medium quite firmly by tamping, or if the cuting beds are in cold frames or outdoors by treading, or if the cuttings are to be accommodated in small pots by jabbing the medium with the ends of the fingers and then leveling its surface. This firming is generally most imporant when sand alone is used as a rooting medium but it is also needed with rooting mixes consisting of sand or perlite and peat moss. Perlite alone is less amenable to packing by these methods; it can be satisfactorily compacted by drenching with a heavy spray of water. Because vermiculite so readily compresses it should never be packed. Watering after the cuttings are planted is all that is needed.

A grid of heating cables underlies the sand of this propagating bench

(b) Leaf cuttings of African-violet

Corylopsis sinensis

Cosmos bipinnatus variety

Corynabutilon vitifolium

Cotinus coggygria

Crataegus laevigata paulii

Crambe cordifolia

Cycnoches ventricosum chlorochilon

Cryptanthus 'It'

A florists' cyclamen

Dutch crocuses, garden varieties

If sand is used as a rooting medium, pack it firmly by tamping

Greenhouse propagating bench, with cuttings

Cuttings of *Taxus* rooting in a greenhouse propagating bench

When planting stem cuttings vertically use a blunt dibber rather than one that tapers to a point and make sure that the base of the cutting rests on the bottom of the hole. Alternatively, draw a heavy knife blade along a straightedge positioned on the surface so that a wide slit of suitable depth is opened in the rooting medium and poke the cuttings into this. Whether the cuttings are planted in individual holes or in slits, unless vermiculite alone is being used, pack the rooting medium against the bases of the cuttings so firmly that a gentle pull on a leaf will not dislodge the cutting. Complete the job by watering copiously with a fine spray.

Humidity for all except deciduous hardwood cuttings (which are completely buried in the rooting medium), and cuttings of cactuses and other succulent plants that need no special attention of this kind, is usually assured either by covering the cuttings with glass or polyethylene plastic film or by keeping them under mist. A familiar example of the first procedure is the old one of inverting a Mason jar over a few rose cuttings planted in a shady spot in the garden. In greenhouses, propagating benches covered with cold frame-like glass

Cuttings in flats rooting in a propagating greenhouse

Rooting cuttings at home is facilitated by such simple devices as these covered with polyethylene plastic film (three figures above)

Greenhouse propagating bench enclosed with polyethylene plastic to maintain humidity

or plastic sash are usual. These ensure local conditions more humid than those in other parts of the structure. In commercial operations involving great numbers of cuttings, entire greenhouses are kept at the desirable humidity. Cold frames serve similarly for rooting cuttings outdoors. In homes, offices, school rooms, and similar places Mason jars, other glass covers, or often and more conveniently polyethylene plastic bags can be used to cover or enclose cuttings during the rooting period.

A propagation unit devised at the Arnold Arboretum makes it simple for home gardeners to root cuttings of a wide variety of deciduous and evergreen hardy shrubs as well as other kinds of plants outdoors. To establish such a unit, take a wooden box approximately 2 feet long by 1 foot wide by 3 to 4 inches deep. Bury it to its rim in the ground, taking care to choose a location not subject to flooding, out of direct sun, but where there is good light. Such places are likely to be found on the north sides of buildings. Line the bottom and sides of the box with a piece of polyethylene plastic film cut big enough to allow about a 2-inch extension above the sides and ends of the box. Next fill the box with a mixture of equal parts coarse sand

Cuttings rooted in a cold frame: (a) Rhododendrons

(b) English holly

Mist systems are highly efficient for rooting cuttings. They assure that a sufficiently high relative humidity is automatically maintained to prevent wilting or drying. For most plants a relative humidity of 90 to 95 percent suits. A further advantage of mist is that the constant or frequent washing of the foliage and stems carries away any fungus spores that light on them and so sharply reduces danger of loss by rotting. Because of the ample amounts of water used it is absolutely essential that

Rooting cuttings under mist: (a) In a greenhouse

(b) In cold frames

(c) A home garden unit, outdoors

and perlite or other favored propagating medium and press it lightly. This is your cutting bed. After the cuttings are planted and watered in, position over the box about 6 inches above the cuttings a flat-topped framework of wire mesh. Cover this and the sides and ends down to the ground with plastic cut to allow a couple of inches all around to be spread outward on the soil surface. Mound soil over the edges of the plastic in contact with the ground. Leave a few small air leaks to forestall excessively high humidity inside the tent.

Aftercare is minimal. It is improbable that further watering will be needed or that diseases or pests will develop. Nevertheless, inspect the unit at about weekly intervals. There is no need to lift the plastic to do this. If signs of trouble appear take off the cover, remove infected or infested cuttings, spray with fungicide or insecticide, and replace the cover. After several weeks check to see whether the cuttings have sufficient roots for them to be potted individually or planted in flats. When this stage is reached, before removal from the propagating unit, gradually condition the cuttings to a harder environment by first partially, and then completely removing the plastic, choosing a humid, cloudy day if possible for this. Or you may take the cover off at night and replace it during the day for about a week.

cutting beds for mist propagation be exceptionally well drained and that a very porous rooting medium is used. Coarse sand, perlite, and vermiculite are generally satisfactory.

Several types of mist systems have been devised and are available from dealers in greenhouse equipment and garden supply houses. They fall into two chief types. Constant mist consists of an uninterrupted rain projected from mist or fog nozzles that break a stream of water into very fine droplets. Intermittent mist is similar, but as the name suggests provides an interrupted instead of a constant bathing of the cuttings with water. Intermittent mist is regulated by a device that turns the flow on and off at predetermined intervals or by one that, by the use of an "electric leaf" or other sensor, opens the flow when the humidity in the vicinity of the cuttings drops below a certain point and turns it off above the same point. Intermittent mist is best in greenhouses. Constant mist, which uses more water, is satisfactory outdoors. There, the cutting bed should be surrounded with a screen or fence of polyethylene plastic film to prevent the mist being blown by wind. Also, economy is achieved by turning the mist off at night.

Different kinds of cuttings are named with reference to whether the tissues of the shoots from which they are made are soft and immature, semimature or "half-ripened" (firm, but not woody), or decidedly woody and fully mature. The first are called softwood cuttings, the second greenwood or half-ripe cuttings, the third hardwood cuttings. Further identification is by the parts of the plants from which they are obtained and the styles in which they are cut or prepared.

Softwood cuttings are most familiar. They are the usual "slips" of home gardeners. Each consists of a section of a leafy shoot generally including two, three, or more nodes (joints) in addition to the basal one. Usually they are terminal pieces, sometimes sections from below the apex with the terminal part taken off as another cutting, or if too soft discarded. Plants such as begonias, chrysanthemums, coleuses, English ivy, fuchsias, geraniums, philodendrons, and wandering jews are commonly increased by cuttings of this type. Numerous others, including many shrubs, can be.

The time to take softwood cuttings often relates to the size the resulting plants desired by a particular date. For example, if you need geraniums in 5-inch pots for planting outdoors in May take cuttings in late August or September, but if specimens in 4-inch pots are big enough for your May planting take cuttings in January or February. Similar adjustments in cutting times are used to control the sizes chrysanthemums, poinsettias, Paris-daisies, and many kinds of plants attain by flowering time.

As a generalization, spring through fall is the best season for taking softwood cuttings, but those of some kinds such as begonias, geraniums, philodendrons, and many others can in suitable environments be rooted with equal facility in winter.

The degree of firmness of the stem tissues, even of softwood cuttings, is a matter of some concern. In only a few cases, with lilacs and Japanese maples for example, do very soft shoots afford the best material for making cuttings. Those of the vast majority should be made from stems that have firmed to the extent that they are no longer obviously sappy and easily squashable.

Greenwood cuttings made of half-ripe shoots provide a reliable means of reproducing many shrubs, such as deutzias, forsythias, mock-oranges, and privets. For these, wait until the shoots of the current season's growth are decidedly firm, but not yet hard and woody. This will be at about the time further extension of the shoots for that season has ceased.

Depending upon the type of plants and the kinds of shoots available, leafy cuttings can be made, as they usually are with ge-

Propagating a shrubby potentilla from greenwood cuttings: (a) Cut terminal shoots from the mother plant

(b) Remove the lower leaves and cut the stems cleanly across just below a node

(c) Plant the cuttings in firmly packed sand or other rooting medium

(d) Drench with a fine spray

(e) Sink the container to its rim in peat moss or sand in a shaded, humid cold frame

raniums and many kinds of begonias, from terminal parts of older shoots or branches, or as is done with chrysanthemums, perennial asters, and other deciduous herbaceous perennials, from the ends of new annual shoots that sprout directly from the crown or rootstock. Other cuttings, notably those of shrubs, including hardy kinds, such as deutzias, forsythias, mock-oranges, and viburnums, and nonhardy ones, such as lantanas and plumbagos, are from young shoots that sprout from older (usu-

(f) After a few weeks, when 1-inch-long roots have developed, pot the plants individually in small pots

(g) Later, when growth is well started cut off the tip of the stem to induce branching

(h) The following year, ready for planting in a nursery bed

ally one-year-old) branches. Such shoots, if not too long and especially if they are a little too soft, are often taken with a "heel" (a thin sliver of the older shoot from which they arise) attached to their bases.

Preventing wilting and drying is essential to success. If some time must elapse between removing the shoots from the parent plant and planting the cuttings, keep the shoots in shade in a polyethylene plastic bag or, if not available, wrapped in moistened newspaper. Do not stand them in water. If this is done some kinds, helio-

Heel cuttings of lantana: (a) Taking the cuttings from a stock plant

(b) Cuttings, with "heel" of older wood at base, prepared for planting

tropes for example, are likely to rot later.

Prepare cuttings for planting by cutting off all leaves from the portion of stem that will be in the rooting medium and by removing all flower buds. With big-leaved cuttings, such as those of some begonias, hydrangeas, and viburnums, the upper parts of the larger leaves that remain may also be removed. Do not, however, denude them too much. Leaves aid by manufacturing food materials needed for root growth. Cut the stem cleanly and squarely across with a sharp knife or a razor blade. Nearly always the best place to make this cut is just beneath a node. It is a good plan to dip all cuttings in a suitably diluted insecticide as soon as they are made and to allow them to dry a little before planting. Except with cactuses and succulents there is no occasion for further delay. With them, allow a period of a few days to ten or so for the cuts to dry and heal somewhat. During this time keep them in a dry, warm, sunny place, or if of kinds known to prefer a little shade, out of direct sun.

Sectional stem cuttings without leaves afford convenient and rapid means of increasing certain tropical and subtropical plants, notably cordylines, dieffenbachias, dracaenas, and others with more or less canelike stems. The technique consists of cutting these into pieces, generally 2 to 3

Softwood cuttings of geranium: (a) Cut off the lower leaves

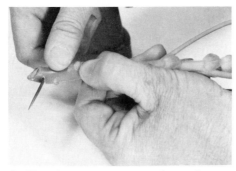

(b) Slice the stem across just beneath a node

(c) Plant the cuttings in sand

(d) A pot of newly planted cuttings

inches long, and setting them, vertically or more usually horizontally, with their tops just showing, in the rooting medium. In a temperature of 70°F, or a few degrees higher, and preferably with a little bottom heat, these soon send roots downward, shoots upward, and become new plants. A variation, often employed with cordy-

Sectional stem cuttings of dieffenbachia: (a) Making the cuttings

(b) Planting the cuttings horizontally

(c) Cuttings planted vertically

(d) Rooted cutting ready for potting

Entire stem of stapelia used as a cutting by coiling it and laying it on the rooting medium

(a) Leaf bud cuttings of geranium ready for insertion in rooting medium

(b) Leaf bud cutting rooted, potted, and sending up a new shoot

Planting leaf cutting of Christmas begonia

Rooted leaf cuttings of peperomia

Portions of leaves of sansevieria rooted as cuttings

Hammer or mallet cutting of geranium

lines and dracaenas, is to lay entire stems, without their leafy tops, horizontally in the rooting medium, not cutting them into pieces until new shoots appear, then making the cuts so that each piece has a growing shoot.

Hammer or mallet cuttings and leaf bud cuttings that succeed with camellias, geraniums, rhododenrons, and many other plants are in effect varieties of sectional cuttings, but not leafless ones. Each consists of a single leaf with a portion of stem attached. With hammer cuttings the attachment is a small section of complete stem resembling a little mallet head at the end of the leafstalk. With leaf bud cuttings it is smaller, consisting only of a scooped-out portion, not the full thickness of the stem, together with the bud in the leaf axil or of a piece formed by splitting a short section of stem longitudinally in half and leaving this attached to the base of the leafstalk. In both types new growth comes from the axillary bud.

Leaf cuttings are made from single leaves or parts of leaves without the axillary bud or any part of stem attached. Success depends upon the ability of some plants to regenerate from leaves alone, in some cases from the stalk, in others from other parts. Prominent among plants that can

readily be increased in this way are African-violets, gloxinias, and most or perhaps all other members of the gesneria family, some kinds of begonias, notably *B. rex-cultorum* varieties and Christmas begonias, peperomias, many kinds of succulents including cotyledons, crassulas, kalanchoes, and sedums, and such other plants as hyacinths, lachenalias, sansevierias, and lewisias. With most succulents no preparation is needed. The leaf is planted as it is when it was pulled off the parent. Most other leaf cuttings are made ready for planting by slicing across their stalks, or if stalkless across the blades, with a razor blade or a sharp knife. Leaves of begonias of the *B. rex-cultorum* group (rex begonias of gardeners) may be cut into wedge-shaped pieces with the junction of two main veins forming the bottom of each portion and these segments used as cuttings, or the veins may merely be cut through just below such meeting points and the entire leaf laid bottom side down on the propagating medium and kept in place by pegging with hairpins or weighting with a few small stones.

Bulb cuttings are less commonly employed as means of increase than many other kinds. This is largely because most gardeners are unfamiliar with this twentieth-century technique. It is most frequently employed in increasing amaryl-

lises (*Hippeastrum*), but is equally as successful with many other bulb plants including *Albuca, Chasmanthe, Cooperia, Haemanthus, Hymenocallis, Lycoris, Narcissus, Nerine, Pancratium, Phaedranassa, Scilla, Sprekelia,* and *Urceolina.* No doubt there are many others susceptible to this method of multiplication.

To make bulb cuttings, take a mature bulb, preferably one large for its kind. Quarter it lengthways making sure each of the four resulting wedges includes a part of the basal plate. Next, cut each of these quarters lengthways into two or more slimmer segments with basal plate tissue attached. Finally, slide a knife blade between every third of fourth piece of the

Bulb cuttings of hippeastrum: (a) Cut the bulb into segments

(b) Slice the segments into smaller pieces

(c) Bulb cuttings planted in a pan (shallow pot)

(d) Roots and a new young bulb sprouting from the bulb cutting

concentric scales that form each wedge, cutting through the piece of basal plate so that each of the ultimate segments, which are the bulb cuttings, has a small piece of plate at its base.

Plant bulb cuttings vertically or at a slight slant with their lower halves buried in the rooting medium, and their tops sticking out. Keep the medium moist, but not wet. Temperatures somewhat higher than normal growing ones for the species are advantageous, and mild bottom heat is helpful, but not essential. Late summer is a good time to make bulb cuttings of many kinds, but the method is relatively new and there is opportunity for testing many sorts of bulb plants as well as for experimenting with different seasons for taking cuttings.

Rooted bulb cuttings: (a) Young bulbs developing from between the scales of the cutting

(b) Cuttings and young bulbs sectioned to show the origin of the roots and new bulbs

Within a few weeks, or in some cases perhaps a few months, each bulb cutting that succeeds will develop between its scales one or sometimes more than one small bulb with leaves. When these are big enough to handle with ease, transplant them to soil of a type favorable to their kind, but much sandier than you would employ for older bulbs. With sorts that can be persuaded to do so, for example, amaryllises, keep them growing, transplanting or repotting as needed, without permitting

them to go dormant until after they have flowered for the first time. With kinds that insist on going dormant, accept this, keep them dry during the period of rest, and replant and start them into growth again at their normal season for such activity.

Evergreen hardwood cuttings made after the season's growth is over and the shoots have firmed and become woody are successful with many trees and shrubs that retain their foliage throughout the year, for example arbor-vitaes, barberries, box-

Evergreen cuttings: (a) Planting a cutting of a false-cypress

(b) Potting the rooted cutting

woods, cypresses, false-cypresses, hollies, junipers, and yews. Such cuttings are made and managed through the rooting period in the same way as half-ripened or greenwood cuttings.

Death by drying is an especial hazard with hardwood cuttings of evergreens. Their often comparatively thin stems are less able than those of softer-wooded cuttings to conduct considerable amounts of water. Because of this, there are two imperatives. Ample supplies of water must be available in the rooting medium at all times and atmospheric relative humidity must be maintained at a consistent high level, 90 percent or above for most kinds. The best way of assuring that the rooting medium is adequately moist without being so wet that needed air is excluded is to make certain that the cutting bed or con-

Rooted cuttings of evergreens: (a) Boxwood

(b) Yew

tainers are extremely well drained and that the rooting medium is so porous that water seeps very rapidly through it. Then drench the rooting medium so frequently that there is never a suspicion of dryness. Cuttings of these kinds do not ordinarily signal, as do those of softer-foliaged sorts, by wilting or more subtle changes in appearance, the need for water. They are likely to be beyond recovery when the first signs of drying or dying are observed. Coarse sand or perlite or either of these mixed with up to one-third part by bulk of peat moss are suitable rooting mediums for hardwood evergreen cuttings.

The double-pan technique is often successful with cuttings, especially evergreens, that are more difficult to root than most sorts. Use two earthenware pans

The double-pan method of rooting cuttings

(shallow flowerpots), one so much larger than the other that when the small pan is set inside the larger one, there is a 2-inch space between their rims. Crock the larger pan in the normal way to assure drainage. Then plug the hole in the bottom of the small pan with cement or with a cork and set the pan inside the bigger one so that their upper rims are level with each other. Pack the space between the two pans with the chosen rooting mix and insert the cuttings in the mix. Water with a fine spray and fill the inner pan with water. Cover with a bell jar or polyethylene plastic. No further overhead watering should be necessary. Sufficient water should pass through the porous sides of the small pan to maintain the rooting mix in a satisfactorily moist condition. Replace the water in the small pan as it is lost in this way. If moisture collects on the inside of the glass or polyethylene cover, raise the cover by propping it up or in any other appropriate way to provide a certain amount of ventilation.

The best time to take the cuttings varies according to kind, from late August to late fall. Those of some sorts, such as hardy false-cypresses, junipers, and thujas, root with greatest facility after the plants from which they are taken have been subjected to two or three sharp frosts. As with greenwood cuttings, protect evergreen hardwood ones from drying from when they are cut from the parent plant until they are made and inserted. Make the cuttings, depending upon kind, 2 to 6 inches in length, without or with heels of older wood at their bases.

Wounding is a technique decidedly helpful with many evergreen hardwood cuttings, particularly those of conifers. It consists of cutting more of the cambium than is done in making the normal cut across the base. The wound may be made by drawing the pointed tip of a sharp knife down the stem from a point about 1½ inches above the base to the end, slicing through the bark to the cambium, but not deeply into the wood beneath. With comparatively thick-stemmed cuttings, such as rhododendrons, the wound may be made by cutting a narrow slice of bark off one

Wounding cuttings of evergreens to encourage root production: (a) A light wound made with the tip of the knife

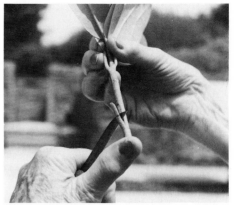

(b) A heavy wound made by removing a narrow slice of bark

side of the cutting for a length of about 1½ inches above the base to the base. Following wounding, the exposed tissues are dusted with hormone powder and the cuttings are planted immediately in a greenhouse or a cold frame shaded from direct sun. The cuttings of some evergreens root with facility and rapidly, others take many weeks or months. Sometimes the latter will form very large basal calluses without roots. It may help then to lift the cuttings out of the rooting medium, pare the calluses with a sharp knife, and replant immediately.

Hardwood cuttings of deciduous shrubs and trees afford a reliable, easy way of raising stocks of many kinds. Although little known to home gardeners, the method of propagation based on the use of these is an ideal one for them. It is much employed by professional nurserymen. Hardwood cuttings are pieces of current season's shoots taken during the season when the plants are without foliage. Here is how to make and manage them.

In fall, as soon after leaf drop as convenient, select vigorous shoots of pencil thickness or somewhat thinner. Section them into uniform lengths of 6 to 10 inches, making the basal cuts squarely across just beneath a node (point on the stem where a leaf was attached), the apical ones on a slant just above a node. If, as with gooseberries, it is not desired to have branches develop from below ground level, rub off all buds except those on the upper inch or two of the cuttings. Tie the prepared cuttings in bundles of up to twenty-five, or even fifty. Bury them, horizontally or butt ends up, outdoors, in a cold frame, or in a box in a cellar or other place where the temperature stays in the 32 to 40°F range, in damp sand, peat moss, or sandy soil. Freezing will not harm hardy kinds, but if they remain solidly frozen for long periods formation of callus tissue is delayed. To guard against this cover the cuttings with a sufficient depth of sand, peat moss, sawdust, or soil. Outdoors, 3 to 4 inches is adequate with, in cold climates, a thick layer of straw, hay, leaves, or other

Hardwood cuttings: (a) Butt the shoots against a backstop so that cuttings are even in length

(b) Tie the cuttings in bundles, label them, and bury them in sand, peat moss, or soil

insulating material put on top of the sand, peat moss, sawdust, or soil after it has frozen to a depth of about 1 inch. In cold frames and cellars, less covering suffices. During the winter callus tissue and often short roots develop from the bottoms of the cuttings, and sometimes roots from their sides.

In spring, at the time deciduous shrubs and trees are about to burst into leaf, dig up the bundles of cuttings, slit the strings that bind them, and without allowing them to dry plant the cuttings in nursery rows 1½ to 2 feet apart in finely pulverized, but not newly fertilized soil. Plant in trenches cut with a spade so that one side is nearly vertical. Strew an inch or so of sand along the bottom of each trench, then set the cuttings 3 to 6 inches apart, upright or slantwise, against the nearly vertical wall with their bases resting on the sand. Let only ½ inch or so of their tips, or with kinds with which it is desirable to have all branches originating above ground, up to 2 inches, protrude above the soil surface. Fill soil into the trench and firm it by treading on it.

During the first summer cuttings handled in this way will produce sturdy bushes from 1 foot to 3 feet or sometimes even taller. Among kinds that respond to this method of propagation are currants, figs, forsythias, gooseberries, grapes, hibiscuses, hydrangeas, jasminums, mock-or-

The next spring: (a) Make a narrow trench with a spade for the cuttings

(b) Dig up and untie the bundles and set the cuttings against the nearly vertical side of the trench

(c) Back-fill the trench with soil, tread it firm, and finish to a level surface

anges, plane trees, poplars, privets, quinces, roses, spireas, weigelas, and willows. Many other sorts of plants can be increased in this way.

Success with hardwood cuttings of kinds difficult to root, such as apples, pears, and peaches, often can be obtained if 2 to 3 inches of the bases of the cuttings have

In fall or spring, after a summer's growth, transplant the young plants to a nursery bed to make further growth: (a) *Stephanandra* roots only from the base of the cutting

(b) *Viburnum* roots along the length of the cutting

been etiolated by keeping light from them throughout the growing season. Nurserymen accomplish this by cutting the stock plants to near ground level in spring and then, as new shoots develop, mounding fine soil around and between them, taking care not to cover the growing tips, and repeating the process at intervals of a few days until the mounds are 4 to 5 inches high. An alternative method is, from the time growth begins, to wrap the lower 3 inches or so of shoots that in fall are going to be used as hardwood cuttings with light-excluding tape.

Root cuttings are pieces of root, generally thick, used to generate new plants. The method is an easy one well suited for use by home gardeners. Among plants commonly raised in this way are anchusas, bouvardias, eryngiums, fatsias, horseradish, Japanese anemones, Oriental poppies, osage-orange, plumbagos, romneyas, sophoras, sumacs, summer phloxes, teco-

Root cuttings: (a) Digging up plant of Cynoglossum for roots to make cuttings

(b) Preparing the cuttings

(c) Planting the cuttings in sand

(d) Cuttings growing vigorously

(e) Well-rooted cutting ready for planting in a nursery bed; cultivating

(f) After the first summer's growth in a nursery bed

mas, trumpet creepers, wisterias, and yuccas. Many others can be.

Early fall is the best season to take root cuttings, about the time the leaves drop or the foliage dies. Dig up stout roots and cut them into pieces 1 inch to 3 inches long. Bury these in damp sand and peat moss, or sawdust, in a cold frame or in boxes in a cold, but frostproof cellar, garage, or similar place or, alternatively, plant them 2 to 3 inches apart in a bed of sand in a cold frame or outdoors.

In spring take the cuttings from their winter storage quarters and plant them in rows 1½ to 2 feet apart in crumbly, but not newly fertilized soil. Allow enough space, usually 4 to 6 inches, between individuals for the young plants to make adequate growth. Plant most root cuttings horizontally, but set the throng-like ones of anchusas, cynoglossums, and horseradish vertically or at an angle. Cover them with soil to a depth of about 2 inches.

This is rather special. To encourage good results they must be afforded environments favorable to their development into sturdy plants. Give particular attention to lifting them from the cutting bed and to

their management immediately afterward. This is a period of transition and adjustment to new conditions and a new way of life.

Do not be impatient. Let the cuttings stay in the cutting bed until they have enough roots to support whatever foliage they have. With most kinds this will not be until the roots are an inch or two long. A frequent question from beginners is "How does one know when the cuttings have rooted?" With a little experience this can be judged fairly accurately by their appearance. Rooted cuttings assume a more jaunty air. They do not show signs of wilting if the atmosphere is a little drier than desirable for newly planted cuttings or if exposed to stronger light than they would have found agreeable at first. They may begin to make new growth. But appearance may not be enough. When you think that rooting has probably taken place, carefully dig out a sample cutting or two, taking care not to damage any roots, and judge from inspection whether the time has come for the next move. If the young plants are to have the tips of their shoots pinched out to induce branching, do this a week or two before or a week or two after the time you transplant. This avoids a double shock.

Avoid checks (setbacks). This is a cardinal rule in the management of rooted cuttings. Do everything possible to promote uninterrupted growth. To this end it is often desirable to harden cuttings somewhat by exposing them to more air, a somewhat drier atmosphere, and increased light before removing them from the rooting medium. This may be done by taking off or partially opening the covers that maintain high humidity and afford shade. To be avoided are extreme transitions from the type of rooting medium in which the cuttings have rooted and the soil into which they are first to be potted or planted. Temper the change from infant to adult diet by using a weaning mix midway in type and texture between that the cuttings have known and the kind of soil agreeable for mature plants of their kind. Leave fertilizer out of the mix and use extra amounts of sand, or perlite, and organic matter such as peat moss or leaf mold. A porous, crumbly soil is needed at the first potting or planting. Do not press or pack it as firmly as you would at later pottings or plantings. And do not let the roots dry during the operation.

Immediately following potting or planting water very adequately with a fine spray, and take whatever measures may be needed to relieve stress and prevent the foliage from wilting. In greenhouses this may best be done by returning the potted plants to an environment similar to that in which the cuttings were rooted and keeping them there until they have recovered from the trauma of planting. Other means of preventing excessive shock are shading from sun and misting the foliage with water from time to time.

CUTWORMS. Closely related to armyworms, but solitary rather than gregarious, the annoying and amazingly destructive cutworms are the larvae of nightflying moths. Repulsive to look at, dingy yellow, bronze, or black, these soft-bodied caterpillars feed on a wide variety of vegetation. Some kinds eat the roots, others climb stems to heights of several feet, but those generally most pestiferous to gardens are the ones grouped as surface cutworms. During the day these remain quiesent, coiled just beneath the soil surface. They feed only at night. In gardens they chiefly damage seedlings and small, newly set out plants, shearing the stems at or just below ground level. When the caterpillars are fully grown they descend a few inches into the soil and pupate, the moths appearing within one to several weeks or the following spring.

Baits purchased from dealers in garden supplies or prepared at home as recommended by Cooperative Extension Agents or other authorities are effective controls. An old-time and still valid method of foiling cutworms is to surround each young plant set out with a collar of stiff paper or thin cardboard. For tomatoes and suchlike plants a section of a milk carton works admirably. Let the base of the collar be buried an inch or so in the ground, the remainder extend for 3 or 4 inches above the surface.

CYANANTHUS (Cyan-ánthus). None of the thirty species of *Cyananthus* is well known in America. Natives of the mountains of southwest China and the Himalayas, they are close relatives of campanulas and like them belong in the bellflower family CAMPANULACEAE. The name, from the Greek *kyanos*, blue, and *anthos*, a flower, refers to the blooms.

Low, summer- and fall-blooming, herbaceous perennials, cyananthuses have more or less trailing stems and small, alternate leaves. Usually solitary, and bell- or funnel-shaped, the flowers are most often blue and are more rarely yellow or white. They terminate the stems and have five-lobed calyxes and corollas and five stamens. The fruits are capsules.

Himalayan *C. lobatus* has a short, sometimes branched, erect rootstock and slender, trailing, leafy, reddish stems 6 to 10 inches long. Its thickish leaves are obovate, narrowing to short stalks, and are lobed or coarsely-toothed. Funnel-shaped, and about 1 inch across, the bright violet-blue flowers have long hairs in their throats. In variety *C. l. albus* the blooms are white. Much more vigorous, *C. l. insignis*, of China, has yellow-green stems and larger flowers. *C. l.* 'Sherriff's Variety' is an improved garden form.

Cyananthus lobatus

Cyananthus lobatus 'Sherriff's Variety'

Another Himalayan, *C. microphyllus* is sometimes misidentified in gardens as *C. integer*, a name that belongs to a species with larger leaves covered with white hairs, and funnel-shaped, scarcely-hairy blooms. The trailing to somewhat erect, reddish stems of *C. microphyllus* are up to 1 foot in length. Its narrowly elliptic to ovate, lobeless and toothless, short-stalked leaves are under ½ inch long. Dark green, they have rolled-under margins. The purple-blue flowers, with cylindrical corolla tubes hairy at their throats, are about 1 inch in diameter.

Garden Uses and Cultivation. Often dif-

Cyananthus microphyllus

ficult to satisfy, these plants challenge skilled growers of alpines. Others should not attempt them. They are unsuited for regions of hot summers. In North America, the Pacific Northwest is most likely to afford the cool, humid conditions they need. Their soil should be deep, gritty, well-drained, and comfortably moist without being wet. A rock garden moraine, lightly shaded from the fiercest heat of summer sun, is a likely place to try cyananthuses. Seeds and summer cuttings are the surest methods of propagation. Careful division in early spring may be successful, but is tricky and uncertain.

CYANASTRACEAE. The characteristics of this family are those of its only genus *Cyanastrum*.

CYANASTRUM (Cyan-ástrum). The only genus of its family, *Cyanastrum* belongs in the cyanastrum family CYANASTRACEAE, a tropical African group of seven species of monocotyledons related to the *Pontederiaceae*. Its name, from the Greek *kyanos*, blue, and *astron*, a star, alludes to the flowers.

Cyanastrums are nonhardy herbaceous perennials that have tuberous rhizomes or corms (bulblike organs) and fleshy roots. Their leaves are oblong-lanceolate to heart-shaped. The flowers, in racemes or panicles, have six perianth parts usually called petals (more correctly, tepals), six stamens, and one style. The fruits are one-seeded capsules.

West African *C. cordifolium* is a handsome ornamental with peculiar, flattened, heart-shaped corms that develop one on top of the other at about ground level. Its glossy leaves have stalks up to 1 foot long, and broadly-heart-shaped blades up to 8 inches long. In bracted, spikelike racemes 2 to 7 inches long that originate from the base of the plant, the fragrant, 1-inch-long, blue-violet flowers are presented. Much shorter racemes of blooms are characteristic of *C. c. compactum*.

Garden and Landscape Uses and Cultivation. In the humid tropics and subtropics, and in greenhouses where a minimum

Cyanastrum cordifolium

Cyanastrum cordifolium (flowers)

night temperature of 60°F or more is maintained and fairly high humidity is assured, the species described is attractive for growing in partial shade. Responding to fertile, always moderately moist, but not wet soil that has a generous organic content, it is readily increased by division and cuttings. Specimens in pots or tubs benefit greatly from weekly or biweekly applications of dilute liquid fertilizer from spring through fall.

CYANELLA (Cyan-èlla). Not frequent in cultivation, the genus *Cyanella* consists of seven species endemic to South Africa. It belongs in the amaryllis family AMARYLLI-DACEAE, or according to botanists who split that group to the tecophilaea family TE-COPHILAEACEAE. Its name, alluding to the flower color of one kind, is a diminutive of the Greek *kyanos*, blue.

Cyanellas have underground bulblike organs called corms with fibrous, matted coats. Their leaves, mostly in basal rosettes, are linear to slender-cylindrical. Generally in loose racemes up to about 1 foot tall or less commonly solitary, the blue, white, or yellow, asymmetrical flowers are without perianth tubes, have six petals (more correctly, tepals), six unequal stamens, one or more distinctly pointing downward, and a style tipped with a minutely-three-lobed stigma. The fruits are capsules. The corms of these plants are eaten by native peoples.

Kinds perhaps cultivated include *C. capensis,* with branched stems, lanceolate, wavy-edged leaves, and lilac blooms. Its stems less freely-branched, *C. odoratissima* has sword-shaped basal leaves, slender-pointed, linear-lanceolate stem leaves, and fragrant, deep rose-pink blooms that fade to pale pink and finally yellow. Fragrant-flowered *C. lutea* has yellow flowers ¾ inch in diameter. In its variety *C. l. rosea* the blooms are pinkish-mauve.

Garden Uses and Cultivation. These are as for *Ixia*.

CYANOPHYLLUM. See Miconia.

CYANOTIS (Cyan-òtis). Creeping, trailing, or weak-stemmed herbaceous perennials, the possibly fifty species of *Cyanotis* belong in the spiderwort family COMMELINACEAE. They inhabit the tropics and subtropics of the Old and the New World. The name, from the Greek *kyanos*, blue, and *ous*, an ear, refers to the petals.

Cyanotises have alternate leaves of small to medium size that with their bases sheathe the stems. The clustered flowers, much like those of *Tradescantia*, are generally purple, blue, or purplish-pink. They have three sepals of nearly equal size, usually joined at their bases into a short tube, three equal-sized petals, generally united at their bases, six stamens, all fertile, and one style. The fruits are capsules.

Distinctly fleshy stems and foliage are typical of *C. kewensis,* a compact sort with prostrate, branching stems. About 1 inch long, its pointed-ovate, nearly triangular, olive-green leaves have purple undersides and like the stems are clothed with brown hairs. The flowers are small and purple. From it, *C. somaliensis* differs in its nar-

Cyanotis somaliensis

rowly-lanceolate to ovate leaves, about 1½ inches in length, being shining green and clothed on both sides with long white hairs. The flowers have bright violet petals, and stamens that are fuzzy with bright blue hairs, and that have golden-yellow anthers.

Less common in cultivation than the sorts described above are *C. arachnoidea* and *C. nodiflora*. A native at high elevations in the Nilgiri Hills of India *C. arachnoidea* has sprawling or semi-erect stems up to 1½ feet long that like the leaves are furnished with cottony or silky white hairs. The fleshy leaves, 2 to 2½ inches long and up to ½ inch wide, taper evenly from base to apex. In terminal clusters ½ to ¾ inch across, the blue flowers have stamens fuzzy with blue hairs and tipped with orange anthers. South African *C. nodiflora* has erect or somewhat sprawling stems up to about 9 inches long and strap-shaped

leaves about 2 inches long. The plants may be bright green and hairless, hairless and glaucous, or distinctly hairy. The flowers, in compact clusters, terminal and from the axils of the upper leaves, range from pale lilac to rich dark blue. Their hairy stamens are tipped with yellow anthers.

Cyanotis arachnoidea

Cyanotis nodiflora

Garden Uses and Cultivation. These are primarily plants for greenhouses and terrariums or for outdoors in the tropics and subtropics. They need minimum temperatures of 60°F and very well-drained soil watered rather sparingly, but not allowed to dry. Constant saturation soon spells disaster. Good light is necessary. If illumination is inadequate the shoots lose their compactness and make weak growth and the foliage lack firmness and good coloring. Pans (shallow pots) are suitable receptacles in which to grown cyanotises. Cuttings are an easy means of propagation. They may also be raised from seed.

CYATHEA (Cyá-thea)—Tree Fern, Sago Fern. This genus as treated here includes ferns previously referred to *Alsophila* and *Hemitelia*, as well as others long familiar as *Cyathea*. Some botanists prefer to recognize the segregate genera. In the broader interpretation *Cyathea* comprises possibly 800 species and is most abundant in mountains in the humid tropics, but extends also to Chile, New Zealand, and South Africa. It is one of several genera of ferns all or some of whose species have conspicuous trunks and are called tree ferns. Others are *Blechnum, Cibotium, Dicksonia,* and *Sadleria.* With one other genus, not known to be cultivated, *Cyathea* comprises the cyathea family CYATHEACEAE. Its name comes from the Greek *kyatheion,* a little cup, in allusion to the cups that hold the spore cases.

Cyatheas have often very tall and frequently spiny trunks. Large to immense, their usually leathery fronds (leaves) are twice-or more-times-pinnate or pinnately-lobed, hairy or not, and may or may not have prickly stalks and ribs. The clusters of spore capsules are on the veins or in the vein axils on the undersides of the fronds. These noble tree ferns are remnants of a flora that covered large parts of the earth's surface in distant geological times. Even now in a few regions they are sufficiently numerous to dominate local floras. Because of the many kinds, their considerable variability, and lack of adequate studies, cyatheas are often difficult to identify as to species. Only a few are known to gardeners.

New World species include *C. arborea, C. insignis* (syn. *Sphaeropteris insignis*), *C. meridensis, C. microdonta,* and *C. suprastrigosus.* The commonest of six or so species of tree ferns native to Puerto Rico, and wild also in other parts of the West Indies, in Trinidad, and from Mexico to northern South America, *C. arborea* has a slender trunk up to 30 feet tall or sometimes taller, and a crown of twice- or thrice-pinnate fronds 6 to 10 feet in length, up to 4 feet wide, and yellowish-green on both surfaces. Native of Jamaica and Cuba, *C. insignis* has a stout, spineless trunk 15 to 20 feet tall. Its leathery fronds are up to 12 feet long and 4 feet wide. They have strongly curved stalks, and blades three-times-pinnately-divided and lobed. They are very dark green on their upper surfaces, and lighter and glaucous beneath. Colombian *C. meridensis* has twice- or thrice-pinnate fronds, green on both sides, and with rather narrow final segments. Tropical American *C. microdonta* (syn. *Alsophila microdonta*) is up to 50 feet tall but mostly lower. It has a thorny trunk and large, thrice-pinnate fronds with thorny stalks and rachises (midribs). The primary divisions of the leaves are 1 to 1½ feet long by up to 8 inches wide. The secondary divisions, stalkless and 3 to 4 inches long, have toothed final segments. Endemic to Costa Rica, *C. suprastrigosus* has a slender trunk, and rather small, thrice-pinnate fronds, dull green above, paler on their undersides.

From New Zealand and Lord Howe Island comes *C. dealbata* (syn. *Alsophila tri-*

Cyathea microdonta

Cyathea suprastrigosus

Cyathea dealbata

color). This noble fern, up to 30 feet tall, has twice- or thrice-pinnate blades 3 to 9 feet in length, and up to 3 feet or more in width. The pointed-oblong primary leaflets are 1 foot to nearly 2 feet long. The ultimate segments, toothed in their upper halves and somewhat sickle-shaped, are approximately ½ inch long. The undersides of young fronds are pale green, those of older ones glaucous to nearly white. The sago fern (*C. medullaris*), the pith of which contains much starch that was eaten by the aborigines, is a native of New Zealand, Australia, and some islands of the Pacific. Nearly black, warty stalks are typical of the fronds of this, a species 40 to 60 feet tall that may have a trunk apparently up to 1 foot wide, but that, except perhaps toward its base, is really narrower. A thick thatch of black roots that covers the trunk is re-

sponsible for the deception. The leaf blades are up to 20 feet long by 6 feet wide. They have pale green undersides and dark green upper ones.

In its native Australia called rough tree fern, *C. australis* (syn. *Alsophila australis*), dark-trunked and about 20 feet tall, has green fronds the stalks of which are slightly roughened with warty protuberances and clothed with straw-colored scales. The blades, three-times-pinnate, 6 to 30 feet in length, and 3 feet or more in width, have slightly glaucous undersides. Similar, but a more rapid grower, *C. cooperi* (syns. *Alsophila cooperi, Sphaeropteris cooperi*), of Australia, has leafstalks much roughened with warty protuberances and covered with glossy, bright brown scales. Its finely-divided blades are metallic-green.

Malaya is home to *C. contaminaris* (syn. *Alsophila glauca*), slender-trunked, and up to 50 feet tall. Stout spines on its leafstalks and midribs and purpling of the midribs with age are characteristic features. Its twice-pinnate leaves, up to 12 feet long by 4 feet wide, are glossy green above and bluish on their undersides.

Garden and Landscape Uses and Cultivation. Among the most graceful of plants of tree proportions, cyatheas are elegant for planting outdoors in climates kindly to them. They can be used singly and in groups at the edges of tree plantings or in open glades and similar places where their tops are assured good light and their trunks and the ground beneath are shaded and humid. They are not plants for cold climates, but most prefer reasonable coolness rather than oppressive tropical heat. Their roots need ample moisture. This can sometimes be assured by planting them on the banks of ponds or streams 2 or 3 feet above the water.

Cyatheas are very amenable to greenhouse cultivation in ground beds and in pots or tubs. They get along surprisingly well in containers comparatively small for the size of the plants, but even so, quite large ones are needed for big specimens. They delight in nourishing, coarse, peaty soil that permits water to drain yet retains enough to be always moist. Dryness even for brief periods is exceedingly harmful. It is well to keep the containers of well-rooted specimens standing in 2 or 3 inches of water. A highly humid atmosphere is requisite, and it is important to keep the trunks constantly moist. This last is most easily done by wrapping around them a 3- or 4-inch layer of sphagnum moss, tied in place with fine wire or nylon thread, or held in a corset of chicken wire, and spraying this with water daily. Shade from strong sun is needed, but as much light as the plants can take without the fronds yellowing or scorching is beneficial. Tropical species do best if the minimum night temperature is 60°F, temperate-region ones, such as those native of New Zealand, five to ten degrees lower. Increases by day of five to fifteen degrees are favorable. Cyatheas do not normally produce offshoots from their bases or trunks and so spores afford the usual means of increase. As young plants these ferns grow quite rapidly, especially *C. medullaris*. For more information see Ferns.

CYATHEACEAE — Cyathea Family. This conglomerate of tree ferns, treated by different authorities as consisting of from two to seven genera and from under 300 to more than 800 species, inhabits high altitude forests from Mexico to Chile and in Malaysia, Australia, and New Zealand. Its members have branched or branchless trunks frequently clothed with matted aerial roots. Prevailingly large to huge, the leaves are in wide-spreading crowns at the tops of the trunks or stems. They are one- or more-times-pinnate and have circular clusters of spore capsules that may or may not have covers (indusia). The only genus ordinarily cultivated is *Cyathea*.

CYATHODES (Cyath-òdes). Epacrids or members of the epacris family EPACRIDACEAE, to which *Cyathodes* belongs, are nearly all natives of Australia, Tasmania, and New Zealand, with a few species in Polynesia, New Guinea, and Malaya. They are, in effect, mainly Southern Hemisphere equivalents of the heath family ERICACEAE, of the Northern Hemisphere. In the wild *Cyathodes* occurs through most of the range of the family. Its members are heathlike, erect or prostrate shrubs, with small, often crowded leaves. The name, from the Greek *kyathos*, a cup, and *odous*, a tooth, alludes to the toothed disks of the flowers. There are fifteen species.

This genus has alternate, stalkless or briefly-stalked, rigid, undivided, toothless leaves and flowers with five sepals, a tubular corolla with five spreading or recurved lobes (petals) densely-bearded to nearly hairless on their bases, as many stamens as corolla lobes and alternate with them, and one style. The fruits are small, berry-like drupes. For the plant some botanists treat as *C. fraseri* see *Leucopogon fraseri*.

Erect and up to 6 feet tall, **C. parviflora** has lanceolate to elliptic-oblong or oblanceolate leaves ½ to ¾ inch long, slightly glaucous on their undersides and with recurved margins. About ⅛ inch long, the flowers are in short, crowded spikes. Black when dried, the ovoid to roundish fruits are about ⅛ inch long.

Low or prostrate and stout-branched, **C. colensoi** is sometimes 6 feet wide. Its leaves, oblong, narrow, and ⅜ inch long, are glaucous on their undersides. The tiny flowers are in terminal racemes of two to five. White, pink, red, or dark crimson, the spherical fruits are less than ¼ inch in diameter. Prostrate **C. empetrifolia** forms

Cyathodes colensoi

patches about 2 feet in diameter. It has slender, wiry branches and linear leaves 3/16 inch long and glaucous on their undersides. The flowers are tiny and in racemes of two to four. The fruits, reddish and egg-shaped, are about ⅛ inch long.

Garden and Landscape Uses and Cultivation. These are suitable for landscaping areas where conditions are agreeable to heaths and heathers (*Erica* and *Calluna*). They need acid, peaty soil and a sunny location. The lower ones are effective in rock gardens. Shearing after flowering or fruiting may be desirable to keep the plants shapely. Propagation is easy by seeds and cuttings and sometimes by division.

CYBISTAX (Cybis-táx). Three species of tropical American trees of the bignonia family BIGNONIACEAE constitute *Cybistax*. The name presumably derives from the Greek *kybistao*, to tumble or summersault, but its application is uncertain.

The leaves of this genus, opposite or somewhat alternate, have long stalks with, spreading from their tops, five to nine short-stalked leaflets. The yellow, two-lipped to strongly five-lobed, bell-shaped flowers are in large branched clusters or panicles. They have five nonprotruding stamens. The fruits are slender capsules.

One of the handsomest tropical flowering trees, **C. donnell-smithii** (syn. *Tabebuia donnell-smithii*), a deciduous native of humid forests in southern Mexico and Central America, is esteemed as a source of lumber called primavera, much used for veneer, furniture, interior trim, and other purposes. Its leaves, mostly of seven oblong to ovate leaflets 2 to 10 inches long and approximately one-half as wide, are downy when young. In loose, pubescent panicles the blooms, also pubescent, are 1½ to 2¼ inches long. The seed capsules are 1 foot to 1½ feet in length by up to a little more than 1 inch in width.

Garden and Landscape Uses and Cultivation. These are as for *Tabebuia*.

CYBISTETES (Cybi-stètes) — Malagas-Lily. Around the only species of this genus there has been great botanical turmoil regarding its correct name. It has repeatedly been confused with *Ammocharis coranica*, which like *Cybistetes* is a deciduous bulb plant of the amaryllis family AMARYLLIDACEAE and is a native of South Africa, although not of the same region as the *Cybistetes*. The reason for the confusion is not far to seek, the habits of growth and appearance of the foliage of the two plants are similar. The name, alluding to the detached ripe umbels of seeds being rolled like tumbleweeds by the wind, is from the Greek *kybisteter*, a tumbler.

The Malagas-lily or St.-Joseph's-lily (**C. longifolia**) is endemic to the southwestern Cape region. It has a bulb up to 9 inches in diameter. The two-ranked leaves, which develop after the flowers, spread in rosette fashion and are up to 1½ feet long by ¾ inch wide. Like those of *Ammocharis* they possess the peculiarity of lengthening from their bases for several years even though their tops die back to the bulbs annually during periods of dormancy. The beautiful blooms, in umbels of six to twenty-four at the tops of thick stems up to 1¼ feet long that arise from one side of the cluster of leaves, are fragrant and pale to deep rose-pink. Each flower, about 3 inches long, has a perianth with a short tube and six petals (or more correctly, tepals) that have a small tuft of hairs within their tips. There are six stamens and a slender style longer than the perianth. The most easily observable differences between *Cybistetes* and *Ammocharis* are that the flowers of the former are placed at a slight angle to the individual stalks by which they are attached to the common stalk of the umbel, whereas those of *Ammocharis* are strictly aligned, and the flower stalks of *Cybistetes*, after the blooms fade and seed capsules develop, lengthen and turn downward, which is not true of those of *Ammocharis*.

Garden and Landscape Uses and Cultivation. These are the same as for *Ammocharis*.

CYCADACEAE—Cycad Family. As treated here the cycad family comprises ten genera and approximately 100 species of gymnosperms. Some authorities limit it to the about twenty sorts of *Cycas* and refer other genera included by conservatives to the segregate families STANGERIACEAE and ZAMIACEAE.

Cycads, which is the group name for plants of the cycad family, are an extremely ancient element in the earth's flora, the present population a mere remnant of the numerous massive sorts that flourished toward the end of the Triassic and the beginning of the Jurassic periods. Cycads are believed to be the most primative seed plants extant. In aspect these plants suggest palms. As natives they are restricted to a few warm parts of the world including species of *Zamia* of the southeastern United States.

Cycads have stout, short or tall, branchless or sometimes branched trunks, those of some low kinds largely subterranean. The leaves are evergreen and of two sorts, the obvious alternate ones that crown the trunks and scalelike ones in spirals on the trunks. The foliage leaves are pinnate or, those of *Bowenia*, twice-pinnate. The sexual organs, primitive flowers, are in cones or loose terminal clusters. Individual plants are unisexual. Genera include *Bowenia, Ceratozamia, Cycas, Dioon, Encephalartos, Lepidozamia, Macrozamia, Microcycas, Stangeria,* and *Zamia.*

CYCADS. This is a collective name for plants of the cycad family CYCADACEAE, the world population of which consists of about one hundred species in ten genera. By some authorities the group is divided into three families to which are given the names CYCADACEAE, to accommodate *Cycas* only, STANGERIACEAE, with the one genus *Stangeria*, and ZAMIACEAE, to which all other cycads are referred.

Geological ages ago, toward the end of the Triassic and the beginning of the Jurassic periods, the ancestors of modern cycads were more numerous and diverse and constituted a much more important part of the earth's flora. They flourished with the dinosaurs.

In appearance these plants suggest palms or tree ferns, but are related to neither. In fact they are primitive seed plants. From present-day seed plants, except *Ginkgo*, they differ in the sexual process being accomplished by self-motile male sperm cells that propel themselves through films of moisture and, when they come into contact with the naked female ovules, fertilize them. This procedure is closely analogous to fertilization in animals. After fertilization the ovules develop into seeds.

Cycads are handsome, slow-growing evergreens well suited for ornamenting outdoor landscapes in warm regions and attractive as conservatory and greenhouse specimens. As a group they are tolerably free from pests and diseases, although scale insects are sometimes bothersome. Outdoors or in, cycads grow best in well-drained, nutritious soil and succeed in partial shade. In caring for them be especially careful not to disturb their surface roots by forking or hoeing near their stems or trunks. Control weeds by hand-pulling rather than hoeing.

Propagation is by careful removal and transplantation of offsets, and by seeds. Sow the latter as soon after they are ripe as practicable in a temperature of 70 to 80°F. Use porous soil containing an abundance of peat moss, leaf mold, or other organic matter and keep it evenly and moderately moist.

CYCAS (Cỳ-cas) — Sago-Palm, Australian Nut-Palm. This genus of the cycad family

CYCADACEAE is sufficiently distinct from other cycads to cause some botanists to regard it as the only one of its family and to classify related genera in separate families. This is not done in this Encyclopedia. Here *Cycas* is regarded as one of ten genera of primitive plants of exceedingly ancient lineage, the remnants of a much more numerous assemblage that populated much of the earth in ancient geological times, and related closely enough to be accepted as one family.

There are about twenty species of *Cycas*, natives of southeastern Asia, Japan, islands of the western Pacific, Australia, East Africa, and Madagascar. The group, imperfectly understood botanically, is much in need of further study. From other cycads *Cycas* differs in its leaves being coiled in the bud stage like those of ferns, in their leaflets having a single thick midrib and no evident side veins, and in the female sporophylls (spore-bearing organs) not being scalelike and in definite cones, but more leaflike and in loose terminal tufts. Also, they fall separately, leaving the shoot that bore them to continue to grow. The name is derived from the Greek *kykas*, a palm tree.

Cycases are strikingly handsome palmlike or tree-fern-like evergreens not botanically related to palms or ferns. They have stout, usually branchless, columnar trunks clothed with persisting bases of old leaves and topped with a crown of spreading, pinnate, leathery foliage. In some kinds the trunks are subterranean. Individual plants are unisexual. Males bear erect, terminal cones, the scales of which bear on their under surfaces numerous pollen sacs. Females develop naked ovules in coarse notches along the margins of the sporophylls. After fertilization these develop into seeds.

In their native regions certain cycases, including *C. revoluta* and *C. circinalis*, are exploited for sago starch, obtained from their trunks. The seeds of *C. circinalis*, poisonous until soaked and washed several times in water to remove the harmful element, in some regions are ground in times of famine into meal for human food. The leaves of *C. revoluta* are used as fresh and dried ornamentals, especially in religious ceremonies and in funeral wreaths.

The sago-palm (**C. revoluta**), native to Java, is most frequent in cultivation. It must not be confused with true palms (species of *Metroxylon*), which also yield sago. From 6 to 10 feet tall, this kind has arching leaves 2 to 7 feet long, with numerous rigid, dark glossy green, downcurved, spine-tipped leaflets with rolled-under margins. Male cones are cylindrical and about 1½ feet long. Female sporophylls, broadly-ovate in outline and densely clothed with a felt of brownish hairs, are cleft in comblike fashion. The flattened-egg-shaped, red, seedlike fruits are 1½

Cycas revoluta

Cycas revoluta, with male cone

Cycas revoluta (female sporophylls)

inches long. From the last, **C. circinalis**, of tropical Africa and islands of the Pacific, differs in the leaflets of its fronds being flat instead of having rolled-under edges. Attaining heights of 10 to 12 feet or sometimes more, it has fronds 5 to 8 feet long and 1½ to 2 feet wide. There are short spines near the bottoms of the leafstalks.

Cycas media, with palms in background and low zamias in foreground

Male cones are cylindrical, woolly-hairy, and up to 2 feet long by 5 inches wide. Each cone-scale is tapered to a long, hooked spine. The spreading, pinnately-notched sporophylls of female plants, clothed with buff hairs, are 6 inches to 1 foot long. The fruits are the size of walnuts.

Kinds less commonly cultivated include the Australian nut-palm (*C. media*), of Australia. This has a trunk up to 15 feet or occasionally 30 feet tall, leaves up to 5 feet in length of flat, linear-lanceolate abruptly spine-tipped leaflets up to 10 inches long and with somewhat incurved margins. Native to Sri Lanka (Ceylon) and parts of Indonesia, *C. rumphii* differs from the last in its leaves having fewer, shorter, lanceolate leaflets and in the scales of its male cones ending squarely or in short points rather than tapering to long points. The leaflets are lighter colored and thinner than those of *C. circinalis*. Sometimes treated as a variety of *C. rumphii*, Australian *C. normanbyana* has a trunk up to 10 feet tall and leaves of flat leaflets up to 8 inches long.

The trunk of *C. micholitzii* (syn. *Pseudocycas micholitzii*), of Vietnam, is largely underground. The leaves, 6 to 10 feet in length, have twice-forked, flat, linear-lanceolate, spine-tipped leaflets up to 1 foot long, with toward their bases short, broad spines. When young the leaflets are glaucous, becoming more or less wavy-edged when adult.

Garden and Landscape Uses. As per-manent outdoor ornamentals these handsome plants can be grown only in tropical and subtropical regions where little or no frost occurs. They associate well with architectural features and are splendid as single specimens and for grouping. In groups they are seen to best advantage when the specimens are of different heights. It is important that sufficient space be left between individuals to allow for the not inconsiderable spread of the foliage at maturity. For best results partial shade and fertile, moist but not wet soil are required.

Cultivation. Cycases are of simple cultivation. They are tenacious to life, often surviving damage by frost and making remarkable recovery from long periods of neglect once better conditions are established. Fine specimens are had only by avoiding such setbacks and ill treatments. It is advantageous to keep the ground immediately around the trunks free of weeds or other growth and to loosen it occasionally without doing damage to the numerous knobby roots near the surface. Container specimens need watering regularly to keep the soil always moderately moist. When they are making new flushes of foliage, weekly applications of dilute liquid fertilizer are very beneficial. In greenhouses cycases can be accommodated in large pots or tubs or in ground beds. A humid atmosphere is appreciated and a minimum winter night temperature of 55 to 60°F, with higher temperatures by day

and at other seasons. Propagation is by fresh seeds sown in sandy soil in a temperature of 70 to 80°F and by offshoots or suckers carefully removed from the bases of old specimens during the season of dormancy. Before potting or planting the suckers all foliage is removed from them. They reestablish themselves most quickly if the pots containing them are sunk to their rims in a bed of peat moss or similar material with a little bottom heat to keep the roots at a temperature of 75 to 80°F. For more information see Cycads.

CYCLAMEN (Cýc-lamen). Similarities between cyclamen and primroses are not at once apparent yet both belong to the primrose family PRIMULACEAE. But then, garden members of that group are of rather widely diverse appearance, including as they do in addition to the two genera just mentioned, shooting stars (*Dodecatheon*), loosestrifes (*Lysimachia*), and pimpernels (*Anagallis*). The genus *Cyclamen* includes fifteen species of the Mediterranean region, adjacent Asia, and central Europe. The name is a classical Greek one.

Cyclamens are hardy and nonhardy, deciduous, herbaceous perennials with corms (solid bulblike or tuber-like organs). They have all basal, long-stalked leaves with usually thickish, kidney- to heart-shaped or angled blades, sometimes toothed or shallowly-lobed, with wavy-toothed or plain margins, those of some kinds prettily patterned or variegated. The solitary flowers, usually plentifully produced, are at the tops of leafless stalks. They have five-parted calyxes and short-tubed corollas with five strongly backward-pointing, twisted petals. There are five nonprotruding stamens, one style, and one stigma. The fruits are capsules, often pulled by the spiral twisting of their stalks to ground level. Sowbread is an old-fashioned name for cyclamens.

The florists' cyclamen is by far the most familiar. It is grown commercially and by amateurs for blooming in winter and spring in greenhouses. Like most florists' flowers it does not occur anywhere wild as gardeners know it. Seemingly not of hybrid

The florists' cyclamen

origin, it represents a triumph of the plant breeders' art. By careful selection from generation after generation of seedlings these modern beauties have been derived from *C. persicum*. Florists' cyclamens are much larger than any wild species. Usually they are had in bloom as ample specimens in 5- or 6-inch pots, but for exhibition purposes magnificent examples in pots 7 or 8 inches in diameter are sometimes achieved. The foliage, the leaves more or less rounded or kidney-shaped, is generally beautifully patterned with silvery zones and markings. The flowers are 2 to 3 inches long and come in a wide range of colors from pale pink to the deepest red and white. They are not noticeably fragrant.

The wild species *C. persicum* (syn. *C. indicum*), native of the eastern Mediterranean region, is a truly charming, nonhardy spring-bloomer. It has a large corm that roots only from its bottom, and heart-shaped-ovate, round-toothed leaves, angled at their bases and more or less marbled with silver. Their stalks are 2 to 6 inches long. Usually delightfully fragrant, the white to rich rose-pink flowers, deepening in color in their throats, are miniatures of those of the grosser florists' cyclamen. Their stalks, longer than the leaves, do not spiral as seed capsules develop.

Other cyclamens, of which only *C. hederaefolium* and *C. purpurascens* have proved reliably hardy in the vicinity of New York City, are for dedicated rock gardeners and others of like interests. In most parts of North America the majority are challenging to grow and for the most part less hardy than could be wished. The application of names to these has been considerably confused. The easiest to grow, at least in the northeast where it not infrequently reproduces from self-sown seeds, is *C. hederaefolium* (syn. *C. neapolitanum*). This fall-blooming native of southern Europe, has flattish corms, rounded beneath, that root only from their tops, and variable foliage. Its leaves, 1½ to 3½ inches across, are obcordate, angled, and shaped like

Cyclamen hederaefolium

those of English-ivy. They have lobed, toothed, or plain margins. They are variously and usually richly patterned with silvery markings. The rose-pink flowers, deeper colored in their throats, have ovate petals ¾ to 1 inch long. Before maturity their anthers are usually pink. The flower stalks spiral conspicuously at fruiting time. In *C. h. roseum* the blooms are lighter pink. Those of *C. h. album* are white. Very like *C. hederaefolium* is *C. africanum*, of North Africa. The chief distinctions are that the latter has larger leaves and flowers and that the anthers of its slightly fragrant blooms are yellowish. The leaves are 3 to 8 inches wide, the petals 1 inch to 1½ inches long. Similar to and often confused with the last, *C. commutatum* has scentless flowers and brownish anthers.

A fall- or late-summer bloomer, *C. purpurascens* (syn. *C. europaeum*) is a hardy species almost as easy to grow in the

Cyclamen purpurascens

northeast as *C. hederaefolium*. In the wild it ranges from southern and central Europe to the Carpathians and Yugoslavia. Often large, its flattened to subspherical corms root from all parts of their surfaces. The smooth-edged or round-toothed, kidney- to heart-shaped leaves, their basal lobes touching or overlapping, are faintly marbled with silver. The carmine-pink very fragrant flowers, their petals, approximately ¾ inch in length, are darker colored near their wide open throats. The anthers are violet-spotted. The styles protrude. The flower stalks spiral as the fruiting capsules form. The blooms of *C. p. album* are white. Plain green foliage is characteristic of *C. p. viridifolium*. The leaves of *C. p. dentatum* are conspicuously toothed.

Spring-blooming hardy cyclamens, some of which in mild regions bloom in winter, include a complex of forms grouped as *C. coum* (syns. *C. hyemale*, *C. ibericum*) occurring wild from the Balkan Peninsula to Turkey, Syria, and the Transcaucasus. These have subspherical to flattish corms rooting from below their middles, and neath, and roundish, kidney- or heart-shaped, lobeless, toothless leaves produced before or with the blooms, silver-

Cyclamen coum

marked or plain green. The small, pink, magenta-pink, or white flowers have constricted throats and round-apexed, ovate petals with at the bottom of each a large purple blotch with a pair of white or pink spots. The flower stalks spiral at fruiting time. Plants cultivated as *C. atkinsii*, originally described as a hybrid between *C. coum* and *C. persicum*, are superior variants of *C. coum*.

Two others that flower in fall are *C. graecum* and *C. cilicicum*. The first inhabits Greece, Macedonia and islands of the Mediterranean, the other Asia Minor. The subspherical to flattish corms of *C. graecum* produce their decidedly fleshy roots only from their bases. The horny-edged leaves are large, obcordate, and irregularly-toothed. Their upper surfaces have whitish veins and markings. When young they are reddish beneath. Pink to deep carmine-pink, the flowers have lanceolate to oblong-lanceolate petals. Pale pink blooms, with darker blotches at the bottoms of their pointed, oblong or obovate petals are typical of *C. cilicicum*. The styles scarcely protrude. The flower stalks spiral as the seed capsules develop. The obscurely-toothed leaves of *C. cilicicum* are roundish-heart-shaped with their short basal lobes often overlapping. Their upper sides are marked with a silvery zone.

Hardy, spring-flowering *C. repandum*, of southern Europe, has subspherical to flattish, hairy corms rooting only from their bottoms. The softish, heart-shaped leaves, dark green, but reddish on their undersides, have a conspicuous gap between their basal lobes. Their margins are shallowly-angular-lobed. Bright carmine-pink deepening in color near their constricted throats, or rarely white, the fragrant flowers have oblong to linear-oblong petals. The style protrudes conspicuously. Variety *C. r. rhodense* has foliage often blotched with yellow and flowers with pink-based, white or light pink petals.

Garden and Landscape Uses and Cultivation. The florists' cyclamen, widely cultivated in greenhouses and usually discarded after its first blooming, is used as an indoor ornamental and in essentially frostless, dryish climates for winter and spring outdoor bedding. If watered enough to keep it from wilting it remains attractive for many weeks in a light window with, late in the season, a little shade from the strongest sun. In warm rooms or if allowed to become so dry that the foliage wilts, it soon deteriorates. The wild form of C. *persicum* is grown in the same way and responds similarly.

Permanent outdoor cultivation of the florists' cyclamen is practicable in California and other mild-climate, dryish regions. There, these plants have survived temperatures down to 22°F. To succeed, it is essential that the site be well drained. Some gardeners recommend planting in beds raised 8 inches or so above the surrounding level in soil containing a large proportion of gravel or coarse sand. Unless drainage is sharp root rot diseases are likely to destroy the plants. Hardy cyclamens are plants of great charm best accommodated in rock gardens, cold frames, and alpine greenhouses.

The florists' cyclamen take somewhat longer than most greenhouse plants raised annually from seeds to reach a sizable maturity. About fourteen months should be allowed. Smaller examples can be had in about a year. In the distant past the tubers were often retained after flowering, gradually dried off, rested through the summer, and repotted and started into growth again in August. This is less satisfactory than growing new plants from seeds each year and is rarely practiced. Sow in pans (shallow pots) or shallow flats from August to December, the later months for smaller-flowering specimens. If possible use heat-sterilized soil. Cyclamens are very subject to infestation by root nematodes and untreated soil can be a source of these. Space the seeds about 1 inch apart and cover them with sandy soil or sand to a depth of about twice their diameter. Keep in a temperature of 60°F, higher of course if the outdoor temperature is higher. Germination is rather slow, normally four to eight weeks.

Transplant when the seedlings have two or three leaves. Do this carefully, paying particular attention to keeping the tops of the developing corms even with the soil surface rather than deeper. Flats are better than small pots. Soil in them is more easily kept in an evenly moist condition. Allow 3 inches between the plants. Crowding causes legginess. Grow in a night temperature of 55°F. Allow, depending upon the brightness of the weather, five to fifteen degrees more by day. A humid, but not dank atmosphere is favorable. Young cyclamens do better near the glass of the greenhouse roof than far beneath it. In early spring pot the plants into 3- or 4-inch pots. In July or August tranfer them to pots 5 or 6 inches in diameter. Do not delay potting until the roots become seriously pot-bound. If you do, the plants are

A flat of young, healthy seedlings of the florists' cyclamen

A florists' cyclamen flowering in a 5-inch pot

likely to become stunted. A potting soil of two parts good topsoil, one part coarse sand or perlite, one to two parts peat moss or leaf mold, and with generous dashes of dried cow manure or sheep manure and bonemeal, suits cyclamens. Pot fairly firmly, always with the tops of the corms showing at the soil surface.

Summer care tests the gardener's ability. Cyclamens like cool, humid conditions. In hot weather they may be kept in a north-facing, shaded cold frame or in a rather heavily shaded greenhouse. Do not crowd the plants. Stand each on an inverted flower pot set on a bench or bed surfaced with gravel or cinders. This makes for better air circulation and permits wetting between the pots to increase humidity. Frequent light overhead spraying with water is beneficial, but not so late in the day that the moisture on the foliage will not dry before night. Water often enough to prevent wilting. If the plants do wilt growth is checked. From the time the final pots are moderately filled with roots give biweekly or weekly applications of dilute liquid fertilizer. Pick off any flower buds that show before October.

In fall, as soon as weather permits, stabilize the night temperature at 50°F, with a daytime increase of five to fifteen degrees. In late September remove all shade

and reduce the frequency of fertilization, but not to the extent that the plants suffer from lack of nutrients. Wetting the foliage overhead should also cease at this time, but continue frequent wetting between pots.

If cyclamens are to be kept for more than one season, a practice not usually followed, when blooming is through gradually reduce the frequency of watering and finally cease. Store dry in a cool, airy place in the soil in which they grew. In August pick away as much of the old soil as possible and repot in new. Second-year plants can be grown into large specimens. It is possible to propagate florists' cyclamens by cutting the corms into pieces, each with a leaf or two attached, in January, and rooting them as cuttings in a greenhouse propagating bench. This is rarely done. The chief pests are aphids, mites, thrips, black vine weevil, and nematodes. Bacterial rot and a stunt disease sometimes afflict these plants.

Hardy cyclamens grow best in deep, well-drained, fairly loose soils that contain an abundance of leaf mold or other agreeable organic matter along with coarse sand, grit, and crushed limestone. They adapt to lightly shaded places beneath deep-rooting trees and shrubs or shielded from sun by a rock garden cliff. They do quite well in shaded cold frames and in pans in alpine greenhouses. Once planted they should be left undisturbed. Care must be taken not to cultivate too closely to them. The chief routine care needed by well-established specimens is an annual top dressing of fresh soil given at the beginning of each growing season. Young corms up to about two years old, but apparently not older ones, may fall victims to the depredations of mice and chipmunks. It is wise to protect against these. The only practicable way of increase is by seeds. These are slow to germinate. Under favorable circumstances they take about a year and may take as long as two. Sow in pans or flats in sandy peaty soil, covering the seeds with a layer of sand or grit. Cover with polyethylene plastic film or in some other way assure that they will be kept constantly damp, not soaking wet, and put the containers in a shaded cold frame protected from disturbance by animals or other creatures. When the seedlings have two or three leaves transplant them with great care to small pots and grow them in a cold frame for a season before setting them in their permanent locations outdoors, or set them directly there.

CYCLANTHACEAE—Cyclanthus Family. The eleven genera of 180 species of stemless or stemmed, sometimes climbing, herbaceous or subshrubby, evergreen perennial monocotyledons belonging here are natives of tropical America including the West Indies. They have long-stalked, leathery leaves in two ranks or spiraled, fanlike or of two often much-cleft lobes or leaflets. Sometimes they are pierced with holes. The inflorescences, much like those of the arum family ARACEAE, are of tiny unisexual flowers in crowded spikes with at their bases several spathes (bracts). The flowers are without or have only rudimentary perianths. The males have numerous stamens, the females four staminodes and one pistil with one to four stigmas. The fruits consist of a fleshy axis with many separate or united berries. The only cultivated genera are Asplundia, Carludovica, Cyclanthus, Dicranopygium, and Sphaeradenia.

CYCLANTHERA (Cyclanth-èra). The name of this genus, derived from the Greek kyklos, a circle, and anthera, an anther, refers to the arrangement of the anthers. The group contains fifteen herbaceous annual and perennial vines, natives of the warmer parts of the Americas including the southern United States. It belongs in the gourd family CUCURBITACEAE. Cyclantheras attach themselves to supports by simple or branched tendrils. Their leaves are opposite and without lobes, or lobed, or they consist of five to nine leaflets that spread like the fingers of a hand. The white, yellowish, or greenish flowers are commonly tiny, the females solitary, the males in racemes or panicles. There are five or occasionally six calyx lobes and a wheel-shaped corolla with five or six lobes (petals). The stamens are united in a column, with their anthers joined in two rings around its apex, so that they form what appears to be a single stamen. Egg- or broad-spindle-shaped, the softish fruits are usually more or less spiny.

The more ornamental of the two kinds cultivated is *C. pedata*, native from Mexico southward. Attaining a height of 10 feet or more, it has hairless leaves with five to seven slightly toothed, narrow leaflets. Its fruits, averaging about 2 inches in length, are oblongish-spindle-shaped and frequently covered with soft spines. When

Cyclanthera pedata, in flower

Cyclanthera pedata, in fruit

young they are pale green, at maturity creamy-white.

Remarkable because of its fruits, which when ripe burst explosively and throw their contents a considerable distance, *C. brachystachya* (syn. *C. explodens*) is native to northern South America. It attains a height of 10 feet or sometimes more and is hairless or nearly so. Its leaves are more or less triangular-ovate to oblong and three-angled or three-lobed. The fruits are green, curved, and about 1 inch long with spines along one side.

Garden and Landscape Uses. Cyclantheras are treated as annuals and are interesting and useful for screening arbors, trellises, fences, and walls across the faces of which strings or wires have been stretched to afford grasps for the tendrils. They are also grown over other suitable supports. In warm weather they grow rapidly. They prosper in fertile, fairly moist soil in full sun.

Cultivation. Seeds are used for propagation. They are handled in the same way as those of cucumbers, melons, and gourds. Where the growing season is long they may be sown directly outdoors, but in the north it is more usual to sow indoors some eight weeks before it is safe to set the plants in the garden (which should not be done before the arrival of settled warm weather). The young plants are grown indoors in a temperature of 65 to 70°F in a fairly humid atmosphere. They are potted singly and at planting out time may occupy pots 4 or 5 inches in diameter. Before planting they should be hardened by standing them for about a week in a cold frame or a sheltered sunny place outdoors. At no time should these plants be permitted to dry out.

CYCLANTHUS (Cyclán-thus). Restricted in the wild to the American tropics, the group to which this genus belongs, the cyclanthus family CYCLANTHACEAE, is most commonly represented in cultivation by *Carludovica*. It is closely related to the palm family PALMAE, and its species are distinctly palmlike in the appearance of their foliage. One species of *Cyclanthus*, an inhabitant of Central and South America and the West Indies, is recognized. Its name comes from the Greek kyklos, a circle, and anthos, a flower, and alludes to the arrangement of the blooms.

Stemless, with leaves springing directly from the rhizome, milky-juiced *C. bipartitus* has clustered, erect leaves with stalks 3 to 6 feet long, and sheathing at their bases. The blades, each with two prominent longitudinal midrib-like veins and wrinkled, leathery surfaces, when young

Cyclanthera pedata

Cyclanthera brachystachya, in fruit

Cyclanthus bipartitus (fruit)

are sometimes undivided, but at maturity are commonly split partway or sometimes to their bases, into two pointed-lanceolate lobes, 1 foot long or longer. The flowers are borne atop long, erect stalks in a spike called a spadix, with a number of large bracts (spathes) at its base. The fragrant blooms are at the sharpened edges of superimposed disks, the alternate ones of which have male flowers, and those between females, or they may be in two parallel spirals, one of only male flowers and the other of only female. The flowers are small, the males with six stamens. The multiple fruits are fleshy aggregations of the stem of the floral spike and the ovaries, with the seeds embedded.

Garden and Landscape Uses and Cultivation. In the tropics and near tropics this species is useful for beds, and for fronting shrubbery, and as underplanting. It is grown in greenhouses and conservatories as a foliage ornamental.

Fertile, porous soil, not lacking for moisture, but not constantly saturated, suits *Cyclanthus*. Part-shade is to its liking. Indoors a minimum winter night temperature of 60°F is suitable, and high humidity favorable. Shade from strong sun is necessary. A generous program of applying dilute liquid fertilizer to well-rooted specimens favors good growth. Repotting, required at intervals of a few years only with largish plants, is done in late winter or spring; at the same time propagation by division may be effected. An alternate method of multiplication is by seeds sown at a temperature of 70 to 80°F.

CYCLOBOTHRA. See Calochortus.

CYCLOPHORUS. See Pyrrosia.

CYCNOCHES (Cyc-nòches)—Swan Orchid. This genus of beautiful and quite astonishing orchids has much in common with *tasetum* and *Mormodes*. It consists of a dozen species, natives of tropical America, and belongs to the orchid family ORCHIDACEAE. The name from the Greek *kyknos*, a swan, and *auchen*, a neck, refers to the prominent columns of the flowers.

Swan orchids are deciduous epiphytes (tree-perchers). They have tall, spindle-shaped to cylindrical pseudobulbs with few to many ribbed leaves, which when they fall leave their persistent bases attached to the pseudobulbs. The flowers are usually unisexual with both sexes on the same plant. Rarely, bisexual blooms are developed. Males and females, depending upon the species, are similar or strikingly dissimilar. When dissimilar, the males have narrower sepals and petals and a lip with finger-like projections from the sides. Bisexual blooms are intermediate. The usually waxy, often fragrant flowers are in arching or pendulous racemes from near the tops of the pseudobulbs. They have

separate, spreading to reflexed sepals and petals and prominent lips that in the normal flower positions are uppermost. In some kinds the lips are fringed and at their bases narrowed to a claw (shaft). The column is long, slender, and curved. Female flowers, more rarely produced than males and not known for all species, are commonly larger and more fleshy, but are fewer in each raceme than males.

Similar male and female blooms are borne by *C. ventricosum*, which ranges in the wild from Mexico to Panama. The flowers, almost 5 inches in diameter, are fragrant. They have light green, spreading or reflexed sepals and petals that become yellow as they age. The lip, at the top of the bloom, is white and has a long claw. The column projects forward in a graceful curve. The popular orchid commonly misnamed *C. chlorochilon* has flowers sometimes 6 inches across, with creamy-white lips with basal claws cylindrical and much shorter than those of *C. ventricosum* is *C. v. warscewiczii*, of Costa Rica and Panama. The base of the lip is usually marked with

Cycnoches ventricosum warscewiczii

a rich black-green semicircular blotch. The stout pseudobulbs are up to 8 inches tall. Rare *C. v. chlorochilon* has much larger blooms than *C. ventricosum*, with lips without claws at their bases. A native of Surinam, *C. loddigesii* is an interesting sort with arching or pendulous spikes of three to eleven, flowers up to 4 inches across that fancifully resemble a swan and have dark-spotted, brownish-green sepals and brownish-green petals, and a darker or lighter colored, fleshy lip.

Dissimilar male and female flowers are borne by *C. egertonianum*, a species wild from Mexico to Brazil. This has rather slender pseudobulbs 3 to 6 inches tall. The male blooms, in usually crowded, pendulous racemes 1 foot to 1½ feet long, are 1½ inches wide or sometimes wider. They have greenish or bronzy-green, sometimes purple-suffused or purple- or brownish-spotted, often reflexed sepals and petals and a white, green, or purplish lip, narrowed to a claw at its base and fringed at

Cycnoches loddigesii

its apex. In *C. e. aureum* (syn. *C. aureum*) the predominently yellow-green, yellow, or nearly white flowers, shaded with green and sometimes sprinkled with pink dots, are larger, the males up to 3½ inches wide. Variety *C. e. dianae* (syn. *C. dianae*) has rosy-pink blooms suffused with white, their lips distinctly clawed and with teeth much shorter than those of *C. egertonianum*.

Garden Uses and Cultivation. Swan orchids are easy to grow. They respond to conditions and care that suit *Catasetum*. Their blooms last fairly well and are splendid and surprising adornments for orchid collections. For more information see Orchids.

CYCNODES. This is the name of bigeneric hybrid orchid the parents of which are *Cychnoches* and *Mormodes*.

CYDISTA (Cy-dísta). About five species are contained in this genus, which belongs in the bignonia family BIGNONIACEAE. They are natives from the West Indies to Brazil. The name refers to the beauty of the blooms; it derives from the Greek *kydistos*, glorious. Cydistas are evergreen vines with opposite leaves consisting of two leaflets and usually a third represented by a tendril. The flowers, in terminal or axillary panicles or more rarely paired, have five-toothed calyxes and bell- to funnel-shaped corollas, with five spreading, rounded lobes and four non-protruding stamens. The winged seeds are in slender podlike capsules.

Commonest in cultivation, the garlic vine *C. aequinoctialis* (syn. *Bignonia aequinoctialis*), is native to Central and South America and the West Indies. This is not to be confused with the yellow-flowered *Anemopaegma chamberlaynii*, or with lavender- to whitish-flowered *Pseudocalymma alliaceum*, which in gardens sometimes masquerade as *Cydista*. True *C. aequinoctialis* is a high climber with lustrous, wavy, stalked, ovate to oblong leaflets 3 to 4 inches long, and white to pink flowers, 2½ to 3 inches long, with pink or purplish veins. The slender seed pods are up to 1 foot long.

Garden and Landscape Uses and Cultivation. These are similar to those of *Clytostoma* and *Allamanda*.

CYDONIA (Cy-dònia)—Quince. One apple-like hardy tree, the common or orchard quince and one other species constitute *Cydonia* of the rose family ROSACEAE. Shrubs called flowering-quince and Japanese-quince are species of *Chaenomeles*. From *Chaenomeles* the genus *Cydonia* differs in its stamens not being joined at their bases. The name is derived from Cydon, now Canea, in Crete.

The common or orchard quince, (**C. oblonga**), native from Iran to Turkestan, is deciduous, broad-headed, and 15 to 25 feet tall. It has stout, crooked branches, and short-stalked, alternate, broad-ovate to broad-elliptic leaves without lobes or teeth. They are 2 to 4 inches long, and gray-woolly on their undersides. The flowers, white or pale pink and solitary at the ends of short leafy shoots, appear in spring when the foliage is well developed. They are 2 inches in diameter and have five sepals, five petals, about twenty stamens, and five styles. The fruits are broadly-pear-shaped, 2½ to 4 inches in diameter, and fuzzy. When ripe they are yellow and very

Cydonia oblonga (fruits)

fragrant. Those of *C. o. maliformis* are apple-shaped and those of *C. o. lusitanica* especially large. Variety *C. o. pyramidalis* is of upright habit. The foliage of *C. o. marmorata* is variegated with yellow and white. The flowers of the last are pink.

Less well known than the common quince and not quite as hardy, **C. sinensis,** of China, up to 20 feet tall, has lanceolate leaves 3 to 4 inches long and flowers about 1½ inches wide with white-based, pink petals. The fruits are oblong-ovoid, deep yellow, and 4 to 6 inches long.

Garden and Landscape Uses. Quince trees are of sturdy, often picturesque appearance and may be used as landscape specimens in the same manner as apple trees. Their fruits, too astringent to be eaten out of hand, are esteemed for jelly, for adding flavor to apple pies, and for other culinary uses. For cultivation see Quince.

CYLINDROPHYLLUM (Cylindro-phýllum). Six species of this nonhardy South African genus of the carpetweed family AIZO-ACEAE, are recognized. Kin of *Mesembryanthemum*, they are low, branching, succulent, desert perennials. Their name, from the Greek *kylindros*, a cylinder, and *phyllon*, a leaf, refers to the form of the leaves. The flowers are solitary, terminal, short-stalked,

yellow, pink, or red, and have five to eight stigmas. The fruits are capsules.

The species commonest in cultivation, *C. calamiforme,* forms short-stemmed or nearly stemless clumps of crowded, slender, subcylindrical, curving, spreading leaves up to 2 or 3 inches long, grayish-green, and joined at their bases. The pale yellow and pinkish, short-stalked flowers, 2 to 3 inches across, have five-lobed calyxes and many narrow petals. Another species, **C. comptonii,** has erect or rather spreading leaves and forms dense patches up to nearly 1 foot across and 4 to 5 inches tall. Its leaves, up to 3½ inches long, are slightly flattened above and have undersurfaces with rounded keels. The white flowers, 3 inches wide, come in summer.

Garden Uses and Cultivation. Cylindrophyllums give little trouble to gardeners with some skill in growing choice succulents. They prosper under conditions that suit most small plants of the *Mesembryanthemum* relationship, and are readily increased from cuttings, and by seeds. Their resting period, during which they should be kept dry and moderately warm, is winter. For additional information see Succulents.

CYLINDROPUNTIA. See Opuntia.

CYMBALARIA (Cymbal-ària)—Kenilworth-Ivy. Closely related to the toadflaxes (*Linaria*), and by some botanists included in that genus, the fifteen species of *Cymbalaria* belong to the figwort family SCROPHU-LARIACEAE. They are natives of temperate parts of the Old World. The name, from the Greek *kymbalon*, cymbals, probably refers to the round leaves of some kinds.

Cymbalarias are trailers or creepers with alternate or opposite, lobed or round-toothed leaves and small, solitary, asymmetrical, snapdragon-like flowers, with five-lobed calyxes and tubular, two-lipped corollas. The upper corolla lip is erect and two-lobed, the lower has three spreading lobes. The throat of the bloom is closed by a prominent palate. There are four stamens and one slim style. The fruits are capsules.

Kenilworth-ivy (**C. muralis** syn. *Linaria*

Cydonia oblonga

Cydonia oblonga (flowers)

Cylindrophyllum, unidentified species

Cymbalaria muralis

cymbalaria) is a native of Europe and Asia, where it frequently grows on old walls. It is naturalized in North America. A pretty annual or evergreen perennial, and hairless, this kind has slender stems up to about 1 foot long. Its often rather distantly spaced, long-stalked leaves, roundish in outline and with three to seven conspicuous lobes, have blades, with purplish undersides, about 1 inch in diameter. The long-stalked blooms, ⅓ inch or a little more in length, and about as much across their faces, vary to some extent in color, but typically are lilac-blue with the palates of their throats yellow and white. They have a lower lip markedly bigger than the upper and a curved spur.

Similar to the last, but somewhat larger, *C. hepaticaefolia* (syn. *Linaria hepaticaefolia*) differs in having mostly opposite, shallowly- or obscurely-three- to five-lobed leaves with pale veins. Like the stems, the foliage is without hairs. The yellow-throated, lilac-blue blooms are a little bigger than those of Kenilworth-ivy. This kind is native to Corsica.

From the Kenilworth-ivy, European *C. globosa* (syn. *Linaria globosa*) differs in minor details only. Its variety *C. g. nana* is very compact. It forms neat clumps of attractive foliage and tiny lavender-pink flowers and is well suited for embellishing rock gardens.

Cymbalaria globosa nana

Cymbalaria pallida

Easily distinguishable from the kinds discussed above by the undersides and stalks of its leaves being clothed with short hairs, *C. pallida* (syn. *Linaria pallida*) has

stems often at first erect to a height of 4 inches, then trailing. The opposite leaves are lobeless or shallowly-three- to five-lobed. The blue-violet flowers with yellow palates are ½ inch long. Their stalks are about as long as the leaves. This species, a native of Italy, is not infrequently cultivated as *C. pilosa*. True *C. pilosa* is more densely-hairy, and has five- to eleven-lobed leaves and smaller, shorter-spurred blooms.

About 1 inch tall, *C. aequitriloba* (syn. *Linaria aequitriloba*) is a little gem. Native to southern Europe, it has three-lobed leaves about ½ inch across, and carried just above its carpet of foliage on stalks not longer than the leaves, mauve-purple blooms. The stems and evergreen leaves are hairy.

Garden and Landscape Uses and Cultivation. Kenilworth-ivy is frequently grown in window gardens and greenhouses in pots and hanging baskets and often springs up as a volunteer in the ground under greenhouse benches. A pretty trailer, it has a rather frail rooting system, so that in regions where it is hardy it can be used effectively as a groundcover over small bulbs in rock gardens and similar places. The other species discussed here are well adapted for rock gardens, for crevices between paving, and other places where low, neat carpeters can serve. The Kenilworth-ivy and *C. pilosa* are hardy in the north, *C. aequitriloba* and *C. hepaticaefolia* are much more tender and do not always survive in the region of New York City. All prosper in gritty, porous, not too rich, moderately moist soil, and except for *C. pilosa*, which does best in sun, they prefer some shade during the heat of the day. These plants are very easily multiplied by division, cuttings, and seed.

CYMBIDIUM (Cym-bídium). With the exception of *Cattleya* and perhaps *Paphiopedilum*, probably no genus of the orchid family ORCHIDACEAE is better known than *Cymbidium*. The pleasing, muted colors of the blooms of this grand group and their amazing lasting qualities endear them to all. Because of the frequency with which they are offered in florists' stores they have become about as familiar to city dwellers as carnations and gardenias. The natural species of *Cymbidium* number about forty. From comparatively few of these, hybridists have evolved literally thousands of horticultural varieties, all beautiful, the best of outstanding loveliness and grace. The name is derived from the Greek *kymbe*, a boat, in allusion to a hollow in the lip of the flower.

The genus occurs natively from Japan, Taiwan, and the Philippine Islands to China, the Himalayas, southeast Asia, Indonesia, and Australia. Among its kinds are ground orchids, tree-perching orchids (epiphytes), and occasionally rock-inhabiting species. As treated here they include

Cymbidiums, in a cool greenhouse

those sometimes segregated as *Cyperorchis*, chiefly on the bases of the flowers having narrower sepals and petals and a nearly straight lip the side lobes of which embrace the column.

Cymbidiums mostly have pseudobulbs, not always large and obvious, sometimes almost completely hidden by the persistent bases of old leaves. The foliage is evergreen and leathery to thin. Generally the leaves are long in proportion to their widths. The flowers are characteristically in spikes or racemes, often long and gracefully arching, sometimes pendulous, from the axils of the lower leaves. They are of very few to as many as eighty blooms, their colors chiefly ranging from white through tones of yellows, greenish-yellows, bronzes, browns, pinks, rose-reds, and various combinations. Each flower has three sepals and two petals of nearly equal style, size, and color and a usually three-lobed lip, botanically representing a third petal. The lip is joined to the bottom of the erect column. There are two pollen masses (pollinia).

The history of these orchids in cultivation dates to 1778 when *C. ensifolium* was brought from China to England, to be followed a decade later by *C. pendulum* from India. Another half century passed before other species began to arrive, a total of twenty between 1837 and 1905. All came first to England and from there were distributed to continental Europe and elsewhere. The first artificial hybrid of which there is record was raised by the famous English nursery of James Veitch. Its parents were *C. eburneum* and *C. lowianum*. This bloomed for the first time in 1889. Nineteen years later in the catalog of one English nursery eighty-eight hybrids were listed. By modern standard these early hybrids had dull-colored blooms. The introduction into cultivation of *C. insigne* in 1904 and of *C. erythrostylum* the following year gave a tremendous fillip to cymbidium breeding, since these fine species pass their good qualities, including desirable growth habits, many-flowered spikes, and white or pink, long-lasting blooms, to their offspring. By 1925 registered hybrids numbered one hundred and seventy.

An award made in England in 1922 to a spectacularly large-flowered white mutant (sport) of *C. alexanderi* named *C. a.* 'Westonbirt' called attention to a new cymbidium of outstanding merit that was to become a parent of a whole race of splendid hybrids. In 1949 it was established that it, as well as a famous mutant of *C. pauwelsii* named *C. p.* 'Compte d'Hemptinne', were tetraploids, that is their cells contained twice the normal number of chromosomes. Itself a hybrid, *C. pauwelsii* has resulted from crossing *C. lowianum* and *C. insigne*.

World War II seriously affected Great Britain's preeminence in raising new hybrid cymbidiums. Much fine breeding stock was sent to the United States. After the war American interest, especially in California, developed rapidly and was greatly stimulated by the formation in 1946 of the Cymbidium Society of America, an organization that has since provided vigorous leadership. In other parts of the world, too, interest in these splendid orchids increased rapidly. By 1972 well over two thousand named hybrids had been registered. Among them were approximately three hundred varieties of the compact plants known as miniatures (taller kinds came to be called standard varieties). The miniatures soon became favorites. Among the species that have contributed to their parentage *C. pumilum* has been frequently employed, with the *C. ensifolium* complex and *C. devonianum* runners-up. Several other species are involved to a lesser extent.

Hybrid cymbidiums: (a) 'Moira', chartreuse-yellow with a cream-yellow lip, spotted with brown

(b) 'Madonna', glistening white with pink lip spotted with maroon and with two yellow ridges in the throat

Cymbidium erythrostylum

Miniature *Cymbidium* 'Flirtation'

Hybrid cymbidiums, in a marvelous array of habits of growth, times of blooming, flower colors and sizes and other variables, are grown today to the near exclusion of the original species. An effort is being made to regenerate interest in the natural species. The hybrids are adequately described, often with illustrations, not infre-

quently in color, in the catalogs of specialists. We shall now consider some of the more important species.

Ivory-white-flowered *C. eburneum* is a handsome Himalayan terrestrial species with clustered, obscure, pseudobulbs mostly concealed by persistent leaf bases. In tufts of up to about fifteen, its two-ranked, pointed-linear leaves, slightly cleft at their apexes, are 1 foot to 2 feet long by ½ inch wide. The flowering stalks, shorter than the leaves, are of one to three well-shaped, strongly fragrant, 3½- to 5-inch-wide, waxy blooms, sometimes slightly pink-tinged. They have petals slightly narrower than the sepals. The broad-ovate-oblong lip, ivory-white with a yellow base with scattered spots of purple around it and a velvety-pubescent crest, has side lobes that curve inward to the column, a crisped-margined center lobe. Much like *C. eburneum* in its style of growth, *C. erythrostylum* has freely-produced, semierect spikes of up to a dozen flowers or more, basically white or pale pink with the lower sepals veined with purple and, in striking contrast, a yellowish-shaded, white lip with strong red lines. The blooms do not open fully and are rather small. This species is native to Vietnam.

A beautiful terrestrial native of Indochina, *C. insigne* has clusters of more or less spherical pseudobulbs 2 to 3 inches long. Its glaucous-green leaves, about 3 feet long, are usually under ½ inch wide. Generally erect and 3 to 4 feet tall, the racemes consist of up to twelve or sometimes more fleshy blooms 3 to 3½ inches wide. Their sepals and petals are pale rose-pink. The lip is suffused with rose-red dotted with purple-red and has a pair of broad yellow lines. Its side lobes stand erectly at the sides of the column. Arching spikes of up to twenty blooms are borne by Burmese *C. tracyanum*. With yellowish-green sepals and petals with purple veinings, and a red-dotted, cream-colored lip with some hairs, the fruity-fragrant flowers are sometimes 5½ inches wide. The blooms of this sort are less long-lasting than those of many other kinds.

Native to Japan, the Ryukyu Islands, and Taiwan, terrestrial *C. lancifolium* has slender pseudobulbs about 5 inches tall with three to five leaves that have pointed-elliptic blades up to about 8 inches long that are minutely toothed toward their apexes. The flowering stalks, up to about 1 foot tall, carry two to six fragrant blooms with pale green sepals and petals, each with a rosy-purple center line and spots of the same color, and a white lip that is reddish purple on its reverse side. About 2 inches wide, the flowers are rather widely

Cymbidium lancifolium

Cymbidium sinense

spaced. Chinese *C. sinense,* a terrestrial species reported to occur also as a native in Japan, the Ryukyu Islands, and Taiwan, has lanceolate-linear leaves and much taller, erect flowering stalks that bear several flowers 2 to 3 inches across with brown and purple sepals and petals and a yellowish-green lip spotted with purple.

Almond-fragrant blooms that, in the fashion of the six species of *Cymbidium,* sometimes classified as *Cyperorchis,* do not open widely, but are somewhat bell-shaped, are typical of Himalayan *C. mastersii* (syn. *Cyperorchis mastersii*). It does not have pseudobulbs. Its tufted, arching, pointed,

Cymbidium mastersii

two-ranked, linear to narrowly-sword-shaped leaves are up to 2½ feet long by ¾ inch wide. The ivory-white flowers, with red-purple spots and an orange-yellow patch on the three-lobed lip, are about 2 inches long. They are in erect racemes up to 1 foot tall, of four to ten. Australian *C. canaliculatum* is a ground orchid with

clusters of poorly developed pseudobulbs and rather thick, dull green, channeled, linear-lanceolate to linear leaves from 9 inches to 1½ feet long by ¾ inch to 1¼ inches wide. Its stiff, arching, freely-produced racemes of crowded blooms are up to 1½ feet long. The flowers, ½ inch to 1½ inches wide and very variable in color, predominantly have green sepals and petals heavily blotched and spotted with deep brown-maroon margined with lemon-yellow, and a crimson-spotted, pink and cream, usually ovate-lanceolate lip with small side lobes.

Pendulous, many-flowered racemes are characteristic of *C. finlaysonianum,* of the Philippine Islands, Indochina, Malaysia, and Indonesia. This robust species has short, stemlike pseudobulbs, and leathery, sword-shaped leaves of substantial texture 2 to 3 feet long by up to 2 inches wide. The racemes are of many rather loosely-arranged, 1½-inch-wide blooms with olive-green-tinted, yellowish-green sepals and petals. The lip has a purple-crimson-tipped, white center lobe. Side lobes are rose-pink streaked with purple.

More tolerant of high temperatures than most cymbidiums, *C. aloifolium,* of Sri Lanka and India, has prospered perched on the branch of a large tree at the Fairchild Botanical Garden, Miami, Florida. This species, practically without pseudobulbs, forms large clumps of arching to pendulous, fleshy, linear to narrowly-strap-shaped leaves 1 foot to 1½ feet long. Its sharply drooping, loose racemes, 2 to 2½ feet long, are of blooms about 1¾ inches across with creamy to pale brownish-yellow sepals and petals and a predominantly reddish-brown, three-lobed lip.

Garden and Landscape Uses. In favored coastal regions of California cymbidiums can be grown permanently outdoors in beds and borders and at least one, *C. aloifolium,* flourishes in southern Florida accommodated as an epiphyte. For locations where drainage is poor it is advisable to plant terrestrial (ground-inhabiting) sorts in beds raised 8 inches to 1 foot above the surface. Except in regions especially suitable for their outdoor cultivation, these orchids need the protection of a greenhouse, or in mild climates a lath house. Their chief uses are as decorative pot plants and as cut flowers in arrangements and corsages. The long-lasting qualities of their blooms, on the plants or cut, especially recommend them for these purposes. The refinement of form and the delicate and unusual colorings of their flowers, as compared for example to those of their more flamboyant cousins, the cattleyas, appeal to the discriminating.

Cultivation. Cymbidiums require cooler growing conditions than many commonly cultivated orchids. True, if air circulation is good and humidity not excessively low, they will endure 100°F on summer days,

but they are more comfortable at lower levels. It is particularly important that there be a considerable drop at night. Unless in summer night temperatures are twenty to twenty-five degrees below daytime ones, flower spikes are not initiated. In winter cymbidiums can withstand three or four degrees of frost, and about 28°F seems to be the critical damaging temperature for blooming specimens, but it is better that they not be so exposed. In greenhouses the most favorable temperatures from fall to spring are 45 to 50°F or for some few kinds 55°F at night, with a daytime rise of five to ten or on sunny days in spring up to fifteen degrees. In summer, from about the time it is safe to plant tomatoes until threat of frost, they do better if removed to a lath house.

Shade from strong sun, what is referred to as filtered sunlight or half-shade, is needed from spring to fall. Too much shade makes for weak, disease-prone plants and inhibits blooming. Too little results in scorched foliage. Outdoors, shade cast during the hottest part of the day by trees nearby but not overhead, or by a building, if it does not keep away the sun all day long, is satisfactory. As much overhead light as possible with exposure to the weaker early morning and evening sun, but not the strong middle-of-the-day sun is ideal. If you must choose between too heavy shade and too sunny a location, settle for the latter. Do not crowd cymbidiums unduly, free access of air between the plants and plenty of side light make for good health. In humid weather air circulation maintained by a fan is likely to be beneficial. Water sufficiently to keep the roots always moist, but drier in winter than at other seasons. When you water give plenty. Saturate the soil thoroughly to flush out accumulated salts. Then give no more until the first signs of approaching dryness are evident. Spraying the foliage with water on bright days from spring to fall is of great benefit. Make sure you do this early so that the leaves dry before nightfall. It is harmful to have water standing in leaf bases for long periods.

Repotting is needed about every two years. Cymbidiums resent root disturbance, and so the job must be done with care. Do this in spring, before the new growths are 3 inches long. Then the danger of damaging new shoots is avoided, and the roots have opportunity to become reestablished well before the plants are called upon to develop flower spikes. Do not use too big containers. Cymbidiums do best when their roots are cozily crowded. Drain the pots very adequately. Take care to keep the bottoms of the pseudobulbs above the soil surface and pack the potting mix quite firmly. After potting keep the plants shaded a little more than usual, and mist their foliage two or three times a day with water.

The soil for cymbidiums may vary as to ingredients so long as it is perfectly drained. Any suspicion of stagnation spells trouble, and eventually disaster. If such conditions prevail the roots soon rot and die. The best soil is a mix containing very generous proportions of organic material, such as leaf mold, coarse peat moss, redwood bark, or fir bark, in addition to a liberal spiking of grit or coarse sand. Be generous with organic fertilizers. A somewhat heavy-handed addition to the soil of rotted or dried cow manure can work wonders. An acidity of pH 4.5 to 5.5 is best.

Propagation of cymbidiums, except commercially where meristems are largely employed or when new varieties are being raised from seed, is by division. This is done at potting time. Let plants that are to be divided become dryish. This facilitates removing the soil from around their roots, and the latter become less brittle and less subject to damage. When dividing, tease out as carefully as possible all decayed and old soil. Then with a sharp knife cut the plants into pieces each of at least three sound pseudobulbs. Do not leave attached old pseudobulbs that are without foliage. Be extremely careful not to bruise incipient buds at the bases of the pseudobulbs. Pot the divisions individually in pots just large enough to hold their roots comfortably. For some time afterward water with special care so as not to get the rooting medium too wet. If you want more plants set the old leafless pseudobulbs or back bulbs as they are called aside in a fairly cool place for a month or six weeks. Then, when their new buds begin to plump up plant them individually in small pots. For more information see Orchids.

CYMBOPOGON (Cymbo-pògon)—Lemon Grass, Citronella Grass, Camel Grass. This group of tropical grasses, some of which are cultivated as sources of essential oils and occasionally as ornamentals, consists of about forty-five species, all natives of the Old World. They belong in the grass family GRAMINEAE. Their name comes from the Greek *kymbe*, a boat, and *pogon*, a beard.

Cymbopogons are tufted aromatic annuals and perennials with linear to lanceolate leaves. Usually their panicles of bloom are much branched. The spikelets are in twos, with one of each pair a stalked male and the other fertile and stalkless.

Lemon grass (*C. citratus*) is unknown in the wild. Most likely it originated in tropical Asia. The commercial source of lemon grass oil, used in perfumery, medicine, and for flavoring is a densely tufted perennial, up to 6 feet tall, with smooth, slender to fairly thick stems, and leaves that may be 3 feet long by ¾ inch wide, with conspicuous mid-veins. The loose, nodding flower panicles are up to 2 feet long. Their spikelets are without awns (bristles).

Cymbopogon citratus

Citronella grass (*C. nardus*), the commercial source of oil of citronella, which is employed as a substitute for attar of roses and as an insect repellent, is a densely-tufted perennial up to 7 or 8 feet tall. Its heavy stems are smooth, and its leaves are up to 3 feet long by ¾ inch wide. The flower panicles, sometimes 2 feet long, are more crowded than those of lemon grass. The spiklets are without awns. A technical difference between citronella grass and lemon grass is that the sessile spikelets of the former are oblong-lanceolate and their lower glumes (chaffy bracts) have backs flat except at their bases, whereas the sessile spikelets of the lemon grass are linear to linear-lanceolate and their lower glumes are shallowly concave below their middles. Citronella grass is unknown in the wild, but probably originated in tropical Asia.

Source of the less well-known camel grass oil, the camel grass (*C. schoenanthus*), is indigenous from North Africa to southwest Asia. A densely-tufted perennial up to 2 feet in height, it differs from the species discussed above in having flower spikelets with awns (bristles) and slender leaves that are semiround rather than flat. It has thin, wiry stems and narrow flower panicles up to 1 foot long.

Garden and Landscape Uses and Cultivation. These oil-producing grasses can be grown outdoors in tropical and subtropical areas in which there is little danger of frost. They succeed in any ordinary fertile soil. They are also sometimes grown in collections of greenhouse plants for the interest of their fragrant foliage and as examples of plants of commercial importance.

In greenhouses they succeed in a winter night temperature of about 55°F, with higher temperatures at other seasons. Full sun is desirable. The soil should be always fairly moist. Dilute liquid fertilizer applied periodically to specimens that have filled their containers with roots stimulates growth. Propagation is by division.

CYMOPHYLLUS (Cymo-phýllus). The only species of this genus is by some botanists included in *Carex*, but by others is believed to be more closely kin of the uncultivated tropical South American *Bisboeckelera* and so is kept distinct from *Carex*. It certainly looks very different from familiar sedges (*Carex*). To the eye untrained in the niceties of botany it vaguely resembles a bromeliad, although it is widely removed from the pineapple family BROMELIACEAE, to which bromeliads belong. In fact *Cymophyllus* is a member of the sedge family CYPERACEAE. Its name comes from the greek *kyma*, a wave, and *phyllon*, a leaf, and was applied because of its undulate-edged leaves.

Native to rich mountain woods from Virginia to West Virginia, South Carolina, and Kentucky, and hardy in sheltered places at least as far north as New York City, *C. fraseri* is a good-looking ornamental well suited for rock gardens, woodland gardens, and gardens devoted to native plants.

Cymophyllus fraseri

It has evergreen, strap-shaped, pale green, leathery, recurving, flat leaves, 1 inch to 2 inches wide and up to 2 feet, but usually less, long. They have fine longitudinal veins, but are without definite midribs. The flower stalks rise from the sheaths formed by the bases of the developing new leaves, one from each. Slender and up to 1¼ feet long, each stalk ends in a solitary, creamy-white spike about 1 inch in length, composed of female flowers in its lower part and males above.

Cultivation. This quite rare and handsome American is entirely amenable to cultivation if given conditions somewhat resembling those it enjoys in the wild. It responds to a soil fat with organic matter, preferably leaf mold or rich compost, that never dries unduly. Constant saturation is not, however, to its liking. Some overhead shade with good side light is favorable. Specimens thrived for twenty years or more in the Thompson Memorial Rock Garden at The New York Botanical Garden. There, they needed the protection afforded by a few pine or other evergreen branches placed over them each fall and left until spring. When the practice of affording that slight aid to temper the dehydrating power of the winter sun was abandoned, the foliage scorched and the plants gradually weakened and eventually died. Propagation of *Cymophyllus* can be effected by seeds sown in sandy peaty soil kept always medium-moist, and by careful division in early spring just as the first signs of new growth are apparent.

CYMOPTERUS (Cym-ópterus). This genus of western North America belongs to the carrot family UMBELLIFERAE. There are about thirty-two species. The name comes from the Greek *kyma*, a wave, and *pteron*, a wing and alludes to the fruits of some kinds.

These are low herbaceous perennials with leaves pinnate or twice-pinnate, or divided in handlike fashion, their margins toothed or toothless. The tiny white, yellow, or purple flowers are in umbels assembled into larger umbels. The fruits are flattened and seedlike.

Stemless or nearly so, *Cymopterus macrorhiza* (syn. *Phellopterus macrorhizus*) has a thick root from which arise at the ground surface a cluster of ovate-oblong, pinnate or twice-pinnate, glaucous, pallid green, somewhat fleshy, leaves up to 3 inches long by nearly 1½ inches wide. The leaflets may be pinnately-lobed or not. On naked stalks longer than the leaves and up to 7 inches tall, the umbels of pinkish flowers are borne in spring. This species is indigenous to Texas and Oklahoma. It enjoys limy soil. Its precise hardiness is not known, but it is capable of withstanding considerable freezing.

Garden Uses and Cultivation. The most appropriate locations for this plant are in rock gardens. It needs gritty, well-drained soil, and a location in sun or part-day shade. Propagation is best effected by seed.

CYNANCHUM (Cyn-ánchum) — Black Swallow-Wort, Mosquito Trap, or Cruel Plant. Presumably alluding, although somewhat far-fetchingly, to the poisonous qualities of some kinds, the name of this genus of more than 100 species is derived from the Greek *kyon*, a dog, and *ancho*, to strangle. The group, belonging to the milkweed family ASCLEPIADACEAE, consists of herbaceous or somewhat woody herbaceous perennials, chiefly twining vines. It is represented in the native floras of Europe, Asia, and Africa and naturalized in North America. Few kinds are cultivated.

Cynanchums contain milky juice, that of at least some kinds, are strongly purgative. Their leaves are opposite and undivided. Their small white, greenish, or purplish, wheel- or shallowly-bell-shaped flowers, are in umbel-like clusters. They are succeeded by seed pods in pairs, much like those of milkweeds. The blooms have five-parted calyxes, deeply-five-lobed corollas with a corona or crown with a scale at the base of each of its five parts, five stamens with their anthers joined around the stigma, and two styles.

Black swallow-wort (*Cynanchum nigrum* syn. *Vincetoxicum nigrum*) is a twining vine, a native of Europe, northern Africa, and western Asia, and naturalized in the northeastern United States. It has stems up to about 4 feet long and nearly stalkless, toothless, narrowly- to broadly-ovate-lanceolate leaves 1½ to 3½ inches long, and hairy along the veins. The heavily scented, dark reddish-purple to black-pur-

Cynanchum nigrum

ple flowers, hairy on their insides, are in short-stalked clusters from the leaf axils. The seed pods are approximately 2 to 3 inches long. Also naturalized in eastern North America from Europe, *C. vincetoxicum* (syns. *Vincetoxicum hirundinaria, V. officinale*) differs from the last in its flowers being greenish-white and not hairy on their insides. Similar *C. fuscatum* (syn. *Vincetoxicum fuscatum*), of the Mediterranean region, is much smaller than *C. nigrum* and has stems and foliage clothed with short hairs. Also native to the Mediterranean region, *C. acutum* is a vine 2 to 12 feet tall with lanceolate leaves about 1½ inches long, with deeply-heart-shaped bases. Its ¼-inch-wide, white to rose-pink, fragrant blooms are in terminal and axillary clusters.

Mosquito trap or cruel plant (*C. ascyrifolium* syns. *C. acuminatifolium, Vincetoxi-*

Cynanchum vincetoxicum

Cynanchum vincetoxicum (flowers)

cum acuminatum) earned its vernacular designations because small insects are often entrapped in the blooms. Native to Japan, Korea, and Manchuria, this sort is erect or nearly so, 1 foot to 2 feet tall, and at most has the ends of its somewhat angular stems twined. Strongly-pinnately-veined and toothless, the pointed-ovate, elliptic, or oblong leaves, 3 to 6 inches long, are hairy on the veins on their undersides. Terminal, and from the upper leaf axils, the flower clusters, shorter than the leaves, are composed of ½-inch-wide, white blooms.

Garden and Landscape Uses and Cultivation. Sometimes planted in flower beds and more informal places for interest and ornament, cynanchums are easy to grow in sunny locations in any ordinary reasonably, well-drained soil. In fact so adaptable is deep-rooted black swallow-wort, and perhaps other kinds, that it can become in places to its liking a serious, invasive pest, difficult to eradicate or control. Propagation is by division, in spring or early fall, and by seed. The species described above are hardy.

CYNARA (Cýn-ara)—Artichoke, Cardoon. The familiar artichoke and less well-known cardoon of vegetable gardens belong in this genus of the daisy family COMPOSITAE. They are the only cultivated ones of fourteen species. Although grown primarily as foods, they are not without decorative appeal. Out of flower some kinds somewhat resemble acanthuses, in flower, giant thistles. Neither of the cultivated kinds is reliably hardy in the north. They hail from warm dryish climates and are discouraged by hard freezing and wet, cold winters.

The genus *Cynara* ranges from the Canary Islands and the Mediterranean region to Kurdistan. It consists of perennial herbaceous plants with large, pinnately- or bipinnately-cleft leaves, stout, erect, branching stems, and large terminal flower heads. Its name is from the Greek *kyon*, a dog, and alludes to a comparison between the spines of the involucre and the teeth of a dog. As is characteristic of the daisy family, each head consists of many individual small flowers called florets backed by an involucre (collar) of leafy bracts. All the florets of *Cynara* are of the disk type (like those of the eyes of daisies). There are no ray florets (those that in daisies look like petals). The involucre, large and conspicuous, is the part that in artichokes is eaten. Artichokes are actually unopen flower heads. The blue, purple, or sometimes white florets have five-parted corollas with slender tubes. The fruits, commonly called seeds, are technically achenes.

The artichoke (*C. scolymus*) is not known in the wild and quite likely was developed

Cynara scolymus

Cynara scolymus (flowers)

from the cardoon by ancient agriculturists. Certainly it has been cultivated since classical times. It is 3 to 5 feet in height and differs from the cardoon in having larger flower heads that terminate each of the main branches and are up to 6 inches in diameter and less divided, almost spineless foliage. Also, the scales of the involucre are broader, more fleshy, and do not end in spines. The flower heads are rich blue.

The cardoon (*C. cardunculus*) attains a height of 6 feet or more and has arching, grayish-green, prominently spiny, pinnately-divided leaves up to 2 feet long, on

Cynara cardunculus

Cynara cardunculus (flowers)

their undersides whitish-pubescent. The violet-purple flower heads have spine-tipped involucral bracts and are approximately 2 to 3 inches in diameter. The edible parts of the cardoon are the young leaves and stalks, which are blanched like those of celery. This species, sparingly naturalized in California, is a widespread introduced weed of the South American pampas. It is a native of southern Europe.

Garden and Landscape Uses and Cultivation. These stately plants are bold features in flower beds and borders and supply strong accents. They associate well with architecture and can be placed with good effect at the bases of walls, near steps, and in similar locations. They need full sun and fertile, well-drained ground. They are propagated by division and by seed. Their flower heads, cut with their stalks and hung upside down in a cool, shaded, well-ventilated place until dry, are useful in dried arrangements. They are also handsome when used as fresh cut flowers. For the cultivation of these plants as vegetables see Artichoke, and Cardoon.

CYNODON (Cý-nodon)—Bermuda Grass. Several species of the about eight of this genus of the grass family GRAMINEAE are useful lawn grasses. One, Bermuda grass, is widely used in North America, especially in the warmer parts. The name, which alludes to the hard, sharp scales on the rhizomes, is derived from the Greek *kyon,* a dog, and *odous,* a tooth. These grasses are mostly natives of Africa, but *C. dactylon* occurs spontaneously in many parts of the world.

Cynodons are perennials that spread by slender runners or underground rhizomes that root from the nodes (joints). They generally have short, erect stems and linear to almost threadlike leaves. Their flower spikes, in groups of two to six or more, are generally in fingered umbels at the tops of the stems. They are composed of small, one-flowered, flattened, stalkless, overlapping spikelets without awns (bristles).

Bermuda grass, (*C. dactylon*), of which there are cultivated many selected varieties and hybrids, forms broad, low mats of leafy stems 3 inches to 1 foot tall that come from underground rhizomes. The leaves, green or grayish-green, are 1 inch to 6 inches long and not over ⅙ inch wide. In clusters of three to six the spreading, slender spikes, up to 2 inches long, have on one side two rows of overlapping spikelets up to ¹⁄₁₀ inch long. For further information see Lawns, Their Making and Renovation.

CYNOGLOSSUM (Cyno-glóssum) — Hound's Tongue, Chinese-Forget-Me-Not, Wild-Comfrey. A widely distributed native of temperate regions, the genus *Cynoglossum* comprises forty or fifty species. Its name derives from the Greek *kyon,* a dog, and *glossa,* a tongue, and alludes to the form and texture of the leaves of some kinds. It belongs to the borage family BORAGINACEAE.

Cynoglossums are herbaceous perennials, biennials, or occasionally annuals, with usually rough-hairy stems and foliage. Their leaves are alternate, undivided, and without lobes or teeth. The blue, purple, or white blooms are in one-sided, somewhat coiled racemes, usually without bracts. They have a five-parted persistent calyx, which commonly increases in size as the fruits develop, and a funnel- to salver-shaped, short-tubed corolla, with five lobes (petals). The throat of the corolla is closed by five conspicuous scales or crests. There are five stamens, which, unlike those of nearly related *Solenanthus,* do not protrude. Their anthers, unlike those of *Borago,* do not form a cone. The ovary is deeply-four-lobed, the style solitary and slender. The four nutlets that form the bur-like fruit are thickly covered with hooked prickles, a characteristic not found in the true forget-me-not (*Myosotis*).

Chinese-forget-me-not (*C. amabile*) is a popular biennial, often grown as an an-

Cynoglossum amabile

Cynoglossum wallichii

Cynoglossum amabile (flowers)

nual, a native of eastern Asia naturalized in parts of North America. It is 1 foot to 2 feet tall and hairy. Usually its stems are without branches. The lower leaves are lanceolate to oblong-lanceolate, have winged stalks, and are 2 to 8 inches long. The upper ones are stalkless and smaller. About ¼ inch long, and with blue, pink, or white corollas somewhat more than ½ inch across and densely-hairy calyxes, the flowers have much the aspect of oversized forget-me-nots.

A European biennial, **C. cheirifolium,** of the Mediterranean region, is occasionally grown. Densely-white-woolly, it is 1 foot to 1½ feet tall and has stalked, lanceolate to spatula-shaped lower leaves 3 to 4 inches long, and smaller upper ones. The flowers, each arising from the axil of a leafy bract, are under ½ inch wide. At first pink to red, they change to purple and finally to blue as they age.

Other biennials or annuals sometimes cultivated are **C. zeylanicum** (syn. *C. furcatum*), of India, and **C. wallichii,** an Asian species. The former attains a height of about 3 feet, the other is one-half to two-

thirds as high. Both have blue flowers, and oblong to lanceolate leaves up to about 10 inches long, those of *C. zeylanicum* silky-hairy.

Two American perennials are wild-comfrey (*C. virginianum*), a native of upland woods from New Jersey to Florida and Louisiana, and *C. grande,* of California and Washington, where it grows on dryish, shaded slopes. Attaining a height of about 2½ feet, **C. virginianum** has branchless stems. Its leaves are chiefly basal and elliptic-oblong. With winged stalks, they are 4 to 8 inches long. Upward on the stems they become progressively smaller and shorter-stalked to stalkless and stem-clasping. The racemes, mostly in threes, are up to 8 inches long and terminate longish, erect stalks. The flowers are blue and ⅓ to ½ inch wide. They come in spring and early summer. Distinct among cynoglos-

sums discussed here because its stems are without hairs and all its leaves are long-stalked, handsome **C. grande** is 1 foot to 3 feet tall and has mostly basal foliage. Its leaves have ovate blades 4 to 8 inches long on stalks often equaling the blades in length. They are slightly-hairy above, more densely-hairy on their undersides. The panicles of blooms are narrow or loose, and widely-branched. The flowers have hairy calyxes and bright blue corollas ⅓ to ½ inch long and about ½ inch wide. They have thick crests in their throats. This kind blooms in spring.

A beautiful Himalayan perennial, **C. nervosum** has branched, hairy stems 2 to 3 feet tall. Its lower leaves are short-stalked, have oblanceolate blades, and are 6 to 10 inches long. The upper leaves are smaller, nearly stalkless, and oblong. All are sparsely-hairy. About ½ inch in diameter, the rich gentian-blue flowers have hairy calyxes. They provide an attractive early summer display.

Garden and Landscape Uses. With few

Cynoglossum nervosum (flowers)

Cynoglossum nervosum

exceptions cynoglossums are a little too weedy in appearance to be really first class garden plants. Notable exceptions are the Chinese-forget-me-not, which provides flowers useful for cutting, and which is attractive in pots for spring display in greenhouses, and *C. nervosum*, a pleasing perennial with rich blue blooms that show to especially fine advantage near those of the plant known in gardens as *Potentilla recta warrenii*. Also meritorious as a flower garden plant is perennial *C. grande*.

Cultivation. Cynoglossums are easy to grow. Where winters are severe the biennials may not be hardy, and it is advisable to treat them as annuals by sowing the seeds in early spring in sunny locations where the plants are to remain. The seedlings are thinned out to stand 4 to 6 inches apart. In regions of fairly cool summers successional sowings ensure a prolonged display. The perennials prefer deep, fertile soil of average moisture content and some shade during the hottest part of the day. They stand full sun if the soil is satisfactorily moist, but not wet. The biennials do well in full sun, and in somewhat dryish locations if the soil is moderately fertile. The perennials are easily propagated by root cuttings and by division as well as by seed.

In greenhouses seeds of Chinese-forget-me-not are sown in September and January. The seedlings are transplanted individually to small pots, from which they are transferred to benches or ground beds or to 4- or 5-inch pots. Throughout they are grown in full sun, where the night temperature is 50°F and by day five to fifteen degrees higher. On all favorable occasions the greenhouse must be ventilated freely. Watering is done to keep the soil moderately moist, but not overly wet. When the plants have filled the available soil with roots regular applications of dilute liquid fertilizer are helpful.

CYNORKIS (Cyn-òrkis). Belonging to the orchid family ORCHIDACEAE, the genus *Cynorkis* consists of about 125 species and is native to tropical and southern Africa, Malagasy (Madagascar), and some other islands of the Indian Ocean. The name is derived from the Greek *Cynosorchis*, used by Dioscorides for an orchid of the Mediterranean region which has two underground tubers thought to resemble a dog's testes in shape.

The sorts of *Cynorkis* include tree-perching (epiphytic), cliff-inhabiting, and ground (terrestrial) species, with some sorts rather indifferent as to habitat. They are deciduous or partially evergreen and have fleshy or tuberous roots. The flowers have concave, subequal sepals, similar or smaller petals, a conspicuous lip, and a short or long spur. Continuous with the column the spreading, three- to five-cleft lip is as long or longer than the sepals and petals.

Cynorchis uncinata

An endemic epiphyte, *C. uncinata* (syn. *Cynosorchis uncinata*) of Madagascar in the wild commonly grows on *Pandanus*. It has a solitary, recurved, pointed-oblong-ovate leaf up to 9 inches long and a flowering stalk nearly as long, with one or more leafy bracts, and terminating in a bracted raceme of up to fifteen showy flowers. The blooms have a slender spur longer than the lip and are rosy-lavender with a white spot on the deeply-four-lobed lip. The lip is up to 1½ inches long by 1 inch wide.

Garden Uses and Cultivation. These are choice orchids for collectors. Success may be had by growing them in a well-drained mix of leaf mold, loamy soil, and pieces of lava cinders or finely broken crocks. In greenhouses most sorts need a winter night temperature of 60°F and a daytime increase of five to fifteen degrees. A humid atmosphere and some shade from strong sun are requisite. When the flowers fade and the foliage of deciduous sorts begins to die, reduce the frequency of watering and finally allow the soil to dry almost completely but not to the extent that the tubers shrivel. At that time store in shade in a humid atmosphere. When new shoots appear, after three or four months, resume watering. Water evergreen sorts through the year. Fertilize lightly during the period of active growth.

CYPELLA (Cyp-élla). South American bulb plants totaling fifteen species belong here. They have iris- or tigridia-like blooms and belong to the iris family IRIDACEAE. From *Iris* and related *Neomarica* they differ in their leaves being longitudinally corrugated and in details of their stamens and stigmas. The name is derived from the Greek *kypellon*, a cup or goblet, and alludes, not very aptly, to the flowers.

Cypellas have elongated bulbs, and linear basal and stem leaves. Solitary or in clusters of up to six, the yellow, orange, or blue blooms sprout from the axils of spathelike bracts at the tops of cylindrical, wiry stems. Those of each cluster open in succession, each bloom lasting up to a day.

The flowers have six separate perianth segmets (petals). The outer three are large, obovate, and spread widely. The others, very much smaller and shorter, are erect, fiddle-shaped, and bend outward at their ends. There are three short, erect, straight stamens, with broad anthers, that are conspicuously overtopped by the branched stigmas. The fruits are capsules containing flattened seeds.

Most familiar to gardeners, *Cypella herbertii* is native to southern Brazil, Uruguay, and Argentina. From 1 foot to 3 feet

Cypella herbertii

tall, it has leaves 6 inches to 1 foot long. Its usually burnished-copper blooms, 3 inches in diameter and with individual stalks 1½ to 2 inches long, are in loose clusters. Each of the outer petals has darker rays or spots at its base. The anthers are pale greenish. The stigmas are purple and have three-branched lobes.

Its flowers dull leaden-blue with yellow in their throats, about 3 inches in diameter, and few together, *C. plumbea* is 2 to 3 feet tall and a native of southern Brazil and Argentina. Its leaves have channeled stalks and are 1 foot to 1½ feet long. The anthers are blue, the styles and stigmas pale green. The branches of the stigmas have one erect and two spreading lobes.

Coming from the Andes of Peru and Bolivia, *C. peruviana* develops its blooms after the foliage has died. The leaves are 6 to 9 inches long. On individual stalks about 1 inch long, the bright yellow flowers are nearly 4 inches across. They have nearly round outer petals marked with reddish-brown on their wedge-shaped bases. Each inner petal has two small purple blotches. The anthers and branches of the stigma are yellow, the latter forked and petal-like.

Garden and Landscape Uses and Cultivation. Cypellas are not hardy in the north. Where there is danger of their bulbs freezing they must be lifted in fall, stored in a temperature of about 40°F and replanted in spring. They are suitable for warm sunny locations in flower gardens, and as novelties they provide interest and variety. They bloom in late summer and fall, but because of the fleeting nature of their flowers make no great or continuous display. A porous, sandy soil that contains fairly liberal amounts of well-decayed organic matter suits cypellas best. Propagation is easy by offsets and seeds.

CYPERACEAE—Sedge Family. Of secondary horticultural importance, this group of monocotyledons consists of about eighty genera and species of grasslike or rushlike herbaceous plants, mostly of damp or swampy habitats. The vast majority are deciduous or evergreen perennials. A few are annuals. Sedges, as the components of this family are called, have a wide natural distribution and are especially abundant in temperate regions of the Northern and Southern hemispheres and in the subarctic, where some sorts such as the cottongrasses (*Eriophorum*) cover great areas.

From grasses sedges are distinguishable by their stems being usually solid and three-angled and the basal, stem-clasping portions of their leaves generally forming closed instead of split sheaths. Most frequently the stems below the flowering parts are without branches. The unisexual or bisexual flowers are tiny, green, greenish, brownish, or less often white. They are in spikelets or spikes, often grouped in heads, umbels, or racemes, and may or may not have perianths, which, if present, consist of bristles or scales. There are two or three stamens and one style. The fruits are seedlike achenes.

Although individual blooms are not showy, the assemblages of them that are the inflorescences are often pleasing and decorative as are the habits of growth of cultivated kinds. Genera in cultivation include *Carex*, *Cymophyllus*, *Cyperus*, *Eleocharis*, *Eriophorum*, and *Scirpus*.

CYPEROCYMBIDIUM. This is the name of bigeneric orchids the parents of which are *Cyperorchis* and *Cymbidium*.

CYPERORCHIS. See Cymbidium.

CYPERUS (Cy-pèrus)—Galingale, Papyrus, Umbrella Plant, Chufa. The ancient Greek name *kypeiros* for certain of the more than 600 species of *Cyperus* is the basis of its scientific one. The genus belongs in the sedge family CYPERACEAE. It is widely distributed throughout most parts of the world.

Mostly herbaceous perennials, some uncultivated kinds annuals, cyperuses have often a rushlike or grasslike aspect, but are quite different in floral structure from rushes, of the family JUNCACEAE, and grasses, of the family GRAMINEAE. More

obvious differences are their characteristically sharply-three-angled stems in contrast to the round ones of rushes and grasses. Cyperuses may have basal, three-ranked leaves or in their place sheathing bracts. Bracts, often leaflike, form a collar (involucre) at the base of the umbels, heads, or panicles of flowers. The insignificant bisexual flowers, enclosed in little two-ranked bractlets, are in flattened spikelets. Each bloom consists of three or fewer stamens and one style with two or three stigmas. There are no sepals or petals. The fruits, technically nuts, are three-angled or flattened.

The papyrus plant from nearly 3,000 years before the birth of Christ was exploited by the Egyptians for making the paper-like papyri they used to write upon. This species grew or perhaps was cultivated along the Nile (it is now extinct there as a wildling). Papyri were prepared by cutting the pith of the stems vertically into thin strips, laying these parallel to each other with another layer at right angles, pressing them together, and then drying the resulting mat in the sun. Irregularities were smoothed by rubbing with the teeth of animals or with shells. The bulrush of the Bible is this species. The chufa is cultivated for its small tubers, which, when roasted, have a delicious nutlike flavor. As tiger nuts they have long been favorites of children in Great Britain.

Papyrus (**C. papyrus**), the largest and most strikingly ornamental cultivated kind,

is a native of river banks and other wet soil areas in the Mediterranean basin. Stately, with jointless, blunt-angled stems 5 to 10 feet tall or taller, it has no basal leaves. A few bracts form collars at the bases of the graceful large terminal mopheads that are the umbels of bloom. The beauty of these is their fifty to one hundred gracefully arching, and drooping, green, hairlike rays of the umbel 9 inches to 1½ feet long or longer.

Umbrella plant (**C. alternifolius**) is a well-known decorative native of Madagascar, naturalized in the West Indies and other warm lands. From 1½ to 4 feet in height, it is without basal leaves. Its rigid stems are erect and finely-grooved. The terminal umbels of bloom rest on involu-

Cyperus alternifolius

Cyperus papyrus

Cyperus alternifolius (foliage and flowers)

Cyperus albostriatus

Cyperus alternifolius variegatus

cres of many firm, leaflike bracts, arranged spirally, that spread outward and slightly downward like the ribs of an umbrella. They are 4 to 8 inches long by under ½ inch wide. The umbels have stalks 1 inch to 3 inches long. Variety *C. a. gracilis*, native to Australia, is smaller and more slender-stemmed than the typical species. Leaves longitudinally striped with white or sometimes all white are characteristic of *C. a. variegatus*. From the umbrella plant **C. adenophorus,** of Brazil, differs in the lower parts of its three-angled, 2-foot-tall stems being leafy and the spikelets of flowers being stalked.

South African, **C. albostriatus** is often misnamed *C. diffusus*. A suckering kind 8 inches to 1½ feet tall, this has long, narrow basal leaves, from a little under to a little over ½ inch wide, with three prominent veins. Beneath the umbels of bloom are five to ten spreading, slightly arched, leafy bracts wider than those of the umbrella plant. The rays are 1 inch to 4 inches long. The light brown spikelets are in clusters of two to three. A variety with leaves and bracts streaked with white is *C. a. variegatus.*

Resembling a miniature papyrus, the plant long known to gardeners as *C. haspan viviparus* is *C. isocladus*, of East Africa and

Cyperus isocladus

Madagascar. It is distinct from *C. haspan*, of coastal North America and tropical and subtropical regions of the New and the Old World. Up to 1 foot tall, **C. isocladus** is without basal leaves. It has stems topped with green mopheads 4 to 8 inches in diameter. The bracts at their bases are reduced to small scales. Each is of thirty to one hundred hairlike rays. The umbels generate plantlets spontaneously.

An inhabitant of swamps and shallow water from Virginia to Florida and Texas and in tropical regions of both the Old and New Worlds, **C. haspan** is 9 inches to 2½ feet tall. It usually has sheathing bracts in place of basal leaves. Its few to many-rayed umbels of loosely spherical flower spikes have two leaflike bracts. They produce a pleasing lacy effect.

The chufa (**C. esculentus sativus**) is a variety of a weedy native of North America, Europe, and Asia, cultivated for food. It has few, narrow basal leaves. At the ends of many underground stolons are small tubers that after roasting have a sweet, nutty flavor. The stems, 4 inches to 2 feet tall, terminate in large, loose, feathery umbels with five to ten often-branched rays

Cyperus haspan

Cyperus esculentus

of various lengths. Below the umbels are a few leaflike bracts.

Garden and Landscape Uses. The chufa is a hardy plant grown for its edible tubers. Other kinds here considered are beautiful moist- and wet-soil ornaments for aquatic gardens, bog gardens, and the margins of pools, as permanent perennials in warm climates, and temporary summer decoratives elsewhere. The lower kinds are also good greenhouse and window pot plants.

Cultivation. As long as the soil is moist or wet cyperuses grow with minimum care in sun or where the strongest light is tempered by a little shade. They prefer fertile earth with a high organic content. Indoors, minimum temperatures may range from 45 to 65°F on winter nights, with increases by day and at other seasons. To assure constant moistness of the soil it is well to keep the bottoms of the pots standing in saucers or other containers in 2 or 3 inches of water, or the pots or tubs may be immersed almost or quite to their rims. Repot annually in spring. Increase is easy by seed, division, and by cutting off the umbels with an inch or two of stem attached and floating them in water. Chufas are propagated by their small tubers.

CYPHOMANDRA (Cyphomán-dra).—Tree-Tomato. Of the thirty species of herbaceous plants, shrubs, and small trees that constitute South American and West Indian *Cyphomandra*, only one is known horticulturally. This genus belongs in the nightshade family SOLANACEAE and is closely related to *Solanum*, from which it differs in technical details of the anthers. The name is from the Greek *kyphos*, a tumor, and *aner*, a male, and refers to a thickening of the anthers.

Cyphomandras are spineless plants with large, alternate, undivided leaves. Those of some species are lobed. The flowers, bell-shaped and in racemes or forked clusters, have five-lobed calyxes, five-lobed, bell- to flatter, more wheel-shaped corollas, and five stamens the anthers of which

form a cone around the small stigma. The fruits, technically berries, are many-seeded.

The tree-tomato (*C. betacea*) is a South American, erect, evergreen shrub, treelike in form, and 8 to 12 feet in height. Its heart-shaped, softly-hairy, prominently-pinnately-veined leaves are up to 1 foot long. When young they often are purplish. The flowers have the fragrance of caramel. They are pinkish, about ½ inch across, and

Cyphomandra betacea

are succeeded by long-stalked, smooth, edible, egg-shaped fruits 2 to 3 inches long, at first gray-green with darker stripes, changing to yellow and finally dull brick-red. Thicker-skinned than tomatoes, when cut open they much resemble those fruits. Slightly acid, yet sweetish, they have a rich, subtle flavor, an appreciation for which has perhaps to be acquired. The unripe fruits are mildly toxic. The whole plant when brushed or bruised has a peculiar, faint, but pervading odor similar to that of some other members of the nightshade family.

Garden Uses. In tropical and subtropical regions the tree-tomato is cultivated outdoors for its fruits and as a curiosity. It is sometimes grown in greenhouses and with minimum care fruits abundantly, usually within one-and-a-half to two years from seed, somewhat sooner from cuttings. The plants are self-fertile, setting fruits freely to their own pollen.

Cultivation. The tree-tomato prospers in any ordinary garden soil that is well-drained and reasonably fertile and in full sun. It is sturdy enough to need no staking or other support. In greenhouses it may be planted in ground beds or accommodated in large pots or tubs. Rich, coarse soil gives best results and when the final containers are filled with roots, regular and generous fertilizing should be practiced from spring through fall. Old specimens may be pruned to shape and repotted or top-dressed with a rich soil mixture in late winter or early spring, just before new growth begins. Except in winter, when the plants are semi-

dormant, the tree-tomato needs copious supplies of water. In winter lesser amounts are needed, but the soil should never be really dry. A minimum winter night temperature of 45 to 50°F is satisfactory, with the day temperature a few degrees higher. These should be raised ten or fifteen degrees when new growth begins at the end of winter. At all times a humid atmosphere is needed. New plants are easy to raise from seeds, and cuttings of lateral shoots can be rooted under mist.

CYPHOSTEMMA. See *Cissus*.

CYPRESS. Without modification this name applies to *Cupressus*. It is more loosely applied to *Taxodium* and sometimes other genera. Common names in which the word cypress occurs are these: bald-cypress (*Taxodium distichum*), cypress-pine (*Callitris*), cypress spurge (*Euphorbia cyparissias*), cypress-vine (*Ipomoea quamoclit*), false-cypress (*Chamaecyparis*), Leland-cypress (*Cupressocyparis leylandii*), Mexican swamp- or Montezuma-cypress (*Taxodium mucronatum*), mourning-cypress (*Chamaecyparis funebris*), pond-cypress (*Taxodium distichum nutans*), standing-cypress (*Ipomopsis rubra*), summer-cypress (*Kochia scoparia culta*), and swamp-cypress (*Taxodium*).

CYPRESS-OF-TULE. This is a giant specimen of *Taxodium mucronatum* located at Tule, Oaxaca, Mexico. It is well known as "El Gigante." See Taxodium.

CYPRIPEDIUM (Cypri-pèdium) — Lady Slipper or Moccasin Flower, Ram's Head Orchid. This name is used as a common one for Old World tropical and subtropical, evergreen lady slipper orchids, frequently grown in greenhouses, and correctly identified as *Paphiopedilum* and *Phragmipedilum*. Botanically the name *Cypripedium* is restricted to a related genus of possibly fifty species that occurs wild north of the equator in both the eastern and the western hemisphere, and includes eleven indigenous to North America. We are dealing with this latter group here. It, and the related genera mentioned above, are recognized as a subtribe of the orchid family ORCHIDACEAE and believed to be the most primitive existing orchids. The name *Cypripedium* derives from the Greek *kypris*, Aphrodite, and *pedilon*, a shoe. It alludes to the form of the lips of the flowers. Cypripediums are ground orchids.

A combination of characteristics differentiates *Cypripedium* from its allied genera. These are the possession of leaves folded or pleated lengthwise and of flowers with one-celled ovaries and corollas that after withering remain attached to the seed capsule. The blooms are of special interest because of their adaptation to pollination by particular insects, chiefly bees. Their lips are pouchlike traps with the only aperture divided into three. Entering the convenient front opening, the insect cannot return by the same route because of inward-directed hairs or an infolded margin. The only way of escape is by one of the other openings, and this requires the insect to force its way past the stigma, and in so doing to deposit any pollen brought on its body from a previously visited bloom. Then it must push by an anther from which pollen is likely to be deposited on its body for transportation to the next flower visited. The floral parts, other than the pouchlike lip, are three widely-spreading sepals, the two lower usually joined, two petals, and a column with two anthers curved over the orifice of the lip. The kinds most likely to be cultivated are hardy natives of North America.

Yellow lady slipper or yellow moccasin flower (*C. calceolus*), the American form sometimes distinguished as *C. c. pubescens* and previously as *C. parviflorum*, occurs in

Cypripedium calceolus

Cypripedium calceolus (flower)

North America, Europe, and Asia. A variable species, it has leafy stems 9 inches to 2½ feet high and ovate-lanceolate leaves up to 8 inches long by about one-half as wide. Solitary or in pairs, the flowers have greenish-yellow to purplish-brown sepals and petals, the latter more or less twisted, and a yellow, generally purple-veined lip.

The upper sepal is ovate-lanceolate and 1¼ to over 3 inches long, the lower sepals, except for a slight division at the apex, are joined, and the petals are about as long as the upper sepal, but are narrower. Yellow lady slipper occurs locally from Newfoundland to the Yukon, British Columbia, South Carolina, Louisiana, and Texas. It inhabits bogs and moist woodlands. Western North American *C. montanum*, similar to the last, but with flowers with a white lip, one to three on a stem, grows as a native in dry and moist places at high elevations. Another kind with flowers with a predominantly white lip, but with a few purple spots on it, and each stem with only one bloom, the sparrow's egg lady slipper (*C. passerinum*) ranges in the wild from the St. Lawrence River to the Arctic Circle.

The small white lady slipper (*C. candidum*), a rare native from Ontario to Nebraska, Pennsylvania, and Missouri, favors limestone bogs, wet meadows, and prairies. It has erect, lanceolate to narrowly-elliptic leaves up to 6 inches long, usually overlapping at their bases, and solitary flowers with often crimson-striped, greenish sepals and petals up to 1½ inches long and a white lip up to 1 inch long. The petals are frequently considerably twisted.

The showy lady slipper (*C. reginae* syn. *C. spectabile*) has blooms with white sepals and petals. This noble orchid is endemic in bogs, swamps, and wet woodlands from Newfoundland to North Dakota, North Carolina, Georgia, and Missouri. Its hairy, leafy stems 1¼ to 3 feet tall, carry one to three blooms about 3 inches wide. The leaves, up to 8 inches long, are elliptic-ovate and have stem-clasping bases. The upper sepal is broadly-ovate, the two lower are completely joined. The wide-spreading petals are almost as long as, but much narrower than, the sepals. The lip is large, white or pink streaked with rose-purple.

Pink lady slipper or pink moccasin flower (*C. acaule*) has no above-ground stem. Its pair of spreading, pubescent, narrow-elliptic leaves are 4 to 8 inches long. From be-

Cypripedium passerinum

Cypripedium acaule

tween them comes an erect, leafless stalk topped by a usually solitary flower with lanceolate, yellowish-green to greenish-brown sepals and petals up to 2 inches long and a large, bladdery, crimson-pink to whitish-pink lip conspicuously veined with deeper pink. In *C. a. albiflorum* the lip is white sometimes veined with pink or green. The pink lady slipper inhabits acid soils in a variety of locations from Newfoundland to Alberta, North Carolina, Alabama, and Kentucky.

The only cypripediums confined in the wild to western North America are *C. californicum* and the brownie or clustered lady slipper (*C. fasciculatum*). Native to California and Oregon, *C. californicum* has blooms similar to those of *C. passerinum*, but differs in the stems having three to twelve blooms from the axils of leaflike bracts. The clustered lady slipper (*C. fasciculatum*) is so called because its several blooms are crowded in terminal clusters. Each has a small, brown-veined, greenish-yellow, globose lip, purplish around its infolded mouth. The sepals and petals are dull purple or greenish, with brown veins. The flower clusters terminate stems up to 1¼ feet long but usually shorter, each with a pair of opposite leaves up to 4½ inches long by about two-thirds as wide.

Ram's head orchid (*C. arietinum*) is so named because its flower in form fancifully suggests the lowered head of a charging ram. Native from Quebec and Manitoba to Minnesota, Massachusetts, and New York, it often inhabits cold swamps as well as other moist, acid soils. It has slender stems up to 1¼ feet tall, with three to five lanceolate to elliptic leaves up to 4 inches long, that terminate in a solitary bloom with a conical, white, deeply purple-veined lip. The sepals, the lower ones not united, and petals are slender, greenish-brown, and up to 1 inch long.

A native of Japan and China sometimes cultivated by enthusiasts, is distinctive and beautiful *C. japonicum.* This has stems up to a little over 1 foot long. Near their tops are a pair of ribbed, nearly opposite, almost orbicular leaves, commonly broader than their 4- to 6-inch lengths. The blooms are solitary, about 2½ inches across. They have red-spotted greenish sepals and petals and a pink or whitish lip spotted with red. Another Japanese, that occurs also in Taiwan, and on the Asian mainland, *C. macranthum* has three to five alternate, elliptic leaves on each stem and solitary flowers with a pink to rose-purple, or rarely white upper sepal, petals, and lip, and united greenish-brown lateral sepals.

Garden Uses and Cultivation. Success in raising cypripediums from seeds has rarely if ever been achieved. For garden plantings dependency is upon plants collected from the wild and the division of cultivated specimens. The first is permissible only when done with the utmost regard for preserving natural environments or when such features are inevitably to be destroyed by building, road-making, and similar developments. Of all native American orchids cypripediums can be transplanted with the best expectation of success. Even so, they need very respectful handling. Great care must be taken not to damage their roots unnecessarily and to take with each clump or division a large mass of undisturbed soil. The location in which they are placed in the garden, soil conditions, amount of light, and other factors that affect growth should be duplicated as closely as possible to those the plants know in the wild, except that they do respond to greater fertility and this is best achieved by mixing with the soil, but not so that it comes into contact with the roots at planting time, a fairly generous amount of well-rotted cow manure and some bonemeal. It is important that the plants never suffer from drought. A mulch of leaf mold applied each fall is beneficial. Early spring is the best time to divide cypripediums. For more information see Orchids.

CYRILLA (Cyríl-la) — Southern-Leatherwood. The lone, but variable representative of *Cyrilla* is a shrub or tree that ranges as a native from Virginia to the West Indies and South America, inhabiting moist and even swampy soils. In North America it rarely exceeds 30 feet in height, but in Puerto Rico, Jamaica, and other parts of the American tropics it may be 50 to 80 feet tall and have a trunk up to 5 feet in diameter. In Puerto Rico the hollowed trunks afford nesting places for rare parrots. They also serve as beehives. The chief use of the wood is as charcoal. The southern-leatherwood (*C. racemiflora*) belongs in the cyrilla family CYRILLACEAE, a small group of New World plants closely related to hollies that includes only two other genera, *Cliftonia* and *Purdiaea*. The name honors Domenico Cirillo, professor of medicine at Naples, Italy, who died in 1799.

Cyrilla racemiflora

Evergreen in the warmer parts of its range, but deciduous where winters are colder, the southern-leatherwood has alternate, short-stalked, toothless leaves that are glabrous, oblong-lanceolate to oblong, lustrous bright green above, and up to 3 inches long or slightly longer. The small white flowers are in slender, cylindrical racemes 3 to 6 inches long that become pendulous in fruit. The fruits are small dry capsules of no decorative merit. Each contains two or three seeds. The leaves of the deciduous form, which is shrubby rather than tree-like, assume warm tones of orange and scarlet in fall. This form of the southern-leatherwood is hardy in sheltered locations as far north as New York City. The evergreen type is much more tender.

Garden and Landscape Uses and Cultivation. An interesting item of interest to collectors of the unusual, this species grows best in moist, sandy, somewhat acid loam in a sunny location. The only pruning needed consists of removing weak twigs and shortening very long shoots in spring, to keep the plants shapely. Propagation is by seeds, and by cuttings under mist in summer.

CYRILLACEAE—Cyrilla Family. Three genera consisting of thirteen species of trees and shrubs constitute this American family of dicotyledons. They have alternate, undivided leaves and racemes of small flowers with calyxes of five lobes or sepals, five-lobed or five-petaled corollas, five or ten stamens, and one style with two to four stigmas. The dry fruits sometimes have two or more wings. Cultivated genera are *Cliftonia* and *Cyrilla*.

CYRTANTHUS (Cyrt-ánthus)—Fire-Lily, Ifafa-Lily. Fire-lily is the name given in South Africa to members of this genus. Its allusion is to the propensity of some kinds in the wild to bloom most abundantly after veld fires. Ordinarily, they often do not flower regularly each year. Belonging to the amaryllis family AMARYLLIDACEAE, the genus *Cyrtanthus* is represented in South Africa by more species than any other group of the amaryllis family there native. By far the greater number of the nearly fifty species are endemic there. A very few occur in tropical Africa. The name, from the Greek *kyrtos*, curved, and *anthos*, a flower, alludes to the curving perianth tubes of the flowers.

Cyrtanthuses are bulb plants. They have narrowly-strap-shaped to linear, deciduous or evergreen leaves, the new ones appearing with or after the flowers. The leafless flower stalks are hollow and at their apexes have an umbel of horizontal to pendulous, waxy, tubular, funnel-shaped, often fragrant flowers. The blooms have six perianth lobes (petals) shorter than the perianth tube, six slender-stalked stamens, and a slender style with a more or less three-lobed stigma. The fruits are oblong capsules containing flat, black, winged seeds. The most common flower colors among these plants are reds and scarlets, but kinds with orange, yellow, and white with pinkish-striped blooms occur.

Best known to gardeners, the Ifafa-lily (*C. mackenii*) has pretty ivory-white, fragrant, waxy blooms in four- to ten-flowered umbels atop stalks 6 inches to 1 foot in length. They are about 2 inches long and have slightly curved perianth tubes and spreading petals. Variants with pink and apricot-pink flowers are known. The up to six leaves are 6 inches to 1 foot long and about ¼ inch wide. In *C. m. cooperi* (syn. *C. lutescens cooperi*) the flowers are cream to yellow. The two to six arching leaves, contemporary with the flowers, are narrow and up to 1¼ feet long. With blooms much like those of *C. mackenii*, but scentless, bright red, and somewhat shorter, *C. o'brienii* has leaves up to about 1 foot long that develop with or after the flowers. They are rich glossy green and about ¼ inch wide. Its leaves, up to 1 foot long and ⅛ inch wide, appearing later than the flowers, *C. ochroleucus* has slender flower stalks up to 1¼ feet tall that end in an umbel of two to four pale yellow blooms 1½ to 2 inches long with spreading petals.

Cyrtanthus o'brienii

Cyrtanthus ochroleucus

Cyrtanthus o'brienii (flowers)

Cyrtanthus mackenii

Outstanding *C. obliquus* has evergreen or nearly evergreen foliage. Its leaves, broadly-strap-shaped, up to 1½ feet long, and 1 inch to 2 inches wide, are twisted near their apexes. The umbels of six to twelve funnel-shaped flowers, with petals that do not spread, are on stalks 1 foot to

1½ feet tall. They are about 3 inches long by 1 inch across their mouths. The base of the perianth tube is yellow, its chief length is flushed with orange, the petals are green.

Attractive, scarlet-flowered *C. parviflorus* differs from *C. o'brienii* in its 1¼-inch-long blooms having tiny petals that, instead of spreading, are in line with the corolla tube. From six to twelve blooms constitute each umbel. The umbels are on 1-foot-tall stems. The leaves of this species, about 1 foot long and ¼ inch wide, come at the same time as the flowers.

Garden and Landscape Uses and Cultivation. Cyrtanthuses do not stand freezing. They are for permanent cultivation outdoors only in regions where the climate

approximates that of their homeland, South Africa. This is true of California and other parts of the west and southwest. It is sometimes recommended that, where winters are too severe for the bulbs' permanent residence outdoors, they be planted in spring, dug up in fall, and stored indoors over winter. This is worth trying, but sufficient knowledge of the results of such treatment is not available to make a recommendation. Outdoors, a site with just a little shade from the strongest sun of the day is probably preferable to full exposure. Fertile, sharply drained soil, kept reasonably moist through the growth period, is needed. Rock gardens and the fronts of flower beds are suitable locations for these rather uncommon plants.

For greenhouse cultivation, pots rather than pans (shallow pots) are best. The bulbs are planted fairly closely together, with their tops just beneath the surface, in a porous, fertile soil, in spring, or for early flowering, in fall. At first water cautiously, more freely as growth develops, and continue as long as foliage is held. When in active growth, ample supplies of moisture are needed. When the leaves begin to die naturally the frequency of watering is reduced, and finally deciduous kinds are kept quite dry through their period of dormancy. Kinds with evergreen foliage are watered throughout the year, but less generously in winter than in summer. In greenhouses, during the part of the year when temperatures can be controlled, a night level of 50°F is adequate, with daytime temperatures five to fifteen degrees higher. If early blooms are desired the night temperature may be maintained at 55°F with proportionate increases by day. At other times, normal outdoor temperatures for the season are satisfactory indoors. In summer the pots may with advantage be moved to a lath house, porch, or lightly shaded place outdoors. Repotting is done in early spring. Propagation is by natural offsets, by seeds, which germinate readily, and by bulb cuttings.

CYRTOMIUM (Cyr-tòmium)—Holly Fern. The species of Old World ferns that constitute *Cyrtomium* belong in the aspidium family ASPIDIACEAE. Native chiefly to warm parts of Asia and Africa, the genus is also indigenous in Hawaii and in the Celebes Islands. Some botanists combine with it the New World genus *Phanerophlebia*. The name, from the Greek *kyrtos*, arched, alludes to the habit of growth. There are about one dozen species.

Cyrtomiums have short, densely-scaly, ascending or erect rhizomes, and once-pinnate, toothed or toothless fronds (leaves) with usually somewhat sickle-shaped, broad leaflets, often with an earlike lobe at one side of the base. The large clusters of spore capsules, scattered over the undersides of the fronds, have centrally supported covers (indusia).

Common holly fern (*C. falcatum*) ranges from Hawaii to Taiwan, Japan, Korea, China, India, and South Africa. It has tufted crowns of rigid, glossy, dark green, very leathery fronds from 1½ to 2½ feet

Cyrtomium falcatum

Cyrtomium falcatum, showing spore clusters

long by up to 8 inches wide. They have shaggy-scaly stalks. Their alternate, seven to twenty-three short-stalked, asymmetrically ovate leaflets are 3 to 5 inches long by 1 inch to 2 inches wide. They have broad, tail-like, pointed apexes and smooth or wavy margins. The undersides of the young leaves are clothed with rust-colored, soft hairs. The spore capsules are in large, round, plentiful clusters. More popular in cultivation than the typical species is *C. f. rochfordianum*, a robust grower, the fronds of which have broader, conspicuously-toothed leaflets. As its name suggests, *C. f. compactum* is dwarfer than the type.

Cyrtomium falcatum compactum

Variable *C. fortunei*, of Japan, Korea, and China, differs from the common holly fern in having fronds with from twenty-four to more than twice that number of duller, broadly-lanceolate to narrowly-ovate-oblong leaflets. The leaflets are 2 to 3 inches long, toothed at their apexes, and usually have an earlike lobe at one side of the base. Native of limestone soils from Hawaii to Japan, China, and India, *C. caryotideum* differs from the kinds discussed above in having lighter green, more drooping fronds, with five to thirteen or rarely up to nineteen oblong-ovate to ovate leaf-

Cyrtomium caryotideum

lets 3 to 6 inches long, usually prominently eared at one side of their bases and with regularly spine-toothed margins.

Garden and Landscape Uses and Cultivation. Holly ferns are excellent ornamentals for outdoors in warm, frostless or nearly frostless regions, for greenhouses, and as houseplants. They succeed without special care in fertile, well-drained soil that contains decomposed organic matter, such as good compost, leaf mold, or peat moss, but is more loamy than that desirable for many ferns. Shade from sun is important, but too heavy shade is likely to result in weak, soft fronds, especially where high temperatures prevail. In tropical and sub-tropical gardens and in large greenhouses these ferns thrive in ground beds. In greenhouses, and as houseplants, they are more often kept in pots, and respond well to this treatment. Indoors, a minimum winter night temperature of 55°F is most favorable, but up to five degrees lower or higher is satisfactory. By day these minimums may be increased by five to fifteen degrees. Watering to keep the soil evenly moist, but not sodden is requisite, and for well-rooted specimens biweekly applications from spring to fall, and monthly applications in winter, of dilute liquid fertilizer are beneficial. Propagation can sometimes be achieved by careful division, but spores afford the most practicable and by far the more rapid means of increase. For more information see Ferns.

CYRTOPODIUM (Cyrto-pódium). The ten species of this handsome New World genus of the orchid family ORCHIDACEAE vary in size from moderately small to quite immense. They are epiphytes (tree-perchers), lithophytes (rock-perchers), and ground orchids with spindle-shaped pseudobulbs and long, deciduous or evergreen, strap- to sword-shaped leaves, strongly ribbed on their undersides and with bases that sheathe the pseudobulbs. Their often highly-colored flowers, from the axils of usually conspicuous bracts, are in generally tall, erect panicles that come from the bases of the new growths. Their prevailing colors are yellow, brown, or red, often with darker markings. Nearly equal in shape and size, the sepals and petals are spreading or more or less reflexed. The lip is three-lobed, with the side lobes erect and the center one spreading. The name, from the Greek *kyrtos*, curved, and *pous*, a foot, alludes to the form of the lip.

Native from southern Florida and Mexico to Argentina, *Cyrtopodium punctatum* has clusters of erect, spindle-shaped pseudobulbs sometimes over 4 feet long, but generally shorter. The deciduous, linear to linear-elliptic, pointed, arching leaves, up to 2½ feet long, drop and leave their spiny bases attached to the pseudobulbs. The sturdy, branched flower stalks may be-

Cyrtopodium punctatum

Cyrtopodium andersonii

from the base of which comes the leaflike or petal-like bract called a spathe, is thickly covered with small, bisexual flowers, each with four to six perianth segments and the same number of stamens. The fruits are usually one-seeded berries.

Probably a native of the Solomon Islands, **Cyrtosperma johnstonii** is a tuberous, short-stemmed species. Its leaves have abundantly spiny, red-spotted stalks 1 foot to 2½ feet long, and arrow-shaped blades

Cyrtosperma johnstonii: (a) Base of plant

come 5 feet tall, but more commonly are shorter. They bear red- or maroon-red-spotted bracts. About 1½ inches in diameter, the blooms are greenish-yellow to clear yellow, with red or red-purple spots or blotches. The three-lobed, yellow-centered, reddish-brown or purplish lip has side lobes that arch over the column, a center lobe waved and toothed at its apex, and a grooved disk.

West Indian and South American *C. andersonii* is much like the last in growth habit. Its pseudobulbs are up to about 4 feet long, and its branched flower stalks may exceed them by a foot or two. The waxy blooms, about 2 inches in diameter and fragrant, are yellow, with the tips of the sepals and petals suffused with green. The lip is citron-yellow, with a nearly orange-yellow blotch and a longitudinally-ridged disk.

The pseudobulbs of *C. virescens* are

shorter and thicker than those of *C. punctatum*, and the greenish-yellow to primrose-yellow flowers, spotted with reddish-brown, rather smaller. The brighter-colored lip is marked with reddish-purple. This is native to South America.

Garden and Landscape Uses and Cultivation. For outdoors in the tropics and warm, frostfree subtropics, cyrtopodiums are excellent ornamentals and are relatively easily cultivated as greenhouse ornamentals. They thrive in rooting mixes appropriate for cymbidiums, but need warmer growing conditions. During their spring-to-fall season of active growth, temperature and humidity levels that suit cattleyas are to their liking. Drier air and cooler surroundings are needed during the winter resting period. At that time, water is withheld, but not to the extent that the pseudobulbs shrivel. When new growth begins watering is resumed, at first cautiously, more generously as leaves become well developed. Throughout the summer, water is given freely and, to well-rooted specimens, occasional applications of dilute, liquid fertilizer. For more information see Orchids.

CYRTOSPERMA (Cyrtó-sperma) Eighteen or perhaps fewer species compose this genus, of wide distribution in the tropics, of the arum family ARACEAE. They are tuberous and rhizomatous, robust herbaceous plants of the style of *Alocasia*. The name, from the Greek *kyrtos*, curved, and *sperma*, a seed, was given because the seeds of some kinds are kidney-shaped. In the Pacific islands and some other places some kinds are cultivated for their taro-like, edible roots.

Cyrtospermas have arrow- or halberd-shaped, rarely somewhat lobed leaves. Their flowers are in inflorescences (commonly called "flowers") typical of the family and constructed like those of calla-lilies. The spadix (spike) of the inflorescence,

(b) Typical leaf

Cyrtopodium andersonii

(c) Inflorescence

almost as long, with long basal lobes. The inflorescences, on spiny stalks shorter than those of the leaves, have brown or purple spathes paler on their insides than on their backs, 4 to 6 inches long, and exceeding the spadixes in length.

Garden and Landscape Uses and Cultivation. These are as for *Alocasia*.

CYRTOSTACHYS (Cyrtós-tachys)—Sealing Wax Palm. Few palms are as colorful as some species of *Cyrtostachys*, with their highly decorative, brilliant red leafstalks. Belonging in the palm family PALMAE, this genus consists of ten species that commonly form dense clumps of many erect, slender, bamboo-like stems topped with graceful feathery foliage. Their flower clusters have two soon-deciduous spathes and flattened and sometimes drooping branches. Individual blooms are unisexual, with both sexes in the same cluster. Male flowers usually have six, rarely twelve or fifteen stamens. The fruits are small. The name is derived from the Greek *kyrtos*, curved, and *stachys*, a spike, and alludes to the curved spikes of flowers. This genus is native to New Guinea and other islands of the Pacific.

Two of the most handsome of the genus, *C. lakka*, of Borneo, and *C. renda*, of Sumatra, are called sealing wax palms. About 15 feet tall, **C. lakka** is somewhat reminiscent of an immense cluster of feather dus-

ters of different lengths and sizes standing duster ends up. Its pinnately-divided leaves, 3½ to 5 feet long, green above and grayish-green beneath, each has about fifty narrow, long-pointed leaflets attached to a rachis (central axis of the leaf) that like the leafstalk is bright red. The fruits, under ½ inch long, and bluntly-conical, contain egg-shaped seeds. They are black with bright red bases. Up to 30 feet tall, **C. renda** is similar to *C. lakka* except that its leafstalks and rachises are brownish-red, its fruits and seeds globular. A variety, *C. r. duvivierianum*, which does not produce seeds, has brilliant red leafstalks and rachises.

Garden and Landscape Uses. These frost-tender palms can be cultivated outdoors in the United States only in southern Florida and Hawaii. They can be grown elsewhere in the tropics and in tropical greenhouses. In the open they are especially decorative when planted as isolated clumps on lawns so that their beauty can be fully appreciated. In their native homes they inhabit moist or even swampy ground, which suggests their needs under cultivation. They are attractive pot and tub plants for indoor decoration.

Cultivation. These palms need fertile, always moist soil and full sun. Most can be propagated by seeds sown in sandy, peaty soil in a temperature of 80 to 90°F, but *C. renda duvivierianum* must be, and

other kinds can be, increased by carefully dividing the clumps. The divisions are potted in sandy, peaty soil, watered thoroughly, kept in a very humid atmosphere, and shaded from bright sun, in a temperature of 70 to 90°F, until they are well rooted. When cultivated in greenhouses these palms should have a minimum winter night temperature of 60 to 65°F, with a five to ten degree increase in the day permitted and at other seasons a minimum night temperature of 65 to 70°F. They prosper in high temperatures and a very humid atmosphere and need some shade from strong summer sun. Established specimens benefit from regular applications of dilute liquid fertilizer. For additional information see Palms.

CYSTOPTERIS (Cyst-ópteris)—Bladder Fern, Brittle Fern. There are up to twelve species of these delightful woodland ferns of the aspidium family ASPIDIACEAE. Mostly natives of temperate regions, they are indigenous to North America, Europe, and Asia. The name, from the Greek *kystis*, a bladder, and *pteris*, a fern, refers to the inflated coverings (indusia) of the spore clusters that are characteristic of *Cystopteris*.

Bladder ferns are of refined appearance. They have slender, creeping rhizomes and erect, spreading, or arching, one- to four-times-pinnately-divided fronds (leaves). The clusters of spore cases are roundish.

Bulblet bladder fern (**C. bulbifera**) is so called because on the undersides of its leaves toward their apexes there are frequently small pea-like bulblets along the midrib. These drop to the ground and generate new plants. This species usually inhabits limestone cliffs in moist woods from Newfoundland to Manitoba, Utah, Georgia, Arkansas, and Arizona. Its dark green, slenderly triangular to linear-lanceolate, hairless leaves are mostly 1 foot to 3 or sometimes 5 feet long, and up to 5 inches wide. They have twenty to forty nearly opposite pairs of primary divisions that gradually diminish in size from base to apex. These are again pinnately-lobed, with the lobes double-toothed.

The brittle fern (**C. fragilis**) is highly variable. Its light green, short-pointed leaves, ovate to ovate-lanceolate, are up to, or sometimes a little over 1 foot long. They are two- or three-times-pinnate; the ultimate segments are toothed. There are nine to fifteen pairs of rather widely-separated, opposite or nearly opposite major leaf divisions, the lowermost pair a little shorter than the next above. The rather few spore-case clusters are irregularly distributed on the under surfaces of the leaves. This species occurs from Labrador to Alaska, Vermont, Michigan, South Dakota, and Texas, in the mountains of North Carolina, western North America, and in Europe and Asia. Variety *C. f. laurentiana* is larger.

Cyrtostachys lakka at Summit Garden, Panama

Garden and Landscape Uses and Cultivation. Bladder ferns are easily cultivated and are suitable for lightly shaded places where the soil contains abundant organic matter and is slightly damp. For *C. fragilis* the earth may be neutral or somewhat acid; *C. bulbifera* prefers limestone soil, but does not demand it. To be seen to best advantage, the last must be located on a cliff or dry wall at an elevation that allows its fronds to arch or hang without reaching the ground. These are ferns for cool locations. They are propagated by spores and division, and *C. bulbifera* by bulblets. For more information see Ferns.

CYTISUS (Cýtis-us)—Broom. Many modern botanists refer plants commonly grown in gardens as *Cytisus* to more than one genus. Because such notable works as *Flora Europaea* does this, it seems inevitable that botanical gardens and perhaps other horticultural establishments and nurseries will come to label their plants accordingly. Such important recent horticultural treatments as those in Bean's *Trees and Shrubs Hardy in the British Isles* and *Hortus Third* accept the broader concept, however, and this is done here. The names that apply under the segregate genera are given in parentheses.

So interpreted, *Cytisus* consists of about fifty species of nonspiny and spiny deciduous and evergreen shrubs and small trees, natives of Europe, the Mediterranean basin, and the Atlantic islands. It belongs to the pea family LEGUMINOSAE. Its name is perhaps derived from the Greek *kytisos*, a kind of clover, probably in allusion to the aspect of the foliage of some kinds. The common name broom applied to members of this genus is used also for *Genista*, *Spartium*, and some other plants.

Cytisuses prevailingly have alternate leaves with three, or less often one leaflet. Their pea-shaped flowers come from the leaf axils or are in leafy terminal clusters. They have a usually bell-shaped calyx with an upper lip with two teeth, a lower one with three, the teeth generally much shorter than the lips. The corollas are yellow, creamy-white, white, or purple, or in some garden varieties various shades of bronze, and crimson. The petals are completely separate. The ten stamens are united. The fruits are linear to oblong pods containing seeds usually with a conspicuous strophiole (wartlike protuberance). The presence of this is the chief distinction between *Cytisus* and *Genista*.

Scotch broom (**C. scoparius** syn. *Sarothamnus scoparius*), an abundant native of western Europe and hardy in fairly sheltered locations in southern New England, is freely naturalized in some milder parts of North America. An erect, freely-branched, green-stemmed, deciduous shrub up to 8 or 9 feet tall or occasionally considerably taller, it has downy young shoots

Cytisus scoparius

the lower leaves of which are stalked and usually have three obovate to oblong-lanceolate leaflets, the upper ones stalkless, and with one lanceolate leaflet. In size the leaflets range from a little under to somewhat over ½ inch long. The bright yellow blooms, their standard or banner petal ¾ inch wide, are solitary or paired. The hairy pods are approximately 2 inches long. Variety *C. s. prostratus* has prostrate stems and densely-silky-hairy young shoots and foliage. Similar to it, *C. s. pendulus* has bigger flowers. Sulfur-yellow blooms are borne by *C. s. sulphureus*, which is lower and more compact than the typical species. As an ancestor of a splendid race of hybrid brooms, *C. s. andreanus* has had great horticultural impact. First discovered in Nor-

Cytisus scoparius andreanus

mandy in 1884, it has flowers with rich brownish-crimson wing petals and a standard or banner petal variously marked with the same hue. This breeds partly true from seeds, but many of its offspring are inferior.

The white Spanish broom (*C. multiflorus*) has long been known in gardens as *C. albus*, a designation that properly belongs to the very much lower Portuguese broom. Native to Spain and Portugal, the white Spanish broom may exceed 10 feet in height. Hardy in sheltered locations in extreme southern New England, it has slender shoots that bear leaves of three leaflets on the lower parts and leaves with single

Cytisus scoparius prostratus

leaflets above. The silvery-hairy leaflets, minute to up to almost ½ inch long, are narrowly-linear. The ⅓-inch-long flowers are produced in abundance from the shoots of the previous year. They are solitary or in twos or threes. The hairy seed pods, ¾ to 1 inch long, contain four to six seeds. This species is averse to limy soils.

The Portuguese broom (*C. albus* syn. *Chamaecytisus albus*) is very different from the Spanish broom, which in gardens not infrequently masquerades under its name.

Cytisus albus

Native of Portugal and Spain, as known in cultivation it is a deciduous, spreading shrub usually not more than 1 foot tall, but in its native range sometimes attaining 3 feet or more. It has downy stems and leaves with three obovate to narrow-elliptic leaflets hairy on both surfaces or only beneath. The white or yellowish-white flowers, with petals that do not spread widely, are ¾ inch long and in terminal clusters of five to ten. The seed pods are hairy, ¾ inch to 1¼ inches in length. This, the hardiest white-flowered broom, is hardy in southern New England.

Purple-flowered *C. purpureus* (syn. *Chamaecytisus purpureus*) is a delightful, distinct, and somewhat variable deciduous native of southeast Europe and the southern European Alps, hardy in sheltered locations in southern New England. Typically 1 foot to 1½ feet tall, with procumbent or ascending branches, this hairless shrub has dark green leaves of three broad-elliptic to obovate leaflets ¼ to 1 inch long. Solitary or in twos or threes from the shoots of the previous year, the flowers are ¾ inch long. In color they vary from pinkish-purple to quite deep purple. Those of *C. p. albus* are white. The seed pods, without hairs and 1 inch to 1½ inches long, contain three or four seeds. This species grafted onto *Laburnum* gave rise to the interesting graft-hybrid (chimera) *Laburnocytisus*.

Other deciduous sorts, hardy as far north as New Jersey or southern New York, and some few where winters are somewhat colder, include these: *C. ardoini*, a rare deciduous native of limestone soils in the French Maritime Alps, is likely to need winter protection in climates harsher than that of southern New Jersey. In gardens rarely exceeding 8 inches in height, in the wild it sometimes attains 2 feet. It has hairy shoots, and leaves, shaggy-hairy when young, of three obovate to oblongish leaflets about ⅓ inch long. Usually up to three, but sometimes more flowers are in each cluster. Golden-yellow and about ½ inch long, they have a nearly round standard or banner petal with incurved margins. The seed pods are hairy, up to 1 inch long, and contain one or two seeds. *C. austriacus* (syn. *Chamaecytisus austriacus*), native from eastern Europe to southcentral Russia, exhibits considerable variation in habit, hairiness, and leaf shape. It is 6 inches to 2 feet tall and has branches procumbent to erect and more or less hairy. The leaves, of three oblong, obovate, or lanceolate leaflets mostly ½ to 1 inch long, have white-hairy undersides and are more or less hairy above. The deep yellow flowers are in heads of usually many. The seed pods are silky-hairy, ¾ to 1 inch long. Variety *C. a. heuffelii* (syns. *C. heuffelii*, *Chamaecytisus heuffelii*), more frequent in cultivation than the species, differs from it in having slenderer, arching, or erect stems and leaves with narrower leaflets. *C. decumbens*, of southern Europe, nearly or quite prostrate, is from 6 inches to 1 foot tall. It has deciduous leaves of one oblong to obovate leaflet ¼ to ¾ inch in length, not more than ⅛ inch in width, and hairy particularly on their undersides. Borne singly or in twos or threes from shoots of the previous year, the rich yellow flowers are ½ inch long or slightly longer. The seed pods are hairy and ¾ to 1 inch long. *C. diffusus* (syn. *C. pseudoprocumbens*) is an attractive native, frequently on limestone soils, of northern Italy and the Balkan Peninsula. Sometimes a little over 1 foot tall, but often lower, it has prostrate stems and more or less erect, slender branches. The leaves have one pointed-oblong-lanceolate leaflet ¼ to ¾ inch in length, and hairless, except perhaps when very young. The bright yellow, ⅜-inch-long flowers are in slender racemes sometimes over 6 inches in length. The hairless or slightly-hairy seed pods are ¾ inch long. *C. emeriflorus* (syn. *Genista glabrescens*) is native to the southern European Alps. It is 1 foot to 2 or occasionally up to 3½ feet tall and has rigid branches. Its leaves are stalked and of three obovate to lanceolate leaflets up to ½ inch long, nearly hairless above, and silky-hairy on their undersides. The yellow flowers, up to ½ inch long, are in densely-leafy clusters of two or four or are solitary. They are borne on shoots of the previous year. The seed pods are hairless and 1 inch to 1½ inches long. *C. glaber* (syns. *C. elongatus*, *Chamaecytisus glaber*), of central and southeastern Europe, is up to 5 feet high, has erect or arching stems, hairy only when young, and leaves with leaflets ¾ to 1 inch long. Two to five together in the leaf axils, the yellow flowers have the banner or standard petal often spotted with reddish-brown. The hairless to hairy seed pods are ¾ to 1 inch long. *C. hirsutus* (syn. *Chamaecytisus hirsutus*) is a very variable native of lime-free soils in central and eastern Europe. Typically it is a more or less decumbent shrub of loose habit, 1 foot to 3 feet tall. Its stems are slender, and when young clothed with hairs that spread outward. Its leaves, shaggy-hairy on their undersides, are of three elliptic to ovate leaflets up to ¾ inch in length by about one-half as wide. The yellow flowers 1 inch long or longer, in axillary clusters of two to four arranged in leafy racemes, have a large standard or banner petal suffused with brown at its middle. The seed pods, more or less hairy, are 1 inch to 1½ inches long. *C. ingrami*, of Spain, is densely-branched and has stalkless or nearly stalkless leaves, those below of three leaflets, the upper ones of one leaflet. The leaflets are elliptic-oblong to obovate, about 1 inch long by ½ to ¾ inch wide. Their upper surfaces are hairless, their undersides clothed with silky hairs. The flowers are large, have a brown-blotched, cream standard or banner petal, and yellow wing and keel petals. The seed pods are straight, hairy, and 1½ inches long. *C. nigricans* (syn. *Lembotropis nigricans*), a native of central and southern Europe and adjacent Asia and hardy in southern New England, is an elegant, late-flowering broom 3 to 6 feet tall. It has downy, cylindrical stems and leaves, sparingly-hairy on their undersides, of three ovate leaflets ½ to 1 inch long. The bright yellow flowers, up to ½ inch long, are crowded in slender racemes up to 1 foot or more in length that terminate shoots of the current season's growth. The seed pods are linear-oblong, hairy, and ¾ inch to 1¼ inches long. *C. procumbens* (syn. *Genista procumbens*) is rather like *C. decumbens*, but somewhat taller. It has decumbent stems and erect, hairy, strongly-grooved and winged branches up to 1½ feet high or sometimes higher. A handsome, free-blooming native of the Balkan Peninsula and east-central Europe, its abundant clustered leaves have each one pointed, linear-oblong to lanceolate leaflet ½ to ¾ inch long by not more than ⅕ inch wide. They are usually hairless on their upper surfaces and hairy beneath. Solitary or in twos or threes, the golden-yellow flowers, crowded in the leaf axils, form leafy, cylindrical racemes 3 to 6 inches in length. The approximately 1-inch-long seed pods are shaggy-hairy. *C. purgans*, of southern Europe, favors lime-free soils. Usually sparsely-foliaged, it has erect, rigid stems and grooved twigs. The upper leaves are of one leaflet, those below of three leaflets. The leaflets are linear-lanceolate, from ¼ to ½ inch long, silky-hairy on their undersides, and nearly hairless above. The deep yellow, vanilla-scented blooms ⅓ to nearly ½ inch long, solitary or in pairs, are produced from shoots of the previous year. The hairy pods are black at maturity. *C. ratisbonensis* (syn. *Chamaecytisus ratisbonensis*), native from central Europe to the southeastern Ukraine, 1 foot to 1½ feet tall or under cultivation sometimes considerably taller, has branches at first hairy, but later hairless. The leaves have three obovate to obovate-lanceolate leaflets, hairless above, silky-hairy on their undersides, and up to a little over ½ inch long. The yellow flowers are solitary or in twos or threes in one-sided racemes. They are 1 inch or more in length. Often the standard or banner petal is spotted orange-red. The seed pods are ¾ to 1 inch long and clothed with hairs. *C. sessilifolius* is a hairless native of southern Europe, not hardy in climates colder than that of southern New Jersey. Its leaves have three elliptic to obovate leaflets up to ¾ inch long. The leaves of nonflowering branches are without stalks, but contrary to the implication of the name (*sessilifolius* means stalkless), those of flowering ones are long-stalked. Rather sparsely produced, the yellow blooms, from a little under to a little over ½ inch wide, are in loose, leafless racemes of three to twelve. The seed pods are hairless and 1 inch to 1½ inches long. *C. supinus* (syn. *Chamaecytisus supinus*), native throughout much of central and southern Europe and the western Ukraine, with rarely procumbent, usually ascending, hairy branches, is usually 2 to 4 feet tall, but occasionally lower. It differs from rather similar *C. austriacus* in that the hairs on its shoots and seed pods spread outward from the surfaces from which they grow. The leaves have obovate to elliptic leaflets ½ to 1 inch long, with very hairy undersides; the upper sides are ultimately hairless. Characteristically bearing heads of two to ten ¾- to 1-inch-long, bright yellow blooms in summer and fall at the ends of shoots of the current season, plants of this occasionally produce additional flowers in racemes in spring. The flowers much resemble those of *C. hirsutus*.

Not hardy in the north, *C. battandieri* of North Africa is a deciduous species distinct because of its vigorous, loose habit of growth, large leaves, and handsome racemes of fragrant blooms. Attaining heights of 15 feet or sometimes more, this has silvery, silky-hairy shoots and foliage. Its long-stalked, laburnum-like leaves have three obovate leaflets 1½ to 3½ inches long by nearly one-half as wide. The erect or upcurved cylindrical racemes of flowers are at the ends of leafy shoots. The flowers have a heart-shaped standard or banner petal about ½ inch wide. The seed pods are erect and hairy, 1½ to 2 inches long.

Deciduous hybrids include the War-

Cytisus battandieri

Cytisus praecox

minster broom (*C. praecox*) which appeared as a chance seedling in an English nursery in 1867. Its parents are *C. purgans* and *C. multiflorus*. From 8 to 10 feet tall, this excellent sort in manner of growth much resembles *C. multiflorus*, but tends to be denser. Its young shoots and leaves, the latter early-deciduous and mostly of one ½-inch-long leaflet, are silky-hairy. The strongly and slightly unpleasantly-scented, sulfur-yellow flowers are borne in great abundance. The Warminster broom produces fertile seeds from which have been raised a number of varieties. These include *C. p.* 'Allgold' and *C. p.* 'Gold Spear', with bright yellow flowers; dwarfer

C. p. luteus, with blooms of deeper yellow; and white-flowered *C. p. albus*.

Other hardy, deciduous hybrids of splendid quality are numerous. Named for its place of origin, *C. kewensis* was discovered as a chance seedling at the Royal Botanic Gardens, Kew, England, in 1891. Its parents *C. ardoini* and *C. multiflorus*, this most excellent kind is a prostrate shrub that bears in great abundance creamy-white to pale yellow flowers about ½ inch long, each with a standard or banner petal ½ inch wide. They are solitary or in twos or threes along shoots of the previous year's growth. The leaves are downy, mostly of three leaflets, sometimes of one. Not more

Cytisus praecox

Cytisus kewensis

than 1 foot high and often lower, this remarkably fine broom may be several feet in diameter. Another chance seedling that originated at Kew, **C. beanii** was discovered in 1900 in a bed of *C. ardoini*, its parents that species and *C. purgans*. Of great charm, this kind is semiprostrate and two or three times as wide as its 6 inch to 1½ feet height. Its slender shoots are hairy when young. Its leaves, most of one linear about ½-inch-long-leaflet, are hairy. Shoots up to 1 foot long in their second year are generously furnished throughout their lengths with, usually in pairs or triplets, occasionally singly, golden-yellow blooms. With the habit of growth of the Scotch broom and 6 to 9 feet tall, **C. dallimorei** originated at the Royal Botanic Gardens, Kew, England in 1900 as the result of a deliberate mating of *C. scoparius andreanus* and *C. multiflorus*. It has downy leaves of mostly three leaflets. The abundant flowers, a little more than ½ inch long, are in tones of rosy- to purplish-pink, with the wing petals rich crimson and the white-keeled standard or banner petal darker on its outside than its face. The first artificially-raised hybrid broom and still one of the best, this kind has played an important part in the percentage of numerous others. A hybrid between *C. purpureus* and presumably *C. elongatus* probably dating to the latter part of the nineteenth century, **C. versicolor** has much the aspect of *C. purpureus* but is a sturdier, fuller, more rounded shrub 2 feet tall or taller. Its flowers, of a shade intermediate between purple and yellow, have densely-hairy calyxes. Named after its raisers, the English nursery Messrs. Hillier, a hybrid between *C. versicolor* and *C. hirsutus hirsutissimus* named **C. versicolor hillieri** is a low shrub with arching branches and large, bronze-flushed, yellow flowers that become buff-pink as they age.

More recent, mostly tall hybrid brooms, many with flowers of colors other than yellow, have ancestries traceable to *C. dallimorei* and *C. praecox*. Others include in their parentage *C. ardoini* and *C. beanii*.

Among these highly esteemed ornamentals, many raised in Europe, some in the United States, and some of which are not reliably hardy in the north, are sorts named 'California', 'Cornish Cream', 'Dorothy Walpole', 'Enchantress', 'Firefly', 'Johnson's Crimson', 'Knaphill', 'Lord Lambourne', 'Pomona', 'Stanford', and 'Zeelandia'. They are described in the catalogs of nurseries offering them.

Not hardy in the north, but suitable for outdoor cultivation in mild, dryish climates such as that of California, and for growing in greenhouses cytisuses that retain their leaves through the winter include natives of the Canary Islands, Madeira, and the Mediterranean region. All now to be described, except *C. filipes* and *C. supranubius*, have yellow blooms: **C. canariensis** (syns. *Genista canariensis, Teline canariensis*) is an abundantly-foliaged, freely-branched, 4- to 6 feet-tall native of the Canary Islands. Like the young shoots, its extremely short-stalked to nearly stalkless leaves, each of three obovate to elliptic leaflets ¼ to ½ inch long by about one-half as wide, are white-hairy. A little over ½ inch in length, the fragrant blooms have a reflexed standard or banner petal hairy on its outside. They are in loose racemes of up to one dozen at the ends of the main branches and numerous short side shoots. The 1-inch-long seed pods are erect. Variety **C. c. ramosissimus** has smaller leaves and more numerous, shorter racemes of bloom. **C. filipes**, of the Canary Islands, has very slender, almost threadlike, sometimes leafless branches. Nearly hairless, its leaves have three narrow-oblanceolate to elliptic leaflets slightly longer than the leafstalk. The white flowers, borne in the leaf axils, are succeeded by seed pods 1 inch to 1½ inches long. **C. maderensis** (syn. *Teline maderensis*), of Madeira, up to 20 feet tall, has young shoots and leaves clothed with brownish-yellow hairs. Older parts except the upper leaf surfaces, which are nearly hairless, are silvery-hairy. The leaves have stalks about ½ inch long and three stalkless elliptic leaflets ½ inch to 1¼ inches long. The flowers are in racemes up to 4 inches long of up to one dozen that terminate the branches and side shoots. The leaves and racemes of *C. m. magnifoliosus* are bigger than those of the typical species. **C. monspessulanus** (syn. *Teline monspessulana*), of southern Europe, North Africa, and Syria, attains heights of 3 to 10 feet and has very upright, leafy branches and markedly grooved branchlets that, when young, are hairy. The short-stalked leaves have three ½- to ¾-inch-long, obovate leaflets sparingly- to densely-hairy on both surfaces, particularly the lower one. The fragrant, ½-inch-long flowers are in short-stalked clusters or racemes from the leaf axils. A lovely hybrid of this species and the next is C. 'Porlock'. **C. racemosus** (syn. *Genista racemosa*) is the correct name of a

Canary Island species that differs but slightly from *C. stenopetalus* and probably should be included in that species. Plants cultivated as *C. racemosus* are usually *C. spachianus*, a sort of hybrid origin. **C. stenopetalus** (syn. *Teline stenopetala*), of the Canary Islands, up to 10 feet tall, has branchlets clothed with white, silky hairs. Its silky-hairy leaves have stalks up to a little over ½ inch long and three narrowly-elliptic to oblanceolate leaflets up to 1½ inches long. The flowers are in somewhat one-sided racemes up to 4 inches long or slightly longer. The seed pods are about 1 inch in length. **C. supranubius** (syn. *Genista fragrans*), of the Canary Islands, is up to 6 feet tall and bushy. It has densely-hairy leaves with stalks slightly longer than the three narrowly-lanceolate to oblanceolate leaflets. In crowded clusters in the leaf axils, the white to pink-tinged, intensely-fragrant flowers are under ½ inch long. The seed pods, hairless, exceed 1 inch in length.

A handsome hybrid, *C. spachianus* is an evergreen, yellow-flowered broom much favored for cultivation as a greenhouse pot

Cytisus spachianus

plant and in mild climates outdoors. Its parents presumably *C. canariensis* and *C. stenopetalus*, in gardens it is commonly misnamed *C. racemosus, Genista racemosa, C. fragrans*, and *Genista fragrans*, the last two names properly synonyms of *C. supranubius*. From *C. canariensis* the hybrid differs in having very decidedly stalked leaves and longer racemes of bloom, from *C. stenopetalus* in having shorter-stalked leaves with smaller leaflets. At maturity 10 to 20 feet tall, abundantly foliaged *C. spachianus* has somewhat hairy, slightly-ribbed young shoots. The leaves have ¼-inch-long stalks and three scarcely-stalked, obovate leaflets ⅓ to ¾ inch long. Their upper surfaces are hairless, their undersides covered with silky down. In slender, hairy-stalked racemes 2 to 4 inches in length, the ½-inch-long flowers are golden-yellow. Rare *C. s. elegans* has grayer leaves with leaflets 1 inch to 2 inches long, and larger flowers.

Garden and Landscape Uses. Brooms afford splendid furnishings for open, sunny locations and well-drained, moderately fertile, but not too-rich soils. Many deciduous sorts succeed in earths too poor in fertility and too dry for the best performance of most shrubs, but those with evergreen foliage appreciate rather more congenial environments. None of the evergreens and not all deciduous sorts are hardy in the north. Among the latter, this is more or less true of *C. ardoini, C. battandieri,* and unfortunately of many of the tall hybrids of the *C. dallimorei* relationship that afford such a remarkably diverse range in the colors of their blooms. These more tender brooms are at their best in climates typical of California and the Pacific Northwest. The hardiest deciduous kinds *C. austriacus* and *C. nigricans* succeed in New England. In the southern part of that region, *C. albus, C. beanii, C. decumbens, C. emeriflorus, C. praecox, C. procumbens, C. purgans,* and *C. scoparius* survive in favorable locations, although *C. scoparius* may suffer some killing back in winter. Slightly more tender, but hardy in sheltered locations in the vicinity of New York City are *C. dallimorei, C. kewensis, C. multiflorus, C. sessilifolius,* and *C. versicolor.*

Cytisuses can be displayed to great advantage in formal and semifloral landscapes and are especially suitable for use near the sea, in heath and chapparel-type plantings, and the lower ones, in rock gardens. Evergreen kinds make good hedges and good greenhouse pot plants. Most sorts prefer somewhat acid to neutral soils. Some decidedly resent alkaline ones.

Cultivation. The chief matters to keep in mind in caring for deciduous cytisuses are that they respond poorly to root disturbance and hard pruning. Grow young plants in pots or other containers until ready for planting in their permanent locations. The best time to plant is spring, just before new growth begins. Needed pruning of deciduous sorts, and a certain amount may be necessary to prevent taller kinds from becoming too tall and straggly, should receive attention as soon as flowering is through. Limit cutting to the removal of unwanted branches and to the shortening of others, but never to the extent that the greater part of the top growth is removed at one time. If severe correctional pruning is needed it is better to spread it over two years than to do it at one time.

Evergreen brooms are less resentful of transplanting and pruning than deciduous sorts, but when transplanting them take care to preserve as large a root ball as practicable and prune them back fairly severely to compensate for an unavoidable loss of some roots. Each year as soon as flowering is through they may be sheared close to the base of the last season's growth. If necessary old specimens may be cut back fairly drastically in early spring just before new growth begins. Such heavy pruning should be followed by fertilizing and sufficient watering to keep the soil evenly moist during the summer, but often specimens so treated produce little or no bloom the following year. Propagation of deciduous cytisuses whenever practicable should be by seeds, but not all sorts produce seeds and hybrids cannot be relied upon to reproduce themselves truly this way. Alternative means of increase are by summer cuttings 2 to 3 inches long of firm shoots planted under mist or in a propagating frame or greenhouse. It is also possible to graft on understocks of *Laburnum,* a procedure sometimes followed with *C. dallimorei.*

As pot and tub plants for late winter and spring display in greenhouses, *C. canariensis, C. c. ramosissimus, C. racemosus,* and *C. spachianus* are especially fine. Often called genistas, they can be grown as bushes or trained as standards (in tree form). These cytisuses are readily propagated from cuttings taken in late summer and rooted in a greenhouse propagating bench or cold frame in sand or perlite or in either of those mixed with peat moss. As soon as roots about 1 inch long have formed, pot the young plants in sandy, peaty soil and keep them growing in a sunny location. When they are well established pinch out their tips to encourage branching. Give subsequent moves into larger pots as growth warrants. Routine care consists of keeping the soil always moderately moist, but not waterlogged and giving well-rooted specimens dilute liquid fertilizer at about two-week intervals from early summer to fall. Night temperature in the greenhouse in winter should be 45 to 50°F, with an increase of five to ten degrees by day. In summer ventilate the greenhouse as freely as possible or at that time stand the plants outdoors plunged nearly to the rims of their pots or tubs in sand, sawdust, peat moss, or some similar material to prevent undue drying and to keep the roots cool. As soon as flowering is through shear the growth of the previous season back from one-third to two-thirds or more of its length and at that time attend to needed repotting.

DABOECIA (Daboèc-ia)—Irish-Heath, St. Dabeoc's-Heath. This horticulturally interesting genus is generally held to comprise only one species, but by some botanists is considered to consist of two. Such botanists segregate the slightly different form that grows in the Azores as *Daboecia azorica*. The genus belongs in the heath family ERICACEAE and thus is related to heather (*Calluna*), heath (*Erica*), and cranberry (*Vaccinium*). The Irish-heath inhabits southwestern Europe, Ireland, and the Azores. Its generic name is derived from its Irish colloquial name, St. Dabeoc's-heath. The genus differs from its relative *Erica* in the flowers not having persistent corollas.

The Irish-heath (*D. cantabrica*) is a very pretty low evergreen shrub that blooms over a long summer period. Its maximum height is about 2 feet, but usually it is considerably lower. It is freely-branched and its stems are glandular-pubescent. Its alternate leaves, unlike those of heathers and heaths, are elliptic, the uppermost ones more narrowly so than those lower on the stems. They are ¼ to ⅝ inch long, have rolled-under margins, rich glossy green upper surfaces, and silvery-tomentose undersides. The flowers are heathlike. Shaped like miniature Japanese lanterns or narrow-mouthed, ovoid bells, they are conspicuously larger than those of hardy heaths and are in erect, terminal spikelike clusters or racemes 3 to 5 inches in length. Each bloom is ⅜ to ½ inch long. As in *Erica*, but not in *Calluna*, the showy part of the bloom is the corolla, but unlike the corollas of *Erica*, the corollas are not persistent, but drop off when the flower fades. Typically the flowers are bright rosy-purple, but in variety *D. c. alba* they are pure white, in *D. c. atropurpurea*, rich reddish-purple, in *D. c. bicolor*, white and rosy-purple on the same plant and some flowers striped with purple and white, and in *D. c. pallida*, pale pink. Variety *D. c. nana* is a low-growing kind with small leaves. The fruits of *Daboecia* are small capsules.

A closely allied plant, sometimes segregated as *D. azorica,* is endemic in the Azores. It differs from the typical Irish-heath in being lower, and having shorter, narrower leaves and squatter flowers, not more than ⅓ inch long, that are without glandular hairs on their outsides and have more revolute lobes. This kind is generally less winter hardy than the common Irish-heath. Hybrids between it and the ordinary Irish-heath are reported.

Garden Uses and Cultivation. Daboecias are charming plants for rock gardens and for use as groundcovers. Their habits of growth and foliage are attractive, and they produce sprays of gay little blooms almost continuously through summer and fall. Plants for acid, peaty soils, they will not tolerate lime. They appreciate sun, but stand part-day shade. The Irish-heath is hardy in sheltered places in southern New England. Planting may be done in early spring or early fall, spacing the plants 6 to 7 inches apart. The soil surface may with advantage be kept mulched with peat moss. The only routine care needed once the plants are established is watering periodically during really dry weather and trimming the plants back in early spring to remove old flowering shoots or others that tend to be straggly, as well as any that have been winterkilled. In cold climates it is very helpful to cover the plants in winter with cut branches of evergreens; those from discarded Christmas trees can be used. Propagation is easy by cuttings and by seeds. Cuttings of firm, current season's shoots, inserted under mist or in a humid cold frame or a greenhouse propagating bench in summer, soon root. A mixture of peat moss and coarse sand is an encouraging rooting medium. Seeds, which are very small, are sown on a finely sifted surface of sandy peaty soil and are only pressed into the surface or at most scarcely covered with fine soil. Sowing may be done in a greenhouse or cold frame in late winter or spring. The seed soil must never dry. To prevent disturbance of the fine seed watering should be done by immers-

ing the seed container in a receptacle and letting the moisture seep up through the drainage hole.

DACRYDIUM (Dacrýd-ium) — Rimu, Mountain Rimu. This genus of evergreen trees and shrubs of the podocarpus family PODOCARPACEAE includes twenty to twenty-five species indigenous to New Zealand, Tasmania, New Caledonia, Fiji, Chile, and Indomalaya. Its name is derived from the Greek *dakrydion*, a small tear, in reference to the tearlike drops of resin that exude from the wood. Dacrydiums are not common in cultivation in North America, although some are adaptable to parts of California and the Pacific Northwest. In their homelands some are useful timber trees.

The genus *Dacrydium* is closely allied to *Podocarpus* and *Phyllocladus*, differing from the former in having two distinct types of leaves, juvenile and adult, and in other botanical details, from the latter being without cladophylls (leaflike branches). In aspect the taller species somewhat resemble spruces, but they rarely are bisexual and their seeds are not in cones. Dacrydiums usually have male and female flowers on separate trees. The females, consisting of a few small scales, one or more of which have a solitary ovule, are at or near the ends of the branches. The males are solitary and in the axils of the upper leaves. The nutlike seeds are seated in usually fleshy, cup-shaped containers.

The only New World representative of the genus, *D. fonkii* of Chile, is probably the smallest of all conifers. Densely-branched and with small scalelike leaves pressed tightly against the stems, it is up to 1 foot tall. Another low species, the mountain rimu (*D. laxifolium*) of New Zealand, is a suberect or prostrate shrub up to 3 feet tall that has been known to bear seeds when only 3 inches high. Its juvenile leaves are up to ½ inch long, its adult ones not over ¹⁄₁₂ inch.

The rimu or red-pine (*D. cupressinum*), of New Zealand, is an important source of

Dacrydium cupressinum in Ireland

DACTYLIS (Dàcty-lis)—Orchard Grass or Cock's Foot. The common orchard grass or cock's foot (**Dactylis glomerata**), the best known member of this genus of five or fewer species of the grass family GRAMINEAE is cultivated for pasture, forage, and hay. Its varieties, *D. g. variegata* and *D. g. aurea*, are grown for ornament. Native of Europe, Asia, and North Africa, *Dactylis* consists of tall, perennial, flat-leaved plants with spikelets of flowers in dense clusters. The name, of no obvious significance, is derived from the Greek *daktylos,* a finger.

Dactylis glomerata

Dactylis glomerata variegata

dark reddish, often beautifully figured, strong, durable lumber that takes a fine polish and is esteemed for cabinet work, buildings, and railroad ties, and in heavy construction. Its bark is used for tanning. The rimu attains a maximum of almost 200 feet, but more commonly is less than one-half that height. Its trunk may be almost 5 feet in diameter. As a young tree it is pyramidal, shapely, and has long drooping branchlets. At maturity its crown is rounded. The juvenile leaves are awl-shaped, up to ½ inch long, spreading, and loosely-arranged. The adult ones are linear, up to ⅛ inch long, and point toward the tips of the branches. They give the impression of being little more than prickles, and the twigs have the appearance of branches of *Lycopodium*. The Hunon-pine (**D. franklinii**) of Tasmania attains a height of 100 feet. The source of fragrant lumber esteemed for cabinetwork and furniture, this species in aspect is reminiscent of a weeping cypress (*Cupressus*). It has slender, drooping branchlets furnished with small, scale-like leaves.

Garden and Landscape Uses. Members of this genus are plants for special collections and specialists. They are not readily available, but should prove worth testing in milder parts of North America.

Dacrydium franklinii (foliage)

Cultivation. Comparatively little is reported regarding the behavior and needs of dacrydiums in cultivation. Some are grown successfully in the south of England in well-drained, light, loamy soils that are always reasonably moist, but, being barely hardy there, they develop very slowly. For additional information see Conifers.

DACTYLEUCORCHIS. This is the name of orchid hybrids the parents of which are *Dactylorhiza* and *Leucorchis*.

The common orchard grass is widely naturalized in North America. A coarse, tufted plant up to 4 feet in height, *D. glomerata* has leaves up to 9 inches long by ¼ inch wide and open, few-branched flower panicles up to 8 inches in length, with few-flowered, nearly stalkless spikelets in crowded, one-sided clusters.

Much less vigorous and smaller than the typical species, pretty *D. g. variegata* has narrower leaves striped longitudinally with green and white. In *D. g. aurea* the leaves are yellow; this variety is less vigorous than the green-leaved type.

Garden Uses and Cultivation. The colored-leaved varieties of orchard grass are suitable for beds and borders, and the low-growing *D. g. variegata* also makes an attractive edging and may be included in rock gardens. No special difficulties attend the cultivation of these grasses. Although they stand light shade, they are at their best in full sun. They appreciate fertile, reasonably moist soils. Propagation is by division in early spring or early fall.

DACTYLOCERAS. This is the name of orchid hybrids the parents of which are *Aceras* and *Dactylorhiza.*

DACTYLOPSIS (Dacty-lópsis). The only representative of this genus is a low succulent perennial of the *Mesembryanthemum* relationship of the carpetweed family AIZOACEAE. Like its kin it has what at casual sight appear to be daisy-like flowers that are very different from daisies because each is a single bloom rather than a collection of florets in a compound head. The name, from the Greek *dactylos,* a finger, and *opsis,* like, alludes to the leaves.

Forming stemless clusters or mats, *Dactylopsis digitata* inhabits saline soils in South Africa. It has soft, alternate, stalkless, finger-like, erect leaves, up to 2½ inches long by up to 1 inch thick, in groups of up to four. Their bases are covered with the papery remains of old leaves. In the summer resting season the leaves wither. New ones appear in fall, and then in early winter the solitary white flowers, ½ inch or slighty more in diameter, are borne. The fruits are capsules.

Garden Uses and Cultivation. Although one of the first South African plants to be cultivated (it was sent to Kew Gardens, England in 1775 by Francis Masson, the first plant collector sent out by that establishment), *Dactylopsis* is extremely rare in gardens. Some gardeners report that it is difficult to grow, others have had success. A treasure for specialist collectors of succulents, it is not hardy. Conditions that suit *Lithops* and other dwarf, extremely fleshy plants of the *Mesembryanthemum* alliance are most likely to agree with *Dactylopsis.* During the summer it should be kept dry, when in growth watered with cautious moderation. There is some evidence that the addition to the water of about two percent common salt is helpful. A dry atmosphere and full sun are essentials. Propagation is by seed and careful division. For additional information see Succulents.

DACTYLORCHIS. See Dactylorhiza.

DACTYLORHIZA (Dactylo-rhìza). Closely akin to *Orchis* and previously included in that genus, *Dactylorhiza,* of the orchid family ORCHIDACEAE, has also been called *Dactylorchis.* Comprising about thirty species,

it is native in Europe, Asia, North Africa, the Atlantic islands, and Alaska. The name, derived from the Greek *dactylos,* a finger, and *rhiza,* a root, alludes to the tubers.

Dactylorhizas are deciduous ground (terrestrial) orchids that differ from *Orchis* in having tubers palmately-lobed or divided into more or less finger-like segments in contrast to the egg-shaped to rounded ones of *Orchis.*

Handsomest of the genus *D. foliosa* (syns. *Orchis foliosa, O. maderensis*), of Madiera, 1½ to 2½ feet tall, has leaves without spots, the lower ones blunt. The rosy-purple blooms are in massive spikes up to about 8 inches long. Favoring moist and wet soils, the European marsh orchid *D. incarnata* (syns. *Orchis incarnata, O. latifolia*) is a variable sort up to 1 foot tall. It has hollow stems and linear-lanceolate,

Dactylorhiza incarnata

unspotted leaves and dense, cylindrical spikes of magenta-red to red-purple flowers. With solid stems and six to nine usually purple-spotted leaves, the lower ones lanceolate, the upper ones narrower, the spotted orchid of Europe (*D. maculata* syn. *Orchis maculata*) attains a height of

Dactylorhiza maculata

about 1 foot. It has dense, conical to cylindrical spikes of whitish to light purple-pink flowers spotted with purple-brown. Much like the last, but with the lips of its flowers with three triangular lobes, the heath spotted orchid (*D. fuchsii* syn. *Orchis fuchsii*) is widespread throughout Europe. Its blooms are strongly marked with more or less continuous, curved reddish lines. Native of Algeria, *D. elata* is much like *D. foliosa.* It has plain green leaves and massive spikes up to 8 inches long of closely-set violet-purple flowers lifted to heights of from 1½ to 2 feet.

Dactylorhiza fuchsii

Dactylorhiza elata

Garden Uses and Cultivation. These are as for *Orchis.* For further information see Orchids.

DAEDALACANTHUS. See Eranthemum.

DAEMONOROPS (Daemón-orops). Little known in cultivation in the Americas, the

genus *Daemonorops* of the palm family PAL-
MAE comprises 100 species of Indomalayan
feather-leaved rattan palms related to *Calamus* and of similar appearance. Its stems
are used for basket work, mats, and wicker
furniture. From the fruits of *D. draco* is obtained a red resin called dragon's blood,
employed as an astringent and a stimulant
and used in varnishes and lacquers (another dragon's blood is the product of *Dracaena draco*). The derivation of the name is
uncertain, but probably is from the Greek
daimon, a demon, and *rhops*, a shrub, in
allusion to the stems of these climbing
feather palms.

Species of this genus are slender-
stemmed, viciously-thorny, mostly climbing forest plants that differ from *Calamus*
in having the outer spathes of the inflorescences boat-shaped and early deciduous, and the flower clusters without
hooked, tail-like whips, although such
whips are present at the ends of the leaves.

Species cultivated, all climbers, may include these: **D. grandis**, of Malaya, has
leaves up to 10 feet long or longer, with
about thirty-six equally-spaced leaflets on
each side of the midrib. **D. margaritae**, of
Hong Kong and the Philippine Islands, has
leaves up to 6 feet long or longer, with fifty
to seventy-five leaflets spaced equally along
each side of the midrib. **D. periacantha**, of
Malaya and Indonesia, has leaves with
many leaflets in groups of three to five.

**Garden and Landscape Uses and Culti-
vation.** These palms are for collectors of
the unusual. They are adapted for cultivation in humid, tropical climates and in
greenhouses in which tropical conditions
are maintained. Little experience is recorded regarding their cultivation, but they
may be expected to thrive under conditions recommended for *Phoenicophorium*.
For additional information see Palms.

DAFFODIL. This is a popular and old vernacular name for large-trumpet varieties of
Narcissi. It is sometimes used loosely and
less accurately for short-trumpet kinds. In
some parts of North America, including
New York, daffodils are often called jon-
quils. This is regrettable because that name
properly should be reserved for *Narcissus
jonquilla* and its varieties and hybrids. The
fall-daffodil is *Sternbergia lutea*, the Peru-
vian-daffodil, *Hymenocallis narcissiflora*.

DAGA, or Teve, is *Amorphophallus campanulatus*.

DAGGER FERN is *Polystichum acrostichoides*.

DAGGER PLANT is *Yucca aloifolia*.

DAHLIA, For the genus *Dahlia* see the next
entry. The climbing-dahlia is *Hidalgoa*. The
dahlia cactus is *Wilcoxia poselgeri*. The sea-
dahlia is *Coreopsis maritima*.

DAHLIA (Dáh-lia). Modern dahlias are hybrid descendants of species native to Mexico and Central America. They are available in a magnificent array of sizes, forms,
and beautiful flower colors and are favorite
garden flowers, ranking with chrysanthemums, irises, peonies, and gladioluses in
popularity, and in the esteem of enthusiasts even nudging the rose itself. This was
not always so. Early in the twentieth century sophisticated gardeners tended to ig-
nore dahlias as coarse and plebeian and
difficult to place with good taste in garden
landscapes. This remarkable change in at-
titude is due, of course, to the splendid
accomplishments of breeders of new varieties, and of dahlia societies in America,
the British Isles, and the continent of Europe. Traditionally dahlia growing has always had a strong appeal. Wild or species
dahlias are less known to gardeners than
the hybrids. A few are occasionally cultivated.

Named to honor the Swedish botanist
Andreas Dahl, who was a pupil of Linnaeus and who died in 1789, *Dahlia* of the
daisy family COMPOSITAE is composed of
twenty-seven species. Of these, two that
occur spontaneously in South America as
well as in Central America and Mexico are
in South America naturalized introductions from further north. Dahlias are chiefly
plants of the highlands. They commonly
are found at altitudes above 4,800 feet.

The genus *Dahlia* consists of herbaceous
and subshrubby perennials mostly with
clusters of root tubers or thickened roots
and usually erect, rarely vinelike, stems.
One species, *D. macdougallii*, not known to
be in cultivation, is an epiphyte (a plant
that perches on trees, but does not extract
nourishment from its hosts) among the
tops of tall trees in rain forests of Oaxaca,
Mexico. The 2-inch-thick stems of this remarkable kind attain lengths of 30 feet.
Species more conventional in habit range
from kinds not much more than 1 foot in
height to the so-called tree dahlias that exceptionally are 30 feet or more tall.

Dahlias have leaves opposite or occasionally in whorls (circles) of three. They
are usually one or more times pinnately-
divided and have terminal as well as lateral
leaflets, most commonly ovate and toothed.
More rarely the leaves are undivided. The
long-stalked flower heads have centers of
bisexual disk florets encircled by female or
sterile ray florets. The involucre (collar behind the flower head) is of two rows of
bracts, the inner ones slightly joined at
their bases, thin and scalelike, and enlarg-
ing as the flowers fade and seed is formed,
the outer more leafy, fleshly, and smaller.
The fruits are compressed, seedlike
achenes. In horticultural varieties, all or
some of the disk florets may be replaced
by ray type florets to give double or semi-
double flower heads. The flower heads

range from white through all colors except
true blues. Sometimes they are so dark
that they appear almost black.

The history of dahlias is interesting. The
first to be grown outside their native lands
were raised from seeds sent to Madrid,
Spain, in 1789. But probably long before
that they had been cultivated by the Aztecs
perhaps as food for animals, almost surely
for medicinal use. Cultivated plants were
probably the source of the seeds sent to
Spain. The oldest written record of plants
native to the Americas, the Badianus Manuscript, an Aztec herbal of 1582, includes
a clearly identifiable colored illustration of
a red-flowered dahlia with the indication
that it was employed for relieving urinary
troubles. The Aztec name cocoxochtl, alluding to the hollow stems resembling
water pipes, suggests the medicinal employment was based in the doctrine of sig-
natures.

The Abbé Cavanillas, Director of the
Royal Botanical Garden at Madrid when
the first dahlias were raised there, recognized and gave names to three kinds. One,
with single red blooms, became *D. coccinea*. The name *D. rosea* was given to one
with single pink blooms. A double-red-
flowered one was named *D. pinnata*. In
1798 dahlias were sent from Spain to Kew
Gardens in England and in 1802 to France.
The first stocks received in England subsequently died and were replenished by
others received from Madrid in 1804. It is
believed that further importations into
England were made from France about
1815. By 1814 fully double-flowered varieties had been developed in Belgium, and
three years later seventy-five horticultural
varieties were being offered by one nur-
seryman. The date of the coming of dahlias
to North America is not recorded. No men-
tion of them is made in *Hortus Elginensis*,
a catalog published in 1811 of the very extensive collections of plants grown by Dr.
David Hossack at the Elgin Botanic Garden
in New York City. Yet certainly by 1821
many varieties were being cultivated in
American gardens. By 1840 great enthusiasm had developed in Europe for dahlias
and high prices were paid for new varieties. Abroad and in America dahlias enjoyed reasonable popularity throughout
the nineteenth century, but it was not until
the twentieth century that the splendid
developments that resulted in modern
dahlias took place. The formation of the
American Dahlia Society in 1915 did much
to stimulate interest in growing dahlias
and breeding new varieties, and in Europe, especially in England, Holland, and
France, enthusiasm remained high and
skillful breeding resulted in great advance-
ments. Among the chief improvements
were the development of varieties that
hold their flowers well above the foliage
instead of nestled among it as was com-

mon with nineteenth-century dahlias, extensions of range of forms and flower colors, and beginning with the introduction of 'Coltness Gem' in 1922, the development of new races of dwarf dahlias that bloom throughout the entire summer.

Species dahlias, that is kinds as they occur in the wild unimproved by horticultural selection and hybridization, are occasionally grown for interest. Of the three kinds of so-called tree dahlias, *D. imperialis* (syn. *D. maxonii*), *D. excelsa*, and *D. tenuicaulis*, the first is most often cultivated. All have distinctly woody, few-branched stems that remain through more than one growing season and flower heads with about eight rays. The stems of *D. imperialis* are 6 to 18 or rarely 35 feet tall and 3 to 4 inches in diameter. The leaves, 2 to 3 feet in length and nearly as broad as long, are two or three times-pinnately-divided, the primary divisions numbering eleven to fifteen. The numerous half erect to slightly nodding or pendulous, slightly bell-shaped flower heads are 4 to 7 inches in diameter. The ray florets are white, whitish-lavender, or rose-purple. From the last, of which it perhaps should be regarded as only a variety, *D. excelsa* may be distinguished by its leaves having only three to seven primary divisions. Its blooms are 4 or 5 inches in diameter. Leaves not exceeding 2 feet in length, and often much smaller, distinguish *D. tenuicaulis*. Also, its stems are much more slender. This attains a height of 3½ to 10 feet. Its pinnate to twice-pinnate leaves have three to seven primary divisions. Its numerous half erect to slightly nodding, 4- to 5-inch-wide flower heads have light purple to rose-purple rays.

Species that have had a considerable part in the ancestry of garden varieties of dahlias include *D. pinnata*, *D. coccinea*, and *D. merckii*. There are probably others. It has been the practice to refer to garden dahlias as varieties of *D. pinnata*, but in view of their undoubted hybrid origin it seems more satisfactory to group them as some authorities do as *D. hybrida*.

As now understood, *D. pinnata* is a vigorous, deciduous, herbaceous perennial 3 to 6 feet tall or taller and with foliage harsh to the touch. Its leaves are pinnately-lobed, pinnate, or more rarely undivided and coarsely-toothed. Its about eight-rayed flower heads, 3 to 4 inches across, vary considerably and sometimes are more or less double. It is not absolutely certain that this, native within a 100-mile radius of Mexico City, is the plant identical with the one originally introduced to Spain and named *D. pinnata*. Native or naturalized in Mexico, Central America, and parts of northern South America, very variable *D. coccinea* is the most widely distributed of all dahlias in the wild. A deciduous herbaceous perennial 2 to 10 feet tall, its stems

Dahlia coccinea

usually are branched only above. Its foliage is less coarse than that of *D. pinnata*. Its hairy or hairless leaves range from undivided to three times-pinnately-divided. The flower heads, about eight-rayed, are lemon-yellow, orange-yellow, scarlet to blackish-scarlet, and variegated. They are 2½ to 3½ inches in diameter. Recognition characteristics of *D. merckii* are its low, compact growth, red stems, hollow leafstalks, practically hairless foliage, and its large number of about eight-rayed flowering heads, up to about 3 inches wide, on branched stalks. Presumably it is the ancestor of the dwarf bedding dahlias of gardens.

Garden varieties of dahlias are classified by the American Dahlia Society and the Central States Dahlia Society in groups according to the forms of the flowers, and separately as to flower size. The groups are as follows.

Single dahlias have blooms with a single outer row of florets surrounding a disk.

Anemone-flowered dahlias have blooms with one or more rows of ray florets surrounding a dense group of tubular florets.

(b) Anemone

(c) Collarette

(d) Peony-flowered

Dahlia types: (a) Single

(e) Formal decorative

(f) Informal decorative

(i) Semi-cactus

at their base and either straight or incurving.

Orchid-flowered dahlias are like single dahlias except that the rays are involute for at least two-thirds of their length.

American Dahlia Society and Central State Dahlia Society classification by flower size is "A" large, over 8 inches in diameter, "B" medium, 6 to 8 inches in diameter, "BB" 4 to 6 inches in diameter, "M" miniature, not more than 4 inches in diameter, "Ball" over 3½ inches in diameter, "Miniature Ball" 2 to 3½ inches in diameter, and "Pompon" not more than 2 inches in diameter.

Height of plant has an obvious bearing on the placement and uses of varieties in gardens. This is given, with other pertinent descriptive information, in the cata-

(g) Ball

(j) Orchid-flowered

A tall dahlia

(h) Cactus

Collarette dahlias have blooms with a single outer row of ray florets, which are usually flat, and one more internal row of small florets surrounding the disk.

Peony-flowered dahlias have blooms with two or more rows of ray florets surrounding a disk.

Formal decorative dahlias have fully double blooms showing no disk. The rays are broad and either pointed or rounded at their tips. The margins of the outer floral rays are flat or slightly revolute and tend

to recurve. The central rays are cupped, and the majority of all rays are in a regular arrangement.

Informal decorative dahlias have fully double blooms showing no disk. The margins of the majority of the floral rays are slightly or not at all revolute. The rays are generally long, twisted or pointed, and usually irregular in arrangement.

Ball dahlias have fully double blooms, ball-shaped or slightly flattened at the top. The ray florets are blunt or rounded at the tips and cupped for more than one-half the length of the florets.

Pompon dahlias have blooms similar to those of ball dahlias, but more globular and of miniature size with florets involute for the whole of their length.

Incurved cactus dahlias have fully double blooms showing no disk. The margins of the majority of the ray florets are fully revolute for at least one-half their length. The pointed rays curve toward the center of the bloom.

Straight cactus dahlias have fully double blooms showing no disk. The majority of the ray florets are fully revolute for at least one-half their length. The pointed rays are straight, slightly incurved, or recurved.

Semi-cactus dahlias have fully double blooms with pointed ray florets revolute for less than one-half their length, broad

logs of specialists. The taller kinds are from 3 to 6 feet tall or taller. A special race, called bedding dahlias, including those called mignon dahlias, are 1 foot to 1½ feet tall. These are especially suited for summer bedding, window and porch boxes, and grouping at the fronts of borders.

Garden and Landscape Uses. Of all fall flowers dahlias are rivaled only by chrysanthemums for garden display and as cut blooms. Varieties can be selected to serve many purposes. In nearly frostless, warm climates, such as that of southern California, the giant tree kinds can be grown permanently outdoors as tall, evergreen subshrubs. Elsewhere they can be accommodated in big tubs, kept outdoors through the summer, and transferred before frost to a conservatory to complete their growth and bloom. But because these need high space they are ill-suited for any, but very large greenhouses, such as are found in botanic gardens and some public parks. Their indoor cultivation is rarely undertaken by amateurs. Other wild species sometimes grown for interest can be displayed in the same ways as garden dahlias of hybrid origin, which are the only

Bedding dahlias: (a) Single-flowered

(b) Double-flowered

kinds with which most gardeners are concerned.

Garden varieties lend themselves for show in beds and borders by themselves and as single specimens or in groups of about three in mixed flower beds. They can also be used as temporary hedges. For cut flowers and as fanciers' collections they can be accommodated in more ample plantings in field or nursery-like areas where they can be set in well-spaced rows to permit ease of manipulation and cultivation. Select types and varieties for ornamental displays on the bases of such factors as height, form, color, profuseness of bloom, and length of flowering season. For low, bedding displays choose kinds that begin blooming early and continue uninterruptedly until frost. For cut flowers varieties that have long flower stalks and blooms that remain in good condition in water for at least several days are best. Consider also flower form, color, and the freedom with which blooms are produced.

Cultivation. As is true of most popular flowers of which there are many highly developed varieties, such as roses, chrysanthemums, irises, and peonies, dahlias are easy to grow, but at the same time respond to understanding care and good treatment. With little trouble almost any-

one can have passing good blooms, considerable skill and attention is needed to produce superb results. Of first importance are site and soil. Dahlias love sun. They need exposure for at least one-half of each day. Longer is better. They also crave good air circulation. Do not locate them in dank corners or places so surrounded with trees or shrubbery that the air is still and damp. Sites that will grow good vegetables are generally satisfactory for dahlias. Dahlias are gross feeders. For good results the soil must be decidedly fertile, but too much nitrogen unbalanced by adequate supplies of phosphorus and potash results in excessive stem and foliage growth at the expense of flower production.

The soil must be well drained. If water stands on the surface, or accumulates below ground nearer than 3 feet to it for long periods, it will not do. In preparation for planting, spade, rototill, or plow to a minimum depth of 10 inches. Deeper is advantageous. Unless the soil has a high organic content, humus-forming material should be added. Take care of this by mixing in compost, well-rotted manure, peat moss, or commercial humus in liberal quantities. Alternatively sow winter rye in fall and turn it under in spring two weeks or more before planting. In addition to compost or its equivalent, in spring two weeks or more before planting broadcast and fork or rototill in a dressing of a complete fertilizer that has a high phosphorus and potash content.

The plants to be set out may be old tubers or pot tubers or green plants raised from cuttings or seeds. Let us consider each in turn. Old tubers are simply those taken from mature clumps dug in fall and stored over winter. Do not plant the clumps intact; divide them with a sharp, heavy knife. In doing this remember that, unlike potatoes, dahlia tubers do not produce sprouts from eyes scattered over their surfaces, but only from the bases of short pieces of old stem attached to the tops of the tubers. When dividing, be sure that each division has a portion of old stem at-

Dividing a clump of tubers

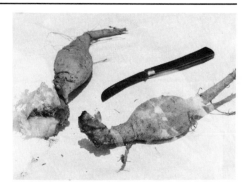

A division ready for planting

tached. If it does not it will not grow. The simplest way of being sure of this is to start the clump of tubers into growth ten days to two weeks ahead of the time of dividing. To do this, set them closely together in flats (shallow boxes) in light compost, peat moss, vermiculite, or similar material kept moist, and put them in a cool, frost-free, light place. Soon growth buds will start and then you can confidently divide the clumps so that each tuber selected for planting has at least one such eye. Pot tubers are usually purchased. They are raised by specialists from propagations made the previous year in time to produce in small pots small tubers that can be wintered in a dormant condition and sold intact the following spring. They are planted without dividing.

Green plants are without tubers at planting time. They are leafy young propagations from seeds sown, or cuttings taken, in late winter or spring. Seeds do not give plants identical to those from which the seeds were taken. Usually there is considerable variation in size of plant, color, form, and size of flowers, and other details. Nevertheless gay floral displays are had from seedlings, and dwarf dahlias raised in this way are especially satisfactory for garden beds and easy and inexpensive to raise. Ordinarily only gardeners interested in developing new varieties raise tall dahlias from seed. For other purposes the results, compared with plants propagated in other ways, are not commensurate with the space required and the effort involved.

Seeds may be sown in the open as soon as it is safe to plant corn, but in the north larger plants and earlier blooms are had from sowings made before this in cold frames, greenhouses, or even in a sunny window. Sow indoors eight to ten weeks before planting out time, using a sandy soil and covering the seeds to a depth of ¼ inch. In a temperature of 60 to 65°F, they germinate in approximately nine days. As soon as their second pair of leaves is fairly developed, transplant individually to 3-inch pots. Later they may need to be moved into containers an inch larger. Keep the young plants growing in a sunny place where the night temperature is 50 to 55°F and that by day five to fifteen degrees higher. To assure bushiness pinch the tips out of the main stems when they are 3 to 4 inches high. For ten days or so prior to planting in the garden harden the plants by transferring them to a cold frame or sheltered, sunny place outdoors.

Cuttings give plants identical with those from which they were taken. To raise stock in this way, plant clumps of tubers in boxes or large pans (shallow pots) in a mixture of sandy soil and large amounts of peat moss or leafy mold in February or March. Put them in a sunny place where the temperature is 55 to 60° at night, and a little higher by day, and keep them moderately moist. As soon as the young shoots are about 3 inches long slice them off with a keen-bladed knife, making sure that a sliver of old stem is attached to the bottom of the cutting. Plant the cuttings in sand, vermiculite, or perlite in a greenhouse propagating bench or approximately similar environment. The essentials are a highly humid atmosphere, good light without exposure to direct sun, a temperature of 60 to 70°F, and a uniformly moist, but not constantly saturated rooting medium.

A flat of green plants, from cuttings

Within a very few weeks the cuttings will have roots 1 inch to 1½ inches long. Then pot them individually in sandy soil in 3-inch pots, return them for about a week to the environment from which they were taken, and from then on treat them as advised above for seedlings.

The time to plant outdoors depends upon whether you are setting unstarted tubers or green plants. Dahlias are very sensitive to frost and must not be planted until the weather is warm and settled. It is normally safe to plant tubers about the time early corn is sown, and green plants about two weeks later. Spacing will depend upon what the plants are expected to achieve. Dwarf dahlias are satisfactory at 1½ to 2 feet between individuals, tall kinds need at least 3 feet between plants, and if in rows these should be 3½ to 4 feet apart.

Tall kinds need staking securely. Special dahlia stakes can be purchased or fashioned out of 1¼-inch square wood. Some growers recommend the use of concrete reinforcing rods ½ or ⅝ inch diameter, coated with rust-proof paint and then with green or brown enamel. Whatever types are used let them be of sufficient length that when they are driven 1½ feet or more into the ground they will be tall enough to support the plants. Drive the stakes before planting.

To plant tubers, excavate holes 6 inches deep and 1 foot wide close to the bases of the stakes. Set the tubers with the eye or sprout toward the stake and cover with 2 to 3 inches of soil. If more than one shoot develops remove all except the strongest. As the stem lengthens, in the normal task of surface cultivation gradually fill the holes to ground level. Green plants are not set as deeply as tubers. Holes for them are made at such a depth that when planted, the top of the root ball is about 2 inches beneath the surface.

A pot-grown green plant, at planting time

Newly planted tubers, staked and labeled according to variety

Subsequent care involves keeping the shoots tied neatly to the stakes to avoid damage by storms or accidents. Dwarf dahlias normally branch freely and need no pinching or disbranching after they are planted out. With taller dahlias pinch the tips out of their stems just above the third pair of leaves. This encourages the development of six side shoots. With smaller-

Secure with soft ties, leaving ample room for the stem to thicken

Disbranching, removing unwanted sideshoots while yet small

A young dahlia making vigorous growth

flowered varieties all may be kept, but with larger-flowered kinds take off all except the four strongest. Subsequent disbranching consists of removing spindly, weak shoots and any that grow toward the center of the plant. Under favorable conditions growth will be rapid, and frequent care in the matters of disbranching and tying retained shoots to the stakes is necessary.

Summer care is not arduous, but must

be frequent. Keep the surface soil shallowly cultivated to keep down weeds and admit air. By the end of July or the beginning of August feeding roots will invade the surface soil. Then cease cultivation, apply a light dressing of a complete garden fertilizer, and mulch with loose compost, salt hay, straw, or half-decayed leaves. The mulch restrains weeds and keeps the earth reasonably cool and uniformly moist. Regular, repeated spraying or dusting will almost surely be needed to control various pests and diseases, the chief ones of which are listed below. During dry spells soak the ground deeply at about weekly intervals. An absolute requisite in producing good dahlias is to keep the plants growing without check. If they cease to make good progress as a result of lack of moisture, inadequate fertility, or other cause, the stems become woody and all hope of a fine display of blooms passes. The largest blooms are had, from large-flowered varieties, by disbranching to the extent that relatively few are produced, and by disbudding, in addition of course, to attention to such other cultural needs as fertilizing, watering, and controlling pests. Disbudding means removing while small all except one of the three flower buds that develop at the end of each stem and all shoots that appear in the axils of the three pairs of leaves below. In climates where frost does not penetrate to the roots, or where they can be protected from freezing by a heavy winter mulch, dahlias can be grown as permanent perennials, but except with the tree kinds, and sometimes the mignon or bedding types, this is rarely done because better results are commonly had by taking the roots up and storing them over winter.

Preparation for winter storage begins after frost has seriously blackened the plants. Then, without delay, dig up the clumps of tubers, being as careful as pos-

Digging dahlia clumps for storage: (a) Cut the stems back to within a few inches of the ground

(b) Use a spading fork to raise the clumps

(c) Lift the clump gently so as not to injure the tubers

(d) Two people, with forks at opposite sides of the clump, can pry the clumps out more easily

sible not to break or damage them, cut the stems back to within about 6 inches of their tops, turn the clumps upside down, and if possible leave them to dry for a few hours. Before nightfall bring them indoors, dust any cut or broken surfaces with sulfur, ferbam or other fungus-inhibiting material, and stand them upside down so that

Label each clump with the name of the variety

Then stand the clumps upside down so that moisture drains from the cut stems

the cut stems can drain. After about a week put them where they are to stay for the winter. Ideal storage is an earth-floored, frostproof cellar where the temperature range is 35 to 50°F. Such accommodations are not always available. Storage where the air is less humid, but not warmer, is satisfactory if measures are taken to prevent drying, with consequent softening and shriveling of the tubers. To do this, cover the clumps with very slightly damp peat moss, vermiculite, or sand or wrap them in several sheets of newspaper and pack them in deep boxes or barrels. Yet another method is to wash the clumps free of soil, allow them to dry, then coat them with wax by dipping them in, and withdrawing them slowly from, hot water with a thick layer of melted paraffin wax floating on its surface. Some gardeners store dahlia clumps in plastic bags containing some vermiculite, and left open for three or four weeks until drops of moisture no longer show on their insides if the bags are closed. In some sections it is possible to winter dahlia tubers in holes about 2 feet deep dug in well-drained ground, covered with a thick layer of straw or hay supported on a framework of stout sticks placed across the hole, and topped with a layer of soil. A bed of straw, dry leaves, or similar material is positioned beneath the tubers. Whichever method of storage is used, it is well to examine the tubers occasionally through the winter for signs of decay and if found to take prompt remedial action.

DAHOON is *Ilex cassine*.

DAIS (Dà-is). One of two shrubs that constitute *Dais* of the mezereum family THYMELAEACEAE is cultivated in frost-free and almost frost-free parts of the United States. Related to *Daphne*, this genus has opposite or alternate leaves. The showy parts of its flowers, which are in erect umbels, consist of five-lobed, corolla-like calyxes. The true corolla is represented by minute scales. There are ten stamens. The name, from the Greek *dais*, a torch, alludes to the form of the flower clusters.

The species of *Dais* here described is a native of South Africa, the other of Madagascar. A neat, deciduous shrub or small tree, *D. cotinifolia* is occasionally up to about 20 feet tall, but usually does not exceed 10 feet. It has glossy, opposite or alternate, obovate or oblong leaves up to 2½ inches long. The fragrant, lilac-pink, starry flowers, ½ to ¾ inch long and about ½ inch across, and hairy on their outsides, are in stalked, terminal clusters, freely borne. From the bark of this species natives in Africa obtained a strong fiber which they used as thread.

Garden and Landscape Uses and Cultivation. This is a general purpose flowering shrub of tidy appearance. It is attractive both in and out of bloom, and succeeds in sun in ordinary soil that is well drained. Plants are easily raised from leafy summer cuttings and from seeds.

DAISY. Without an adjective the word daisy, if longest usage is to determine its application, properly belongs to *Bellis perennis*, the old-time day's eye of England. In the United States this is known as English daisy and the word daisy without qualification is generally used for the ox-eye-daisy (*Chrysanthemum leucanthemum*). Many other plants of the daisy family COMPOSITAE have daisy as part of their colloquial names. Among such are African-daisy (*Arctotis stoechadifolia, Lonas annua*), Barberton-daisy (*Gerbera jamesonii*), blue-daisy (*Felicia amelloides*), camphor-daisy (*Haplopappus phyllocephalus*), Carmel-daisy (*Scabiosa prolifera*), crown-daisy (*Chrysanthemum coronarium*), Dahlberg-daisy (*Dyssodia tenuiloba*), globe-daisy (*Globularia*), gloriosa-daisy (*Rudbeckia hirta*), kingfisher-daisy (*Felicia bergeriana*), lazy-daisy (*Aphanostephos*), Livingston-daisy (*Dorotheanthus bellidiformis*), Michaelmas-daisy (*Aster*), Nippon-daisy (*Chrysanthemum nipponicum*), orange-daisy (*Erigeron aurantiacus*), painted-daisy (*Chrysanthemum coccineum*), Paris-daisy (*Chrysanthemum frutescens*), poached-eggs-daisy (*Myriocephalus stuartii*), Shasta-daisy (*Chrysanthemum superbum*), sleepy-daisy (*Xanthisma texanum*), Swan-river-daisy (*Brachycome iberidifolia*), Tahoka-daisy (*Machaeranthera tanacetifolia*), Transvaal-daisy (*Gerbera jamesonii*), turfing-daisy (*Tripleurospermum tchihatchewii*), and white-daisy (*Layia glandulosa*).

DAISY BUSH or DAISY TREE. See Olearia.

DAKUA WOOD is *Agathis vitiensis*.

DALBERGIA (Dal-bérgia)—Rosewood, Sissoo. This genus is one of the many tropical and subtropical ones of the pea family LEGUMINOSAE. Its name commemorates the Swedish naturalist Nils Dalberg, who died in 1820, and his brother Carl Gustaf Dalberg, who collected plants in South America. The group consists of 300 species of trees, shrubs, and woody vines and occurs in both the Western and Eastern hemispheres. Certain of its kinds are the sources of the true rosewoods used extensively for furniture and other purposes. Few of its species are cultivated.

Dalbergias have alternate, pinnate leaves, with the leaflets arranged alternately, and terminal or axillary panicles of small, pea-shaped, white or purplish blooms with united stamens. The fruits are flat pods that do not open to shed the seeds but, after they fall, in contact with the ground soon rot. Two species, *D. sissoo* and *D. latifolia*, are among the most valuable lumber trees of India and are planted in forests, as well as for shade and ornament. The former, and to a lesser extent *D. lanceolaria*, are grown to some extent in Florida. These and occasionally other species, are used similarly elsewhere in the tropics and subtropics.

The sissoo (*D. sissoo*) is deciduous and up to 80 feet tall. Its leaves have a zig-zagged central axis and usually five, but sometimes three, pointed, broadly-elliptic or ovate leaflets 1 inch to 3 inches long, which are pubescent when young, but are without hairs later. The terminal leaflet is the largest. The yellowish-white, very fragrant flowers, about ⅓ inch long, are in axillary panicles. The pods, each containing one to four seeds, are 1½ to 4 inches long and up to ½ inch wide. This species is naturalized in Florida.

More graceful and elegant than the sissoo is *D. lanceolaria* (syn. *D. frondosa*). This native of India is a tall, deciduous tree with slender, somewhat drooping branches and leaves 3 to 8 inches long, of nine to seventeen blunt, ovate or obovate leaflets 1 inch to 2 inches long and often notched at their ends. The flowers are in big terminal and axillary panicles of small one-sided racemes. They are pale mauve and up to ⅓ inch long. The pliable pods, 1½ to 4 inches in length, narrow at their ends and contain one to three seeds.

Native to southern India, *D. latifolia* is there a tall deciduous tree, but further north in India is usually much lower. Its leaves are 4 to 6 inches long and have three to seven broad-elliptic to nearly round, blunt leaflets, sometimes notched at their apexes. The leaflets are ½ inch to 3 inches long and comparatively long-stalked. The whitish flowers, ¼ inch long, are in short panicles from the leaf axils. The oblong-lanceolate, strap-shaped pods have one to four seeds.

Other tree species worthy of attention include *D. paniculata,* a semievergreen native of India, with an open crown of slender branches and leaves of eleven to thirteen roundish leaflets. It is adaptable to dry soils as are semievergreen *D. mammosa* and *D. cochinchinensis*, both of Vietnam and both open-headed and with slender branches. The twice-pinnate leaves of feathery appearance of *D. mammosa* consist of numerous, inch-long, narrow leaflets. The leaves of *D. cochinchinensis* have eleven to thirteen leaflets and are almost or quite 1 foot long.

Garden and Landscape Uses and Cultivation. These are attractive shade and ornamental trees for warm, frost-free climates. They prefer soils not deficient in moisture and are propagated by seed.

DALEA (Dà-lea)—Smoke Tree, Indigo Bush. Its name commemorating the English botanist Dr. Samuel Dale, who died in 1739, *Dalea* comprises possibly 200 species of the pea family LEGUMINOSAE. From nearly related *Petalostemon*, which some botanists include, *Dalea* in its narrower sense differs in its flowers having more than five stamens.

Daleas are herbaceous plants, shrubs, and small trees of the Americas. They have alternate, rarely undivided, more often pinnate leaves, with a terminal and lateral toothed or toothless leaflets. The small flowers, purple, pink, white, or yellowish, are in terminal spikes or racemes. They have five-toothed, persistent calyxes. Their corollas are pea-shaped with their lower pair of petals united as a keel or separate. There are nine or ten stamens, united at least part way from their bases into a tube, and one style. The fruits are small pods.

The smoke tree (*D. spinosa* syn. *Psorothamnus spinosus*) is a much-branched, spiny shrub or tree, up to 25 feet tall, native to desert regions in the southwestern United States and adjacent Mexico. Ashy-gray throughout with fine pubescence, it has quickly deciduous foliage and bright blue-purple flowers. The leaves, oblong-wedge-shaped, toothed, and up to 3 inches long, are dotted with conspicuous dark glands. The abundant blooms, ⅓ to nearly ½ inch

Dalea spinosa, in the Mojave Desert

Dalechampia roezliana

long, are in spikes ¾ to 1¼ inches long. The name smoke tree is also applied to *Cotinus*.

A shrub 1 foot to 3 feet tall, **D. frutescens,** native from Texas to Mexico, has glaucous-green leaves up to 1 inch long of thirteen to seventeen obovate leaflets, and violet-colored blooms in spikes ½ to ¾ inch long.

Herbaceous perennial kinds, often woody at their bases and occasionally cultivated, include those now to be described. Native from South Dakota to Texas, **D. aurea** is pubescent and 1 foot to 1½ feet tall. It has silvery foliage. The leaves, up to 2 inches long, have five to rarely nine oblong-oblanceolate to narrowly-obovate leaflets. The yellow flowers are in dense spikes up to 2 inches long. White-flowered **D. enneandra,** indigenous from North Dakota to Texas, attains heights up to about 3½ feet. Its hairless leaves have five to eleven long, linear to narrow-oblong leaflets ¼ to ½ inch long. The loose flower spikes are 2 to 4½ inches long.

Garden and Landscape Uses and Cultivation. Daleas, best adapted for native plant gardens and naturalistic and informal areas, are occasionally accommodated in flower beds. They are likely to respond best to conditions approximating those under which they grow in the wild. Good soil drainage and exposure to sun are essen-

tials. Transplanting is likely to be hazardous, therefore it is recommended that plants raised from seeds be grown in pots until they are set in their permanent locations.

DALECHAMPIA (Dalechámp-ia). Few non-botanists would relate the cultivated members of this genus to the castor-bean (*Ricinus*), yet a rather close affinity exists. Both belong to the spurge famiy EUPHORBIACEAE and are kin to poinsettias and numerous cactus-like succulents of parts of Africa and other desert regions. They look like neither. Widespread in the tropics, *Dalechampia* consists of about 110 species, very few of which are known in cultivation. The name honors James Dalechamp, a French physician and botanist, who died in 1588. The genus consists of mostly vines, together with a few shrubs. They have alternate, lobeless leaves with large stipules (basal appendages) and dense clusters of small, bisexual flowers without petals, but with two large petal-like bracts backing each flower cluster.

Best known is **D. roezliana,** a stiff, erect, branched, leafy, evergreen shrub, 3 to 4 feet tall, and a native of Mexico. Its firm, obovate-oblanceolate, stalkless leaves, up to 6 inches in length and often coarsely toothed above their middles, narrow to heart-shaped bases. The two large, showy,

long-lasting floral bracts are broad-ovate and 2 to 2½ inches long. Rich rose-pink or rose-red and conspicuously veined, they are short-stalked and, at their margins, toothed. Among the tiny yellow flowers a number of much smaller bracts develop. Variety *D. r. alba* has white or creamy-white bracts. Much rarer is **D. scandens,** which in the wild ranges from Mexico through Central America and the warmer parts of South America and in the West Indies. This is a twining woody vine,

Dalechampia scandens

softly-hairy and with deeply three- to five-lobed leaves with toothed margins and stalks about as long as their blades. Where the leafstalk joins the blade are two erect hornlike appendages. The tiny rich green flowers are accompanied by a pair of tri-lobed bracts that spread by day and close at night. They are broadly-ovate and up to 1½ inches across. The bracts, usually green, are sometimes creamy-white tipped with green.

Garden Uses and Cultivation. Only in the humid tropics and in greenhouses providing a similar environment can these plants be expected to give satisfaction. They grow in any reasonably fertile soil, well-drained and moderately moist, and appreciate a little shade from strong direct sun. In greenhouses they should be watered freely from spring through fall, more sparingly in winter. A minimum winter night temperature of 55 to 60°F is satisfactory, at other seasons night temperatures may be higher. At all times day temperatures may exceed those maintained at night by five to ten degrees. Shrubby *D. roezliana*, which grows rather slowly, may begin blooming when less than 1 foot high. Take care not to use pots too large for its roots. More rampant than the last, *D. scandens* needs more generous root room. Both are easy to propagate from cuttings in a humid greenhouse propagating case, and by seeds, sown in sandy peaty soil in 60 to 70°F.

DALIBARDA (Dalibàrd-a)—Dewdrop. Inhabiting moist woods and swamps from Nova Scotia and Quebec to New Jersey, Ohio, and Minnesota, one small member of the rose family ROSACEAE is the only species of *Dalibarda*. Its name commemorates T. F. Dalibard, a French horticulturist, who died in 1779.

A herbaceous perennial with slender, creeping stems and violet-like, round-toothed, ovate, slightly pubescent leaves with blades up to 2 inches long and wide, and downy stalks up to 4 inches long, *D. repens* has two distinct kinds of flowers. The more obvious petaled ones are mostly sterile, their function being display rather than fruit production. The others, coming later than the petaled ones, short-stalked and without petals, are succeeded by fruits that are dry, achene-like, one-seeded drupes. The sterile blooms are solitary or in pairs atop downy stalks 2 to 4 inches long that overtop the foliage. Resembling the flowers of strawberries (*Fragaria*) and cinquefoils (*Potentilla*), they are about ½ inch across and have five spreading, toothed, white petals and a conspicuous cluster of numerous stamens. Although resembling the strawberry in aspect, this plant is botanically much more closely related to blackberries (*Rubus*), to which it exhibits little obvious likeness.

Garden Uses and Cultivation. A neat grower of minor importance horticulturally, this sort is likely to be of interest only to students and collectors of our native flora. It is best adapted for inclusion in native plant gardens and rock gardens, but is not always easy to establish. Like many acid-soil plants, it tends to be finicky and capricious. Perhaps it needs some unknown mycorrhiza associated with its roots. Who knows? In any case it should be set in very acid, peaty or woodsy soil that is always moist, and in shade. It is propagated by division and by summer cuttings planted in acid leaf mold or in a mixture

of acid peat moss and sand, in a humid cold frame. Seeds may also be used to secure increase, but these do not seem to be produced abundantly.

DAME'S-VIOLET is *Hesperis matronalis*.

DAMMER-PINE is *Agathis dammara*.

DAMNACANTHUS (Damna-cánthus). To the madder family RUBIACEAE belongs this genus of evergreen, much-branched, often spiny shrubs. Indigenous from Japan to northern India, it comprises six species. The name, derived from the Latin *damno*, to injure, and *acanthus*, a spine, alludes to its thorny armament.

These plants have rigid branches and opposite, very short-stalked, ovate, leathery leaves in the axils of which the small white blooms, solitary or in pairs, are borne. The flowers have four- or five-toothed calyxes, funnel-shaped corollas with four or five spreading lobes (petals), four or five stamens, and a slender style ending in a two- to four-lobed stigma. The fruits are berry-like drupes (fruits of the plum type) containing one to four stones.

Variable, and native to Japan, the Ryukyu Islands, Korea, China, Thailand, and northern India, *Damnacanthus indicus* is up to 5 feet in height, but often lower. Its shoots, leafstalks, and sometimes the mid-

Damnacanthus indicus

ribs of its leaves are shortly-hairy. The slender, forking branches spread horizontally and have needle-like spines ⅓ to ¾ inch long that are from one-half as long to as long as the leaves. The leaves, pointed at their tips, are lustrous, broad-ovate to broad-elliptic, and up to 1 inch in length. Solitary or in pairs, the four-petaled flowers are ½ inch long or longer. Spherical and up to ¼ inch in diameter, the very long-lasting berries are bright red. Variety *D. i. microphyllus* has leaves under ½ inch long. In *D. i. ovatus* the leaves are ovate, and the spines prominent.

Differing from *D. indicus* in having spines only one-fourth to one-half as long as the

leaves, *D. major* (syn. *D. i. major*) is a native of Japan. Its ovate leaves are ¾ inch to 1½ inches long. Variety *D. m. lancifolius* has broad-lanceolate leaves 1¼ to 2½ inches long, *D. m. parvifolius*, of Japan, the Ryukyu Islands, and Korea, leaves smaller than those of *D. major*.

Garden and Landscape Uses and Cultivation. These quite beautiful shrubs are not hardy in the north, but in regions of mild winters are suitable for the fronts of shrub beds, rock gardens, and hedges. They succeed in sun or part-shade in ordinary fertile soils, not excessively dry, and are propagated by summer cuttings under mist or in a greenhouse propagating bed, and by seed.

DAMPING DOWN. More commonly used in Great Britain than America, the term damping down means wetting greenhouse floors, walls, and other surfaces including the benches on which pot plants are grown. It is done to increase humidity and in hot weather to lower temperatures somewhat.

An important greenhouse procedure, damping down may be needed two, three, or more times a day where tropical plants are grown, less often for most other kinds, and in greenhouses devoted to succulent plants rarely or not at all. When damping down with a hose or watering can, take care not to wet soil or foliage. If this is needed, it is better to damp down first, then water plants needing that attention, and finally to spray the foliage. Such orderly procedure minimizes danger of soaking the soil of plants not in need, and of wetting foliage best left dry. Damping down floors and walls can be done by a system of water pipes fitted with mist nozzles that can be turned on at intervals, either manually or by an automatic timer.

DAMPING OFF. The killing of seedlings while they are germinating, or very shortly afterward, by soil fungi is called damping off. A common trouble, especially in greenhouses, and when the weather is warm and humid outdoors, it can take tremendous toll, to the extent of wiping out entire populations of young seedlings. Several different fungi cause damping off. Some kill the germinating seeds before above-ground growth appears, others cause the collapse or withering of the seedlings after they are well up. There is no cure for infected plants, but effective precautionary measures greatly reduce the probability of infection and limit its spread if it occurs.

Sowing in sterilized soil or other sterile medium, such as vermiculite, perlite, milled sphagnum moss, or baked sand, almost eliminates the chance of damping off, although it is possible even then for the causal fungi to be introduced by soil-contaminated hands or the use of dirty pots, and in more unlikely ways. In lieu of using sterile soil good results maybe had by watering the prepared seed pots, flats, or beds immediately after sowing and again a week after the seedlings appear above ground with an anti-damping off preparation based on oxyquinoline. Such preparations are sold by dealers in garden supplies. If nonsterilized, nontreated soil is used it is useful to coat the seeds before sowing with a protectant dust. These, based on various chemical fungicides, are available from dealers in garden supplies.

There are predisposing causes to damping off. To avoid these, make sure that containers in which seeds are sown are well drained and clean. Avoid using soil that as a result of watering will lose its porosity and become compact. Do not sow too thickly. Do not, by growing at too high temperatures or in too little light, permit seedlings to become drawn (thin, leggy, and weak). Maintain fairly even temperatures. Wide fluctuations tend to encourage damping off. In greenhouses sturdy growth and hence resistance to damping off can be encouraged by keeping the seedlings near the glass rather than at long distances from it. Ventilate to assure good air circulation and discourage dank, "dead" air, but avoid cold drafts. Water with care, especially in dull and humid weather. Do not have the soil surface or plants wet for long periods. With this in mind, as far as possible confine watering to bright days and early enough in the day that surface moisture dries before nightfall. Do not water unless the soil is definitely approaching a stage of dryness that unless water is given the seedlings may wilt. When you water, soak the soil thoroughly. If water must be given in dull or humid weather or so late in the day that the drying of surface moisture within about an hour is unlikely, rather than allow seedlings to suffer from lack of moisture, avoid if possible using a sprinkling can or similar device. Instead, immerse the pots or flats halfway to their rims in water until moisture reaches the surface by capillarity. Then remove them promptly.

If damping off appears, prompt early action may prevent its spread. First correct predisposing causes discussed above. Then water with an anti-damping off preparation with an oxyquinoline base. It may even be helpful to sprinkle powdered sulfur, powdered charcoal, or baked sand among patches of affected seedlings and in their immediate surrounding areas. Another rescue procedure that often works at the first sign of trouble is to transplant unaffected seedlings immediately to other pots or flats. This results in correcting some of the predisposing conditions to attack, such as overcrowding, and consequent shading, and makes possible an environment more favorable to the seedlings and less favorable to the attacking fungus. Because transplanting may be necessary before the seedlings reach the stage when it would normally be done handle the young plants with particular care.

DAMSON, or damson plum, is *Prunus insititia.*

DANAE (Dà-nae)—Alexandrian-Laurel. The name of this genus of the lily family LILIACEAE is that of the daughter of King Acrisius of Argos. There is only one species, an evergreen shrub related to butcher's-broom (*Ruscus*). It is called Alexandrian-laurel and is native to Asia Minor and Iran.

Alexandrian-laurel (*Danae racemosa* syn. *Ruscus racemosus*) is 2 to 4 feet tall and branched. It has slender, erect or arching, smooth, evergreen stems. What appear to be leaves are really cladodes; they are flattened branches that look like and function as leaves. The true leaves are the tiny scales in the axils of which the cladodes develop. The cladodes are alternate, short-stalked, oblong-lanceolate, pointed, 1½ to 4 inches long, and bright green above and below. The tiny yellowish to white flowers are bisexual (in this way differing from those of *Ruscus*). They are stalked, and in terminal racemes of four to six. Each has six perianth lobes (usually called petals), six stamens, with their filaments (stalks) joined, and a short, knobbed style. The fruits are red berries, each with a paler, saucer-shaped disk at its bottom, ¼ inch in diameter, and containing one or rarely two seeds. Alexandrian-laurel is hardy about as far north as Virginia.

Garden and Landscape Uses. An exceedingly graceful, lively green shrub, with long arching stems that produce a distinctly bamboo-like effect, and bright red berries, Alexandrian-laurel is an admirable ornamental for partially shaded locations. It succeeds in ordinary soils not excessively dry, and its leafy shoots are excellent for cutting to mix with flowers in arrangements.

Cultivation. No particular care is required. If the plants tend to become too dense, cut out some of the older stems in spring. A mulch of compost or other organic, moisture-conserving material maintained around the plants is all to the good. Propagation is by division in spring or early fall and by seeds removed as soon as they are ripe from their surrounding pulp and sown immediately in a cold frame.

DANDELION. Unmodified this word alludes to *Taraxacum,* dealt with in the next entry. The Cape-dandelion is *Arctotheca,* the desert-dandelion, *Malacothrix glabrata,* the dwarf-dandelion, *Krigia,* the mountain-dandelion, *Agoseris.*

DANDELION. The dandelion (*Taraxacum officinale*) of the daisy family COMPOSITAE is well known as an ubiquitous weed of lawns, cultivated lands, waysides, and waste places. Were it rare, it would doubtless be prized for its showy golden blooms. In fact its ornamental qualities make no appeal to gardeners. Dandelions are grown, however, as healthful, edible greens for use in salads and for eating boiled. In many places in North America wild dandelions are gathered in spring for these uses, especially by immigrants from southern Europe. Cultivated dandelions are superior to wild ones. There are several varieties or strains, selected for the size, abundance, and succulence of their foliage. Some have cut or frilled leaves.

As a crop dandelions are treated as an annual or a biennial. As may be imagined, they are tolerant of a wide variety of sites and soils. But the best results are had from deep, fertile soil and adequate moisture. Prepare the land by spading, rototilling, or plowing. Mix in humus-forming material, such as compost or manure, and a dressing of a complete garden fertilizer. Sow the seeds thinly in early spring in rows 1 foot apart. Thin the seedlings to 9 inches to 1 foot apart. Summer care consists of keeping weeds down by frequent shallow cultivation or by mulching. Watering deeply at intervals in dry weather helps.

Harvesting may be done in fall, but because dandelion greens are usually more appreciated in spring than later, it is usual if fall harvesting is done to pick only leaves and to leave the roots in the ground over winter to produce spring foliage. Spring harvesting consists of cutting off the entire heads of foliage just below the lowest leaves. Tenderness is induced and the greens are less bitter if they are blanched in the manner of escarole. This is done by excluding light for two or three weeks either by lifting the leaves upward and tying their tops together or by straddling the rows with a pair of broad boards leaned so that their top edges meet to form a tentlike covering.

Excellent winter salading can be had by digging the roots in fall, storing them packed in slightly damp soil, sand, or peat moss in a celler or other place where the temperature is a little above freezing, and forcing them into growth in successive batches in the manner whitloof chicory is forced. To do this, plant the roots a few inches apart in deep boxes in soil or a mixture of sand and peat moss kept moist, but not saturated. Keep in the dark in a temperature of 55 to 60°F. In a very short time a crisp, succulent crop will be ready for harvesting.

DANEWORT is *Sambucus ebulus.*

DANGLEBERRY is *Gaylussacia frondosa.*

DAPHNE (Dáph-ne)—Garland Flower, Spurge-Laurel, Mezereon. Restricted in the wild to temperate and subtropical Europe and Asia, *Daphne,* of the mezereum family THYMELAEACEAE, comprises seventy species of deciduous and evergreen shrubs. The name is the ancient Greek one of the unrelated sweet bay or laurel (*Laurus nobilis*).

Daphnes, erect or prostrate, have usually tough, flexible, young branches, and alternate or rarely opposite, short-stalked, undivided leaves without lobes or teeth. Their small, usually fragrant flowers are in clusters or short racemes with sometimes bracts immediately beneath them. The showy part of the bloom is a tubular, corolla-like calyx with four petal-like lobes, often hairy on the outside, generally glistening within. There is no corolla. The eight very short-stalked stamens in two rows, are included in or sometimes slightly protrude from the calyx tube. There is a short style, or none, and a large headlike stigma. The fruits are one-seeded, berry-like, fleshy or leathery drupes (fruits constructed like plums).

Deciduous daphnes commonly 1 foot tall or taller include the mezereon (*D. mezereum*), of Europe and western Asia, *D. genkwa*, of China and Korea, and *D. giraldii*, of China. With stiffly erect branches 3 to 5 feet tall, **D. mezereum** has alternate wedge-shaped, oblong to oblanceolate, thin,

Daphne mezereum alba

hairless leaves 1½ to 3½ inches long and ⅓ to 1 inch wide. Its numerous, fragrant, ½-inch-wide, rather harsh purplish-pink flowers, in stalkless clusters of two to four strung along shoots of the previous season's growth, appear before the leaves. Up to ½ inch in length they are silky-hairy on their outsides. About ¼ inch long, the scarlet, roundish-egg-shaped fruits make a good display. This kind is hardy through most of New England. Variety *D. m. alba* has pure white flowers and yellow fruits. The flowers of *D. m. grandiflora* are bigger than those of the typical species, those of *D. m. rosea* are clear pink. About 3 feet tall, lovely **D. genkwa** is of loose, sometimes rather straggly habit. It has mostly opposite, oblong-lanceolate leaves 1 inch to 2 inches long, up to a little more than ½ inch wide, and silky-hairy beneath, at least on

Daphne genkwa

Daphne genkwa (flowers)

the veins. Appearing before the foliage and not fragrant, the charming lavender-blue flowers are in loose clusters of up to seven along shoots of the previous year's growth. Densely-silky-hairy on their outsides, they are about ½ inch long. The fruits are white. This is hardy in southern New England. About as hardy, **D. giraldii** is erect and up to 2 feet tall. Crowded at the ends of the branches, its hairless, oblanceolate leaves are 1¼ to 2¼ inches long. Slightly fragrant, the flowers, not hairy and about ⅓ inch long, are in terminal, bractless clusters of up to eight. The fruits are bright red.

Deciduous in cold climates, semievergreen in mild ones, **D. altaica** hails from the Altai Mountains. Freely-branched and up to 4½ feet in height, this has hairless, narrow-oblong to oblanceolate leaves 1¼ to 2½ inches long by up to a little over ½ inch wide. Its flowers, up to ⅓ inch in length, are in bractless clusters of up to ten. The egg-shaped fruits are yellowish-red. Beautiful, free-flowering **D. burkwoodii** and very similar *D. b.* 'Somerset' are hybrids between *D. caucasica* and *D. cneorum*. Hardy in southern New England, in mild climates they are evergreen or par-

Daphne burkwoodii

Daphne burkwoodii 'Somerset'

Daphne burkwoodii 'Somerset' (flowers)

tially so, in harsher ones deciduous. Rounded, up to 6 feet tall, they have narrow-oblanceolate leaves ¾ inch to 1¼ inches wide. Variety 'Somerset' tends perhaps to be lower, broader, and more compact than some forms of this hybrid. Very freely produced, the beautiful, sweetly fragrant flowers are in crowded, 2-inch-wide terminal clusters. About ½ inch wide, they are creamy-white sometimes slightly tinged with lilac-pink. The fruits are red. A slow-growing, Canadian-raised hybrid between *D. burkwoodii* and *D. retusa*, evergreen **D.**

Cyrtopodium andersonii

Cytisus spachianus

Delonix regia

Cymbidium hybrid

Seedling bedding dahlias with marigolds
behind

Daphne burkwoodii 'Somerset'

A well-bloomed, bedding dahlia

A cactus-flowered dahlia

Dasylirion, unidentified species

mantensiana is from 1½ to 4 feet tall and has a spread of 3 feet or more. Its leaves are about 1½ inches long. The small clusters of fragrant, soft pink blooms are produced intermittently over a long season.

Evergreen daphnes ordinarily 2 feet tall or taller include popular *D. odora*, its white-flowered variety *D. o. alba*, and *D. o. marginata*, which has creamy-white margins to its leaves and, unusual for a variegated variety, is somewhat hardier and more robust than its green-leaved type. Native of Japan and China, **D. odora** is a hairless, upright shrub 3 to 6 feet tall or sometimes taller. It blooms in winter or early spring.

Daphne laureola

Daphne odora

Its blunt-pointed-elliptic-oblong leaves, 1½ to 3½ inches long, are dark green. The very sweetly fragrant, rosy-purple blooms, in stalked, crowded heads with lanceolate bracts, are ½ inch or a little more in diameter. They are not hairy. This is hardy about as far north as Washington, D.C. Resembling *D. odora* and as fragrant, somewhat hardier **D. hybrida** is a good hybrid between that species and *D. collina.* Its pink-tinged to purple-lilac flowers, in evidence from fall to spring, differ from those of *D. odora* in being, like those of its other parent, hairy on their outsides.

Spurge-laurel (**D. laureola**), a hairless, bushy, evergreen 3 feet or so tall, is hardy about as far north as Philadelphia. Native to western and southern Europe, it has lustrous, pointed-oblanceolate leaves 2 to 3½ inches long. Scentless or sometimes fragrant, its ⅓-inch-long, yellowish-green flowers are in nearly stalkless, axillary racemes of five to ten. The egg-shaped fruits are bluish-black. Endemic to the Pyrenees, *D. l. philippii* is lower, more spreading, and has smaller flowers. A hybrid between this and *D. mezereum*, erect, semievergreen **D. houtteana** has purplish leaves similar in shape to those of *D. laureola*. Its paler lilac-

colored to dark red-purple flowers are in short-stalked, lateral clusters of up to five.

Other evergreens include compact, slow-growing, much-branched, Chinese **D. retusa,** up to 3 feet tall. This kind has densely-bristly- hairy young shoots. Its oblong to oblong-lanceolate, hairless leaves, 1 inch to 3 inches long, usually notched at their apexes, have rolled-under margins. White inside and rosy-purple outside, the ¾-inch-wide, lilac-scented blooms are in many-flowered, terminal, bracted heads about 3 inches wide. The fruits are red. This is hardly about as far north as Philadelphia. Uncommon **D. oleoides,** not hardy in the north, is native from the Mediterranean region to the Himalayas. A pubescent evergreen 2 to 3 feet tall, it has thick,

Daphne oleoides

elliptic, obovate or lanceolate, bristle-tipped leaves up to 1½ inches in length that become hairless on their upper sides at maturity. About ½ inch long, the fragrant, creamy-white to pale pinkish-lilac blooms are in terminal clusters. From 2 to 3 feet tall and wide, evergreen **D. collina** forms a neat mound. Its lustrous, dark green, obovate leaves, paler, and very hairy on their undersides, are ¾ inch to 2 inches long and up to a little more than ½ inch wide. The purplish-rose-pink blooms are in bracted clusters of ten to fifteen at the branch ends. They are fragrant, ½ inch in

diameter, and felted with hairs on their outsides. The plant identified as *D. c. neapolitana* may be a natural hybrid between *D. collina* and *D. cneorum* or possibly between *D. oleoides* and *D. cneorum*. Smaller looser, and more spreading than *D. collina*, it is perhaps easier to grow.

Garland flower (**D. cneorum**) and its varieties are popular for rock gardens, fronts of flower beds, and similar locations. This hardiest of evergreen daphnes, a native of dry or stony places, usually in limestone

Daphne cneorum

soils, from Spain to the Ukraine, survives throughout most of New England. The typical species is up to 1 foot tall by 3 feet wide. It has trailing branches and crowded, alternate, wedge-shaped-oblanceolate leaves up to 1 inch long. Lustrous green above, they are somewhat glaucous on their undersides. The very fragrant blooms, in many-flowered, flattish clusters without bracts, are pink. They are about ½ inch long and densely-hairy on their outsides. The fruits are yellowish-brown. Variety *D. c. alba*, with white flowers, is less vigorous, *D. c. major* is larger in all its parts, *D. c. eximia* resembles the last and is superior to it, *D. c. pygmaea* is very dwarf, and *D. c. variegata*, its leaves margined with white, is not particularly attractive. Variety *D. c.* 'Silver Leaf' is similar to or identical with the last. Lower, of less sturdy habit, and less attractive than *D. cneorum* to which it is allied, **D. striata,** of western France, is

Daphne striata

a prostrate or nearly prostrate evergreen with stalkless, linear-lanceolate, hairless, blunt leaves. Its flowers, in terminal clusters of up to twelve, are rosy-pink to magenta-pink, with pink-streaked corolla tubes. Choice and rare, *D. s. alba* has white blooms.

Other low daphnes particularly suited for rock gardens and similar locations include deciduous European **D. alpina.** Neat and from under 6 inches to about 1½ feet tall, it has obovate to oblong-obovate,

Daphne alpina

Daphne alpina (flowers)

grayish-green leaves up to 1½ inches long, hairy on both surfaces, and usually crowded toward the branch ends. In bractless clusters of up to ten, the sweetly-scented flowers, ⅓ inch long, are white, and on their outsides silky-hairy. The orange-red fruits are slightly pubescent. This is hardy in southern New England. Evergreen **D. arbuscula** inhabits rocky, limestone regions in eastern Czechoslovakia. A rounded shrublet up to 6 inches tall, it somewhat resembles *D. cneorum.* Crowded near the ends of its red branches, the blunt, linear-lanceolate leaves, sometimes hairy on their undersides, are about ¾ inch in length.

Their margins are rolled under, their upper surfaces are longitudinally grooved. A little over ½ inch long, the downy, fragrant, rosy-pink blooms are in clusters of up to eight. This kind is hardy in southern New England. Evergreen and about as hardy as the last, **D. blagayana,** an inhabitant of screes in the mountains of the Balkan Peninsula and northward, is a creeping kind up to 1 foot tall. It has long, sparingly-branched stems, leafless except near their ends. Loosely-branched and hairless, it has 1- to 2-inch-long, blunt, wedge-shaped, obovate to oblong-obovate leaves. Its white, yellowish-white, or pinkish-white flowers, fragrant and somewhat under ½ inch across, are in silky-bracted heads about 2 inches wide, of ten to fifteen or sometimes more, that are somewhat under ¾ inch long. The spherical fruits are pinkish-white. Choice **D. petraea** (syn. *D. rupestris*), endemic to a very limited area in northern Italy, is a spreading evergreen alpine shrublet up to 6 inches in height. It forms an intricate mat of short, thick, tortuous, procumbent branches. Its stalkless, narrowly-spatula-shaped to oblongish, blunt to pointed, leathery, lustrous leaves are ¼ to ½ inch long. The fragrant, rose-pink blooms, in terminal clusters of three to six, are up to ⅓ inch long. They are downy on their outsides. Challenging to grow, in the wild this kind inhabits crevices in sunny limestone cliffs at high altitudes. It is hardy in southern New England.

Garden and Landscape Uses. Among the most beautiful and choicest shrubs, daphnes are used variously in gardens. With few exceptions the evergreens, not hardy in the north, are admirable for milder climates, and *D. odora* for winter flowering in cool greenhouses and sunrooms. Deciduous daphnes are mostly hardy. With special reference to height and growth habit the various kinds may be used at the fronts of shrub borders, in foundation plantings and rock gardens, and as edgings. The smallest are generally choice, often challenging subjects for alpine greenhouse cultivation. Daphnes are often unpredictable, thriving in some gardens, living less happily or failing completely in others under apparently similar conditions. Not infrequently they die unexpectedly and unexplainably. It may well be that some kinds are naturally short-lived. Daphnes as a group require well-drained, moderately moist soil. In the wild many, including most of European origin, favor limestone, but some are certainly satisfied in cultivation with neutral or slightly acid soils. Here belong *D. blagayana, D. cneorum, D. genkwa, D. hybrida, D. laureola, D. mezereum,* and *D. odora.* Gardeners hold different opinions about the need for alkaline soils for several kinds, including *D. cneorum.* Trial and error seems the best way of determining the importance of this as it applies to particular gardens. Califor-

nian experience emphasizes need for care in watering. Too frequent applications are likely to result in disaster with *D. odora.* Light or part-day shade is favorable to some kinds, including the spurge-laurel and *D. odora.*

Cultivation. Daphnes do not transplant readily. Once established they are best left undisturbed. Young specimens well rooted in containers are better for setting out than are larger field-grown ones. A mulch of limestone grit mixed with leaf mold or peat moss helps keep soil temperatures and moisture content more uniform than without such covering. No regular pruning is required. Occasional straggly shoots may need to be cut back. Propagation is by seeds, cuttings, layering, and, the deciduous kinds, by grafting onto *D. mezereum.* A simple way of increasing *D. genkwa* is by driving the blade of a spade vertically into the ground some little distance away from the base of the bush. From the ends of major roots severed in this way new shoots will develop. When large enough these can be dug up and transplanted. Seeds of daphnes often germinate slowly and erratically. Their performance is improved if they are freed from the surrounding pulp, mixed with slightly damp peat moss or vermiculite, and stored in a closed jar or polyethylene plastic bag in a temperature of 40°F for three months before they are sown.

In greenhouses *D. odora* and its varieties succeed where the winter night temperature is 40 to 50°F and that by day ranges, relative to the brightness of the weather, from five to fifteen degrees higher. They benefit from being stood outside in summer in partial shade. Watering is done to keep the soil always moist but not sodden. Repotting is needed at intervals of two to four years only. It is a mistake to put them in containers excessively large relative to the size of the plants. Do not overfertilize, but applications of mild liquid nourishments may be given at intervals of two weeks when new foliage is developing and monthly afterward, except in winter.

DAPHNIPHYLLUM (Daphni-phýllum). Ten species or more of *Daphniphyllum,* of eastern and southeastern Asia, are recognized. They belong to the spurge family EUPHORBIACEAE, or according to some authorities to a family of its own, the DAPHNIPHYLLACEAE. Its name is derived from that of the unrelated genus *Daphne* and the Greek *phyllon,* a leaf. It alludes to the similarity in appearance of the foliage of some species to that of some daphnes.

Daphniphyllums are evergreen trees and shrubs with alternate, long-stalked, leathery, hairless leaves, glaucous on their undersides, and without lobes or teeth. Their little unisexual flowers, in racemes from the leaf axils, are without petals. They may have three to six plainly visible or scarcely evident sepals, or these may be lacking. Male blooms have five to eighteen short-stalked stamens, females five or more staminodes (nonfunctional stamens) and two to four short, reflexed styles. The fruits are drupes (the technical name for fruits structured like plums).

A handsome tree up to 30 feet in height, or a tall, broad, rounded shrub, **D. macropodum,** native to Japan and southern Korea, is hardy about as far north as Virginia. It has smooth, often reddish, young shoots.

Daphniphyllum macropodum

Its rhododendron-like leaves, oblong to obovate and 6 to 8 inches long by 1½ to 3 inches wide have sixteen to nineteen pairs of lateral veins. They are lustrous green above, conspicuously glaucous beneath, and have reddish stalks 1½ to 2½ inches long. The yellowish-green, stalked flowers, the sexes on separate plants, have fairly well-developed sepals. The males have about ten stamens, the females usually two styles. The dark blue roundish to egg-shaped fruits are not quite ½ inch long. Variety *D. m. variegatum* has broad, irregular, creamy-white margins to its leaves. A shrub up to about 5 feet, but quite often not more than 2 to 3 feet tall, **D. humile** (syn. *D. macropodum humile*), usually broader than high, branches freely from its base. Its leaves are 2 to 5 inches long by ¾ inch to 2 inches wide. They have twelve to sixteen pairs of lateral veins. This is native to Japan and Korea.

Garden and Landscape Uses and Cultivation. Daphniphyllums are among the most handsome broad-leaved evergreens. Their beauty is that of their foliage; neither their flowers nor fruits contribute any display. They are partial to filtered light; they succeed in part-shade better than in sun. For their best comfort the soil should be deep, acceptably fertile, and moist, but not wet. Limestone soils are especially to the liking of daphniphyllums. The only pruning needed is any deemed desirable to thin out shoots that crowd and to keep the

plants shapely. Spring is the best time to do this. Propagation may be by seeds, but more commonly is effected by cuttings, about 3 inches long, of firm young shoots taken in July or August and rooted under mist or in a greenhouse propagating bed, preferably with slight bottom heat.

DARLING-RIVER-PEA is *Swainsona greyana.*

DARLINGTONIA (Darling-tònia)—California Pitcher Plant, Cobra Plant. Insectivorous plants, kinds that get part of their sustenance from the bodies of live prey, are fascinating. The wild flora of North America includes a good many different kinds, some of genera unknown elsewhere in the wild, and highly restricted in their distribution even on their own continent. With these belongs the California pitcher plant (*Darlingtonia californica* syn. *Chrysamphora californica*), one of the oddest and most remarkable plants in the world. Even to a nonbotanist this curious species exhibits similarities to the other American pitcher plants *Sarracenia* and the two, together with the exceedingly rare South American *Heliamphora*, constitute the sar-

Darlingtonia californica

Darlingtonia californica (flowers)

racenia family SARRACENIACEAE. There is only one species of *Darlingtonia*, a genus named in honor of the American botanist and writer William Darlington, who died in 1863. It is a herbaceous perennial with a cluster of erect pitchers that are modified leaves, 8 inches to 2 feet long. They are yellowish-green and conspicuously veined. From their bases they widen upward in trumpet fashion and end in a sinister-looking, broad hood from which hangs a two-lobed crimson and green appendage, the whole fancifully suggesting the head of a cobra. The hood, translucent with opaque spots, admits filtered light to the interior of the pitcher which is partially filled with a digestive fluid usually containing, in various stages of decay, the bodies of unwary insects attracted by the carrion odor. The creatures enter the opening beneath the hood, step on the smooth interior surface, and slip to their doom in the fluid that fills the bottom of the pitcher, there to contribute to the sustenance of their captor. Even if they struggle out of the liquid they are prevented from climbing upward by down-pointing hairs.

The pitchers are not flowers. The latter are carried singly on stalks with a few scalelike leaves that arise from the bottoms of the plants and are taller than the pitchers. They develop from spring to summer, are more unusual than beautiful, and dark purple and nodding. There are five sepals, five heavily veined petals ¾ inch to 1½ inches long, twelve to fifteen stamens, and a single style. The seed pods are inversely egg-shaped capsules.

Contrary to common belief Darlingtonias are not dependent upon their animal prey for nutrients. They obtain most of the elements they require from the soil and air in the manner of more ordinary plants, and can get along quite nicely without insects and like creatures.

Although generally rare in the wild, the California pitcher plant is abundant locally. It inhabits sphagnum bogs and similar wet places in California and Oregon up to an altitude of about 8,000 feet.

Garden Uses. This fascinating plant is not difficult to grow, but many who attempt it fail because of ignorance of its needs. A common cause of trouble is putting into the pitchers pieces of meat and suchlike tidbits. It is better not to do this. Animal food is not necessary for the plant's well-being and, in any case, it is better to rely on chance entrapments than forced feeding. The California pitcher plant is much hardier than is generally supposed; it has been grown outdoors successfully in New England. Its most appropriate location is in a bog garden or a wet place by pond side or stream. It thrives in full sun or with slight shade. It is also satisfactory in cool greenhouses.

Cultivation. Large specimens collected from the wild rarely survive, which accounts for the nearly one hundred percent failure with plants obtained from dealers who flamboyantly advertise this rarity from time to time in the public press. Small well-rooted starts should be acquired, or plants may be raised from seeds. The best starts are shoots 1 inch to 3 inches tall, taken from the outsides of clumps and carefully potted. Larger divisions are much less successful. A potting mixture of coarse sand, leaf mold, and live sphagnum moss is best, and the potted plants are stood with the bottoms of the containers in an inch or so of water, and are covered with a glass jar or covering of polyethylene plastic film to assure a humid atmosphere. Seeds are handled by sowing on the surface of rather tightly packed live sphagnum moss in pots or pans (shallow pots) stood in saucers of water under a bell jar or polyethylene cover. Older plants are satisfied with a somewhat acid rooting mixture consisting largely of peat moss or leaf mold and coarse sand, with a little topsoil mixed in. The water supplied must not be alkaline. Rain water is preferred. Where winters are cold a protective covering of branches of pines or other evergreens is advisable.

In greenhouses a cool, humid environment is essential. In winter the night temperature should not be above 40 to 50°F, and by day not more than five to ten degrees higher. In summer light shade is necessary. Repotting, needed about every second year, is done about midsummer when the plants are not actively making new growth.

DARNEL is *Lolium temulentum.*

DARWINIA (Dar-wínia). Related to more familiar *Chamaelaucium,* but quite different in aspect, *Darwinia,* comprising thirty-five species of the myrtle family MYRTACEAE, occurs wild only in Australia. Its name honors the physician and poet Dr. Erasmus Darwin, who died in 1802. He was grandfather of the distinguished English naturalist and author of the *Origin of Species* Charles Darwin.

Darwinias are medium-sized to small or prostrate, heathlike shrubs. They have opposite or nearly opposite, small, undivided leaves and tiny flowers enclosed by persistent petal-like bracts. Each bloom has a five-lobed calyx, five petals, ten stamens, and alternating with them, the same number of tiny staminodes (nonfunctional stamens). The style protrudes and is bearded at its summit. The fruits, consisting of the ovary and hardened receptacle of the flower, and thin-skinned and leathery, usually contain one seed. They do not open to release the seeds.

A broad shrub about 2 feet tall with distinctive bluish-green foliage, *D. citriodora* has almost opposite, oblong-lanceolate to ovate-lanceolate leaves ¼ to ½ inch in length, that when crushed smell of lemon.

The flowers, usually in terminal, drooping heads surrounded by reddish or greenish-white bracts, are not especially conspicuous. Prostrate and freely-branched *D. homoranthoides* has three-angled, linear leaves about ⅜ inch long in two ranks. Its flowers, solitary from the leaf axils of the branchlets, are white, with red anthers. Low and much-branched *D. thymoides* has linear to lanceolate leaves up to ⅜ inch long, with rolled-back margins. The stalkless white flowers are in terminal heads. From 1½ to 2 feet tall, *D. hookeriana* has narrow-oblong leaves up to ¾ inch long and drooping clusters of flowers hidden by pink bracts 1 inch or somewhat more in length.

Garden and Landscape Uses and Cultivation. Darwinias are adapted for rock gardens and other selected sites in California and elsewhere with frost-free or nearly frost-free winters, and warm, dry summers. They need very well-drained, preferably sandy soil, and appreciate mild applications of organic fertilizer. Some kinds are benefited and kept from becoming straggly by occasional trimming. Most seem to appreciate slight shade from the strongest sun. Propagation is by cuttings and seeds.

DASHEEN or TARO. This large-growing relative of the ornamental elephant's ear (*Colocasia antiquorum*) is cultivated widely in the tropics and to some extent in the far south for its edible tubers. Much resembling the elephant's ear in appearance, dasheen or taro is *Colocasia esculenta* of the arum family ARACEAE.

For its sucessful cultivation dasheen requires a very warm, frostless growing season of seven months or longer, a humid atmosphere, and deep, fertile, moist soil with good subsurface drainage. Planting may be done two to three weeks before the date of the last expected frost or a jump may be gained on the season by starting tubers in pots indoors and setting the young plants out after all danger of frost is over. For planting use small tubers weighing 2 to 5 ounces. Set these like potatoes are planted 2 feet apart in rows 3 to 3½ feet apart. In fall, when fully mature, dig the tubers, allow them to dry on the soil surface, and store in a well-ventilated building where the temperature is about 50°F.

DASYLIRION (Dasy-lírion)—Sotol, Desert Spoon, Bear-Grass. There are eighteen species of *Dasylirion* of the lily family LILIACEAE, or, according to those who accept the segregation, of the agave family AGAVACEAE. They are natives of the southwestern United States and Mexico. The name, from the Greek *dasys,* thick, and *lirion,* a lily, alludes to the tufted foliage.

Dasylirions have woody trunks up to a few feet tall or sometimes subterranean, topped with large, symmetrical, dome-

Dasylirion, unidentified species blooming at Buenos Aires, Argentina

shaped to globular heads of numerous long, narrow, usually prickly-margined, leathery, evergreen leaves. In some kinds the leaves end in a small brush of dry fibers. The little greenish-white, unisexual flowers, the sexes on different plants, are in slender panicles of crowded racemes or spikes that rise several feet above the foliage. Bell-shaped, the blooms have six toothed petals or more correctly tepals, nearly equal in size and shape, six protruding stamens, and a short style ending in three stigmas. The fruits are dry, three-winged, one-seeded capsules. From *D. wheeleri* the Indians prepared a food and an intoxicating beverage called sotol. Its leaves are collected and sold as "cactus spoons" and "desert spoons" for use as decorations. Good conservation calls for the discouragement of this practice. For the plant once named *D. hartwegianum* see Calibanus.

Lustrous green, flat, prickly-margined leaves shredded into little brushes at their apexes are characteristic of *D. texanum*, of Texas, and of *D. acrotriche* and *D. graminifolium*, of Mexico. Sometimes called beargrass, **D. texanum** has a short or wholly subterranean trunk. Its 2- to 3-foot-long leaves, ½ inch wide, have yellow prickles that turn brown with age. The panicles of bloom are up to 15 feet tall. Leaves about 3 feet long by ½ inch wide, margined with very short, yellowish-white prickles are typical of **D. graminifolium.** Its trunk up to 3 feet or more high, **D. acrotriche** has leaves 2 to 3 feet long and up to ½ inch wide. The pale yellow prickles have brown apexes. The panicles attain heights of 9 to 15 feet.

Dull, glaucous, flat leaves are typical of the desert spoon (*D. wheeleri*), of Texas, Arizona, and Mexico, and of *D. serratifolium*, of Mexico. The short-trunked **D. wheeleri** has nearly smooth leaves 2 to 3 feet long by 1 inch broad, with brown-tipped, yellow prickles. The flower panicles are 9 to 15 feet tall. The leaves of **D.**

Dasylirion acrotriche, its trunk with a skirt of old leaves, at the Royal Botanic Gardens, Kew, England

serratifolium are 2 to 3 feet long by 1 inch to 1½ inches wide. Their margins are furnished with rather widely spaced, long prickles.

Dull, glaucous foliage not shredded at the leaf tips characterizes some kinds, among them short-trunked **D. glaucophyllum,** of Mexico. The leaves of this kind are 2 to 3 feet long by 1 inch to 1½ inches wide. Their prickles are yellowish-white. The panicles of bloom are 12 to 18 feet tall. Its trunk up to 6 feet tall, Mexican **D. longissimum** is distinct from others discussed here because its leaves are four-sided instead of flat and do not have prickly edges. They are very numerous, 4 to 6 feet long, and up to ⅜ inch wide. Usually they are not frayed into miniature brushes at their ends. The flower panicles are 6 to 18 feet tall.

Garden and Landscape Uses. Dasylirions, striking in bloom and handsome at all times, are appropriate for a variety of landscape uses in warm, dry regions with climates similar to those they know in the

Dasylirion longissimum, at Huntington Botanic Garden, San Marino, California

wild. They can be displayed to advantage in large rock gardens and similar places. In containers they are attractive for decorating terraces, patios, steps, and other architectural features. They are suitable for including among collections of succulents and other desert plants in conservatories and large greenhouses.

Cultivation. Of simple cultivation, dasylirions revel in very porus, well-drained soil and sunny locations. They are easily propagated by cuttings and by seeds. In greenhouses winter night temperatures of 50 to 55°F are adequate with an increase of five to fifteen degrees during the day permitted. Watering may be done with fair freedom from spring to fall, always allowing the earth to approach dryness between applications, and with much greater restraint in winter.

DATE. The date palm (*Phoenix dactylifera*) is cultivated as a commercial fruit crop in parts of California and the southwest where the climate is suitable, but except as an ornamental it is without home garden im-

Dasylirion acrotriche, at Mexico City, Mexico

Date palm plantation, in California

portance. Elsewhere in desert and semi-desert regions where hot, dry summers prevail, notably in North Africa and to a much lesser extent locally in Baja California, date palms are an important source of nutritious food.

Date palms are unisexual. It is usual to ensure pollination of the females by, at flowering time, tying small branches of male blooms among the clusters of female ones. For the best results and to influence the time of ripening, the choice of pollinating trees is important. There is much difference also in the quality and abundance of the fruits borne by different females. Seedling trees are usually markedly inferior. For fruit production, only propagations by suckers taken from varieties of proven merit should be planted. The plants are set about 30 feet apart. Under good conditions suckers removed when they are three to six years old will begin fruiting in five to six years. The fruits are generally harvested by cutting entire bunches, which are then ripened or cured by keeping them for a period in a warm place.

The species called Canary Island date-palm is *Phoenix canariensis*, the one known as wild date-palm, *P. sylvestris*. Other plants that have common names including the word date are the Chinese-date (*Zizyphus jujuba*), the date-plum (*Diospyros lotus*), the Jerusalem-date (*Bauhinia monandra*), the Trebizonde-date (*Elaeagnus angustifolia*), and the date-yucca (*Samuela*).

DATIL is *Yucca baccata*.

DATISCA (Datís-ca). This rarely cultivated genus of the datisca family DATISCACEAE consists of two species, one native of California and Baja California and one from the Mediterranean region to India. The name is of Greek derivation but its meaning is obscure.

Datiscas are stout, herbaceous, branching perennials with alternate, pinnate or pinnately-lobed, toothed leaves and tiny unisexual flowers in clusters and racemes. The sexes are on separate plants. The blooms have no corollas. The males have calyxes with four to nine lobes, and eight or more stamens, fertile blooms have three-lobed calyxes and sometimes two to four stamens. The fruits are capsules.

Very much resembling hemp (*Cannabis sativa*) in general appearance, although not botanically related, **Datisca cannabina** is a hardy herbaceous perennial up to 7 feet tall. It has pinnate leaves with seven to eleven leaflets about 2 inches long by ½ inch wide and small yellow flowers, the males in clusters, the females in long loose racemes from the leaf axils. Male flowers have from eight to twenty-five stamens, females have three-cleft styles. This species is native from Asia Minor to India.

Garden Uses and Cultivation. The ornamental value of this species lies in its

Datisca cannabina

graceful foliage. It may be displayed effectively, either singly or in groups in beds, at the fronts of shrub plantings, and similar places. For its best development it needs deep, fertile, well-drained ground and a sunny location. Female plants that have been pollinated are most decorative, therefore it is advantageous to plant mostly females, but to have at least one male nearby as a pollinator. Propagation is by division and seed. Established specimens should be fertilized each spring.

DATISCACEAE—Datisca Family. The characteristics of this family of dicotyledons of one genus are given under Datisca.

DATURA (Da-tùra) — Angel's Trumpet, Thorn-Apple, Jimson-Weed. Some authorities include *Brugmansia* in *Datura* of the nightshade family SOLANACEAE, but here they are treated as separate genera. So interpreted, *Datura* consists of eight species, natives of tropical, subtropical, and warm-temperate regions chiefly of the Western Hemisphere. The name is believed to be derived from its East Indian one, *dhatura* or *dutra*, or possibly from the Arabic one *tatorah*.

Daturas are coarse annuals and short-lived herbaceous perennials. They have large, mostly alternate, undivided, smooth-edged or wavy-toothed leaves, usually more or less ill-scented when bruised. The stalked, up-facing, often fragrant, short-lived flowers are solitary from the leaf axils. They have long-tubular calyxes, spathelike or with five lobes or teeth, and splitting lengthwise or falling away in such a manner that the base is left as a cup cradling the seed capsule. The trumpet-shaped, wide-mouthed corolla, white, white and purple, purple, violet, red, or yellow, has five shallow lobes or is pleated and has ten teeth. There are five stamens and a slender style with two stigmas. The dry or sometimes fleshy, often prickly or spiny fruits are capsules containing large seeds.

From daturas are obtained the alkaloids, atropine, hyoscyamine, and hyoscine, used medicinally. In the Old and the New World the intoxicating, narcotic, and hypnotic effects resulting from eating or drink-

Datura stramonium (fruits)

ing parts or preparations of daturas, especially of their seeds, have been known since ancient times. This knowledge was widely applied in religious ceremonies, divination, and prophecy, as well as medicinally. In India thugs employed preparations of *Datura* to stupefy their victims, and in South America slaves and wives were conditioned for burial along with their dead masters and husbands by feeding them large quantities of *Datura* mixed with tobacco, to deaden their senses. Jimson-weed, Jamestown weed, or thorn-apple (**D. stramonium**), commonly found in waste places in North America and elsewhere, is one of the most virulently poisonous daturas. It somewhat resembles *D. metel,* but has smaller, white blooms. Two of the common names of this allude to the experiences of British soldiers sent in 1676 to quell a rebellion in Jamestown, Virginia. After cooking and eating young jimson weeds they became insane for a period of eleven days and then recovered.

Annual **D. metel** (syns. *D. chlorantha, D. fastuosa*) originally of Asia, is now much more widely dispersed both in the wild and in cultivation. Downy or hairless, and

Datura metel

3 to 5 feet tall, it has narrowly-ovate leaves up to 8 inches in length, with asymmetrical bases and more or less wavy-toothed margins. Its fragrant flowers are typically white, less commonly yellow or purple. They are 5 to 6 inches long and in some cultivated sorts have double or triple, five- or ten-lobed corollas nested inside each other in hose-in-hose fashion. There are five or six stamens. The fruits are pendulous, short-spiny, nearly spherical, and about 1½ inches in diameter. Varieties are *D. m. alba,* with white flowers, *D. m. aurea,* with yellow flowers, *D. m. caerulea,* with blue flowers, *D. m. cornucopaea,* with double flowers, and *D. m. huberana,* with blue, red, and yellow flowers.

Differing from the last in its stems and foliage being softly-gray-pubescent, **D. inoxia,** of the southwestern United States

and Mexico, is a beautiful annual or herbaceous perennial up to 3 feet tall or taller. It has forked branches and asymmetrically-ovate, slightly-wavy-toothed leaves up to about 2¼ inches in length, the upper ones often opposite. The fragrant blooms are 6 to 8 inches long by 4 to 6 inches wide. White, faintly tinged with pink or lavender, they have ten teeth and are succeeded by subspherical, long-spined, pubescent fruits 2 inches in diameter. More frequent in cultivation than the typical species, *D. i. quinquecuspida* (syn. *D. wrightii*) is often misnamed *D. meteloides,* which name is properly a synonym of the species *D. inoxia.* The variety differs from the species in

Datura inoxia quinquecuspida

being in mild climates always perennial, more widely spreading, and in having the corollas of its white to lavender flowers with five instead of ten teeth.

Garden and Landscape Uses and Cultivation. The chief employment of daturas as ornamentals is as flower garden annuals. Even the perennials bloom well the first year. All are easy to manage. They prosper in ordinary, well-drained soils in sunny locations and are raised from seeds sown where the plants are to remain or are started early indoors and the young plants transplanted to the garden after all danger of frost has gone. They must be spaced widely enough apart to avoid overcrowding. The blooming season is prolonged if, by removing the flowers as soon as they fade, the development of seed pods is prevented.

DAUBENTONIA. See Sesbania.

DAUCOPHYLLUM. See Musineon.

DAUCUS (Daù-cus)—Carrot, Queen Anne's Lace. Of the genus *Daucus,* the name of which comes from *daukos,* the ancient Greek name of some plant of the same family, only the cultivated carrot is of garden importance. The pretty weed of waste places called Queen Anne's lace is of interest be-

cause it is the species from which the cultivated carrot was developed.

The genus *Daucus* belongs in the carrot family UMBELLIFERAE. It comprises sixty species of annuals and biennials of North America, Europe, Asia, and Africa. They have pinnately-dissected leaves and tiny, mostly white flowers in umbels of smaller umbels. The small fruits are ovoid or oblong, flattened and ribbed.

Queen Anne's lace (**D. carota**) is an erect and branching biennial, hairy or hairless. It is 6 inches to 3 feet tall and has finely-dissected, twice- or thrice-pinnate leaves. Its tiny five-petaled, white flowers are in flat, concave, or slightly rounded umbels.

Daucus carota

They are white or faintly purplish, usually with one or more of the central flowers dark red-purple. This native of Europe and Asia is freely naturalized in North America. It may be used to a limited extent in meadow-like parts of wild gardens. It is easily raised from seeds sown where the plants are to remain and bloom. For the cultivated carrot see Carrot.

DAVALLIA (Da-vállia)—Squirrel's Foot Fern or Ball Fern, Rabbit's Foot Fern. To those botanists who accept the segregation, *Davallia* is the type genus of the davallia family DAVALLIACEAE. A more conservative accounting includes it in the polypody family POLYPODIACEAE. The genus consists of forty species of epiphytic (tree-perching), epipetric (rock-perching), and terrestrial ferns, natives from southwestern Europe to the Canary Islands, tropical and subtropical Asia, Madagascar, Australia, and islands of the Pacific. Its name commemorates the Swiss botanist Edmond Davall, who died in 1798. Some ferns previously known by this name are included in *Humata* and *Sphenomeris.*

Davallias are mostly of small to medium size. They have creeping rhizomes, generally furnished with abundant chaffy scales or hairs, and finely divided, firm to leathery foliage. The fronds (leaves), commonly ovate to triangular in outline, are one- to four- or sometimes five-times-pinnate. The clusters of spore capsules, along or close to the margins of the ultimate segments, have slender, semitubular coverings (indusia) that open toward the margins of the leaflets.

A plant often misidentified as *D. canariensis* is Malayan **D. trichomanoides.** This kind has rhizomes with light brown scales. Its evergreen fronds have blades 6 to 9 inches in length, four times divided into overlapping, strap-shaped ultimate segments. It rarely produces spore clusters. True **D. canariensis** is native to Spain, Portugal, the Canary Islands, and Madeira. It has rhizomes with bright chestnut-brown scales. The leaves are evergreen. The stalks and blades are of about equal length. The latter are triangular, hairless and three- or four-times-pinnate into usually blunt, not sharp-toothed or horned, lanceolate to ovate-oblong ultimate segments 1/16 inch wide or wider. The leaf blades are 1 foot to 1½ feet long by about two-thirds as wide as long.

Squirrel's foot fern or ball fern (**D. mariesii** syn. *D. bullata*), deciduous in cool environments, may be evergreen if kept warm and moist. It has rhizomes densely covered with whitish or pale brown hairlike scales. The blades of its fronds are usually under 1 foot long. Ovate and one-half or more as broad as long, they are four

Davallia mariesii: (a) New growth beginning after domant period

(b) In full foliage

times divided into sharp-toothed ultimate segments. This is a native of Japan, Taiwan, Formosa, and China.

Rabbit's foot fern (**D. fejeensis**), as its botanical name implies, is native to the Fiji Islands. This quite variable evergreen has fronds with triangular, lacy blades about 1 foot long, four- or five-times-pinnate. The ultimate segments are linear, up to ¼ inch in length. Polynesian *D. f. plumosa* is a dwarf variety with more finely cut, four-times-pinnate, dainty, evergreen, pendu-

Davallia fejeensis

lous fronds. Coarser than *D. fejeensis* and its variety, **D. solida** (syn. *D. lucida*), of Malaya to Queensland, has rhizomes clothed with brown scales or hairs. Its glossy, broad-bladed, 1- to 2-foot-tall evergreen fronds are three-times-pinnate, their apexes broad and undivided. The slightly round-toothed ultimate segments are rhomboid-ovate.

Davallia solida

Other cultivated sorts include these: **D. denticulata,** a variable native of the Old World tropics that occurs in the wild both on the ground and perched on trees, has densely-fibrous, thick, creeping rhizomes and shining, leathery fronds 1 foot to 2 feet long or sometimes longer and up to 1¼

feet wide. Their blades are triangular and lacily three- or four-times-pinnate. Several horticultural varieties have been named. **D. divaricata,** of tropical Asia, has stout, creeping, scaly rhizomes and rather short-stalked, finely-divided, thrice-pinnate fronds 2 to 3 feet long. Their lower leaflets are up to 1 foot long by one-half as wide. **D. pentaphylla,** of Java and Polynesia, is a dwarf sort with rhizomes covered with fine black, hairlike scales. The evergreen, once-pinnate fronds, up to 8 inches long

Davallia pentaphylla

Davallia pentaphylla

by about 4 inches wide, have two to six pairs and a terminal one, of lustrous, linear, wavy-toothed leaflets. **D. pyxidata,** Australian, has stout, densely-scaly, creeping rhizomes and fairly short-stalked fronds 1 foot to 2 feet long by under 1 foot wide. They have glossy, lacey, triangular blades three- to four-times-pinnately-lobed. The segments of the lower leaflets are lanceolate, 2 to 3 inches long by about 1 inch wide.

Garden and Landscape Uses. Among the best nonhardy ferns, davallias are simple to grow and very decorative. Most often they are accommodated in pans (shallow pots) or hanging baskets in greenhouses and conservatories, or in the humid tropics and warm subtropics outdoors. The squirrel's foot fern is especially suitable for hanging baskets, and the Japanese use rhizomes of it to make spheres or balls that are suspended in the manner of baskets. They also train the rhizomes into more fanciful shapes such as monkeys and other animals, dolls, bells, and pillars.

Cultivation. Davallias need a porous, sharply drained rooting medium that contains an abundance of organic matter. One of a little turfy loam mixed with much course leaf mold or peat moss, with the addition of some fir bark such as is used for potting orchids, and a generous dash of peanut-size pieces of charcoal is satisfactory. It should be kept evenly moist from spring to fall, for evergreen kinds drier in winter, for deciduous kinds quite dry then. Good light with only sufficient shade to keep the leaves from scorching is ideal. The atmosphere should be humid. Repotting is best done in late winter or early spring. At that time and generally subsequently it is necessary to peg the rhizomes to the rooting medium. This can be done with pieces of wire bent like hairpins. In intervening years in spring just pick out a little of the old rooting compost from between the rhizomes and replace it with new. Well-rooted specimens benefit from biweekly applications of dilute liquid fertilizer from spring to fall. Night temperatures of 55 to 60°F in winter with a daytime increase of five to fifteen degrees and higher temperatures at other seasons are appropriate for most kinds. For *D. mariesii* and *D. canariensis* temperatures ten degrees lower suit. Propagation can be by spores, but more often is done by dividing the rhizomes in spring into sections and pinning them onto the soil surface. In a temperature of 60 to 70°F these soon root and become established. For more information see Ferns.

Rhizomes of *Davallia* pinned to soil surface for propagation

DAVALLIACEAE — Davallia Family. This family of one dozen genera of small to large ferns, by some botanists included in the POLYPODIACEAE, consists chiefly of tree- or rock-perching sorts. Its more than 2,500 species usually have creeping, frequently extensive rhizomes, much less commonly suberect ones. Their fronds (leaves) are undivided and lobeless or are once- or more-times-pinnately-lobed or -divided. Located at the vein ends, the clusters of spore capsules have usually pouch-shaped covers (indusia). Genera in cultivation include *Arthropteris*, *Davallia*, *Humata*, and *Nephrolepsis*.

DAVIDIA (Davíd-ia) — Dove Tree, Handkerchief Tree. Probably the main reason this beautiful deciduous, flowering tree is not planted as frequently as it would appear to deserve is its tardiness in reaching the blooming stage. Were it as precocious as flowering cherries, crab apples, or even flowering dogwoods, it would doubtless more commonly ornament our landscapes. Although it usually takes a long time for the dove tree to first bloom, it subsequently flowers regularly. It may well be that by grafting scions of flowering trees onto young seedlings, instead of waiting for the seedlings to reach flowering size, blooming would be achieved much sooner. The dove tree or handkerchief tree (*Davidia involucrata*) is the only species of its genus. It belongs to the nyssa family NYSSACEAE and is native to China. The name honors the French missionary and distinguished student of Chinese plants, Abbé Armand David. He died in 1900.

In aspect much like a linden, ***D. involucrata*** rarely attains 65 feet in height and may be almost as wide. In cultivation it is commonly smaller. Densely and attractively foliaged, it has alternate, slender-stalked, prominently-veined, pointed-heart-shaped, coarsely-toothed leaves with blades 3 to 6 inches long by about two-thirds as wide as they are long, and densely pubescent on their undersides. The young foliage is slightly ill-scented. In bloom the dove tree is quite a wonderful sight. The flowers are in numerous small tight clusters. From the base of each cluster dangles a large white or creamy-white bract, above which is a smaller, erect hoodlike one. Rarely there are three bracts. It is the bracts that make the show. They flutter in the breeze and give reason for the names dove

(b) In bloom

Davidia involucrata: (a) Tree

(c) A cluster of flowers and showy bracts

(d) Fruit

tree and handkerchief tree. The bracts are ovate to oblong-obvate, the larger one up to 6½ inches in length, the smaller up to one-half that size. The globular heads of true flowers are ¾ inch in diameter. They terminate pendulous stalks 3 inches long. Without sepals and petals, the blooms are not conspicuous. Each cluster consists of one bisexual flower and numerous males. The solitary pear- to olive-shaped, fleshy fruits, are drupes about 1½ inches long by 1 inch wide, and green with a purplish bloom. Variety *D. i. vilmoriniana* differs from the species in having the undersides of its leaves hairless or at most sparsely-hairy along the veins.

Garden and Landscape Uses. The dove tree is splendid as a lawn specimen. It can also be included in mixed plantings of trees and shrubs, but should not be crowded because it needs generous space to spread its branches and adequately present its charms. It is most strikingly displayed against backgrounds of dark-foliaged evergreens. The variety *D. i. vilmoriniana* is hardy in sheltered locations in southern New England, the typical species is a little more tender. Dove trees appreciate deep, fertile soil and sunny locations.

Cultivation. Once established, the dove tree requires little or no attention. No

pruning is ordinarily needed. Because young specimens are more tender to cold than older ones, toward the northern limits at which this tree is hardy it is well to wrap them in burlap or in other ways give slight winter protection during their first few years. Propagation can be by seed. It is advisable not to allow these to dry. If they do, they may take a year or more to germinate. As soon as the fruits are ripe free the seeds from surrounding pulp and sow them in sandy peaty soil in a cool greenhouse or cold frame; or mix them with damp sand or peat moss, put them in polyethylene plastic bags, store them for five months at about 60°F and three months at 40°F, and then sow. They will germinate in a few weeks. When the seedlings are large enough, and after all danger of frost is over, transplant them to an outdoor nursery bed. Summer cuttings can be rooted under mist or in a greenhouse propagating bed. Layering affords another means of increase.

DAVIESIA (Dav-ièsia). More than sixty species of shrubs of the pea family LEGUMINOSAE comprise *Daviesia*. The name honors the British Reverend Hugh Davies, who died in 1821. All are natives of Australia and Tasmania.

Daviesias have leathery, alternate leaves or in place thereof sharp spines or tiny scales. The small pea-like flowers, yellow, orange, red, or purplish-red, are in dense, axillary racemes or clusters. They have short calyxes with the two upper of the five lobes often united. The standard or upper petal is roundish and approximately the same length as the lateral petals and keel. The stamens are not joined. The fruits are small, triangular, purselike pods with sharp-pointed apexes. They contain one or two seeds.

A hairless shrub 2 to 5 feet tall, *D. corymbosa* has slightly angular branches and alternate, usually lanceolate-linear, rigid leaves 2 to 5 inches long and up to ½ inch wide. The flowers are in dense, stalked racemes, of which the blooms occupy ½ inch to 1½ inches, from the leaf axils. They are red and orange and quite showy. From 2 to 6 feet tall, *D. latifolia* is a hairless shrub with ovate-oblong, ovate-lanceolate, to obovate leaves 2 to 4 inches long, ¾ inch to 1½ inches wide, and with very evident veins. From the leaf axils develop the small orange-yellow flowers in fairly loose racemes 1 inch to 2½ inches long. The upper or standard petal of each is darker colored at its base.

Garden and Landscape Uses and Cultivation. Daviesias are suited for cultivation outdoors in climates as mild as that of southern California, and in cool greenhouses. They stand little frost. They have the same uses and need the same care as *Chorizema*. Before sowing, their seeds should

be treated by pouring boiling water over them, and allowing them to soak for twelve to twenty-four hours.

DAWN-REDWOOD is *Metasequoia glyptostroboides.*

DAY FLOWER. See Commelina. False-day-flower is *Commelinantia.*

DAY-LILY. See Hemerocallis.

DDT. This synthetic chlorinated hydrocarbon came into prominence as a pesticide in World War II and afterward was used in ever increasing amounts and with marked success in controlling plant pests. Later it was ruled dangerous, and its use in many countries prohibited, because of well-based concern about its harmful effects on the environment related to its virtual indestructability. It is not biodegradable, but remains unchanged as it passes from one organism to another in the food chain. Minute amounts taken in are accumulated and stored in plants, animals, and man, especially in fatty tissues, often with extremely damaging, and even fatal results, and with not yet fully understood harmful effects to reproductive systems.

DEADLY NIGHTSHADE is *Atropa belladonna.* Woody nightshade (*Solanum dulcamara*) is frequently misnamed deadly nightshade.

DEAMIA (Dèam-ia). Two species constitute *Deamia* of the cactus family CACTACEAE, a genus many botanists include in nearly related SELENICEREUS. Its members are natives of Mexico to Colombia. Its name is based on that of the botanist who first collected it, the American Charles Clemon Deam. He died in 1953.

Deamias clamber in trees, on cliffs, or over rocky ground. They have long, slender, three- to eight-angled or winged stems with clusters of bristles and spines and often aerial roots. The large funnel-shaped white flowers open by day. The outsides of the perianth tubes, ovary, and spiny fruits have little scales with bristles in their axils.

Of unusual appearance, *D. testudo* (syn. *Selenicereus testudo*) varies considerably in the shape and other details of its long stems, which cling snugly to tree trunks or other supports and then become raised or humped at their centers like the shell of a tortoise, for which reason *testudo*, the Latin name for a tortoise, is used in the name of this species. The largest segments of stems are up to 1 foot long and are 4 or sometimes up to 8 inches wide. They have three to eight wings. The smaller segments are narrower and are winged or triangular. Ten brownish spines or more, from nearly ½- to about ¾-inch-long, come from each areole. Bristly when young, they are needle-like later. Up to 1 foot in length, the flowers are white or yellowish-white.

Garden and Landscape Uses and Cultivation. Unlike the great majority of cactuses, the one described here needs for its best development high temperatures, abundant humidity, and shade from strong sun. It prospers under conditions that suit tropical orchids. It should be provided with a trunk or branch of locust or other long-lasting wood, a tree fern trunk, or other suitable support to which its aerial roots can cling. For further information see Cactuses.

DEATH-CAMAS is the common name of *Zigadenus nuttallii* and *Z. venenosus*.

DEATH TREE, SHEBA VALLEY. See *Synadenium cupulare*.

DEBREGEASIA (Debreg-eàsia). The nettle family URTICACEAE contains this African, Asian, and Malaysian genus of five species of shrubs and small trees, the name of which commemorates the early nineteenth-century sea captain explorer Prosper Justin de Bregeas.

Debregeasias have alternate, short-stalked leaves with wrinkled upper surfaces, three chief veins, and toothed margins. Their insignificant flowers are unisexual, with the sexes on the same or separate plants. They are in usually clustered, spherical heads. Each minute bloom has a generally four-parted, more rarely three- or five-parted, perianth, those of the females urn-shaped or obovoid, becoming fleshy as the fruits form. The males have four short stamens, the females a knoblike stigma. The edible, raspberry-like fruits are conglomerates of small berry-like drupes, each developed from a single female flower.

Native to China, Japan, the Ryukyu Islands, and Taiwan, *Debregeasia edulis* (syn. *D. japonica*) is a deciduous shrub about 6 feet tall. It has hairy shoots, and lanceolate to linear, finely-toothed leaves up to 6 inches long, up to 1 inch wide, densely-white-woolly beneath, and on their upper sides pubescent along the veins. The heads of flowers are solitary or two or three together in the leaf axils. The fruits are orange-yellow.

Indigenous from low altitudes in the Himalayas to Java, *D. longifolia* (syn. *D. velutina*) is occasionally a tree up to 20 feet in height, more commonly a tall shrub. It has lanceolate to oblong-lanceolate, toothed leaves 3 to 6 inches long, clothed on their undersides with snow-white hairs, and on their top sides with short, rough ones. The flower heads are two or more together on short, branching stalks. The orange to red fruits are the size of peas.

Garden and Landscape Uses. The hardiest debregeasia is *D. edulis*. This has roots that survive outdoors in southern New England, even though in that severe climate the tops are killed to the ground and new shoots develop from the base each spring. Other species are suitable only for regions of little or no frost, and for greenhouses. The striking contrast between the white undersides of the leaves and their upper surfaces forms the chief attraction. Debregeasias show to advantage in front of dark-foliaged shrubbery.

Cultivation. Cuttings 2 to 3 inches long made from shoot ends in late summer root readily under mist or in greenhouse propagating beds. New plants are also easily raised from fresh seeds. Ordinary soil is satisfactory, and established specimens need little routine care other than thinning out old and unwanted branches in winter or early spring. In greenhouses *D. longifolia* grows satisfactorily in fertile, porous soil, where the temperature in winter is about 55°F at night and by day five to fifteen degrees higher. Moderate humidity and watering are needed, and in summer light shade from strong sun.

DECAISNEA (Decaìs-nea). Named in commemoration of the French botanist Joseph Decaisne, onetime Director of the Jardin des Plantes in Paris, who died in 1882, one or two species of deciduous shrubs 10 to 15 feet tall, constitute *Decaisnea* of the lardizabala family LARDIZABALACEAE.

Decaisneas have erect stems with thick pith and big, pointed winter buds. Their opposite, odd-pinnate leaves are clustered toward the ends of the branches. The lateral panicles of loose racemes are of drooping, unisexual and bisexual flowers without petals. They have six yellowish-green sepals that resemble petals, are narrow-lanceolate, 1 inch long or longer, and taper to long points that curve gracefully outward. Male flowers have six long-stalked stamens joined as a column. In bisexual blooms the stamens are shorter and not united, and there are three separate pistils. The curious and attractive cylindrical fruits have numerous constrictions that make them look like huge caterpillars. They are 3 to 4 inches long by ½ to ¾ inch wide and although insipid are eaten in the Orient. They contain a slimy, white pulp in which numerous black seeds are embedded.

Chinese *D. fargesii* has dull blue fruits. The plant called *D. insignis* is an Himalayan variant that has yellow fruits, but scarcely differs otherwise. It is more tender than the blue-fruited plant. The latter is hardy in southern New England.

Garden and Landscape Uses and Cultivation. Decaisneas are interesting and attractive shrubs for collectors of the unusual. They prosper in fertile, loamy soil and require little care. They are propagated by seeds, which may be sown in a

Decaisnea fargesii, in bloom

greenhouse, cold frame, or sheltered place outdoors in spring.

DECARYIA (Dec-áryia). The only species of this interesting endemic of Madagascar belongs to the didierea family DIDIEREACEAE. Its name honors the twentieth-century plant collector and botanist R. Decary.

A tree up to 25 feet tall, *Decaryia madagascariensis* is remarkable because of its very markedly zigzag twigs furnished with spreading thorns up to ⅓ inch long or a little longer. The fleshy, reverse-heart-shaped leaves, about ¼ inch long, sprout singly from below a pair of thorns. The tiny white bisexual and female flowers are in umbel-like clusters 2 to 3 inches long. The blooms have two each sepals and petals, eight or ten stamens or in female flowers staminodes (nonfunctioning stamens), and one style ending in a scarcely enlarged stigma. The dry fruits contain a single seed.

Garden Uses and Cultivation. This species, scarce in cultivation, is of interest to collectors of choice succulents and other desert plants. It can be grown outdoors in warm, desert and semidesert regions and in sunny greenhouses. Well-drained, very porous soil, for specimens grown indoors kept moderately moist, but allowed to become nearly dry between waterings from spring to fall, and in winter kept dry, suits. A night temperature in winter of 55 to 60°F, with a few degrees increase by day is satisfactory. Higher temperatures at other seasons are in order. Increase is by cuttings and by seeds.

DECEMBER, GARDENING REMINDERS FOR. So far as outdoor gardening is concerned northern gardeners can now to a large extent relax and take their ease. But not in all regions are all tasks completed. It is never wise to install protective winter coverings of salt hay, leaves, and suchlike

Salt hay, with a few brushwood branches to hold it in place, being positioned over herbaceous perennials

To renovate overgrown deciduous shrubs, prune now and as opportunities arise through the winter

(c) Croton (*Codiaeum*)

materials until the ground is frozen to a depth of a couple of inches or so, and this in milder parts of the north should not be until after Christmas. This gives opportunity to employ branches of discarded Christmas trees as winter covering, and very satisfactory they are. You may be able to obtain unsold trees without cost. Christmas tree branches are excellent for covering the roots of shrubs of borderline hardiness and for herbaceous perennials including bulb plants.

Very serious damage can result from snow sliding off roofs onto evergreens (deciduous shrubs are less likely to suffer severely). Snow guards affixed to roofs prevent this. Where there are no guards it may be well to afford protection by installing temporary wooden roofing over prized specimens.

Relaxation from outdoor work in December is only justified if you attended earlier to all tasks necessary to bring the garden into condition for winter. If you failed to do this procrastinate no longer. Consult the entry in this Encyclopedia titled November, Gardening Reminders For to check tasks you may have overlooked.

Grape vines may be pruned in all regions other than where cold is so severe that the vines may be subject to a certain amount of winterkilling. In such regions delay pruning until February or even March. Renovative pruning of old overgrown deciduous shrubs may be started in December and continued whenever weather permits until spring growth begins.

Propagation by hardwood cuttings of a great many kinds of deciduous shrubs is an easy and successful way of securing new plants. Make the cuttings now, bury them under sand, sawdust, peat moss, or similar material and in spring take them out and plant them in lines in a nursery bed.

Now catch up with your horticultural reading. Enjoy your books and take advantage of public and specialist libraries. The literature of gardening is vast. Do not neglect the older classics. The newest offerings are not necessarily the best. Enjoy

the works of such great gardeners as Liberty Hyde Bailey, E. A. Bowles, Reginald Farrer, Gertrude Jekyll, Mrs. Francis King, Elizabeth Lawrence, William Robinson, Louise Beebe Wilder, John Wister, and Richardson Wright and of horticultural explorers such as E. H. M. Cox, David Douglas, George Forrest, F. Kingdon-Ward, and E. H. Wilson. Appraise newer offerings. A fair proportion of horticultural books pub-

Tropical greenhouse plants stand lower temperatures now than from spring through fall: (a) Acalypha

(b) African-violet

lished each year are decidedly worthwhile. A gardener without at least a modest library forgoes much of the fun and interest of his avocation or vocation.

Greenhouse temperatures may with advantage (and fuel economy) by kept somewhat lower during this period of short days and relatively weak sunshine than earlier. Most tropical plants get along well with a night level of 58 to 60°F rising by day five to ten, or in very sunny weather

(d) *Clerodendrum*

(e) Chinese hibiscus

(f) Philodendrons

(c) Poinsettia

(c) Cineraria

as much as fifteen degrees. About the beginning of the month reduce the night temperature for poinsettias to 55°F, but not lower.

Cool greenhouse plants such as calceolarias, carnations, Christmas cherries, Christmas peppers, cinerarias, freesias, primulas, and stocks, for which winter night temperatures of 45 to 50°F are recommended, will be sturdier if at this time the lower of these levels is maintained. Take care not to overwater. Soil kept con-

stantly soggy is likely to cause roots to rot. Little or no fertilizing is needed this month.

Potting most plants is best deferred until days are longer and roots are more active, but annuals, including calceolarias, cinerarias, and primulas, may be moved into larger containers in which they will bloom in spring.

(d) Freesia

Greenhouse plants decorative in December: (a) Christmas cherry

Greenhouse plants that need cool conditions now to bloom later: (a) Calceolaria

(e) *Primula obconica*

(b) Christmas pepper

(b) Carnations

Chrysanthemums cut back after flowering, and to be used to provide cuttings later, should be watered sparingly and kept in a light place where the night temperature is 40 to 45°F.

Bulbs of Easter lilies may be planted in December, the date determined by the needs of the variety, the temperature at which they are to be grown, and the date of Easter. Examine dormant amaryllis (*Hippeastrum*) bulbs at two-week intervals. As soon as new growth is detected, saturate

(f) Stock

(b) To flower later

Bulbs of Easter lilies may be potted now

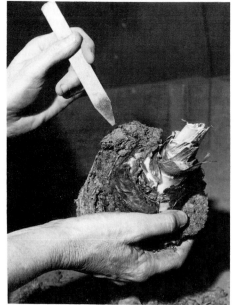

Amaryllis (*Hippeastrum*): (a) Topdress or repot old bulbs and start them into growth

vent temperatures, on sunny days, from rising to harmful levels.

Houseplant needs at this season are much like those suggested for greenhouse plants, but because temperatures in dwellings are frequently higher and the relative humidity is lower than in greenhouses more frequent watering is likely to be necessary. In December maximum light benefits all houseplants, even ferns, African-violets, and others generally satisfied with low intensities. Keep your plants as near windows as practicable, but on cold nights move them away or pull a shade or insert a few sheets of paper between them and the panes to prevent chilling or more serious damage by cold. Sponge the leaves of foliage plants, such as dracaenas, English ivies, fatshederas, philodendrons, and rubber plants, occasionally to remove grime

the soil with water, top-dress or repot according to the needs of individual bulbs, and put them in a sunny place where the temperature is about 70°F and the atmosphere is humid, and resume regular watering.

Cold frames containing plants likely to be harmed by very low temperatures must, unless they are blanketed by a thick covering of snow, be protected by covering them at nights and on very cold days with mats. Snow is excellent insulation. Do not brush it off the frames so long as the weather is cold. Ventilate cold frames during spells of mild weather by opening one end of each sash an inch or two or more to dissipate excessive humidity and pre-

In cold frames, protect cold-sensitive plants by spreading insulating mats over them at night

Ventilate cold frames on mild days

and to allow more light through as well as to improve their appearance.

In the south outdoor work appropriate for December varies with latitude and altitude. Winter tasks in the upper south do not differ greatly from those practicable in northern gardens except that milder weather makes it possible to continue until later planting, construction work, and readying land for spring sowing and planting. If dahlias have not yet been dug and stored attend to this now.

In warmer parts of the south deciduous and evergreen trees, shrubs and vines, including fruits trees, and roses may be planted in December. Be sure to water newly planted evergreens thoroughly and then to mulch them. Plant tulip bulbs that have been kept in cold storage. If you are unable to plant immediately after you receive them from the dealer store them in a refrigerator at about 40°F until you can. Sowing hardy annuals of practically all sorts, as well as cool-season vegetable garden crops, is now in order in the lower south, but not in colder areas. Keep lawns of rye grass mown fairly closely and particularly rye grass that has been sown over permanent summer grasses.

Other December tasks for southern gardeners include pruning deciduous shrubs that flower in summer or later on current season's shoots, but not, unless they are much overgrown and are to be rehabilitated, kinds that produce their blooms in spring on shoots of the previous year. Pruning them in winter sacrifices part of the spring floral display. Apply dormant sprays to trees and shrubs in need of them to control scale insects and certain other pests. Do this only when there is no danger of freezing and in accordance with recommendations of State Agricultural Experiment Stations or other reliable authorities.

West Coast gardens in December differ in the care they need and the opportunities they afford according to region. In northern sections and at higher elevations, few tasks are compelling, but further south there is no dearth of work to be done.

Planting deciduous and evergreen trees and shrubs may be undertaken successfully in practically all regions. If you are not yet ready to plant you can perhaps prepare sites for planting them a little later. Now is the time to plant calla-lilies, and it is about the last opportunity to set out bulbs of true lilies, poppy anemones, and Persian ranunculuses. Where winters are mild you may set out a wide selection of annuals, biennials, and herbaceous perennials, including calceolarias, Canterbury bells, carnations, cinerarias, columbines, primulas, snapdragons, and stocks. In regions of little or no frost make sowings of annuals and cool-season vegetables.

Other tasks appropriate now include pruning fruit trees as soon as they become dormant and pruning deciduous trees and shrubs other than spring bloomers. Hardwood cuttings of deciduous shrubs and cuttings of evergreens are likely to root well and give good results if made and planted at this time.

DECIDUOUS. As applied to such plant parts as leaves and petals deciduous means falling or shedding naturally at a particular season or stage of growth. As applied to entire plants it alludes to kinds that normally lose their leaves and are completely without foliage for a portion of each year. In cold regions deciduous trees and shrubs drop their foliage in fall. In desert and semidesert regions many plants lose their foliage at the beginning of the dry season.

DECKENIA (Deckèn-ia) One enormous tree of the palm family PALMAE, endemic to the Seychelles Islands, is the only member of the genus *Deckenia*. Its name commemorates the nineteenth-century African explorer Baron von der Decken.

Sometimes 150 feet in height, *D. nobilis* has a slender trunk that when mature commonly has a few roots exposed above ground. A unisexual species, it has long, pinnate leaves with sixty pairs or more of leaflets cleft at their tips. The long, drooping flower clusters have short spathes that soon fall. The female flowers are very small. Each is flanked by two males each with nine stamens.

Garden and Landscape Uses and Cultivation. This species is for collectors of the rare and unusual. Very little information is available about its cultivation. It may be expected to respond to conditions favorable to *Phoenicophorium*. For more information see Palms.

DECODON (Dé-codon)—Water-Willow or Swamp-Loosestrife. The only member of this genus of the loosestrife family LYTHRACEAE is a native of swamps and watersides from Massachusetts to Ontario, Minnesota, Florida, and Louisiana. Its name derives from the Greek *decas*, ten, and *odous*, a tooth. It refers to the top of the calyx.

The water-willow or swamp-loosestrife (*Decodon verticillatus* syn. *Nesaea verticillata*) is a subshrubby, wet soil or aquatic, often partly submerged perennial, with slender, angled, recurved or arching wandlike stems 3 to 8 feet long. Its nearly stalkless, toothless leaves are opposite or in threes or fours. They are elliptic-lanceolate to lanceolate, 1½ to 8 inches long, and are often waved at their margins. Their undersides are more or less pubescent. The rose-purple flowers, borne in summer in the upper leaf axils in dense clusters of two to eight, are about 1 inch across. They have five to seven triangular sepals with a linear appendage between each pair, five lanceolate, creped petals, and ten, or rarely only eight, stamens one-half of which are

longer than the others and protrude. The fruits are capsules. The tips of the arching stems of the water-willow root spontaneously in contact with soil or water, and give rise to new arching stems. As a result, the plant spreads rapidly.

Garden and Landscape Uses. Although unsuitable for small gardens and intimate plantings, this native American can serve usefully as a waterside plant in landscapes of broader scope. It is hardy and useful for holding banks. Its long stems, furnished with willow-like leaves, arch gracefully over the water and are congruous there. The flowers are pleasing, but not highly showy. An attractive feature is the fine red foliage hues the plant assumes in fall.

Cultivation. Provided the soil is moist and the location sunny no difficulty is had in cultivating this vigorous plant. It is easily propagated by seeds sown in soil kept constantly moist, by removing and transplanting branches that have layered (self-rooted) themselves, by division in spring, and by summer cuttings.

DECUMARIA (Decum-ària). Decumarias are less well known to gardeners than their relatives, the hydrangeas. There are only two species, one native of eastern North America, the other of China. Only the former is at all common in cultivation. Much hardier than its Asian relative, it may be grown outdoors in southern New England. Decumarias are slender woody vines that attach themselves to their supports by aerial rootlets in the fashion of the climbing hydrangea (*Hydrangea petiolaris*) and *Schizophragma*. They belong in the saxifrage family SAXIFRAGACEAE. Their generic name is from the Latin *decumus* (*decimus*), ten, and refers to the parts of the flowers, which are often, but by no means always, in tens.

These vines have opposite, stalked leaves, smooth-edged or sparingly toothed, and fertile, bisexual, white, fragrant flowers in small, dense, terminal clusters. Each flower has seven to ten minute sepals, the same number of oblong petals, twenty to thirty stamens, and united styles with a seven- to ten-lobed stigma. The fruits are ribbed, urn-shaped capsules containing numerous tiny seeds. The absence of large, sterile, marginal blooms in the flower clusters distinguishes this genus from *Schizophragma* and from most sorts of *Hydrangea*. From both genera *Decumaria* differs in its flowers having fewer petals and stamens and from *Hydrangea* is not having separate styles.

The American deciduous or nearly deciduous *D. barbara* climbs to a height of 30 feet. Its pointed or sometimes blunt, ovate to elliptic leaves, 2 to 4 inches long and about one-half as wide as long, are hairless, lustrous on their upper sides, and sometimes slightly hairy on the veins beneath. The round-topped clusters of flowers, 2 to 4 inches across, appear in late

Decumaria barbara

Decumaria barbara (flowers)

spring or early summer. This species inhabits wet woods and swamps, in shade or partial shade, from New Jersey to Pennsylvania, Florida, and Louisiana.

The evergreen **D. sinensis,** of central China, is a less robust climber with usually blunt, oblong leaves up to 3 inches long and about one-half as wide as long, and with pubescent leafstalks. It is not likely to be hardy north of Virginia.

Decumaria sinensis

Garden and Landscape Uses. Although by no means showy in bloom, decumarias are interesting for clothing walls, tree trunks, and other surfaces that afford their rootlets foothold. They produce attractive foliage patterns and grow neatly. They are excellent in shade and thrive in full sun provided the soil is not dry. The American *D. barbara* is a very satisfactory groundcover. At The New York Botanical Garden a patch in full sun and without access to support has remained not more than a few inches high and has completely covered the soil with good-looking foliage from spring through fall for thirty years or more. Its growth is sufficiently dense to inhibit weeds. Planted in shadier locations, this species is less compact and its leaves darker green.

Cultivation. Decumarias grow luxuriently in moist, fertile soils in shade or sun. They require but little care other than any pruning or shearing to restrain their spread or keep them tidy. Propagation is by seeds and by leafy summer cuttings planted in a greenhouse propagating bed or under mist.

DECUMBENT. Prostrate on the ground with the tip upturned; said of stems and branches.

DEEPLY PREPARED. Gardeners apply this term to the ground. When they speak of deeply-prepared soil they mean earth that has been conditioned for planting by spading, rototilling, or plowing to a depth of at least 8 inches and by incorporating with it organic matter such as manure, compost, or leaf mold. Other amendments, for instance lime and fertilizer, may be used to further improve the quality of the soil.

DEER. These beautiful animals are capable of causing considerable damage to gardens not only in fairly sparsely populated regions but, where nearby woodlands afford shelter, in suburban areas also. They are particularly partial to apple trees, yews, and tulips, and are not averse to browsing on vegetables. Spraying trees and shrubs attractive to them with a commercial deer repellent may bring relief, but to be effective this must be repeated at intervals. Where these animals cause serious damage the best recourse is to bring the matter to the attention of local game wardens or other wild life authorities.

The word deer forms parts of the colloquial names of certain plants. Examples include deer bush (*Ceanothus integerrimus*), deer-cabbage (*Lupinus diffusus*), deer fern (*Blechnum spicant*), deer foot (*Achlys triphylla*), and deer-grass (*Rhexia*).

DEERBERRY is *Vaccinium stamineum.*

DEERINGIA (Dee-ríngia). Native of Saxony, Dr. Karl Deering, who lived in Eng-

land and died in 1749, is commemorated by the name of this genus. It belongs in the amaranth family AMARANTHACEAE and consists of a dozen species, native from Indo-Malaya to Australia and Madagascar.

Deeringias are vining herbaceous plants, subshrubs, and shrubs. They have alternate leaves, and inconspicuous, unisexual or bisexual, small flowers in terminal, spikelike panicles. The blooms have five-parted perianths, and five stamens united in a ring. The fruits are small and berry-like.

A woody vine 10 to 15 feet tall, **Deeringia amaranthoides** (syn. *D. baccata*) is hairless and has slender-pointed, ovate-lanceolate, toothless leaves. Its greenish-white flowers are in panicles of interrupted spikes up to 1 foot long. The red fruits are about ¼ inch in diameter. This kind is native to Australia. Variety *D. a. variegata* has variegated foliage.

Garden and Landscape Uses and Cultivation. In climates free or nearly free of frost *D. amaranthoides* and its variety are interesting vines for planting outdoors against suitable supports. They grow in ordinary soils in sunny locations and are increased by seeds and cuttings.

DEGARMOARA. This is the name of hybrid orchids the parents of which include *Brassia, Miltonia,* and *Odontoglossum.*

DEINANTHE (Deinán-the). It is a pity that this close relative of *Hydrangea* is not better known to gardeners. Both its kinds are beautiful. They belong in the saxifrage family SAXIFRAGACEAE. The name, alluding to the large size of the fertile blooms, is from the Greek *deinos*, extraordinary, and *anthos*, a flower. In addition to this striking feature, *Deinanthe* differs from *Hydrangea* in its species being herbaceous perennials, instead of shrubs and woody vines.

Deinanthes have woody rhizomes, and erect stems with opposite, quite large, stalked, toothed, broadly-ovate or elliptic leaves, those of one kind being deeply cleft at their apexes. The flowers, displayed in

Deinanthe bifida, with flower buds

summer, are in terminal clusters that contain several bisexual blooms and fewer, marginal, usually sterile ones. Each has a short-tubed, toothed calyx. In the sterile flowers there are usually two or three rather large, persistent, petal-like calyx lobes (sepals). There are five petals, many stamens, and a solitary style, the latter distinguishing *Deinanthe* from *Cardiandra*, the blooms of which have three styles. The fruits are many-seeded capsules.

Native of Japan, white-flowered **D. bifida** is 1¼ to 2 feet in height. Its lower leaves are small and scalelike. Those above are 4 to 8 inches long, about two-thirds as wide, and coarsely toothed. At their apexes they are very distinctly two-lobed in fishtail fashion. The flowers are in clusters of up to twenty. The sterile blooms are up to ¾ inch or slightly more in diameter.

From the above, Chinese **D. caerulea** is easily distinguished by its coarsely-toothed leaves not being lobed at their apexes and by its flowers being violet-blue. Up to 1½ feet in height, this attractive species has nodding bisexual flowers up to 1¼ inches in diameter and smaller sterile ones with three petal-like calyx lobes.

Deinanthe caerulea

Garden Uses and Cultivation. Deinanthes succeed outdoors in sheltered locations near New York City. They enjoy peaty, fertile soil, well drained but never lacking for moisture, and locations sheltered from strong winds. Scorching heat is not to their liking; they appreciate light shade. Propagation is by division in spring and by seed.

DEKENSARA. Trigeneric orchids the parents of which include *Brassavola*, *Cattleya*, and *Schomburgkia* are named *Dekensara*.

DELONIX (Del-ónix)—Royal Poinciana, Flamboyant, Flame Tree. Perhaps the most gorgeous of flowering tropical trees, the royal poinciana, of Madagascar, is commonly planted throughout the tropics. None who visits those regions during its

flowering season can fail to be impressed, even overwhelmed, by the magnificence of its display. This is especially true where wet and dry seasons alternate. There, the royal poinciana is neither stingy nor sporadic in its blooming; enough of the tens of thousands of flowers a sizable specimen flaunts, open at one time to transform the tree into a flaming brilliant parasol that lights the landscape from afar in much the same way, but even more intensely, as a gorgeous sugar maple does in a New England fall. In constantly humid parts of the tropics the floral display of this tree is likely to be less concentrated and to extend over a longer season with never quite such an all-out display at one time. In Florida the main display is in May and June, in Hawaii in June and July.

The royal poinciana is the only one of the three species of *Delonix*, native of Madagascar and Africa, commonly cultivated. The group belongs in the pea family LEGUMINOSAE. Its name derives from the Greek *delos*, evident, and *onyx*, a claw, and alludes to the long-clawed petals. These are wide-topped trees with twice-pinnate, fernlike leaves the final divisions of which are quite small. Their brilliant crimson, scarlet or orange-yellow flowers are not pea-shaped, but are formed similarly to those of the honey-locust (*Gleditsia*) and the Kentucky coffee-tree (*Gymnocladus*). In loose, terminal or axillary clusters, each

has five sepals and petals, the latter with round blades narrowed to basal long claws. There are ten stamens. The fruits, flat, woody pods, resembling gigantic lima beans are solid between the individual seeds and hang on the trees practically throughout the year.

The royal poinciana (**D. regia** syn. *Poinciana regia*) is a wide-spreading, horizontal-branched, flattish-topped tree 20 to 50 feet in height. Its head of foliage suggests a giant lacy parasol. It grows rapidly and in most lands is deciduous, at least for a brief season. Usually while the branches are without leaves the flowers begin to open and continue their display for several weeks. Individual blooms are 4 to 5 inches wide and are many together in huge panicles. The petals are intense scarlet or crimson-scarlet with one marked with white or yellow. The sepals are bright green with scarlet lines. From 1 foot to 2 feet long, the seed pods turn blackish-brown at maturity.

Garden and Landscape Uses and Cultivation. The royal poinciana may be grown in southern California and southern Florida, but does not there reach the height it does under more tropical conditions. For its best display it needs ample room, preferably a site where it can be viewed from a little distance away. It is sometimes planted as a street tree, but for that use has disadvantages. Its roots are aggressive

Delonix regia

and likely to split pavements and masonry, and its brittle branches break easily and are much subject to storm damage. Because of this the tree should not be set closer than 30 feet to buildings. At its best in deep, well-drained soil of reasonable fertility, this tree develops with astonishing speed; under favorable conditions it attains a height of 25 feet in four years. It needs full sun. It is easily raised from seeds; other methods of propagation are less practicable. Injuries to trunk and branches through breakage or other cause are likely to attract termites, other destructive insects, and harmful fungi. To deter these, smooth the broken surfaces promptly and cover them with tree wound paint or asphaltum paint, but not with tar.

DELOSPERMA (Delo-spérma). Among the easiest to grow of the *Mesembryanthemum* relationship of the carpetweed family AI-ZOACEAE is *Delosperma*. This is a genus of 120 species, natives of South Africa. Its name, derived from the Greek *delos*, visible, and *sperma*, seed, alludes to the compartments of the seed capsules being without or having only rudimentary covers. Plants previously segregated as *Corpuscularia* are included in *Delosperma*.

Mostly succulent subshrubs, a few species of delospermas are tuberous-rooted herbaceous perennials or biennials. Comparatively few are of special beauty or interest, and most have rather inconsequential blooms. There are exceptions. The leaves of delospermas are cylindrical or three-angled with sometimes recurved apexes, or are nearly flat. Except in a few species, the older leaves of which are smooth, the foliage is covered with little projections called papillae. The white, yellow, or red flowers, in aspect daisy-like, but unlike daisies, single blooms instead of heads of many florets, have five or six sepals of which two are commonly longer than the others. The blooms, solitary or in branched clusters, mostly open about midday or in the afternoon. The fruits are capsules, the seeds of which are exposed and thus lend themselves in the native habitats of delospermas to being washed out and distributed by rain.

One of the most popular and attractive species, *D. echinatum* is a compact subshrub up to 1 foot tall, but often lower. It is much branched and has thick, sausage-shaped leaves a little less than ½ inch long beset, like the stems, in porcupine-like fashion with white-bristly papillae that on young shoots glisten in sunshine. The solitary, short-stalked, whitish to yellowish, long-lasting blooms range in diameter from a little under to a little over ½ inch. Shrubby and freely-branched, *D. cooperi* has nearly cylindrical, recurved, soft leaves up to 2 inches long, with somewhat flattened upper surfaces. Because the papillae are in longitudinal rows the leaves

have the appearance of being striped with grayish-green. About 2 inches wide, and mostly in groups of three to seven, the flowers are violet-red. Resembling the last, but with much shorter leaves and very much smaller flowers, is *D. aberdeenense.* Of similar aspect but smaller, *D. abyssinicum* is much like *D. cooperi*. It has light pink flowers, sometimes solitary, more often in groups of three to seven.

Delosperma abyssinicum

The plant formerly named *Corpuscularia lehmannii* is *D. lehmannii*. This kind has prostrate stems up to nearly 1 foot long from which sprout short shoots consisting of a few pairs of ¾-inch-long leaves whitened with incrustations of calcium oxalate. The light yellow flowers, 1½ to 2 inches wide and with 1-inch-long stalks have calyxes of six sepals. Similar, but considerably smaller than the last, *D. taylori*

Delosperma lehmannii

usually blooms more freely. The flowers of *D. t. albanense* are pale pink. Much resembling the wandering jew types of *Tradescantia* and related genera in habit, *D. tradescantioides* has long, freely-branched, prostrate, rooting stems, with rather distantly spaced pairs of pointed-elliptic to ovate leaves 1 inch long or a little longer

Delosperma tradescantioides

by somewhat under ½ inch wide. Their upper surfaces are often slightly concave. Solitary and white, the flowers are somewhat over ½ inch across.

A herbaceous perennial, *D. sutherlandii* has thick roots and produces each year stems up to ½ inch long, with pairs of finely rough-hairy, lanceolate leaves slightly united at their bases. From 2 to 3½ inches long by ¾ inch wide, the leaves have slightly channeled upper surfaces. The showy, violet-pink blooms, on stalks 2 to

Delosperma sutherlandii

4 inches long, are solitary and 2¼ to 2¾ inches wide. Another herbaceous perennial is turnip-rooted *D. brunnthaleri*. This compact kind has pointed-lanceolate to pointed-ovate, somewhat recurved leaves about 1½ inches long and slightly united at their bases. They are covered with minute papillae, and have hair-fringed margins. The violet-pink flowers are ¾ inch in diameter.

Garden and Landscape Uses. Delospermas can be grown permanently outdoors in warm dry climates, in less favorable climates outdoors in summer and in a sunny greenhouse or similar location through the winter, or they can be kept indoors permanently. They are satisfactory for rock gardens and similar locations and for pots and pans (shallow pots). Trailing *D. tradescantioides* is suitable for hanging baskets.

Cultivation. Excellent soil drainage and fairly nutritious, porous earth of ordinary quality suit delospermas. Full sun is necessary for best growth. New plants are easily had from cuttings, and seeds also germinate readily. Unless cross pollinated with another plant of different seedling stock, the flowers of some delospermas do not set seeds.

DELOSTOMA (Delós-toma). Five species of trees and shrubs of Andean South America and the West Indies constitute *Delostoma* of the bignonia family BIGNONI-ACEAE. They are rarely cultivated. The wide mouths of their flowers give reason for the generic name, derived from the Greek *delos*, evident, and *stoma*, a mouth. Delostomas have opposite, undivided, elliptic to nearly round leaves and terminal racemes or panicles of a few large, slightly two-lipped, funnel-shaped blooms each with five spreading corolla lobes (petals). There are four normal stamens and one aborted thread-like one, a staminode. They do not protrude. The fruits are woody, many-seeded capsules.

Native of Ecuador and Colombia, *D. roseum* is occasionally grown in mild, essentially frost-free climates. It has ovate to roundish leaves, hairy beneath and up to 6 inches long, and panicles up to 8 inches in length of pink flowers about 1½ inches long. The dark purple fruits are about 2 inches in length.

Garden Uses and Cultivation. These are as for *Clytostoma* and *Allamanda*.

DELPHINIUM (Del-phínium) — Larkspur. Modern hybrid delphiniums are among the stateliest and most gorgeous flowers gardens produce. They have quality. They are choice. In temperate regions they are easy to grow, yet not so easy that good care does not bring superior results, not so easy that as with sweet alyssum and California-poppies a few seeds dribbled into a likely spot and left to fend for themselves will bring results, and certainly they are not take-over plants of a kind that some vigorous perennials are, always ready to invade and appropriate neighbors' territory. Delphiniums are well behaved, responsive to intelligent attention. In addition to the magnificent hybrids there are a few species that connoisseurs of the worthwhile and the unusual cultivate, most of these natives of North America.

The genus *Delphinium* as now understood consists of more than 200 species, natives of many parts of the Northern Hemisphere. It belongs in the buttercup family RANUNCULACEAE. To the nonbotanist it is difficult to conceive of flowers more generally unlike in aspect than delphiniums and buttercups, but there it is; relationship is based on technical details of floral structure, not gross appearance. As now understood by botanists specializing in the floras of those parts of Europe, North Africa, and adjacent Asia where they are natives, and as accepted here, *Delphinium* no longer includes the garden annual larkspurs. They and kin of them are recognized as constituting the genus *Consolida*. This is not likely to disturb gardeners, who never really thought of their annual larkspurs and perennial delphiniums as being one. Some botanists, but not most American ones, segregate the native American species as *Delphinastrum*. This is not done here. The name *Delphinium* comes from the Greek *delphinion*, which, derived from *delphis*, a dolphin, alludes to the shape of the flower of some kinds.

Delphiniums include herbaceous perennials some with tubers and some, in cultivation at least, rather short-lived biennials, and annuals. They have leaves cleft or divided in hand-fashion (palmately). Their flowers are in spikes or racemes, branchless or branched in panicle fashion. Most are in shades of blue, varying to lavender and violet, and sometimes white. A few have red or yellow blooms. The flowers are highly asymmetrical. Each has five petal-like sepals one of which is lengthened into a spur into which project the spurs of the two uppermost of the usually four petals. The other pair of petals, if present, are small. There are numerous stamens arranged spirally and generally two to five separate pistils each becoming a small pod-like fruit called a follicle. Some if not all delphiniums are poisonous to cattle and horses, but apparently not to sheep. Most destructive to cattle in California, probably because of its prevalence, is *D. menziesii*.

The history of hybrid delphiniums is comparatively recent yet its roots go back a long way. They trace to the introduction into England of Siberian *D. elatum* perhaps in the early years of the seventeenth century. This, according to Parkinson, a famous herbalist of the time, was being grown in diverse colors, white, blush, and purple or violet and in double-flowered forms. However, a serious student of the group opines that Parkinson's plant was some other and that the first perennial delphinium to reach British gardens came nearly two centuries later from the seed house of Vilmorin in France. Be that as it may, the Siberian delphinium, as well as *D. grandiflorum* and *D. cheilanthum*, was being grown early in the nineteenth century. The first record of a hybrid delphinium is of one that bloomed about 1839 in a garden near Manchester, England. Its parents are believed to have been *D. grandiflorum* and *D. cheilanthum*. About the middle of the century breeding of other hybrids began and about that time the first American delphinium reached England. It was a form of *D. virescens* named *D. vimineum* obtained from Texas, but so far as we know played no part in hybridizing. Systematic breeding of superior hybrids started with the famous Victor Lemoine of Nancy, France, about 1850, and his firm remained prominent and active in delphinium breeding for nearly a century. It seems that the Lemoine hybrids were the results of interbreeding *D. elatum*, *D. cheilanthum*, and *D. tatsienense*. Other breeders soon entered the field. In England James Kelway took the lead, followed later by Charles Langdon, of Blackmore and Langdon, and subsequently by Watkins Samuel and Frank Bishop, the latter working for the nursery firm of Bakers at Wolverhampton. A decided benchmark was the appearance as a chance seedling, sometime before 1880, that was named *D. belladonna*.

Meanwhile, following World War I, interest in raising new hybrid delphinium stirred and swelled in the United States. This led to the foundation of the American Delphinium Society, which had a profound effect in stimulating interest in delphiniums and the production of races especially suited to American conditions. Unfortunately the society went out of existence in the 1950s. American leaders in breeding and growing delphiniums include such outstanding ones as Newell F. Vanderbilt, of California, Colonel Joel E. Spingarn, of New York, Leon H. Leonian, of West Virginia, Charles F. Barber, of Oregon, and the California firm of Vetterle and Reinelt. Vanderbilt worked with among other kinds the American species *D. scopulorum*, *D. nudicaule*, and *D. coccineum*. Except in parts of North America favored by cool summers the multiplication of hybrid delphiniums by division and cuttings is not practicable. While English gardeners, in a climate that permits such practices were filling their flower beds and borders with unbelievably splendid horticultural varieties, envious Americans had to content themselves with much inferior plants raised from seeds, and for the most part these were not very successful perennials. But American breeders met the challenge and developed new strains of such merit and beauty that they were sought and grown by Europeans. Most notable of these were the Pacific hybrids of Vetterle and Reinelt. The virtue of these American strains is that they come surprisingly true from seeds and even plants that vary from the type of the variety are rarely less beautiful. True, they are rarely long-lived, but are so easy to raise that in North America at least, vegetative propagation of delphiniums is rarely practiced.

Modern blue-, lavender-blue, purple-, and white-flowered hybrids fall into two groupings. There are those with erect, massive, cylindrical spires of quite densely arranged, single and semidouble flowers

Modern hybrid delphiniums: (a) 6-foot-tall specimen, with doronicums in front

(b) A 4-foot-tall, two-year-old plant

Delphinium belladonna, hybrid

that under good conditions attain heights of 6 to 8 feet or more. Most popular of these are the 'Pacific Hybrids'. Even under favorable conditions these are short-lived and are most conveniently treated as biennials or annuals. Their handsome spikes are of large, well-formed flowers that come in a wide spectrum of colors and in pure white. Similar English hybrid strains, best suited to regions of fairly cool summers and moderate winters, are under favorable conditions more successful as perennials. A second grouping, the belladonna hybrids, grow only 3 to 4 feet tall, but produce more flower spikes from each plant than the 'Pacific Hybrids' or English hybrids and tend to be more permanent in regions of hot summers. Compared with their taller relatives their single or sometimes semidouble flowers are smaller and more loosely arranged. Because of these characteristics the flower spikes are more airy and less massive than those of the stately 'Pacific Hybrids' and the English hybrids. They are splendid in less formal flower arrangements than those for which

the others are so well suited. The color range of the belladonna hybrids is more restricted than those of the taller kinds. Outstanding among the belladonna hybrids are 'Bellamosa', with dark blue flowers, 'Casa Blanca', the flowers of which are white, and 'Cliveden Beauty', with clear azure-blue or turquoise-blue blooms.

Chinese delphiniums or Chinese larkspurs, as they are often called, are popular for growing as annuals or short-lived perennials, good perhaps for two or three years. Usually listed as *D. chinense,* or *D. sinense,* they are forms of **D. grandiflorum.** Attractive and quite different in habit from other commonly grown delphiniums they are generally 1 foot to 2 feet tall and taprooted. They branch freely and widely and have leaves cut into narrow segments. In loose sprays, the generously produced wiry-stalked blooms, with wide-spreading sepals and straight or slightly hooked spurs of the same color as the sepals, stand well apart from each other. They come in rich blue, light blue, blue and purple, mauve, and white. Very similar, but taller,

with leaves less finely divided and flowers that do not open quite fully and have longer spurs, which are curved, than those of the common *D. grandiflorum* varieties is Chinese **D. tatsienense.**

Red- and pink-flowered hybrids have been the goal of several breeders of delphiniums, but results so far have been disappointing. As parents, the breeders have relied upon *D. cardinale* and *D. nudicaule* to infuse the desired hues. One such hybrid developed in Holland in 1933 from crosses made earlier between *D. nudicaule* and a blue-flowered hybrid delphinium was named *D. ruysii.* The first of its kind was offered for sale in 1937 as *D. r.* 'Pink Sensation'. A refinement of this was named 'Rose Beauty'. Unfortunately *D. ruysii* hybrids failed to live up to early expectations and after a short time were no longer cultivated. No doubt breeders will persist in their quest for good red and pink flowered hybrids and perhaps will use *D. semibarbatum* as a starting point for yellow-flowered ones. In time they may succeed, but as yet they have not.

Asian species, ancestors of modern hybrid delphiniums and others, are not in general cultivation. Occasionally grown in botanical gardens and by plant breeders, they are so excelled in floral display value by their hybrid offspring that they make little appeal to gardeners concerned only with that aspect of plant cultivation. Here are descriptions of the progenitors of the hybrid races: **D. cheilanthum,** a native of Siberia where it occupies slopes, tundra, and river valleys even into the arctic, is normally 1 foot to 3 feet tall and has branchless stems, free of hairs or nearly so. Its more or less rounded-heart-shaped leaves are grayish-hairy on their undersides and have three chief lobes cleft nearly to the base of the blade. The flowers are blue and from 1½ to 2 inches across. **D. elatum,** of Siberia, inhabits clearings, woodland margins, and river valleys in

Delphinium elatum

thin birch and aspen forests. A perennial 3 to 7 feet tall or taller, it has hollow, somewhat hairy or hairless stems and rounded leaves deeply-heart-shaped at their bases and cleft two-thirds their depth into three main lobes. The flowering portion of the stem is branched or branchless, and rather loosely flowered. The blooms are blue, up to 1 inch across. *D. grandiflorum,* native in Siberia, Mongolia, and China favors dry meadows and stony slopes. Perennial, it has usually branched stems 1 foot to 2 feet

Delphinium grandiflorum

tall. Its leaves are hairy, divided into five lobes. The few bright blue flowers are in loose, broad racemes. They are 1 inch to 1½ inches wide.

Bright yellow flowers are characteristic of rarely cultivated *D. semibarbatum* (syn. *D. zalil*), native to semidesert regions of central temperate Asia. A perennial 1 foot to 2½ feet tall, this has five-lobed leaves that because of the peculiar arrangement of the lobes appear to be three-lobed. The 1-inch-wide flowers are in fairly dense racemes. This is most likely to succeed under conditions favorable to *D. cardinale* and *D. nudicaule.*

Scarlet-flowered *D. cardinale,* called scarlet larkspur, is endemic to dry openings in chaparral and woodlands in California and Baja California. It is not reliably hardy in the north. From 3 to 6 feet tall, this has thick, deep, woody roots and erect, branched or branchless, hollow stems. The basal leaves, withered by the time the plant is in full bloom and 2 to 8 inches in diameter, are divided into five wedge-shaped segments each shallowly- to deeply-three-lobed. The stem leaves are of five to seven deeply-cleft lobes. They are sometimes hairy along the veins. Usually brilliant scarlet, rarely yellow, the slender-stalked flowers are in loose racemes or panicles. They have spurs up to ¾ inch and petals up to over ½ inch long. There are three pistils. The follicles (seed pods) are erect. Its cornucopia-shaped flowers less brilliant than those of scarlet larkspur, they are prevailingly dull red to orange-red with red-tipped, yellow upper petals, less commonly yellow, *D. nudicaule* is native from Oregon to California. It survives, but

Delphinium nudicaule

does not thrive in the vicinity of New York City. This kind has clusters of thin roots and erect or ascending, often glaucous stems 9 inches to 2 feet tall or occasionally taller. The stalked leaves, up to 4 inches in diameter, are cleft into three or five shallowly-round-lobed lobes. The blooms have spurs ½ to ¾ inch long and sepals up to ½ inch long. The follicles are spreading. Natural hybrids between *D. nudicaule* and *D. decorum* have flowers from coral-pink to yellow, lavender, and purple.

North America, especially the western part of the continent, is rich in species of perennial delphiniums, some of great loveliness, but alas most not very amenable to cultivation. They range from tall to diminutive. Most are tuberous-rooted and although likely to have scanty foliage, their blooms are frequently large, brilliantly colored, and abundant. They are well described in floras and wild flower books of the region. Beside red-flowered *D. cardi-*

nale and *D. nudicaule* discussed earlier these may tempt the adventurous: *D. depauperatum,* slender-stemmed and usually up to 1 foot high, has small, dark but bright, blue blooms. *D. geyeri,* from 1½ to 6 feet tall, has slender, sometimes branched racemes of showy, lively rich blue or occasionally light blue flowers. *D. hesperium,* 1 foot to 3 feet in height, has flowers dark blue to purple-blue, whitish or pinkish. *D. leucophaeum* (syn. *D. menziesii ochroleucum*), 1 foot to 2½ feet high, has creamy-white to yellowish flowers with bright cyan-blue upper petals. *D. menziesii,* 9 inches to 1½ feet tall, has showy, rich dark blue blooms in few-flowered racemes. *D. nuttallianum,* 6 inches to 1 foot tall or a little taller, has nodding, bright dark blue or purplish-blue flowers. *D. scopulorum* has slender, branchless stems 1½ to 3 feet tall, and loose racemes of indigo-blue blooms. *D. tricorne,* 1 foot to 1½ feet tall or sometimes taller, has racemes of blue or

Delphinium tricorne

sometimes white flowers. *D. trolliifolium,* stout-stemmed and 3 to 6 feet tall, bears large bright dark blue blooms.

Garden and Landscape Uses. Native American species may with propriety be used in native plant gardens and other naturalistic areas, and the dwarfer kinds may be given places in rock gardens. A few species from other lands too may serve this latter purpose. Except as they interest breeders or are represented in botanical garden and suchlike collections, the taller foreigners are unlikely to attract gardeners. They are so thoroughly outshone and superseded by the magnificence of the hybrids that they simply are not worthwhile.

Hybrid delphiniums belong in flower garden beds and borders and in cut flower gardens. The stately spires of bloom of many and the looser ones of others, such as those of *D. belladonna* varieties are not only unexcelled but are unequaled by any

Hybrid delphiniums grown for cut flowers

other hardy plants in their color range. Quite different *D. grandiflorum* varieties, although perennials, are most commonly grown as annuals to decorate flower beds and the fronts of borders. Their delicate sprays of flowers are useful for cutting. Sites for delphiniums, at least taller ones, must be sheltered from wind. Exposure, even though the plants are staked, is likely to cause much breakage. Delphiniums are sun-lovers, but will stand a little part-day shade. Ground for them should be well-drained, deep, and thoroughly tilled. Good soil texture is important. Apart from these strictures the characteristics of the soil may vary considerably. Good delphiniums can be produced in sandy, loamy, or even clayey earths from slightly acid to alkaline. These plants respond well to limestone soils.

Cultivation. As pointed out earlier, although cultivated delphiniums are herbaceous perennials the common practice in America is to treat them as biennials, annuals, or at most short-lived perennials. American gardeners seldom, and in most parts of the continent certainly not successfully, engage in propagating delphiniums by division and cuttings as is the practice in Europe. They depend upon seeds. One of the first elements in the successful cultivation of the hybrids is to obtain the finest possible strain of seed. Go to a reliable source for this. Expect to pay a relatively high price. High that is in relation to the packet cost of inferior strains, low in comparison to the magnificent results obtainable.

Sow the seeds in a cold frame or a sheltered, nicely prepared seed bed outdoors in May to give flowers the following year. A few plants from this sowing may develop blooming stalks the first summer. If they do, cut these off early so that the plant can concentrate its energies on building a strong root system and crown to flower handsomely the following year. To have blooms the first year sow indoors in February in a temperature of 60°F. Transplant the seedlings to flats or small pots and grow them in a sunny greenhouse in a night temperature of 50°F, with day temperatures five to ten degrees or so higher, until danger of frost is passed. Then, after hardening them in a cold frame for a week or so, plant them where they are to bloom. Allow a minimum of 2 feet between plants of vigorous hybrid kinds. More is better. About 1 foot is enough between varieties of *D. grandiflorum*.

In preparation for planting, spade, rototill, or plow deeply and incorporate a fair amount of compost or other decayed organic material. Avoid using fresh manure. Well-decayed manure is excellent. Mix in also a dressing of a complete garden fertilizer and if the soil is acid rake in lime. Delphiniums prefer a soil acidity approaching pH 7. Routine care during their first season involves keeping weeds down by frequent shallow surface cultivation or by mulching. The latter is of special benefit after warm weather arrives. Water deeply at approximately weekly intervals in dry weather. A light application of a complete fertilizer made about a month after the young plants are set out stimulates sturdy growth. Plants from seeds sown early indoors will bloom the first year beginning in August. Taller-growing kinds, but not *D. grandiflorum chinense* varieties, need careful staking. Do not allow seedlings from late spring or summer sowings to bloom the first year. To do this reduces their chances of surviving the winter. Cut off incipient flower spikes early. Do not discard hybrid delphinium plants after their first season of bloom. A proportion of them are likely to survive and bloom the next season, but not all can be relied upon to do this.

Second-year plants, survivors of those that flowered well the previous year and those raised from late spring or summer sowings that were not permitted to bloom the first year, benefit in cold regions where there is little snow cover from being given some winter protection. A light covering of salt hay, branches of evergreens, or other loose material that will moderate the ill effects of freezing and thawing that heaves the soil and breaks their roots, yet permit a free circulation of air around the crowns of the plants, is what is needed. In late winter gradually remove this covering and spread a dressing of a complete fertilizer and stir it shallowly into the surface soil. If you want the tallest and finest flower spikes reduce the number of shoots that come from the base of plants that send up many. Do this while they are small. Leave only three to five, removing the weakest and retaining the strongest. As spring advances delphiniums grow rapidly. Stake tall kinds before their stems are damaged by winds, storm, or other cause. Bamboo canes, fairly stout and pushed well into the ground to gain good anchorage, are very suitable supports. Place them with care, close to the bases of the plants, but not so that they pierce the crowns. Install them to be as inconspicuous as possible. Tie the stems securely but not overtightly to the canes. If the plants are being grown for cut flowers only, it will do to stick three or four canes around each and link them with a few circles of twine. An application made when the first flower buds are clearly evident of a side dressing or top dressing of fertilizer fairly rich in nitrogen encourages the production of tall flowering stalks and large blooms. As an alternative some gardeners apply dilute liquid fertilizer about once a week from the time the buds show until the first flowers open. In dry weather weekly deep soakings with water are needed. These are better given by flooding the ground than by overhead sprinkling. Delphiniums grown without adequate moisture are little more than ghosts of what they can be when more generously treated. Weeds must be kept down. This can be done by frequent shallow cultivations or by mulching. Where summers are hot, the latter is preferable.

When cutting flowers for use in arrangements or faded spikes from displays, and these latter should be removed promptly to prevent exhausting seed production, make the cuts above the main leaves. The removal of substantial amounts of foliage is debilitating. As soon as the first blooming of second-year and older plants is over scratch a light dressing of fertilizer into the soil, keep the plants adequately watered in dry weather, and you will be rewarded with a second blooming.

When cutting delphinium flowers, leave the plants with ample foliage

In greenhouses perennial delphiniums can be forced to bloom from March onward. For the best results set in outdoor nursery beds in early spring young plants raised from seeds sown indoors in winter or outdoors the previous year. In the nursery bed allow 2 to 3 feet between the rows of plants and 1 foot to 1½ feet between individuals in the rows. Throughout the growing season give your best attention to such matters as weeding, cultivating, and watering. Cut off early any flowering stalks that appear or at most allow one spike to develop until it reaches a stage where you can judge the quality of the bloom to determine whether the plant will be worth greenhouse space. In fall, after the first hard frost, dig the plants with as big root balls as practicable and pot them in containers just large enough to accommodate them comfortably. These ordinarily will be 6 to 8 inches in diameter. Water the newly potted plants thoroughly and put them in a well-ventilated cold frame buried to the rims of their containers in peat moss, sand, or similar loose material. In January or February bring them into a sunny greenhouse with a night temperature of 45 to 50°F, and daytime temperatures approximately five to ten degrees higher. Water moderately at first, more freely as the amount of foliage increases. Thin out the young shoots so that from one to three, or if a larger number of spikes rather than securing the biggest possible flower spikes interests you most, up to five, shoots remain. When growth is reasonably well advanced begin a program of weekly or semiweekly applications of dilute liquid fertilizer. Stake and tie the stems effectively and neatly. On all favorable occasions ventilate the greenhouse freely. Too high temperatures and excessive humidity are detrimental.

DEMAZERIA. See Desmazeria.

DENDROBIUM (Den-dròbium). A formidable 900 is the estimated number of spe-

cies that compose *Dendrobium*, an Old World genus of the orchid family ORCHI-DACEAE. A representative selection of species and numerous hybrids are popular with orchid collectors, but the vast majority of kinds that exist in widely differing environments, from Korea and Japan to Australia and New Zealand, and tropical Asia, Indonesia, and Pacifica have never been cultivated. The name, alluding to the plants living in trees, is derived from the Greek *dendron*, a tree, and *bios*, life.

Dendrobiums are tree-perching orchids. They are epiphytes, which unlike parasites and saprophytes do not abstract nourishment from the trees they inhabit. In growth habit and aspect the various sorts differ widely. They may be evergreen, partially deciduous, or deciduous. Their pseudobulbs, which complete their growth in one season and are clustered or spaced at intervals along rhizomes, are usually long, slender, jointed, and more or less canelike; less commonly they are short and fat. The leaves are mostly thick and comparatively short. They may be flat or cylindrical, but are never longitudinally-pleated. The long or short racemes of flowers terminate stalks that arise from near the tops of the stems or from the apexes of the pseudobulbs. The blooms show much variation in form and color according to species. They have three nearly equal-sized sepals, of which two are united to the base of the column to form a spur or small pouch. The other sepal is similar to the two petals, except that it may be wider or narrower. The lip is three-lobed or lobeless. There are four pollen masses (pollinia).

The descriptions that follow are of a selection of popular kinds: *D. aggregatum* has clustered, four-angled evergreen, pseudobulbs 2 to 3 inches tall each with

Dendrobium aggregatum

one leaf about as long. In arching to pendulous racemes up to 6 inches in length, of up to fifteen blooms, the honey-scented, golden-yellow flowers are about 1¾ inches wide. This is native to tropical Asia. *D. atroviolaceum*, of New Guinea, has furrowed, club-shaped pseudobulbs about 8 inches tall that taper to their bases and have near their apexes two to four ever-

green, elliptic leaves up to 7 inches long by 2 inches wide. Fragrant and 1½ to 2½ inches across, the flowers have purple-blotched, creamy-white petals and sepals and a three-lobed lip rich violet-purple inside. *D. bigibbum* is an evergreen native of Australia. It has cylindrical or slightly spindle-shaped pseudobulbs 1 foot to 1½ feet tall that, mostly toward their tops, have up to six narrowly-oblong-lanceolate leaves up to about 4 inches long. The flowers, somewhat variable in color, are commonly rosy-mauve with a deeper-colored lip with a white or whitish crest in the throat. *D. capituliflorum*, of New Guinea, has tapered pseudobulbs 2 to 6 inches long and

Dendrobium capituliflorum

lanceolate leaves often rich purple on their undersides. Crowded in heads 1 inch or a little more across, the many about ½-inch-long flowers are white or very pale green with a green column and a green base to the lip. *D. chrysanthum* ranges from Thailand to the Himalayan region. Its lax, slender, canelike pseudobulbs are about 2 feet in length. Deciduous, its slightly twisted, ovate-lanceolate leaves are up to a little more than 6 inches long. The fragrant, fleshy blooms, about 2 inches across, are in twos or threes from first-year psuedobulbs. They are golden-yellow marked on

Dendrobium chrysanthum

the toothed lip with crimson or red-purple blotches. **D. chrysotoxum,** of China, Indochina, Thailand, and Burma, has clusters of club- to spindle-shaped pseudobulbs up to 1 foot tall or taller with at their apexes two, three, or more deciduous leaves 4 to 6 inches long by about 1½ inches wide. The flowers are in rather loose, arching to pendulous racemes up to 1 foot in length. Golden to orange-yellow and very fragrant, they are 1½ inches or sometimes a little more in diameter. The lip is fringed, hairy on its upper surface, and has a curved orange to dark maroon band. **D. clavatum,** of southern China to the Himalayas, has drooping, cylindrical pseu-

Dendrobium cuthbertsonii

Dendrobium chrysotoxum

dobulbs up to 3 feet long and evergreen leaves. Its 2- to 3-inch-wide flowers, in racemes of four to six, are of orange-yellow flowers each with a toothed, bright yellow lip with a maroon blotch. **D. crumenatum,** of southeast Asia and Indonesia, has spindle-shaped pseudobulbs topped with a long, slender, often branched, stemlike portion that bears the leaves and flowers. The leaves are stalkless, deciduous, and up to about 5 inches long by ¾ inch wide. The solitary or paired blooms are pure white or tinged pink with, on the lip, a five-ridged disk or patch of yellow. **D. cuthbertsonii** of New Guinea, a delightful kind about 2 inches high, has miniature pseudobulbs each with two or three linear to narrow-elliptic leaves not more than 1 inch long. The solitary flowers, approximately 1½ inches across, come in a considerable range of colors and combinations chiefly in the orange-red, scarlet, and red range, but including yellow and white. In favorable environments individual blooms remain fresh for five to six months. **D. dearei** inhabits the Philippine Islands. It has stemlike pseudobulbs up to 3 feet long and strap-shaped leaves, notched at their ends and

about 3 inches long. The uppermost ones remain, those lower on the stems are deciduous. The green-throated, white flowers in racemes of up to six are 2 to 3 inches wide. **D. densiflorum** (syn. *D. thyrsiflorum*) is a native of the Himalayas, Burma, and Indochina. Strikingly beautiful, it has sometimes angled, conical pseudobulbs each with three to five rigid, evergreen leaves up to 6 inches in length. Sprouting from the upper parts of the pseudobulbs,

Dendrobium densiflorum

the very crowded, drooping racemes are cylindrical, up to 9 inches long by one-third as wide. The lightly fragrant, 2-inch-wide, bright yellow blooms have a velvety-hairy, but not fringed, orange lip. **D. devonianum** is a beautiful species with slender, drooping, stemlike, often branched pseudobulbs up to 4 feet long or longer. A deciduous native from the Himalayas to Burma, Thailand, and Indochina, its leaves are linear-lanceolate and up to 5 inches long. The short-stalked, fragrant blooms, solitary or in pairs, come from the joints of the pseudobulbs after the leaves have fallen. They are 2 to 3 inches wide, creamy-white to yellowish, sometimes tinged with pink and sometimes with magenta tips to petals and fringed lip. There are two orange blotches on the lip. **D. farmeri** in

habit of growth is much like *D. densiflorum* from which it differs in its flowers being pale lilac-mauve to white with a hairy, yellow lip tipped with rose-pink. This sort is native from the Himalayas to Malaya. **D. fimbriatum** ranges in the wild from the Himalayas to Burma, Malaya, Thailand, and Indochina. It has slender, stemlike, erect or arching pseudobulbs, up to 6 feet in length, thickened at their bases. Its pointed-lanceolate leaves are about 6 inches long. Up to about fifteen together in loose racemes up to 8 inches long, usually developed when the pseudobulbs are without foliage, the curiously-scented, 3-inch-wide blooms have a conspicuously fringed

Dendrobium fimbriatum

lip. They are bright yellow to orange-yellow with, on the lip, a deeper orange-yellow, velvety patch. **D. f. oculatum** is more robust and has slightly bigger blooms with a pair of rich-velvety, blackish-brown blotches on the lip. **D. findlayanum,** of Burma to Thailand, has generally erect, stemlike pseudobulbs with much-swollen joints. The leaves are deciduous, oblong-lanceolate, up to about 4 inches long. Sprouting from the pseudobulbs, usually in pairs, the up to 3-inch-wide fragrant blooms are from blush-white to lilac deepening in hue toward the tips of the sepals and petals. The lip has a large yellow disk with behind it a blotch of magenta-purple. **D. formosum** is wild from the Himalayas to Burma, Thailand, and the Andaman Islands. Its slightly spindle-shaped pseudobulbs, up to 1½ feet in length, are furrowed when old and have persistent leaf sheaths with black hairs. The oblong to ovate, evergreen leaves, notched at their apexes, are up to 5 inches long. The blooms, from at or near the apexes of the pseudobulbs and from two to four together, 4 inches wide or a little wider, are snow-white with yellow or orange-yellow throats. **D. f. giganteum** has shorter, stouter pseudobulbs, with flowers 5 to 6 inches across. **D. gouldii** hails from the Solomon

Dendrobium formosum giganteum

Dendrobium kingianum

Dendrobium nobile

Dendrobium johnsoniae

Dendrobium loddigesii

Dendrobium nobile virginale, white flowers, with a cream lip

Islands. Its thick, stemlike pseudobulbs are up to 4 feet tall or taller. The two-ranked, evergreen, ovate-elliptic leaves are up to 6 inches long. In erect or arching racemes of sometimes as many as twenty-five, the flowers are yellowish with purple-brown staining and with dark purple or rich brown on the petals. The yellowish, three-lobed lip has five conspicuous, purple-stained ridges. **D. heterocarpum** (syn. *D. aureum*) is a very variable native from the Philippine Islands to southeast Asia, India, Sri Lanka (Ceylon), and Indonesia. It has upright to pendulous, stout, stemlike pseudobulbs and deciduous foliage. The leaves are pointed-oblong-lanceolate and up to 5 inches long. Borne from the leafless pseudobulbs, the fragrant, 2½- to 3-inch-wide blooms are in clusters. They are cream to yellow and have a lip white with a velvety-hairy, golden-yellow disk usually veined with crimson or sometimes the entire lip is yellow. **D. johnsoniae**, a magnificent native of New Guinea, has

slender, club-shaped pseudobulbs under 1 foot long, each with two or three evergreen leaves, elliptic to oblong, and up to 6 inches long. From at or close to the tips of the pseudobulbs are produced loose racemes up to 1 foot long each with up to one dozen, but often fewer, snow-white, 4- to 5-inch wide, fragrant blooms with the

three-lobed lip marked with purple. **D. kingianum** is an Australian that exhibits considerable variability. Its club-shaped pseudobulbs not over 1 foot tall, may be thick or stout below and slender in their upper parts. Often they branch at their tops. The leaves are evergreen, lanceolate, up to 4 inches long, with six or fewer from each pseudobulb. The dainty, fragrant, 1-inch-wide blooms do not open fully. They

are delicate pink to deep mauve with the lip usually with a deeper blotch. Rarely the flowers are white. **D. loddigesii**, of China and the island of Hainan, has branched, stemlike, prostrate or pendulous pseudobulbs up to about 8 inches long. Its deciduous, lustrous oblong-lanceolate leaves are approximately 3 inches in length. From 1½ to 2 inches wide, the fragrant solitary flowers come from the sides of the leafless

pseudobulbs. They are rose-purple or white suffused with rose-purple, have a fringed orange lip margined with white and tipped with purple. **D. nobile,** native from the Himalayas to Thailand, Indochina, China, and Taiwan, is the commonest species in cultivation. It has clustered, erect, stemlike pseudobulbs up to about 2 feet in length, often more or less thickened in their upper

parts, that bear two-ranked, deciduous, oblong leaves, notched at their apexes and up to 4 inches long. The very fragrant, usually not widely-expanded flowers, variable in size and up to about 3 inches wide, are in twos or threes from the upper parts of the leafless pseudobulbs. They vary considerably in color, and many horticultural varieties are largely based on such variations. Typically they have white sepals and wavy-edged petals tipped with rose-pink. The creamy-white lip, tubular

toward its lower part, has a deep crimson to crimson-purple throat, is tipped with rose-pink. **D. phalaenopsis** (syn. *D. bigibbum phalaenopsis*), of New Guinea to Australia and Timor, is similar to *D. bigibbum*, but larger in its parts and more robust. It has canelike pseudobulbs up to 3 feet tall or taller and two-ranked, evergreen leaves. The flowers, in long, pendulous racemes, are variable in color. They range from pure white to deep purple with usually a darker lip and are 3 inches wide or sometimes wider. The petals are much bigger than the sepals. *D. p. schroederianum* is a splendid variety with larger blooms. **D. pierardii**

Dendrobium pierardii

Dendrobium sanderae

Dendrobium superbum dearei

ranges from China to the Himalayas, Burma, Malaya, and Thailand. It has drooping to arching stemlike pseudobulbs and deciduous, pointed-lanceolate leaves up to 4 inches long. The 2-inch-wide, fragrant, thin-textured flowers are solitary or in twos or threes from the leafless pseudobulbs. They are delicate pink with a yellowish lip veined with rose-purple. *D. primulinum* has approximately the same geographical range as the last. This has clustered, slender, pendulous or arching pseudobulbs up to 1 foot in length and

Dendrobium speciosum

Dendrobium victoriae-reginae

Dendrobium primulinum

deciduous leaves, the latter oblongish, up to 4 inches long. Borne on leafless pseudobulbs, usually single and up to 2½ inches in diameter, the fragrant flowers are white with pink-tipped sepals and petals and with a sulfur- to primrose-yellow lip, its lower part tubular and streaked with red-purple. **D. sanderae,** of the Philippine Islands, is very like *D. dearei*. Its flowers tend to be bigger, up to 4 inches across. They are pure white, and striped and shaded purple in the throat. **D. speciosum,** of Australia, has club-shaped, ribbed pseudobulbs 1 foot to 2 feet long or somewhat longer, with five, leathery leaves 8 to 10 inches long. The flowers, many together

in erect or arching racemes that may be 2 feet long, never spread their narrow, 1½-inch-long sepals and petals widely. They are cream to yellow with red or violet spots on the very short, three-lobed lip. **D. superbum** (syn. *D. macrophyllum*), of the Philippine Islands, Indonesia, Indochina, and Malaya, is a splendid kind. Deciduous, it has pendulous, stemlike pseudobulbs up to 4 feet long or considerably longer. The lustrous, two-ranked, oblong-lanceolate leaves are up to 5 inches in length. Usually in pairs, the blooms, 3 to 4 inches across and often not expanding widely, come from the sides of the pseudobulbs. Most commonly they are rosy-purple with a pubescent lip with a pair of deeper purple blotches, but they vary considerably in coloring. Often they are fragrant. *D. s. album* has white flowers with a pale purple lip. *D. s. anosmum* has white flowers with the lip pale purple at its base. *D. s. dearei* has blooms pure white except for a suggestion of pale yellow on the lip. *D. s. giganteum* has bigger blooms than those of the typical species. *D. s. huttonii* has pure white flowers with a pair of purplish blotches on the disk. **D. victoriae-reginae,** of the Philip-

pine Islands, has erect, slender, cylindrical pseudobulbs 6 inches to 1 foot tall, and narrow, strap-shaped leaves 2½ to 3½ inches long. In clusters of seven, the 1½-inch-wide flowers are violet to blue-purple, veined with deeper tones of the same color and grading to white at the bases of the sepals and petals.

Mostly less well known in cultivation than the sorts described above are these: **D. arachnites** of Burma has more or less cylindrical pseudobulbs 3 to 4 inches tall, each with usually two or three pseudo-linear-lanceolate leaves 2½ to 5 inches long. Its somewhat fragrant flowers, from the upper parts of the pseudobulbs, are 2 to 2½ inches across. They have very narrow orange-scarlet to cinnabar-red sepals and petals and a darker lip sometimes veined with purple. The fiddle-shaped lip is recurved somewhat at its apex. **D. chrysoglossum** of New Guinea is a slender and attractive sort with erect or suberect stemlike pseudobulbs up to 2 feet long. Borne in short racemes toward the apexes of the older, leafless pseudobulbs, and strapshaped, the leaves are 2 to 3 inches long. The 1 inch-long flowers have pleasing pink

Dendrobium arachnites

Dendrobium delacouri

Dendrobium teretifolium fairfaxii

Dendrobium chrysoglossum

Dendrobium delicatum

Dendrobium uncinatum

Dendrobium infundibulum

sepals and petals and a bright orange lip. **D. delacouri** of Indochina and Thailand has narrow-cylindrical-club-shaped pseudobulbs up to about 1 foot long each with a few long ovate leaves. The flowers, 1 inch or slightly more across, have buff-yellow sepals and petals and a brown-lined, deep yellow lip fringed with yellow-tipped brown hairs. **D. delicatum** of Australian origin resembles **D. kingianum** and may be a hybrid between that species and *D. speciosum*. Quite variable, it has more flowers on taller spikes than *D. kingianum*. They are cream, tinged on the outsides of the sepals with pink or peppered with fine brown-purple spots. **D. infundibulum** of Burma has erect, slender pseudobulbs 1 foot to 1½ feet long and with black hairs. Its 3 to 4 inch-wide flowers, from the upper parts of the pseudobulbs, are ivory-white with a patch of yellow on the lip. They have an inch-long, funnel-shaped spur. **D. teretifolium** of Australian has very slender, pendulous stems up to 10

feet long but often much shorter, and similar leaves from 6 inches to 1 foot long. In racemes of three to twelve, the 1 to 2 inch-wide flowers have slender white sepals and petals and a red lip. The flowers of *D. t. fairfaxii* have the bases of their sepals and lip suffused with reddish-brown or purple. A native of New Guinea, **D. uncinatum** is 4 to 6 inches tall. Each short pseudobulb has three or four leaves about 2 inches

long. The fan-shaped flowers are about 1¼ inches long. They are lavender-pink with a vinous-purple lip that lies tightly against the column and has a bright orange tip. **D. undulatum** of Australia forms clumps of stout, 2 to 4 feet-long pseudobulbs and somewhat wavy 3 to 4 inch-long leaves. The flowers, from the apexes of the pseudobulbs, are in racemes of ten to twenty. From 2 to 2½ inches wide, they have very wavy sepals and petals that vary from dullish brown to light lavender with cream margins and the reverse sides spotted purplish.

Garden Uses and Cultivation. Dendrobiums of many kinds, including numerous fine hybrids, are popular and have various uses. In Hawaii and other tropical places they are esteemed as outdoor ornamentals and for supplying blooms for flower arrangements, boutonnieres, and similar purposes. They are also grand for greenhouses. In the main easy to grow, they are well suited to the needs of beginners as well as experienced gardeners. As is to be expected of a group with such a wide nat-

Dendrobium undulatum

ural distribution, the needs of different kinds, especially as to temperature, vary considerably. Dendrobiums are classed as cool, intermediate, or tropical according to their temperature preferences. In greenhouses a winter night temperature of 45 to 50°F suits cool dendrobiums. Those classed as intermediate require 55°F at night in winter. For tropical dendrobiums the minimum winter night temperature should be 60°F. A little higher is advantageous for some. Day temperatures, depending upon the brightness of the weather, may be five to fifteen degrees above those maintained at night. From spring to fall higher temperatures than those recommended for winter are in order. Of the kinds described here, *D. primulinum* is rated as needing cool conditions, *D. loddigesii* and *D. nobile* respond to slightly more warmth. Intermediate dendrobiums include *D. aggregatum*, *D. densiflorum*, *D. devonianum*, *D. fimbriatum*, *D. johnsoniae*, and *D. kingianum*. Suitable for intermediate or somewhat warmer conditions are *D. chrysanthum*, *D. chrysotoxum*, *D. dearei*, *D. farmeri*, *D. findlayanum*, *D. formosum*, *D. heterocarpum*, *D. pierardii*, *D. sanderae*, and *D. superbum*. Decidedly tropical dendrobiums include *D. atroviolaceum*, *D. bigibbum*, *D. crumenatum*, *D. gouldii*, and *D. phalaenopsis*.

Another point for dendrobium growers to bear in mind is their varied requirements regarding a need for a dormant or resting season. Those with evergreen foliage need no period of definite dormancy. Their roots must be kept moderately moist throughout the year, somewhat drier when not growing than during their active development of new stems or pseudobulbs. For the most part evergreen kinds favor tropical or near tropical environments. Dendrobiums that lose their foliage completely for a period each year need cool, fairly airy conditions at that time. Also, at the onset of dormancy when they begin to lose their foliage, the amount of water sup-

plied must be gradually reduced by increasing the intervals between applications and finally withholding it completely. This ripens the pseudobulbs in preparation for flowering. When blooming is through, it may be desirable to cut out all or some of the old pseudobulbs. Because of their different habits of growth dendrobiums, according to kind, are adapted to various containers, such as pots, pans (shallow pots), and hanging baskets. One condition all must have is a freely-drained rooting medium that may be any of those commonly employed for epiphytic orchids such as orchid peat and bark chips. These orchids are responsive to fertilizer given in small amounts at frequent intervals during the growing season. Excessively large containers are not to their liking. Only sufficient shade is needed to keep the foliage from being scorched. For more information see Orchids.

DENDROCALAMUS (Dendrocál-amus)— Giant Bamboo. Twenty species of Indo-Malaysia and China belong in *Dendrocalamus* of the grass family GRAMINEAE. The name derives from the Greek *dendros*, a tree, and *kalamos*, a reed, and refers to the gigantic size of some of these grasses. They have short, thick rhizomes and form dense, imposing clumps of thick-walled, hollow or solid stems. Their flowers, in globular heads grouped in branching clusters, have six stamens.

The largest known bamboo is the giant bamboo (*D. giganteus*), of southeastern Asia. Under favorable circumstances it grows to a height of 100 feet or more and has stems up to 1 foot in diameter. This noble species develops its new shoots with great rapidity. They may elongate at the rate of 1 foot to 1½ feet in twenty-four hours. It is a strictly tropical kind. Not as tall, but just as beautiful, the male bamboo (*D. strictus*), of India, Burma, and Java, is the most important source of good quality paper pulp in India. Unlike most bamboos, this species withstands dry conditions surprisingly well, surviving drought better than any other bamboo. In parts of its natural range it is occasionally subjected to temperatures as low as 22°F. Attaining a height of 40 to 60 feet, the male bamboo has arching or erect stems 4 or 5 inches in diameter. There is considerable variation in geographical strains and in different seedlings in the abundance of branches, size of leaf, and other details. Some plants have hollow stems, whereas in others they are nearly or quite solid. Some strains, at least under certain conditions, are deciduous. The flowering phase of this bamboo is believed to occur every 25 to 35 years. After blooming the plant dies. The male bamboo has proved satisfactory in southern California and Florida and is common in Puerto Rico. The colloquial name does not refer to the sex of the plant. Native

from India to Cochin-China, *D. latifolius*, grows to a height of 60 to 70 feet. Its stems are erect and it has tapering, short-stalked leaves up to 10 inches long by 2 inches wide. This species succeeds in climates such as that of southern California.

Garden and Landscape Uses and Cultivation. For information see Bamboos.

DENDROCEREUS (Dendro-cèreus). One species endemic to Cuba constitutes *Dendrocereus* of the cactus family CACTACEAE. Its name is from the Greek *dendron*, a tree, and *Cereus*, the name of another cactus genus. Britton and Rose, who first described the genus, selected the name because in their opinion *D. nudiflorus* (syn. *Cereus nudiflorus*) in outline more resembled a tree than did any other cactus. They presented it as having from a distance the aspect of an apple tree.

This impressive cactus has a stout, erect trunk that above develops many erect, spreading or drooping, dull green branches with three to five thin, flangelike, very spiny ribs. The trees, up to 35 feet tall, have trunks up to 2 feet or more in diameter and wide-spreading heads. The areoles (places on cactus stems from which spines and flowers grow) are covered with a felt of short hairs. They have two to fifteen black-tipped spines up to 1¾ inches long or sometimes they are spineless. The short-lived flowers open at night. They are 4 to 5 inches long and have many spreading, perianth parts (petals), the inner ones white and 1¾ inches long, the outer greenish-yellow and shorter. The numerous stamens are shorter than the petals. The style, very thick and 1 inch long, is capped with a many-lobed stigma. Greenish, smooth, and 4 to 6 inches long, the tough-skinned fruits are more or less pear-shaped.

Garden and Landscape Uses and Cultivation. These are the same as for other massive columnar cactuses. See Cactuses.

DENDROCHILUM (Dendro-chìlum)—Chain Orchid. These attractive, small-flowered "collection" orchids are frequently in cultivation mislabeled with the obsolete name *Platyclinis*. They number 100 species or more, natives from Indomalaysia to the Philippine Islands and New Guinea. Their name, from the Greek *dendron*, a tree, and *cheilos*, a lip, alludes to their habitats, and to their flowers being lipped. They are members of the orchid family ORCHIDACEAE.

Predominantly epiphytic (perching on trees), but sometimes growing on rocks, and rarely in the ground, dendrochilums have small pseudobulbs, clustered or spaced along the rhizomes, or none, and solitary leaves. The flowers are in terminal, pendulous racemes from the ends of erect stalks. They have spreading, narrow sepals and petals, and a lip nearly as long. The column has two hornlike appendages.

The golden chain orchid (*Dendrochilum filiforme* syn. *Platyclinis filiformis*), of the Philippine Islands, has clusters of furrowed, long-egg-shaped pseudobulbs 1 inch tall or a little taller. The leaves are erect or nearly so, long-stalked, broadly-linear to narrowly-elliptic, and 5 to rarely 8 inches long. The arching, slender racemes, with their stalks up to 1 foot or 1½ feet long, hang vertically. They come mostly

Dendrochilum filiforme

in spring and summer. The fragrant, canary-yellow to yellowish-white blooms, up to 100 of them to each raceme, are in two ranks. The individuals are little more than ¼ inch across and have broad-elliptic sepals and petals that exceed the lip. Much like *D. filiforme*, but less robust and mostly blooming in fall, **D. uncatum** (syn. *Platyclinis uncata*) has clustered pseudobulbs, and arching and drooping racemes with stalks 6 to 9 inches in length. The scented blooms, about ⅜ inch wide, are pale green to yellowish-white, and have brown or yellow-brown lips. This kind is from the Philippine Islands. Differing from *D. filiforme* in having somewhat translucent, greenish-yellow flowers that develop later in the season, **D. cucumerinum** was named because of the supposedly cucumber-like odor of its blooms. Native of the Philippine Islands, this blooms in fall or early winter.

Flowers about ½ inch across are displayed by spring-blooming **D. glumaceum** (syn. *Platyclinis glumacea*), also of the Philippine Islands, in two-ranked chains that with their stalks are 1 foot or more in length. They are whitish to pale yellow, and have lips suffused with green at their bases. The crowded, egg-shaped pseudobulbs, the younger ones with a pair of reddish scales, and inside them a large red-tipped bract, are up to about 2 inches long. The long-stalked, narrowly-lanceolate, frequently wavy-edged leaves have blades up to 10 inches long by 1½ inches wide. Similar, but stronger, larger, and with broader foliage, **D. longifolium** (syn. *Platyclinis longifolia*), of Sumatra and Java, flowers in

spring. It has greenish-yellow flowers with very long-pointed sepals and brown-marked lips.

Sturdy **D. cobbianum** (syn. *Platyclinis cobbiana*) has tightly clustered, furrowed pseudobulbs 1 inch to 2 inches long. Its softish, long-stalked leaves are narrowly-lanceolate to elliptic-lanceolate and wavy. The strictly pendulous, twisted or spiraled racemes of flowers, up to 8 inches long or longer, have stalks up to 1 foot in length. Whitish to pale yellow, the fragrant blooms are ½ to ¾ inch across. They have yellow or orange-yellow lips. This native of the Philippine Islands blooms mostly in fall.

Garden Uses and Cultivation. Dendrochilums are at their best as comparatively big specimens, hence it is a mistake to divide them too frequently. They are displayed advantageously in pans (shallow pots) and pots and grow well in a variety of rooting mediums commonly used for orchids. One that finds favor with many growers is a mixture of osmunda fiber and chopped sphagnum moss. Dendrochilums are very intolerant of poor conditions at their roots. Repotting must be done before the rooting medium decays and becomes stale. These orchids seem to do best when confined in containers smallish with respect to the sizes of the specimens. They enjoy tropical warmth and humidity and need partial shade. No season of rest is required, but in winter they are kept just a little drier than at other times, and fertilizing is discontinued. Intermediate or tropical greenhouse temperatures are appropriate. For further information see Orchids.

DENDROMECON (Dendromè-con)—Bush-Poppy, Tree-Poppy. As here interpreted *Dendromecon* consists of one variable species. Some botanists recognize two or three species and some have divided it into as many as twenty. The genus is native to the southwestern United States and Mexico and is closely related to the Matilija-poppy (*Romneya*). From this it is distinguished by its lobeless leaves and yellow flowers. It is a member of the poppy family PAPAVERACEAE. The name is from the Greek *dendron*, tree, and *mekon*, poppy.

Dendromecons are open-branched, evergreen shrubs with alternate, undivided, yellowish or grayish-green leaves and solitary flowers with two sepals, four petals, numerous stamens, and two stigmas. Their sap is not milky. The fruits are linear capsules.

Of stiff, rounded habit, the bush-poppy or tree poppy (**D. rigida**), of California and Baja California, is usually 1 foot to 6 feet tall, but sometimes it attains 10 feet. It has whitish bark and linear-lanceolate or lanceolate-oblong, rigid leaves 1 inch to 4 inches long that have minutely-toothed edges. The golden-yellow flowers 1 inch to 2 inches across, at the ends of short

Dendromecon rigida

branches, are on stalks longer than the leaves. Proportionately broader leaves, perfectly smooth at their margins, and flower stalks no longer than the leaves distinguish *D. r. harfordii*, a native of islands off the coast of California. This sort sometimes attains a height of 18 feet. Variety *D. r. rhamnoides* has paler green, more distantly-spaced, sharply-tapered leaves. It is a native of islands off the coast of California.

Dendromecon rigida harfordii

Garden Uses. These plants are rather less hardy than the Matilija-poppy (*Romneya*) and are not adaptable for cultivation outdoors in eastern North America. They are at their best in gardens in the west and southwest where warm, dry summers are to their liking. There, they are handsome for shrub plantings, setting near buildings, and other landscape purposes. They flower attractively over a long period from early summer through fall, the individual blooms, under favorable conditions, lasting for several days.

Cultivation. Dendromecons need a sunny location and well-drained soil. Pruning, done in spring, is restricted to cutting out weak and unwanted branches and shortening any that have become bare at their bases. The cut back branches soon develop new shoots. Propagation is by cuttings in summer in a greenhouse or a cold frame, by layering, and by seeds, which take a long time to germinate.

DENDROPANAX (Dendró-panax). To the aralia family ARALIACEAE belongs this genus of about seventy-five species of shrubs and trees of the tropics and subtropics of the Old and the New Worlds. The name is from the Greek *dendron,* a tree, and the name of the related genus *Panax.* Without spines, dendropanaxes have undivided, sometimes lobed, leaves. Their small greenish flowers are in umbels that are sometimes themselves disposed in umbels. They have tiny calyxes, five petals, and the same number of stamens. The fruits are berries.

A hairless, evergreen shrub or small tree, *Dendropanax trifidus* (syns. *Gilibertia japonica, Hedera japonica*) has stoutish branches and lustrous, leathery, long-stalked leaves, the largest with blades up to 8 inches long by 6 inches wide. They are wedge-shaped at their bases and have two or three lobes at their tops or may be lobeless and smaller. Three or five major veins spread from the bottoms of the leaves. In young specimens the leaves are often deeply-three- or -five-lobed. The small, greenish-yellow flowers, of little decorative merit, are in mostly solitary, terminal umbels. The fruits are black, and about ⅓ inch long. This is endemic to Japan.

Garden and Landscape Uses and Cultivation. Not hardy in the north, this quite beautiful evergreen is excellent for partially shaded locations in mild climates. It is a good general-purpose tall shrub or small tree that thrives in ordinary soil and needs no special care. It is also useful as a pot or tub plant for greenhouses and indoor ornament. For such usages it should be given fertile, well-drained, loamy soil, kept moderately moist. A reasonably humid atmosphere and shade from strong sun are required. In greenhouses a winter night temperature of 45 to 50°F is satisfactory, with an increase of five to fifteen degrees by day. Propagation is by cuttings and by seeds.

DENDROPHYLAX (Dendro-phỳlax). An extraordinary native orchid is one of four species of *Dendrophylax,* of the orchid family ORCHIDACEAE, a genus endemic to the West Indies and southern Florida. The name, derived from the Greek *dendron,* a tree, and *phylax,* a guard or watchman, alludes to the habitats of these plants.

These orchids are leafless epiphytes (tree-perchers). Each plant consists of a short stem, a tangled mass of grayish roots, and in season, blooms. The roots contain chlorophyll and perform the photosynthetic food-building functions usually carried out by leaves. The blooms are large and handsome. They have somewhat similar

spreading sepals and petals and a complex lip with a large center lobe, two small side ones, and a long spur. The column is short.

Native to the West Indies and southern Florida, where it is most common on live oaks and royal palms, *D. lindenii* (syns. *Polyrrhiza lindenii, Angraecum lindenii*) has a brief stem, and roots mostly clinging to trunks and branches of trees, up to several feet in length. Its long-lived, fragrant, white, waxy flowers, about 5 inches long and in racemes of several, open successively over a long period. The upper (dorsal) sepal and the petals are reflexed, the other sepals curve downward and outward. The pure white, large lip, concave at its base, has two short side lobes and a center one that develops into a pair of long, outward- and downward-spreading lobes with curled apexes. Endemic to the West Indies, *D. funalis* (syn. *Polyrrhiza funalis*) has brief stems and rather few slender roots up to about 1 foot in length, many of which hang free from the tree on which the plant grows. On stalks 1 inch to 2 inches tall, the solitary or paired, long-lasting flowers have ¾-inch-long, greenish-yellow, reflexed sepals and petals and a big pure white lip. The spur is 2 inches long or longer.

Garden Uses and Cultivation. The cultivation of these plants is attended with difficulties. Rarely is it possible to reestablish them after removal from their host tree although possibly they can be persuaded to attach themselves to slabs of tree fern trunk. Probably the best plan is to leave them attached to the piece of branch or trunk on which they originally grew and suspend this in a humid, intermediate-temperature or warm-temperature greenhouse in bright light, with shade only from the strongest sun. Water must be supplied in abundance at all seasons. For more information see Orchids.

DENMOZA (Den-mòza). The genus *Denmoza* of the cactus family CACTACEAE contains two species, natives of Argentina. Its name is an anagram of that of the Argentinean province of Mendoza.

Denmozas much resemble *Echinopsis,* but are readily distinguished by their flowers being markedly asymmetrical.

As a young specimen *D. rhodacantha* is more or less spherical but as it ages its stems lengthen and become approximately columnar and up to 5 feet tall. They are 3½ to 6 inches, or when aged 1 foot in diameter and have fifteen to thirty ribs with gray-woolly areoles (cushions from which spines originate) spaced about ¾ inch apart. Six to twelve spines spring from each areole. They are white to blood-red and about 1½-inches long. The 2- to 2½-inch-long blooms appear at the tops of the plants and have scarlet perianth segments (petals) and carmine protruding sta-

mens and styles. Very similar, but with thinner, more numerous spines and bristly hairs is *D. erythrocephala.*

Garden and Landscape Uses and Cultivation. These are those detailed as appropriate for low, columnar cactuses under Cactuses.

DENNSTAEDTIA (Denn-staédtia)—Hay-Scented Fern, Cup Fern, Lacy Ground Fern. The precise limits of this group of ferns are not well established. They merge through a few species that might as well be allotted to one genus as the other with *Microlepia.* Chiefly tropical and subtropical, but represented in the native floras of Canada, the United States, Japan, Chile, and Tasmania, *Dennstaedtia* consists of seventy species. It belongs in the pteris family PTERIDACEAE. The name commemorates August W. Dennstaedt, an early nineteenth-century German botanist.

Dennstaedtias are medium- to large-sized, extensively creeping, often thicket-forming ferns, with branched, hairy rhizomes and hairy or hairless fronds (leaves) one- to three-times-pinnately-divided. The clusters of spore capsules, at the edges of the frond segments, are not concealed by scalelike coverings (indusia). It is not always easy to distinguish the species one from another.

The hay-scented fern (*D. punctilobula*)

Dennstaedtia puntilobula: (a) Foliage

(b) In a naturalistic setting

inhabits rocky slopes, cleared places, open woodlands, and sandstone cliffs from Newfoundland to Ontario, Minnesota, Georgia, Alabama, and Arkansas. It has deciduous fronds up to 3 feet tall, with ovate, twice-pinnate blades, at their bases up to 10 inches wide, with their divisions deeply-pinnately-lobed. The seventeen to twenty-five pairs of primary divisions of the fronds are lanceolate and have fourteen to twenty-five pairs of deeply-cleft secondary divisions. The minute clusters of spore capsules are on or at the bases of little recurved teeth. This is the only New World species with gland-tipped hairs on the undersides of its fronds.

Native from southern Florida to Mexico, Central and South America, the Greater Antilles, and Trinidad, *D. bipinnata* (syn. *D. adiantoides*) occupies wet forests, slopes and banks, and open places. It has creeping rhizomes and glossy, hairless, thrice-pinnate fronds with sharply-toothed ultimate segments. The blades are up to 5 feet long by 3 feet wide. The leafstalks are light brown and up to 3 feet long. From this, *D. globulifera* differs most obviously in having soft, rather than leathery or semileathery, firm fronds, with the apexes of their ultimate segments at most shallowly and rather bluntly, instead of more conspicuously sharply, toothed. Also, the basal subdivisions of the primary divisions near the middles of the fronds are opposite rather than alternate. This kind occurs in habitats similar to those favored by *D. bipinnata*, from Mexico to Argentina and in the Greater Antilles. A variable species, native from Central America to the Greater and Lesser Antilles, Trinidad, and Paraguay, *D. obtusifolia* includes plants previously segregated as *D. ordinata*. Closely similar *D. dissecta* is perhaps only a variant of the same species differing in its clusters of spore cases being about one-half as thick as they are broad. From its rhizomes *D. obtusifolia* sends hairy-stalked fronds with triangular, thinly-hairy, much-divided blades 3 to 6 feet long. Its clusters of spore capsules are about as broad as they are thick.

Ranging in the wild from Mexico to Brazil and the Greater Antilles, the common cup fern (*D. cicutaria*) favors humid forests, hillsides, open areas, and river and stream banks. It has thrice-pinnate fronds 3 to 8 feet tall and up to 3 feet wide at the bases of their blades. They have alternately arranged primary divisions and final divisions pinnately-lobed and toothed. The cup- or urn-shaped spore-case clusters are at the bases of indentations in the leaf margins. Australian *D. davallioides,* sometimes called lacy ground fern, has thrice-pinnate fronds with lower leaflets up to 9 inches long by 4 inches wide. The pointed-linear segments have toothed edges.

Garden and Landscape Uses and Cultivation. The limits of their natural distri-

Dennstaedtia cicutaria

butions indicate the relative hardiness of the kinds discussed. Where hardy, dennstaedtias may be planted to adorn semiwild and naturalistic landscapes. They tend to be too rampant for admission to areas reserved for choice, less vigorous plants. The nonhardy kinds are sometimes grown in greenhouses, where they succeed without special care where the winter night temperature is about 55°F and that by day is five to fifteen degrees higher. Good soil drainage and nourishing earth always fairly moist that has a generous organic content suit these ferns. Let the soil be coarse and aerated rather than composed of finer particles and compact. Shade from strong sun is necessary, although as long as the earth is satisfyingly moist dennstaedtias will take more intense light than most ferns. Certainly specimens in greenhouses should not be shaded from fall to spring. Propagation is easy by division and spores. For additional information see Ferns.

DENTARIA (Den-tària)—Toothwort, Milk Maids. Often pretty and dainty, spring-blooming hardy herbaceous perennials of Northern Hemisphere woodlands constitute *Dentaria* of the mustard family CRUCIFERAE. Its name, alluding to toothlike projections on the rhizomes of some kinds, comes from the Latin *dens*, a tooth. There are possibly thirty species, but some that have been described as such may be natural hybrids. The botany of the group is by no means well understood, and specific limits are to some extent subject to interpretation.

Much like *Cardamine,* and sometimes included in that genus, *Dentaria* generally differs in having fleshy rootstocks and in the foliage of nearly all species being divided or deeply lobed in palmate (hand-like) patterns, but these identifying characteristics do not always hold true. The two or three, opposite or alternate stem leaves are high on erect, otherwise naked stems. The few longer-stalked basal leaves sometimes die before the flowers are fully expanded. Dentarias have evenly continuous or sharply constricted, horizontal rhi-

zomes. The constrictions mark the ends of successive seasons' growth, and the rhizomes break readily at those points. The flowers are rather few together in terminal racemes. White, pink, or purplish, and fairly large, they have four sepals, four petals, six stamens, of which two are shorter than the others, and a slender style. The fruits are narrow capsules with one row of flattened seeds.

Eastern North American toothworts likely to be cultivated are *D. diphylla, D. laciniata, D. heterophylla,* and *D. anomala.* All have flowers with petals from a little under to a little over ½ inch long. From the other three *D. diphylla* is readily distinguished by its rhizomes not being constricted at intervals. From 8 inches to over 1 foot tall, its basal and usually two opposite or nearly opposite stem leaves each have three coarsely-toothed leaflets not more than twice as long as wide. This inhabits rich, moist woodlands from New Brunswick to Minnesota, and southward in the mountains to Georgia and Alabama. About as tall, and with similar flowers, *D. laciniata* has foliage divided into much more slender segments. Its basal leaves, which have

Dentaria laciniata

usually died by flowering time, are similar to the stem ones, of which there is commonly a whorl (circle) of three. They are divided or deeply lobed into three linear to lanceolate, toothed or nearly toothless segments that are four times as long or longer than wide. This species, highly variable in foliage, occurs from Quebec to Minnesota, Florida, Arkansas, and Kansas. The basal leaves of **D. heterophylla,** usually present at flowering time, are much like those of *D. diphylla,* but its two or rarely three, opposite or nearly opposite stem leaves resemble those of *D. laciniata.* Its blooms are similar to those of the others. This is native from New Jersey to Indiana, Georgia, and Alabama. Quite probably a hybrid between *D. laciniata* and *D. diphylla* or *D. heterophylla,* and often found growing with those species, **D. anomala** occurs in Connecticut and Kentucky. Unlike those of *D. diphylla* its rhizomes are

A Pacific hybrid delphinium

Dendrobium chrysotoxum

Dendrobium fimbriatum oculatum

A hybrid delphinium

Dalea species

Dendromecon rigida (flowers)

Dendrobium nobile

Diabroughtonia 'Alice Hart'

Dianthus, a garden variety

Dianthus, a garden variety

Dentaria diphylla

constricted at intervals. From *D. laciniata* and *D. heterophylla* it differs in having stem leaves with ovate segments not more than three times as long as wide.

Western American **D. californica** is a variable species with several varieties recognized by botanists. Commonly called milk maids, it inhabits shaded banks and slopes from Oregon to Baja California. It has ovoid rhizomes, and hairless aboveground parts. Its basal leaves, typically of three broadly-ovate leaflets up to 2 inches wide, are toothed. The two, three, or rarely more stem leaves have three lanceolate to ovate, toothed or toothless leaflets or lobes. The slender, erect stems, 6 inches to nearly 1½ feet tall, terminate in many-flowered racemes of white or pale pink blooms with petals from a little shorter to slightly longer than ½ inch long. Variety *D. c. integrifolia* has foliage more fleshy than that of the species. The basal leaves of *D. c. cuneata* have five to seven leaflets and are generally twice-pinnate, those of *D. c. sinuata* (syn. *D. integrifolia tracyi*) and *D. c. cardiophylla* are generally undivided, and so usually are the stem leaves of *D. c. sinuata*, but not those of *D. c. cardiophylla*.

Native of moist meadows from Oregon to British Columbia, **D. tenella** differs from *D. californica* and its varieties in that any tubers it may have do not exceed ⅛ inch in diameter instead of being at least nearly twice that thickness. Hairless and up to about 8 inches tall, this species has usually undivided basal leaves and stem leaves of three or sometimes five, usually toothless, narrowly-lanceolate to narrowly-ovate leaflets. The pink blooms are in few-flowered racemes. Variety *D. t. pulcherrima* is more robust and has thicker leaves, the basal ones of three to five leaflets.

Garden and Landscape Uses and Cultivation. Easily colonized in woodlands and other shaded places where the soil is rich with organic matter and fairly moist, and suitable for growing under similar conditions in rock gardens and native plant gardens, toothworts make pleasing spring displays of bloom and have attractive foliage. They are increased by dividing the rhizomes, which may be done as soon as the foliage dies, or in very early spring, and by seeds sown in a cold frame or a sheltered, shady place outdoors, in sandy soil well supplied with leaf mold or peat moss, and kept uniformly moist, but not saturated.

DEODAR CEDAR is *Cedrus deodara*.

DERMATOBOTRYS (Dermáto-botrys). Alluding to the presentment of the blooms, the name of the genus *Dermatobotrys* is derived from the Greek *dermatos*, skin or bark, and *botrys*, a cluster. There is only one species, a South African somewhat doubtfully assigned to the figwort family SCROPHULARIACEAE, by some thought to belong in the nightshade family SOLANACEAE.

An epiphyte (plant that perches on trees without taking nourishment from its host), **D. saundersii** is a deciduous shrub with long, thick, leafless stems and four-angled branches with foliage at their ends. The somewhat pointed leaves are ovate to oblong, coarsely-toothed, and 2 to 6 inches long by approximately one-third as wide as long. From 1½ to 2 inches long, the flowers hang in clusters from below terminal tufts of foliage. They have a small, five-parted calyx, a tubular, light-red corolla with a yellow interior, five fertile stamens, and one style. The fruits are many-seeded, egg-shaped berries ¾ inch long.

Garden Uses and Cultivation. This very interesting plant is fairly easy to grow in a greenhouse in which a dry, airy atmosphere, and a winter night temperature of 55°F are maintained. It succeeds in well-drained, coarse, peaty soil kept moderately moist during the growing season, nearly dry when the plant is dormant. Increase is easy by seeds and by cuttings.

DERRIS (Dér-ris). Three species of *Derris*, but most especially *D. elliptica*, are the chief sources of the insecticide rotenone, a product of their roots. The roots are also used as native fish poisons. Another species is cultivated as an ornamental. The genus comprises trees, shrubs, and lianas of the tropics and subtropics of the Old and the New Worlds. They belong in the pea family LEGUMINOSAE. There are about eighty species. The name, from the Greek *derris*, a leather covering, has reference to the fruits.

Members of the group have alternate, pinnate leaves with an odd number of leaflets. They are without tendrils. The pea-like white, pink, or purple flowers are in racemes, panicles, or clusters from the leaf axils. They are succeeded by pods that do not split open, but rot after they fall to the ground to release the seeds.

The Malay jewel vine (**D. scandens**) is an attractive, tall-growing, woody vine, native to China, Malaya, and Australia. Its leaves have nine to fifteen oblongish, paired leaflets 1 inch to 2 inches in length, with the terminal leaflet larger than the others. The small, pale pink flowers are in slender, axillary racemes. The pods, flat, winged on one side and about 2 inches long, contain one to three seeds.

Garden and Landscape Uses and Cultivation. The Malay jewel vine is a good-looking, warm-country climber adapted for use where a strong-growing vine can be employed advantageously. It succeeds under ordinary garden conditions and is increased by seed.

DESCHAMPSIA (Des-chámpsia) — Hair Grass. The genus *Deschampsia* belongs in the grass family GRAMINEAE and includes about forty species. Its name honors the eighteenth-century French physician and naturalist M. Deschamps.

Natives of temperate and cold parts of the Northern and Southern hemispheres and of mountains in the tropics, deschampsias are tufted grasses with slender, usually branchless stems and narrow or hairlike, flat, rolled, or folded leaves. Their panicles of flowers are loosely-branched or compact. The flattened spikelets are small and two- or rarely three-flowered. The common name hair grass is applied also to *Aira*.

Crinkled hair grass (**D. flexuosa**), formerly known to gardeners as *Aira flexuosa*, inhabits North America from coast to coast as far south as North Carolina and Arkansas, great parts of Europe and Asia, and mountains in Africa. Perennial, it usually forms dense tufts of slender, wiry stems and tightly rolled, bristle-like leaves not more than 1/25 inch wide and with blades up to 8 inches long. Mostly its leaves are basal or come from near the bottoms of the stems. In bloom this grass is 1 foot to 3 feet tall. Its glistening flower panicles, 1½ to 6 inches long and about one-half as wide, are loose, open, and usually partly nodding. The spikelets that compose them are brownish, silvery, or purplish, up to ¼ inch long, with bristles as long or slightly longer. The bristles are characteristically bent or twisted below their middles. The crinkled hair grass inhabits dry woods, fields, sand hills, and moorlands.

The tufted hair grass (**D. caespitosa**), is another quite beautiful grass, but is less often cultivated than *D. flexuosa*. It grows in wet soils and bogs from Greenland to Alaska, New Jersey, West Virginia, and Arizona and in Europe and Asia. It is a variable species and botanists distinguish several varieties. From *D. flexuosa* it differs in having flat or folded leaves 1/12 inch broad or broader, and the bristles of its flower spikelets are almost or quite straight. Perennial, it attains a height of 2 to 5 feet.

Garden and Landscape Uses and Cultivation. These grasses grow without difficulty in any ordinary garden soil, the crinkled hair grass preferring dryish conditions and the other wetter. They prosper in full sun, but stand a little shade. Propagation is easy by division in spring or early fall, and by seed. These are attractive in garden beds and borders, supply attractive panicles for use in fresh or dried flower arrangements.

DESERT. The common names of various plants include the word desert. Examples are Australian desert-kumquat or Australian desert-lime (*Eremocitrus*), desert candle (*Caulanthus inflatus* and *Eremurus*), desert-dandelion (*Malacothrix glabrata*), desert gum (*Eucalyptus*), desert-holly (*Atriplex hymenelytra*), desert-hollyhock (*Sphaeralcea ambigua*), desert-honeysuckle (*Anisacanthus* *thurberi*), desert-ironwood (*Olneya tesota*), desert-lily (*Hesperocallis undulata*), desert-mallow (*Sphaeralcea ambigua*), desert-marigold (*Baileya*), desert-parsley (*Lomatium*), desert-rose (*Adenium obesum*, *Hibiscus farragei*, and *Alyogyne huegelii*), desert spoon (*Dasylirion wheeleri*), desert star (*Monoptilon*), desert-sunflower (*Geraea canescens*), desert-willow (*Chilopsis*), and Sturt's desert-pea (*Clianthus formosus*).

DESERT GARDENS. Largely in response to the amenities that accrue from air conditioning, more and more Americans are making their homes in the deserts and semideserts of the Southwest. The region is vast, occupying sizable portions of southern California, southern Nevada, Arizona, New Mexico, and Texas. There Americans live and there they garden, and very successfully.

Desert gardening is completely different from humid-climate horticulture, yet its underlying principles are the same. No less than gardeners elsewhere, those who tend plants in regions of scanty rainfall must work with their environments, not against them. To do otherwise courts disappointment if not disaster. Gardening in deserts presents challenges that successfully met can bring great rewards.

What are the peculiarities of deserts as homes for plants? Obvious are the scarcity of water in soil and atmosphere, high temperatures, at least at times, with often very great differences between those of day and night, and brilliant sunshine. Limiting factors less likely to be thought of by those unfamiliar with deserts are winds that can shred and sand-blast unsheltered vegetation, soils low in organic matter and high in soluble salts, high alkalinity or salinity of available irrigation water, and high soil temperatures. All can be tempered or turned

(b) Fairchild Tropical Garden, Miami, Florida

to advantage by those who work sympathetically with the environment.

Deserts are not all alike. Elevation as well as longitude has a profound influence. In some high and medium-high deserts frost and winter freezing sets limits to the kinds of plants that survive. In low, subtropical deserts frosts are infrequent, freezing rare. In most deserts exceptional cold at intervals of several years determines the kinds of plants that are actually permanent, but optimistic gardeners tend to base their plantings more on average or usual years than on exceptional and destructive ones. As with gardeners everywhere, selecting the right plants, and especially with short-season crops, sowing and planting at the most advantageous times, are of critical importance in desert gardening.

Styles of desert gardens are broadly two, naturalistic developments that emphasize and take advantage of the desert scene, and oasis-like creations that introduce the softer elements of gardens in more humid regions. Both can be charming and effective, but the former may be more desirable because it presents unique garden landscapes impossible to have elsewhere. Naturalistic desert gardens have the added advantage of needing less attention in the matter of irrigation. A combination of both types is perhaps ideal; verdant areas surrounding the home gradually merging into more typical, naturalistic desert plantings beyond can be a happy arrangement.

Soil management is of special importance in establishing and maintaining desert gardens. Desert soils are characterized by their usually very low organic content. At every planting mix with the earth to a depth of 6 inches at least a 2- to 3-inch layer of peat moss, sawdust, ground tree bark, or other available organic amendment fortified with a nitrogenous fertilizer (sulfate of ammonia at 4 to 5 ounces to each 10 square feet is satisfactory). Build-up of soluble salts to harmful concentrations in the soil must be guarded against. All desert soils have comparatively high percentages of such salts and, application of fertilizers, and irrigation practices, can increase these beyond the level of tolerance. This stalls seed germination, stunts growth, and may cause leaf margins to wither and turn brown. The handling of this problem must be adjusted to particular conditions. If water low in total salts is available leach the soil thoroughly before planting by pouring into it great amounts of water, sufficient to carry the excess salts to lower levels. If drainage is adequate and when irrigation is necessary ample water is supplied, no further leaching may be needed. But in regions where irrigation water has a high salt content, as in southern California, periodic leachings after the initial one are likely to be necessary as indicated by simple soil tests or by judgment

Desert garden views: (a) The Desert Botanical Garden, Phoenix, Arizona

(c) Huntingdon Botanical Gardens, San Marino, California

(d) Los Angeles State and County Arboretum, Arcadia, California

based on experience. If sodium salts are present in substantial amounts in the water the clay content of the soil is deflocculated to the extent that the ground gradually becomes more and more pasty and impervious to the passage of water and air, to the great detriment of plants. With soils low in calcium treat this condition by applying gypsum. Those with a high calcium content, such as caliche earths, respond to the application of sulfur or sulfate of iron.

Caliche itself can be a problem in places where only a thin layer of soil covers a thick, perhaps impermeable layer of caliche. Then the best solution is to dig out the caliche to a depth of at least 2 feet and replace it with a mix containing 25 to 30 percent organic matter and 1 to 2 pounds of sulfate of iron to each cubic yard. Be sure that water will drain freely from the bottoms of holes dug in caliche and be certain they are of sizes that will afford enough root room for the plants at maturity. Alternatively you may plant in beds raised well above the caliche.

Windbreaks are necessary for the best results. Without such, the kinds of plants that succeed are likely to be severely limited. The most effective breaks in open country are wide ones of shrubs and trees in four or five rows spaced about 15 feet apart. The row to windward may be of Russian-olive, the next tamarisk, the third Arizona cypress, then a row of lombardy poplar, and finally a leeward planting of Aleppo pines. Quite obviously, on many sites such protection is clearly impracticable and narrower windbreaks must do. In such places a row each of tamarisk and Arizona cypress, with the former to windward will provide good shelter. In built-up communities, other buildings, plantings around them, and street trees will do much to moderate wind. Walls, fences, and louvered baffles strategically placed give additional protection.

Mulching is of tremendous help in desert gardens. Success with many plants

(e) The Botanical Garden, Mexico City, Mexico

largely depends upon it. Not only does 2 to 4 inches of loose organic material spread over the ground conserve moisture and gradually nourish the earth by adding organic matter and nutrients, it also prevents the upper few inches becoming so hot under the blazing sun that roots are killed. Maintain mulches permanently around trees, shrubs and other perennial plants, adding more to the surface as that beneath decays. Coarse materials, such as chunks of tree bark, look well and are not likely to blow away. In windy places anchor finer-textured mulches by surfacing them with bark of this type or with coarse gravel or crushed rock.

Watering, as well may be imagined, except for cactuses, succulents and a few other plants native to desert regions, is of first importance in desert gardens. Unless it is done regularly do not expect results. Given adequate water, a wide variety of trees, shrubs, lawn grasses, flowers, vegetables, and fruits flourish amazingly. The frequency of watering will depend upon the time of the year and other variables including type of soil. The basic rule, when water is given is to soak the earth to the full depth of the roots, then give no more until there is need. Shallower sprinklings encourage surface rooting, and desert plants not anchored deeply suffer more severely from dryness than deeper-rooted ones and are more subject to being blown over by wind. When it becomes necessary to leach hazardous accumulations of salts from the soil apply five to six times as much water as normal at one application. Annuals, vegetables, newly planted lawns, and container plants may need watering once or twice a day during the hottest and driest parts of the year, but established lawns and shrubs and trees will likely get along well with deep waterings at intervals of three or four days to two weeks. Fertilize plants other than cactuses and succulents, in desert gardens frequently, but with small amounts at each application.

Plants especially suited for gardens in deserts include of course a wide selection of cactuses and many other succulents, lists of which are given under the entries Cactuses and Succulents. Plantings of these can be extraordinarily beautiful as well as appropriate. But desert gardeners are by no means restricted to these. There are a wide variety of other plants, evergreen and deciduous, including cycads, palms, pines, yuccas, and others of noble aspect. Here is a selection of kinds other than cactuses and succulents. Not all will do well in all desert places, but most will if they are watered adequately.

DECIDUOUS TREES: almond (*Prunus amygdalus*), black locust (*Robinia pseudoacacia*), blue palo verde (*Cercidium*), Bolleana poplar (*Populus alba pyramidalis*), Carolina poplar (*Populus canadensis eugenei,*) fig (*Ficus carica*), fruitless white mulberry (*Morus*), honey-locusts (*Gleditsia triacanthos*

varieties), jacarandas [in warm climates], Japanese pagoda tree (*Sophora japonica*), Jerusalem-thorn (*Parkinsonia*), jujube (*Zizyphus jujuba*), lemon bottlebrush (*Callistemon*), Lombardy poplar (*Populus nigra italica*), 'Modesto' ash (*Fraxinus velutina glabra*), orchid tree (*Bauhinia*), pistache (*Pistacia chinensis*), Siberian elm (*Ulmus parviflora*), silk tree (*Albizia*), sycamores (*Platanus acerifolia, P. occidentalis, P. racemosa, P. wrightii*), white poplar (*Populus alba*), willows (*Salix babylonica* and some others), and *Zelkova serrata*.

EVERGREEN TREES: Aleppo pine (*Pinus halepensis*), Athel tamarisk (*Tamarix aphylla*), California pepper tree (*Schinus molle*), citruses in variety, *Cupressus glabra*, digger pine (*Pinus sabiniana*), *Eucalyptus camaldulensis, E. polyanthemos*, and *E. viminalis*, Italian stone pine (*Pinus pinea*), Japanese black pine (*Pinus thunbergii*), *Magnolia grandiflora*, and olive (*Olea europaea*).

EVERGREEN SHRUBS: *Abelia grandiflora, Cocculus laurifolius, Cotoneaster lacteus* and *C. pannosus, Elaeagnus pungens, Euonymus japonicus*, evergreen privet (*Ligustrum japonicum*), firethorn (*Pyracantha*, several sorts), hop bush (*Dodonaea viscosa* and its purple-leaved variety *D. v. purpureus*), Junipers (*Juniperus*, many kinds), lavender-cotton (*Santolina chamaecyparissus*), myrtle (*Myrtus*), Natal-plum (*Carissa grandiflora*), ocotilla (*Fouquieria*), oleander (*Nerium*), Oriental arbor-vitae (*Thuja orientalis* varieties), *Photinia serrulata*, pineapple-guava (*Feijoa*), rosemary (*Rosmarinus officinalis*), smoke tree (*Dalea*), sugar bush (*Rhus*), yaupon (*Ilex vomitoria*), yellow-oleander (*Thevetia*), and xylosmas.

DECIDUOUS SHRUBS: Apache plume (*Fallugia*), bird-of-paradise flower (*Caesalpinia*), butterfly bush (*Buddleia*) [in some places evergreen], California privet (*Ligustrum*) [in some places evergreen], chuparosas (*Anisacanthus* and *Beloperone*), crape-myrtle (*Lagerstroemia*), desert-honeysuckle (*Anisacanthus*), fairy duster (*Calliandra*), Japanese barberry (*Berberis*), and varieties *Lysiloma microphylla*, pomegranate (*Punica*), rose-of-Sharon (*Hibiscus*), Spanish-broom (*Spartium*) [the stems give an evergreen effect], and yellow bells (*Tecoma*) [in some places is evergreen].

VINES: Boston-ivy (*Parthenocissus*), bougainvilleas [in warm climates], cape-honeysuckle (*Tecomaria*), coral vine (*Antigonon*), grape vine (*Vitis*), Hall's honeysuckle (*Lonicera*), orange trumpet vine (*Pyrostegia*), trumpet vine (*Campsis*), Virginia creeper (*Parthenocissus*), and wisterias.

Herbaceous plants including annuals and biennials and perennials including bulbs in considerable variety can be grown in desert gardens if water is available for irrigation. So can a good selection of vegetables and some fruits.

DESFONTAINIA (Desfont-aìnia). One of the handsomest of temperate South Amer-

ican flowering shrubs, the only cultivated *Desfontainia* is one of possibly five species of the Andes of Chile and Peru. It belongs in the logania family LOGANIACEAE. Its name, often spelled *Desfontainea*, commemorates the French botanist René Louiche Desfontaines, who died in 1833.

Hardy only in mild climates, such as that of California, *Desfontainia spinosa* when out of bloom looks much like an evergreen English holly (*Ilex aquifolium*), but is im-

Desfontainia spinosa

mediately distinguishable because its leaves are opposite instead of alternate. The shrub is bushy and attains a maximum height of about 10 feet, but is often smaller. It has pale, glossy branches and broad-elliptic to ovate, lustrous, spiny leaves 1 inch to 2½ inches long. The flowers, which come in summer and fall, are quite astonishing on such a holly-like plant. They are in terminal clusters and because of their shape and striking colors have a decided fire-cracker or decorative candy-like appearance. They are tubular-funnel-shaped, about 1½ inches long, bright crimson-scarlet, and tipped with five small yellow corolla lobes (petals). Each has a five-lobed, green calyx with its margins fringed with hairs. There are five stamens. The fruits are spherical to egg-shaped, many-seeded berries.

Garden and Landscape Uses. This choice evergreen is admirable for displaying prominently in shrub borders, foundation plantings, and other landscape settings, and as an individual specimen. When well placed and thriving it is a splendid addition to almost any garden. For its best satisfaction it needs a little broken shade as protection from the hottest sun, and deep, moderately fertile, encouraging soil, never excessively dry.

Cultivation. Desfontainias can be raised from cuttings, about 3 inches long, taken in summer and rooted under mist or in a greenhouse propagating bed, but the best results are had from seeds sown in sandy peaty soil kept moderately moist. The seedlings should be shaded lightly from strong sun. Established specimens are grateful for an organic mulch maintained around them and for watering thoroughly and regularly during dry weather. They need no pruning, except the occasional shortening of an unruly shoot to keep them shapely, and any cutting necessary to limit their size. Spring is the season to attend to this.

DESMANTHUS (Des-mánthus). Chiefly natives of tropical and subtropical America, but with some in North America, the thirty species of *Desmanthus* belong in the section of the pea family LEGUMINOSAE that includes the sensitive plant (*Mimosa*), the silk tree (*Albizia*), and *Acacia*. Accordingly, the flowers are not pea-like, but are in fuzzy heads or spikes, a characteristic accounted for in the name which comes from the Greek *desme*, a bundle, and *anthos*, a flower, and alludes to the heads of bloom.

Of minor garden importance, the members of this genus are herbaceous perennials and shrubs with twice-pinnate, mimosa-like foliage. The tiny white or greenish flowers, clustered in tight heads, have five-lobed calyxes, five petals, and five or ten usually much-protruding stamens. A hardy herbaceous perennial, *D. illinoensis* is 3 to 6 feet tall and has conspicuously angled, hairless or minutely-hairy stems. Its leaves, 2 to 4 inches long, have six to twelve pairs of major divisions, each divided into twenty to thirty pairs of oblong leaflets up to ⅕ inch long and often hairy along their margins. The flower stalks, up to 1¼ inches long, terminate in solitary small heads of bloom, succeeded by short, strongly curved pods up to 1 inch long, in dense, nearly spherical heads. A succession of flowers is produced through the summer. This species ranges in the wild from Ohio to Colorado, Florida, Texas, and New Mexico. Very similar, but with more rigid seed pods up to 2¾ inches long, *D. leptolobus* is indigenous from Missouri to Kansas and Texas.

Garden Uses and Cultivation. These plants have little to recommend them except for inclusion in collections of native plants and for occasional use in naturalistic plantings. They grow without difficulty in ordinary garden soil, moist or dry, in sunny places, and are raised from seed.

DESMAZERIA (Des-mazèria)—Spike Grass. The only species that belongs here is native of the western Mediterranean region. South African plants previously included, but not of horticultural importance, are segregated as *Plagiochloa*. The name, sometimes spelled *Demazeria*, honors Jean Baptiste Henri Joseph Desmazieres, a French botanist, who died in 1862. The genus belongs in the grass family GRAMINEAE.

An annual 4 inches to 1 foot tall, spike grass (*D. sicula*) has solitary or tufted stems and linear, flat leaf blades up to 10 inches long by ⅒ inch wide. Stems and leaves are hairless. The flower spikes are flattened, somewhat one-sided or with the spikelets distinctly in two ranks. They are 2 to 3 inches long and composed of many-flowered, flattened, narrowly-oblong to ovate- or elliptic-oblong, nearly stalkless spikelets without bristles. In the wild this grass favors sandy soils.

Garden Uses and Cultivation. An attractive ornamental for flower borders and fresh or dried bouquets, this species is easily grown in ordinary soil in full sun. Seeds are sown thinly in spring where the plants are to remain, and if necessary to prevent crowding, the seedlings are thinned out at an early stage.

DESMODIUM (Des-mòdium)—Tick-Clover, Tick-Trefoil, Telegraph Plant. Estimates of the number of species of this nearly cosmopolitan genus (it is not native of Europe) of the pea family LEGUMINOSAE range from 150 to 450. They include herbaceous plants, subshrubs, and shrubs, mostly of the tropics and subtropics, but including some natives of temperate regions. The name, alluding to the jointed seed pods, is from the Greek *desmos*, a chain. From closely related *Lespedeza* desmodiums differ in their pods containing more than one seed.

The leaves of desmodiums rarely have a single blade, more commonly they are of three, or sometimes five leaflets. The pea-like blooms, in axillary or terminal racemes or panicles, are small, and usually purple. They have ten stamens most often with one separate and the others joined. The fruits are pods, constructed in scallop fashion between the seeds and when ripe usually easily separable into segments, those of many kinds given to attaching themselves to clothing in the manner of little burrs. This is reason the colloquial names tick-clover and tick-trefoil. Species of *Lespedezia* and *Rhynchosia* have in the past been named *Desmodium*.

As a group desmodiums find little favor with gardeners. A few hardy kinds are occasionally planted. Among such is ***Desmodium canadense,*** a herbaceous perennial 2 to 7 feet tall with leaves, pubescent on their undersides, and three blunt-oblong to ovate-lanceolate leaflets 2 to 4 inches long. Coming in late summer, the racemes of pink to purple flowers, each bloom ½ inch long, are in quite showy, loose panicles. The 1-inch-long seed pods are three- to five-jointed. This native from Nova Scotia to Alberta, South Carolina, and Arkansas favors moist soils.

Not hardy in the north, *D. tiliaefolium,* of the Himalayas, is somewhat shrubby and 3 to 6 feet tall. It has pubescent shoots and leaves of three broad-ovate leaflets 2 to 4 inches long, which are nearly hairless above and slightly or densely-hairy on their undersides. In large terminal panicles its slender-stalked, pale lilac to pink blooms, about ½ inch long, are borne in late summer. The slender seed pods, 2 to 3 inches long, contain six to nine seeds.

The telegraph plant (**D. *motorium* syn. D. *gyrans*) is one of the most remarkable and intriguing of plants. Interest lies in that by day, whenever the temperature is above 72°F, and especially in sunshine, the two small side leaflets of each leaf move continuously, and slowly but visibly, up and down, their tips describing narrow ellipses in the air. This, a shrub 2 to 4 feet tall, is a native of tropical Asia. It has no merit as an ornamental. Its leaves have three leaflets, the central one much larger and broader than the nearly linear side ones. It is linear to broadly-elliptic and has a blade up to 4½ inches long and 1½ inches wide, or smaller. The purple to violet flowers, a little under ½ inch long, are in racemes and raceme-like panicles. The pods are 1 inch to 1½ inches long.

Garden and Landscape Uses and Cultivation. The telegraph plant is grown as a botanical curiosity, other kinds for decorating informal areas and flower borders. The decoratives thrive in sun or part-day shade in well-drained soils. They are propagated by division and by seed. The seeds should be sown as soon as ripe, or if stored for any length of time, be treated by soaking for fifteen minutes, before sowing, in concentrated sulfuric acid.

The telegraph plant needs high temperatures and humidity. In greenhouses the minimum at night should be 70°F, and by day five to fifteen or more degrees higher. Although perennial, it is usual to treat this as an annual, raising it from seeds sown in late winter in a temperature of 70 to 75°F. Cuttings taken in spring provide an alternative method of increase. These root readily in a greenhouse propagating bed with a little bottom heat. When well established the tips of the young plants are pinched out to encourage branching. Ordinary fertile potting soil watered to keep it moderately moist is satisfactory. Good light with only slight shade from strong summer sun is needed.

DESMONCUS (Desmón-cus). This group of sixty-five prickly, slender-stemmed, climbing palms of the American tropics superficially resembles the rattans (*Calamus* and *Daemonorops*) of the Old World tropics, but the botanical relationship is not close. The similarities are due to parallelism in development rather than consanguinity, if that term can be applied to bloodless organisms. The genus *Desmoncus* is nearly allied to *Bactris*. It belongs in the palm family PALMAE. The name is from the Greek *desmos*, a band or chain, and *onkos*, a hook. It alludes to the tail-like extensions of the midribs of the leaves, which bear sharp, backward-pointing, fishhook-like thorns of menace to travelers, but useful to the plant in its climb upward in forests to the light.

These palms have slender, prickly or unarmed stems and alternate leaves that are pinnate or pinnatisect (pinnately-cleft, but not deeply enough to form entirely separate leaflets). The flower clusters, which appear in the leaf axils, are branched and bear greenish flowers followed by pea-like fruits.

Native of Trinidad, Tobago, and nearby South America, **D. orthacanthos** (syns. *D. horridus, D. major*) is a vigorous climber with stems up to 60 feet long and leaves that may exceed 6 feet in length. The latter have fifteen to twenty-three pairs of narrow leaflets and often a few spines on the midribs beneath. The terminal whiplike tail of the leaf has five to eight pairs of straight hooks. The orange-red fruits are about ⅜ inch in diameter.

Less robust **D. polyacanthus,** of Trinidad and eastern South America, has stems up to 45 feet and leaves up to 4½ feet long with eight to twelve pairs of lanceolate leaflets and a tail of four to six pairs of reflexed hooks. The red fruits are a little over ½ inch in diameter.

Garden and Landscape Uses and Cultivation. These plants are appropriate for fanciers of palms who can provide humid, tropical conditions outdoors or in greenhouses. They are little known in cultivation, but may be expected to thrive under conditions favorable to *Phoenicophorium.* For further information see Palms.

DETARIUM (De-tàrium)—African-Mahogany. Four species of the pea family Leguminosae constitute *Detarium.* They are native trees of tropical Africa that have pinnate leaves with an even number of few opposite leaflets. Their small flowers are in panicles from the leaf axils or are on lateral one-year-old branchlets. They have four sepals, no petals, ten separate stamens, and a slender style. The stalkless, nearly orbicular, compressed seed pods do not split to release the seeds. The meaning of the name is not apparent.

In its native haunts a large-headed forest tree from 30 to 120 feet tall, African-mahogany (**D. senegalense**) has pinnate leaves 5 to 9 inches long, with six to twelve alternate, oblong-elliptic leaflets 1 inch to 2½ inches in length. Its subspherical seed pods, 2 to 2½ inches across, contain a sweet, edible pulp that is offered for sale in African markets. This tree is the source of commercial lumber called African-mahogany, and of a fragrant resin.

Garden and Landscape Uses and Cultivation. As an ornamental, and as an example of a tree of economic importance, African-mahogany is planted to some extent in tropical and warm subtropical regions. It succeeds under humid conditions in ordinary soil, and is propagated by seed.

DEUTEROCOHNIA (Deutero-còhnia). One of the horticulturally lesser known genera of the pineapple family Bromeliaceae, this consists of seven species of arid regions in the southern Andes. The name, honoring Ferdinand Julius Cohn and alluding to this being the second genus dedicated to him, is from the Greek *deuteros,* second, and the name of the person commemorated.

In vegetative aspect, but not in bloom, *Deuterocohnia* resembles *Puya* and, like it, inhabits dry, sunny places. Deuterocohnias usually grow on sun-baked cliffs. From puyas they are readily distinguishable by their flowering stalks carrying loose panicles instead of tight heads of flowers and originating not in the center, but from one side of the rosette of foliage.

Deuterocohnias, unlike most bromeliads, are not tree-perchers (epiphytes). They grow in the ground or on rocks. They are evergreen perennials with short woody stems and rosettes of tough, thick leaves well adapted to minimize loss of water by transpiration. The flowering stalks are remarkable in that they contain, unusual for monocotyledons, a layer of cambium-type tissue that continues to grow and so makes it possible for the stalk to put forth new branches over a period of several years. The individual blooms are small and not showy. They have three each sepals and petals, the latter erect so the bloom is cylindrical. There are six stamens. The fruits are capsules.

Occasionally cultivated **D. schreiteri** has rosettes of many spirally-arranged, narrow-triangular, coarsely-spine-toothed, rigid leaves up to about 1 foot long, clothed with a dense covering of tiny gray scales. The slender flowering stalks, 2 to 3 feet tall and loosely-branched have many tubular, 1-inch-long, yellow blooms. From the last, **D. meziana** differs in its rosettes being of strongly recurved, flexible, light waxy-green, narrowly-triangular leaves with spines at their margins pointing in opposite directions. Up to 6 or 7 feet tall, the branched flowering stalk has red-orange blooms.

Deuterocohnia schreiteri

Deuterocohnia schreiteri (flowers)

Garden and Landscape Uses and Cultivation. These are as for dyckias, puyas, and pitcairneas.

DEUTZIA (Deùt-zia). Consisting of fifty species of deciduous and, rarely, evergreen shrubs, *Deutzia* belongs to the saxifrage family Saxifragaceae, It chiefly inhabits eastern Asia and the Himalayas, but is represented in the native flora of the Philippine Islands and by one to three species endemic to Mexico. The name commemorates Johan van der Deutz, a Dutch patron of botany, who died in 1788.

Deutzias have hollow shoots with generally brown, peeling bark. These features, as well as the five rather than generally four petals distinguish them from mock-oranges (*Philadelphus*), which have shoots with solid pith and close or flaky bark. Their leaves are opposite, short-stalked, and toothed. Rarely solitary, the flowers are more commonly in panicles or clusters, usually at the ends of short branchlets from shoots of the previous year's growth. Prevailingly white, they are sometimes pink or purplish. Each has a five-toothed calyx, five petals, ten or rarely more stamens, shorter than the petals and with usually winged stalks, and three to five styles. The fruits are capsules containing numerous tiny seeds.

Compact **D. gracilis** is a favorite. Usually not exceeding 3 feet in height, it sometimes is taller. It has many slender, erect or arching branches, and oblong-lanceolate leaves, generally under 2 inches in length, occasionally bigger, hairless or nearly so on their undersides. The white flowers, up to ¾ inch across, are in erect racemes or panicles 1½ to 3½ inches long. This native of Japan is hardy throughout most of New England.

The largest-flowered deutzia and also the earliest to bloom, **D. grandiflora,** native to northern China, is hardy in southern New England. About 6 feet in height, this kind has ovate leaves ¾ inch to 2 inches long, white-hairy on their undersides, with short, harsh hairs above. Its white flowers,

Deutzia gracilis

Deutzia scabra (flowers)

Deutzia scabra candidissima

singly, in twos or threes, are 1 inch to 1¼ inches across. They have long, slender calyx lobes, stamens with stalks with big, spreading, recurved teeth, and three styles longer than the stamens.

From the deutzias described above **D. scabra,** of Japan and China, differs in having both sides of its dull leaves covered with stellate (star-shaped) hairs. Its flowers, in slender, erect panicles 2 to 4½ inches long, have erect petals up to ¾ inch long. They are white or white with pinkish or purplish outsides. The stamens, nearly as long as the petals, are toothed near their apexes. This is hardy in southern New England. Pure white, double blooms are characteristic of *D. s. candidissima.* Double flowers rosy-purple on the outsides of the petals are borne by *D. s. plena.* The flowers of popular *D. s.* 'Pride of Rochester' are much like those of *D. s. plena,* but come earlier in the season. Hardy throughout most of New England, Chinese **D. parviflora** is about 6 feet high. It has elliptic to ovate-lanceolate, pointed, unequally-toothed leaves with scattered, stellate hairs on both surfaces. The ½-inch-wide, white flowers are many together in flat clusters 1½ to 3 inches across. In the bud stage the petals overlap.

Evergreen or nearly so **D. pulchra** is a large shrub native of mountain regions in the Philippine Islands and Taiwan. It has pointed-ovate to pointed-lanceolate leaves 2 to 3 inches long, and white to yellowish-white flowers about ¾ inch wide in terminal panicles. It is hardy in California and other frostless or nearly frostless places.

Hybrid deutzias are fairly numerous and include some of the best for garden display. Here belongs *D. candelabrum,* the parents of which are *D. gracilis* and *D. scabra.* Very attractive in bloom, **D. candelabrum** is about 6 feet tall. It has erect panicles of white flowers, broader than those of *D. gracilis.* It is hardy in southern New England. A hybrid of *D. lemoinei* and *D. scabra,* named **D. candida,** has ovate to ovate-oblong leaves up to 2 inches long, green, and with scattered hairs on both surfaces. The slender-stalked flowers, nearly ¾ inch across, are white. The parents of **D. excellens** are *D. vilmorinae,* of China, and the hybrid *D. rosea grandiflora.* It has ovate-oblong leaves up to a little over 2 inches long. They are grayish-white on their undersides with stellate (star-shaped hairs). The ¾-inch-wide white blooms have stamens with stalks with large spreading teeth. The name **D. hybrida** is applied to a complex of varieties resulting from hybridizing *D. longifolia* and *D. discolor,* both of China. Among the best are *D. h.* 'Contraste' and similar *D. h.* 'Magicien', with big pink flowers, and *D. h.* 'Mont Rose', with large lavender-pink flowers late in the deutzia flowering season. These are hardy in southern New England. Hybrid **D. kalmiaeflora** has as parents *D. parviflora* and *D. purpurascens,* the last native of western China. Hardy in southern New England, about 6 feet tall *D. kalmiaeflora* has cup-shaped flowers ¾ inch wide. They have petals white on their insides, carmine-pink without. The hybrid between *D. gracilis* and *D. parviflora* is **D. lemoinei,** which is hardy through most of New England. Free-flowering and handsome, this is up to 8 feet in height. It has long-pointed, sharply-toothed leaves, with scattered hairs on their under surfaces. The flowers, in erect panicles 1½ to 3 inches tall or a little taller, are white and ½ to ¾ inch in diameter. Variety *D. l. compacta* is dwarf and compact, *D. l. erecta* has denser panicles of bloom, *D. l.* 'Boule de Neige' has larger flowers in dense clusters. Hardy in south-

Deutzia scabra

Deutzia lemoinei

Deutzia purpurascens

ern New England, **D. magnifica** is a hybrid offspring of *D. scabra* and Chinese *D. vilmorinae*. A variable deutzia of which there are several named varieties, *D. magnifica* is generally similar to *D. scabra,* but its panicles of double, white flowers are shorter and broader. The blooms have usually four styles. The flowers of *D. m. erecta* are in more crowded panicles. Variety *D. m. latiflora* has single, open blooms 1½ inches wide. A hybrid between *D. gracilis* and *D. purpurascens,* the last native of western China, *D. rosea* differs from *D. gracilis* in its generally pinkish flowers being in broader, looser panicles. Several variants of this hybrid have been named. One of the best, *D. r. eximia* has blooms pink or purplish on their outsides. The few panicles of *D. r. grandiflora* are looser than those of typical *D. rosea.* A little less hardy

Deutzia rosea

in winter than the other deciduous species described above, Chinese **D. purpurascens** is up to 6 feet tall. It has slender, arching stems and toothed, oblong-ovate to lanceolate leaves 2 to 2½ inches long and hairy on their undersides. In clusters of up to ten, the ¾ to 1 inch-wide flowers are white with purplish outsides to the petals and with purple, winged filaments (stalks of the stamens).

Garden and Landscape Uses. Despite the freedom with which they bloom, deutzias are less highly regarded as garden and landscape furnishings than many other spring- and early-summer-blooming shrubs. The reasons are twofold. Except when in flower, they lack character and are weak rather than strong elements in the landscape. They present no fall display of foliage color or fruits. Unless pruned regularly they tend to become ragged and nondescript. In cold climates branch ends are likely to be killed back in winter, adding to the untidy appearance of uncared-for specimens. They have the virtues of being adaptable to all ordinary garden soils, of succeeding in sun or part-day shade, of being generally free of pests and diseases, and of growing quickly. They are seen at their best when grouped in front of evergreens.

Cultivation. Deutzias are very easily propagated by summer leafy cuttings planted under mist or in a cold frame or greenhouse propagating bed and by hardwood cuttings. Some, such as *D. gracilis,* can be increased by division. The species can be raised from seed. Routine care consists of drastically thinning out old shoots after flowering and judiciously shortening those that remain. In cold localities late winter or early spring pruning to remove winterkilled portions of shoots is necessary for good appearance. An annual mulch of compost or other organic material applied in late spring is beneficial. For forcing into early bloom in greenhouses pot *D. gracilis* in early fall. Water thoroughly, then bury the pots nearly to their rims in a bed of peat moss or similar material in a cold frame. In February or March bring the plants into a sunny greenhouse where the night temperature is about 50°F and that by day five to fifteen degrees higher. Keep well watered, and until the flowers begin to open spray the tops two or three times a day to encourage the buds to develop. After flowering keep the plants growing under cool conditions until all danger of frost has passed, then remove them from

the pots and plant them outdoors. The same plants should not be forced in successive years.

DEVIL or DEVIL'S. These words are employed in the common names of various plants, among them these: blue devil (*Echium vulgare*), creeping devil (*Machaerocereus eruca*), devil flower (*Tacca chantrieri*), devil-in-a-bush (*Nigella damascena*), devil tree (*Alstonia scholaris*), devil wood (*Osmanthus americanus*), devil's bit (*Chamaelirium luteum*), devil's-bit-scabious (*Succisa pratensis*), devil's club (*Oplopanax horridus*), devil's hand (*Chiranthodendron pentadactylon*), devil's head (*Homalocephala texensis*), devil's-ivy (*Epipremnum aureum*), devil's paintbrush (*Hieracium aurantiacum*), devil's pins (*Hovea pungens*), devil's-potato (*Echites umbellata*), devil's tongue (*Amorphophallus rivieri*), and devil's walking stick (*Aralia spinosa*).

DEW PLANT. See Aptenia.

DEWBERRY. See Blackberry.

DEWDROP. See Dalibarda. Golden dewdrop is *Duranta repens.*

DEWOLFARA. This is the name of hybrid orchids the parents of which include *Ascocentrum, Ascoglossum, Euanthe, Renanthera,* and *Vanda.*

DIABROUGHTONIA. This is the name of bigeneric orchids the parents of which are *Broughtonia* and *Diacrium.*

DIACATTLEYA. See Caulocattleya.

DIACRIUM. See Caularthron.

DIALAELIA. This is the name of bigeneric orchid hybrids the parents of which are *Diacrium* and *Laelia.*

DIALAELIOCATTLEYA. This is the name of trigeneric orchids the parents of which are *Cattleya, Diacrium,* and *Laelia.*

DIALAELIOPSIS. This is the name of orchid hybrids the parents of which are *Diacrium* and *Laeliopsis.*

DIAMOND FLOWER is *Ionopsidium acaule.*

DIANELLA (Dian-élla)—Flax-Lily. To the lily family LILIACEAE belongs this genus of evergreen, fibrous-rooted, herbaceous plants of Australia, New Zealand, islands of the Pacific including Hawaii, and tropical Asia. There are about thirty species. From the berries of native ones the Hawaiians extracted a dye for tapa cloth, and used the leaves for thatching. The name is a diminutive of Diana, goddess of the hunt.

Dianellas, or flax-lilies as they are called in Australia, have usually branched, slender rhizomes, and two-ranked strap-shaped leaves, sheathing at their bases and resembling those of some irises and grasses. In some the foliage is all basal, in others leaves are clustered at the tops of the stems. The small, usually nodding, starry, blue, bluish, or white flowers, are numerous and in loosely-branched panicles. They have six spreading perianth segments (petals, or more correctly tepals) that often bend downward as the blooms age, six stamens with thickened stalks, and a slender style tipped with a very small stigma. The fruits are fleshy, few-seeded berries.

A fine Australian, *Dianella tasmanica* has rigid purplish-green or grayish-green, broadly-sword-shaped, long-pointed, all basal leaves 2 to 4 feet long and about 1 inch wide, distinctly keeled and channeled below, and with their edges with small spines. Its dainty blue blooms, in lax, many-branched panicles longer than the leaves, in California begin opening about February and continue in succession for many weeks. They are ½ to ¾ inch across, and are succeeded in summer by beautiful, long-lasting, slender-stalked, bright blue berries up to ¾ inch in diameter. Another beautiful Australian, in the wild favoring dry soils, **D. revoluta** has all basal leaves 2 to 3 feet long with rolled-under margins, and lax, angled flower stalks 2 to 4 feet long. Its deep purple-blue flowers, about 1 inch across, and in large, loose panicles, are succeeded by rather small blue berries.

Native in Hawaii, Australia, China, and other parts of Asia, **D. ensifolia** (syn. *D. nemerosa*) is up to 6 feet tall. Its leaves,

Dianella ensifolia

about 1 foot long by 1 inch wide, are both basal and on the lower parts of the stems. The whitish to bluish flowers are ⅓ inch long. Its berries are blue.

A New Zealander, **D. intermedia,** known in its homeland as blueberry, inhabits woods and more open areas. It has linear leaves, 1 foot to 4 feet long, and panicles, 1 foot to 2 feet long and carried above the foliage, of numerous white or delicate blue

flowers, followed by blue, ½-inch-long berries.

Garden and Landscape Uses. In warm-temperate and subtropical climates dianellas are agreeable plants for flower gardens and other places where clumps of grassy foliage and displays of small blooms and brilliant colored berries can be shown to advantage. They prosper in ordinary garden soil in sunny locations. They are also satisfactory for ground beds and pots in greenhouses and conservatories.

Cultivation. These plants pose no special problems for gardeners. They may be planted in fall or spring and remain undisturbed for many years. Container specimens need repotting every two or three years. In greenhouses winter night temperatures of 45 to 50°F are satisfactory, with a rise of five to ten degrees by day permitted. The soil should be kept moderately moist. More water is required when the plants are making active growth than at other times. Then, too, applications of dilute liquid fertilizer are beneficial, especially to specimens that have filled their containers with roots. Propagation is easy by seed and by division.

DIANTHUS (Dian-thus)—Carnation, Pinks. This group of mostly Northern Hemisphere plants is of considerable interest to gardeners. It includes carnations, sweet williams, garden or border pinks and rainbow or Indian pinks, as well as perennial alpine and rock garden sorts. Because of the horticultural importance of several of these groups and because of their different cultural requirements they are dealt with separately in this Encyclopedia under the headings Carnations, Pinks, and Sweet Williams, respectively. Here we describe and discuss other sorts.

The genus *Dianthus* consists of three hundred species and belongs in the pink family CARYOPHYLLACEAE. Its name, derived from the Greek *dios*, divine, and *anthos*, flower, was applied by Theophrastus in allusion to the beauty and fragrance of the blooms. Dianthuses are mostly low evergreen plants that form broad, more or less compact, basal clumps or mounds of stems and foliage above which rise sparsely-leaved flowering stems, but some have stems leafy throughout and are without basal concentrations of foliage. The stems are usually conspicuously thickened at the nodes and have opposite, generally narrow-linear, grasslike leaves, which in many kinds are glaucous-gray-green. The lower parts of the stems of perennial sorts are often subshrubby. The flowers are in dense clusters of usually few, or are loosely arranged or sometimes solitary. Their prevailing colors are white, pink, and red, more rarely lavender or yellow. The sepals are united to form a tube for one-half or more of their lengths, with at its base two or more bracts. Natural species charac-

istically have five petals, but some garden varieties have more and often their flowers are fully double. The petals are usually toothed or fringed at their edges and narrow sharply at their bases to form slender claws. Except in double blooms there are ten stamens and two styles. The many seeds are in cylindrical capsules that open at their tops.

Dianthuses in the wild are frequently very variable and many when grown in proximity to other kinds, as they often are in gardens, hybridize freely, so that plants raised from their seeds are not identical with the seed parent. Because of this, much confusion exists about identifying and naming dianthuses correctly, and often plants in cultivation carry wrong names. In some cases species appear under the name of other species, frequently hybrids masquerade as good species, and often mongrel seedlings are labeled with the names of species or of horticultural varieties to which they have little or no likeness. A perusal of nurserymen's catalogs will reveal, in addition to the dianthuses described here, offerings of named horticultural varieties, many of undetermined species relationship. Most of these are excellent garden plants and can be relied upon to prosper and bloom satisfactorily.

The cottage pink (**D. plumarius**), in its wild form native from central Europe to

Dianthus plumarius

Siberia, is ancestor to the garden varieties called border pinks, dealt with in this Encyclopedia in the entry Pinks. The natural species forms mats of glabrous and glaucous stems and foliage and in bloom is 6 to 12 inches in height. Its leaves are very narrow, the lower ones, up to 4 inches long, have prominent midribs and are keeled beneath. The very fragrant blooms, pink, red, or more rarely white, with darker centers, grow one to three or rarely up to five, atop each stem and are 1 inch to 1½ inches across. The petals are dissected so that their margins are fringed to one-third or more of their depth and there is a beard of long fine hairs at the throat of the bloom.

Much like *D. plumarius* but commonly much smaller and with grass-green, or rarely gray-green, narrow, sharp-pointed leaves, **D. arenarius** is a quite variable native from northern Europe to Siberia. Except for their centers, which are greenish or grayish, its lacy blooms are always white and have deeply-fringed petals. They

Dianthus arenarius

are fragrant, but less so than those of *D. plumarius*. One of the earliest species to bloom, in the vicinity of New York City *D. arenarius* is in full flower in mid-May. Some forms of it are very low, compact, and cushion-like. The flower stems 6 inches to more than 1 foot tall, may have a few long branches each with one to several blooms 1 inch to 1½ inches across and with slender calyx tubes.

The maiden pink (**D. deltoides**) is one of the most reliable and long-lived dianthuses. A vigorous grower, this makes

Dianthus deltoides

wide-spreading, flat mats of prostrate, barren stems and grass-green foliage. Its leaves are linear to linear-lanceolate. The flower stems, rising to heights of 6 inches to 1 foot, are usually forked above with each stem or branch topped with a solitary flower about ¾ inch across. The blooms are light to dark rose or magenta-pink, more or less spotted or speckled, and have sharply-toothed petals. Common in cultivation, this is a variable species. An es-

pecially low, compact form is *D. d. serpyllifolius*, one with glaucous-blue foliage is *D. d. glaucus*, whereas white flowers are characteristic of *D. d. albus*. The maiden pink, native from Great Britain to Japan, is naturalized in parts of North America.

The cheddar pink (**D. gratianopolitanus** syn. *D. caesius*) is a glabrous, glaucous, blue-green, loose mat-forming sort 3 inches to 1 foot high in bloom. Native from south-

Dianthus gratianopolitanus

ern England to southern France, this has often branchless flowering stems with one or two, rarely three, fragrant flowers, rose-pink and 1 inch to 1¼ inches across, more or less bearded in their throats and with toothed petals. The soft, narrow leaves are up to 2 inches long. This excellent garden plant has produced many good varieties, some with semidouble and some with double blooms.

Deliciously fragrant, **D. superbus** is usually green and 9 inches to 2 feet tall. Its stems, woody toward their bases, above are mostly branched. Its few basal leaves are up to 2 inches long by ⅜ inch wide. The stem leaves are shorter and more tapering. From 1½ to 2 inches or sometimes more in diameter, the blooms are very pale

Dianthus superbus

to deeper lilac or rose-pink or rarely pure white. The petals are deeply slashed into numerous narrow segments. Fine red hairs line the throat of each flower. Individual stems carry from two to twelve blooms fairly closely together. This species is often short-lived, but its charm is such that it is well worth the trouble of propagating it fairly frequently.

Akin to *D. superbus*, but commonly lower, its flower stems with longer branches and its fragrant flowers with shorter calyx tubes, **D. monspessulanus** is native from Spain to the Caucasus. A good garden plant, this sort is hairless and forms loose mounds or cushions of soft grass-green to somewhat glaucous foliage. Its flower stems rise to heights of 8 inches to 1 foot or sometimes more and are sparingly forked above or branchless. The blooms, 1 inch to 1½ inches or sometimes more in diameter, are rose-pink to carmine or white, and have deeply-lobed petals. Variety *D. m. sternbergii* is lower, glaucous, with solitary flowers.

Dianthus monspessulanus sternbergii

Clear lemon-yellow, scentless flowers distinguish cluster-headed **D. knappii**, of Hungary and Yugoslavia. Unfortunately this suffers from the disadvantage that some of the flowers in the heads fade and turn brown before others, thus marring the full beauty of the display before blooming is finished. Also, there is no great profusion of bloom at any time. A lowland species that in bloom attains a height of 8 inches to 1½ feet, this has wiry stems branchless or scarcely forked except at their tops. Its leaves are green or somewhat glaucous and up to 2½ inches long and ¼ inch wide. Each compact cluster consists of eight to ten blooms ¾ inch across.

Cushion- or mat-forming **D. petraeus** (syn. *D. kitaibelii*) is a native of eastern and central Europe. It has green foliage and slender, forked flower stems 6 inches to 1¼ feet tall bearing white blooms about ¾ inch wide. These differ from those of *D.*

Dianthus knappii

when happily located, a loose cushion 2 to 4 inches tall of stems and glaucous, broadly-linear, long-pointed leaves on top of which sit the solitary blooms, full, round, with toothed petals, and 1 inch to 1½ inches in diameter. The blooms of this are bright rose-pink with conspicuous darker centers spotted with white and are bearded in their throats. The undersides of the petals are rose-pink. A high mountain native of the European Alps, *D. glacialis* out of bloom is not more than 3 to 4 inches in height and makes a tight mound of glabrous, green foliage bespangled with usually solitary, very short-stemmed, fragrant blooms about ¾ inch in diameter and flesh-colored or deeper pink with whitish centers and the undersides of its petals yellowish or greenish. This has much the appearance and habit of growth of *D. alpinus,* but its petals are not as flat.

Other dianthuses likely to be cultivated include these: *D. brevicaulis,* a somewhat glaucous, compact alpine from Mount Taurus in Asia Minor, has many rosy-mauve flowers about ½ inch in diameter. The blooms just top the 2- to 3-inch high foliage mass. Its basal leaves are fringed with hairs. *D. carthusianorum* is a very variable cluster-head with erect, mostly branchless stems up to 1½ feet tall, and usually green foliage. Its small flowers, pale red to dark red or purple-red, are in tight heads. This is a widely distributed native of Europe. *D. corymbosus* (syn. *D. tymphresteus*), native of the Balkan Peninsula, is tufted, green, clammy-hairy, and up to 1 foot tall or taller. Its leaves are linear to linear-lanceolate. Its ½- to ¾-inch-wide, pink blooms have bearded, toothed petals. *D. fragrans,* of the Caucasus, forms low clumps above which rise to a height of above 1 foot branched stems bearing very fragrant, white to pink blooms often spotted with deeper pink, up to 1 inch in diameter. *D. freynii,* closely related to *D. glacialis,* is a high mountain native of mountains in Yugoslavia that forms tight clumps of glaucous foliage, in bloom is only 1½ to 2½ inches in height. Its solitary flowers are bright pink. *D. furcatus,* low

plumarius, which they otherwise somewhat resemble in being beardless in their throats and their petals not deeply dissected. Also, its very narrow leaves are shorter than those of *D. plumarius.* Frequent in gardens, especially rock gardens, *D. p. noeanus* (syn. *D. noeanus*) is attractive in habit and foliage and easily makes itself at home in a variety of soils and situations. Native from the Balkan Peninsula to Iran, this becomes a low, very dense cushion of rigid, sharp-pointed, green or grayish-green foliage distinctly prickly to the touch, above which rise to a height of 6 inches to 1 foot very slender, arching flowering stems. These branch above and bear mostly five to nine much-fringed, fragrant white blooms of rather flimsy texture, about ¾ inch in diameter, and with six bracts at the bottom of the calyx tube. Blooming later than most kinds, *D. p. noeanus* is often misidentified as *D.*

corymbosus and as *Acanthophyllum spinosum,* but it is very different from the plants to which those names rightly belong. It is also sometimes confused with *D. spiculifolius,* a much looser-growing species with flat rather than needle-like leaves and flowers 1 inch or more in diameter that have four bracts at the bottom of the calyx tube.

Dwarf and lovely *D. alpinus* is native from the European Alps to Greece and arctic Russia. It forms low cushions of short stems and grass-green blunt leaves above which rise on 2- to 4-inch stems solitary, scentless flowers 1 inch to 1½ inches in diameter. They are pink to crimson, with darker spots on a white ground forming an eye, which often fades with age, at the center of the bloom. The undersides of the petals are buff or greenish-white. The throats of the flowers are bearded. Variety *D. a. albus* has white flowers. Similar to *D. alpinus,* but more difficult to satisfy in cultivation, *D. callizonus* is perhaps the most lovely of all alpine dianthuses and forms,

Dianthus petraeus noeanus

Dianthus alpinus

Dianthus arvernensis

and glaucous, has flower stems up to 8 inches tall. Its purplish-pink blooms about ¾ inch across are one to three on a stem. It is a native of the mountains of northern Italy. *D. gallicus,* closely related to *D. monspessulanus,* native from France to Portugal, forms mounds of hairy, glaucous foliage and has stems up to about 1 foot tall, each with one to three pink to white, fragrant flowers about 1 inch in diameter. *D. graniticus,* of southern France, is related to *D. scaber* (syn. *D. hirtus*). It has low clumps of stiffish, narrow, green leaves and flower stems 4 inches to 1¼ feet in height that bear, singly or few together, reddish-purple flowers up to ⅛ inch in diameter. *D. hispanicus* (syn. *D. requienii*), of the Pyrenees, is a glaucous, tufted plant, in bloom 6 inches to somewhat over 1 foot tall. This has branchless or branched stems with solitary, rose-purple flowers with purple anthers. *D. japonicus,* of Japan and Manchuria, resembles the wild form of the sweet william (*D. barbatus*), but has shorter, blunter, ovate-lanceolate leaves. It is up to about 1½ feet in height and has red flowers with paler centers, about eight together in close clusters. *D. microlepis* is a cushion-forming species about 2 inches high with soft leaves and small, solitary, rose-pink to white flowers ½ to ¾ inch across among its foliage. It is native at high elevations in the Balkans. *D. myrtinervius,* native of Macedonia, is closely related to *D. deltoides,* but much smaller. It forms a compact tuft up to 2 inches high and has many solitary, rose-purple blooms. Its leaves are about ¼ inch long. *D. nitidus,* from the mountains of central Europe, is a green, loose, cushiony plant with flower stems rising to a height of 1 to 2 feet, and each with one to three rose-pink-spotted white blooms about 1 inch in diameter and with shallow-toothed petals. *D. pavonius* (syn. *D. neglectus*), a rare native of the European Alps, is similar to *D. glacialis,* but somewhat larger and usually has solitary flowers, bright crimson-pink with lighter centers and shallowly-toothed margins to the petals. This species rarely exceeds 6 inches in height and has flowers 1 inch to 1¼ inches across. *D. pinifolius,* of southeastern Europe, is a cluster-head with narrow leaves and stems 6 inches to 1½ feet tall. Its rose-pink blooms are five to eight together in dense heads. *D. roysii* is a hybrid between *D. callizonus* and *D. gratianopolitanus.* From 4 to 6 inches tall, it has large, showy, pink flowers. *D. seguieri* is closely related to *D. chinensis* (treated in this Encyclopedia under Pinks), but is reliably perennial and more slender. Native to southern Europe, it has broad, green leaves and bright pink flowers in loose or tight clusters. *D. serotinus* is a glaucous, cushion-forming kind with leafy flowering stems up to about 1½ feet tall and white, fragrant blooms with fringed petals. It is a native of eastern Europe. *D. subacaulis*

forms dense cushions of foliage up to 3 inches high and has one or two pink blooms up to ½ inch across on each short flowering stem. It is native to the mountains of southern Europe. *D. sylvestris,* the wood pink, forms clumps of usually green basal leaves up to 4 inches long above which are lifted to heights of 6 inches to 1 foot flowering stems each with one to three rose-pink blooms up to 1 inch in diameter. It is native from Spain to eastern Europe.

Garden and Landscape Uses. The dianthuses dealt with here include many kinds of great value for display in rock gardens, dry walls, and as edgings to flower beds, and for similar uses. Some are suitable for alpine greenhouses. Of great importance is that they bloom in late spring and early summer after the great flush of low, spring-blooming perennials is over. Deprived of dianthuses, most rock gardens, at least most sunny ones, would lose much charm. It would be hard to replace them with plants as lovely and as easy to grow. They are attractive in and out of bloom. The blue-green foliage masses of many kinds harmonize splendidly with rocks and provide neat and pleasing foils for other plants in bloom. Rock garden dianthuses include true aplines, species that in their native homes ascend well above tree line, as well as kinds from lower elevations. Characteristically they are low, compact plants.

Cultivation. Three important points to remember about dianthuses are that they are sun lovers, they will not tolerate wet soils, and the earth in which they grow must be porous and the subsurface drainage adequate, and they are not acid-soil plants. They prosper in neutral earths and some need and most tolerate alkaline ones. Fairly frequent propagation, easy to accomplish, is usually necessary for best results. Some kinds are naturally rather short-lived and others tend to become straggly and unkempt with age.

Except for a few tricky-to-grow high alpine species favored by connoisseurs of rock garden plants, dianthuses are not difficult to manage. True species are easily raised from seeds sown in pots of gritty soil in a cold frame in May or earlier in a greenhouse. Transplant the seedlings as soon as they are large enough to handle to other pots or to nursery beds where they will remain until early fall or the following spring when they are moved to their final locations. Seeds do not, of course, provide means of increasing choice varieties true to type. Even those sold under species names are often hybrids that give variable progeny. The offspring may bear a general or even close resemblance to the parents, but often are not identical.

Vegetative propagation is usually the only way of raising plants that are identical to the desired ones. This can be done by

division, cuttings, and layers. The first is the least practicable. Few dianthuses lend themselves well to division. All may be grown from cuttings. These should be made from the ends of nonflowering shoots and be from 1 inch to 3 inches long or sometimes longer, according to kind. The cutting should include at least three or four pairs of leaves. The basal leaves are cut off and the stem sliced squarely across just beneath a node with a sharp knife. The cuttings are then inserted in a propagating bed of coarse sand or perlite in a cold frame, cool greenhouse, or under a Mason jar or bell jar outdoors, or they may be set under mist without any glass or polyethylene covering. Layering is suitable only for kinds that have fairly long stems. These are bent to touch the ground at a point some distance from their tips and there are wounded on the underside by making a cut about one-third of the thickness of the stem and ¼ to ½ inch long in the direction of the tip. The wounded part is then pegged into a slight hollow made in the soil and is covered with sandy soil kept moist. After rooting has taken place the new plant is severed from its parent and transplanted.

Care of established plants is minimal. It includes the removal of old flower stalks after the blooms have faded and at that time a light cutting back of sorts that show a tendency to become too loose and straggly. A generous sprinkling of bonemeal, but not of fertilizers that supply rapidly available nitrogen, stirred into the soil in early spring is beneficial. If the surface soil around the plants has been eroded by washing, mix the bonemeal with soil and apply it as a top-dressing. The maintenance around most dianthuses of a mulch of stone chips, especially of limestone chips, is beneficial and aesthetically appropriate.

Diseases and Pests. Although dianthuses are subject to a number of diseases, especially rots, wilt, and rust, outdoor kinds grown under good conditions and not in wet, unkindly soils are not too susceptible. Should disease attack plants it is usually simplest to destroy the victims promptly and start with clean, fresh stock, although at the first incidence of infection with rots it may be helpful to spray several times at 7- to ten-day intervals with a fungicide; this practice is recommended as a preventative of rust. The chief pests of dianthuses are slugs, cutworms, aphids, caterpillars, thrips, and red spider mites.

DIAPENSIA (Dia-pènsia). Of the four species of this genus of the diapensia family DIAPENSIACEAE, only one is likely to be cultivated. It is *Diapensiua lapponica,* a far northern native of North America, Europe, and Asia. The other kinds inhabit high altitudes in the Himalayas and western China. Diapensias are relatives of *Galax,* but differ from that genus markedly in appearance and in being shrubs, or perhaps shrublets would be a better word, since they are so tiny that they belie the ordinary conception of a shrub. They resemble another North American representative of the family, the pyxie (*Pyxidanthera barbulata*). The name *Diapensia* is an ancient Greek one of uncertain application.

The group consists of cushiony or prostrate, tightly-matted, evergreen plants with hairless, mostly opposite, undivided leaves, and small, white, or more rarely rose-purple or yellow, stalked flowers, solitary at the ends of the stems. The blooms have a calyx deeply divided into five slightly overlapping segments (sepals), a short-tubed corolla with five broad, spreading lobes (petals), five stamens joined to the corolla, and a slender style with a three-lobed stigma. The fruits are capsules.

Chiefly arctic, occurring in North America northward from Newfoundland, but also inhabiting the summits of high mountains in northern New England and New York, **D. lapponica** is about 3 inches high, compact, and has matted stems 2 to 4 inches long. The overlapping, spatula-shaped leaves are ¼ to ½ inch long. The white, bell-shaped blooms have wide-spreading petals. The flowers are about ⅓ inch long, ½ to ¾ inch wide, and have stalks about an inch long when the flowers are expanded but longer later.

Garden Uses and Cultivation. It is to be regretted that these arctic-alpine gems do not take kindly to cultivation. They are not for casual gardeners. Only avid collectors of the choicest and most elusive alpines are likely to be steeled to the almost inevitable disappointments that attend attempts to tame these northerners. But if success comes, the alpine garden enthusiast is well rewarded. There is no likelihood of succeeding where summers are hot and humid. Cool nights are a must for diapensias, and they will not tolerate scorching days. The recommendation is to plant in acid peaty soil containing a great abundance of coarse sand, grit, and stone chips and to arrange for constant lime-free moisture to seep from beneath. A rock garden moraine, or cultivation in pans (shallow pots) in cold frames afford the best chances of success. Propagation is by seed, by very careful division in spring, and by summer cuttings. It is not easy.

DIAPENSIACEAE—Diapensia Family. Six or seven genera of dicotyledons totaling about twenty species of low shrublets and stemless perennials that inhabit arctic, subarctic, and north-temperate regions compose the diapensia family. They have opposite or alternate, evergreen, undivided leaves and, solitary or in racemes, flowers with five-lobed calyxes and five-lobed or five-parted corollas, five stamens, and one style with a three-lobed stigma. The fruits

are capsules. The best known genera are *Diapensia, Galax, Pyxidanthera,* and *Shortia.*

DIASCIA (Di-áscia)—Twinspur. Of the approximately forty known species of *Diascia* only one is cultivated. The group is entirely South African and belongs in the figwort family SCROPHULARIACEAE. Most of its members are annuals, a few herbaceous perennials. The name is from the Greek *di,* two, and *askos,* a pouch. It refers to the corollas having two spurs.

These are erect or spreading plants with usually opposite leaves and racemes of pink or violet blooms in which the flowers are interspersed with leaves or leafy bracts. The calyx is five-lobed. The two-lipped corolla is flat or shallowly-concave with only a suggestion of a tube. Its lower lip is three-lobed, its upper two-lobed. The lower one has twin spurs. There are four stamens and a slender style. The fruits are rounded or elongated capsules containing numerous seeds.

The twinspur (**D. barberae**) is an erect annual of elegant appearance and up to

Diascia barberae, variety

about 1 foot tall. It has ovate, toothed leaves up to 1½ inches long, and slender-stalked flowers with the central lobe of the lower lip much larger than the others. The blooms are about ½ inch in diameter and in racemes about 6 inches long. They are rosy-pink with a yellow spot in the throat. Their anthers are yellow.

Garden Uses. Where it thrives outdoors (it does not tolerate extremely hot summers), this sort is splendid for edgings and for the fronts of flower borders. It may also be used in rock gardens, especially in mild climates where such plantings are not limited to alpines and similar plants. It is quite pretty. The twinspur also makes a pleasing greenhouse pot plant for blooming in spring and early summer.

Cultivation. For outdoors, seeds are usually sown inside early and the seedlings transplanted to flats in which they

grow for eight to ten weeks before being planted where they are to flower. This is done after all danger of frost is past. The newly sown seeds are placed in a temperature of 60 to 65°F. They germinate in about two weeks. In flats the young plants are spaced 2 inches apart. They are grown in a sunny greenhouse having a night temperature of 55 to 60°F, and day temperatures five or ten degrees higher. Their tips should be pinched out when the plants are 2 to 2½ inches tall and the tips of the resulting branches when they attain a length of 3 inches. Prior to planting in the garden they are hardened by being placed in a cold frame or sheltered place outdoors for about a week. In the garden such plants are set about 6 inches apart. An alternative method is to sow the seeds in early spring where the plants are to bloom and thin the young plants to 4 to 5 inches apart. After the first bloom is over the plants are cut back to a height of 2 or 3 inches to induce a second crop of flowers, and this may be repeated when that crop begins to fail.

For blooming in greenhouses seeds are sown in September. The plants are grown as detailed above and when they begin to crowd in the flats, are transplanted individually to 4-inch pots or three in a 6-inch pot. Alternatively, seeds may be sown directly in 5-inch pots and the seedlings thinned to leave three or four plants in each. The potted plants are grown to flowering size in a sunny greenhouse with a night temperature of 55 to 60°F and daytime temperatures a few degrees higher. When well rooted they need watering freely and weekly or biweekly applications of dilute liquid fertilizer. When flowering is nearly over they may be cut back, top dressed, and kept growing to encourage a second blooming.

DIASTEMA (Dia-stéma). This genus of the gesneria family GESNERIACEAE is closely related to *Achimenes* and *Kohleria.* It consists of about forty species of herbaceous plants of Central and South America. Those in cultivation are 4 to 6 inches tall. They have scaly rhizomes and opposite leaves, usually coarsely-toothed, and generally furnished with stiffish hairs. The flowers have cylindrical corolla tubes and five spreading corolla lobes (petals) that form a slightly two-lipped face to the bloom. There are four stamens in two pairs. The fruits are capsules, not beaked at their ends. The name is derived from the Greek *di,* two, and *stemon,* a stamen.

Among cultivated kinds *D. quinquevulnerum,* from Colombia and Venezuela, is popular. It has ovate-elliptic, bright green leaves with indented veins. The flowers, about ¾ inch in diameter, are white with yellow throats and, in those that live up to the specific name, have the five corolla lobes each marked with a pinkish-purple spot; more usually only the three lower are

Diastema quinquevulnerum

so ornamented. The blooms are in loose clusters.

Another commonly grown kind is *D. vexans,* from Colombia. This has slender, hairy, spreading stems and velvety, ovate to lanceolate leaves from the axils of which, as well as from the terminations of the stems, the rather long-stalked white flowers are borne. Their corolla tubes are funnel-shaped and each spreading corolla lobe has a conspicuous purple spot. Lavender corolla lobes with a purple spot on each, and crinkled, lanceolate, dark green, toothed leaves distinguish *D. maculatum* from other cultivated kinds. Its leaves are reddish on their undersides and are covered with stiff hairs. The corolla tube is straight, yellowish, and about ¾ inch long. The face of the flower is about ¾ inch across. Having flowers the same size as those of the last named, *D. longiflorum* is a native of Venezuela. Despite its specific name, its blooms are not longer than those of some other species in cultivation. Its bright green, oblong-ovate leaves are tinged with brown along their double-toothed margins. Distantly spaced stiff hairs cover the upper leaf surface. The flowers have corolla tubes ¾ inch long and are in terminal clusters. The corolla lobes and undersides of the straight corolla tubes are white, except for a magenta mark on each lobe. The upper side of the tube is bright pink. The throat of the flower is yellow. The face of the bloom is about ½ inch across. Having flowers about one-half the size of the species discussed above, *D. rupestre* is native of southern Mexico. Its white blooms, on comparatively long stalks, are unspotted. They have yellow throats and sometimes the top of the corolla tube is marked with red lines. The leaves are bright green and have stiff hairs.

Garden Uses and Cultivation. These are plants for the collector of gesneriads (plants belonging in the gesneria family). They grow happily in a humid greenhouse, shaded from strong sun, and in terrariums. They are harmed by dry air and are less suited than African violets and some

other gesneriads for use as room plants. They do well when grown under fluorescent lights and need a minimum winter night temperature of about 60°F. Their soil should be coarse, loose, and contain an abundance of organic matter. Throughout their growing season it should be kept uniformly moist, dry only when the plants are resting, as they do each year. When they have filled their containers with healthy roots they may, during their season of growth, be given occasional applications of dilute liquid fertilizer. Increase is easily secured from seeds, by separating rhizomes from a mother plant and either planting them whole or breaking them into pieces first, by stem cuttings, and by leaf-cuttings.

DIBBER or DIBBLE. Dibbers or dibbles are used to plant seedlings of small to medium size. The best are fashioned of hard wood. Commercial types usually have a pistol-grip handle formed by about a forty-five degree bend in the shaft and their lower parts are commonly sheathed with steel. Many gardeners make their own dibbers and dispense with metal coverings. Those of a size suitable for setting out young cabbages, broccoli, and similar plants are easily fashioned from the handle of an old spade or spading fork. Saw off the upper

Dibbles: (a) One fashioned from the handle of an old spading fork

(b) Smaller dibbles, the sharp-pointed one in the center is less satisfactory than the blunt-ended ones

foot or so, and sharpen the sawed end to a blunt point. Smaller dibbers in assorted sizes are useful for transplanting seedlings into flats and cold frames. These can be made from straight pieces of oak or other hard woods. They need no handles. They may be from about 4 to 6 inches long, ½ to 1 inch thick, and their lower parts gradually tapered to blunt points. A pencil with its eraser-tipped end used to make the holes and firm the soil about the roots serves well with small seedlings.

In use, dibbers are pushed into the soil to make holes large enough to accommodate the roots without harmful crowding. After the hole is made, the seedling is held in position with one hand, the dibber is used with the other to push or press soil against the roots. Do not make the holes so deep or press the soil against the seedlings in such a way that they are left hanging with the lower parts of their roots in a cavity. In very clayey soils dibbers compact earth along the sides of the hole to an extent that may make it difficult for new young roots to penetrate. On soils of this type planting with a trowel is generally to be preferred.

DICENTRA (Di-céntra)—Bleeding Heart, Dutchman's Breeches, Squirrel-Corn, Golden Eardrops, Steer's Head. Bleeding hearts, Dutchman's breeches, and their generic kin are mostly delightful, easily grown garden plants. They constitute *Dicentra* of the fumitory family FUMARIACEAE and number eighteen species of which eight are natives of North America and ten of eastern Asia. In the wild all except two American kinds inhabit open locations and lightly shaded fringes of woodlands. Shadier humid woods are favored by *D. canadensis* and *D. cucullaria.* One Asian species, *D. peregrina,* rare in cultivation, has as its home the summits and slopes of extinct volcanoes. The name, alluding to the floral spurs, is from the Greek *dis,* twice, and *kentron,* a spur.

Dicentras are hardy herbaceous perennials and annuals. They have rhizomes, tubers, or taproots and are stemless or with erect or climbing stems. The much-dissected, ferny leaves, usually long-stalked, are all basal, or basal and alternate on the stems. They are without hairs and are twice- or sometimes thrice-divided into threes, or more rarely, are pinnately divided. The ultimate leaflets are ovate-lanceolate, and lobed or toothed. The flowers, solitary, in racemes, or panicles, have minute, scalelike sepals, and four weakly united petals that form a usually flattened, more or less heart-shaped corolla. The two outer petals, bigger than the others, are pouched or spurred at their bases. There are six stamens in two series and a slender style with a two- to four-crested or horned stigma. The fruits are linear or oblong, several-seeded capsules.

One of the most dependable and lovely spring-blooming herbaceous perennials, common bleeding heart (**D. spectabilis**) is of bushy appearance. It is 1½ to 2 feet tall and has brittle, leafy stems. Its leaves, mostly up to 1 foot long, are divided into many broad-obovate or wedge-shaped toothed or cleft leaflets and overtopped by gracefully arching, branchless racemes of pendulous, beautifully heart-shaped, typically rose-pink to rosy-red, or rarely white flowers, from the ends of which the two white inner petals protrude. The blooms

Dicentra spectabilis

Dicentra spectabilis (flowers)

are 1 inch long or longer. This favorite kind is a native of Korea, China, and eastern Siberia.

Two American species and a variety of one, *D. eximia*, *D. formosa*, and *D. f. oregana* (syn. *D. oregana*), constitute a closely related natural group of pretty plants with essentially all-basal, finely-divided, ferny leaves, broadly triangular in outline, and panicles of nodding flowers, with occasionally, but rarely, one small leaf on their stalks. Supposed hybrids between *D. f. oregana* and *D. eximia* include the vigorous and attractive patented horticultural varieties named 'Bountiful', 'Debutante', 'Paramount', and 'Silversmith'. Some of these, as well as *D. nevadensis* and forms of *D.*

eximia, have been hybridized with *D. peregrina* to produce a series of delightful intermediates that are excellent garden plants.

Ranging in the wild from New York to the mountains of North Carolina and Tennessee, charming **D. eximia** blooms more or less continuously through spring and summer, making its most abundant display early. From 1 foot to 1¼ feet tall, it has green or slightly glaucous foliage and rose-pink, pink, or rarely white, narrowly-heart-shaped flowers under 1 inch in length. Variable **D. formosa** differs from *D. eximia* in its rose-pink to pink, or rarely white flowers being proportionately broader just below the point at which the outer petals bend outward, in the reflexed portions

being not over ⅕ inch long, and in the crests of the inner petals not protruding. This species is indigenous from California to British Columbia. Variety *D. f. oregana*, smaller than the species, has leaves distinctly glaucous on both surfaces. Its creamy-yellow flowers have pink-tipped inner petals. This variety is restricted in the wild to a small area that spans the border between California and Oregon. A related Californian species, **D. nevadensis** (syn. *D. formosa nevadensis*) has smaller flowers and narrower, more sharply-toothed leaves.

Dicentra eximia

Dicentra eximia (flowers)

Dicentra formosa oregana

Dutchman's breeches (**D. cucullaria**) well deserves its colloquial name. Its pleasant little flowers strung four to ten or rarely more together in one-sided racemes that exceed the foliage, indeed call to mind inverted pairs of the traditional pantaloons of Hollanders. They are white, often tinged with pink, with the extremeties of the petals yellow or orange-yellow. From 6 inches to over 1 foot tall, this native from Nova Scotia to North Dakota, Georgia, and Kansas, and of Washington, Oregon, and Idaho, has underground clusters of small, grainlike, white, tear-drop-shaped tubers. Its broadly-triangular leaves, three-times-thrice-divided, with the ultimate segments toothed, are mostly about 6 inches in length and have glaucous undersides. Dutchman's breeches blooms in early spring and dies down completely by midsummer.

Dicentra cucullaria

Squirrel-corn (**D. canadensis**) much resembles Dutchman's breeches, but has fewer, larger, spherical, yellow tubers and spurs to its flowers that instead of being prolonged into typical "pants legs" are very short. This kind is native from Quebec and Nova Scotia to Minnesota, North Carolina, Tennessee, and Missouri.

Golden eardrops (**D. chrysantha**), a wildling of dryish soils in California and Baja California, differs from all the kinds described above in having an erect, rigid stem 3 to 4½ feet tall. It is sparsely furnished with twice-pinnate, glaucous leaves, mostly up to 1 foot long, and with the ultimate leaflets pinnately-lobed. The narrow panicle is of few to numerous upfacing, slightly unpleasantly scented, golden-yellow, ½-inch-long flowers. Much resembling the last, but coarser and sometimes 10 feet tall, **D. ochroleuca**, of California, has thrice-pinnate leaves. Its cream- to off-white blooms have purple-tipped inner petals.

Steer's head (**D. uniflora**) is appropriately tagged colloquially and botanically, since its flowers are indeed of a shape rem-

Dicentra canadensis

iniscent of a steer's head, and are solitary. It is native of California and Oregon. Stemless, and with a horizontal rhizome or cluster of small tubers, this species has one to three twice- or thrice-three-times-divided leaves with glaucous undersides, 1½ to 3 inches in length. The nodding to erect, pink, lilac-pink, or white flower, stained with light brown, has curved, very wide-spreading ends (the "horns" of the "steer's head") to its outer petals, and purple-tipped, crestless inner petals (the wedge-shaped "head" of the "steer"). A variable species, **D. peregrina** (syn. *D. pusilla*) is native from Japan to Siberia. It differs from *D. uniflora* is not being tuberous, in having leaves up to 4 inches long or longer, and in its usually two or three white to puplish flowers on each stalk.

Garden and Landscape Uses. Common bleeding heart (*D. spectabilis*) is preeminently for flower beds and is first rate for early forcing in greenhouses. Its blooms lend themselves to elegant use in arrangements. It grows well in ordinary fertile soil, but unfortunately dies down in early summer leaving gaps often difficult to hide. Setting temporary plants as fillers too close to the bleeding hearts is likely to damage their thick fleshy roots. The best solutions are to locate in front of the bleeding hearts such plants as chrysanthemums or phloxes that gain sufficient height as the season advances to hide the empty spaces, or to sow seeds of accommodating annuals, such as California-poppies or sweet alyssum, around the failing tops of the bleeding hearts.

Not presenting the problem of losing their foliage in early summer, *D. eximia* and *D. formosa* are useful for planting in the fronts of flower beds and colonizing in informal areas and rock gardens. They are best in partial shade, but in not excessively dry soil succeed in sunny places. They are likely to spread by self-sown seeds. Golden eardrops and less colorful *D. ochroleuca* can be used to the rears of flower beds, and can colonize less formal environments, especially in climates that approximate those

of their native California. They prosper in dryish soil.

The other dicentras discussed are at their best strewn plentifully in rich, light woodland soil beneath trees, or in other shaded areas.

Cultivation. As a group dicentras make no extravagant demands upon gardeners' skills. Given half a chance they settle into comfortable permanence and enhance their environments yearly with attractive foliage and charming flowers. They are grateful for being allowed to develop undisturbed. Frequent transplanting is neither needful nor desirable. Propagation is easy by division in early spring, or after the foliage has died, and by seeds. Root cuttings, sometimes recommended as a means of propagating *D. spectabilis*, rarely prove successful. White-flowered variants of normally pink-flowered species are commonly weaker than the more typical kinds and are more difficult to maintain in cultivation.

Greenhouse forcing is easy, and with *D. spectabilis, D. eximia, D. formosa*, and their varieties and hybrids, results in specimens with much appeal. In preparation for forcing, the plants are dug up in fall or, in the case of *D. spectabilis*, better still as soon as the foliage dies away, care being taken to preserve the roots as undamaged as possible, and are potted in porous, fertile soil in well-drained pots just big enough to accommodate the roots without too much crowding. After potting and watering, the plants are placed in a cold frame where they remain until early the following year. Beginning in January they are brought in successive batches, to provide a long season of bloom, into a sunny greenhouse. A fairly humid atmosphere and a night temperature not higher than 50 to 55°F, increased by day by five to fifteen degrees depending upon the brightness of the weather, soon stimulates growth and quite rapid development to flowering results. Watering, done with some caution at first, is done more frequently as the foliage mass increases.